T0344993

Effective Field Theories

Effective field theories are a widely used tool in various branches of physics. This book provides a comprehensive discussion of the foundations and fundamentals of effective field theories of quantum chromodynamics (QCD) in the light quark sector with an emphasis on the study of flavor symmetries and their realizations. In this context, different types of effective field theories pertaining to various energy scales are considered and selected applications are devised. This text also covers the formulation of effective field theories in a finite volume and its application in the analysis of lattice QCD data. *Effective Field Theories* is intended for graduate students and researchers in particle physics, hadron physics and nuclear physics. Exercises are included to help readers deepen their understanding of the topics discussed throughout, with solutions available to lecturers.

Ulf-G. Meißner is Professor in Theoretical Physics at Bonn University, Germany, and Director at the Institute for Advanced Simulation, Forschungszentrum Jülich, Germany. His research focuses on strong interaction physics, especially nuclear and particle physics. He was awarded the Lise Meitner Prize of the European Physical Society and the Distinguished Scientist Award of the Chinese Academy of Sciences.

Akaki Rusetsky works at the Helmholtz Institute for Radiation and Nuclear Physics at Bonn University, Germany. His research focuses on strong interaction physics, effective field theories and lattice field theories. He received the Visiting Scientist award under the President's International Fellowship Initiative of the Chinese Academy of Sciences.

Effective Field Theories

ULF-G. MEIßNER

Bonn University
and
Forschungszentrum Jülich

AKAKI RUSETSKY

Bonn University

CAMBRIDGE
UNIVERSITY PRESS

CAMBRIDGE
UNIVERSITY PRESS

University Printing House, Cambridge CB2 8BS, United Kingdom

One Liberty Plaza, 20th Floor, New York, NY 10006, USA

477 Williamstown Road, Port Melbourne, VIC 3207, Australia

314–321, 3rd Floor, Plot 3, Splendor Forum, Jasola District Centre, New Delhi – 110025, India

103 Penang Road, #05–06/07, Visioncrest Commercial, Singapore 238467

Cambridge University Press is part of the University of Cambridge.

It furthers the University's mission by disseminating knowledge in the pursuit of
education, learning, and research at the highest international levels of excellence.

www.cambridge.org
Information on this title: www.cambridge.org/9781108476980
DOI: 10.1017/9781108689038

© Cambridge University Press 2022

First published 2022

A catalogue record for this publication is available from the British Library.

ISBN 978-1-108-47698-0 Hardback

Additional resources for this publication at www.cambridge.org/effectivefieldtheories.

Contents

Preface

Effective field theories (EFTs) have become a premier tool in many branches of theoretical physics, in particular in nuclear, hadronic and particle physics. It is the aim of this book to present the foundations and applications of such an approach to the strong interactions within the very successful Standard Model of particle physics. Although we cover a wide variety of topics, we concentrate on applications in hadronic and particle physics. Furthermore, our aim is not to consider all possible EFTs in these fields, but rather to discuss the physics related to the many fascinating phenomena emerging in the light quark sector of Quantum Chromodynamics. We stress that we put a strong emphasis on the basics and the foundations of these EFTs and only work out a few assorted applications. All this is based on a thorough discussion of the pertinent symmetries and their realizations, which is one fundamental cornerstone of any EFT. Notable omissions are heavy quark EFTs, the wide field of nuclear physics (with the exception of the so-called pionless nuclear EFT covered here) and also the Standard Model EFT, which has gained prominence in recent years. Whenever possible, we refer to appropriate books, reviews or papers. In any case, this book should enable the reader to work out an appropriately tailored EFT for any physical system under consideration. Furthermore, whenever necessary, we explain the underlying quantum field theoretical (QFT) tools that are required to master EFTs. However, we have not made any attempt to provide a self-contained QFT book, as many excellent books on this subject are available.

The genesis of this book is quite interesting. It was originally started by one of the authors (UGM) at Forschungszentrum Jülich more than two decades ago but came to a halt when he moved to Bonn University and took over a number of administrative positions. In a way, this was fortunate, as over the years we frequently lectured together on courses in Theoretical Hadron Physics, Advanced Theoretical Hadron Physics, Effective Field Theories and the like, which allowed us to sharpen our view on these topics, ultimately leading to the book in its present form.

Many colleagues have contributed to our understanding of the topics discussed here, in particular Véronique Bernard, Bugra Borasoy, Gilberto Colangelo, Michael Döring, Gerhard Ecker, Evgeny Epelbaum, Jambul Gegelia, Feng-Kun Guo, Hans-Werner Hammer, Martin Hoferichter, Barry Holstein, Misha Ivanov, Norbert Kaiser, Joachim Kambor, Bastian Kubis, Valery Lyubovitskij, Maxim Mai, Ferenc Niedermayer, Jin-Yi Pang, Fernando Romero-López, Jacobo Ruiz de Elvira, Chien-Yeah Seng and Jia-Jun Wu. We are especially grateful to Jürg Gasser and Heiri Leutwyler for sharing their deep insights.

We especially thank Feng-Kun-Guo and Bernard Metsch for a careful reading of the manuscript, Hans-Werner Hammer for useful remarks on Efimov physics and Fabian Müller for help with the exercises. They should not, however, be made responsible for any error.

We are grateful for the support and patience of Simon Capelin, Sarah Lambert, Henry Cockburn, Vince Higgs and Arya Thampi from Cambridge University Press who were instrumental in making this book possible.

We are also grateful for the hospitality extended by the Institute for Theoretical Physics and the Institute of High-Energy Physics of the Chinese Academy of Sciences, the School of Physics of Peking University in Beijing, China, and the Institute for Nuclear Theory at Seattle, Washington, USA, where parts of this book were written.

This book could not have been completed without generous funding from the Deutsche Forschungsgemeinschaft, the Chinese Academy of Sciences, Volkswagen-Stiftung, European Research Council and Shota Rustaveli National Science Foundation of Georgia.

Basic Concepts

1.1 Introduction

In the description of any physical phenomenon, there always arises the question of effectively separating the few relevant degrees of freedom from the myriad irrelevant ones. One could even state that the very existence of physics as an exact science directly hinges on the *possibility* of such a separation because, at any time, only a limited amount of information about a given physical system is available.

Consequently the question arises, which criteria should be used to achieve such a separation? A universal criterion is based on the comparison of the length (momentum) scales that are inherent to any complex system. In the presence of multiple scales, the physics at a low momentum (large-distance) scale is insensitive to the dynamics at high momenta (short distances). This is called *scale separation* and should be considered as of one the cornerstones of the concept of effective field theory. A trivial example is provided by Newtonian mechanics. In order to describe the free fall of a stone, the knowledge of the structure of the stone (molecules, atoms, quarks and gluons, etc.) is not needed. Another example is provided by the well-known multipole expansion in electrodynamics. The electrostatic potential produced by an arbitrary static distribution of charges that are located in a small area near the origin (see Fig. 1.1) is given *at large distances* by

$$V(\mathbf{r}) = \sum_{i=1}^{N} \frac{q_i}{|\mathbf{r} - \mathbf{d}_i|} = \sum_{i=1}^{N} \frac{q_i}{r} + \sum_{i=1}^{N} \frac{q_i (\mathbf{d}_i \cdot \mathbf{r})}{r^3} + \sum_{i=1}^{N} \frac{q_i \left[3(\mathbf{d}_i \cdot \mathbf{r})^2 - \mathbf{d}_i^2 \mathbf{r}^2 \right]}{2r^5} + \cdots . \quad (1.1)$$

Here, $r_i = |\mathbf{r} - \mathbf{d}_i|$ is the distance between the ith charge and the observer located at the point P. Introducing the total charge, the dipole moment and the quadrupole moment, in order,

$$Q = \sum_{i=1}^{N} q_i, \qquad \mathbf{P} = \sum_{i=1}^{N} q_i \mathbf{d}_i, \qquad Q_{\alpha\beta} = \frac{1}{2} \sum_{i=1}^{N} q_i \left[3 d_{i\alpha} d_{i\beta} - \delta_{\alpha\beta} \, \mathbf{d}_i^2 \right], \quad (1.2)$$

one obtains

$$V(\mathbf{r}) = \frac{Q}{r} + \frac{\mathbf{P} \cdot \mathbf{r}}{r^3} + \frac{Q_{\alpha\beta} r_\alpha r_\beta}{r^5} + \cdots . \quad (1.3)$$

This expansion converges if the distance r between the observer and the center of the charge distribution is much larger than the size of the charge distribution itself, that is, $|\mathbf{d}_i| \ll r$.

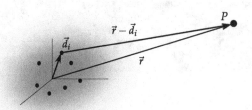

Figure 1.1 An electrostatic potential produced by an arbitrary localized static distribution of charges. The observer is located at point P far away from the charge distribution. Figure courtesy of Serdar Elhatisari.

Equation (1.3) demonstrates that by choosing the appropriate degrees of freedom, or variables, for describing the problem at large distances (i.e., choosing $\mathbf{r}$ instead of the individual distances $\mathbf{r}_i$), the solution of the problem is considerably simplified and can be described by a few parameters, here $Q, \mathbf{P}, Q_{\alpha\beta}$. These characterize the system as a whole rather than its individual components. Equation (1.2) can be considered as a *matching condition,* giving the expressions of these parameters in terms of the underlying physics at short distances.

This separation of scales is encountered in any field of physics. In this chapter, we consider the application of this idea in quantum field theory and demonstrate how the physics at the heavy scales (at short distances) can be consistently integrated out from the theory, leading to an effective theory, which contains the light degrees of freedom only.

1.2 Warm-up: Effective Theory for Scattering on the Potential Well

1.2.1 Effective Range Expansion

Before addressing effective field theories, we would like to start from a more familiar example and consider constructing an effective theory for quantum-mechanical scattering on a short-range potential. This allows us to explain many fundamental concepts and notions of effective field theories in an intuitive and transparent fashion.[1] Namely, we shall consider a spherical potential well, depicted in Fig. 1.2. The potential of the well is given by

$$U(r) = \begin{cases} -U_0 & \text{for} & r \le b\,, \\ 0 & \text{for} & r > b\,. \end{cases} \qquad (1.4)$$

[1] A similar problem has been addressed, for example, in the beautiful lectures given by Lepage [1], which we strongly recommend for further reading.

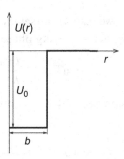

The spherical potential well. U_0 and b are the depth and the range of the potential, respectively.

This choice has the advantage that the Schrödinger equation (note that throughout we work in natural units $\hbar = c = 1$),

$$\left(-\frac{1}{2m}\nabla^2 + U(r)\right)\psi(\mathbf{r}) = E\psi(\mathbf{r}), \tag{1.5}$$

is exactly solvable, thus rendering our arguments explicit. Here, m denotes the mass of the particle that moves in the potential $U(r)$. Note that any short-ranged potential of an arbitrary form, but with the same range b, should lead to a similar behavior of the wave function at large distances $r \gg b$, since the details of the potential at short range should not matter.

The solution of this quantum-mechanical problem is well known from textbooks, and we display only those results here, which will be needed in the following. Due to rotational symmetry, the wave function in Eq. (1.5) can be factorized into the radial part $R_\ell(r)$ and the angular part, given by the spherical harmonics with the angular momentum ℓ and projection m:

$$\psi(\mathbf{r}) = R_\ell(r)Y_{\ell m}(\theta, \phi), \tag{1.6}$$

with θ and ϕ the polar and the azimuthal angle, respectively. For large values of $r \gg b$, the asymptotic form of the solution with $E > 0$ is given by

$$R_\ell(r) \to \frac{A_\ell}{r}\sin\left(kr - \frac{\pi\ell}{2} + \delta_\ell(k)\right). \tag{1.7}$$

Here, A_ℓ is a constant, k is the wave vector, related to the energy by $E = k^2/(2m)$, and $\delta_\ell(k)$ is the *scattering phase shift,* which encodes all information about the behavior of the wave function at asymptotically large distances. It is given by the expression

$$\tan\delta_\ell(k) = \frac{k j'_\ell(kb) j_\ell(Kb) - K j_\ell(kb) j'_\ell(Kb)}{k n'_\ell(kb) j_\ell(Kb) - K n_\ell(kb) j'_\ell(Kb)}, \tag{1.8}$$

where $j_\ell(x)$ and $n_\ell(x)$ denote the spherical Bessel functions of the first and second kind, respectively, and $K^2 = k^2 + 2mU_0$. Further, the prime denotes differentiation with respect to the argument.

In the following, we restrict ourselves to S-wave scattering with $\ell = 0$ and thus drop the index ℓ. The S-wave scattering phase is given by

$$\tan \delta(k) = \frac{k \tan(Kb) - K \tan(kb)}{K + k \tan(kb) \tan(Kb)}.$$ (1.9)

Using this expression, we may write down the *effective-range expansion* (ERE) for the phase shift:

$$k \cot \delta(k) = -\frac{1}{a} + \frac{1}{2} r k^2 + v_4 k^4 + O(k^6).$$ (1.10)

Here, a is the *scattering length*, r is called *effective range,* and the higher coefficients $v_4, v_6, \ldots$ are known under the name of *shape parameters.* Generally, $a, r, v_4, v_6, \ldots$ are referred to as *effective-range parameters.* The explicit expressions for these parameters are obtained by Taylor-expanding Eq. (1.9). It is convenient to express the results in terms of b and the dimensionless parameter $x = b\sqrt{2mU_0}$:

$$a = bf_0(x),$$

$$r = bf_2(x),$$

$$v_{2n} = b^{2n-1} f_{2n}(x).$$ (1.11)

Next, we display the first two coefficients explicitly:

$$f_0(x) = 1 - \frac{\tan x}{x},$$

$$f_2(x) = \frac{3 \tan x - 3x + 3x \tan^2 x - 6x^2 \tan x + 2x^3}{3x(x - \tan x)^2},$$ (1.12)

and so on.

Next, let us consider the limit $x \to \pi/2 + \pi n$. We can easily convince ourselves that $f_2(x), f_4(x), \ldots$ stay finite in this limit. Concerning the first coefficient, matters are, however, different. As seen from Eq. (1.12), $f_0(x) \to \infty$ as $\tan x \to \infty$. Thus, we have two distinct possibilities (we consider the magnitudes of the various parameters or scales):

1. All effective-range parameters are of *natural size,* which is determined by the interaction range b. Namely, $a \sim b$, $r \sim b$, $v_{2n} \sim b^{2n-1}$.
2. We have an *unnaturally large scattering length,* namely, $a \gg b$. All other parameters are of natural size, that is, we still have $r \sim b$, $v_{2n} \sim b^{2n-1}$.

Note also that the convergence of the effective range expansion is in both cases controlled by the parameter b. In other words, the effective range expansion converges when $kb \ll 1$. Physically, this means that for the large distances $r \sim 1/k \gg b$, the scattering on the short-ranged potential irrespective of its shape can be parameterized by the first few coefficients in this expansion (just as any static charge distribution in classical electrodynamics at large distances can be characterized by the first few coefficients in the multipole expansion of the electric field). At distances $r \sim 1/k \sim b$ this expansion does not make sense anymore, since all terms become equally important.

Finally, let us find out what the condition $x \to \pi/2 + \pi n$ means. To this end, note that the bound-state spectrum in the potential well is determined by the equation

$$\sqrt{2mU_0 + \kappa^2}\cot(b\sqrt{2mU_0 + \kappa^2}) = \kappa, \qquad (1.13)$$

where κ denotes the bound-state momentum, which is related to the energy by $E = -\kappa^2/(2m)$. When the bound state emerges exactly at threshold (i.e., at $\kappa = 0$), from Eq. (1.13) one gets $\cot(b\sqrt{2mU_0}) = 0$, that is, $b\sqrt{2mU_0} = x = \pi/2 + \pi n$. This shows that the scenario with the unnaturally large scattering length is realized when the parameters of the potential are fine-tuned so that a very shallow bound state emerges. The existence of a such zero-energy bound state does not affect the other effective range parameters.

1.2.2 Construction of the Effective Theory

At large distances $r \gg b$ (or, equivalently, at small momenta $kb \ll 1$), one cannot resolve the details of the potential. All short-ranged potentials at this distance should look pretty much the same and similar to the potential with zero range. In the first approximation, one can replace the exact potential with a local δ-function potential, that is,

$$U(r) \to C_0 \delta^{(3)}(\mathbf{r}), \qquad (1.14)$$

and adjust the single available coupling C_0, so that the lowest-order term in the effective-range expansion of the scattering phase (the scattering length) is the same in the original theory and in the effective theory. Since the effective range expansion encodes all physical information about the system at low energies, it is then intuitively clear that the zero-range potential in Eq. (1.14) – in the lowest-order approximation – is equivalent to the initial potential $U(r)$ with a finite range at low energies. The procedure of adjusting C_0 goes under the name of *matching*.

The issues, considered in what follows, have been addressed in the literature; see, for example, Refs. [2, 3]. In order to find the scattering phase shift in the effective theory, one writes down the Lippmann–Schwinger equation for the S-wave scattering amplitude $T(p,k)$:

$$T(p,k) = V(p,k) + \frac{2}{\pi} \int \frac{q^2 dq}{q^2 - k^2 - i\varepsilon} V(p,q) T(q,k), \quad \varepsilon \to 0^+, \qquad (1.15)$$

with $V(p,k)$ the potential in momentum space. Further, k and p denote the magnitudes of the incoming and outgoing relative three-momenta, respectively. The on-shell amplitude $T(k,k) \doteq T(k)$ obeys elastic unitarity,

$$\mathrm{Im}\, T(k) = k|T(k)|^2, \qquad (1.16)$$

and can be expressed through the phase shift

$$T(k) = \frac{1}{k} e^{i\delta(k)} \sin\delta(k). \qquad (1.17)$$

We may also introduce the scattering R-matrix, which obeys the Lippmann–Schwinger equation,

$$R(p,k) = V(p,k) + \frac{2}{\pi} \text{P.V.} \int \frac{q^2 dq}{q^2 - k^2} V(p,q) R(q,k). \tag{1.18}$$

Here, unlike Eq. (1.15), the integral is equipped with the principal value (P.V.) prescription. The T- and R-matrices are related. On shell, this relation takes the form

$$T(k) = \frac{R(k)}{1 - ikR(k)}, \qquad R(k) = \frac{1}{k} \tan \delta(k). \tag{1.19}$$

In the following, we prefer to work with the R-matrix. The Fourier transform of the potential in Eq. (1.14) is given by

$$V(p,k) = C_0. \tag{1.20}$$

The solution of the Lippmann–Schwinger equation, (1.18), is

$$R(k) = \frac{C_0}{1 - C_0 I_2(k^2)}, \qquad I_2(k^2) = \frac{2}{\pi} \text{P.V.} \int \frac{q^2 dq}{q^2 - k^2}. \tag{1.21}$$

Here, we encounter the problem of an ultraviolet (UV) divergence. Since the δ-function potential is singular at short distances, the integral I_2 diverges at the upper limit and should be regularized. The most straightforward way to do this is to introduce a momentum cutoff, Λ. Then, the integral is equal to

$$I_2(k^2) = \frac{2}{\pi} \text{P.V.} \int^\Lambda \frac{q^2 dq}{q^2 - k^2} = \frac{2}{\pi} \Lambda + O(1/\Lambda). \tag{1.22}$$

In the following, the terms of order $1/\Lambda$ are always neglected and never displayed explicitly. The R-matrix at leading order, which is given by Eq. (1.21), turns out to be constant. The matching condition then reads

$$R(0) = -a, \qquad C_0 = -\frac{a}{1 - aI_2(0)}. \tag{1.23}$$

It follows from Eq. (1.23) that C_0 should be Λ-dependent in order to ensure that the observable (the scattering length a) does not depend on the cutoff.

In order to illustrate the matching, let us do a simple numerical exercise. We arbitrarily choose the parameters of the square well as $b = 1$ and $x = b\sqrt{U_0} = \pi/4$. In Fig. 1.3, we display the exact phase shift, given by Eq. (1.9), as well as the phase shift, obtained in the effective field theory with a zero-range potential given in Eq. (1.14), where the parameter C_0 is determined from the matching to the exact scattering length. The parameter Λ is set equal to $1/b$ (i.e., to the inverse of the short-range scale of the model). It is seen that, up to the momenta $k^2/\Lambda^2 \leq 0.5$, the phase shift is reproduced in the effective theory rather well.

We are not going to stop here, however: we ask ourselves whether it is possible to *systematically* improve the description of the phase shift. To this end, note that, using the leading-order potential (with no derivatives), it is possible to adjust only the scattering length in the effective theory. The higher-order coefficients of the effective-range

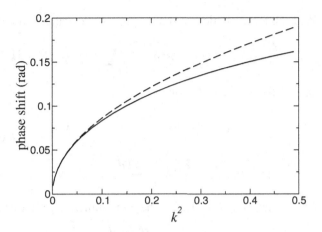

Figure 1.3 The phase shift obtained at leading order in the effective theory (dashed line). For comparison, we plot the exact solution for the potential well given by Eq. (1.9) (solid line).

expansion at this order are all zero (because the R-matrix is constant at this order). To reproduce these as well, we need more adjustable parameters. For example, the effective range also can be tuned, if we add a term with two derivatives to the potential, and so on.

Let us now explicitly demonstrate how this can be done. We modify the potential in Eq. (1.14):

$$U(r) \rightarrow C_0 \delta^{(3)}(\mathbf{r}) + C_2 \nabla^2 \delta^{(3)}(\mathbf{r}). \tag{1.24}$$

We note that the first term in this expansion is referred to as leading order (LO), whereas the second gives the next-to-leading order (NLO) contribution. Calculating the Fourier transform, we get

$$V(\mathbf{p} - \mathbf{k}) = \int d^3\mathbf{r}\, e^{-i(\mathbf{p}-\mathbf{k})\mathbf{r}} \left(C_0 \delta^{(3)}(\mathbf{r}) + C_2 \nabla^2 \delta^{(3)}(\mathbf{r}) \right)$$

$$= C_0 - C_2 (\mathbf{p} - \mathbf{k})^2. \tag{1.25}$$

Projecting onto the S-wave gives

$$V(p,k) = \frac{1}{2} \int_{-1}^{+1} d\cos\theta\, V(\mathbf{p} - \mathbf{k})$$

$$= C_0 - C_2 (p^2 + k^2) = \sum_{i,j=1}^{2} v_i(p) C_{ij} v_j(k), \tag{1.26}$$

where

$$v_1(k) = 1, \quad v_2(k) = k^2, \quad C = \begin{pmatrix} C_0 & -C_2 \\ -C_2 & 0 \end{pmatrix}. \tag{1.27}$$

We look for a solution of the Lippmann–Schwinger equation using the ansatz:

$$R(p,k) = \sum_{i,j=1}^{2} v_i(p) D_{ij}(k^2) v_j(k). \tag{1.28}$$

The matrix D obeys the following equations:

$$D(k^2) = C + C\Gamma(k^2)D(k^2), \qquad \Gamma(k^2) = \begin{pmatrix} I_2(k^2) & I_4(k^2) \\ I_4(k^2) & I_6(k^2) \end{pmatrix},$$

$$I_{2n}(k^2) = \frac{2}{\pi}\text{P.V.}\int \frac{q^{2n}dq}{q^2 - k^2}. \tag{1.29}$$

$I_2(k^2)$ is given by Eq. (1.22). The remaining integrals are equal to

$$I_4(k^2) = \frac{2}{\pi}\text{P.V.}\int^{\Lambda} \frac{q^4 dq}{q^2 - k^2} = \frac{2}{\pi}\left(\frac{1}{3}\Lambda^3 + \Lambda k^2\right),$$

$$I_6(k^2) = \frac{2}{\pi}\text{P.V.}\int^{\Lambda} \frac{q^6 dq}{q^2 - k^2} = \frac{2}{\pi}\left(\frac{1}{5}\Lambda^5 + \frac{1}{3}\Lambda^3 k^2 + \Lambda k^4\right). \tag{1.30}$$

The solution for the R-matrix is then given by

$$R(k) = \frac{C_0 + C_2^2 I_6(k^2) - 2k^2 C_2(1 + C_2 I_4(k^2)) + k^4 C_2^2 I_2(k^2)}{(1 + C_2 I_4(k^2))^2 - I_2(k^2)(C_0 + C_2^2 I_6(k^2))}. \tag{1.31}$$

This expression can be expanded in powers of k^2, and the first two coefficients can be matched to the effective-range expansion for the scattering phase. Introducing the dimensionless variables,

$$x_0 = \frac{\Lambda C_0}{\pi}, \qquad x_2 = \frac{\Lambda^3 C_2}{\pi}, \tag{1.32}$$

we obtain two coupled equations for x_0, x_2:

$$h_0 \doteq \frac{\Lambda a}{9\pi} = \frac{5x_0 + 2x_2^2}{18(5x_0 + 2x_2^2) - 5(3 + 2x_2)^2},$$

$$h_2 \doteq \frac{27\pi\Lambda r}{100} = \frac{x_2(3 + x_2)(3 + 2x_2)^2}{(5x_0 + 2x_2^2)^2}. \tag{1.33}$$

It is possible to find an explicit solution to these equations:

$$x_0 = \frac{1}{5}(\alpha y - 2x_2^2), \ x_2 = \frac{1}{2}(\sqrt{y} - 3), \ y = \frac{9}{1 - 4\alpha^2 h_2}, \ \alpha = -\frac{5h_0}{1 - 18h_0}. \tag{1.34}$$

Substituting this solution in Eq. (1.32), one arrives at the values of C_0 and C_2 that reproduce the first two terms in the effective-range expansion. It is clear that the description can be improved systematically, adding terms with more derivatives to the potential. All this makes sense for $k^2 \ll 1/b^2$, for which the effective range expansion is justified.

In Fig. 1.4, the numerical solution for the phase shift at next-to-leading order is depicted. It is seen that there is a systematic improvement as compared to the leading order. The next-to-leading order phase shift almost follows the exact curve. Here, it

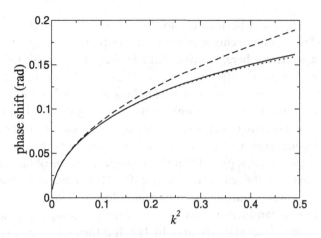

Figure 1.4 The same as in Fig. 1.3. Shown is the phase shift in the effective theory obtained at the leading (dashed line) and the next-to-leading order (dotted line), respectively, in comparison to the exact solution (solid line).

is worth mentioning that the effective range expansion is convergent very fast in the simple case considered here: keeping only the first two terms in this expansion, one obtains a phase shift which is indistinguishable from the exact solution by bare eye (in the region where the effective theories are applicable).

1.2.3 Regularization

Let us now address the question how the results depend on the regularization. Up to now, we have used cutoff regularization to tame the ultraviolet divergences. Here, we consider the use of dimensional regularization [4, 5], where the expressions look particularly simple.[2] This happens because all "no-scale" integrals vanish by definition in this scheme, that is,

$$\int d^d\mathbf{q}\,(\mathbf{q}^2)^\alpha = 0 \quad \text{for all } \alpha, \tag{1.35}$$

with d the number of space dimensions, which is set to three after the integral has been evaluated. It is then immediately seen that in this regularization $I_2 = I_4 = I_6 = 0$ and, hence,

$$R(k) = C_0 - 2C_2 k^2. \tag{1.36}$$

Matching to the effective-range expansion is then straightforward:

$$C_0 = -a, \quad C_2 = \frac{1}{4}ra^2, \tag{1.37}$$

and so on. It is seen that, in this scheme, the couplings $C_0, C_2, \ldots$ do not depend on any scale, except the one implicitly present in the effective-range expansion parameters, that is, the scale b.

---

[2] A good introduction to this method is given in the textbook [6].

There has been discussion in the literature, whether certain *physical* results may depend on the choice of the regularization in a *non-perturbative* setting, which we are considering here; see, for example, Refs. [2, 3, 7, 8]. Of course, all results obtained within perturbation theory should be strictly regularization independent. To understand the argument, it suffices to look at Eq. (1.34). From this equation, it is clear that the limit $\Lambda \to \infty$ exists only in the case $h_2 \le 0$ (and, hence, the effective range $r < 0$), whereas there is no restriction on the sign of the effective range in dimensional regularization, and in nature this quantity obviously may have either sign. Further, it has been shown in Refs. [9, 10] that this restriction can be obtained from the so-called Wigner bound on the scattering phase shift [11], assuming that the interaction has zero range.

However, putting the argument differently, one could state that there is no justification to consider the limit $\Lambda \to \infty$ in the preceding expansion, since the physical cutoff scale for the system is given by $1/b$. It is then seen that no problem emerges, if the low-energy expansion is carried out in the presence of a finite cutoff on the order of $1/b$. We may also conclude that the physical results indeed do not depend on the regularization chosen, as long as the regularization parameter is chosen within a reasonable range.

1.2.4 Counting Rules

From the very beginning, our method is aimed at a systematic improvement of the description of the scattering phase (any physical observables, in general) in the low momentum region. This means that, for instance, the contributions to the phase shift from the terms with more derivatives, which could be added to Eq. (1.24), will be suppressed by a factor $(kb)^2 \ll 1$, where k is a small momentum. Let us ask, however, what is the meaning of the effective potential in the cutoff regularization. Of course, since all couplings depend on the cutoff, one has to consider the potential at a single cutoff. Following our intuition, we expect that there exists a string of effective couplings, determining the effective potential to all orders, so that adding more terms affects the observables less and less. The situation is, however, more subtle. If one adds higher-order terms, one has also to readjust the lower-order terms because this is required by matching. For example, comparing Eq. (1.34) to the solution at the lowest order $x_0 = -9h_0$ and $x_2 = 0$, we see that the change in the values of the dimensionless parameters x_0, x_2 is of order one, since h_0, h_2 are of order one for $\Lambda \sim 1/b$. The same pattern holds at higher orders as well. Consequently, albeit a systematic improvement of the precision in the description of physical observables can be achieved, the terms in the effective potential, strictly speaking, cannot be ordered according to their relative importance. Adding a formally higher-order term leads to a renormalization in the lower-order terms as well. In other words, no consistent *power counting* scheme can be defined in this case.

The situation is different in the dimensional regularization scheme where, for example, C_0 can be fixed through the scattering length and stays put for all orders. The same is true for the higher-order couplings. The question about the *size* of these couplings is, however, more subtle. If the scattering length is of natural size, we have, as expected,

$C_0 = O(b)$, $C_2 = O(b^3)$ and so on, so that the Taylor expansion for $R(k)$ converges for $k^2 \ll 1/b^2$. As seen from Eq. (1.37), this is no more true in the case of the unnaturally large scattering length, which renders the radius of convergence of this expansion smaller.

The reason for the difference between the cutoff and dimensional regularization is easy to understand. Consider, for simplicity, the case of natural-sized couplings. In dimensional regularization, there is only one short-distance parameter b, the various couplings have different dimensions (they are proportional to different powers of b) and do not talk to each other. In contradistinction to this, there exist two short-distance parameters in the cutoff regularization, and a dimensionless variable Λb can be constructed. The couplings at different orders can then mix in the matching condition, and this mixing differs with increasing orders. An analog to this phenomenon emerges when a heavy particle is present in the loops of the effective theory (e.g., relativistic baryon Chiral Perturbation Theory and the breakdown of the counting rules). We shall address this issue in the following.

We postpone the discussion of other regularization schemes to subsequent chapters.

1.2.5 Error Estimates

An important issue for any theoretical prediction is the precision that can be achieved. Stated differently, similar to experimental measurements, theoretical predictions also carry an uncertainty, also called the *theoretical error*. One of the main advantages of effective theories is the possibility of estimating this uncertainty. Here, we briefly discuss the so-called *naive dimensional analysis* (NDA) and refer to later sections for more precise and refined methods to deal with error estimates. Consider some observable $\mathcal{O}$ that is expanded in a small parameter Q, with Q the ratio of small (soft) momenta to the hard scale Λ (or a collection of small parameters as encountered in later sections),

$$\mathcal{O} = \mathcal{O}_{LO} + \mathcal{O}_{NLO} + \cdots = \sum_{i=0} c_i Q^i , \qquad (1.38)$$

where the coefficients c_i are assumed to be of order one, which is called *naturalness*. Here, LO, NLO, ..., denote the leading, next-to-leading, ... order in the expansion in Q. Note that the lowest order can start with some other power of Q, but this does not invalidate the following considerations. If, for example, $Q = 0.1$, we expect that the corrections at NLO to be on the order of 10% and at NNLO on the order of 1%. Two remarks are relevant here. First, such NDA estimates have always to be taken with a grain of salt. Often one encounters the situation that at a given order, some symmetry might suppress the contributions or in a very different case, there can be a remarkable enhancement of a given order due to some close-by state not accounted for explicitly (like, e.g., the Delta resonance in pion–nucleon scattering). In such cases, the NDA obviously will fail. Second, NDA does not allow us to estimate the sign of the corrections. Note further that sometimes only even powers appear in the expansion; the model just discussed is an expansion in powers of k^2/Λ^2, where the hard scale was identified with $1/b$, and the observable under consideration is $k\cot\delta$. This important issue will be taken up later in various places.

1.2.6 Renormalization Group Equations

As mentioned already, the couplings in the cutoff scheme should run with the cutoff parameter, Λ, in order to ensure that the observables are cutoff-independent. The renormalization group (RG) equations at leading order can be easily obtained by differentiating Eq. (1.23) with respect to Λ:

$$\Lambda\frac{d}{d\Lambda}C_0(\Lambda) = -\frac{2}{\pi}\Lambda C_0(\Lambda)^2. \tag{1.39}$$

The equations at next-to-leading order can be obtained by differentiating the solutions given in Eq. (1.34) with respect to the scale Λ and taking into account the fact that a and r are Λ-independent. The right-hand side of Eq. (1.39) gives the so-called β-function that determines the running of the coupling $C_0(\Lambda)$ with respect to Λ. Recalling the discussion in the previous section, we may expect that the beta functions in cutoff regularization at different orders are different merely by a quantity of order one.

1.2.7 What Did We Learn from This Example?

- At low momenta ($k \ll 1/b$) a scattering process can be characterized by a small set of effective range expansion parameters.
- The interaction range is implicitly encoded in the size of the effective range expansion parameters. Namely, if the scattering length is of natural size, then we have $a \sim b$, $r \sim b$, $v_4 \sim b^3$ and so on. In case of an unnaturally large scattering length, only the first of these relations is not valid.
- An unnaturally large scattering length is related to the formation of a near-threshold bound state (or a virtual state).
- One may construct a low-energy effective theory, approximating the square well potential by a series of the δ-function potential and derivatives thereof. The couplings in front of these potentials are adjustable parameters and are used to reproduce the effective-range expansion parameters order by order. This procedure goes under the name of matching.
- Albeit the matching conditions may look different in different regularizations, the resulting scattering amplitude, expressed in terms of the effective range parameters, is the same in all regularizations up to terms of higher orders.
- Last but not least, it is interesting to mention that the matching fixes not only the scattering amplitude at small momenta, but the spectrum of the shallow bound states as well. To see this, it suffices to note that, according to Eq. (1.19), the poles of the T-matrix (corresponding to the bound states) emerge for purely imaginary values of k, corresponding to the solution of the equation

$$R^{-1}(k) - ik = -\frac{1}{a} + \frac{1}{2}rk^2 + \cdots - ik = 0. \qquad (1.40)$$

If the effective theory reproduces the values of $a, r, \ldots$, then the solution of this equation will also be reproduced up to higher-order terms.

- Theoretical uncertainties can and should be estimated. A first estimate can be based on naive dimensional analysis, but in general more sophisticated methods should be used.

1.3 Integrating out a Heavy Scale: a Model at Tree Level

1.3.1 Matching at Tree Level

After this warm-up example, let us proceed with the construction of an effective theory in a simple field-theoretical model. This model is described by the following Lagrangian,[3]

$$\mathcal{L} = \frac{1}{2}(\partial\phi)^2 + \frac{1}{2}(\partial\Phi)^2 - \frac{m^2}{2}\phi^2 - \frac{M^2}{2}\Phi^2 - \frac{g}{2}\phi^2\Phi, \qquad (1.41)$$

with ϕ, Φ denoting the light and heavy scalar fields with masses m, M, respectively, and $m \ll M$. Further, the coupling constant g has dimension [mass] and ∂ is a shorthand notation for ∂_μ, where μ enumerates the space-time indices.

Let us consider processes involving light particles only (that means we only have external legs made from the light particle species), with energies E much smaller that the heavy mass M (i.e., assume that $E \sim m \ll M$). For such energies we expect that the presence of the heavy particle cannot lead to any observable consequence and the system can be described by an effective Lagrangian containing the light degrees of freedom only.[4] The inverse of the heavy mass $1/M$ plays the role of the short-range scale b in our case of the square well potential, and in the limit $M \to \infty$, a local potential emerges.

Consider, in particular, the $2 \to 2$ scattering process $\phi\phi \to \phi\phi$ at low energies. The momenta of the initial (final) particles are p_1 and p_2 (p_3 and p_4). The scattering amplitude in perturbation theory is a series in the coupling constant g, so that the matching to the effective field theory can be performed at each order independently. At tree level, the scattering amplitude is given by the diagrams depicted in Fig. 1.5 and is equal to

---

[3] In general, there will be an additional linear term $c\Phi$ present in the Lagrangian in Eq. (1.41), which is needed to cancel tadpole diagrams with one external Φ-leg. Here, however, we work in tree approximation, where it is possible to put $c = 0$.

[4] There is a well-known example, which exactly follows the path outlined in this toy model. Namely, in the Standard Model, the interactions between left-handed charged currents are mediated by the $W^\pm$ bosons with a mass $M_W \simeq 80$ GeV. If the momentum transfer in a process is much smaller than M_W, the flavor-changing weak interactions are described by the local four-fermion Fermi Lagrangian. The Fermi coupling G_F, which appears in the effective theory at tree level, is inversely proportional to M_W^2.

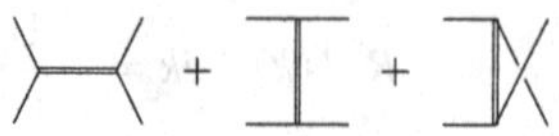

Figure 1.5 The tree-level scattering amplitude for the process $\phi\phi \to \phi\phi$ in the model described by the Lagrangian given in Eq. (1.41). Single and double lines correspond to the light and heavy fields, respectively. Shown are the s-, t- and u-channel contributions, in order.

$$T_{\text{tree}} = \frac{g^2}{M^2 - s} + \frac{g^2}{M^2 - t} + \frac{g^2}{M^2 - u}, \tag{1.42}$$

where $s = (p_1 + p_2)^2$, $t = (p_1 - p_3)^2$, $u = (p_1 - p_4)^2$ are the usual Mandelstam variables. On the mass shell, these variables obey the relation $s + t + u = 4m^2$.

In the limit of a large mass M, the amplitude in Eq. (1.42) can be expanded in a Taylor series:

$$T_{\text{tree}} = \frac{3g^2}{M^2} + \frac{g^2}{M^4}(s + t + u) + \frac{g^2}{M^6}(s^2 + t^2 + u^2) + \cdots. \tag{1.43}$$

At low energies, each subsequent term in this expansion is suppressed by a factor E^2/M^2 with respect to the previous one, where E is the characteristic energy of the light particles.

Our aim is to find a Lagrangian that contains only ϕ-fields, and which reproduces the expansion of the amplitude in Eq. (1.43). In general, such an *effective* Lagrangian must contain an infinite tower of quartic terms in the field ϕ. By analogy with Eq. (1.24) we may try to use the Lagrangian of the following form:

$$\mathcal{L}_{\text{eff}} = \frac{1}{2}(\partial\phi)^2 - \frac{m^2}{2}\phi^2 - C_0\phi^4 - C_1\phi^2\Box\phi^2 - C_2\phi^2\Box^2\phi^2 + \cdots, \tag{1.44}$$

with $\Box = \partial^\mu\partial_\mu = \partial\partial$. Note that at tree level the mass parameters in both the underlying and effective Lagrangians are equal. As we shall see, this is no more the case at one loop order.

The tree-level amplitude, obtained from this Lagrangian, takes the form

$$T_{\text{tree}}^{\text{eff}} = -24C_0 + 8C_1(s + t + u) - 8C_2(s^2 + t^2 + u^2) + \cdots. \tag{1.45}$$

This amplitude is shown in Fig. 1.6. Demanding $T_{\text{tree}}^{\text{eff}} = T_{\text{tree}}$ leads to *matching conditions* which enable one to express the couplings of the effective theory $C_0, C_1, C_2, \ldots$ in terms of the parameters of the underlying theory g, m and M.

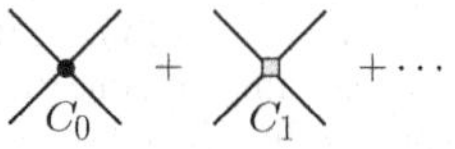

Figure 1.6 The tree-level scattering amplitude for the process $\phi\phi \to \phi\phi$ in the effective theory described by the Lagrangian given in Eq. (1.44). This amplitude can be obtained from the amplitude shown in Fig. 1.5 by contracting all heavy lines to a point.

1.3.2 Equations of Motion

The matching condition is imposed on *observables*, that is, in our case, on the scattering amplitude defined on shell, $p_i^2 = m^2$. As is known, the Mandelstam variables on shell obey the constraint

$$s + t + u = 4m^2, \tag{1.46}$$

and the tree-level amplitudes in the full theory and in the effective theory are given by the expressions

$$T_{\text{tree}} = \frac{3g^2}{M^2} + \frac{4g^2 m^2}{M^4} + \frac{g^2}{M^6}(s^2 + t^2 + u^2) + \cdots \tag{1.47}$$

and

$$T_{\text{tree}}^{\text{eff}} = -24C_0 + 32m^2 C_1 - 8C_2(s^2 + t^2 + u^2) + \cdots . \tag{1.48}$$

The matching conditions, which enable one to express the couplings of the effective theory in terms of the parameters of the underlying theory, take the form

$$-24C_0 + 32m^2 C_1 = \frac{3g^2}{M^2} + \frac{4g^2 m^2}{M^4}, \quad -8C_2 = \frac{g^2}{M^6}, \tag{1.49}$$

and so on.

Note that the mass-shell matching does not allow one to determine the couplings C_0 and C_1 separately. According to Eq. (1.49), only the combination $-24C_0 + 32m^2 C_1$ can be determined from the matching condition. This is related to the fact that (accidentally in this model) all second-order terms can be eliminated by using the equations of motion (EOM). In order to prove this, note that on the one hand

$$\phi^2 \Box \phi^2 = 2\phi^3(\Box + m^2)\phi - 2m^2\phi^4 + 2\phi^2(\partial\phi)^2, \tag{1.50}$$

and on the other hand,

$$\phi^2(\partial\phi)^2 = \underbrace{\frac{1}{3}\partial^\mu(\phi^3\partial_\mu\phi)}_{=\text{total derivative}} - \frac{1}{3}\phi^3(\Box + m^2)\phi + \frac{m^2}{3}\phi^4. \tag{1.51}$$

Using the EOM

$$(\Box + m^2)\phi = -4C_0\phi^3 + \cdots, \tag{1.52}$$

it is seen that $\phi^3(\Box + m^2)\phi$ is transformed into a sum of operators containing more than four fields and, therefore, does not contribute to the tree-level amplitude. Finally, the term proportional to ϕ^4 can be lumped together with the similar term in the Lagrangian. To summarize, the second-order terms can be completely eliminated from the Lagrangian. Thus, without losing generality, one may set $C_1 = 0$ everywhere.

1.3.3 Unitarity Bound

We have constructed an effective theory that is equivalent to the underlying theory at tree level. However, one does not stop at tree level. The effective field theory is a full-fledged field theory, so one has to consider loop diagrams, generated by the Lagrangian in Eq. (1.44) as well. Here a question arises naturally: The underlying theory is a superrenormalizable theory (the single coupling constant g has the dimension of mass), whereas the resulting effective theory contains a tower of non-renormalizable vertices. How should one deal with these divergences? Or stated differently, how can one interpret the equivalence of these two theories beyond the tree level?

Moreover, it can be seen that the tree-level amplitude in the effective field theory necessarily violates unitarity. In order to see this, it is convenient to define the partial-wave amplitudes

$$T_\ell^{\text{eff}}(s) = \frac{1}{32\pi\sqrt{s}} \int_{-1}^{+1} d\cos\theta \, T^{\text{eff}}(s, \cos\theta) P_\ell(\cos\theta),$$

$$T^{\text{eff}}(s, \cos\theta) = 16\pi\sqrt{s} \sum_{\ell=0}^{\infty} (2\ell+1) T_\ell^{\text{eff}}(s) P_\ell(\cos\theta), \tag{1.53}$$

where $P_\ell(\cos\theta)$ denote the conventional Legendre polynomials and θ is the scattering angle in the center-of-mass system. The unitarity relation for the partial-wave amplitudes gives

$$\text{Im}\, T_\ell^{\text{eff}}(s) \geq p |T_\ell^{\text{eff}}(s)|^2, \qquad p = \sqrt{\frac{s}{4} - m^2}, \tag{1.54}$$

where the inequality turns into the equality below the first inelastic threshold, $s_{\text{thr}} = (4m)^2$, where processes like $\phi\phi \to \phi\phi\phi\phi$ are not allowed energetically.

Transforming Eq. (1.54), we get

$$p(\text{Re}\, T_\ell^{\text{eff}}(s))^2 + p\left(\text{Im}\, T_\ell^{\text{eff}}(s) - \frac{1}{2p}\right)^2 - \frac{1}{4p} \leq 0. \tag{1.55}$$

Now, it is immediately seen that the real part of the amplitude obeys the so-called *unitarity bound*:

$$|\text{Re}\, T_\ell^{\text{eff}}(s)| \leq \frac{1}{2p}. \tag{1.56}$$

This bound is violated by the tree amplitude given in Eq. (1.45). For example, in the partial wave with $\ell = 0$ the tree-level amplitude is equal to

$$\text{Re}\, T_0^{\text{eff}}(s) = \frac{1}{16\pi\sqrt{s}} \left(-24C_0 + 32m^2 C_1 - 8C_2\left(\frac{2}{3}(s - 4m^2)^2 + s^2\right) + \cdots\right). \tag{1.57}$$

Substituting this expression into Eq. (1.56), it is seen that the left-hand side grows with increasing s, whereas the right-hand side decreases. Using the values of the coupling constants, determined by the matching condition given in Eq. (1.49), and assuming $s \gg m^2$, it is seen that the unitarity bound is saturated at

$$s_M = M^2 \sqrt{\frac{16\pi - 3\tilde{g}^2}{5\tilde{g}^2/3}} + O(1), \qquad \tilde{g} = \frac{g}{M}. \tag{1.58}$$

Note that the large-M limit in the underlying theory is performed so that the dimensionless quantity $\tilde{g}$ stays finite. Otherwise, the leading coupling C_0 could not be finite. Consequently, the quantity s_M is of order of M^2. If $s > s_M$, loops are necessary in order to render the tree-level amplitude unitary. In turn, this means that the loops must be of the same order of magnitude as the tree amplitude, heralding trouble in the perturbative expansion.

In reality, if s is on the order of $s_M \sim M^2$, the effective theory cannot be applied any more, and one should resort to a perturbative expansion in the underlying theory, which is superrenormalizable and where the amplitude decreases as s^{-1} at large values of s. It is said that the underlying theory provides a *Wilsonian ultraviolet (UV) completion* of the effective theory at scales of order M.

1.4 The Model at Tree Level: Path-Integral Formalism

Consider the generating functional of the theory described by the Lagrangian in Eq. (1.41):

$$Z(j,J) = \int d\phi d\Phi \exp\left\{ i \int d^4x (\mathcal{L}(\phi,\Phi) + j\phi + J\Phi) \right\}, \tag{1.59}$$

where $j(x), J(x)$ denote classical external sources for the fields $\phi(x)$ and $\Phi(x)$, respectively. The Green's functions are obtained by functional differentiation of Z with respect to these sources (once per each external leg) and by putting $j = J = 0$ at the end.

Since we are interested here in the Green's functions of the light field only, we may put $J = 0$ and consider the quantity $Z(j) \doteq Z(j, J = 0)$. Performing a shift of the integration variable,

$$\Phi \to \Phi - \frac{g}{2}(\Box + M^2)^{-1}\phi^2, \tag{1.60}$$

it is possible to rewrite the generating functional in the following form:

$$Z(j) = \int d\phi d\Phi \exp\left\{ i \int d^4x \left(-\frac{1}{2}\Phi(\Box + M^2)\Phi + \frac{g^2}{8}\phi^2(\Box + M^2)^{-1}\phi^2 \right.\right.$$
$$\left.\left. -\frac{1}{2}\phi(\Box + m^2)\phi + j\phi \right) \right\}. \tag{1.61}$$

The integration over the variable Φ in the first term gives a constant that can be included into the normalization. Expanding now the second term in the exponential, we get

$$\frac{g^2}{8} \phi^2 (\Box + M^2)^{-1} \phi^2 = \frac{g^2}{8M^2} \left(\phi^4 - \phi^2 \frac{\Box}{M^2} \phi^2 + \phi^2 \frac{\Box^2}{M^4} \phi^2 + \cdots \right). \qquad (1.62)$$

Comparing this expansion with Eq. (1.44), we may immediately read off

$$C_0 = -\frac{g^2}{8M^2}, \qquad C_1 = \frac{g^2}{8M^4}, \qquad C_2 = -\frac{g^2}{8M^6}, \qquad \cdots, \qquad (1.63)$$

and the result in Eq. (1.49) is reproduced. Of course, as we already know, C_0 and C_1 are not independent, as on the mass shell only a linear combination thereof survives. One could use the EOM in Eq. (1.62) in order to reduce the number of the independent matching conditions. There is, however, nothing wrong in using an overcomplete set of independent couplings.

It is legitimate to ask why this result is valid only at tree level, even if no approximation has been made so far. The answer to this question is that the Taylor expansion of the integrand in the path integral is not justified, since the value of the integral changes as a result of this expansion. On the other hand, at tree level, the path integral is equal just to the value of the integrand along the classical trajectory. Consequently, in this case, the expansion is justified, since an integration over ϕ is no longer performed.

A final remark is in order. It is easy to see that before Taylor-expanding, the theory with the effective Lagrangian, which contains only ϕ fields, is formally equivalent to the underlying theory to all orders in perturbation theory. The effective theory contains a vertex, $\phi^2 (\Box + M^2)^{-1} \phi^2$, and is thus nonlocal. Its high-energy behavior is, however, damped by the inverse D'Alembertian and corresponds to that of the original super-renormalizable theory. The expansion makes a local effective Lagrangian out of a nonlocal one, but at the cost of a worse behavior at high momenta. It is clear that the expansion breaks down at momenta of the order of M, and we are back to the underlying theory.

1.5 Equations of Motion and Field Redefinitions

In the previous sections, those terms in the Lagrangian, which vanish by using the EOM, have been dropped. In what follows, we shall prove that these terms do not contribute to the S-matrix and thus to physical observables, and hence the procedure is justified. Moreover, we shall prove that the two Lagrangians, which differ from each other by field redefinition, lead to the same S-matrix and thus the theories described by these Lagrangians are equivalent.[5]

In the framework of field theory, the S-matrix for a generic process $n \to m$ is obtained by using the well-known Lehmann–Symanzik–Zimmermann (LSZ) rule for the Green's function with $n + m$ external legs.[6] (For simplicity, we consider here the case of a real scalar field with a mass m, but the argument applies without modifications

[5] In what follows, we mainly follow the arguments given in Ref. [12].
[6] The LSZ formalism is considered in detail in most of the field theory textbooks; see, e.g., Ref. [13].

to other cases as well.) The S-matrix is related to the T-matrix through $S = 1 + iT$. If none of the external momenta are equal, the relation between the T-matrix element and the Green's function is given by

$$iT(p_1, \cdots, p_m; q_1, \cdots, q_n) = (iZ^{-1/2})^{n+m}$$

$$\times \lim_{p_i^2, q_j^2 \to m^2} \prod_{i=1}^{m} \theta(p_i^0)(m^2 - p_i^2) \prod_{j=1}^{n} \theta(q_j^0)(m^2 - q_j^2) G(p_1, \cdots, p_m; q_1, \cdots, q_n), \quad (1.64)$$

where G is the Fourier transform of the $n + m$-point Green's function:

$$(2\pi)^4 \delta^{(4)}(p_1 + \cdots + p_m - q_1 - \cdots - q_n) G(p_1, \cdots, p_m; q_1, \cdots, q_n)$$

$$= \int \prod_{i=1}^{m} d^4 x_i e^{ip_i x_i} \prod_{j=1}^{n} d^4 y_j e^{-iq_j y_j} \langle 0 | T\phi(x_1) \cdots \phi(x_m)\phi(y_1) \cdots \phi(y_m) | 0 \rangle. \quad (1.65)$$

Here, the symbol "T" denotes the conventional time-ordering, $TA(x)B(y) = \theta(x^0 - y^0)A(x)B(y) + \theta(y^0 - x^0)B(y)A(x)$, and Z stands for the wave function renormalization constant which is given by the residue of the two-point function at the one-particle pole:

$$D(p^2) = i \int d^4 x e^{ipx} \langle 0 | T\phi(x)\phi(0) | 0 \rangle,$$

$$D(p^2) \to \frac{Z}{m^2 - p^2}, \quad \text{as } p^2 \to m^2. \quad (1.66)$$

In other words, the Green's function contains poles in all external momenta, when the latter approach the mass shell. The generic S-matrix element is obtained from the Green's function by extracting the residue on the mass shell and multiplying by a factor $iZ^{-1/2}$ for each external leg.

The amputated Green's function Γ is defined as

$$G(p_1, \cdots, p_m; q_1, \cdots, q_n) = \prod_{i=1}^{m} D(p_i^2) \prod_{j=1}^{n} D(q_j^2) \Gamma(p_1, \cdots, p_m; q_1, \cdots, q_n), \quad (1.67)$$

and the T-matrix element can be determined from the amputated function as

$$iT(p_1, \cdots, p_m; q_1, \cdots, q_n) = (iZ^{1/2})^{n+m} \lim_{p_i^2, q_j^2 \to m^2} \Gamma(p_1, \cdots, p_m; q_1, \cdots, q_n). \quad (1.68)$$

Below, we shall demonstrate that the T-matrix element does not depend on the choice of the interpolating field. To this end, let us consider a general nonlinear local field transformation of the type

$$\phi'(x) = F[\phi(x)] = \phi(x) + a_2 \Box \phi(x) + \cdots + b_0 \phi^2(x) + b_1 \partial_\mu \phi(x) \partial^\mu \phi(x)$$

$$+ b_2 \phi(x) \Box \phi(x) + \cdots + c_0 \phi^3(x) + \cdots. \quad (1.69)$$

The single requirement is that the matrix element of the field ϕ' between the vacuum and the one-particle state $\langle 0 | \phi'(x) | p \rangle$ is different from zero. Note also that, to ease the notations, the coefficient in front of $\phi(x)$ in the r.h.s. of Eq. (1.69) is set equal to 1, since an arbitrary constant coefficient can be removed by a mere rescaling of the field that does not change the S-matrix elements.

It is clear that the Green's functions for the fields ϕ and ϕ' differ off shell. In order to extract the T-matrix element, however, we need the behavior only in the vicinity of the mass shell. Only the diagrams that are one-particle reducible in all external particles can possess poles in all external momenta and therefore contribute to the T-matrix elements. Diagrammatically, this corresponds to the situation when the particle described by the field ϕ escapes the connected part of the diagram and then turns into ϕ' without interacting with other external legs. Hence, the $n+m$ point Green's function of the fields ϕ' is given by

$$G'(p_1,\cdots,p_m;q_1,\cdots,q_n)$$
$$=\prod_{i=1}^{m}\Pi(p_i^2)D(p_i^2)\prod_{j=1}^{n}\Pi(q_j^2)D(q_j^2)\Gamma(p_1,\cdots,p_m;q_1,\cdots,q_n)+\cdots. \qquad (1.70)$$

Here, $\Pi(p^2)$ denotes the sum of all one-particle irreducible diagrams, which describe the transition of ϕ into ϕ' (see Fig. 1.7b), and the ellipses denote the regular terms. (These terms do not contain one-particle reducible diagrams in at least one of the external lines.) These regular terms do not contribute to the T-matrix, and we shall consistently omit them in the following. The key point is that the amputated Green's function Γ is the same in the two cases, simply because it contains the same set of diagrams in both cases (cf. Figs. 1.7a and 1.7b).

Moreover, the two-point function of field ϕ' is given by

$$D'(p^2) = \Pi^2(p^2)D(p^2)+\cdots, \qquad (1.71)$$

where, again, the ellipsis denotes the regular terms.

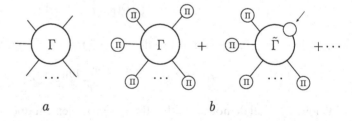

$a \qquad\qquad\qquad b$

Figure 1.7 (a) The representation of the $n+m$ point Green's function of the fields ϕ through the amputated Green's function Γ. Solid lines denote the dressed propagator of a single particle. (b) The same for the $n+m$ point Green's function of the fields ϕ'. The singular part is given by a set of the diagrams, where the single particle lines corresponding to the field ϕ are emanating from the connected part of the diagram and then turn into ϕ' without interacting with other lines. The blocks denoted by Π contain the set of all one-particle irreducible diagrams that describe this transition. An example of the regular part that does not contribute to the T-matrix also is shown. In this part the vertex indicated by the arrow cannot be separated from the rest of the diagram by cutting just one internal line. The amputated function Γ is the same in both cases, and $\tilde{\Gamma}$ does not contribute on shell.

Now, taking into account the fact that $\Pi(p^2)$ is regular in the variable p^2 because it is a sum of the one-particle irreducible diagrams, we obtain the relation between the wave function renormalization constants of the fields ϕ and ϕ':

$$Z' = Z\Pi^2(m^2). \tag{1.72}$$

Finally, using Eq. (1.70), we see that the T-matrix elements, obtained from the Green's functions of the fields ϕ and ϕ' with the use of the LSZ formula, are exactly the same, since the factors $\Pi(m^2)$, obtained for each external leg, are canceled by the same factor emerging in $(Z')^{-1/2}$. This proves the statement that the T-matrix elements do not depend on the choice of the interpolating field.

Based on this result, we can now show that the field redefinitions leave the S-matrix elements invariant. The generating functional for the Green's functions in the path integral formulation is given by

$$Z(J) = \int d\phi \exp\left(i \int d^4x \left[\mathcal{L}(\phi(x)) + J(x)\phi(x)\right]\right). \tag{1.73}$$

The Green's functions are obtained by differentiating this expression with respect to the external sources $J(x)$ and setting them to zero at the end.

Next, within this path integral, let us perform the field transformation given in Eq. (1.69). The Jacobian of this transformation is

$$\left|\frac{d\phi'}{d\phi}\right| = \exp\left\{\text{Tr}\ln\left(\frac{d\phi'}{d\phi}\right)\right\}. \tag{1.74}$$

Here, in order to simplify the notations, we shall carry out the calculation of the determinant, when the field transformation has the following form (cf. Eq. (1.69)),

$$\phi'(x) = \phi(x) + a_2 \square \phi(x) + b_0 \phi^2(x), \tag{1.75}$$

albeit the treatment is, of course, completely general. Then,

$$\frac{d\phi'}{d\phi} = \delta^{(D)}(x-y) + a_2 \square \delta^{(D)}(x-y) + 2b_0 \phi(x)\delta^{(D)}(x-y) \doteq 1 + r. \tag{1.76}$$

Here, we have anticipated that an UV regularization will be needed to calculate the determinant and write down the expression in D dimensions, setting $D \to 4$ at the end of calculations. Using $\ln(1+r) = r - r^2/2 + \ldots$, we obtain

$$\text{Tr}\ln\left(\frac{d\phi'}{d\phi}\right) = \int d^D x\, r(x,x) - \frac{1}{2}\int d^D x\, d^D y\, r(x,y) r(y,x) + \cdots,$$

$$\int d^D x\, r(x,x) = \int d^D x \left(a_2 \square \delta^{(D)}(0) + 2b_0 \phi(x)\delta^{(D)}(0)\right) = 0,$$

$$\int d^D x\, d^D y\, r(x,y) r(y,x) = \int d^D x \left(a_2^2 \square^2 \delta^{(D)}(0) + 4b_0^2 \phi(x)^2 \delta^{(D)}(0)\right.$$

$$\left. + 4a_2 b_0 \phi(x)\square \delta^{(D)}(0)\right) = 0, \tag{1.77}$$

and so on. Here, we have used the fact that, in dimensional regularization, $\delta^{(D)}(0) = \partial_\mu \delta^{(D)}(0) = \ldots = 0$. Hence, the Jacobian of the transformation is equal to unity in dimensional regularization. Of course, the physical results do not depend on the regularization. Using another regularization can be accounted for by using a different renormalization prescription.

The rest is then straightforward. Under the field transformation the generating functional turns into

$$Z(J) = \int d\phi \exp\left(i \int d^4x \left[\mathcal{L}(F[\phi(x)]) + J(x)F[\phi(x)] \right] \right), \qquad (1.78)$$

where we already took into account the fact that the Jacobian is equal to unity. This generating functional produces the Green's functions of the operator $F[\phi(x)]$ in a theory that is described by the Lagrangian $\mathcal{L}(F[\phi(x)])$. However, since the S-matrix elements do not depend on the choice of the interpolating field, the same S-matrix elements will be obtained from the generating functional

$$\tilde{Z}(J) = \int d\phi \exp\left(i \int d^4x \left[\mathcal{L}(F[\phi(x)]) + J(x)\phi(x) \right] \right). \qquad (1.79)$$

Comparing this with the original expression in Eq. (1.73), we may conclude that the theories, described by the Lagrangians $\mathcal{L}(\phi(x))$ and $\mathcal{L}(F[\phi(x)])$ (before and after the field transformations), lead to the same S-matrix and are thus equivalent to each other.

Finally, we can easily show that in the Lagrangian it is possible to consistently drop the terms that vanish by using the EOM [12]. For illustration, let us consider a theory described by the Lagrangian

$$\mathcal{L} = \frac{1}{2}\partial_\mu \phi \partial^\mu \phi - \frac{m^2}{2}\phi^2 - \frac{\lambda}{4}\phi^4. \qquad (1.80)$$

The classical EOM for this theory is

$$E[\phi(x)] = \frac{\delta S}{\delta \phi(x)} = -(\Box + m^2)\phi(x) - \lambda \phi^3(x) = 0, \qquad (1.81)$$

where S denotes the action functional.

Let us now consider an arbitrary local functional $H[\phi]$ and amend the initial Lagrangian:

$$\mathcal{L}(x) \to \mathcal{L}(x) + \varepsilon H[\phi(x)]E[\phi(x)]. \qquad (1.82)$$

In other words, the additional term vanishes for the solutions of the classical EOM. We shall now demonstrate that the amended theory leads to the same S-matrix as the original one. In order to do this, note that the infinitesimal transformation

$$\phi(x) \to \phi(x) + \varepsilon H[\phi(x)] \qquad (1.83)$$

leads to the new Lagrangian:

$$\mathcal{L}(x) \to \mathcal{L}(x) + \varepsilon H[\phi(x)]\frac{\delta S}{\delta \phi(x)} + O(\varepsilon^2) = \mathcal{L}(x) + \varepsilon H[\phi(x)]E[\phi(x)] + O(\varepsilon^2). \qquad (1.84)$$

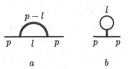

The self-energy of the light particle at one loop in the model described by the Lagrangian given in Eq. (1.41). Single and double lines correspond to the light and heavy fields, respectively.

Thus, the Lagrangians $\mathcal{L}(x)$ and $\mathcal{L}(x) + \varepsilon H[\phi(x)]E[\phi(x)]$ differ by the field transformation and hence lead to the same S-matrix. Using this procedure, we may eliminate all terms in the Lagrangian that vanish as solutions of the classical EOM.

1.6 Light Particle Mass at One Loop

Let us start the loop calculations for the simplest case of the two-point function of the light field, which is described by the Lagrangian given in (1.41). This two-point function can be written down in the following form:

$$D(p^2) = i \int d^4x e^{ipx} \langle 0|T\phi(x)\phi(0)|0\rangle = \frac{1}{m^2 - p^2 - \Sigma(p^2)}. \tag{1.85}$$

The self-energy of the light particle, $\Sigma(p^2)$, in the underlying theory at one loop is described by two diagrams shown in Fig. 1.8. We shall calculate these using dimensional regularization. The contribution of the diagram in Fig. 1.8a is given by

$$\Sigma_a(p^2) = g^2 \int \frac{d^D l}{(2\pi)^D i} \frac{1}{m^2 - l^2} \frac{1}{M^2 - (p-l)^2}. \tag{1.86}$$

In (1.86) D denotes the number of space-time dimensions. In addition, in all propagators the usual causal prescription mass$^2 \to$ mass$^2 - i\varepsilon$, with $\varepsilon \to 0^+$, is implicit.

Performing the integral with the help of the Feynman trick,

$$\frac{1}{AB} = \int_0^1 \frac{dx}{(xA + (1-x)B)^2}, \tag{1.87}$$

as $D \to 4$, we obtain

$$\Sigma_a(p^2) = -2g^2 L - \frac{g^2}{16\pi^2} \int_0^1 dx \ln \frac{xm^2 + (1-x)M^2 - x(1-x)p^2}{\mu^2}, \tag{1.88}$$

where μ denotes the scale of dimensional regularization, and

$$L = \frac{\mu^{D-4}}{16\pi^2} \left(\frac{1}{D-4} - \frac{1}{2}(\Gamma'(1) + \ln 4\pi) \right). \tag{1.89}$$

Here, $\Gamma(z)$ is the Γ-function and $\Gamma'(1) = -\gamma$, where $\gamma = 0.577215665\ldots$ denotes Euler's constant. Integrating over the variable x, we obtain

$$
\Sigma_a(p^2) = -2g^2 L - \frac{g^2}{16\pi^2}\left\{\frac{1}{2}\left(1 - \frac{M^2 - m^2}{p^2}\right)\ln\frac{m^2}{\mu^2} + \frac{1}{2}\left(1 + \frac{M^2 - m^2}{p^2}\right)\ln\frac{M^2}{\mu^2}\right.
$$

$$
\left. - \frac{\lambda^{1/2}}{2p^2}\left(\ln\frac{\frac{1}{2}\left(1 - \frac{M^2-m^2}{p^2}\right) - \frac{\lambda^{1/2}}{2p^2}}{\frac{1}{2}\left(1 - \frac{M^2-m^2}{p^2}\right) + \frac{\lambda^{1/2}}{2p^2}} - \ln\frac{-\frac{1}{2}\left(1 + \frac{M^2-m^2}{p^2}\right) - \frac{\lambda^{1/2}}{2p^2}}{-\frac{1}{2}\left(1 + \frac{M^2-m^2}{p^2}\right) + \frac{\lambda^{1/2}}{2p^2}}\right) - 2\right\}, \quad (1.90)
$$

where

$$
\lambda \doteq \lambda(p^2, m^2, M^2), \qquad \lambda(x, y, z) = x^2 + y^2 + z^2 - 2xy - 2yz - 2zx \quad (1.91)
$$

denotes the Källén triangle function. Expanding this expression for large M, we obtain

$$
\Sigma_a(p^2) = -2g^2 L - \frac{g^2}{16\pi^2}\left(-1 + \ln\frac{M^2}{\mu^2}\right)
$$

$$
- \frac{g^2}{16\pi^2 M^2}\left(m^2\ln\frac{M^2}{\mu^2} - m^2\ln\frac{m^2}{\mu^2} - \frac{p^2}{2}\right) + O(M^{-4}), \quad (1.92)
$$

where the notation $O(M^{-4})$ includes also the terms of the form $O(M^{-4}\ln^k M^2)$. We shall consistently adhere to this notation in the following.

The calculations in case of the diagram in Fig. 1.8b (the "tadpole") can be done analogously. The result is given by

$$
\Sigma_b(p^2) = \frac{g^2}{2M^2}\int\frac{d^D l}{(2\pi)^D i}\frac{1}{m^2 - l^2} = \frac{g^2 m^2}{M^2}L - \frac{g^2 m^2}{32\pi^2 M^2}\left(1 - \ln\frac{m^2}{\mu^2}\right). \quad (1.93)
$$

Adding these two expressions, we finally obtain

$$
\Sigma_a(p^2) + \Sigma_b(p^2) = -2g^2 L - \frac{g^2}{16\pi^2}\left(-1 + \ln\frac{M^2}{\mu^2}\right) + \frac{g^2 m^2}{M^2}L
$$

$$
- \frac{g^2}{16\pi^2 M^2}\left(m^2\ln\frac{M^2}{\mu^2} - \frac{3m^2}{2}\ln\frac{m^2}{\mu^2} - \frac{1}{2}(p^2 - m^2)\right) + O(M^{-4}). \quad (1.94)
$$

Next, let us answer the following question. We know that the effective Lagrangian in Eq. (1.44) reproduces all Green's functions of the underlying theory in the tree approximation. Are the results of the *loop calculations* in the underlying theory also reproduced by the loops in the effective theory, if one is using the same Lagrangian? The answer to this question is *no*, as will become clear from our calculations at one loop using the effective Lagrangian given in Eq. (1.44). Note also that, since we are calculating the Green's function and not the S-matrix element, one cannot use the EOM and eliminate the operator, which is multiplied by the coupling C_1.

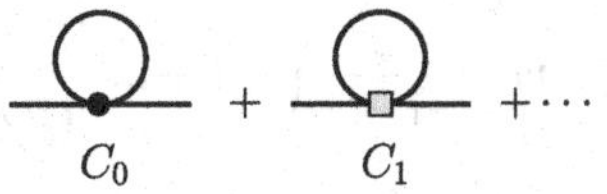

$$C_0 \qquad\qquad C_1$$

Figure 1.9 The self-energy of the light particle in the effective theory described by the Lagrangian given in Eq. (1.44). C_0 multiplies the vertex with no derivatives, C_1 the vertex with two derivatives and so on.

In the effective theory up to and including $O(M^{-2})$, only the first diagram in Fig. 1.9, which is proportional to $C_0 = -g^2/(8M^2)$, contributes. The result is given by

$$\Sigma_{\text{eff}}(p^2) = -12C_0 \int \frac{d^D l}{(2\pi)^D i} \frac{1}{m^2 - l^2} + O(M^{-4})$$

$$= -24C_0 m^2 L_{\text{eff}} + \frac{3C_0 m^2}{4\pi^2}\left(1 - \ln\frac{m^2}{\mu_{\text{eff}}^2}\right) + O(M^{-4}). \tag{1.95}$$

In this expression, μ_{eff} denotes the scale of dimensional regularization in the effective theory, which need not be the same as the one in the underlying theory, and L_{eff} is determined from Eq. (1.89) with the replacement $\mu \to \mu_{\text{eff}}$.

As clearly seen from the preceding equations, $\Sigma_a + \Sigma_b \neq \Sigma_{\text{eff}}$ at one loop. One may now ask the question, how could one modify the effective theory so that the Green's functions are the same? It can be seen that for the two-point Green's function, expanded up to the terms of order M^{-4}, this can be achieved by supplementing the effective Lagrangian by counterterms that correspond to mass and wave function renormalization:

$$\mathcal{L}_{\text{eff}} \;\to\; \mathcal{L}_{\text{eff}} + \frac{A}{2}(\partial\phi)^2 - \frac{B}{2}\phi^2,$$

$$A = \frac{g^2}{32\pi^2 M^2} + O(M^{-4}),$$

$$B = g^2\left(2L_{\text{eff}} + \frac{1}{16\pi^2}\left(\ln\frac{M^2}{\mu_{\text{eff}}^2} - 1\right)\right)$$

$$+ \frac{g^2 m^2}{M^2}\left(2L_{\text{eff}} + \frac{1}{16\pi^2}\left(\ln\frac{M^2}{\mu_{\text{eff}}^2} - 1\right)\right) + O(M^{-4}). \tag{1.96}$$

Note that A and B do not depend on the scale μ_{eff}, as well as the quantities $\Sigma_a, \Sigma_b, \Sigma_{\text{eff}}$.

The quantities of interest are, however, not the (ultraviolet divergent) coefficients of the Lagrangian, but the renormalized quantities. Using explicit expressions for the two-point function, we are now in a position to perform the matching of the renormalized masses. In order to do this, let us note that the physical mass m_P of the light particle, which is determined by the position of the pole in the two-point Green's function, should be the same in both theories. In the underlying theory, at one loop, the pole is located at

$$m^2 - m_P^2 - (\Sigma_a(m_P^2) + \Sigma_b(m_P^2)) = 0. \tag{1.97}$$

From this equation we obtain, to lowest order in g,

$$m_P^2 = m_r^2 + \frac{g^2}{16\pi^2}\left(-1 + \ln\frac{M^2}{\mu^2}\right) + \frac{g^2 m_r^2}{16\pi^2 M^2}\left(\ln\frac{M^2}{\mu^2} - \frac{3}{2}\ln\frac{m_r^2}{\mu^2}\right)$$

$$+ O(M^{-4}),\tag{1.98}$$

where m_r denotes the *running (or renormalized) mass* in the underlying theory in the modified minimal subtraction ($\overline{\text{MS}}$) scheme, which is defined through the subtraction of the divergent piece proportional to L:

$$m_r^2(\mu) = m^2 + 2g^2 L - \frac{g^2 m^2}{M^2} L.\tag{1.99}$$

Here, m denotes the bare mass in the underlying theory. Note that, at this order, it is still not necessary to consider the loop corrections of other parameters of the theory.

Since we have modified the effective Lagrangian to ensure that the Green's functions in the underlying and the effective theories coincide, the poles in both theories will be at the same place. The physical mass calculated in the effective theory is given by the solution of the following equation:

$$m^2 + B - (1 + A)m_P^2 - \Sigma_{\text{eff}}(m_P^2) = 0,\tag{1.100}$$

and takes the form

$$m_P^2 = m_{r,\text{eff}}^2 + \frac{3g^2 m_{r,\text{eff}}^2}{32\pi^2 M^2}\left(1 - \ln\frac{m_{r,\text{eff}}^2}{\mu_{\text{eff}}^2}\right) + O(M^{-4}),\tag{1.101}$$

where we used the matching condition for the constant C_0 at tree level. Further, $m_{r,\text{eff}}$ denotes the running mass in the effective field theory, which is related to the bare mass in the following manner:

$$m_{r,\text{eff}}^2(\mu_{\text{eff}}) = m_{\text{eff}}^2 - \frac{3g^2 m_{\text{eff}}^2}{M^2} L_{\text{eff}}.\tag{1.102}$$

The bare mass in the effective theory, m_{eff}, which appears in Eq. (1.102), can be read off from the effective Lagrangian,

$$\mathcal{L}_{\text{eff}} = \frac{1}{2}(\partial\phi)^2 - \frac{m^2}{2}\phi^2 + \frac{A}{2}(\partial\phi)^2 - \frac{B}{2}\phi^2 + \text{quartic terms}$$

$$= \frac{1}{2}Z_{\text{eff}}(\partial\phi)^2 - \frac{m_{\text{eff}}^2}{2}Z_{\text{eff}}\phi^2 + \text{quartic terms},$$

$$Z_{\text{eff}} = 1 + A, \qquad m_{\text{eff}}^2 = \frac{m^2 + B}{1 + A}.\tag{1.103}$$

Since observables (here the physical mass) should be the same in the underlying theory and in the effective theory, this finally gives the relation between the running mass in both theories:

$$m_{r,\text{eff}}^2(\mu_{\text{eff}}) = m_r^2(\mu) + \frac{g^2}{16\pi^2}\left(-1 + \ln\frac{M^2}{\mu^2}\right)$$

$$+ \frac{g^2 m_r^2(\mu)}{16\pi^2 M^2}\left(\ln\frac{M^2}{\mu^2} - \frac{3}{2}\left(1 + \ln\frac{\mu_{\text{eff}}^2}{\mu^2}\right)\right) + O(M^{-4}).\tag{1.104}$$

As we see, the *running masses* in both theories are not the same beyond tree approximation.[7] Moreover, these masses *run differently* with respect to the scale variations:

$$\mu \frac{dm_r^2(\mu)}{d\mu} = \frac{g^2}{8\pi^2} - \frac{g^2 m_r^2(\mu)}{16\pi^2 M^2}$$

$$\mu_{\text{eff}} \frac{dm_{r,\text{eff}}^2(\mu_{\text{eff}})}{d\mu_{\text{eff}}} = \frac{3g^2 m_{r,\text{eff}}^2(\mu_{\text{eff}})}{16\pi^2 M^2} + O(M^{-4}). \tag{1.105}$$

The above RG equations can be obtained by differentiating the expression for the physical mass with respect to the scale and setting this derivative to zero, because the physical mass does not depend on the scale. Moreover, it should be pointed out that even if the scale μ is present in Eq. (1.104), the running mass in the effective theory, $m_{r,\text{eff}}(\mu_{\text{eff}})$, in fact, does not depend on this scale. This statement can be straightforwardly checked by using the first of the equations in Eq. (1.105). This happens because Eq. (1.104) was obtained from the matching to the physical observable, which has to be scale-independent.

A few concluding remarks:

i) As we have seen, matching the two Lagrangians at tree level does not mean that the loops calculated with these Lagrangians also will match. The difference, however, can be taken away completely by renormalization. This means that both theories are physically equivalent. This is a particular case of the decoupling theorem [14], as detailed in what follows.

ii) Matching enables us to express the parameters of the effective theory in terms of the parameters of the underlying theory. What makes sense is the relation between the finite quantities, for example, between the running masses and the couplings.

iii) Both sets of the running parameters depend on their own scales (μ and μ_{eff}, respectively). The parameters of the effective theory do not depend on the underlying scale μ, if they can be determined from the matching to physical observables.

iv) Note that in the relation given by Eq. (1.104), all logarithms containing the light mass cancel. This is the manifestation of the general pattern, which states that the couplings of the effective theory do not have a nonanalytic behavior that emerges at the light scales. All of this nonanalytic behavior has to be reproduced by the loops in the effective theory. On the contrary, the parameters of the effective theory encode the short-distance dynamics and thus depend on the light mass, at most, in a polynomial form. For consistency, here we assume that the scales μ, μ_{eff} are also "hard." On the other hand, reducing μ_{eff} down to the "light" scale, the couplings will no more depend analytically on this scale. We shall observe this phenomenon explicitly in Chiral Perturbation Theory (ChPT).

---

[7] Strictly speaking, only the matching of *observables* in two theories (i.e., the masses and the S-matrix elements) is required. The two-point function is not an observable. So, in principle, one could leave the wave function renormalization constant Z_{eff} free. However, not much will change in our discussion of the physical mass if we lift the restriction on this constant.

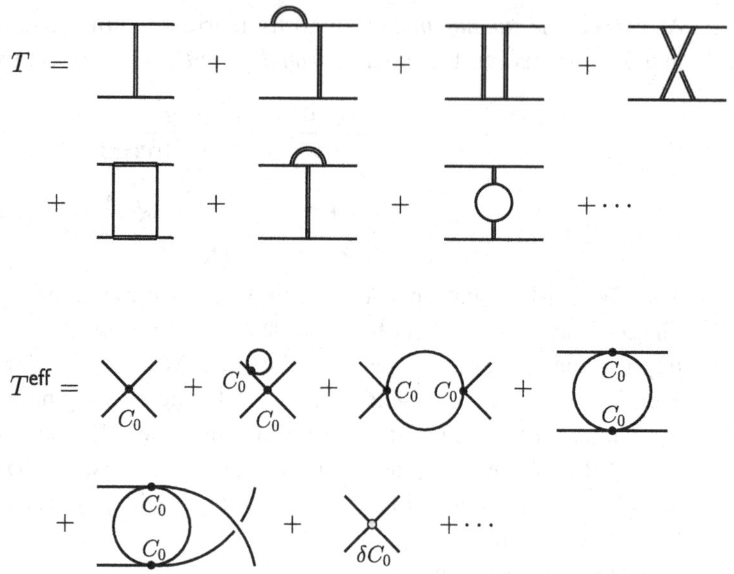

Figure 1.10 Representative set of the diagrams that contribute to the $\phi\phi \to \phi\phi$ amplitudes in the underlying (upper panel) and the effective theory (lower panel).

v) As we know, the dimensionful coupling constant g is on the order of the heavy mass M in the large-M limit. As one sees from Eq. (1.104), the running mass in the underlying theory is not protected from large loop corrections,[8] and it is driven up to the heavy scale, unless some fine-tuning is enforced. This phenomenon is closely related to the *hierarchy problem* in the Standard Model.

1.7 Matching of the Quartic Coupling at One Loop

After matching the two-point function, we turn to the Green's functions with more external legs. Matching of the $\phi\phi \to \phi\phi$ scattering amplitudes at one loop proceeds analogously. First of all, we have to calculate the scattering amplitude in the underlying theory and in the effective theory. A representative set of the diagrams is shown in Fig. 1.10. The matching condition is

$$T = T^{\text{eff}}. \tag{1.106}$$

It is seen that as a result of this matching condition, the quartic couplings in the tree-level effective Lagrangian, given by Eq. (1.44), are modified according to $C_i \to C_i + \delta C_i$. This is shown schematically in Fig. 1.10.

---

[8] The protection might arise due to the symmetries, e.g., the chiral symmetry in case of fermions. However, in the case that we are considering, there are no such symmetries.

Since the one-loop contributions to the scattering amplitude in the effective theory (see Fig. 1.10) are divergent, the modified C_i should also contain divergent parts,

$$C_i = \nu_i L_{\text{eff}} + C_i^r(\mu_{\text{eff}}),\qquad(1.107)$$

where the coefficients ν_i determine the running of the renormalized couplings $C_i^r(\mu_{\text{eff}})$ with respect to the scale μ_{eff}:

$$\mu_{\text{eff}}\frac{dC_i^r(\mu_{\text{eff}})}{d\mu_{\text{eff}}} = -\frac{\nu_i}{16\pi^2},\qquad(1.108)$$

where the ν_i are the pertinent β-functions whose explicit values are not needed here. Matching enables us to express the *renormalized* couplings $C_i^r(\mu_{\text{eff}})$ in terms of the fundamental parameters of the underlying theory. Comparing with Eq. (1.63), which contains matching at tree level, and using the fact that g has dimension of mass, we get

$$C_i^r(\mu_{\text{eff}}) = (-)^{i+1}\frac{g^2}{8M_r^{2(i+1)}(\mu_{\text{eff}})}\left\{1+\kappa_i\frac{g^2}{16\pi^2 M_r^2(\mu_{\text{eff}})}\right\},\qquad(1.109)$$

where $M_r(\mu_{\text{eff}})$ is the renormalized heavy mass in the underlying theory, and the dimensionless constants κ_i can depend only on the dimensionless arguments m_r/M_r and μ_{eff}/M_r. (Without loss of generality and in order to ease the notation, we used here $\mu = \mu_{\text{eff}}$.) In Eq. (1.109) we further took into account the fact that in the underlying (superrenormalizable) theory the coupling g is not renormalized, and we used g instead of g_r everywhere. Moreover, as became clear from the discussion in Section 1.6, the coupling constants determined from the matching cannot contain infrared singularities at $m_r \to 0$, since these singularities are the same in the underlying and in the effective theory, canceling each other in the matching condition. An example of this is the cancellation of all $\ln(m_r^2)$-terms in the matching of the two-point functions; see Section 1.6. Consequently, the κ_i are a polynomial in the variable m_r^2/M_r^2, and the dependence on this variable can be traded for the derivative terms by using the EOM in the Lagrangian;[9] see Section 1.3. On the contrary, the dependence on the second variable μ_{eff}/M_r is nonanalytic, as in perturbation theory logarithms $\ln(\mu_{\text{eff}}/M_r)$ usually appear.

Carrying out the matching at one loop is straightforward but not very enlightening, since a large number of Feynman diagrams have to be calculated. In what follows, we shall demonstrate how the same goal can be achieved within the path-integral formalism with considerably less effort. To this end, we evaluate the generating functional given in Eq. (1.61) at one loop by using the saddle-point technique. In the beginning, we carry out the integration over the field Φ. (This integration gives an uninteresting constant, which can be included in the normalization of the path integral.) Further, we expand the action functional in this integral around the *classical* solution for the

---

[9] Using the EOM is justified, since the S-matrix elements, which are used in the matching condition, do not change. One should bear in mind, however, that the off-shell behavior of the Green's function changes, if the EOM are used.

field ϕ, setting $\phi = \phi_c + \xi$. Here, the field ξ denotes a quantum fluctuation around the classical solution ϕ_c, which obeys the following EOM:

$$0 = (\Box + m^2)\phi_c(x) + j(x) + \frac{g^2}{2} \int d^4y\, \phi_c(x) D_M(x-y)\phi_c^2(y)$$

$$= (\Box + m^2)\phi_c(x) + j(x) + \frac{g^2}{2M^2} \phi_c^3(x) + \cdots, \qquad (1.110)$$

with

$$D_M(x-y) = \langle x|(\Box + M^2)^{-1}|y\rangle = \int \frac{d^4p}{(2\pi)^4} \frac{e^{-ip(x-y)}}{M^2 - p^2}$$

$$= \frac{1}{M^2} \int \frac{d^4p}{(2\pi)^4} e^{-ip(x-y)} \left(1 + \frac{p^2}{M^2} + \cdots\right)$$

$$= \frac{1}{M^2} \delta^{(4)}(x-y) - \frac{1}{M^4} \Box \delta^{(4)}(x-y) + \cdots. \qquad (1.111)$$

Retaining terms up to second order in the expansion over ξ, and taking into account the fact that $d\phi = d\xi$, the generating functional in Eq. (1.61) can be rewritten as

$$Z(j) = \int d\xi \exp\left\{i \int d^4x \left(-\frac{1}{2}\phi_c(\Box + m^2)\phi_c + \frac{g^2}{8}\phi_c^2(\Box + M^2)^{-1}\phi_c^2 + j\phi_c\right)\right\}$$

$$\times \exp\left\{i \int d^4x\, d^4y \left(-\frac{1}{2}\xi(x)H(x-y)\xi(y) + O(\xi^3)\right)\right\}, \qquad (1.112)$$

with

$$H(x-y) = (\Box + m^2 + S(x))\delta^{(4)}(x-y) - \Lambda(x-y),$$

$$S(x) = -\frac{g^2}{2}(\Box + M^2)^{-1}\phi_c^2(x),$$

$$\Lambda(x-y) = g^2\phi_c(x)\langle x|(\Box + M^2)^{-1}|y\rangle\phi_c(y). \qquad (1.113)$$

Note that there are no terms linear in ξ, because ϕ_c is the solution of the EOM that makes the action functional stationary.

Evaluating the Gaussian integral over ξ in a standard manner, we obtain

$$Z(j) = \exp\left\{i \int d^4x \left(-\frac{1}{2}\phi_c(\Box + m^2)\phi_c\right.\right.$$

$$\left.\left. + \frac{g^2}{8}\phi_c^2(\Box + M^2)^{-1}\phi_c^2 + j\phi_c\right) + iS_{\text{eff}}\right\}, \qquad (1.114)$$

where

$$S_{\text{eff}} = \frac{i}{2}\text{Tr}\ln\left((\Box + m^2 + S) - \Lambda\right) = \frac{i}{2}\text{Tr}\ln(\Box + m^2) + \frac{i}{2}\text{Tr}\left((\Box + m^2)^{-1}S\right)$$

$$- \frac{i}{4}\text{Tr}\left((\Box + m^2)^{-1}S(\Box + m^2)^{-1}S\right) - \frac{i}{2}\text{Tr}\left((\Box + m^2)^{-1}\Lambda\right)$$

$$+ \frac{i}{2} \operatorname{Tr}((\Box + m^2)^{-1} S (\Box + m^2)^{-1} \Lambda) - \frac{i}{4} \operatorname{Tr}((\Box + m^2)^{-1} \Lambda (\Box + m^2)^{-1} \Lambda) + \cdots$$

$$= T_0 + T_1 + T_2 + T_3 + T_4 + T_5 + O(g^6). \tag{1.115}$$

Here, "Tr" denotes the trace of an operator in coordinate space, that is,

$$\operatorname{Tr} A = \int d^4 x \langle x | A | x \rangle. \tag{1.116}$$

Note that T_0 is an uninteresting constant, which can be included in the normalization of the path integral. T_1 and T_3 are quadratic in the field ϕ_c and contribute to the renormalization of the two-point function of the light field. We have studied this issue in detail in Section 1.6. The remaining terms T_2, T_4 and T_5, which contribute to the renormalization of the quartic couplings, can be rewritten as

$$T_2 = \frac{g^4}{16} \int d^4 x d^4 y d^4 u d^4 v \left(-i D(u - v) D(v - u) D_M(u - x) D_M(v - y) \right) \phi_c^2(x) \phi_c^2(y),$$

$$T_4 = \frac{g^4}{4} \int d^4 x d^4 y d^4 u d^4 v (-i D(v - u) D(u - y) D_M(y - v) D_M(u - x))$$

$$\times \phi_c^2(x) \phi_c(y) \phi_c(v),$$

$$T_5 = \frac{g^4}{4} \int d^4 x d^4 y d^4 u d^4 v (-i D(v - u) D_M(u - x) D(x - y) D_M(y - v))$$

$$\times \phi_c(x) \phi_c(y) \phi_c(u) \phi_c(v), \tag{1.117}$$

where $D(x - y)$ is a light scalar propagator with a mass m. Schematically, the three quantities T_2, T_4, T_5 are depicted in Fig. 1.11.

Let us now consider the strategy for matching at one loop. First, we recall that the matching condition is altered by loop corrections, because the heavy particles are present in the loops and the Taylor expansion in the inverse powers of the heavy mass cannot be straightforwardly carried out in the Feynman integrals. Namely, let us denote $T_M\{T_i\}$, $i = 2, 4, 5$, the quantities T_i, evaluated from the same diagrams shown in Fig. 1.11, but with the Taylor-expanded heavy particle propagator,

$$\frac{1}{M^2 - l^2} \rightarrow \frac{1}{M^2} + \frac{l^2}{M^4} + \cdots. \tag{1.118}$$

Here and in what follows, the symbol "T_M" stands for the procedure of Taylor-expanding in inverse powers of the heavy mass.[10] Then, the difference,

$$\Delta T = \sum_{i=2,4,5} (T_i - T_M\{T_i\}), \tag{1.119}$$

should be compensated by adjusting the quartic coupling constants. This gives us the desired matching condition for these couplings.

[10] Note that, graphically, the operation T_M amounts to contracting the heavy propagators to one point. Consequently the diagrams describing $\phi\phi \rightarrow \phi\phi$ scattering in the effective theory arise from the diagrams T_2, T_4 and T_5, shown in Fig. 1.11.

Figure 1.11   A schematic representation of T_2, T_4 and T_5. The solid and double lines denote the light and heavy fields, respectively. The arrow points toward the one loop graph in the effective theory that is obtained from T_2, T_4, T_5 by contracting the heavy propagators.

From Fig. 1.11 we immediately conclude that T_2 will not affect the matching condition, because it does not contain heavy particles in the loops. Consequently,

$$T_2 - T_M\{T_2\} = 0 \,. \tag{1.120}$$

T_4 and T_5 will, however, affect the matching condition. Let us start with the quantity T_4. The vertex diagram, which is part of T_4 (see Fig. 1.11), is given by

$$- iD(v-u)D(u-y)D_M(y-v)$$

$$= \int \frac{d^4 p_1}{(2\pi)^4} \frac{d^4 p_2}{(2\pi)^4} \, e^{-ip_1(v-u)-ip_2(u-y)} \Gamma_v(p_1,p_2) \,, \tag{1.121}$$

with

$$\Gamma_v(p_1,p_2) = \int \frac{d^D l}{(2\pi)^D i} \frac{1}{(m^2-(p_1+l)^2)} \frac{1}{(m^2-(p_2+l)^2)} \frac{1}{(M^2-l^2)} \,. \tag{1.122}$$

Note that the second heavy propagator $D_M(u-x)$, which is outside the loop, can be expanded in inverse powers of M without much ado.

We are interested in the quantity $R_v(p_1,p_2) = \Gamma_v(p_1,p_2) - T_M\{\Gamma_v(p_1,p_2)\}$. Since the quantity $R_v(p_1,p_2)$ should be a low-energy polynomial in the small momenta p_1, p_2, one may expand it in a Taylor series:

$$R_v(p_1,p_2) = R_v(0,0) + p_1^\mu \frac{\partial}{\partial p_1^\mu} R_v(p_1,p_2) \bigg|_{p_1,p_2=0}$$

$$+ p_2^\mu \frac{\partial}{\partial p_2^\mu} R_v(p_1,p_2) \bigg|_{p_1,p_2=0} + \cdots . \tag{1.123}$$

Note that in the effective Lagrangian this expansion translates into the derivative expansion in the light fields. In order to perform matching at lowest order in the inverse heavy mass M, it suffices to retain the first term in this expansion. Generalization to higher orders is straightforward.

Calculating $\Gamma_v(0,0)$, we get, on the one hand,

$$\Gamma_v(0,0) = \int \frac{d^D l}{(2\pi)^D i} \frac{1}{(m^2-l^2)^2} \frac{1}{M^2-l^2} = \frac{1}{16\pi^2} \int_0^1 \frac{dx\, x}{xm^2+(1-x)M^2}$$

$$= \frac{1}{16\pi^2} \left(\frac{1}{m^2-M^2} - \frac{M^2}{(m^2-M^2)^2} \ln \frac{m^2}{M^2} \right)$$

$$= -\frac{1}{16\pi^2 M^2} \left(1 + \ln \frac{m^2}{M^2} \right) + O(M^{-4}) \,. \tag{1.124}$$

On the other hand,

$$T_M\{\Gamma_v(0,0)\} = \int \frac{d^D l}{(2\pi)^D i} \frac{1}{(m^2 - l^2)^2} \left\{ \frac{1}{M^2} + O(M^{-4}) \right\}$$

$$= \frac{2}{M^2} L_{\text{eff}} - \frac{1}{16\pi^2 M^2} \ln \frac{m^2}{\mu_{\text{eff}}^2} + O(M^{-4}). \tag{1.125}$$

Subtracting these two expressions gives

$$R_v(0,0) = \frac{2}{M^2} L_{\text{eff}} - \frac{1}{16\pi^2 M^2} \left(1 + \ln \frac{\mu_{\text{eff}}^2}{M^2} \right) + O(M^{-4}). \tag{1.126}$$

As expected, the nonanalytic terms proportional to $\ln m^2$ cancel in this difference. Substituting now this expression into Eqs. (1.121) and (1.117), we finally obtain

$$T_4 - T_M\{T_4\} = \frac{g^4}{4M^2} R_v(0,0) \int d^4x \, \phi_c^4(x) + O(M^{-6}). \tag{1.127}$$

The quantity T_5 can be treated analogously. Here, we need the expression of the box integral at zero momenta (see Fig. 1.11):

$$\Gamma_b(0,0) = \int \frac{d^D l}{(2\pi)^D i} \frac{1}{(m^2 - l^2)^2} \frac{1}{(M^2 - l^2)^2} = \frac{1}{16\pi^2} \int_0^1 \frac{dx \, x(1-x)}{(xm^2 + (1-x)M^2)^2}$$

$$= \frac{1}{16\pi^2} \frac{-2(M^2 - m^2) + (M^2 + m^2)\ln(M^2/m^2)}{(M^2 - m^2)^3}$$

$$= \frac{1}{16\pi^2 M^4} \left(-2 + \ln \frac{M^2}{m^2} \right) + O(M^{-6}). \tag{1.128}$$

The same integral, with the Taylor-expanded heavy propagator, is equal to

$$T_M\{\Gamma_b(0,0)\} = \int \frac{d^D l}{(2\pi)^D i} \frac{1}{(m^2 - l^2)^2} \frac{1}{(M^2)^2} = -\frac{2}{M^4} L_{\text{eff}} - \frac{1}{16\pi^2 M^4} \ln \frac{m^2}{\mu_{\text{eff}}^2}.$$
$$\tag{1.129}$$

From these equations we obtain

$$R_b(0,0) = \Gamma_b(0,0) - T_M\{\Gamma_b(0,0)\}$$

$$= \frac{2}{M^4} L_{\text{eff}} + \frac{1}{16\pi^2 M^4} \left(-2 - \ln \frac{\mu_{\text{eff}}^2}{M^2} \right) + O(M^{-6}). \tag{1.130}$$

Finally, from Eq. (1.117) we have

$$T_5 - T_M\{T_5\} = \frac{g^4}{4} R_b(0,0) \int d^4x \, \phi_c^4(x). \tag{1.131}$$

Eqs. (1.127) and (1.131) allow one to read off the matching of the low-energy constant C_0 at one loop:

$$C_0 = -\frac{g^2}{8M^2} - \frac{g^4}{4M^2} R_v(0,0) - \frac{g^4}{4} R_b(0,0) + O(M^{-6})$$

$$= -\frac{g^2}{8M^2} - \frac{g^4}{M^4} L_{\text{eff}} + \frac{g^4}{64\pi^2 M^4} \left(3 + 2\ln \frac{\mu_{\text{eff}}^2}{M^2} \right) + O(M^{-6}). \tag{1.132}$$

Figure 1.12 Renormalization of the heavy mass at one loop. The solid and double lines denote the light and heavy fields, respectively.

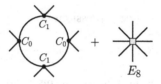

Figure 1.13 The renormalization of the Green's function with eight external legs in the effective theory. In order to cancel the divergence, a local term with eight ϕ-fields is introduced in the Lagrangian. The pertinent coupling is denoted by E_8.

In order to arrive at the final result, one has to express everything in Eq. (1.132) in terms of the renormalized couplings. As already mentioned, g is not renormalized. The quantity M^2 should, however, be renormalized (see Fig. 1.12):

$$M^2 = M_r^2(\mu_{\text{eff}}) - g^2 L_{\text{eff}}, \qquad (1.133)$$

where, without loss of generality, one may assume that the scales in the underlying and effective theories coincide, $\mu = \mu_{\text{eff}}$.

Substituting this expression into Eq. (1.132), we finally obtain

$$C_0 = -\frac{g^2}{8M_r^2} - \frac{9g^4}{8M_r^4}L_{\text{eff}} + \frac{g^4}{64\pi^2 M_r^4}\left(3 + 2\ln\frac{\mu_{\text{eff}}^2}{M_r^2}\right) + O(M_r^{-6})$$

$$= v_0 L_{\text{eff}} + C_0^r(\mu_{\text{eff}}). \qquad (1.134)$$

It is seen that $C_0^r(\mu_{\text{eff}})$ can be written in the form of Eq. (1.109). Reading off the coefficient κ_0, we get

$$\kappa_0 = -6 - 4\ln\frac{M^2}{\mu_{\text{eff}}^2} + O(M_r^{-2}). \qquad (1.135)$$

The coefficient κ_0 does not depend on the light mass m at this order. This is, however, not true in general, that is, to all orders in the expansion in the inverse powers of M_r, unless the EOMs are used.

Finally, from Eq. (1.134) we can straightforwardly ensure that the renormalized coupling constant at this order obeys the well-known RG equation in the ϕ^4 theory:

$$\mu_{\text{eff}}\frac{dC_0^r}{d\mu_{\text{eff}}} = \frac{9}{2\pi^2}(C_0^r)^2. \qquad (1.136)$$

Last but not least, it should be noted that the effective Lagrangian beyond tree level contains terms with $6, 8, \ldots \phi$-fields as well. These are needed, in particular, to cancel the divergences in the loop diagrams of the effective theory of the type shown in Fig. 1.13. Such terms emerge as a result of using the EOM in the quartic terms as well.

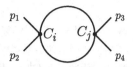

Figure 1.14 Insertion of two irrelevant vertices into the one-loop diagram. The incoming and outgoing momenta are p_1, p_2 and p_3, p_4, respectively. The total momentum is $P = p_1 + p_2 = p_3 + p_4$.

1.8 Dependence of the Effective Couplings on the Heavy Mass

In the simple model considered in the previous sections, the heavy mass M sets the hard scale at which the structure of the theory changes. For this reason, it is interesting to find out how the parameters of the low-energy theory depend on the heavy mass. First, let us consider the effective Lagrangian at tree level. We can judge about the leading behavior of these couplings in the limit $M \to \infty$ on the basis of the mass dimensions of these couplings alone. Only the effective mass of the light particle has a positive mass dimension. The coupling C_0 is dimensionless, and the couplings C_i with $i > 0$ have negative mass dimension. On dimensional grounds, the leading behavior in M in the latter couplings should be proportional to $g^2/M^{2(i+1)} \propto M^{-2i}$. Consequently, the couplings C_i, $i > 0$, fall off as negative powers of M in the limit $M \to \infty$. The dimensionless couplings are defined as $\tilde{C}_i = C_i M^{2i}$. The couplings $\tilde{C}_i$ are said to be of *natural size* if they are of order one. The dimension of the operators in the Lagrangian, which are multiplied by these couplings, is correlated with the preceding counting, in order to ensure that the Lagrangian has the correct mass dimension.

According to the mass dimension, the couplings are referred to as *relevant* (positive mass dimension), *marginal* (dimensionless) and *irrelevant* (negative mass dimension). It is seen that at low energies corresponding to the limit of a very large M, the contribution from the irrelevant couplings to the Green's functions is suppressed by powers of the large mass M.

Does the situation change beyond the tree level? Let us consider the insertion of irrelevant couplings in the loops. For simplicity, consider one loop in the effective theory with the insertion of two irrelevant vertices multiplied by the couplings C_i and C_j; see Fig. 1.14. The product of these two couplings falls off as $M^{-2(i+j)}$. Further, the mass dimension of the diagram in Fig. 1.14 is equal to zero. So, in order to obtain the required mass dimension, the preceding factor should be multiplied by $mass^{2(i+j)}$, where *mass* denotes any available mass scale in the effective theory: external momenta, effective mass or the regulator mass in the loops.

The discussion is particularly simple in dimensional regularization. The diagram in Fig. 1.14 is given by the expression

$$I_{ij} = \frac{\tilde{C}_i \tilde{C}_j}{M^{2(i+j)}} \int \frac{d^D l}{(2\pi)^D i} \frac{N(l; p_1, p_2, p_3, p_4)}{(m^2 - l^2)(m^2 - (P-l)^2)}, \qquad (1.137)$$

where the tree-level couplings $\tilde{C}_i = C_i M^{2i}$ are dimensionless and stay finite as $M \to \infty$. Further, $p_1, \ldots, p_4$ are the external momenta with $P = p_1 + p_2$, and the numerator N, which has the mass dimension $2(i+j)$, depends on the integration momentum l, the external momenta and the light mass m (at this order, one may replace the running effective mass of the light particle $m_{r,\text{eff}}$ by m). After integration, the dependence on the scale μ_{eff} appears. However, in dimensional regularization the dependence on the scale μ_{eff} is logarithmic and thus safe (i.e., the power of M in front of the integral is not changed through the multiplication by a logarithm). Loops with insertions of the irrelevant couplings are also irrelevant in the limit $M \to \infty$. Thus, irrelevant couplings can be eliminated from the theory at one loop level as well. Moreover, as seen in this example, the naive power counting is respected in dimensional regularization. The insertion of two couplings that scale as M^{-i} and M^{-j} yields a result that scales like $M^{-(i+j)}$.

The argumentation is a bit more subtle in arbitrary regularization (say, cutoff regularization), where the powers of the large regulator scale Λ_{cut} can appear. This situation also emerges if we have a multi-scale problem, with heavy particles appearing in the effective field theory loops together with light particles (one prominent example being pion–nucleon scattering in Chiral Perturbation Theory). According to the dimensional counting, the maximal power of Λ_{cut} is contained in the maximally UV-divergent piece of the integral I_{ij} in Eq. (1.137), which does not depend on the external momenta $p_1, \cdots p_4$. Denoting this maximally divergent piece by $\tilde{I}_{ij}$, we have

$$\tilde{I}_{ij} = \frac{\tilde{C}_i \tilde{C}_j}{M^{2(i+j)}} \int^{\Lambda_{\text{cut}}} \frac{d^4 l}{(2\pi)^4 i} \frac{l^{2(i+j)}}{(m^2 - l^2)^2} \sim \frac{\tilde{C}_i \tilde{C}_j \Lambda_{\text{cut}}^{2(i+j)}}{M^{2(i+j)}}. \tag{1.138}$$

In other words, this term is no more suppressed since $\Lambda_{\text{cut}} \sim M$. Note, however, that the above contribution does not depend on the external momenta and has exactly the same form as the contribution coming at tree level from the marginal vertex with the coupling C_0. Consequently, the whole contribution $\tilde{I}_{ij}$ can be removed by renormalization of C_0, which we are free to perform. We arrive at the same conclusion as earlier within dimensional regularization: the contributions from the irrelevant couplings are irrelevant at one loop as well. Thus, our results, as expected, do not depend on the regularization used, only the naive power counting holds no more in the cutoff regularization (cf. the effective theory of the potential well, Section 1.2.4).

The above arguments can be readily generalized for any number of insertions in diagrams with an arbitrary number of loops and external legs. The picture particularly simplifies in the limit $M \to \infty$, where the contributions from the irrelevant operators vanish. One arrives at the so-called decoupling theorem by Appelquist and Carazzone [14], which, loosely spoken, states that the whole contribution of the infinitely heavy degrees of freedom can be included in the renormalization of the masses and coupling constants, and the wave function renormalization constants of the light degrees of freedom.[11] In the context of the simple model considered in this section, the statement amounts to the claim that arbitrary renormalized Green's functions in the underlying and the effective theories in the limit $M_r \to \infty$ are related according to

$$G_r(p_1, \cdots, p_n; m_r, M_r, g, \mu) = Z^{n/2} G_{r,\text{eff}}(p_1, \cdots, p_n; m_{r,\text{eff}}, C_0^r, \mu) + \cdots, \quad (1.139)$$

where, for simplicity, we took $\mu_{\text{eff}} = \mu$, Z denotes the wave function renormalization constant and the ellipses stand for the terms that vanish in the limit $M_r \to \infty$.

To summarize: at low energies the contributions from the irrelevant operators in the Lagrangian to all physical observables are suppressed by inverse powers of the large scale and are thus small. Therefore, the physics at low energies is governed by a few operators with nonpositive mass dimension. This is obviously true at tree level but, as we have seen, holds as well when insertion of effective vertices in the loops is considered. This property of the effective theories has important implications when we try to establish the limits of existing theories and look for physics beyond them. As a simple example, consider QED, which is described by a Lagrangian containing only the relevant and marginal operators. Whatever the physics beyond QED is, it can be described by a string of the effective interactions in the Lagrangian, constructed with the use of the electron and photon fields. The operators with the smallest mass dimension will be least suppressed and, hence, more important at low momenta. Recalling the engineering dimensions of the fermion and photon fields, $[\psi] = [\bar{\psi}] = \frac{3}{2}$ and $[\mathscr{A}_\mu] = 1$, one may conclude that, at leading order, the effective Lagrangian can contain a single operator of dimension 5, the well-known Pauli term,

$$\delta \mathcal{L} = \frac{e}{M} \bar{\psi} \sigma_{\mu\nu} \mathscr{F}^{\mu\nu} \psi, \quad (1.140)$$

where e is the electric charge, $\mathscr{F}^{\mu\nu} = \partial^\mu \mathscr{A}^\nu - \partial^\nu \mathscr{A}^\mu$ denotes the electromagnetic field strength tensor, and $\sigma_{\mu\nu} = \frac{i}{2} [\gamma_\mu, \gamma_\nu]$. According to dimensional counting, the operator of dimension 5 in the Lagrangian should be accompanied by an inverse power of a large scale M and thus suppressed at low energies. Further, note that the operator given in Eq. (1.140) contributes to the anomalous magnetic moment of the electron, and this contribution is not contained in QED. Consequently, measuring the anomalous magnetic moment very precisely in the experiment and confronting the result with the sufficiently accurate calculations in QED, one obtains a lower bound on the scale M of the physics beyond QED; see, for example, Ref. [15]. The search of new physics at the precision frontier generally proceeds along similar patterns.

---

[11] Note that in the theories with spontaneous breaking of the symmetry, the masses of some particles are equal to the product of the vacuum expectation value of the Higgs field and the coupling constant of a given particle with the Higgs field (a nice example is the Standard Model). The heavy mass limit in such theories can mean: a) the vacuum expectation value becomes large, or b) some of the couplings become large. The decoupling theorem applies in the first case only, whereas in the second case we are dealing with the theory in the strong coupling limit, where the perturbative arguments cannot be used.

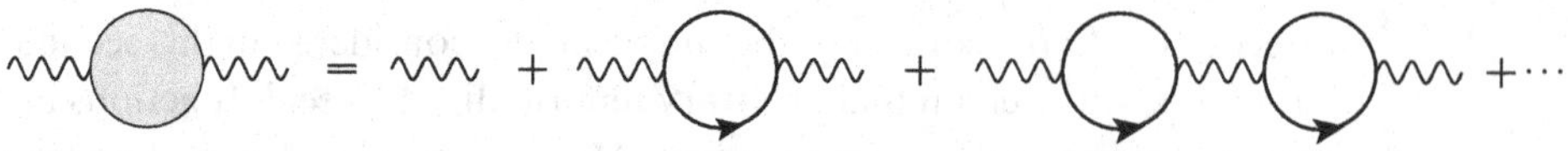

Figure 1.15 The dressed photon propagator by summing up the self-energy insertions. The wiggly and solid lines denote photons and electrons, respectively.

1.9 Decoupling in Different Renormalization Schemes

Next, we shall discuss the choice of the renormalization scheme that will prove to be very instructive. As we have seen in the previous sections, the decoupling of a heavy scale in the theory proceeds differently, if different regularizations and renormalization schemes are used (e.g., the $\overline{\text{MS}}$ scheme in dimensional regularization versus cutoff regularization). In this section, we wish to elaborate on this issue.

Consider the well-known example of charge renormalization in QED at one loop. Summing up all self-energy insertions in the photon propagator (see Fig. 1.15), it is seen that the latter obeys the Dyson–Schwinger equation:

$$D_{\mu\nu}(p) = D^0_{\mu\nu}(p) + iD^0_{\mu\lambda}(p)e^2\Pi^{\lambda\rho}(p)D_{\rho\nu}(p). \qquad (1.141)$$

For simplicity, we choose the Feynman gauge, where the free photon propagator is given by $D^0_{\mu\nu}(p) = -g_{\mu\nu}/p^2$. Furthermore, the one-loop self-energy is given by

$$\begin{aligned}
e^2\Pi_{\mu\nu}(p) &= i(p_\mu p_\nu - p^2 g_{\mu\nu})e^2\Pi(p^2) \\
&= \frac{ie^2}{2\pi^2}(p_\mu p_\nu - p^2 g_{\mu\nu})\left\{-\frac{16\pi^2}{3}L - \int_0^1 dx\, x(1-x)\ln\frac{m_e^2 - p^2 x(1-x)}{\mu^2}\right\},
\end{aligned} \qquad (1.142)$$

where m_e and e denote the electron mass and charge, respectively, and μ is the scale of dimensional regularization. The UV-divergent quantity L is defined in Eq. (1.89).

The solution of the Dyson–Schwinger equation takes the form

$$D_{\mu\nu}(p) = -\left(g_{\mu\nu} - \frac{p_\mu p_\nu}{p^2}\right)\frac{1}{p^2(1 + e^2\Pi(p^2))} - \frac{p_\mu p_\nu}{p^4}. \qquad (1.143)$$

As seen from Eq. (1.142), the quantity $\Pi(p^2)$ is ultraviolet-divergent. This divergence has to be "eaten up" by the charge renormalization. Namely, the bare charge e is also divergent. The renormalized charge and renormalized self-energy are defined by requiring that

$$\frac{e^2}{1 + e^2\Pi(p^2)} = \frac{e_{\text{ren}}^2}{1 + e_{\text{ren}}^2\Pi_{\text{ren}}(p^2)}. \qquad (1.144)$$

The method to remove the divergences in $\Pi_{\text{ren}}(p^2)$ and, hence, the definition of e_{ren}, depends on the renormalization prescription used. In what follows we shall compare two different renormalization schemes. (The indices "r" and "R" are used to distinguish between these schemes):

a) *The $\overline{MS}$ scheme:* In this scheme, the renormalized self-energy is obtained by just dropping the term proportional to L:

$$e^2\Pi_r(p^2;\mu^2) = -\frac{e^2}{2\pi^2}\int_0^1 dx\, x(1-x)\ln\frac{m_e^2 - p^2x(1-x)}{\mu^2}. \qquad (1.145)$$

b) *The MOM scheme:* In this scheme, the renormalized self-energy is obtained by a subtraction at $p^2 = \mu_0^2$ with $\mu_0^2 < 0$:

$$e^2\Pi_R(p^2;\mu_0^2) = e^2\Pi(p^2) - e^2\Pi(\mu_0^2) = -\frac{e^2}{2\pi^2}\int_0^1 dx\, x(1-x)\ln\frac{m_e^2 - p^2x(1-x)}{m_e^2 - \mu_0^2x(1-x)}. \qquad (1.146)$$

In both cases, the divergent quantity, which is subtracted, is a constant independent of p^2. Writing down $\Pi(p^2) = \Pi_{\text{div}} + \Pi_{\text{ren}}(p^2)$, we get

$$e_{\text{ren}}^2 = e^2(1 - e^2\Pi_{\text{div}} + O(e^4)). \qquad (1.147)$$

Furthermore, Π_{div} and $\Pi_{\text{ren}}(p^2)$ both depend on the renormalization scale (μ in the $\overline{MS}$ scheme, μ_0 in the MOM scheme), whereas their sum does not. Hence, e_{ren} also depends on this scale, since the bare parameter, e, does not. The dependence on the scale is described by the RG equations:

(a) *$\overline{MS}$ scheme:*

$$\mu\frac{de_r(\mu)}{d\mu} = \beta_r(e_r(\mu)),$$

$$\beta_r(e) = \frac{e}{2}\mu\frac{d}{d\mu}\Pi_r(p^2;\mu^2) = \frac{e^3}{12\pi^2}. \qquad (1.148)$$

(b) *MOM scheme:*

$$\mu_0\frac{de_R(\mu_0)}{d\mu_0} = \beta_R(e_R(\mu_0)),$$

$$\beta_R(e) = \frac{e}{2}\mu_0\frac{d}{d\mu_0}\Pi_R(p^2;\mu_0^2) = -\frac{e^3}{4\pi^2}\int_0^1 \frac{dx\, x^2(1-x)^2\mu_0^2}{m_e^2 - \mu_0^2x(1-x)}. \qquad (1.149)$$

It is instructive to study two limiting cases, with m_e much larger and much smaller than the renormalization scale. We have:

$$\beta_r(e) = \frac{e^3}{12\pi^2}, \quad \text{all values of } m_e,$$

$$\beta_R(e) = \frac{e^3}{12\pi^2}, \quad \text{if } m_e \ll \mu_0, \quad \text{the same result as in the } \overline{MS} \text{ scheme,}$$

Figure 1.16   Vacuum polarization correction to the one-photon exchange diagram in electron–electron scattering. The shaded circle represents the full propagator.

$$\beta_R(e) = -\frac{e^3 \mu_0^2}{60\pi^2 m_e^2} \to 0, \quad \text{if } m_e \gg \mu_0, \quad \text{decoupling, approaches a free theory}.$$

$$(1.150)$$

Does this result mean that the decoupling of the heavy scale occurs only within the MOM renormalization scheme? Of course not, as can be seen from the discussion in Section 1.6. The lesson to be learned here is different. The decoupling is explicit, if everything is expressed in terms of *low-energy quantities*. Such a low-energy quantity is, for example, the physical charge, which can be defined as follows. Consider the scattering amplitude of two electrons, with the one-photon exchange diagram modified by an electron loop;[12] see Fig. 1.16:

$$T_{ee \to ee} = \bar{u}(p_1', s_1') \gamma^\mu u(p_1, s_1) \frac{e_{\text{ren}}^2}{-q^2(1 + e_{\text{ren}}^2 \Pi_{\text{ren}}(q^2))} \bar{u}(p_2', s_2') \gamma_\mu u(p_2, s_2), \quad (1.151)$$

where p_i, s_i and p_i', s_i' ($i = 1, 2$) denote the momenta and the spins of the electrons in the initial and in the final state, respectively, and $q = p_1' - p_1 = p_2' - p_2$. To simplify the discussion, we do not consider the second diagram, which is obtained by a permutation of the two electrons in the initial or in the final state.

Consider now these expressions at very low momenta, $\mathbf{p}_i^2 \ll m_e^2$ and $\mathbf{p}_i'^2 \ll m_e^2$. It is easily seen that

$$p_i^0 = \sqrt{m_e^2 + \mathbf{p}_i^2} = m_e + \frac{\mathbf{p}_i^2}{2m_e} + \cdots = m_e + O(m_e^{-1}). \quad (1.152)$$

Similar relations hold for $p_i'^0$. Further, $q^2 = (p_i'^0 - p_i^0)^2 - (\mathbf{p}_i' - \mathbf{p}_i)^2 = -(\mathbf{p}_i' - \mathbf{p}_i)^2 + O(m_e^{-1})$. The nonrelativistic reduction of the Dirac spinors takes the form

$$\bar{u}(p', s') \gamma^\mu u(p, s) = \bar{u}(0, s') \frac{\not{p}' + m_e}{\sqrt{p'^0 + m_e}} \gamma^\mu \frac{\not{p} + m_e}{\sqrt{p^0 + m_e}} u(0, s)$$

$$= (2m_e) g^{\mu 0} \delta_{s's} (1 + O(m_e^{-1})). \quad (1.153)$$

According to this, the amplitude at low momenta is given by

$$T_{ee \to ee} = (2m_e)^2 \delta_{s_1' s_1} \delta_{s_2' s_2} \frac{e_{\text{ren}}^2}{\mathbf{q}^2(1 + e_{\text{ren}}^2 \Pi_{\text{ren}}(-\mathbf{q}^2))} (1 + O(m_e^{-1})). \quad (1.154)$$

---

[12] According to the Ward identity, only vacuum polarization contributes to charge renormalization.

On the other hand, in the Born approximation, this amplitude is proportional to the interaction potential between two electrons,[13] which takes into account the vacuum polarization effect. Dropping the overall normalization factor and spin indices, we get

$$V(r) = \int \frac{d^3\mathbf{q}}{(2\pi)^3} e^{i\mathbf{q}\mathbf{r}} \frac{e_{\text{ren}}^2}{\mathbf{q}^2(1 + e_{\text{ren}}^2 \Pi_{\text{ren}}(-\mathbf{q}^2))}. \tag{1.155}$$

Neglecting the correction in the denominator, the static Coulomb potential is obtained from this expression.

We fix the parameters of the theory (here, the electric charge) at large distances, that is, by measuring the force acting on small charged oil droplets (Millikan-type experiment). The elementary charge measured in this manner corresponds to $\alpha = e_{\text{phys}}^2/(4\pi) \simeq 1/137$, the fine-structure constant. Since the distances in such an experiment are much larger than the Compton wavelength of the electron, in momentum space we are focusing on the region $\mathbf{q}^2 \to 0$. In this region, the modified potential (1.155) asymptotically coincides with the Coulomb potential at large distances. This gives

$$V(r) = \int \frac{d^3\mathbf{q}}{(2\pi)^3} e^{i\mathbf{q}\mathbf{r}} \frac{e_{\text{phys}}^2}{\mathbf{q}^2(1 + e_{\text{phys}}^2(\Pi_{\text{ren}}(-\mathbf{q}^2) - \Pi_{\text{ren}}(0)))},$$

$$e_{\text{phys}}^2 = \frac{e_{\text{ren}}^2}{(1 + e_{\text{ren}}^2 \Pi_{\text{ren}}(0))}. \tag{1.156}$$

For the different renormalization schemes we have the following:

(a) $\overline{MS}$ scheme:

$$e_r^2(\mu) = \frac{e_{\text{phys}}^2}{1 + \dfrac{e_{\text{phys}}^2}{12\pi^2} \ln \dfrac{m_e^2}{\mu^2}}. \tag{1.157}$$

(b) MOM scheme:

$$e_R^2(\mu_0) = \frac{e_{\text{phys}}^2}{1 + \dfrac{e_{\text{phys}}^2}{2\pi^2} \displaystyle\int_0^1 dx\, x(1-x) \ln\left(1 - \dfrac{\mu_0^2}{m_e^2} x(1-x)\right)}. \tag{1.158}$$

Differentiating $e(\mu)$ and $e(\mu_0)$ with respect to μ and μ_0, respectively, and taking into account that the quantity e_{phys} is a physical observable that is scale-independent, we again arrive at the RG equations, (1.148) and (1.149). Finally, expressing everything in terms of e_{phys}, the modified Coulomb potential takes the form

$$V(r) = \int \frac{d^3\mathbf{q}}{(2\pi)^3} e^{i\mathbf{q}\mathbf{r}} \frac{e_{\text{phys}}^2}{\mathbf{q}^2(1 + e_{\text{phys}}^2 F(\mathbf{q}^2))},$$

---

[13] Strictly speaking, this is the potential energy of two electrons and not the static electromagnetic potential. We shall, however, use this short term in the following and hope that it does not lead to any confusion.

$$F(\mathbf{q}^2) = \frac{1}{2\pi^2} \int_0^1 dx\, x(1-x) \ln\left(1 + \frac{\mathbf{q}^2}{m_e^2} x(1-x)\right). \tag{1.159}$$

Note that the quantity $F(\mathbf{q}^2)$ is scale-independent and is the same in both regularizations. The decoupling is explicit as $F(\mathbf{q}^2) \to 0$ for $m_e^2 \to \infty$. Thus, the whole difference between the two regularizations is hidden in Eqs. (1.157) and (1.158), which describe how the renormalized charge $e_r(\mu)$ and $e_R(\mu_0)$ behave at $m_e \to \infty$, when e_{phys} is fixed. This behavior is different. Namely, $e_R(\mu_0) \to e_{\text{phys}}$, meaning that $e_R(\mu_0)$ stays a perfect low-energy quantity in this limit. On the contrary, the limit $m_e \to \infty$ cannot be performed at a fixed e_{phys} and μ in the quantity $e_r(\mu)$, because of the large logarithms $\ln(m_e^2/\mu^2)$ in perturbation theory. In order to suppress these logarithms, one has to take $\mu \sim m_e$, meaning that one is fixing the charge at a scale of order m_e. Thus, $e_r(\mu)$ is not a quantity defined at low energy, and the decoupling is not explicit if the expressions are written in terms of $e_r(\mu)$ instead of e_{phys}.

1.10 Floating Cutoff

In Section 1.8 we gave arguments in favor of the conclusion that at low energies, only superrenormalizable and renormalizable interactions, described by relevant and marginal operators in the Lagrangian, survive, whereas the contributions from irrelevant operators are suppressed by the powers of a large mass. In this section, we address this issue from a different point of view, using a method that is based on the ideas of Wilson's renormalization group [16]. In this method the high-frequency modes in the generating functional of the theory are systematically integrated out. The discussion here closely follows Polchinski's original paper [17]; see also [18].

We do not want to focus on any particular model. To this end we shall interpret M merely as some hard scale of the theory, after which the unknown physics starts, be this a new particle with a mass M, nonlocal effects, or whatever. Further, in order to make the arguments maximally transparent, here we shall use a momentum cutoff instead of dimensional regularization. Consider, for simplicity, a theory with a single scalar field ϕ. The Euclidean generating functional for the *renormalized Green's functions* in momentum space is given by

$$Z(J) = \int [d\phi]_M \exp\left\{S(\phi, C_i(M)) + J\phi\right\}, \tag{1.160}$$

where $[d\phi]_M$ denotes the path integral measure with a cutoff on the high-frequency modes with $p \sim M$. This shorthand notation should be interpreted as follows: calculating Eq. (1.160) in perturbation theory, a cutoff at the momentum scale of order M is introduced in all Feynman graphs. Note that the explicit form of the cutoff does not play a role. Further, $S(\phi, C_i(M))$ is the action functional that contains the *bare* coupling constants $C_i(M)$ corresponding to the cutoff at a scale M. In addition, in a renormalizable theory (in a conventional sense), it is possible to choose $C_i(M)$ so that the generating functional remains finite as $M \to \infty$. This is not possible in a non-renormalizable theory, but everything is perfectly well defined if the cutoff stays finite.

Now, let us ask the question how the action depends on the cutoff when the latter varies from M to some lower value Λ_{eff}, where our effective field theory is defined. This scale obeys the inequality $m \ll \Lambda_{\text{eff}} \ll M$. (Here, m denotes the light scale in the theory, for example, the mass of a light particle described by the field ϕ.) More precisely, we consider a smooth change of a cutoff from M to Λ_{eff}, defining a scale $\Lambda_{\text{eff}} \leq \Lambda \leq M$. The Euclidean path integral is given by Eq. (1.160), with M replaced by Λ.

Here, we are only interested in the low-frequency modes; thus, we assume that

$$J(p) = 0 \qquad \text{for} \quad p^2 > \Lambda_{\text{eff}}^2. \tag{1.161}$$

The crucial point is that, in order to ensure that the renormalized Green's functions do not depend on Λ, the effective action $S(\phi, C_i(\Lambda))$ should obey certain *RG flow equations*. In other words, the effective couplings that enter $S(\phi, C_i(\Lambda))$ should depend on Λ in a manner that compensates the explicit Λ-dependence coming from the cutoff. For example, the mass parameters at two scales in the theory with the interaction Lagrangian $\mathcal{L}_I = C_0 \phi^4$ at one loop are related by

$$m^2(\Lambda) = m^2(M) - 12C_0 \int^M \frac{d^D l}{(2\pi)^D} \frac{1}{m^2 + l^2}\bigg|_{\text{Eucl.}}$$

$$+ 12C_0 \int^\Lambda \frac{d^D l}{(2\pi)^D} \frac{1}{m^2 + l^2}\bigg|_{\text{Eucl.}}$$

$$= m^2(M) - \frac{3C_0}{4\pi^2}\left(M^2 - \Lambda^2 - m^2 \ln \frac{M^2}{\Lambda^2} + \cdots\right),$$

$$\Lambda \frac{d}{d\Lambda} m^2(\Lambda) = \frac{3C_0}{2\pi^2} \Lambda^2 \left(1 + O\left(\frac{m^2}{\Lambda^2}\right)\right), \tag{1.162}$$

where the momentum integrals are evaluated in Euclidean space. Note that, in order to obtain these equations, the tadpole diagram (the first diagram in Fig. 1.9) has been evaluated. In general, at a scale Λ, the effective Lagrangian includes the contributions from all momenta $\Lambda < p < M$, which emerge through the loops. Thus, the Lagrangian at lower scales necessarily contains all derivative vertices, even if the theory did not contain non-renormalizable operators at $\Lambda = M$ at all. The RG flow equation tells us that these will be generated at lower scales. For dimensional reasons these operators $\phi^2 \Box \phi^2$, $\phi^2 \Box^2 \phi^2$, $\cdots$ will be suppressed by the respective powers of Λ if $m \ll \Lambda$ still holds. The first-order differential equations, analogous to one given in Eq. (1.162), emerge for the couplings $C_i(\Lambda)$. These are nothing but the conventional RG equations.

In the following it will be useful to consider the mass as one of the couplings $C_i(\Lambda)$. These couplings define a point $C = \{C_i(\Lambda)\}$ in the (infinite-dimensional) parameter space, which moves along some trajectory when Λ changes from M to Λ_{eff}. This trajectory is defined by the RG equations and the boundary conditions at $\Lambda = M$.

It is convenient to define the dimensionless couplings, $\tilde{C}_i(\Lambda)$, according to

$$C_i(\Lambda) = \Lambda^{\Delta_i} \tilde{C}_i(\Lambda). \tag{1.163}$$

If Δ_i is positive, zero or negative, the pertinent operators are referred to as relevant, marginal and irrelevant couplings, respectively.

The RG equations for the dimensionless couplings are written as follows:

$$\Lambda\frac{d}{d\Lambda}\tilde{C}_i(\Lambda) + \Delta_i\tilde{C}_i(\Lambda) = \beta_i(\tilde{C}(\Lambda)), \qquad (1.164)$$

where the β_i denote the pertinent β-functions. In what follows, we shall demonstrate that, if the couplings are sufficiently small (i.e., allowing for a perturbative treatment), for a completely arbitrary boundary condition at $\Lambda = M$, the system at a lower scale $\Lambda = \Lambda_{\text{eff}}$ lives on a surface with a dimension given by a total number of the relevant and marginal operators only. In other words, systems with very different values of the irrelevant couplings at a high scale behave similarly at lower scales. Stated differently, the information about the high scale is erased during the RG evolution. Only the relevant and marginal couplings survive.

In order to understand how this happens, consider a toy example with two couplings: a marginal one, $C_0(\Lambda)$, and an irrelevant one, $C_1(\Lambda)$, with a mass dimension equal to -2. We set, for simplicity, all other couplings to zero. The RG equations for the couplings C_0, C_2 take the form

$$\Lambda\frac{d\tilde{C}_0}{d\Lambda} = \beta_0(\tilde{C}_0, \tilde{C}_1),$$

$$\Lambda\frac{d\tilde{C}_1}{d\Lambda} - 2\tilde{C}_1 = \beta_1(\tilde{C}_0, \tilde{C}_1). \qquad (1.165)$$

The boundary conditions are fixed at $\Lambda = M$:

$$\tilde{C}_i(\Lambda)\big|_{\Lambda = M} = \tilde{C}_i^{(0)}, \qquad i = 0, 1. \qquad (1.166)$$

Let us now assume that the pair $(\bar{C}_0, \bar{C}_1)$ is some solution of the preceding equations. Consider a small deviation $\tilde{C}_i^{(0)} \to \tilde{C}_i^{(0)} + \delta\tilde{C}_i^{(0)}$ in the boundary conditions. Then, the solutions will also change. We denote $\varepsilon_i = \tilde{C}_i - \bar{C}_i$ and note that the RG equations linearize for ε_i:

$$\Lambda\frac{d\varepsilon_0}{d\Lambda} = \overline{\frac{\partial\beta_0}{\partial\tilde{C}_0}}\varepsilon_0 + \overline{\frac{\partial\beta_0}{\partial\tilde{C}_1}}\varepsilon_1,$$

$$\Lambda\frac{d\varepsilon_1}{d\Lambda} - 2\varepsilon_1 = \overline{\frac{\partial\beta_1}{\partial\tilde{C}_0}}\varepsilon_0 + \overline{\frac{\partial\beta_1}{\partial\tilde{C}_1}}\varepsilon_1, \qquad (1.167)$$

where the bar means that the partial derivatives are evaluated at $(\bar{C}_0, \bar{C}_1)$. The term $-2\varepsilon_1$ in Eq. (1.167) is crucial in our discussion, since it will cause a damping of the variation of the deviations in the ε_1-direction.

The equations for $\varepsilon_0, \varepsilon_1$ can be decoupled by introducing a new variable,

$$\zeta_1 = \varepsilon_1 - \varepsilon_0\frac{d\bar{C}_1/d\Lambda}{d\bar{C}_0/d\Lambda}. \qquad (1.168)$$

The physical meaning of the variable ζ_1 is the distance between two neighboring trajectories in the $\tilde{C}_0, \tilde{C}_1$ plane along the $\tilde{C}_1$ axis; see Fig. 1.17. As already mentioned, the

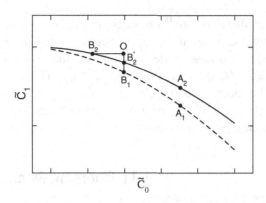

Figure 1.17 Two neighboring trajectories in the $\tilde{C}_0, \tilde{C}_1$ plane. Let the points A_1, A_2 correspond to a pair $(\tilde{C}_0, \tilde{C}_1)$ at some Λ. The distance between the curves is just the length of $|A_1 A_2|$. When $\Lambda \to \Lambda + \Delta\Lambda$, then A_1 moves to B_1 and A_2 to B_2. The vertical distance between two curves in this case is, however, given by the length of $|B_1 B_2'| = |OB_1| - |OB_2'|$, which coincides with variable ζ_1 defined in Eq. (1.168).

RG equation for this quantity decouples. Indeed, using Eqs. (1.165) and (1.167), it can be straightforwardly shown that ζ_1 obeys the following equation:

$$\Lambda \frac{d\zeta_1}{d\Lambda} - 2\zeta_1 = \left(\overline{\frac{\partial \beta_1}{\partial \tilde{C}_1}} + \overline{\frac{\partial \beta_0}{\partial \tilde{C}_0}} - \Lambda \frac{d}{d\Lambda} \ln \bar{\beta}_0 \right) \zeta_1 . \tag{1.169}$$

In perturbation theory, the r.h.s. of this equation is small, and the RG evolution of ζ_1 is governed by the term $-2\zeta_1$ in the l.h.s. of the equation, determined by the mass dimension of the irrelevant coupling. Hence, the solution of the equation at $\Lambda = \Lambda_{\text{eff}}$ to lowest order in perturbation theory is given by

$$\zeta_1(\Lambda_{\text{eff}}) = \zeta_1(M) \left(\frac{\Lambda_{\text{eff}}^2}{M^2} \right) . \tag{1.170}$$

This means that all RG trajectories approach each other in the infrared, and there is one essential parameter left instead of two: The value of $\tilde{C}_1$ is predicted, given the value of $\tilde{C}_0$. Moreover, the value of the parameter $\tilde{C}_0$ is also not independent, it just marks the place where we are on a single trajectory in the infrared. In other words, $\tilde{C}_0$ can be traded for the scale Λ_{eff}. Let us stress once more that the validity of this result is restricted to the perturbative regime, where the small corrections are supposed not to change the leading-order behavior. Strong non-perturbative effects might invalidate some arguments used in the proof and lead to the modification of the results.[14]

Imagine now that we arbitrarily change input values of couplings $\tilde{C}_0, \tilde{C}_1$ at the scale $\Lambda = M$. The curves will still converge in the infrared. Moreover, the resulting curve will be defined by the same RG equation for a single coupling $\tilde{C}_0$, since $\tilde{C}_1$ is a function of $\tilde{C}_0$ in the infrared limit. Thus, adjusting $\tilde{C}_0(M), \tilde{C}_1(M)$ so that $\tilde{C}_0(\Lambda)$ at $\Lambda = \Lambda_{\text{eff}}$ takes the

---

[14] A nice example for this is provided by the Thirring model; see, e.g., the discussion in Ref. [19].

same value as before,[15] one ends up with the same curve below Λ_{eff}. Stated differently, different theories at the scale M converge to the same theory with a single coupling $\tilde{C}_0(\Lambda_{\text{eff}})$ at a lower scale, which means that some information, which was present at the scale M, gets erased at the scale Λ_{eff}. This is exactly the essence of decoupling. Furthermore, note that $C_1(\Lambda_{\text{eff}}) = \tilde{C}_1(\Lambda_{\text{eff}})/\Lambda_{\text{eff}}^2$ is still power-suppressed in the low-energy region, if the cutoff Λ_{eff} is taken much larger than the light particle mass and the characteristic momenta.

1.11 Emergent Symmetries

The fact that only a few couplings survive in the infrared region can have far-reaching implications. Namely, if we set all couplings that become irrelevant at low energies to zero, the resulting Lagrangian might have a higher symmetry than the original one, containing, for example, a whole string of operators with increasing mass dimension. Consequently, one speaks of an *emergent symmetry*. This means that the theory at low energies exhibits a symmetry that was not present in the original Lagrangian at high energies. We note that the content of this section is somewhat speculative and it might be skipped upon first reading.

Below, we will demonstrate this effect within a simple model for a global symmetry. We shall follow the perturbative approach outlined in Refs. [20, 21]. The Lagrangian of the model is given by

$$\mathcal{L} = \frac{1}{2}(\partial\phi_1)^2 + \frac{1}{2}(\partial\phi_2)^2 - \frac{m^2}{2}(\phi_1^2 + \phi_2^2) - \frac{\lambda}{4}(\phi_1^4 + \phi_2^4) - \frac{g}{2}\phi_1^2\phi_2^2$$

$$+ \text{counter terms}. \tag{1.171}$$

Here, $\phi_{1,2}$ are real scalar fields. The preceding Lagrangian possesses a discrete symmetry with respect to $\phi_{1,2} \to -\phi_{1,2}$ and $\phi_1 \leftrightarrow \phi_2$. If $\lambda = g$, however, the theory becomes invariant under the continuous $O(2)$ group

$$\phi_1 \to \phi_1 \cos\theta - \phi_2 \sin\theta, \qquad \phi_2 \to \phi_1 \sin\theta + \phi_2 \cos\theta, \tag{1.172}$$

where θ is an angle. Let us now consider the renormalization group running for the couplings of the Lagrangian. To this end, we shall first calculate the pertinent β-functions at one loop, using the saddle-point method in the path integral. Expanding the action functional in the vicinity of the classical solution of the EOM,

$$\phi_{1,2}(x) = \phi_{1,2}^c(x) + \xi_{1,2}(x), \tag{1.173}$$

we get

$$\int d^4x \mathcal{L}(x) = \int d^4x \left(\mathcal{L}^c(x) + \frac{1}{2}\Xi^T\hat{D}\Xi + O(\xi^3)\right), \tag{1.174}$$

---

[15] We have two couplings $\tilde{C}_0(M), \tilde{C}_1(M)$ at our disposal, to adjust a single coupling $\tilde{C}_0(\Lambda_{\text{eff}})$. It is evident that some residual freedom at the scale M remains after matching, performed at a lower scale.

where $\mathcal{L}^c(x)$ is the Lagrangian evaluated with the classical solution $\phi_{1,2}^c(x)$, $\Xi^T = (\xi_1, \xi_2)$, and

$$\hat{D} = \begin{pmatrix} -(\Box + m^2) - 3\lambda(\phi_1^c)^2 - g(\phi_2^c)^2 & -2g\phi_1^c\phi_2^c \\ -2g\phi_1^c\phi_2^c & -(\Box + m^2) - 3\lambda(\phi_2^c)^2 - g(\phi_1^c)^2 \end{pmatrix}$$

$$= \hat{D}_0 + \hat{D}_1, \quad \hat{D}_0 = \begin{pmatrix} -(\Box + m^2) & 0 \\ 0 & -(\Box + m^2) \end{pmatrix}. \tag{1.175}$$

Note also that the linear terms in $\xi_{1,2}$ are absent in Eq. (1.174) because the $\phi_{1,2}^c(x)$ obey the EOM. The path integral can now be evaluated in the semiclassical approximation:

$$\int d\phi_1 d\phi_2 \exp\left[i\int d^4x \mathcal{L}(x)\right] = \int d\xi_1 d\xi_2 \exp\left[i\int d^4x\left(\mathcal{L}^c(x) + \frac{1}{2}\Xi^T\hat{D}\Xi\right)\right]$$

$$= (\det \hat{D})^{-1/2} \exp\left[i\int d^4x \mathcal{L}^c(x)\right]$$

$$= \exp\left[i\left(\int d^4x \mathcal{L}^c(x) + \frac{i}{2}\operatorname{tr}\ln\hat{D}\right)\right]. \tag{1.176}$$

Expanding the logarithm in powers of the fields $\phi_{1,2}^c$,

$$\frac{i}{2}\operatorname{tr}\ln\hat{D} = \frac{i}{2}\operatorname{tr}\ln(\hat{D}_0 + \hat{D}_1)$$

$$= \frac{i}{2}\operatorname{tr}\ln\hat{D}_0 + \frac{i}{2}\operatorname{tr}(\hat{D}_0^{-1}\hat{D}_1) - \frac{i}{4}\operatorname{tr}(\hat{D}_0^{-1}\hat{D}_1\hat{D}_0^{-1}\hat{D}_1) + \cdots, \tag{1.177}$$

it is immediately seen that the renormalization of the quartic couplings can be read off from the last term in this equation. This term can be rewritten in the following form:

$$-\frac{i}{4}\operatorname{tr}(\hat{D}_0^{-1}\hat{D}_1\hat{D}_0^{-1}\hat{D}_1) = -\frac{i}{4}\int d^4x d^4y D(x-y)D(y-x)\operatorname{tr}(\hat{D}_1(y)\hat{D}_1(x)), \tag{1.178}$$

with

$$D(x-y) = \langle x|(\Box + m^2)|y\rangle = \int \frac{d^4k}{(2\pi)^4}\frac{e^{-ik(x-y)}}{m^2 - k^2}. \tag{1.179}$$

Then, using translational invariance, one may write

$$D(x-y)D(y-x) = i\int \frac{d^4k_1}{(2\pi)^4}e^{-ik_1(x-y)}\int \frac{d^4k_2}{(2\pi)^4 i}\frac{1}{(m^2 - (k_1+k_2)^2)(m^2 - k_2^2)}. \tag{1.180}$$

The integral over k_2 is UV-divergent. For clarity, let us use cutoff regularization. Then,

$$\int \frac{d^4k_2}{(2\pi)^4 i}\frac{1}{(m^2 - (k_1+k_2)^2)(m^2 - k_2^2)} = \frac{1}{16\pi^2}\ln\frac{\Lambda^2}{m^2} + \cdots, \tag{1.181}$$

where the ellipsis denotes the terms that do not depend on Λ in the limit of large cutoff $\Lambda \gg m$. Hence,

$$D(x-y)D(y-x) = i\delta^{(4)}(x-y)\frac{1}{16\pi^2}\ln\frac{\Lambda^2}{m^2} + \cdots, \tag{1.182}$$

and

$$-\frac{i}{4}\text{tr}\,(\hat{D}_0^{-1}\hat{D}_1\hat{D}_0^{-1}\hat{D}_1) = \frac{1}{4}\frac{1}{16\pi^2}\ln\frac{\Lambda^2}{m^2}\int d^4x\,\text{tr}\,(\hat{D}_1(x)\hat{D}_1(x))$$

$$= \frac{1}{16\pi^2}\ln\frac{\Lambda^2}{m^2}\int d^4x\left(\frac{9\lambda^2+g^2}{4}((\phi_1^c)^4+(\phi_2^c)^4)+(3\lambda g+2g^2)(\phi_1^c)^2(\phi_2^c)^2\right). \tag{1.183}$$

The coefficients in front of the operators $(\phi_1^c)^4 + (\phi_2^c)^4$ and $(\phi_1^c)^2(\phi_2^c)^2$ in the effective action (that includes $\mathcal{L}^c$) should not depend on the cutoff Λ. This condition gives

$$0 = \Lambda\frac{d}{d\Lambda}\left(-\frac{1}{4}\lambda+\frac{1}{16\pi^2}\ln\frac{\Lambda^2}{m^2}\frac{9\lambda^2+g^2}{4}\right),$$

$$0 = \Lambda\frac{d}{d\Lambda}\left(-\frac{1}{2}g+\frac{1}{16\pi^2}\ln\frac{\Lambda^2}{m^2}(3\lambda g+2g^2)\right). \tag{1.184}$$

Hence, the RG equations at one loop take the following form:

$$\Lambda\frac{d}{d\Lambda}\lambda = \frac{3}{8\pi^2}\left(3\lambda^2+\frac{1}{3}g^2\right),$$

$$\Lambda\frac{d}{d\Lambda}g = \frac{3}{8\pi^2}\left(2\lambda g+\frac{4}{3}g^2\right). \tag{1.185}$$

The quantity of interest is $\eta = g/\lambda$. The RG equation for this quantity is readily obtained from Eq. (1.185):

$$\Lambda\frac{d}{d\Lambda}\eta = \frac{3}{8\pi^2}(-\lambda\eta)\left(\frac{1}{3}\eta^2-\frac{4}{3}\eta+1\right). \tag{1.186}$$

As seen, the β-function for the coupling η vanishes at $\eta = 0, 1, 3$; these are *fixed points*. The point $\eta = 0$ (i.e., $g = 0$) corresponds to two mutually noninteracting scalar fields. The case $\eta = 3$ (i.e., $g = 3\lambda$) is reduced to the former by the substitution $\phi_1 = (\psi_1 + \psi_2)/\sqrt{2}$ and $\phi_2 = (\psi_1 - \psi_2)/\sqrt{2}$. Finally, the case $\eta = 1$ (i.e., $g = \lambda$) corresponds to the $O(2)$ symmetric case. Note also that the derivative of the β-function with respect to η is positive for $\eta = 1$ (infrared attractor) and negative for $\eta = 0, 3$ (infrared repulsor). Here, we assume that λ is positive so that the Hamiltonian is bound from below.

Let us now consider the running of the coupling η in the vicinity of the fixed points. If $\eta \simeq 1$, the RG equation linearizes:

$$\Lambda\frac{d}{d\Lambda}\eta \simeq c(\eta-1), \qquad c = \frac{3\lambda}{8\pi^2} > 0. \tag{1.187}$$

The solution of this differential equation takes the form

$$\eta(\Lambda) - 1 = (\eta(M) - 1)\left(\frac{\Lambda}{M}\right)^c, \tag{1.188}$$

where Λ and M are two different scales with $M \gg \Lambda$. Now, it is seen that starting from some high scale M and evolving down to the scale Λ, the coupling $\eta(\Lambda)$ tends to one irrespective of the value $\eta(M)$ at the high scale. In other words, the $O(2)$ global symmetry emerges at low energies, even if at high energies there was no such symmetry and the couplings λ and g were independent. Note also that with the fixed points $\eta = 0, 3$ the situation is the opposite: the sign of the pertinent c's is now negative, and the coupling $\eta(\Lambda)$ is repelled from the fixed points as $\Lambda \ll M$.

Next, we turn to the case of *local* symmetries. As is well known, theories with (massive) vector fields are, in general, non-renormalizable. Only when these fields describe gauge bosons of some local symmetry group (unbroken or broken) can the theory be made renormalizable. On the other hand, as we have learned from the previous discussions, the non-renormalizable interactions that are described by the operators with higher dimensions become irrelevant at low energies, that is, only renormalizable interactions survive. This raises the intriguing question of whether gauge symmetries are also emergent at low energies and need not be postulated from the beginning. In a loose language, one could even speak of "the emergence of light from chaos" (i.e., massless photons out of massive vector fields) at low energies.

The preceding question is difficult. In Refs. [20, 21], a one-loop perturbative study has been carried out in different models, including those with supersymmetry. The results are inconclusive. In some cases, we encounter infrared attractors, whereas in other cases we do not. On the other hand, in Ref. [22], very plausible *non-perturbative* arguments were given in favor of the statement that gauge theories should emerge in the infrared limit. In what follows, we shall illustrate these arguments in a simple example.

Let us start from a theory that contains a vector field $G_\mu = -iG_\mu^a T^a$ that belongs to the adjoint representation of some *compact* group $\mathcal{G}$. The quantities T^a denote the generators of the group $\mathcal{G}$, normalized as $\mathrm{tr}\,(T^a T^b) = \delta^{ab}/2$. Except G_μ, there can be other fields (matter fields) in the theory, which transform as basis vectors of some irreducible representations of $\mathcal{G}$. In the following, we do not need to specify these fields explicitly. Note that the symmetry considered here is *global*.

The most prominent signature of an (unbroken) gauge symmetry is the emergence of a vector particle with zero mass. This manifests itself as the pole in the two-point function of two vector fields, located at $p^2 = 0$. It is, however, more convenient to consider the two-point function of currents $J_\mu = -iJ_\mu^a T^a$, which couple to the field. For example, if $\mathcal{G}$ is the group $SU(N)$, and the vector field G_μ interacts with the fermions in the fundamental representation, ψ, then a convenient choice for J_μ^a by analogy with QED is

$$J_\mu^a(x) = \bar{\psi}(x)\gamma_\mu T^a \psi(x). \tag{1.189}$$

The reason for considering currents instead of fields will become clear in what follows.

Next, using the invariance under global transformations, it can be seen immediately that the two-point function

$$\Delta_{\mu\nu}(x-y) = \int dG_\mu \cdots \text{tr}\big(J_\mu(x)J_\nu(y)\big)\exp\Big(i\int d^4x\mathcal{L}[G_\mu,\cdots]\Big) \qquad (1.190)$$

develops a pole at the physical mass of the vector particle. Here, the ellipses stand for the matter fields, which are not shown explicitly.[16]

Up to now, the theory did not possess gauge invariance at all, as $\mathcal{G}$ is a global group. In Ref. [22], Förster, Nielsen and Ninomiya have proposed an elegant trick to formally elevate the global symmetry to a local one. To this end, one *defines* the gauge transformation of G_μ in a standard manner:

$$G_\mu(x) \mapsto G_\mu^\Lambda(x) = \Lambda(x)G_\mu(x)\Lambda(x)^{-1} - \partial_\mu\Lambda(x)\Lambda(x)^{-1}, \qquad (1.191)$$

where $\Lambda(x)$ is an element of a *local* group $\mathcal{G}$. The matter fields transform under the gauge group in the standard manner as well. Note also that the current, which was introduced in Eq. (1.189), under the gauge transformation transforms as

$$J_\mu(x) \mapsto \Lambda(x)J_\mu(x)\Lambda(x)^{-1}, \qquad (1.192)$$

so that the trace in Eq. (1.190) is invariant under the transformations. This is the rationale for considering the two-point function of currents instead of the fields G_μ. Further, the path integral measure is invariant under local transformations. Performing the transformations on the integration variables, one may write

$$\Delta_{\mu\nu}(x-y) = \int d\Lambda \int dG_\mu \cdots \text{tr}\big(J_\mu(x)J_\nu(y)\big)\exp\Big(i\int d^4x\mathcal{L}[G_\mu,\cdots]\Big)$$

$$= \int d\Lambda \int dG_\mu \cdots \text{tr}\big(J_\mu(x)J_\nu(y)\big)\exp\Big(i\int d^4x\mathcal{L}[G_\mu^\Lambda,\cdots]\Big).$$

$$(1.193)$$

The action functional in Eq. (1.193) consists of two parts: the one that preserves gauge invariance, and the one that does not; that is,

$$S[G_\mu^\Lambda,\cdots] = \int d^4x\mathcal{L}[G_\mu^\Lambda,\cdots] = S_0[G_\mu,\cdots] + S_1[G_\mu^\Lambda,\cdots]. \qquad (1.194)$$

Defining next

$$\exp\big(i\delta S[G_\mu,\cdots]\big) = \int d\Lambda\exp\big(iS_1[G_\mu^\Lambda,\cdots]\big), \qquad (1.195)$$

we finally get

$$\Delta_{\mu\nu}(x-y) = \int dG_\mu \cdots \text{tr}\big(J_\mu(x)J_\nu(y)\big)\exp\big(iS_0[G_\mu,\cdots] + i\delta S[G_\mu,\cdots]\big). \qquad (1.196)$$

---

[16] Two disclaimers should be immediately made here. First, we implicitly assume that the theory is regularized on a finite space-time lattice (see the following), and hence there is no need for adding a gauge-fixing term. Second, for simplicity, we assume that there is no confinement of the vector field; otherwise, the physical mass cannot be defined.

At first glance, a paradoxical result emerges. Starting from a gauge non-invariant expression, we obtain an expression where everything is written in terms of gauge-invariant quantities. Performing the gauge transformation $G_\mu \rightarrow G_\mu^\Lambda, \Lambda \in \mathcal{G}$ in Eq. (1.195), it can be easily seen that the quantity $\delta S[G_\mu, \cdots]$ is explicitly gauge-invariant. However, the gauge invariance of the latter does not yet guarantee that it is given in form of the integral over a local Lagrangian, and thus one gets a gauge-invariant effective theory at the end. In fact, the variables Λ in Eq. (1.195) can be considered as *dynamical fields*, living on a compact group manifold and interacting with *external* fields $G_\mu, \ldots$. An analog is a system of spins that are localized on the sites of a regular lattice in space and that interact with an external magnetic field. It is known that, for varying parameters of the model, two different phases can be realized:

- High-temperature phase, no spontaneous magnetization:

  In this case, the variables $\Lambda(x)$ are not correlated at distances larger than the inverse UV cutoff. This corresponds to the situation where S_1 is "small" and the path integral can be calculated by expanding the exponent $\exp(iS_1) = 1 + iS_1 + \ldots$. In this case, δS can be written as

$$\delta S[G_\mu, \cdots] = \int d^4x \sum_i C_i \mathcal{O}_i(x), \tag{1.197}$$

  where the $\mathcal{O}_i(x)$ are *local, gauge-invariant operators* and the C_i are couplings. At low energies, only relevant and marginal operators survive and we end up with a gauge theory with massless gauge bosons.[17] So, light from chaos indeed emerges.[18]

- Low-temperature phase, spontaneous magnetization:

  In this case, the correlation length of variables Λ is large and a local, gauge-invariant effective Lagrangian cannot be derived. The field G_μ is in general massive.

A nontrivial part of the statement is that there exists a region in the parameter space where the transition to the disordered phase takes place and, hence, an effective low-energy gauge-invariant theory emerges from the theory, which was not gauge-invariant originally. The statement has been backed, for example, by the numerical results of Monte-Carlo simulations in the two-dimensional *XY* model on the lattice [23]. Even being tempting, we do not claim here that the fundamental gauge interactions in Nature necessarily emerge from a theory that does not exhibit gauge symmetry at high energies. Although it is an interesting option to entertain, a further discussion goes beyond the scope of this book.

---

[17] These can still acquire mass through the Higgs mechanism in the interactions with matter fields.

[18] A gauge theory must be compact in order to be dynamically stable. A simple counter-example is provided by a single free massive vector field, described by the Lagrangian

$$\mathcal{L} = -\frac{1}{4}(\partial_\mu G_\nu - \partial_\nu G_\mu)^2 + \frac{m^2}{2} G_\mu G^\mu. \tag{1.198}$$

This is a free theory, and the vector meson there has a nonzero mass [22].

1.12 Triviality of the ϕ^4-Theory

Let us consider a theory that is described by the Lagrangian

$$\mathcal{L} = \frac{1}{2}(\partial\phi)^2 - \frac{m^2}{2}\phi^2 - \frac{\lambda}{4}\phi^4, \qquad (1.199)$$

where ϕ denotes a one-component real scalar field. It is well known that this theory is perturbatively renormalizable in a strict sense. To all orders in perturbation theory, the divergences arising in the Green's functions can be removed by the renormalization of the mass m, the coupling constant λ and the multiplicative renormalization of the field $\phi \to Z^{1/2}\phi$, where Z is the wave function renormalization constant. For a detailed discussion, see, for example, the textbook [24].

In this section, we use cutoff regularization and denote the cutoff scale by Λ. Although dimensional regularization is more simple and elegant, it lacks physical transparency. In the following, m and λ are the bare parameters. The renormalized parameters are denoted by m_R and λ_R, respectively.

Suppose that one calculates the Green's function of the fields ϕ with n external legs, the n-point function. The unrenormalized Green's function is given by

$$(2\pi)^4 \delta^{(4)}\left(\sum_{i=1}^{n} p_i\right) G(p_1, \cdots, p_n; m, \lambda, \Lambda)$$

$$= \int \prod_{i=1}^{n}(dx_i\, e^{ip_i x_i})\, \langle 0|T\phi(x_1)\cdots\phi(x_n)|0\rangle. \qquad (1.200)$$

The renormalized Green's function follows as

$$G_R(p_1, \cdots, p_n; m_R, \lambda_R, \Lambda) = Z^{-n/2}(m_R, \lambda_R, \Lambda)G(p_1, \cdots, p_n; m, \lambda, \Lambda), \qquad (1.201)$$

with m_R, λ_R functions of m, λ and Λ and vice versa. Renormalizability means that, for fixed values of m_R, λ_R, one may choose the dependence on Λ in the parameters m, λ and Z so that all Green's functions $G_R(p_1, \cdots, p_n; m_R, \lambda_R, \Lambda)$ stay finite in the limit $\Lambda \to \infty$. If the Green's functions are expanded up to k loops in perturbation theory, then

$$G_R(p_1, \cdots, p_n; m_R, \lambda_R, \Lambda) = G_R(p_1, \cdots, p_n; m_R, \lambda_R) + O(\Lambda^{-2}\ln^k\Lambda), \qquad (1.202)$$

where $G_R(p_1, \cdots, p_n; m_R, \lambda_R)$ is the renormalized Green's function in the limit $\Lambda \to \infty$, and the remainder describes the so-called *scaling violation*.

In order to set the framework unambiguously, one in addition should fix the renormalization prescription, that is, one should define how the quantities m_R, λ_R, Z are expressed through the pertinent bare quantities. In the following, we choose the renormalization at vanishing external momenta (a particular choice within the so-called MOM scheme). Consider first the two-point Green's function:

$$G_R(p, -p; m_R, \lambda_R, \Lambda) = Z^{-1}(m_R, \lambda_R, \Lambda)G(p, -p; m, \lambda, \Lambda)$$

$$\doteq -i\big[\Gamma_R^{(2)}(p^2; m_R, \lambda_R, \Lambda)\big]^{-1}. \qquad (1.203)$$

In order to fix m_R and Z, we may require

$$\Gamma_R^{(2)}(0; m_R, \lambda_R, \Lambda) = m_R^2, \tag{1.204}$$

and

$$\left. \frac{d}{dp^2} \Gamma_R^{(2)}(p^2; m_R, \lambda_R, \Lambda) \right|_{p^2=0} = -1. \tag{1.205}$$

Note that m_R is *not* equal to the physical mass denoted by m_P. Also, m_R is different from the renormalized mass m_r in the $\overline{\text{MS}}$ scheme, which was considered in the previous sections.

In order to fix the coupling constant λ_R, one considers the four-point function:

$$T_R(p_1, p_2, p_3, p_4; m_R, \lambda_R, \Lambda)$$

$$= -iZ^{-2}(m_R, \lambda_R, \Lambda) \prod_{i=1}^4 (m_P^2 - p_i^2) G(p_1, p_2, p_3, p_4; m, \lambda, \Lambda). \tag{1.206}$$

The renormalized coupling λ_R is defined as

$$T_R(0, 0, 0, 0; m_R, \lambda_R, \Lambda) = -6\lambda_R. \tag{1.207}$$

The preceding three relations indeed determine the bare quantities m, λ, Z in terms of the renormalized ones, m_R, λ_R, and the cutoff Λ.

In order to see how these relations work in practice, let us perform calculations at one loop. The calculation of the two-point function at this order yields

$$\left[G(p, -p; m, \lambda, \Lambda) \right]^{-1} = m^2 - p^2 + 3\lambda I(m^2, \Lambda),$$

$$I(m^2, \Lambda) = \int^\Lambda \frac{d^4 k}{(2\pi)^4 i} \frac{1}{m^2 - k^2}. \tag{1.208}$$

Note that, at this order, $I(m^2, \Lambda)$ does not depend on p^2. Consequently, differentiating the relation

$$\Gamma_R^{(2)}(p^2; m_R, \lambda_R, \Lambda) = Z(m_R, \lambda_R, \Lambda) \left[m^2 - p^2 + 3\lambda I(m^2, \Lambda) \right] \tag{1.209}$$

with respect to p^2 and using Eq. (1.205), we obtain

$$Z(m_R, \lambda_R, \Lambda) = 1. \tag{1.210}$$

Further, Eq. (1.204) at this order yields

$$m_R^2 = m^2 + 3\lambda I(m^2, \Lambda), \qquad m^2 = m_R^2 - 3\lambda_R I(m_R^2, \Lambda). \tag{1.211}$$

Here, we have used the fact that $m_R = m$ and $\lambda_R = \lambda$ at lowest order. Substituting the explicit expression for the self-energy integral

$$I(m^2, \Lambda) = \frac{1}{16\pi^2} \left[\Lambda^2 - m^2 \ln \frac{\Lambda^2}{m^2} + O\left(\frac{1}{\Lambda^2} \right) \right], \tag{1.212}$$

we get

$$m^2 = m_R^2 - \frac{3\lambda_R}{16\pi^2}\left[\Lambda^2 - m_R^2\ln\frac{\Lambda^2}{m_R^2} + O\left(\frac{1}{\Lambda^2}\right)\right].\tag{1.213}$$

Note that m_R is equal to the physical mass m_P at this order.

Next, calculating the four-point function at one loop, we get

$$T(p_1, p_2, p_3, p_4; m_R, \lambda_R, \Lambda) = -6\lambda$$

$$+(6\lambda)^2\left(J(s, m^2, \Lambda) + J(t, m^2, \Lambda) + J(u, m^2, \Lambda)\right),\tag{1.214}$$

where $s = (p_1 + p_2)^2$, $t = (p_1 + p_3)^2$, $u = (p_1 + p_4)^2$, and

$$J(s, m^2, \Lambda) = \frac{1}{2}\int^\Lambda \frac{d^4k}{(2\pi)^4 i}\frac{1}{(m^2 - k^2)(m^2 - (k - p_1 - p_2)^2)}$$

$$= \frac{1}{32\pi^2}\int_0^1 dx\left[\ln\frac{\Lambda^2}{m^2} + 1 - \ln\frac{m^2 - x(1-x)s}{m^2} + O\left(\frac{1}{\Lambda^2}\right)\right].\tag{1.215}$$

Using Eq. (1.207), we finally obtain

$$\lambda_R = \lambda - \frac{9\lambda^2}{16\pi^2}\left[\ln\frac{\Lambda^2}{m^2} + 1\right].\tag{1.216}$$

Fixing λ_R to a given value (independent of Λ) leads to

$$0 = \Lambda\frac{\partial}{\partial\Lambda}\lambda_R = \Lambda\frac{\partial}{\partial\Lambda}\lambda - \frac{9\lambda^2}{8\pi^2}.\tag{1.217}$$

This yields the RG equation that determines the dependence of the bare coupling λ on the cutoff Λ:

$$\Lambda\frac{\partial}{\partial\Lambda}\lambda = \beta(\lambda, m, \Lambda).\tag{1.218}$$

The arguments of the β-function on the right-hand side of this equation are the bare coupling λ, the bare mass m and the cutoff Λ. The quantities m and Λ can be further expressed through λ and the renormalized parameters m_R, λ_R. Since the latter are fixed, one could consider the explicit dependence on λ only, suppressing all other variables. Expanding now the β-function in powers of the coupling constant, we get

$$\beta(\lambda) = -\beta_0\lambda^2 - \beta_1\lambda^3 - \cdots,\tag{1.219}$$

where, in our case,

$$\beta_0 = -\frac{9}{8\pi^2} < 0.\tag{1.220}$$

Let us consider the solution of the RG equation to lowest order. Integrating the equation

$$\Lambda\frac{\partial}{\partial\Lambda}\lambda = -\beta_0\lambda^2\tag{1.221}$$

gives

$$\frac{1}{\lambda(\Lambda)} - \frac{1}{\lambda_0} = \beta_0\ln\frac{\Lambda}{\Lambda_0}, \qquad \lambda_0 = \lambda(\Lambda)\big|_{\Lambda=\Lambda_0}.\tag{1.222}$$

This equation can be rewritten as

$$\lambda(\Lambda) = \frac{\lambda_0}{1 + \beta_0 \lambda_0 \ln \frac{\Lambda}{\Lambda_0}}. \tag{1.223}$$

Recall that λ_0 is the bare coupling constant corresponding to the (fixed) renormalized constant at the value of the cutoff $\Lambda = \Lambda_0$. It has a certain numerical value that depends on λ_R, Λ_0 and m_R. Renormalizability of the theory is then equivalent to the statement that, when Λ is varied continuously from $\Lambda = \Lambda_0$ to infinity, we can always choose a value of the bare coupling $\lambda(\Lambda)$, which would correspond to the *same* values of the renormalized parameters. Let us explicitly check whether this statement is valid at one loop.

The answer to this question depends on the sign of the coefficient of β_0. As seen from Eq. (1.223), if β_0 is positive, $\lambda(\Lambda)$ decreases monotonically and approaches zero, as Λ increases from Λ_0 to infinity. (Recall that λ_0 has to be positive, otherwise the Hamiltonian of the model is not bound from below.) This means that the limit $\Lambda \to \infty$ exists and, moreover, the first-order perturbative result can be trusted at large values of Λ (asymptotic freedom). However, if β_0 is negative, as in our example (see Eq. (1.220)), the situation changes dramatically. The bare coupling constant becomes infinite at

$$\Lambda = \Lambda_0 \exp\left(-\frac{1}{\beta_0 \lambda_0}\right). \tag{1.224}$$

This phenomenon is called a *Landau pole* [25–27]. In this case, the limit $\Lambda \to \infty$ cannot be performed, unless $\lambda_R = 0$, that is, in a *trivial* theory. The latter statement, however, comes with a grain of salt. Namely, approaching the critical value of Λ, the coupling constant λ grows, and the applicability of perturbation theory is questionable. It is therefore necessary to find out whether higher-order corrections might invalidate the statement.

Below, we shall formulate the condition under which the limit $\Lambda \to \infty$ exists. To this end, let us integrate the first-order differential equation (1.218):

$$\ln \frac{\Lambda}{\Lambda_0} = \int_{\lambda_0}^{\lambda} \frac{d\lambda'}{\beta(\lambda')}. \tag{1.225}$$

If $\Lambda \to \infty$, the l.h.s. of this equation is positive and diverges. The integral on the r.h.s. of the equation should thus behave in the same way. There are several alternatives:

- The β-function has a zero,

$$\beta(\lambda) = 0 \quad \text{at } \lambda = \lambda^*, \tag{1.226}$$

where the quantity λ^* defines the location of a so-called *fixed point*. Let us assume, for illustrative reasons, that the zero in the β-function is of first order. (Other cases can be treated similarly.) In this case, we can perform a Taylor expansion in the vicinity of the fixed pole:

$$\beta(\lambda) = C(\lambda - \lambda^*) + O((\lambda - \lambda^*)^2), \tag{1.227}$$

where C is the derivative of the β-function at the fixed point. We assume that $C < 0$ (ultraviolet fixed point). We also assume that $\lambda_0 > \lambda^*$ (the case $\lambda_0 < \lambda^*$ can be considered analogously). Suppose now that $\lambda(\Lambda)$ for some Λ is located in the vicinity of λ^*. Then, Eq. (1.225) can be rewritten as

$$\ln\frac{\Lambda}{\Lambda_0} = \int_{\lambda_0}^{\lambda^*+\varepsilon} \frac{d\lambda'}{\beta(\lambda')} + \int_{\lambda^*+\varepsilon}^{\lambda} \frac{d\lambda'}{C(\lambda'-\lambda^*)} . \tag{1.228}$$

Here, ε is a small quantity and, in the second integral, the function $\beta(\lambda)$ can be replaced by the first term in the Taylor expansion. Denoting the first integral by A (it does not depend on λ), we get

$$\ln\frac{\Lambda}{\Lambda_0} = A + \frac{1}{C}\ln\frac{\lambda-\lambda^*}{\varepsilon} , \tag{1.229}$$

so that

$$\lambda - \lambda^* = \varepsilon\left(\frac{\Lambda}{\Lambda_0 e^A}\right)^C . \tag{1.230}$$

If $C < 0$, $\lambda(\Lambda)$ converges toward λ^* as $\Lambda \to \infty$. For this reason, we speak of an *ultraviolet fixed point,* or an *attractor.* On the contrary, there is no attraction to this point in the ultraviolet if $C > 0$.

- The function $\beta(\lambda)$ has a definite sign. For $\Lambda \to \infty$, the l.h.s. of Eq. (1.225) diverges, and the r.h.s. should do so as well. This divergence can arise only at $\lambda \to \infty$, that is, the asymptotic behavior of $\beta(\lambda)$ is restricted by

$$\left|\frac{\beta(\lambda)}{\lambda}\right| \leq \text{constant}, \quad \text{as } \lambda \to \infty. \tag{1.231}$$

In this case, the limit $\Lambda \to \infty$ can be also performed. There are no singularities at finite values of Λ.

- Finally, consider the situation when $\beta(\lambda)$ grows faster than $\lambda^{1+\delta}$, with $\delta > 0$, as $\lambda \to \infty$. In this case, the integral on the r.h.s. of Eq. (1.225) converges, whereas the logarithm on the l.h.s. diverges, as $\Lambda \to \infty$. Consequently, there exists a critical value of Λ, called Λ_{crit}, determined from the equation

$$\ln\frac{\Lambda_{\text{crit}}}{\Lambda_0} = \int_{\lambda_0}^{\infty} \frac{d\lambda'}{\beta(\lambda')} , \tag{1.232}$$

where the coupling constant blows up, and thus performing the limit $\Lambda \to \infty$ is not possible. This situation resembles the previously discussed Landau pole in perturbation theory.

Thus, the question whether the limit $\Lambda \to \infty$ can be performed in the renormalized ϕ^4-theory boils down to the study of the asymptotic behavior of the β-function at large values of λ and its zeros. Clearly, non-perturbative methods should be used in order to solve this problem.

In the series of papers [28–30] Lüscher and Weisz have demonstrated that, indeed, the ϕ^4-theory is trivial, that is, the limit $\Lambda \to \infty$ *cannot* be performed at $\lambda_R \neq 0$. Schematically, the argument in these papers goes as follows. (For simplicity, we consider a case of a single scalar field in the phase with unbroken symmetry.) In the unrenormalized theory, one has two dimensionless quantities: $\bar{m} = m/\Lambda$ and λ. In the renormalized theory, these are $\bar{m}_R = m_R/\Lambda$ and λ_R. Further, one can express the renormalized parameters in terms of the bare ones in a wide range of bare parameters, including the physically interesting domain $m_R/\Lambda \to 0$. These relations take the form

$$\bar{m}_R = \bar{m}_R(\lambda, \bar{m}), \qquad \lambda_R = \lambda_R(\lambda, \bar{m}). \tag{1.233}$$

Now, it is possible to invert the second relation, expressing $\bar{m}$ through λ and λ_R. Substituting this into the first relation, one gets

$$\bar{m}_R = F(\lambda, \lambda_R). \tag{1.234}$$

In Refs. [28–30], the trajectories of $\bar{m}_R$ were studied for a fixed λ_R, using a combination of the high-temperature expansion of the Green's functions and a numerical solution of the RG equations. It was shown that the dependence on λ is monotonic, and the minimum is achieved when $\lambda \to \infty$. This result means that the value of $\bar{m}_R^{-1}$ is bounded *from above* at every fixed λ_R, and one may write

$$\ln \bar{m}_R^{-1} = \ln \frac{\Lambda}{m_R} \leq f(\lambda_R), \tag{1.235}$$

where $f(\lambda_R)$ can be evaluated numerically, using the methods just mentioned. In other words, for each given $\lambda_R \neq 0$, there exists an upper bound on Λ, that is, the theory is trivial. Note further, as shown in Ref. [28], the function $f(\lambda_R)$ can be well approximated by an expression of the form $f(\lambda_R) = A/\lambda_R + B \ln \lambda_R + O(1)$. If λ_R is not very large, the second-order perturbation expression works very well, yielding

$$\ln \frac{\Lambda}{m_R} \leq -\frac{1}{\beta_0 \lambda_R} - \frac{\beta_1}{\beta_0^2} \ln(-\beta_0 \lambda_R) + C(\lambda_R), \tag{1.236}$$

where $-1.7 \leq C(\lambda_R) \leq 1.3$ for $\lambda_R \leq 1.5$.

At this point, one has to discuss what the obtained result means in practice. The fact that the ultraviolet cutoff cannot be moved to infinity does not a priori invalidate the results that can be obtained from this theory at low momenta, that is, momenta much smaller than the cutoff Λ. In other words, the ϕ^4-theory perfectly makes sense as an *effective theory* if one does not insist that it should be valid at all energies. The contributions coming from high momenta, which can be characterized by the higher-dimensional operators in the Lagrangian, are small at the momenta much less than the cutoff.

This statement has important implications, if one insists that, for consistency, the mass of a scalar particle must be smaller than the ultraviolet cutoff. This results in the existence of an upper bound on the mass of the scalar particle(s), the so-called *triviality bound*, in theories, where the interactions in the scalar sector are described by the ϕ^4 Lagrangian. A prominent example is the triviality bound set on the mass of the Higgs particle in the Standard Model before the actual discovery of the Higgs. Below we shall give a simple, intuitive derivation of the bound from the one-loop running of the coupling constant.

Consider the Lagrangian of the Higgs sector of the Standard Model that contains one complex doublet field Φ:

$$\mathcal{L}_H = \frac{1}{2} \partial_\mu \Phi^\dagger \partial^\mu \Phi - \frac{m^2}{2} \Phi^\dagger \Phi - \frac{\lambda}{4} (\Phi^\dagger \Phi)^2 \,. \tag{1.237}$$

For a rough estimate one can use the lowest-order perturbative results for the running of the renormalized coupling constant. To this end, by analogy with Eq. (1.216), one can define the scale-dependent renormalized constant $\lambda_R(\mu)$:

$$\lambda = \lambda_R(\mu) - \frac{1}{2} \beta_0 \lambda_R(\mu)^2 \left[\ln \frac{\Lambda^2}{\mu^2} + 1 \right] \,, \tag{1.238}$$

where β_0 in the theory with two complex scalar fields, described by the Lagrangian (1.237), is given by

$$\beta_0 = \frac{3}{2\pi^2} \,, \tag{1.239}$$

and the quantity λ_R, which was defined previously, corresponds to the choice of the scale $\mu = m_R$; see Eq. (1.216). The RG running of the renormalized quartic coupling is given by

$$\mu \frac{\partial}{\partial \mu} \lambda_R(\mu) = -\beta_0 \lambda_R^2(\mu) \,. \tag{1.240}$$

Integrating this differential equation leads to

$$\lambda_R(\mu_0) = \frac{\lambda_R(\mu)}{1 + \frac{3}{2\pi^2} \lambda_R(\mu) \ln \frac{\mu}{\mu_0}} \,, \qquad \mu \geq \mu_0 \,. \tag{1.241}$$

Taking into account the fact that $\lambda_R(\mu)$ is positive, the following upper bound on $\lambda_R(\mu_0)$ emerges:

$$\bar{\lambda}_R \doteq \lambda_R(\mu_0) \leq \frac{2\pi^2}{3 \ln \frac{\mu}{\mu_0}} \,. \tag{1.242}$$

Further, at the order of perturbation theory we are working, we can use tree-level results for the masses from the Standard Model:

$$M_H^2 = 2 \bar{\lambda}_R v^2 \,, \qquad M_W^2 = \frac{1}{4} g^2 v^2 \,, \qquad g = \frac{e}{\sin \theta_W} \,. \tag{1.243}$$

Here, M_H and M_W are the masses of the Higgs and W-bosons, respectively, $v \simeq 246 \, \text{GeV}$ in the vacuum expectation value of the Higgs field, e and g are the electromagnetic and the $SU(2)$ gauge couplings, respectively, and $\sin^2 \theta_W \simeq 0.23$, where θ_W denotes the weak mixing (Weinberg) angle. Using now Eq. (1.243), we obtain

$$\left(\frac{M_H}{M_W} \right)^2 = \frac{8 \bar{\lambda}_R}{g^2} \,. \tag{1.244}$$

Finally, substituting the value of the coupling g, the following rough estimate can be obtained:

$$\frac{M_H}{M_W} \leq \frac{4\pi}{g\sqrt{3}} \frac{1}{(\ln(\mu/\mu_0))^{1/2}} \simeq \frac{900 \, \text{GeV}}{M_W} \frac{1}{(\ln(\mu/\mu_0))^{1/2}} \,. \tag{1.245}$$

In the spontaneously broken phase, μ_0 should be chosen on the order of the Higgs mass. Further, the quantity μ defines the scale up to which the theory is consistent. (It does not make sense to move the scale μ beyond the cutoff Λ.) If μ is as small as $2\mu_0$ (the smaller values are barely consistent with the requirement $m_H \ll \Lambda$), the preceding equation yields $M_H \leq 1070$ GeV. If μ is taken up to the Planck mass $m_{Pl} = 10^{19}$ GeV, the upper limit on the Higgs mass goes down to $\simeq 140$ GeV. Combining the results of lattice calculations with higher-order calculations in perturbation theory, it is possible to arrive at more refined constraints on the Higgs mass.

Very interesting further questions emerge in connection with the problem considered in the present section. How is the bound affected in the Standard Model when the interactions with the gauge bosons and fermions are taken into account? How would the results change, if the Higgs particle is composite? The answer to these intriguing questions is, however, beyond the scope of the present book. Further information on this subject can be found, for example, in Refs. [31–35].

1.13 Relevant Degrees of Freedom at Low Momenta

In this chapter we have considered theories, where both the heavy and light degrees of freedom corresponded to the fields represented in the Lagrangian, and perturbation theory was assumed to work at both high and low energies. At low energies, the heavy degrees of freedom could be neatly integrated out, and a perturbative matching could be performed. Conceptually, this is the most clean and transparent case; but in Nature the separation of the low- and high-energy modes can proceed along many different patterns. In what follows, we list a few of them.

The construction of the low-energy effective theory of QCD is perhaps the most important example, which will be considered in much detail here. The degrees of freedom of the underlying theory are quarks and gluons, none of them surviving in the low-energy limit. Formally, the (light) quarks and gluons have very small (even zero) masses. However, the strong non-perturbative interactions between them lead to confinement (no quarks and/or gluons in the asymptotic states) and to the creation of a mass gap of order of 1 GeV determined by the mass of the proton, which is the lightest *stable* particle in QCD, whose mass does not vanish in the chiral limit. It turns out that at low energies only the Goldstone bosons, which emerge as a result of the spontaneous chiral symmetry breaking in QCD, have masses lighter than the heavy scale of QCD on the order of 1 GeV, and thus represent the only relevant degrees of freedom at these energies. Perturbative matching is, of course, not possible.

Another interesting possibility emerges, for example, in nonrelativistic effective theories, or in heavy quark (heavy baryon) effective field theories, which will also be considered here in detail. In this case, the role of the heavy fields is played by the heavy components of the *same* field. These heavy components correspond to the antiparticles, whose contribution is now relegated to the effective couplings. The heavy scale of

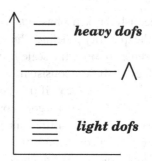

Figure 1.18 Degrees of freedom (dofs) in a typical EFT, which operates at small momenta corresponding to the light particle masses. Beyond some separation scale Λ, heavy particles appear. Their effect on the low-energy EFT is only indirect by providing the strengths of multiparticle operators in the light particle sector.

the theory is determined by the mass gap separating the particles and the antiparticles, that is, by the mass of the particle itself.

Finally, note that, in the context of the many-body problems, effective fields often describe collective excitations, a prominent example being the Landau–Ginsburg theory. The energies of the excitations should be much smaller than the natural hard momentum scale in the problem, given by the inverse of the lattice spacing in the crystal.

To summarize, there exist different scenarios in nature for how the low-energy degrees of freedom emerge from the dynamics at short distances. A universal rule to find these degrees of freedom is to examine the low-energy spectrum of the system, that is, to investigate the low-energy singularity structure of the S-matrix. This allows us to choose the appropriate variables for describing the system at low energies without any reference to the short-distance dynamics.

1.14 Construction Principles of an EFT

Here, we briefly summarize what we have learned, or, stated differently, what the principles are behind the construction of an effective field theory. These are:

- *Scale separation:* This is arguably the most basic concept underlying any EFT. The EFT is operative at small momenta/energies (at large distances), and the physics in this regime is insensitive to what happens at large momenta (small distances) as depicted in Fig. 1.18. The heavy particles can have only an indirect effect at low energies by providing the strengths of certain multiparticle operators in the light particle sector. However, the EFT can be formulated without any knowledge of these heavy degrees of freedom, fitting the appearing LECs to some data at low energies. This scale separation further implies that renormalizability in the strict sense is not applicable. In fact, it should be obvious that any quantum field theory is indeed an effective field theory.

- *Relevant degrees of freedom:* Related to the scale separation are the active particles in the low-energy EFT, which either decouple from the heavy degrees of freedom or are generated through some symmetry breaking, as discussed in detail in Section 1.13. These relevant degrees of freedom are intimately connected to the low-energy singularity structure of the pertinent S-matrix.
- *Symmetries:* Symmetries play an important role in the construction of any EFT. First, discrete and continuous symmetries constrain the possible interactions of the particles. This is equivalent to the construction of any QFT. Second, the realization of symmetries plays an important role, as spontaneous breaking can generate the pertinent low-energy degrees of freedom. As will be discussed in detail in later chapters, this exactly happens for QCD in the confinement regime (at small momenta).
- *Power counting:* With the relevant particles and the constraints from symmetries, one can write down an infinite tower of allowed interactions in the EFT. Power counting is the tool to order all these terms according to their relevance. Symbolically, any matrix element can be perturbatively expanded in powers of energies/momenta Q over the breakdown scale Λ as

$$\mathcal{M} = \sum_{\nu} \left(\frac{Q}{\Lambda}\right)^{\nu} f_{\nu}(Q/\mu, g_i) \,, \tag{1.246}$$

where μ is a renormalization scale related to the required renormalization of loop diagrams and the g_i are coupling constants, often called low-energy constants (LECs). In case of an EFT, the index ν is bounded from below, which allows for a systematic and controlled expansion. Furthermore, f_{ν} is a function of $O(1)$, and this property is called "naturalness." Note also that describing bound states requires some type of non-perturbative resummation.

- *Matching:* Matching is not universal to EFTs but appears very often. One example we already encountered was the matching of the EFT two-body scattering phase to the effective range expansion, which allows us to express the LECs in terms of physical parameters. Another example that will prominently appear in the next chapter is the matching of a nonrelativistic EFT to its relativistic counterpart. Yet another example of matching is the reduction of a theory at high energies down to the low-energy EFT, which might proceed in steps, and matching is performed at the boundaries between the different EFTs to guarantee a smooth transition when lowering the resolution scale. A classical example is the weak $\Delta S = 1$ Hamiltonian, where the W-bosons and the t, b and c quarks are integrated out successively; see, for example, Ref. [36].

1.15 Literature Guide

Most of the material presented in this chapter is standard and overlaps with many textbooks and lecture notes. In this context we mention here the recent books on effective theories by Petrov and Blechman [37] and Burgess [38], as well as a selection of lecture notes and reviews [12, 19, 39–43].

As already mentioned, the use of the effective theory approach in the problems of nonrelativistic quantum mechanics was discussed in the extremely instructive lectures by Lepage [1]. The two-body scattering with pointlike potentials and the issue of the equivalence of different regularizations and renormalization schemes as well as the relation to the Wigner bound have been extensively discussed in the literature. Apart from the papers, which were already cited [2, 3, 8], important aspects of the same problem have been considered in Refs. [7, 44].

The decoupling theorem was first proved by Appelquist and Carazzone [14]. For related work, see, for example, Refs. [45–47]. A very detailed introduction to the issue is contained in a book by Collins [6], to which the reader is referred for further references. The decoupling in the case of two scalar fields has been considered in detail in Refs. [48, 49].

The Wilson renormalization group approach was first described in Ref. [16]; see also Ref. [50]. It is discussed, in particular, in the textbook by Peskin and Schroeder [24]. The dependence of the effective couplings on the floating cutoff was first considered in Polchinski's paper [17]; see also the discussion in Weinberg's textbook [18].

The Landau pole was introduced in the seminal papers [25–27]. A thorough discussion of this issue from different points of view is given in the review article [51]. In a series of papers [28–30], Lüscher and Weisz give an extremely clear and concise discussion of triviality in the ϕ^4 theory. The triviality bound on the Higgs mass is considered in Refs. [52–56].

Various aspects of the emerging symmetries at low energy have been considered in Refs. [20–23]; see also Ref. [57].

References

[1] G. P. Lepage, "How to renormalize the Schrödinger equation," nucl-th/9706029.

[2] D. R. Phillips, S. R. Beane and T. D. Cohen, "Nonperturbative regularization and renormalization: simple examples from nonrelativistic quantum mechanics," Annals Phys. **263** (1998) 255.

[3] S. R. Beane, T. D. Cohen and D. R. Phillips, "The potential of effective field theory in *NN* scattering," Nucl. Phys. A **632** (1998) 445.

[4] C. G. Bollini and J. J. Giambiagi, "Dimensional renormalization: the number of dimensions as a regularizing parameter," Nuovo Cim. B **12** (1972) 20.

[5] G. 't Hooft and M. J. G. Veltman, "Regularization and renormalization of gauge fields," Nucl. Phys. B **44** (1972) 189.

[6] J. C. Collins, Renormalization. An Introduction to Renormalization, the Renormalization Group, and the Operator Product Expansion. Cambridge University Press (1984).

[7] K. A. Scaldeferri, D. R. Phillips, C. W. Kao and T. D. Cohen, "Short range interactions in an effective field theory approach for nucleon-nucleon scattering," Phys.
Rev. C **56** (1997) 679.

[8] J. Gegelia, "About the equivalence of cutoff and conventionally renormalized
effective field theories," J. Phys. G **25** (1999) 1681.

[9] D. R. Phillips and T. D. Cohen, "How short is too short? Constraining contact
interactions in nucleon-nucleon scattering," Phys. Lett. B **390** (1997) 7.

[10] D. B. Kaplan, "More effective field theory for nonrelativistic scattering," Nucl.
Phys. B **494** (1997) 471.

[11] E. P. Wigner, "Lower limit for the energy derivative of the scattering phase shift,"
Phys. Rev. **98** (1955) 145.

[12] A. V. Manohar, "Introduction to Effective Field Theories," arXiv:1804.05863
[hep-ph].

[13] C. Itzykson and J. B. Zuber, Quantum Field Theory, Mcgraw-Hill (1980) 705 pp.
(International Series in Pure and Applied Physics).

[14] T. Appelquist and J. Carazzone, "Infrared Singularities and Massive Fields," Phys.
Rev. D **11** (1975) 2856.

[15] D. Djukanovic, J. Gegelia and U.-G. Meißner, "Triviality of quantum electrodynamics revisited," Commun. Theor. Phys. **69** (2018) 263.

[16] K. G. Wilson, "Renormalization group and critical phenomena. 1. Renormalization group and the Kadanoff scaling picture," Phys. Rev. B **4** (1971) 3174,
3184.

[17] J. Polchinski, "Renormalization and effective Lagrangians," Nucl. Phys. B **231**
(1984) 269.

[18] S. Weinberg, The Quantum Theory of Fields. Vol. 1: Foundations, Cambridge
University Press (1995).

[19] A. V. Manohar, "Effective field theories," arXiv:hep-ph/9606222.

[20] J. Iliopoulos, D. V. Nanopoulos and T. N. Tomaras, "Infrared stability or anti
grand unification," Phys. Lett. **94B** (1980) 141.

[21] I. Antoniadis, J. Iliopoulos and T. Tomaras, "On the infrared stability of gauge
theories," Nucl. Phys. B **227** (1983) 447.

[22] D. Förster, H. B. Nielsen and M. Ninomiya, "Dynamical stability of local gauge
symmetry: creation of light from chaos," Phys. Lett. **94B** (1980) 135.

[23] E. Poppitz and Y. Shang, "'Light from chaos' in two dimensions," Int. J. Mod.
Phys. A **23** (2008) 4545.

[24] M. E. Peskin and D. V. Schroeder, "An introduction to quantum field theory,"
Addison-Wesley (1995) 842 pp.

[25] L. D. Landau, A. A. Abrikosov and I. M. Khalatnikov, "The electron mass in
quantum electrodynamics," Dokl. Akad. Nauk Ser. Fiz. **95** (1954) 1177.

[26] L. D. Landau and I. Y. Pomeranchuk, "On point interactions in quantum
electrodynamics," Dokl. Akad. Nauk Ser. Fiz. **102** (1955) 489.

[27] L. D. Landau, Theoretical Physics in the 20th Century: A Memorial Volume to
W. Pauli, Interscience (1960).

[28] M. Lüscher and P. Weisz, "Scaling laws and triviality bounds in the lattice ϕ^4 theory. 1. One component model in the symmetric phase," Nucl. Phys. B **290** (1987) 25.

[29] M. Lüscher and P. Weisz, "Scaling laws and triviality bounds in the lattice ϕ^4 theory. 2. One component model in the phase with spontaneous symmetry breaking," Nucl. Phys. B **295** (1988) 65.

[30] M. Lüscher and P. Weisz, "Scaling laws and triviality bounds in the lattice ϕ^4 theory. 3. N-component model," Nucl. Phys. B **318** (1989) 705.

[31] M. J. Dugan, H. Georgi and D. B. Kaplan, "Anatomy of a composite Higgs model," Nucl. Phys. B **254** (1985) 299.

[32] I. Brivio and M. Trott, "The standard model as an effective field theory," Phys. Rept. **793** (2019) 1.

[33] G. Panico and A. Wulzer, "The composite Nambu-Goldstone Higgs," Lect. Notes Phys. **913** (2016) 1.

[34] R. Contino, "The Higgs as a composite Nambu-Goldstone boson," arXiv:1005.4269 [hep-ph].

[35] C. T. Hill and E. H. Simmons, "Strong dynamics and electroweak symmetry breaking," Phys. Rept. **381** (2003) 235 Erratum: [Phys. Rept. **390** (2004) 553].

[36] F. J. Gilman and M. B. Wise, "Effective Hamiltonian for delta s = 1 weak nonleptonic decays in the six quark model," Phys. Rev. D **20** (1979), 2392.

[37] A. A. Petrov and A. E. Blechman, Effective Field Theories, World Scientific Publ. Co. (2016).

[38] C. P. Burgess, Introduction to Effective Field Theory, Cambridge University Press (2021).

[39] A. Pich, "Effective field theory: course," hep-ph/9806303.

[40] D. B. Kaplan, "Effective field theories," nucl-th/9506035.

[41] H. Georgi, "Effective field theory," Ann. Rev. Nucl. Part. Sci. **43** (1993) 209.

[42] D. B. Kaplan, "Five lectures on effective field theory," nucl-th/0510023.

[43] J. Polchinski, "Effective field theory and the Fermi surface," In Boulder 1992, Proceedings, Recent directions in particle theory 235–274 [hep-th/9210046].

[44] S. K. Adhikari and A. Ghosh, "Renormalization in nonrelativistic quantum mechanics," J. Phys. A **30** (1997) 6553.

[45] S. Weinberg, "Effective gauge theories," Phys. Lett. **91B** (1980) 51.

[46] E. Witten, "Heavy quark contributions to deep inelastic scattering," Nucl. Phys. B **104** (1976) 445.

[47] Y. Kazama and Y. P. Yao, "Decoupling, effective Lagrangian, and gauge hierarchy in spontaneously broken nonabelian gauge theories," Phys. Rev. D **25** (1982) 1605.

[48] B. A. Ovrut and H. J. Schnitzer, "A new approach to effective field theories," Phys. Rev. D **21** (1980) 3369.

[49] B. A. Ovrut and H. J. Schnitzer, "Decoupling theorems for effective field theories," Phys. Rev. D **22** (1980) 2518.

[50] K. G. Wilson, "The renormalization group: critical phenomena and the Kondo problem," Rev. Mod. Phys. **47** (1975) 773.

[51] D. J. E. Callaway, "Triviality pursuit: can elementary scalar particles exist?," Phys. Rept. **167** (1988) 241.

[52] R. F. Dashen and H. Neuberger, "How to get an upper bound on the Higgs mass," Phys. Rev. Lett. **50** (1983) 1897.

[53] L. Maiani, G. Parisi and R. Petronzio, "Bounds on the number and masses of quarks and leptons," Nucl. Phys. B **136** (1978) 115.

[54] N. Cabibbo, L. Maiani, G. Parisi and R. Petronzio, "Bounds on the fermions and Higgs boson masses in grand unified theories," Nucl. Phys. B **158** (1979) 295.

[55] D. J. E. Callaway, "Nontriviality of gauge theories with elementary scalars and upper bounds on Higgs masses," Nucl. Phys. B **233** (1984) 189.

[56] M. A. B. Beg, C. Panagiotakopoulos and A. Sirlin, "Mass of the Higgs Boson in the canonical realization of the Weinberg-Salam theory," Phys. Rev. Lett. **52** (1984) 883.

[57] C. Wetterich, "Gauge symmetry from decoupling," Nucl. Phys. B **915** (2017) 135.

2 Nonrelativistic Effective Theories

2.1 Introduction

The material of Chapter 1 can be summarized in a few words: effective theories become truly effective when they are applied to systems with multiple momentum scales. Using an effective field theory (EFT) framework enables one to concentrate on *one particular* energy scale (or region) and integrate out others, including them in the couplings. Usually (but not always), this is a heavier scale.

In this chapter, we consider processes where the typical three-momenta of the involved particles are much smaller than the mass of these particles. Physics can be described by nonrelativistic quantum mechanics in this limit, treating particles and antiparticles separately. For example, one could forget about the antiparticles and build a nonrelativistic theory with particles only as particle number is a conserved quantity in nonrelativistic quantum mechanics. Just the opposite happens in relativistic quantum field theory. There, particles and antiparticles are both described by the same relativistic field and cannot be separated from each other. Further, in relativistic field theory, the particles and antiparticles can annihilate, so the number of particles and of antiparticles is not conserved.

The usefulness of the nonrelativistic description is mainly caused by the following fact. If the three-momenta of particles become much smaller than their masses, defining a soft momentum scale, the relativistic description necessarily leads to the violation of the power-counting rules in these small momenta. This happens because the relativistic theory, which treats the particles and antiparticles simultaneously, contains a built-in hard scale, namely the mass gap between particles and antiparticles, which is exactly equal to twice the particle mass. In the presence of such a hard scale, the momenta from different regions start to contribute in the Feynman integrals, and the power-counting rules break down.[1] We shall see later that in nonrelativistic theories such a breaking can be avoided.

An important aspect of the preceding discussion often remains forgotten. It is exactly the separating of particles and antiparticles that revives counting rules and leads to a consistent effective field theory. If one has reason to believe that particle/antiparticle creation and annihilation do not play an important role in the

---

[1] This issue will be taken up in Chapter 4 in the context of Chiral Perturbation Theory with baryons (and other matter fields).

description of the processes one is interested in, one may integrate out the antiparticles, even if the three-momenta of the particles are of the same order as their masses.[2] A nice example of this is provided by kaon decays into three pions. The momenta of the final-state pions are generally not that small as compared to the pion mass. Still, it is plausible that the contributions of the five-pion intermediate states do not play any significant role for the description of the $K \to 3\pi$ decay amplitude (recall that the five-pion threshold lies sufficiently above the kaon mass), and a nonrelativistic setting can be applied for the description of this decay. In this setting the pion energy is given by the full relativistic expression $p^0 = \sqrt{M_\pi^2 + \mathbf{p}^2}$ rather by its nonrelativistic expansion. We shall consider this example in great detail in this chapter.

To summarize, the notion of a nonrelativistic effective theory (NREFT) has "kinematical" (square root versus nonrelativistic expansion) as well as "dynamical" (conservation of the particle number) aspects. The latter aspect is more important. For instance, it is possible to construct a systematic nonrelativistic setting respecting power counting, where the energy of a particle is given by the relativistic dispersion relation. In contrast, we are not aware of a consistent formulation of a nonrelativistic framework that includes particle creation and annihilation. In what follows, we will mostly use scalar field theories to develop the formalism of NREFTs and only touch upon certain aspects of nonrelativistic quantum electrodynamics and quantum chromodynamics, NRQED and NRQCD, respectively.

2.2 Foldy–Wouthuysen Transformation

A reduction of a relativistic field theory to a NREFT can be best understood using the example of the well-known Foldy–Wouthuysen transformation [1], where the separation of particle and antiparticle degrees of freedom is made explicit. Consider first the case of a free Dirac field of mass m, which is described by the Lagrangian

$$\mathcal{L} = \bar{\Psi}(x)(i\slashed{\partial} - m)\Psi(x) = \Psi^\dagger(x)(i\partial_t - H)\Psi(x). \tag{2.1}$$

Here, $\Psi(x)$ denotes the quantized four-component Dirac field operator with

$$H = \boldsymbol{\alpha}\mathbf{p} + \beta m, \tag{2.2}$$

where $\mathbf{p} = -i\nabla$ is the three-momentum operator, and

$$\beta = \gamma^0 = \begin{pmatrix} \mathbb{1} & 0 \\ 0 & -\mathbb{1} \end{pmatrix}, \qquad \boldsymbol{\alpha} = \gamma^0\boldsymbol{\gamma} = \begin{pmatrix} 0 & \boldsymbol{\sigma} \\ \boldsymbol{\sigma} & 0 \end{pmatrix}, \tag{2.3}$$

where the γ^μ and $\boldsymbol{\sigma}$ are the 4×4 Dirac and 2×2 Pauli matrices, respectively, and $\mathbb{1}$ denotes the 2×2 unit matrix.

---

[2] Some technical modifications could be necessary in order to restore the counting rules, e.g., amending the procedure for the evaluation of the Feynman integrals; see what follows.

With the use of a unitary transformation (field redefinition), the above Lagrangian can be reduced to a Lagrangian that contains two noninteracting Pauli spinors instead of one Dirac spinor. This transformation has the following form:

$$\Psi(x) \mapsto U\Psi(x), \qquad H \mapsto U(H - i\partial_t)U^\dagger, \qquad (2.4)$$

where $U = e^{iS}$, with S an Hermitean operator, and $U^\dagger = U^{-1}$.

In order to achieve the separation, one has to choose the operator U so that the transformed Hamiltonian H does not contain the matrix $\boldsymbol{\alpha}$, which is non-diagonal, thus connecting the "large" and the "small" components of the Dirac spinor. This can be achieved by the following choice:

$$U = \cos\theta(|\mathbf{p}|) + \frac{\beta \boldsymbol{\alpha} \mathbf{p}}{|\mathbf{p}|} \sin\theta(|\mathbf{p}|), \qquad U^\dagger = U^{-1}, \qquad (2.5)$$

where the function $\theta = \theta(|\mathbf{p}|)$ fulfills the following relation:

$$\tan 2\theta(|\mathbf{p}|) = \frac{|\mathbf{p}|}{m}. \qquad (2.6)$$

The transformed Hamiltonian takes the following form:

$$H = \beta\sqrt{\mathbf{p}^2 + m^2}. \qquad (2.7)$$

Introducing now two Pauli spinors, ψ and χ,

$$\Psi(x) = \begin{pmatrix} \psi(x) \\ \chi(x) \end{pmatrix}, \qquad (2.8)$$

the original Lagrangian can be rewritten as

$$\mathcal{L} = \psi^\dagger(i\partial_t - \sqrt{\mathbf{p}^2 + m^2})\psi + \chi^\dagger(i\partial_t + \sqrt{\mathbf{p}^2 + m^2})\chi. \qquad (2.9)$$

Note that the square root of an operator is a nonlocal quantity. What is implicitly meant here is the expansion of this operator in inverse powers of the mass:

$$\sqrt{\mathbf{p}^2 + m^2} = m + \frac{\mathbf{p}^2}{2m} - \frac{\mathbf{p}^4}{8m^3} + \cdots. \qquad (2.10)$$

Of course, in the present (trivial) case, the results of this expansion can be summed up in the final expressions to all orders.

The Lagrangian, written down in terms of the Pauli fields ψ and χ, describes a free particle and a free antiparticle. It is important to note that these two fields do not talk to each other and hence, for example, barring the field $\chi(x)$ from the Lagrangian does lead to a consistent theory that describes particle degrees of freedom only. The EOM of the free particle field takes the following form

$$(i\partial_t - \sqrt{\mathbf{p}^2 + m^2})\psi(x) = 0. \qquad (2.11)$$

The Fourier transform of the solution of this EOM does not contain negative-energy modes:

$$\psi_\alpha(x) = \int \frac{d^3\mathbf{k}}{(2\pi)^3} e^{-ik^0x^0 + i\mathbf{k}\mathbf{x}} b_\alpha(\mathbf{k}), \qquad (2.12)$$

with $k^0 = \sqrt{\mathbf{k}^2 + m^2}$. Further, $b_\alpha(\mathbf{k})$ denotes the annihilation operator for a particle with three-momentum $\mathbf{k}$ and spin projection α, $\alpha = 1,2$. From the canonical anticommutation relations for the field $\psi(x)$ one obtains

$$\{b_\alpha(\mathbf{p}), b_\beta^\dagger(\mathbf{k})\} = \delta_{\alpha\beta}(2\pi)^3 \delta^{(3)}(\mathbf{p} - \mathbf{k}). \tag{2.13}$$

The propagator of the free field takes the form

$$i\langle 0|T\psi_\alpha(x)\psi_\beta^\dagger(y)|0\rangle = \int \frac{d^4k}{(2\pi)^4} e^{-ik(x-y)} S_{\alpha\beta}(k), \tag{2.14}$$

where

$$S_{\alpha\beta}(k) = \frac{\delta_{\alpha\beta}}{\sqrt{\mathbf{k}^2 + m^2} - k^0 - i\varepsilon}. \tag{2.15}$$

Note that the propagator of this nonrelativistic particle has one pole, in contrast to the relativistic propagator, which contains the particle as well as antiparticle poles. In this approach, an antiparticle described by the field $\chi(x)$ can be considered analogously to the particle, only some signs will be changed.

We conclude the consideration of this trivial example by noting that our notion of a "free particle" is in fact arbitrary and is based on the splitting of the Lagrangian into the "free" and "interaction" parts. Above, we have opted to include everything into the "free" part. This is a valid, but not the only option. An alternative could be to use Eq. (2.10) and to split the Lagrangian in the following way:

$$\mathcal{L} = \mathcal{L}_{\text{free}} + \mathcal{L}_{\text{int}} = \psi^\dagger\left(i\partial_t - m - \frac{\mathbf{p}^2}{2m}\right)\psi + \psi^\dagger\left(\frac{\mathbf{p}^4}{8m^3} + \cdots\right)\psi + \text{antiparticles}. \tag{2.16}$$

Such a splitting has advantages when discussing the counting rules. If one adopts such a splitting, the Fourier-transform of the field still has the form given in Eq. (2.12), but with $k^0 = m + \mathbf{k}^2/(2m)$ instead of the full relativistic expression. Likewise, the propagator, instead of Eq. (2.15), is now given by

$$S_{\alpha\beta}(k) = \frac{\delta_{\alpha\beta}}{m + \mathbf{k}^2/(2m) - k^0 - i\varepsilon}. \tag{2.17}$$

On the other hand, the vertices collected in $\mathcal{L}_{\text{int}}$ in Eq. (2.16) lead to insertions in the fermion line as shown in Fig. 2.1. Summing up all these insertions, one again arrives at the full relativistic propagator given in Eq. (2.15).

After the above trivial example, let us consider a fermion moving in an external electromagnetic field. The Lagrangian of this system is given by

$$\mathcal{L} = \bar{\Psi}(x)(i\slashed{D} - m)\Psi(x), \qquad D_\mu = \partial_\mu + ie\mathscr{A}_\mu^{\text{ext}}(x). \tag{2.18}$$

Here, $\mathscr{A}_\mu^{\text{ext}}(x) = (\mathscr{A}_0, -\mathscr{A})$ is a c-number external electromagnetic field.

The diagonalization of the above Hamiltonian cannot be performed to all orders explicitly. However, writing down the transformation (2.4), one may choose the operator S so that the transformed Hamiltonian H does not contain a non-diagonal matrix

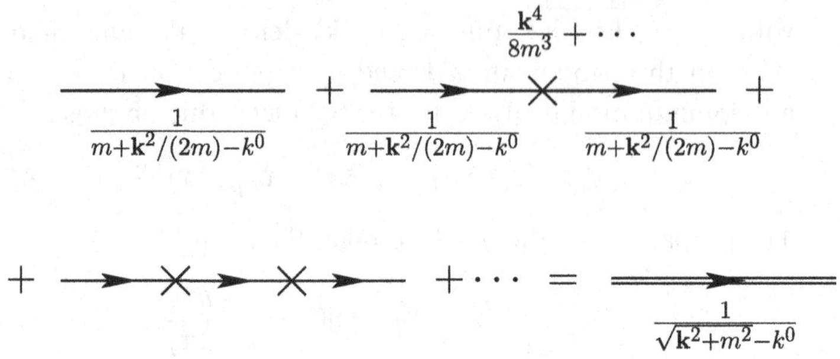

Figure 2.1 Summing up insertions (represented by the cross) in the nonrelativistic propagator, shown by the solid line. The double line represents the result of this summation.

α up to a given order in the expansion in $1/m$. An iterative procedure for constructing S is described in many textbooks. We refer the reader to Ref. [2] for details. Here, we merely display the final result for the transformed Hamiltonian, which is diagonal up to and including the terms of order $1/m^2$:

$$H = \beta \left(m - \frac{\mathbf{D}^2}{2m} - \frac{\mathbf{D}^4}{8m^3} \right) + e\mathscr{A}^0 - \frac{e}{2m} \beta \, \boldsymbol{\sigma} \mathbf{B} - \frac{ie}{8m^2} \, \boldsymbol{\sigma} [\boldsymbol{\nabla} \times \mathbf{E}]$$

$$- \frac{e}{4m^2} \, \boldsymbol{\sigma} [\mathbf{E} \times \mathbf{p}] - \frac{e}{8m^2} \, \boldsymbol{\nabla} \mathbf{E} + \cdots, \tag{2.19}$$

where $\mathbf{E} = -\boldsymbol{\nabla}\mathscr{A}^0 - \partial_t \mathscr{A}$ and $\mathbf{B} = [\boldsymbol{\nabla} \times \mathscr{A}]$ denote the electric and magnetic fields, respectively. The construction can be straightforwardly carried out to higher orders.

The Lagrangian in terms of the two-component Pauli spinors takes the following form:

$$\mathcal{L} = \psi^\dagger \left(iD_t - m + \frac{\mathbf{D}^2}{2m} + \frac{\mathbf{D}^4}{8m^3} + \frac{e}{2m} \boldsymbol{\sigma} \mathbf{B} + \frac{ie}{8m^2} \boldsymbol{\sigma} ([\mathbf{D} \times \mathbf{E}] - [\mathbf{E} \times \mathbf{D}]) \right.$$

$$\left. + \frac{e}{8m^2} (\mathbf{D}\mathbf{E} - \mathbf{E}\mathbf{D}) + \cdots \right) \psi + \text{antiparticles}. \tag{2.20}$$

Here, everything is expressed in terms of the covariant derivative D_μ. The Lagrangian for the antiparticles looks similar, only the signs of the terms that multiply the matrix β in Eq. (2.19) should be changed. Note also that the resulting Lagrangian is explicitly gauge-invariant. This does not come as a surprise. The original Lagrangian is gauge-invariant, and this property has nothing to do with the $1/m$ expansion. Hence, the result should be gauge-invariant to all orders in this expansion.

It is very instructive to carry out a similar nonrelativistic reduction of a scalar field as some new aspects arise. Here, for simplicity, we restrict ourselves to the free Lagrangian of a single relativistic scalar field $\varphi(x)$:

$$\mathcal{L} = \frac{1}{2} \dot{\varphi}^2 + \frac{1}{2} \varphi \nabla^2 \varphi - \frac{M^2}{2} \varphi^2. \tag{2.21}$$

Introducing a two-component field,

$$\Phi = \begin{pmatrix} \varphi_+ \\ \varphi_- \end{pmatrix}, \qquad \varphi_\pm = \frac{1}{2}\left(\varphi \pm \frac{i}{M} \dot{\varphi} \right), \tag{2.22}$$

with $\dot{\varphi} = \partial \varphi / \partial t$, one can rewrite the original Lagrangian in the following form (up to a total time derivative):

$$\mathcal{L} = M \Phi^\dagger \sigma_3 (i\partial_t - H) \Phi, \qquad H = (\sigma_3 + i\sigma_2)\frac{\mathbf{p}^2}{2M} + M\sigma_3. \tag{2.23}$$

Here, $\sigma_1, \sigma_2, \sigma_3$ are the usual 2×2 Pauli matrices, and $\mathbf{p}$ again denotes the pertinent momentum operator. This Lagrangian can be diagonalized by performing the following linear transformation:

$$\Phi = M^{-1/2}\left(\cosh\frac{\alpha}{2} - \sigma_1 \sinh\frac{\alpha}{2} \right)\Theta, \qquad \tanh\alpha = \frac{\mathbf{p}^2/(2M)}{M + \mathbf{p}^2/(2M)}. \tag{2.24}$$

The resulting Lagrangian in terms of the new two-component field $\Theta = (\theta_+, \theta_-)^T$ is given by

$$\mathcal{L} = \Theta^\dagger \sigma_3 (i\partial_t - w\sigma_3)\Theta = \theta_+^\dagger (i\partial_t - w)\theta_+ + \theta_-^\dagger (-i\partial_t - w)\theta_-, \tag{2.25}$$

with $w(\mathbf{p}) = \sqrt{M^2 + \mathbf{p}^2}$.

In order to obtain the Green's functions in the relativistic theory, one has to supplement the relativistic Lagrangian with the source term $j\varphi$. The path integral can be trivially performed in the free theory. Namely, expressing the relativistic field φ in terms of $\theta_\pm$, it can be shown that

$$j\varphi = j\frac{1}{2\sqrt{w}}(\theta_+ + \theta_-) + (\theta_+^\dagger + \theta_-^\dagger)\frac{1}{2\sqrt{w}} j. \tag{2.26}$$

Treating the fields $\theta_\pm$ in the path integral as independent variables, one gets the following expression for the generating functional:

$$Z(j) = \int d\theta_+ d\theta_- \exp\left\{ i \int d^4 x \left(\theta_+^\dagger (i\partial_t - w)\theta_+ + \theta_-^\dagger (-i\partial_t - w)\theta_- \right. \right.$$

$$\left. \left. + j\frac{1}{2\sqrt{w}}(\theta_+ + \theta_-) + (\theta_+^\dagger + \theta_-^\dagger)\frac{1}{2\sqrt{w}} j \right) \right\}$$

$$= \exp\left\{ -\frac{i}{2} \int d^4 x d^4 y\, j(x) D(x-y) j(y) \right\}, \tag{2.27}$$

where

$$D(x-y) = \int \frac{d^4 p}{(2\pi)^4} \frac{e^{-ip(x-y)}}{2w(\mathbf{p})} \left\{ \frac{1}{w(\mathbf{p}) - p^0 - i\varepsilon} + \frac{1}{w(\mathbf{p}) + p^0 - i\varepsilon} \right\}$$

$$= \int \frac{d^4 p}{(2\pi)^4} \frac{e^{-ip(x-y)}}{M^2 - p^2 - i\varepsilon}. \tag{2.28}$$

It is clear that the propagation of the fields $\theta_\pm$ gives rise to the poles at $p^0 = \pm w(\mathbf{p})$, which sum up to the relativistic propagator. So, integrating out the negative-energy component θ_-, one gets the propagator with one particle pole only, given by the first term in the curly brackets in Eq. (2.28).

We would like to finish this discussion with two remarks. First, let us establish the equal-time commutation relations for the nonrelativistic fields $\theta_\pm$. To this end, we consider the expansion of the free relativistic field into creation and annihilation operators:

$$\varphi(x) = \int \frac{d^3\mathbf{k}}{(2\pi)^3 2w(\mathbf{k})} \left(a(\mathbf{k})e^{-ikx} + a^\dagger(\mathbf{k})e^{ikx} \right).$$ (2.29)

Using Eqs. (2.22) and (2.24), one obtains straightforwardly

$$\theta_+(x) = \int \frac{d^3\mathbf{k}}{(2\pi)^3 \sqrt{2w(\mathbf{k})}} a(\mathbf{k})e^{-iw(\mathbf{k})t + i\mathbf{k}\mathbf{x}},$$

$$\theta_-(x) = \int \frac{d^3\mathbf{k}}{(2\pi)^3 \sqrt{2w(\mathbf{k})}} a^\dagger(-\mathbf{k})e^{iw(\mathbf{k})t + i\mathbf{k}\mathbf{x}}.$$ (2.30)

In Eq. (2.30), reinterpreting $a^\dagger(-\mathbf{k}) = b(\mathbf{k})$ as the annihilation operator of an anti-particle, one may verify once more that $\theta_\pm$ correspond to the positive/negative energy solutions of the free Klein–Gordon equation (or, equivalently, to particles and antiparticles). Further, the equal-time commutation relations are given by

$$[\theta^\dagger_\pm(\mathbf{x},t), \theta_\pm(\mathbf{y},t)] = \pm\delta^{(3)}(\mathbf{x} - \mathbf{y}).$$ (2.31)

This result was expected from the beginning: since the nonrelativistic Lagrangian contains only one power of the time derivative, the canonically conjugated variables are given by the pairs $\theta_\pm, \theta^\dagger_\pm$ and do not involve the time derivative of the scalar fields, as in the relativistic case.

Our second remark concerns the number of independent degrees of freedom. In the beginning, we had two, namely the field and its canonical momentum, or a and $a^\dagger$. At the end, we have two fields, θ_+ and θ_-, each of them containing only a or $a^\dagger$. The canonical momenta are given by $\theta^\dagger_+$ and $\theta^\dagger_-$, respectively, as is manifest in the commutation relations (2.31). These, however, cannot be considered as independent from $\theta_\pm$; otherwise, the number of the independent variables would double. This fact has implications; for example, in the path integral formulation, when the Gaussian integrals over the variables $\theta_\pm$ are evaluated. The calculation yields the pertinent determinant to the power $(-1/2)$ and not -1 that agrees with the relativistic result.

The preceding discussion can be extended to the interacting cases. For example, a scalar particle interacting with the external electromagnetic field is considered in detail in the textbook [2] (more references are given in the literature guide at the end of this chapter). We, however, want to stop our discussion here, because we are not going to use the Foldy–Wouthuysen-type transformations for the derivation of the nonrelativistic Lagrangians anyway. Rather, this was needed just to get a feeling of the general structure of the nonrelativistic field theory before starting to work with them. Such types of transformations will, however, resurface when we consider the low-energy effective theory of QCD in the presence of matter fields (in particular nucleons); see Chapter 4.

In conclusion, we briefly summarize the lessons learned in these simple examples. Using field transformations, we were able to separate the particle and antiparticle degrees of freedom in the original relativistic Lagrangian. The propagators of the particle and antiparticle fields contain one pole each, in contrast to the original relativistic propagator. Finally, the resulting Lagrangian inherits the symmetries of the original theory except, of course, Lorentz-invariance, which is lost in the inverse mass expansion. By summing up the relativistic corrections, however, it is possible to arrive at the original, relativistically invariant expressions.

One may use the hints from these examples and consider the nonrelativistic reduction of the generic relativistic theories. However, new and nontrivial aspects also arise. In the following sections, we shall address these fundamental issues in much detail.

2.3 Construction of the Nonrelativistic Lagrangian for a Scalar Field

Consider first a theory with a single relativistic scalar field $\Phi(x)$, which is described by the Lagrangian

$$\mathcal{L} = \frac{1}{2}(\partial \Phi)^2 - \frac{M^2}{2}\Phi^2 - \frac{\lambda}{4}\Phi^4 + \text{counterterms}. \tag{2.32}$$

The Lagrangian has the symmetry $\Phi \to -\Phi$ and, hence, it describes processes with an even number of external legs. Processes with an equal number of particles in the initial and final states (e.g. $1+2 \to 3+4$, $1+2+3 \to 4+5+6$, and so on) play a special role. In such a case, all external three-momenta can simultaneously be made much smaller than the mass M. Consequently, such processes can be described within a nonrelativistic framework that can be obtained from the preceding relativistic theory. On the other hand, reactions with an unequal number of particles in the initial and final states always involve a mass gap proportional to M. In this case, a nonrelativistic theory cannot be used.

At this point, one might try to move forward along the path discussed in the previous section; that is, to carry out the expansion of the Lagrangian in powers of $1/M$. However, as we already know, radiative corrections will modify the coefficients of this expansion. For this reason, we shall act differently here: we merely write down the most general effective nonrelativistic Lagrangian and then *match* its couplings to the parameters of the underlying relativistic theory perturbatively. At tree level, this will yield the same result as the expansion of the Lagrangian in powers of $1/M$. In addition, we will not consider antiparticles at all. In contrast to a relativistic field theory, this is a consistent procedure here.

We start from the kinetic part that describes the freely moving nonrelativistic particle. We opt to perform the $1/M$ expansion also there. Then, the kinetic Lagrangian takes the following form

$$\mathcal{L}_{\text{kin}} = \phi^\dagger \left(i\partial_t - M + \frac{\nabla^2}{2M} \right)\phi + \phi^\dagger \left(\frac{\nabla^4}{8M^3} + \cdots \right)\phi. \tag{2.33}$$

Here, $\phi(x)$ denotes a nonrelativistic field, which is *not Hermitean,* other than $\Phi(x)$. Namely, $\phi(x)$ contains only an annihilation operator:

$$\phi(x) = \int \frac{d^3\mathbf{k}}{(2\pi)^3} \, e^{-ik^0x^0+i\mathbf{k}\mathbf{x}} \, a(\mathbf{k}), \qquad k^0 = M + \frac{\mathbf{k}^2}{2M}. \tag{2.34}$$

The nonrelativistic propagator has only one pole in the complex k^0-plane:

$$i\langle 0|T\phi(x)\phi^\dagger(y)|0\rangle = \int \frac{d^4k}{(2\pi)^4} \frac{e^{-ik(x-y)}}{M + \mathbf{k}^2/(2M) - k^0 - i\varepsilon}. \tag{2.35}$$

As we have already seen, summing up all relativistic insertions in the two-point function, one arrives at the propagator, which has a pole at $k^0 = \sqrt{\mathbf{k}^2 + M^2}$.

Next, let us consider scattering in the two-particle sector. The simplest local Lagrangian, which can be constructed from the nonrelativistic fields, contains no derivatives:

$$\mathcal{L}_0^{(2)} = C_0 \phi^\dagger \phi^\dagger \phi \phi. \tag{2.36}$$

Note that the Lagrangian has to contain two ϕ and two $\phi^\dagger$ fields; otherwise, it would not be Hermitean. Since ϕ contains only an annihilation operator (and hence $\phi^\dagger$ contains only a creation operator), the preceding vertex will contain an equal number of creation and annihilation operators (i.e., it will conserve particle number).

In order to write down the most general effective Lagrangian, one has to define an appropriate power-counting scheme. In the relativistic theories, which were considered in Chapter 1, the pertinent small parameter was given by a typical small four-momentum, divided by the mass of a heavy particle, which is integrated out. The situation is different here: only three-momenta are small, and a small parameter is given by $|\mathbf{p}|/M$, where M is the mass of the particle itself. So, the expansion is carried out in the three-momenta (spatial derivatives) only. Concerning the time component, as seen from the kinetic part of the Lagrangian, Eq. (2.33), the quantity $p^0 - M$ counts at $O(p^2)$. Here, p denotes a generic small quantity, which for the case at hand is $|\mathbf{p}|/M$. In the Lagrangian, however, we do not need this counting explicitly, as all time derivatives can be systematically removed everywhere except from the kinetic term using the EOM and field redefinitions.

Next, let us construct the most general two-particle Lagrangian at order p^2. It is more convenient to work in momentum space. At tree level, the scattering amplitude is given merely by the matrix element of the Lagrangian between free states; thus, one could equivalently consider the construction of the most general scattering amplitude for the process $\phi(p_1)+\phi(p_2) \to \phi(q_1)+\phi(q_2)$ at $O(p^2)$. Since the total three-momentum is conserved, one may define three linearly independent three-momenta $\mathbf{p} = \frac{1}{2}(\mathbf{p}_1 - \mathbf{p}_2)$, $\mathbf{q} = \frac{1}{2}(\mathbf{q}_1 - \mathbf{q}_2)$ and $\mathbf{P} = (\mathbf{p}_1 + \mathbf{p}_2) = (\mathbf{q}_1 + \mathbf{q}_2)$ in an arbitrary frame. As in the nonrelativistic theory, Lorentz invariance reduces to space rotations, so the amplitude at $O(p^2)$ can depend on the following six invariants:

$$\mathbf{p}^2, \quad \mathbf{q}^2, \quad \mathbf{P}^2, \quad \mathbf{p}\mathbf{P}, \quad \mathbf{q}\mathbf{P}, \quad \mathbf{p}\mathbf{q}. \tag{2.37}$$

One can directly write down the operators in the effective Lagrangian, which yield contributions to the tree amplitude exactly of this type. Further, it can be easily seen that the operators corresponding to $\mathbf{pP}$, $\mathbf{qP}$ and $\mathbf{pq}$ are absent due to the Bose symmetry (they will generally be present in the case of nonidentical particles). Indeed, under the replacement $p_1 \leftrightarrow p_2$ and/or $q_1 \leftrightarrow q_2$ the momenta change as follows: $\mathbf{p} \to -\mathbf{p}, \mathbf{q} \to -\mathbf{q}$ and $\mathbf{P} \to \mathbf{P}$. Consequently, the last three invariants are odd under this replacement and thus cannot be present in the amplitude. Further, Hermiticity ensures that the first two invariants can occur only in the combination $\mathbf{p}^2 + \mathbf{q}^2$. Hence, the effective Lagrangian at $O(p^4)$ contains only two operators:

$$\mathcal{L}_2^{(2)} = C_1 \left((\phi^\dagger \overset{\leftrightarrow}{\nabla}{}^2 \phi^\dagger)(\phi\phi) + \text{h.c.} \right) + C_2 \left((\phi^\dagger \phi^\dagger)\nabla^2(\phi\phi) + \text{h.c.} \right), \qquad (2.38)$$

where $\overset{\leftrightarrow}{\nabla} = \frac{1}{2}(\overset{\leftarrow}{\nabla} - \overset{\rightarrow}{\nabla})$.

The construction of the Lagrangian at higher orders proceeds along similar lines. In order to reduce the number of independent terms, one can use the EOM. This concerns, for example, the so-called off-shell terms, which give vanishing contributions to the S-matrix elements at tree level. An example of such a term at $O(p^4)$ is given by

$$O_{\text{off-shell}} = (\phi^\dagger \overset{\leftrightarrow}{\nabla}{}^4 \phi^\dagger)(\phi\phi) - (\phi^\dagger \overset{\leftrightarrow}{\nabla}{}^2 \phi^\dagger)(\phi \overset{\leftrightarrow}{\nabla}{}^2 \phi) + \text{h.c.} \qquad (2.39)$$

In momentum space, the matrix element of this operator is proportional to

$$(\mathbf{p}^2 - \mathbf{q}^2)^2 = \frac{1}{4}(\mathbf{p}_1^2 + \mathbf{p}_2^2 - \mathbf{q}_1^2 - \mathbf{q}_2^2)^2. \qquad (2.40)$$

It is immediately seen that this matrix element vanishes on the energy shell. Indeed, using the lowest-order EOM for the fields ϕ and $\phi^\dagger$,

$$\nabla^2 \phi = -2M(i\partial_t - M)\phi, \qquad \nabla^2 \phi^\dagger = 2M(i\partial_t + M)\phi^\dagger, \qquad (2.41)$$

one can bring the off-shell operator to the following form:

$$O_{\text{off-shell}} = M^2(i\partial_t)^2(\phi^\dagger \phi^\dagger \phi\phi). \qquad (2.42)$$

Thus, the off-shell term is reduced to a total derivative and can be dropped from the Lagrangian.

The sectors that contain three and more particles can be treated similarly. For example, the lowest-order, nonderivative Lagrangian in the three-particle sector takes the following form:

$$\mathcal{L}_0^{(3)} = D_0 \phi^\dagger \phi^\dagger \phi^\dagger \phi\phi\phi. \qquad (2.43)$$

At this stage, one might start worrying about the proliferation of different Lagrangians that all come with the unknown couplings. The situation is, however, not that bad. As will be seen later, due to the particle number conservation in the nonrelativistic theory, the Lagrangians from the higher sectors do not contribute to processes with lower numbers of particles. For example, if the scattering amplitude in the two-particle sector is considered, the Lagrangians $\mathcal{L}^{(3)}, \mathcal{L}^{(4)}, \dots$ do not matter at all. If processes in the three-particle sector are studied, only $\mathcal{L}^{(2)}$ and $\mathcal{L}^{(3)}$ are relevant, and so on. Such a clear hierarchy simplifies the use of nonrelativistic theories considerably.

Finally, note that nonrelativistic theories containing spinor, vector or other fields can be constructed in the similar fashion. Moreover, one could consider theories in which some particles are heavy (nonrelativistic) and some are light (relativistic). For example, a lowest-order Lagrangian describing a heavy spin-1/2 particle with a mass M and a light scalar with a mass m takes the following form:

$$\mathcal{L} = \psi^\dagger \left(i\partial_t - M + \frac{\nabla^2}{2M} + g\phi \right) \psi + \frac{1}{2}(\partial\phi)^2 - \frac{m^2}{2}\phi^2 + \cdots . \tag{2.44}$$

Here, ψ is the nonrelativistic fermion field (Pauli spinor), whereas ϕ describes the relativistic particle.

2.4 Symmetries

2.4.1 Lorentz Invariance

In the nonrelativistic setting, manifest Lorentz invariance is apparently lost. However, if the nonrelativistic theory is derived from a relativistic one, it still remembers its ancestor. Namely, the S-matrix elements calculated in the resulting nonrelativistic theory are still relativistically invariant (because they are the same as in the relativistic theory, up to a given order in the expansion in the inverse heavy mass). Hence, the matching to the relativistic theory automatically takes care of the relativistic invariance, so that certain relations between the effective nonrelativistic couplings emerge, which guarantee that the S-matrix elements are invariant under Lorentz transformations.

At lowest order in perturbation theory, the nonrelativistic amplitudes are given by the matrix elements of the interaction Lagrangian. From this, one may conclude that the interaction Lagrangian must be relativistically invariant, albeit the nonrelativistic fields, unlike their relativistic counterparts, transform in a very complicated, nonlinear way with respect to Lorentz boosts (see, e.g., Ref. [3]), and the invariance is not explicit. As already mentioned, Lorentz invariance manifests itself in relations between the various couplings of the effective Lagrangian, which appear at different orders. A very convenient way to systematically derive these constraints is provided by the so-called reparameterization invariance [4], which we consider in what follows. For simplicity, we limit our consideration to scalar fields. We start by noting that, introducing the nonrelativistic field ϕ, say, in Eq. (2.33), instead of the relativistic one, Φ, one has implicitly fixed the reference frame, treating the time t and the space coordinates $\mathbf{x}$ differently. This choice of the frame can be characterized by a timelike unit four-vector $v_\mu = (1, \mathbf{0})$ with $v^2 = 1$. The nonrelativistic Lagrangian in Eq. (2.33) can then be rewritten as follows:

$$\mathcal{L}_{\text{kin}} = \phi_v^\dagger \left(iv\partial - M - A\frac{\partial_v^2}{2M} + \cdots \right) \phi_v . \tag{2.45}$$

Here, we made explicit the dependence of the field ϕ_v on the choice of the unit vector v and introduced the notation $\partial_v^2 = \partial^2 - (v \cdot \partial)^2$. Note also that A in the Eq. (2.45) denotes an arbitrary constant. Of course, if the kinetic part of the effective Lagrangian is obtained through the expansion of the exact expression of the relativistic energy, then $A = 1$;, see, for example, Refs. [4, 5]. However, we want to keep this constant free in the beginning and show that the constraint $A = 1$ can be obtained by using the reparameterization invariance, which can be directly related to the requirement of the relativistic invariance of the Lagrangian.

We start our argument by noting that there is nothing special about the choice of the unit vector v in the direction of the time axis. Performing a generic Lorentz boost, we obtain another timelike unit vector w and can work with the nonrelativistic fields ϕ_w instead of ϕ_v. It is very convenient to interpret the transformation $v \to w$ in the following way. In momentum space, the introduction of the vector v is equivalent to the splitting of the four-momentum of the particle as

$$p_\mu = M v_\mu + k_\mu, \tag{2.46}$$

where the residual four-momentum k_μ is small, $v_\mu k^\mu \ll M$. This splitting is not unique since the pair $(v + q/M, k - q)$, with $q = O(1)$, corresponds to the same momentum p_μ, as the original pair (v, k). Further, since the vector $w_\mu = v_\mu + q_\mu/M$ should be a timelike unit vector again, we have

$$w^2 = \left(v + \frac{q}{M} \right)^2 = v^2 + \frac{2v \cdot q}{M} + \frac{q^2}{M^2} = 1. \tag{2.47}$$

This yields $2Mv \cdot q = -q^2$. It is easy to see now that (a) each transformation, corresponding to some vector q_μ, which obeys this constraint, is an (infinitesimal) Lorentz boost with the parameters of order of $1/M$, and, vice versa, (b) each infinitesimal Lorentz transformation with parameters of order of $1/M$ can be represented by such a transformation with some q_μ. Hence, the invariance of the Lagrangian under these transformations (reparameterization invariance [4]) is equivalent to the statement about the Lorentz invariance of the Lagrangian.

Obviously, the fields ϕ_v and ϕ_w are related as

$$\phi_w(x) = e^{iqx} \phi_v(x), \qquad w = v + \frac{q}{M}. \tag{2.48}$$

The kinetic part of the Lagrangian after this transformation becomes

$$\mathcal{L}_{\text{kin}} \to \phi_w^\dagger \left(v \cdot (i\partial + q) - M - \frac{A}{2M} \left((\partial - iq)^2 - (v \cdot \partial - iv \cdot q)^2 \right) + \cdots \right) \phi_w. \tag{2.49}$$

Expressing v through w and expanding in the inverse powers of M, up to the terms of order M^{-2}, one gets

$$\mathcal{L}_{\text{kin}} \to \phi_w^\dagger \left(i \left(w - \frac{q}{M} \right) \cdot (\partial - iq) - M \right.$$

$$\left. - \frac{A}{2M} \left((\partial - iq)^2 - \left(w \cdot \partial - \frac{q \cdot \partial}{M} - iw \cdot q + i\frac{q^2}{M} \right)^2 \right) + \cdots \right) \phi_w$$

$$= \phi_w^\dagger \left(iw \cdot \partial - M - A\frac{\partial_w^2}{2M} + (A - 1) \left(\frac{iq \cdot \partial}{M} - \frac{q^2}{2M} \right) + \cdots \right) \phi_w. \qquad (2.50)$$

Here, the relation $2Mw \cdot q = q^2$ was used. It is now seen that the invariance of the Lagrangian requires $A = 1$. Of course, we had this result before (at tree level). Our present result goes beyond this; it tells us that, if a Lorentz-invariant regularization is used in the path integral, $A = 1$ holds at all orders.

One may pursue this method to higher orders in $1/M$, as well as for the particles with nonzero spins. In particular, many useful constraints on the couplings of NRQED and NRQCD were established; see, for example, Refs. [3, 4, 6, 7]. At higher orders in the inverse mass expansion, the simple reparameterization transformations, considered previously, need to be amended [7]. Still, the main advantage of using reparameterization invariance is that it allows to restrict the number of independent couplings in the Lagrangian *without any specific reference to the underlying relativistic theories and prior to the matching.* On the other hand, matching to the relativistic theories always adjusts these couplings so that the constraints are automatically obeyed [5].

We end this section by briefly mentioning the role of Galilean invariance. In the non-relativistic limit Lorentz transformations reduce to Galilei transformations. Hence, the kinetic term of the lowest-order nonrelativistic Lagrangian is invariant under Galilei transformations. Moreover, the derivative $\overleftrightarrow{\nabla} = (\overleftarrow{\nabla} - \overrightarrow{\nabla})/2$, introduced in the previous section, is Galilean-invariant and, hence, all interaction terms, which contain only powers of $\overleftrightarrow{\nabla}$, are Galilean-invariant as well. Note also that the terms describing the motion of the center of mass of two particles are *not* Galilean-invariant. As we shall see in what follows, the couplings in front of such terms are suppressed by inverse powers of M. Thus, in the two-body sector, a Galilean-invariant theory is obtained from the general nonrelativistic Lagrangian by dropping all relativistic kinetic terms ($O(p^4)$) and higher, as well as all interaction terms that are related to the motion of the center of mass).

The Galilean invariance in a theory, which describes several particles with different masses, requires further scrutiny. We refer here to the situation that occurs when the transitions between different channels with slightly different masses are allowed. (It is assumed that the mass gap between different channels is of the order of the small three-momenta and is much smaller than the masses of individual particles.) An example is given by specific transitions between charmed mesons [8]. More precisely, consider the transitions $D^{*0} \leftrightarrow D^0\pi^0$. The masses of the involved particles are $M_{D^{*0}} = 2006.85\,\text{MeV}$, $M_{D^0} = 1864.84\,\text{MeV}$ and $M_{\pi^0} = 134.98\,\text{MeV}$. The mass gap is $\delta = M_{D^{*0}} - M_{D^0} - M_{\pi^0} = 7.03\,\text{MeV}$, which is much smaller than the masses of all particles participating the reaction. Galilean invariance, however, *requires* that the mass gap δ exactly vanishes in the kinetic term, since the mass is a conserved quantity in the nonrelativistic theory. Hence, in order to impose Galilean invariance, one has to treat the parameter δ as a perturbation. In this case, the theory is invariant only at the lowest order in δ [8].

2.4.2 Discrete Symmetries

The discrete symmetries are parity transformations, $\mathbf{x} \to -\mathbf{x}$, charge conjugation and time reversal,[3] $t \to -t$. Let us work out how these are realized in an NREFT.

Parity: The realization of parity in the nonrelativistic framework is straightforward. The transformation rules for the nonrelativistic fields under parity can be directly derived from those for their relativistic counterparts.

Charge conjugation: In the relativistic theory, particles and antiparticles are described by the same local relativistic field. This is different in the nonrelativistic framework, where one has separate fields for a particle and an antiparticle. A nice example of this is provided by the triplet of pions. Choosing the Cartesian basis, the relativistic Lagrangian can be written as (assuming one common mass for the pion fields)

$$\mathcal{L} = \frac{1}{2} (\partial \mathbf{\Phi})^2 - \frac{M_\pi^2}{2} \mathbf{\Phi}^2 + \cdots, \tag{2.51}$$

where the ellipsis denotes the interaction terms. Defining the physical fields as

$$\Phi_\pm = \frac{1}{\sqrt{2}} (\mp \Phi_1 + i \Phi_2), \qquad \Phi_0 = \Phi_3, \tag{2.52}$$

we get

$$\mathcal{L} = -\partial \Phi_+ \partial \Phi_- + \frac{1}{2} (\partial \Phi_0)^2 + M_\pi^2 \Phi_+ \Phi_- - \frac{M_\pi^2}{2} \Phi_0^2 + \cdots. \tag{2.53}$$

Charge conjugation corresponds to $\Phi_\pm \to -\Phi_\mp$ and $\Phi_0 \to \Phi_0$. The relativistic Lagrangian is invariant under this transformation. Note that the charged fields contain both particle and antiparticle degrees of freedom; it is impossible to separate them without violating locality.

Now the nonrelativistic Lagrangian in terms of the Cartesian fields is given by

$$\mathcal{L} = \boldsymbol{\phi}^\dagger \left(i \partial_t - M_\pi + \frac{\nabla^2}{2 M_\pi} + \cdots \right) \boldsymbol{\phi} + \cdots. \tag{2.54}$$

Defining the physical fields $\phi_\pm$, ϕ_0 in analogy with Eq. (2.52), we get

$$\mathcal{L} = \sum_{i=\pm,0} \phi_i^\dagger \left(i \partial_t - M_\pi + \frac{\nabla^2}{2 M_\pi} + \cdots \right) \phi_i + \cdots. \tag{2.55}$$

The nonrelativistic fields $\phi_\pm$, ϕ_0 are separated in the kinetic part. This means that, for example, one may even completely discard ϕ_- while retaining ϕ_+ (if one is interested in those two-particle processes, where a negatively charged pion never appears in the

---

[3] As time is directional, it would be better to speak of motion reversal. We adhere, however, to the commonly used notation.

intermediate as well as in the initial or final state). Obviously, such a theory will not be invariant under charge conjugation. Still, it is a valid nonrelativistic theory equivalent to the initial (invariant) relativistic theory in the sector with fixed total charge Q.

Reversing the argument, one may write down the nonrelativistic Lagrangian, containing all the fields $\phi_\pm, \phi_0$, and multiply each term in the Lagrangian by an arbitrary coupling. The invariance under charge conjugation can then be imposed by hand, assuming that the couplings, multiplying similar terms with ϕ_+ and ϕ_-, are the same. The same result will, of course, emerge if the nonrelativistic couplings are determined from matching to the (invariant) underlying relativistic theory. To summarize, the invariance under charge conjugation is realized differently in the two theories: while the relativistic Lagrangian is constructed from invariant operators from the beginning, invariance of the nonrelativistic Lagrangian is achieved by adjusting the couplings for the operators that contain charge-conjugated fields.

Time reversal: In field theory, time reversal is realized through an antiunitary operator T. Invariance with respect to time reversal implies that the matrix elements of the S-matrix between arbitrary asymptotic states α, β obey the following relation:

$$\langle\beta|S|\alpha\rangle = \langle\alpha|T^{-1}ST|\beta\rangle. \tag{2.56}$$

Taking into account the antiunitarity of the operator T, we get

$$\langle\alpha|S|\beta\rangle = \langle\alpha|T^{-1}TST^{-1}T|\beta\rangle^* = \langle\beta|TST^{-1}|\alpha\rangle^*. \tag{2.57}$$

This leads to $TST^{-1} = S^\dagger$. Hence, the Lagrangian of a theory, which is invariant under time reversal, should obey the following relation:

$$T\mathcal{L}(\mathbf{x},t)T^{-1} = \mathcal{L}^\dagger(\mathbf{x},-t). \tag{2.58}$$

Note that one usually considers Hermitean Lagrangians in field theory, and no Hermitean conjugate appears explicitly on the right-hand side of the above equation. We wish, however, to keep the formalism as general as possible. One particular case where this might be useful will be considered here. Let us start from a theory that contains a triplet of pions, interacting through both strong and electromagnetic interactions. Let us further restrict ourselves to the scattering in the two-pion sector with total electric charge $Q = 0$. We wish to write down the nonrelativistic Lagrangian, which describes such a process with small three-momenta. For example, the lowest-order nonderivative Lagrangian, which describes the elastic process $\pi^+\pi^- \to \pi^+\pi^-$, is given by

$$\mathcal{L}_{+-,+-} = C_0\phi_+^\dagger\phi_-^\dagger\phi_+\phi_- + \cdots. \tag{2.59}$$

Further, under time reversal, the nonrelativistic fields transform as

$$T\phi_\pm(\mathbf{x},t)T^{-1} = \eta_T\phi_\pm(\mathbf{x},-t), \qquad \eta_T^2 = 1. \tag{2.60}$$

It is now seen that the above Lagrangian is invariant under time reversal, *even if* $\mathrm{Im}\,C_0 \neq 0$.

It is interesting to discuss where the imaginary part of C_0 arises from, rendering the nonrelativistic Lagrangian non-Hermitean. In Fig. 2.2 some diagrams based on the relativistic theory are displayed. These correspond to two-photon annihilation of a $\pi^+\pi^-$

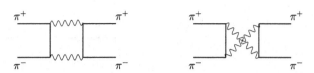

Figure 2.2 Diagrams with two photons in the intermediate state, which contribute to the process $\pi^+\pi^- \to \pi^+\pi^-$ in the relativistic theory.

pair in the intermediate state. Since the mass gap of such a process is of order of $2M_\pi$, it cannot be included in the nonrelativistic Lagrangian explicitly. However, since in this theory one is anyway interested only in the process $\pi^+\pi^- \to \pi^+\pi^-$, one may choose to include the diagrams shown in Fig. 2.2 into the low-energy effective couplings. To this end, one has to evaluate the Feynman integrals in the relativistic theory and expand the result in Taylor series in the external three-momenta of the pions. The presence of the mass gap guarantees that this is possible, and the higher-order terms in the expansion are suppressed by the inverse powers of the pion mass. The coefficients of the expansion can then be identified with the contributions of the local four-pion terms (the lowest-order term of this type is shown in Eq. (2.59)). Note that the pertinent contribution to, for example, the coupling C_0 is *complex,* because the diagrams in Fig. 2.2 have nonzero imaginary part at threshold. This example shows that: a) invariance under time reversal can be implemented in the nonrelativistic theory in a standard way, b) non-Hermitean Lagrangians in the effective theories are not exotic, they emerge after "shielding" some physical intermediate states and relegating their contribution to the effective couplings, and c) one may "integrate out" not only heavier, but also lighter degrees of freedom as well. In this case unitarity, which is observed in the underlying theory, is restored by the contributions from the imaginary parts of the effective couplings.

2.5 Perturbation Theory, Matching and the Effective Range Expansion

2.5.1 Two-Point Function of a Massive Field

NREFT is a full-fledged field theory in which the amplitudes are given by a sum of all pertinent Feynman diagrams. However, the structure of the nonrelativistic propagator (the presence of a single pole instead of two poles in the relativistic theory) and of the nonrelativistic vertices (particle number conservation in each vertex) leads to dramatic simplifications, as many of the diagrams vanish. Indeed, from Eq. (2.35) one may conclude that the particles propagate only forward in time. The propagator vanishes, when $x^0 < y^0$, since in this case the integration contour over the variable k^0 can be closed in the upper half plane, where the integrand has no poles. Hence, all diagrams except those where all particles propagate forward in time can be dropped.

The only mathematical complication that might potentially arise here is related to the fact that the propagator is undefined at $x^0 = y^0$. Here, in a few examples, we shall

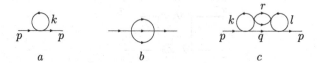

Figure 2.3 Loop corrections to the two-point Green's function in the nonrelativistic theory: (a) the tadpole, (b) three particles in the intermediate state (this diagram is absent in the nonrelativistic theory due to particle number conservation) and (c) a four-loop diagram, which is allowed by the particle number conservation.

show how this problem can be handled by using dimensional regularization (of course, physical results are regularization-independent). Let us start from the one-particle sector of the theory and calculate the two-point function of the heavy field. As we have already seen (see, e.g., Fig. 2.1), summing up tree-level contributions from the kinetic term leads to the relativistic dispersion law $k^0 = \sqrt{M^2 + \mathbf{k}^2}$. We would like to show now that there are no other contributions to the two-point function. The simplest contribution might arise from the tadpole diagram, shown in Fig. 2.3a. Its contribution is proportional to

$$I_a \propto \int \frac{d^D k}{(2\pi)^D i} S(k), \qquad S(k) = \frac{1}{E_\mathbf{k} - k^0 - i\varepsilon}, \qquad (2.61)$$

where $E_\mathbf{k} = M + \mathbf{k}^2/2M$ is the nonrelativistic energy of a particle with a three-momentum $\mathbf{k}$. Note that we have used dimensional regularization in Eq. (2.61). Performing the integral over the variable k^0 first, one gets $i\pi$ as the result. Performing the integration over space components afterwards yields a vanishing result, that is, $I_a = 0$ in dimensional regularization.[4]

A similar pattern is realized in the diagrams containing more loops, like those shown in Fig. 2.3b,c. It can be shown that all such diagrams vanish after performing the Cauchy integral over the fourth components of the loop momenta. We do not consider these calculations in detail here.

2.5.2 Perturbation Theory in the Two-Particle Sector

The same method can be applied to study the Green's functions in the two-, three- and more particle sectors. It is now clear why the sectors are separated from each other (e.g., why the three-particle coupling in Eq. (2.43) does not contribute to the two-particle scattering amplitude). The reason is that such a contribution would necessarily involve some lines in the Feynman diagrams with particles going backwards in time,

---

[4] Note that here we cannot use the argument of closing the integration contour over k^0 in the upper half-plane to avoid a pole. This is seen immediately because, by the same token, one might close the contour in the lower half-plane and include the pole that leads to a different result. The reason for this is that the integrand does not decrease sufficiently fast to justify neglecting the integral over the semicircle at infinity. In fact, the integral diverges logarithmically there. Using the method given here allows us to demonstrate that $I_a = 0$ in dimensional regularization without resorting to Cauchy's theorem and closing the integration contour at infinity.

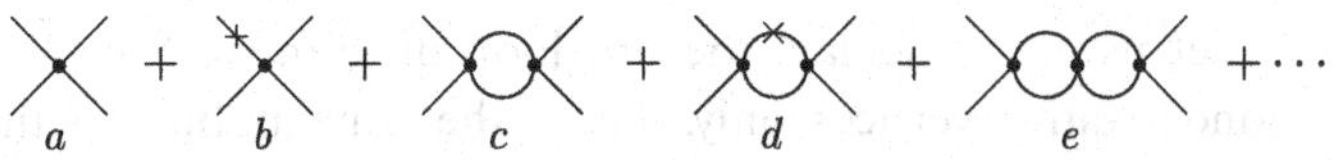

Figure 2.4
Bubble graphs that contribute to the nonrelativistic, two-particle scattering amplitude. The filled circle denotes the local four-particle vertices, with and without derivatives, and the crosses represent the relativistic insertions in the external and internal lines.

and these can be shown to vanish with the use of the arguments just presented. Hence, only certain classes of loop diagrams can contribute. In the case of the two-particle scattering amplitude, which will be considered next, these are solely the so-called bubble diagrams, shown in Fig. 2.4. Summing up these diagrams, we arrive at the familiar Lippmann–Schwinger equation for the amplitude, which is known from nonrelativistic quantum mechanics.

Let us now consider the matching procedure in the two-particle sector in more detail. Assume, for simplicity, that our theory contains a single massive scalar field $\phi(x)$. One starts with the calculation of the four-point function in the non-relativistic theory:

$$G(p_1, p_2; q_1, q_2) = \int d^4x_1 d^4x_2 d^4y_1 d^4y_2 \, e^{ip_1x_1 + ip_2x_2 - iq_1y_1 - iq_2y_2}$$

$$\times \langle 0|T\phi(x_1)\phi(x_2)\phi^\dagger(y_1)\phi^\dagger(y_2)|0\rangle. \tag{2.62}$$

Next, we could sum up all tree-level relativistic insertions in the external lines. (As we already know, there are no more contributions in the two-point functions in the nonrelativistic theory containing massive fields only.) Consequently, the poles in the external momenta move to the right place, determined by the relativistic dispersion law. Applying the LSZ formalism and taking into account the fact that the wave function renormalization constant for the nonrelativistic field is, obviously, equal to one, for the scattering amplitude one gets

$$i(2\pi)^4\delta^{(4)}(p_1 + p_2 - q_1 - q_2)T(p_1, p_2; q_1, q_2)$$

$$= \lim_{p_i^0 \to w(\mathbf{p}_i), q_j^0 \to w(\mathbf{q}_j)} \prod_{i=1}^{2}(w(\mathbf{p}_i) - p_i^0)\prod_{j=1}^{2}(w(\mathbf{q}_j) - q_j^0)G(p_1, p_2; q_1, q_2). \tag{2.63}$$

The external momenta are on the mass shell, that is, $p_i^2 = q_i^2 = M^2$, with $i = 1, 2$. Note, however, that here the relativistic insertions in the internal lines are not resummed, but are treated at a given order in perturbation theory.

Next, we shall consider various contributions to the scattering amplitude, arising from the individual diagrams in Fig. 2.4 (note that the external legs are already removed). The contributions from the tree diagrams Fig. 2.4a,b can be read off from the Lagrangians (2.36) and (2.38):

$$T_{\text{tree}} = 4C_0 - C_1\left((\mathbf{p}_1 - \mathbf{p}_2)^2 + (\mathbf{q}_1 - \mathbf{q}_2)^2\right) - 8C_2\mathbf{P}^2 + O(p^4), \tag{2.64}$$

where $P = p_1 + p_2 = q_1 + q_2$ is the total four-momentum.

Let us now calculate the one-loop diagram in Fig. 2.4c, assuming that we have nonderivative vertices only. Then, the contribution to the scattering amplitude in dimensional regularization is

$$
T_{1-\text{loop}}^{\text{non}-\text{der.}} = \frac{1}{2} (4C_0)^2 \int \frac{d^D k}{(2\pi)^D i} S(k) S(P-k)
$$

$$
= 8C_0^2 \int \frac{d^d k}{(2\pi)^d} \frac{1}{E_{\mathbf{k}} + E_{\mathbf{P-k}} - P^0 - i\varepsilon} = (4C_0)^2 \frac{iM\tilde{q}_0}{8\pi}, \qquad (2.65)
$$

where $d = D - 1$ and

$$
\tilde{q}_0^2 = M \left(P^0 - 2M - \frac{\mathbf{P}^2}{4M} \right). \qquad (2.66)
$$

A very important property of the loop integral in dimensional regularization is now explicit: The quantity $\tilde{q}_0$ is of order p and vanishes at threshold. Moreover, the contribution of the two-loop diagram with nonderivative vertices shown in Fig. 2.4e can be immediately written down:

$$
T_{2-\text{loop}}^{\text{non}-\text{der.}} = (4C_0)^3 \left(\frac{iM\tilde{q}_0}{8\pi} \right)^2. \qquad (2.67)
$$

The generalization to higher orders is straightforward. Each loop adds an $O(p)$ factor $\tilde{q}_0$ that is consistent with the power counting. Indeed, counting the kinetic energy $k^0 - M$ as $O(p^2)$, three-momenta as $O(p)$ and each propagator as $O(p^{-2})$, for a loop we get $O(p^{2+3\cdot1-2\cdot2}) = O(p)$. Consequently, in order to carry out the calculations at a given order in the nonrelativistic theory, we may truncate the perturbative expansion at a certain maximal number of loops. This property of the nonrelativistic theory is extremely useful, since it allows us to obtain relations between the effective couplings and the experimentally measurable quantities in the two-body scattering (the parameters of the effective range expansion), which hold exactly to all orders in perturbation theory.

In order to complete the discussion of the power counting, it suffices to consider in addition relativistic insertions and derivative vertices. The relativistic contribution, depicted in Fig. 2.4d, leads to the following Feynman integral:

$$
T_{1-\text{loop}, 1-\text{ins.}}^{\text{non}-\text{der.}} = (4C_0)^2 \int \frac{d^D k}{(2\pi)^D i} \frac{\mathbf{k}^4}{8M^3} S(k)^2 S(P-k)
$$

$$
= (4C_0)^2 \frac{i\tilde{q}_0}{64\pi M} \left(5\tilde{q}_0^2 + \frac{5\mathbf{P}^2}{2} + \frac{\mathbf{P}^4}{16\tilde{q}_0^2} \right). \qquad (2.68)
$$

It is immediately seen that the counting rules are obeyed, as an insertion of an $O(p^4)$ operator gives a result at $O(p^{2+3\cdot1+4-3\cdot2}) = O(p^3)$. Moreover, relativistic insertions lead to the correct placement of the cut of the one-loop function in the complex plane. The shift of the position of the cut is signaled by the presence of the nonanalytic term $\mathbf{P}^4/(16\tilde{q}_0^2)$ in the brackets. Indeed, expanding the fully relativistic expression

$$
q_0^2 = \frac{\mathbf{P}^2}{4} - M^2 = \frac{1}{4}\left(P^0 - \sqrt{\mathbf{P}^2 + 4M^2}\right)\left(P^0 + \sqrt{\mathbf{P}^2 + 4M^2}\right)
$$

$$
= \tilde{q}_0^2\left(1 + \frac{\mathbf{P}^2}{8M^2} + \frac{\tilde{q}_0^2}{4M^2} + \frac{\mathbf{P}^4}{64M^2\tilde{q}_0^2} + \cdots\right), \tag{2.69}
$$

one gets

$$
(4C_0)^2\frac{iMq_0}{8\pi} = (4C_0)^2\frac{iM\tilde{q}_0}{8\pi}\left(1 + \frac{\mathbf{P}^2}{16M^2} + \frac{\tilde{q}_0^2}{8M^2} + \frac{\mathbf{P}^4}{128M^2\tilde{q}_0^2} + \cdots\right). \tag{2.70}
$$

It is seen that the nonanalytic contribution is exactly reproduced. Later, in Section 2.7, we shall explicitly demonstrate that the insertions in the internal lines can be resummed to all orders, and the result indeed reproduces the correct relativistic expression.

Next, we consider the insertion of the derivative vertices in the loops. Take, for instance, the diagram in Fig. 2.4c and replace one of the vertices by the vertex with two derivatives. The pertinent Feynman diagram is given by

$$
T_{1-\text{loop}}^{\text{two-der.}} = -\frac{1}{2}(4C_0)\int\frac{d^Dk}{(2\pi)^Di}\,S(k)S(P-k)
$$

$$
\times\left(16C_2\mathbf{P}^2 + 2C_1(2\mathbf{k}-\mathbf{P})^2 + C_1((\mathbf{p}_1-\mathbf{p}_2)^2 + (\mathbf{q}_1-\mathbf{q}_2)^2)\right)
$$

$$
= -\frac{iM\tilde{q}_0}{8\pi}(4C_0)\left(16C_2\mathbf{P}^2 + 8C_1\tilde{q}_0^2 + C_1((\mathbf{p}_1-\mathbf{p}_2)^2 + (\mathbf{q}_1-\mathbf{q}_2)^2)\right).
$$

$$
\tag{2.71}
$$

Again, the power counting holds: Adding two powers of momenta to the integrand increases the power of the result by two. It is now clear that the method works at higher orders exactly in the same way. However, we would like to stress here once more that the nice power counting emerges in dimensional regularization only. Using, for example, cutoff regularization, we would arrive at a far messier picture.

2.5.3 Matching in the Two-Particle Sector

The S-matrix elements in the relativistic and nonrelativistic theories should be equal to each other. This leads to the following matching condition between the amplitudes:

$$
T(p_1, p_2; q_1, q_2) = \prod_{i=1}^{2}\frac{1}{(2w(\mathbf{p}_i))^{1/2}(2w(\mathbf{q}_i))^{1/2}}A(p_1, p_2; q_1, q_2), \tag{2.72}
$$

where $A(p_1, p_2; q_1, q_2) = A(s, t)$ denotes the relativistic amplitude, and the kinematic factor on the right-hand side (r.h.s.) stems from the different normalization of the one-particle states in the relativistic and the nonrelativistic theories:

$$\langle p|q\rangle = (2\pi)^3 2w(\mathbf{p})\delta^{(3)}(\mathbf{p}-\mathbf{q}) \quad \text{(relativistic)},$$

$$\langle p|q\rangle = (2\pi)^3 \delta^{(3)}(\mathbf{p}-\mathbf{q}) \quad \text{(nonrelativistic)}. \tag{2.73}$$

The matching condition Eq. (2.72) should be understood in the following way: Both the left-hand and right-hand sides should be expanded in a series of all three-momenta, and the nonrelativistic couplings should be adjusted so that these expansions coincide up to a given order. Note, however, that there are terms on both sides that are *not* low-energy polynomials in three-momenta, namely the terms that contain odd powers of q_0. A freedom in choosing the values of the effective constants allows one to match the polynomial parts of the amplitudes. Should it turn out that the non-polynomial parts on both sides are not the same, the matching would fail. However, as expected, this is not the case. The whole non-polynomial part of the relativistic amplitude in the vicinity of the elastic threshold $s = 4M^2$ (i.e., in the region, where the nonrelativistic approach is applicable) is reproduced automatically at a given order, once the polynomial part is matched at the pertinent order.

From the preceding discussion one may also conclude that it would be very convenient to find a formulation of the matching condition, which does not include such non-polynomial terms at all. In the two-particle sector, this is indeed possible. To this end, one has to define the so-called K-matrix, which is related to the scattering amplitude through the following equation:

$$A(p_1, p_2; q_1, q_2) = K(p_1, p_2; q_1, q_2) + \frac{i}{4} \int \frac{d^3\mathbf{k}_1}{(2\pi)^3 2w(\mathbf{k}_1)} \frac{d^3\mathbf{k}_2}{(2\pi)^3 2w(\mathbf{k}_2)}$$

$$\times K(p_1, p_2; k_1, k_2)(2\pi)^4 \delta^{(4)}(q_1 + q_2 - k_1 - k_2) A(k_1, k_2; q_1, q_2). \tag{2.74}$$

Performing the partial-wave expansion in the CM frame,

$$A(p_1, p_2; q_1, q_2) = 4\pi \sum_{\ell=0}^{\infty} Y_{\ell m}(\hat{p}) A_\ell(s) Y_{\ell m}^*(\hat{q}),$$

$$K(p_1, p_2; q_1, q_2) = 4\pi \sum_{\ell=0}^{\infty} Y_{\ell m}(\hat{p}) K_\ell(s) Y_{\ell m}^*(\hat{q}), \tag{2.75}$$

where $Y_{\ell m}$ are the spherical harmonics and $\hat{p}, \hat{q}$ denote unit vectors in the direction of the relative three-momenta in the outgoing and incoming states, respectively. On the energy shell, we obtain an algebraic relation between T_ℓ and K_ℓ:

$$A_\ell(s) = \frac{16\pi\sqrt{s}}{q_0 \cot \delta_\ell(q_0) - iq_0}, \qquad K_\ell(s) = \frac{16\pi\sqrt{s}}{q_0} \tan \delta_\ell(q_0). \tag{2.76}$$

Here, $\delta_\ell(q_0)$ denotes the phase shift in the ℓth partial wave, and $q_0^2 = s/4 - M^2$.

The effective-range expansion (ERE) takes the form

$$q_0 \cot \delta_\ell(q_0) = q_0^{-2\ell} \left(-\frac{1}{a_\ell} + \frac{1}{2} r_\ell q_0^2 + O(q_0^4) \right).$$

(2.77)

Here, a_ℓ and r_ℓ are the scattering length and the effective range, respectively.

Next, note that the nonrelativistic scattering amplitude T, which was considered in the previous section, obeys the Lippmann–Schwinger equation. In order to derive this equation, we first obtain the canonical Hamiltonian from the given nonrelativistic Lagrangian density and then split it into the free and interaction terms, $\mathbf{H} = \mathbf{H}_0 + \mathbf{H}_I$. In order to ease the bookkeeping, it is convenient to include all relativistic corrections in the free part $\mathbf{H}_0$. Then, the interaction part contains only four-particle vertices:

$$\mathbf{H}_I = - \int d^3\mathbf{x} \left(C_0 \phi^\dagger(\mathbf{x},0)\phi^\dagger(\mathbf{x},0)\phi(\mathbf{x},0)\phi(\mathbf{x},0) + \cdots \right).$$

(2.78)

The full Green's function $\mathbf{G}(E)$ and the T-operator $\mathbf{T}(E)$ are defined through the equations

$$\mathbf{G}(E) = \frac{1}{E - \mathbf{H}} = \mathbf{G}_0(E) + \mathbf{G}_0(E)\mathbf{H}_I\mathbf{G}(E),$$

$$\mathbf{G}_0(E) = \frac{1}{E - \mathbf{H}_0},$$

$$\mathbf{T}(E) = \mathbf{H}_I + \mathbf{H}_I\mathbf{G}_0(E)\mathbf{T}(E),$$

$$\mathbf{G}(E) = \mathbf{G}_0(E) + \mathbf{G}_0(E)\mathbf{T}(E)\mathbf{G}_0(E).$$

(2.79)

Sandwiching the third equation between plane-wave two-particle states, one gets (using again dimensional regularization)

$$T(p_1, p_2; q_1, q_2; E) = V(p_1, p_2; q_1, q_2) + \frac{1}{2} \int \frac{d^d k_1}{(2\pi)^d} \frac{d^d k_2}{(2\pi)^d}$$

$$\times \frac{V(p_1, p_2; k_1, k_2)(2\pi)^d \delta^{(d)}(\mathbf{q}_1 + \mathbf{q}_2 - \mathbf{k}_1 - \mathbf{k}_2)T(k_1, k_2; q_1, q_2; E)}{w(\mathbf{k}_1) + w(\mathbf{k}_2) - E - i\varepsilon},$$

(2.80)

with

$$T(p_1, p_2; q_1, q_2; E) = -(2\pi)^d \delta^{(d)}(\mathbf{p}_1 + \mathbf{p}_2 - \mathbf{q}_1 - \mathbf{q}_2)\langle p_1, p_2 | \mathbf{T}(E) | q_1, q_2 \rangle,$$

$$V(p_1, p_2; q_1, q_2) = -(2\pi)^d \delta^{(d)}(\mathbf{p}_1 + \mathbf{p}_2 - \mathbf{q}_1 - \mathbf{q}_2)\langle p_1, p_2 | \mathbf{H}_I | q_1, q_2 \rangle.$$

(2.81)

Equations (2.80) and (2.81) are written down in d dimensions, anticipating the need for regularization, and the negative signs in Eq. (2.81) are chosen in accordance to the usual sign convention $S = 1 + iT$ for field-theoretical amplitudes. On the energy shell, $E = w(\mathbf{p}_1) + w(\mathbf{p}_2) = w(\mathbf{q}_1) + w(\mathbf{q}_2)$, the amplitude T coincides with the nonrelativistic T-matrix considered in the previous section. Note that, owing to our choice of the free Hamiltonian, the free particles obey the exact relativistic dispersion law. Calculating the scattering amplitude term by term in the Born series, one should first expand the energy denominators in powers of $1/M$ and then evaluate the integrals in dimensional regularization. This prescription, which is implicit in Eq. (2.80), guarantees that the results obtained in this way are consistent with our findings in the previous section.

The argument now goes as follows. The denominator in Eq. (2.80) can be split into two pieces:

$$\frac{1}{w(\mathbf{k}_1) + w(\mathbf{k}_2) - E - i\varepsilon} = \frac{\text{P.V.}}{w(\mathbf{k}_1) + w(\mathbf{k}_2) - E - i\varepsilon} + i\pi\delta(w(\mathbf{k}_1) + w(\mathbf{k}_2) - E).$$

(2.82)

Then, for $E = w(\mathbf{q}_1) + w(\mathbf{q}_2)$, Eq. (2.80) can be rewritten as a system of two equations:

$$R(p_1, p_2; q_1, q_2; E) = V(p_1, p_2; q_1, q_2) + \frac{1}{2} \text{P.V.} \int \frac{d^d\mathbf{k}_1}{(2\pi)^d} \frac{d^d\mathbf{k}_2}{(2\pi)^d}$$

$$\times \frac{V(p_1, p_2; k_1, k_2)(2\pi)^d\delta^{(d)}(\mathbf{q}_1 + \mathbf{q}_2 - \mathbf{k}_1 - \mathbf{k}_2)R(k_1, k_2; q_1, q_2; E)}{w(\mathbf{k}_1) + w(\mathbf{k}_2) - E}$$

(2.83)

and

$$T(p_1, p_2; q_1, q_2; E) = R(p_1, p_2; q_1, q_2; E) + \frac{i}{4} \int \frac{d^3\mathbf{k}_1}{(2\pi)^3} \frac{d^3\mathbf{k}_2}{(2\pi)^3}$$

$$\times R(p_1, p_2; k_1, k_2; E)(2\pi)^4\delta^{(4)}(q_1 + q_2 - k_1 - k_2)T(k_1, k_2; q_1, q_2; E).$$

(2.84)

In the second equation we have already put $d \to 3$, since no divergences appear. Comparing now Eq. (2.84) with Eq. (2.74), it is seen that the only difference consists in the normalization factor $1/(2w(\mathbf{k}_1)2w(\mathbf{k}_2))^{1/2}$. Using the matching condition given in Eq. (2.72), one may easily read off the matching condition for the relativistic K-matrix:

$$K(p_1, p_2; q_1, q_2) = \prod_{i=1}^{2} \frac{1}{(2w(\mathbf{p}_i))^{1/2}(2w(\mathbf{q}_i))^{1/2}} R(p_1, p_2; q_1, q_2; E),$$

(2.85)

where $E = w(\mathbf{p}_1) + w(\mathbf{p}_2) = w(\mathbf{q}_1) + w(\mathbf{q}_2)$.

Furthermore, let us have a closer look at Eq. (2.83). Expanding the denominators and performing the principal-value integrals in dimensional regularization, it is straightforward to ensure that all these integrals vanish, yielding on shell

$$R(p_1, p_2; q_1, q_2; E) = V(p_1, p_2; q_1, q_2).$$

(2.86)

The meaning of the matching condition (2.85) becomes now crystal-clear. This equation does not contain non-polynomial terms at all. The left-hand side is written in terms of the effective-range expansion parameters $a_\ell, r_\ell \ldots$, whereas the right-hand side is equal to the matrix element of the interaction Hamiltonian and is given in terms of $C_0, C_1, C_2, \ldots$. The values for these constants can then directly be read off.

Up to and including $O(p^2)$ the matching condition in our model gives

$$C_0 = -\frac{2\pi a_0}{M}, \qquad C_1 = \frac{\pi a_0^2}{2M}\left(r_0 - \frac{1}{a_0 M^2}\right), \qquad C_2 = -\frac{\pi a_0}{4M^3}.$$

(2.87)

Note that, at this order, there are only two independent couplings. The matching to the relativistic amplitude expresses C_0, C_1, C_2 in terms of two parameters a_0 and r_0. This is

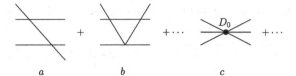

Diagrams that contribute to the three-particle scattering amplitude. The diagrams (a) and (b) contain vertices that describe the interaction of particle pairs, while the third one is a spectator. The diagram (c) corresponds to the contact three-particle interaction, which is described by a low-energy polynomial.

a nice demonstration of the constraints imposed by relativistic invariance. (It is easy to check that the second term in C_1, as well as C_2 vanish in the nonrelativistic limit, so these can be regarded as relativistic corrections.)

Two concluding remarks are in order:

- The matching at lower orders is not affected at higher orders. For example, Eq. (2.87) stays intact at all orders.
- All the beautiful properties just presented imply the use of dimensional regularization. If a different regularization is used (say, a cutoff regularization), many of the properties get lost. For example, Eq. (2.86) is not valid anymore, and the matching for the low-energy couplings should be carried out at higher orders anew.

2.5.4 Matching in the Three- (and More) Particle Sectors

The matching condition in the N-particle sector can be written by analogy to the two-particle sector:

$$T(p_1,\cdots,p_N;q_1,\cdots,q_N)$$

$$= \prod_{i=1}^{N} \frac{1}{(2w(\mathbf{p}_i))^{1/2}(2w(\mathbf{q}_i))^{1/2}} A(p_1,\cdots,p_N;q_1,\cdots,q_N). \tag{2.88}$$

However, contrary to the two-particle sector, here it is generally impossible to easily separate the polynomial and non-polynomial pieces. Also, in the two-particle sector, the sole source of the non-polynomial behavior is the bubble diagram. There are more such sources if $N > 2$. For example, in the three-particle amplitude, there are diagrams of two types, shown in Fig. 2.5. Namely, there are diagrams, where two particles interact and the third is a spectator (e.g., the diagrams in Figs. 2.5a,b), and the diagrams that all three particles come to one point (see, e.g., Fig. 2.5c). In the nonrelativistic theory, such diagrams are described by the Lagrangian given in Eq. (2.43). The diagrams of the first type contain, in general, both non-polynomial and polynomial pieces, whereas in the diagrams of the second type, the non-polynomial pieces are absent. Hence, in order to carry out the matching, one has to carefully separate polynomial and non-polynomial pieces first. With a particularly simple example, we shall demonstrate how this can be done.

Let us assume that the relativistic theory is described by the Lagrangian

$$\mathcal{L} = \frac{1}{2}(\partial\Phi)^2 - \frac{M^2}{2}\Phi^2 - \frac{\lambda}{4}\Phi^4. \tag{2.89}$$

This relativistic Lagrangian does not contain a six-particle coupling, which would correspond to the topology shown in Fig. 2.5c. However, such a coupling is induced in the nonrelativistic theory. Indeed, the leading-order three-particle amplitude, shown in Fig 2.5a, is given by

$$A(p_1,p_2,p_3;q_1,q_2,q_3) = \sum_{i=1}^{3}\sum_{l=1}^{3} \frac{(6\lambda)^2}{M^2 - (p_j + p_k - q_l)^2}. \tag{2.90}$$

Here, outgoing and incoming momenta are labeled by p_i, p_j, p_k and q_l, q_m, q_n, respectively, and $(ijk), (lmn) = (123), (231), (312)$.

We can now expand the relativistic propagators for small three-momenta:

$$\frac{1}{M^2 - (p_j + p_k - q_l)^2} = \frac{1}{2w(\mathbf{p}_j + \mathbf{p}_k - \mathbf{q}_l)}$$

$$\times \left(\frac{1}{w(\mathbf{p}_j + \mathbf{p}_k - \mathbf{q}_l) - p_j^0 - p_k^0 + q_l^0 - i\varepsilon} + \frac{1}{w(\mathbf{p}_j + \mathbf{p}_k - \mathbf{q}_l) + p_j^0 + p_k^0 - q_l^0 - i\varepsilon} \right)$$

$$= \frac{1}{2M(E_{\mathbf{p}_j + \mathbf{p}_k - \mathbf{q}_l} - p_j^0 - p_k^0 + q_l^0 - i\varepsilon)} + \cdots + \frac{1}{4M^2} + \cdots. \tag{2.91}$$

The first part of (2.91) contains the nonrelativistic propagator. The same nonpolynomial contribution arises in the nonrelativistic theory, and these two contributions exactly cancel in the matching condition. The second, polynomial part remains, however, unbalanced and should be taken care of by adjusting the nonrelativistic three-particle effective coupling D_0. Dropping the non-polynomial pieces and setting all external three-momenta to zero, the matching at $O(\lambda^2)$ yields

$$D_0 = \frac{9\lambda^2}{32M^5}. \tag{2.92}$$

The matching at higher orders and for more particles follows the same pattern. Moreover, in analogy to the effective-range expansion, one could define, for example, the threshold three-particle amplitude to all orders after subtracting the diagrams that are sufficiently singular at threshold (there is a finite number of such diagrams in the nonrelativistic theory). However, due to the more complicated singularity structure of the three (and more) particle amplitudes, such kind of definitions are, in general, not very useful.

2.6 Nucleon–Nucleon Scattering: the Case of a Large Scattering Length

The discussion, which was carried out in the previous sections, tacitly assumed that the effective couplings in the nonrelativistic Lagrangian are of natural size. In particular, if Λ is a heavy scale in the theory (e.g., if the interaction between two particles is mediated through the exchange of a particle with a mass Λ, like the interaction between two nucleons, which arises through the pion exchange), then the couplings C_0, C_1, C_2 in the nonrelativistic Lagrangian are on the order of

$$C_0 = \frac{\tilde{C}_0}{\Lambda^2}, \qquad C_{1,2} = \frac{\tilde{C}_{1,2}}{\Lambda^4}, \cdots, \tag{2.93}$$

where the $\tilde{C}_{0,1,2}, \ldots$ are numbers of order one. However, as shown in Section 1.2, this nice picture may break down under certain circumstances. Namely, the couplings in Nature may conspire such that a bound (or virtual) state emerges very close to the elastic threshold. Then, as one knows, the scattering length becomes very large. (In fact, it tends to infinity, when the bound state moves closer and closer to the threshold.) Other coefficients in the ERE stay finite and are determined by the actual range of interactions.

Let us consider the S-wave elastic amplitude. In the following, we shall limit ourselves to the case of particles with no spin, even though we sometimes shall explicitly refer to nucleon–nucleon (NN) scattering. In order to ease the notation, we do not explicitly display the index $\ell = 0$ and the overall normalization factor. Also, for simplicity, we drop the relativistic corrections and limit ourselves to the CM frame. This, in particular, leads to $\tilde{q}_0 = q_0$. (We shall use the symbol q_0 in the subsequent formulae.) The expansion of the amplitude is given by

$$T(s) \propto \frac{1}{-1/a + \frac{1}{2}rq_0^2 + O(q_0^4) - iq_0} = -a + ia^2 q_0 + a^3 q_0^2 - \frac{1}{2}a^2 rq_0^2 + \cdots. \tag{2.94}$$

It is seen that the radius of convergence of this expansion is given by the condition $|q_0 a| < 1$. In other words, if a becomes very large, the radius of convergence shrinks. Note that such a situation does not represent something very exotic; for example, the singlet and triplet S-wave NN scattering lengths are $-23.8\,\text{fm}$ and $5.4\,\text{fm}$, respectively. This corresponds to the scales $\Lambda = 8\,\text{MeV}$ and $\Lambda = 36\,\text{MeV}$ in the singlet and triplet channels, respectively, which are both much smaller than the pion mass, $M_\pi \simeq 140\,\text{MeV}$. The reason for the emergence of such small scales also is well known: there is a shallow bound state in the triplet channel (the deuteron), and a shallow virtual state in the singlet channel. Such shallow bound states exist in many other physical systems as well. Hence, adapting the nonrelativistic approach to this new situation is interesting also for practical reasons.

We start by noting that the amplitude, given in Eq. (2.94), can be re-expanded differently by (formally) promoting a from $O(1)$ to $O(p^{-1})$:

$$T(s) \propto \frac{1}{-1/a - iq_0} + \frac{rq_0^2/2}{(-1/a - iq_0)^2} + \cdots. \tag{2.95}$$

The subsequent terms in this expansion are of order $p^{-1}, 1, p, \ldots$, and the large coefficients, proportional to the positive powers of a, do not appear anymore. However, justifying such regrouping of terms in the effective theory is not an easy task. Indeed, say, in the first term of the above expression, all loops with the nonderivative vertices are summed up, yielding

$$-a(1 - iaq_0 + (iaq_0)^2 + \cdots) = \frac{1}{-1/a - iq_0}. \tag{2.96}$$

However, the radius of convergence of this expansion is given by the inequality $|aq_0| < 1$, and the summation cannot be justified for the values of q_0 above that point.

In Refs. [9–12], a simple and elegant way out of this dilemma was found. As mentioned there, the particular way of expanding the amplitude in Eq. (2.94) is linked to the use of the minimal subtraction prescription in the calculation of a loop in Eq. (2.65) that leads to the result proportional to iq_0. Changing the renormalization prescription will change this result as well. However, since the amplitude does not depend on the renormalization prescription, this change will be compensated by the change of the effective couplings $C_0, C_1, C_2, \ldots$. As shown in Refs. [9–12], it is possible to choose a prescription so that the EFT expansion of the amplitude becomes consistent, and the answer is given by (2.95).

We shall demonstrate the above statement, following Refs. [9–11] (note that Ref. [12] arrives at the same result without explicitly referring to dimensional regularization). The starting point is Eq. (2.65). In the CM frame, the loop function is given by

$$I = \int \frac{d^D k}{(2\pi)^D i} \frac{1}{(E_\mathbf{k} - k^0 - i\varepsilon)(E_\mathbf{k} - P^0 + k^0 - i\varepsilon)} = \int \frac{d^d k}{(2\pi)^d} \frac{1}{2E_\mathbf{k} - P^0 - i\varepsilon}$$

$$= \frac{M(-q_0^2 - i\varepsilon)^{d/2-1}\Gamma(1 - d/2)}{(4\pi)^{d/2}}. \tag{2.97}$$

Next, one introduces the scale of dimensional regularization, μ, in a standard manner, rendering the variable q_0^2 dimensionless. In the minimal subtraction scheme, one can perform the limit $d \to 3$ after analytic continuation in d, so that no divergences arise and the dependence on the scale μ drops out. One can, however, postulate a new subtraction scheme (the so-called power-divergence subtraction scheme or PDS), noting that (2.97) has a pole in d at $d \to 2$:

$$I = \frac{M\mu^{d-2}(-q_0^2 - i\varepsilon/\mu^2)^{d/2-1}\Gamma(1 - d/2)}{(4\pi)^{d/2}} = \frac{M\mu^{d-2}}{4\pi(2 - d)} + O(1). \tag{2.98}$$

Subtracting the pole term from I and performing the limit $d \to 3$, in the PDS scheme one finally gets

$$I = \frac{M}{4\pi} (\mu + iq_0).$$ (2.99)

Note that, putting $\mu = 0$, one arrives at the old result obtained within the minimal subtraction scheme.

As already mentioned, the effective couplings should also depend on μ, in order to compensate the μ-dependence of the amplitude, which is given by the following expression:

$$T(s) \propto \frac{C(q_0^2, \mu)}{1 + \frac{M}{8\pi} C(q_0^2, \mu)(\mu + iq_0)},$$

$$C(q_0^2, \mu) = 4C_0(\mu) - 8C_1(\mu)q_0^2 + O(q_0^4).$$ (2.100)

The RG equations for the couplings can be easily obtained from the differentiation of (2.100):

$$\mu \frac{d}{d\mu} \left(\frac{8\pi}{MC(q_0^2, \mu)} + \mu \right) = 0.$$ (2.101)

Each coefficient of Eq. (2.101) in the expansion in powers of q_0^2 should vanish. This gives

$$\mu \frac{d}{d\mu} C_0(\mu) = \frac{M\mu}{2\pi} C_0(\mu)^2,$$

$$\mu \frac{d}{d\mu} C_1(\mu) = \frac{M\mu}{\pi} C_0(\mu)C_1(\mu),$$ (2.102)

and so on.

The above RG equations can be solved, taking into account the boundary conditions at $\mu = 0$, which can be obtained from Eq. (2.87) in the nonrelativistic limit:

$$C_0(0) = -\frac{2\pi a}{M}, \qquad C_1(0) = \frac{\pi a^2 r}{2M}.$$ (2.103)

The solution is

$$C_0(\mu) = -\frac{2\pi}{M} \frac{1}{\mu + 1/a}, \qquad C_1(\mu) = \frac{\pi r}{2M} \frac{1}{(\mu + 1/a)^2}.$$ (2.104)

Now, we see that if $a \sim \Lambda^{-1}$ is of natural size, we can safely choose $\mu = 0$ (standard minimal subtraction) in all the preceding expressions, getting $C_0 \sim (M\Lambda)^{-1}$, $C_1 \sim (M\Lambda^3)^{-1}$ and so on. However, even for $a \to \infty$, a consistent power counting is possible for $\mu \neq 0$. In order to demonstrate this, it is convenient to choose $\mu = O(p)$. Then, $C_0 \sim (M\mu)^{-1}$, $C_1 \sim (M\Lambda\mu^2)^{-1}$ and so on. (Note that, still, $r \sim \Lambda^{-1}$.) In this power counting, all bubble diagrams containing the nonderivative coupling $C_0(\mu)$ are of the same order in p and should be resummed. Otherwise, as can be easily verified, the insertions of the derivative-type vertices like $C_1(\mu)$ and higher are perturbative. The modified expansion of the amplitude, written in Eq. (2.95), is thus justified.

To summarize, in case of an unnaturally large scattering length, the standard loop expansion of the amplitude in the effective theory has a very small radius of convergence. One could, however, partially resum the perturbative series that leads to a larger radius of convergence, set by the effective range and not by the scattering length. Moreover, such a resummation can be formally justified within EFT, using the freedom to choose the renormalization prescription and to reshuffle contributions between the loops and effective couplings.

Finally, note that in case of the NN interactions we can pursue the approach, described in this section, even further and add explicit pion exchanges, which are treated perturbatively [10, 11] (although such a perturbative treatment is doubtful for the tensor force, as long known in nuclear physics). An alternative approach to the problem promotes pion exchanges to the leading order and treats them non-perturbatively, on an equal footing with the lowest-order contact interactions [13, 14], the so-called Weinberg counting. We, however, do not consider the above approaches in any detail and refer the reader to the recent reviews of the status of the theory of nuclear forces [15, 16], where all these questions are addressed. Finally, we note that the Weinberg counting can be justified in large-N_c QCD [17], with N_c the number of colors.

2.7 Relativistic Kinematics: Two-Particle Processes

In Section 2.5, the inclusion of the relativistic corrections has been addressed. It is true that these can be taken into account systematically, order by order, in the calculation of the amplitude. However, a proper bookkeeping turns out to be very cumbersome at higher orders. One has seen an example of this already in Eqs. (2.68) and (2.70). Albeit it is clear that, after including the relativistic corrections, the cut in the loop function will gradually move to the right position, the polynomial factors, multiplying the quantity q_0, cannot be written down in a compact form to all orders, rendering the matching condition (in an arbitrary moving frame) rather clumsy.

An even simpler example can be found at tree level in the $\lambda \varphi^4$-theory. The lowest-order relativistic amplitude is just a constant, $A(s,t) = A$. From the matching condition (2.72), however, it is clear that the constant A will appear in all effective couplings $C_0, C_1, C_2, \ldots$. The reason is purely kinematical, all powers of momenta emerge after the expansion of the square roots that account for the fact that the normalization of the one-particle states in the relativistic and nonrelativistic states is chosen differently. To summarize, albeit in the framework just considered the relativistic corrections can be evaluated up to a given order, the result is not very transparent and cannot be easily rewritten in an explicitly Lorentz-invariant form. Fortunately, in the two-particle sector, a slightly modified approach exists, which enables one to solve all these problems with a surprising ease [18, 19]. We shall briefly review this approach in what follows. As in the previous sections, we shall restrict ourselves to particles without spin. The generalization of the approach to the case of particles with spin is possible (see, e.g., [20].)

We shall, however, treat an extension of the approach to the scattering of particles with different masses M_1 and M_2 explicitly.

As becomes clear from the preceding discussion, in order to obtain an explicitly covariant version of the nonrelativistic theory, two conditions should be fulfilled:

- The relativistic insertions in the *internal* lines should be summed up to all orders. However, as a result of such a summation, the hard scale M explicitly appears in the propagator. This fact leads to the breaking of the counting rules that should be taken into account.
- The normalization of the nonrelativistic one-particle states should be changed so that the matching condition becomes explicitly Lorentz-invariant.

It turns out that the preceding two modifications essentially suffice to achieve the goals. The only nontrivial part here is the prescription used to calculate the individual Feynman integrals. This new prescription is needed to rectify the power counting, which was destroyed by the appearance of the heavy scale in the propagators. The proper procedure is described in what follows.[5]

The change of the normalization is achieved by rescaling of the nonrelativistic fields $\phi_i(x) \to (2W_i)^{1/2} \phi_i(x)$ $(i = 1, 2)$ and W_i is the differential operator $W_i = \sqrt{M_i^2 - \nabla^2}$ (such a differential operator should always be understood as a Taylor expansion in inverse powers of a large scale). Summing up all relativistic insertions in the kinetic term, we arrive at the following Lagrangian:

$$\mathcal{L} = \sum_{i=1}^{2} \phi_i^\dagger (2W_i)(i\partial_t - W_i)\phi_i + \mathcal{L}^{(0)} + \mathcal{L}^{(2)} + \cdots, \qquad (2.105)$$

where $\mathcal{L}^{(0)}, \mathcal{L}^{(2)}, \ldots$ denote the four-particle Lagrangians at leading order, next-to-leading order and so on. The leading-order Lagrangian can be written immediately:

$$\mathcal{L}^{(0)} = C_0 \phi_1^\dagger \phi_2^\dagger \phi_1 \phi_2. \qquad (2.106)$$

Note that the constant C_0 here is different from the one that appears in Eq. (2.36). (The two have even different mass dimensions.) We, however, want to avoid a proliferation of different symbols in the notation and hope that this will not lead to any confusion. Also, we postpone the discussion of the higher-order terms in the Lagrangian and focus first on the bubble sum produced by the leading-order term. The loop function emerging in this theory in an arbitrary reference frame, defined by the total CM momentum P^μ, is given by

$$I = \int \frac{d^D k}{(2\pi)^D i} \frac{1}{4w_1(\mathbf{k})w_2(\mathbf{P}-\mathbf{k})} \frac{1}{(w_1(\mathbf{k}) - k^0 - i\varepsilon)(w_2(\mathbf{P}-\mathbf{k}) - P^0 + k^0 - i\varepsilon)}$$

$$= \int \frac{d^d k}{(2\pi)^d} \frac{1}{4w_1(\mathbf{k})w_2(\mathbf{P}-\mathbf{k})} \frac{1}{w_1(\mathbf{k}) + w_2(\mathbf{P}-\mathbf{k}) - P^0 - i\varepsilon}, \qquad (2.107)$$

---

[5] The situation closely resembles the situation with Chiral Perturbation Theory in the one-nucleon sector, where the breaking of the power counting is cured by the use of similar methods, see Section 4.10.2.

with $w_i(\mathbf{k}) = \sqrt{M_i^2 + \mathbf{k}^2}$. According to the naive power counting, $w_i(\mathbf{k}) - M_i = O(p^2)$, $P^0 - M_1 - M_2 = O(p^2)$ and $\mathbf{P} = O(p)$. For consistency, one should count the mass difference $M_1 - M_2$ also as $O(p^2)$. In this counting, the loop function counts as $O(p)$. However, carrying out a straightforward calculation, it can be verified that the pertinent integral is a quantity of order one. Most easily, this is seen in the equal mass case and in the rest frame $\mathbf{P} = 0$ and $P^0 = \sqrt{s}$. Adding and subtracting the same integral, evaluated at $s = 4M^2$, one ensures that the difference is ultraviolet finite and proportional to $\sqrt{s - 4M^2}$; that is, this part obeys the naive power counting. The integral evaluated at $\sqrt{s} = 2M$, however, counts as $O(1)$. Indeed, in this integral, one can rescale the integration momentum $\mathbf{k}$ by M (there are no other momenta or scales available). As a result, we arrive at a pure number that does not vanish in dimensional regularization. Of course, we could set a particular renormalization prescription to make the power-counting violating term vanish. The problem, however, resurfaces at higher loop orders and requires readjusting the renormalization prescription order by order. In what follows, it will be shown how this problem can be solved in one shot at all orders.

It is clear why the same problem does not emerge in the "standard" formulation of the nonrelativistic approach, which operates with propagators that are Taylor-expanded in inverse powers of the heavy scale M. For example, in the case that we have just considered, the expansion of the loop function, evaluated at threshold, $\sqrt{s} = 2M$, gives no-scale integrals only, which vanish in dimensional regularization. Hence, it is clear that our goal would be achieved, if we would postulate an amended rule for calculating Feynman integrals: given a generic Feynman integral, one first Taylor-expands the integrand in inverse powers of M, integrates in dimensional regularization and sums up the result to all orders again. Such a procedure is equivalent to setting a particular renormalization prescription, with the advantage that it can be formulated universally for all loops in the two-particle scattering.

Technically, the procedure can be carried out very easily, if the following algebraic identity is used:

$$\frac{1}{4w_1 w_2} \left\{ \frac{1}{w_1 + w_2 - P^0} - \frac{1}{w_1 + w_2 + P^0} + \frac{1}{w_1 - w_2 + P^0} - \frac{1}{w_1 - w_2 - P^0} \right\}$$
$$= \frac{1}{2P^0} \frac{1}{\mathbf{k}^2 - (\mathbf{k} \cdot \mathbf{P}/P^0)^2 - q_0^2}, \tag{2.108}$$

where $q_0 = \lambda^{1/2}(s, M_1^2, M_2^2)/(2\sqrt{s})$ and λ denotes the conventional Källén triangle function, $\lambda(x, y, z) = x^2 + y^2 + z^2 - 2(xy + yz + zx)$. Note also that a shift in the integration variable $\mathbf{k} \to \mathbf{k} + \mu_1 \mathbf{P}$ has been performed, where

$$\mu_{1,2} = \frac{1}{2}\left(1 \pm \frac{M_1^2 - M_2^2}{s}\right). \tag{2.109}$$

Now, one can express the integrand in Eq. (2.107) from Eq. (2.108) and afterwards expand it in inverse powers of M_i. The term that originally appears on the r.h.s.

of Eq. (2.108) is not expanded. The expansion of other terms gives polynomials in momenta, because the denominators are of order one. For example,

$$\frac{1}{w_1 + w_2 + P^0} = \frac{1}{2(M_1 + M_2)}$$

$$- \frac{1}{4(M_1 + M_2)^2} \left(\frac{(\mathbf{k} + \mu_1 \mathbf{P})^2}{2M_1} + \frac{(\mathbf{k} - \mu_2 \mathbf{P})^2}{2M_2} + P^0 - M_1 - M_2 \right) + \cdots. \quad (2.110)$$

The integration of the low-energy polynomial at all orders gives zero. Consequently, in our prescription, from Eq. (2.107) we obtain

$$I = \frac{1}{2P^0} \int \frac{d^d k}{(2\pi)^d} \frac{1}{\mathbf{k}^2 - (\mathbf{k} \cdot \mathbf{P}/P^0)^2 - q_0^2 - i\varepsilon}. \quad (2.111)$$

In order to evaluate this integral, we split the integration momentum into the components parallel and perpendicular to the direction of the momentum $\mathbf{P}$:

$$\mathbf{k} = \frac{\mathbf{P}}{|\mathbf{P}|} k_{\parallel} + \mathbf{k}_{\perp}, \qquad k_{\parallel} = \frac{\mathbf{k} \cdot \mathbf{P}}{|\mathbf{P}|}. \quad (2.112)$$

The integration measure transforms to $d^d k = dk_{\parallel} d^{d-1} \mathbf{k}_{\perp}$, and the integral takes the form

$$I = \frac{1}{2P^0} \int \frac{dk_{\parallel} d^{d-1} \mathbf{k}_{\perp}}{(2\pi)^d} \frac{1}{\left(\sqrt{s}/P^0\right)^2 k_{\parallel}^2 + \mathbf{k}_{\perp}^2 - q_0^2 - i\varepsilon} = \frac{i q_0}{8\pi \sqrt{s}}. \quad (2.113)$$

Note that, in order to arrive at the final result, we have rescaled the integration variable $k_{\parallel} \to k_{\parallel} P^0/\sqrt{s}$.

The most important property of the above expression is that it is Lorentz-invariant, even though we have started with Eq. (2.107), where the invariance is not explicit. It can be seen that the particular renormalization prescription that is imposed on the loop function guarantees this invariance.

Let us now consider higher orders. Since within the given renormalization prescription all principal-value integrals vanish in dimensional regularization, the matching condition takes a particularly simple form (cf. Eqs. (2.85) and (2.86)):

$$K(p_1, p_2; q_1, q_2) = V(p_1, p_2; q_1, q_2). \quad (2.114)$$

At this stage, we can consider the higher-order interaction Lagrangians $\mathcal{L}^{(2)}, \ldots$. In order to carry out the expansion of the relativistic K-matrix in an arbitrary frame, it is convenient to introduce the CM and relative four-momenta in the following manner:

$$P = p_1 + p_2 = q_1 + q_2, \qquad p = \mu_2 p_1 - \mu_1 p_2, \qquad q = \mu_2 q_1 - \mu_1 q_2, \quad (2.115)$$

where $s = P^2$. Note that on shell $p^2 = q^2 = q_0^2$. Moreover, in the CM frame,

$$p^\mu = (0, \mathbf{p}), \qquad q^\mu = (0, \mathbf{q}), \qquad \cos\theta = \frac{\mathbf{p} \cdot \mathbf{q}}{|\mathbf{p}||\mathbf{q}|}, \quad (2.116)$$

where θ is the scattering angle. One sees that in an arbitrary frame, one could define

$$\cos\theta = -\frac{p \cdot q}{q_0^2}. \quad (2.117)$$

The partial-wave expansion can be readily generalized in an arbitrary frame:

$$K(p_1, p_2; q_1, q_2) = \tilde{K}_0(s) - 3(p \cdot q)\tilde{K}_1(s) + \frac{5}{2}(3(p \cdot q)^2 - q_0^4)\tilde{K}_2(s) + \cdots. \tag{2.118}$$

Here, $\tilde{K}_0, \tilde{K}_1, \tilde{K}_2, \ldots$ can be expressed through the ERE parameters in different partial waves, namely

$$8\pi\sqrt{s}\tilde{K}_\ell^{-1}(s) = -\frac{1}{a_\ell} + \frac{1}{2}r_\ell q_0^2 + \cdots. \tag{2.119}$$

Further,

$$p \cdot q = \frac{1}{4}(p_1 - p_2) \cdot (q_1 - q_2) - \frac{(M_1^2 - M_2^2)^2}{4s}. \tag{2.120}$$

The last term is $O(p^4)$ in our counting. Hence, up to and including $O(p^2)$-terms, we get

$$K(p_1, p_2; q_1, q_2) = -8\pi a_0(M_1 + M_2) - \frac{4\pi a_0(M_1 + M_2)}{M_1 M_2}q_0^2 - 4\pi a_0^2 r_0(M_1 + M_2)q_0^2$$

$$+ 6\pi a_1(M_1 + M_2)(p_1 - p_2) \cdot (q_1 - q_2) + O(p^4), \tag{2.121}$$

where

$$q_0^2 = \frac{s}{4} - \frac{1}{2}(M_1^2 + M_2^2) + \frac{(M_1^2 - M_2^2)^2}{4s} = \frac{1}{2}p_1 \cdot p_2 - \frac{1}{2}M_1 M_2 + O(p^4). \tag{2.122}$$

The generalization of these expressions to higher orders is straightforward. Now, looking at Eq. (2.121), it is easy to write the Lagrangian $\mathcal{L}^{(2)}$, which reproduces the $O(p^2)$ terms in this equation:

$$\mathcal{L}^{(2)} = C_1\left((\phi_1^\dagger)^\mu(\phi_2^\dagger)_\mu \phi_1 \phi_2 - M_1 M_2 \phi_1^\dagger \phi_2^\dagger \phi_1 \phi_2 + \text{h.c.}\right)$$

$$+ C_2\left(\phi_1^\dagger(\phi_2^\dagger)^\mu - (\phi_1^\dagger)^\mu \phi_2^\dagger\right)\left((\phi_1)_\mu \phi_2 - \phi_1(\phi_2)_\mu\right), \tag{2.123}$$

where

$$(\phi_i)_\mu = (W_i, -i\nabla)\phi_i, \qquad (\phi_i^\dagger)_\mu = (W_i, i\nabla)\phi_i^\dagger. \tag{2.124}$$

The matching to the relativistic K-matrix finally yields

$$C_1 = -\frac{\pi a_0(M_1 + M_2)}{M_1 M_2} - \pi a_0^2 r_0(M_1 + M_2),$$

$$C_2 = -6\pi a_1(M_1 + M_2). \tag{2.125}$$

Here, a_0, r_0 are the leading S-wave ERE parameters, and a_1 is the P-wave scattering length. One should mention that, going to the higher orders, only one new parameter emerges in each partial wave at each step. Hence, in this approach, there are no "superfluous" couplings obeying constraints. Due to Lorentz invariance, these constraints are implemented automatically.

Next, inserting the Lorentz-invariant derivative vertices in the loops, it can be verified that the result is also Lorentz-invariant. Indeed, one could rewrite the analog of the integral in Eq. (2.107) as

$$I = \int \frac{d^d \mathbf{k}_1}{(2\pi)^d 2w_1} \frac{d^d \mathbf{k}_2}{(2\pi)^d 2w_2} N \frac{(2\pi)^d \delta^{(d)}(\mathbf{k}_1 + \mathbf{k}_2 - \mathbf{P})}{w_1 + w_2 - P^0 - i\varepsilon}, \qquad (2.126)$$

where N denotes a Lorentz-invariant polynomial, which emerges from the derivative vertices. (This polynomial is set to unity in Eq. (2.107).) Now, recall that, after expansion of the integrand in inverse powers of the heavy scale, only the absorptive part of the integral survives that is equivalent to the replacement:

$$\frac{1}{w_1 + w_2 - P^0 - i\varepsilon} \to i\pi \delta(w_1 + w_2 - P^0). \qquad (2.127)$$

From the preceding expression it is immediately seen that the loop function is given by an explicitly Lorentz-invariant expression, and our previous statement is thus verified.

2.8 Cusps in Three-Particle Decays

In this section we describe one example where the use of the nonrelativistic approach allows us to achieve the result in the most efficient and transparent way. In our opinion, this example serves as a nice demonstration of both the beauty and the rich potential of the method of nonrelativistic Lagrangians.

In the years 2003 to 2005, the NA48 experiment at CERN collected a vast data sample of about 10^8 decays $K^+ \to \pi^0 \pi^0 \pi^+$ (neutral mode) and of about $4 \cdot 10^9$ decays $K^+ \to \pi^+ \pi^+ \pi^-$ (charged mode) that allowed them to achieve a remarkable statistical precision [21, 22]. The measured $\pi^0 \pi^0$ invariant mass distribution in the neutral mode is shown in Fig. 2.6, where a pronounced unitary cusp at $s_{12} = 4M_{\pi\pm}^2$ is clearly visible.[6] The parameters of this cusp can be determined very accurately because of the huge amount of data available.

Already in 1961, Budini and Fonda [23] have shown that a cusp in the invariant mass distribution emerges due to the $\pi\pi$ rescattering. Moreover, they provided an analytic formula for the cusp behavior and have pointed out that the study of charged kaon decays allows us to measure the particular isospin combination $a_0 - a_2$ of the $\pi\pi$ scattering lengths, with $I = 0, 2$ the total isospin of the S-wave pion pair. Forty-five years later, this idea was rediscovered and developed further in Refs. [24, 25]. A simple, intuitive justification of the idea can be given by looking at Fig. 2.7. The K^+ meson can decay (weakly) in either one of the two (charged or neutral) channels, and the pions emerging from such decays may rescatter (strongly) in the final state. Suppose now that the kaon decays and pion interactions in the final state are described in some (unspecified) relativistic field theory. In such a theory, the cusp emerges from the interference

---

[6] We assign the labels 1,2,3 to the decay products so that the label 3 is always reserved for the "odd" pion (π^+ in the neutral mode and π^- in the charged mode). The kinematical variables are defined as $s_{ij} = (p_i + p_j)^2$, where $(ij) = (12), (23), (31)$.

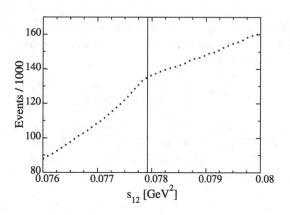

Figure 2.6 $K^+ \to \pi^0\pi^0\pi^+$ decay: Invariant mass distribution of the $\pi^0\pi^0$ pair in the vicinity of the cusp. The vertical line denotes the position of the charged pion threshold. The threshold cusp is clearly visible. This figure is made by using the data from Ref. [21]. We thank Brigitte Bloch-Devaux for providing us with these data.

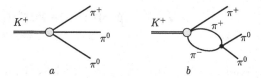

Figure 2.7 Diagrams that contribute to the decay $K^+ \to \pi^0\pi^0\pi^+$: (a) the tree-level diagram, and (b) the one-loop diagram, which describes the process $\pi^+\pi^- \to \pi^0\pi^0$ in the final state. The interference of these two diagrams leads to the cusp structure in the invariant mass distribution of the $\pi^0\pi^0$-pair.

of the diagrams a and b shown in Fig. 2.7. Indeed, the diagram in Fig. 2.7a displays a tree-level process. In field theory, this vertex is a regular function of the kinematical variables (a polynomial in the external momenta) and, at the lowest order in the Taylor expansion, it can be approximated by a constant that we denote by A_0. The diagram shown in Fig. 2.7b behaves differently. It is given by a product of a decay amplitude A_+, the $\pi^+\pi^- \to \pi^0\pi^0$ scattering amplitude T_x and the loop function of charged pions $I(s_{12})$. (The cusp in the scattering amplitude T_x was worked out in Ref. [26] in the framework of Chiral Perturbation Theory.) The first quantity is again a smooth function and can be approximated by a constant. Further, exactly at the cusp, the $\pi\pi$ scattering amplitude is given by the difference of the S-wave scattering lengths $a_0 - a_2$ (modulo a normalization factor). Hence, in the vicinity of the cusp it can be replaced by a constant up to terms that vanish as $s_{12} \to 4M_\pm^2$. The loop function, however, is singular and is given by

$$I(s_{12}) = \operatorname{Re} I(s_{12}) + \frac{i}{16\pi}\sqrt{1 - \frac{4M_{\pi^\pm}^2}{s_{12}}}, \quad \text{if } s_{12} > 4M_\pm^2,$$

$$I(s_{12}) = \operatorname{Re} I(s_{12}) - \frac{1}{16\pi}\sqrt{\frac{4M_{\pi^\pm}^2}{s_{12}} - 1}, \quad \text{if } s_{12} < 4M_\pm^2. \tag{2.128}$$

In this expression, the real part of the relativistic loop is a smooth function, which can be approximated by a polynomial of a low order in the variable s_{12} a renormalization prescription can be used in which this function is equal to zero at threshold. Thus, the singularity at the cusp is generated solely from the square root contained in the imaginary part.

Next, adding the diagrams in Fig. 2.7a,b and squaring, we get

$$|M_0|^2 = |M_a + M_b|^2 = \left| A_0 + A_+ T_x \left(\mathrm{Re}\, I + iC \sqrt{s_{12} - 4M_{\pi^\pm}^2} \right) \right|^2$$

$$= \left(A_0 + A_+ T_x \mathrm{Re}\, I \right)^2 + \left(CA_+ T_x \right)^2 \left(s_{12} - 4M_{\pi^\pm}^2 \right).$$

(2.129)

Here, it is assumed that $s_{12} > 4M_{\pi^\pm}^2$ and C denotes some irrelevant normalization constant. On the contrary, if $s_{12} < 4M_{\pi^\pm}^2$, we get

$$|M_0|^2 = |M_a + M_b|^2 = \left(A_0 + A_+ T_x \mathrm{Re}\, I \right)^2 + \left(CA_+ T_x \right)^2 \left(s_{12} - 4M_{\pi^\pm}^2 \right)$$

$$- 2CA_+ T_x \left(A_0 + A_+ T_x \mathrm{Re}\, I \right) \sqrt{4M_{\pi^\pm}^2 - s_{12}}.$$

(2.130)

The decay rate for the decay of a kaon into three pions is given by

$$d\Gamma = (2\pi)^4 \delta^{(4)}(P - p_1 - p_2 - p_3) |M_0|^2 \frac{d^3\mathbf{p}_1}{(2\pi)^3 2p_1^0} \frac{d^3\mathbf{p}_2}{(2\pi)^3 2p_2^0} \frac{d^3\mathbf{p}_3}{(2\pi)^3 2p_3^0}.$$

(2.131)

Integrating over the angles, we get

$$d\Gamma = \frac{1}{(2\pi)^3} \frac{1}{32M_K^3} |M_0|^2 ds_{12} ds_{23}.$$

(2.132)

In the preceding equation, $P = (M_K, \mathbf{0})$ denotes the momentum of the kaon. Further, for a fixed s_{12}, the variable s_{23} varies in the interval $s_{23}^{\min} \le s_{23} \le s_{23}^{\max}$, where

$$s_{23}^{\min} = (E_2^* + E_3^*)^2 - \left(\sqrt{E_2^{*2} - M_2^2} + \sqrt{E_3^{*2} - M_3^2} \right)^2,$$

$$s_{23}^{\max} = (E_2^* + E_3^*)^2 - \left(\sqrt{E_2^{*2} - M_2^2} - \sqrt{E_3^{*2} - M_3^2} \right)^2,$$

(2.133)

and E_2^*, E_3^* are the energies of particles 2 and 3 in the (12) rest frame:

$$E_2^* = \frac{s_{12} - M_1^2 + M_2^2}{2\sqrt{s_{12}}}, \qquad E_3^* = \frac{M_K^2 - s_{12} - M_3^2}{2\sqrt{s_{12}}}.$$

(2.134)

The invariant mass distribution is proportional to

$$\frac{d\Gamma}{ds_{12}} = \frac{1}{(2\pi)^3} \frac{1}{32M_K^3} \int_{s_{23}^{\min}}^{s_{23}^{\max}} ds_{23} |M_0|^2.$$

(2.135)

Using Eqs. (2.129) and (2.130) in this expression, it is seen that the last term in Eq. (2.130) leads to a cusp (square-root singularity), which is visible in Fig. 2.6 exactly at $s_{12} = 4M_{\pi^\pm}^2$.

In order to extract the quantity $a_0 - a_2$ from the experimentally measured cusp parameters, the knowledge of A_0 and A_+ is needed. (All other quantities that enter the expression of $|M_0|^2$ are known.) These can be determined from the separate measurements of the charged and neutral channels (say, fitting the total decay rates in the two channels). Thus, the goal of measuring $a_0 - a_2$ indeed can be achieved.

The problem of the intuitive approach, which was just described, consists in the fact that it matches neither the accuracy with which we would like to measure $a_0 - a_2$ nor the very high experimental precision. The diagrams just considered are not the only singular ones that are possible. Taking this fact into account and using two- and three-body unitarity, the preceding approach has been extended to include all singularities up to two loops [25]. However, methods based on unitarity do not allow for a straightforward inclusion of the electromagnetic effects (which are also relevant at the level of precision achieved in the experiment). Moreover, as it turned out later, Ref. [25] is based on certain assumptions about the singularity structure of the three-body amplitudes, which do not account for, for example, the singularities away from the cusp region, or the Landau-type singularities [18, 19]. On the other hand, in a conventional relativistic field-theoretical framework, one could easily solve all the preceding problems. For example, in this approach, there is no need to assume any singularity structure of the amplitudes a priori; they are generated automatically from the calculation of the Feynman diagrams that enter the perturbative expansion of an amplitude at a given order. Real and virtual photons also can be treated in a standard manner in this approach. Unfortunately, the relativistic approach produces its own problems that render its use impossible in the present case. The (very severe) problem is that the relativistic field-theoretical approach, unlike the approach based on unitarity, does not yield a regular expansion in terms of the physical scattering lengths (the ERE parameters, in general). Hence, the relations that one gets between the cusp parameters and the scattering lengths in this approach can be valid only perturbatively, up to a given order. Based on the previous experience, it cannot be seriously expected that these perturbative calculations can be carried out at the accuracy that matches the experimental precision.

The nonrelativistic effective theories, which were introduced in the previous section, are tailored for solving such problems. First, these represent a full-fledged field theory, meaning that all low-energy singularities of the relativistic amplitudes are automatically reproduced by calculating the pertinent Feynman diagrams. Further, the electromagnetic interactions can be straightforwardly included. On the other hand, unlike the relativistic theory, here the role of the couplings is played by the ERE parameters, which are exactly the quantities one wants to extract from the experiment. This circumstance drastically reduces the theoretical error in the relation between the scattering lengths and the cusp parameters, and renders the nonrelativistic approach ideally suited for the problem in question.

In the nonrelativistic framework, the Lagrangian consists of two parts. The weak part describes the decay $K \to 3\pi$ into different channels. Up to and including $O(p^4)$, this part of the Lagrangian is given by

$$\mathcal{L}_K = K^\dagger 2W_K(i\partial_t - W_K)K + \frac{G_0}{2}\left(K^\dagger\phi_+\phi_0^2 + \text{h.c.}\right)$$

$$+ \frac{G_1}{2}\left(K^\dagger(W_\pm - M_{\pi^\pm})\phi_+\phi_0^2 + \text{h.c.}\right) + \frac{G_2}{2}\left(K^\dagger(W_\pm - M_{\pi^\pm})^2\phi_+\phi_0^2 + \text{h.c.}\right)$$

$$+ G_3\left(K^\dagger\phi_+(W_0^2\phi_0\phi_0 - W_0\phi_0 W_0\phi_0) + \text{h.c}\right) + \frac{H_0}{2}\left(K^\dagger\phi_-\phi_+^2 + \text{h.c.}\right)$$

$$+ \frac{H_1}{2}\left(K^\dagger(W_- - M_{\pi^\pm})\phi_-\phi_+^2 + \text{h.c.}\right) + \frac{H_2}{2}\left(K^\dagger(W_- - M_{\pi^\pm})^2\phi_-\phi_+^2 + \text{h.c.}\right)$$

$$+ H_3\left(K^\dagger\phi_-(W_\pm^2\phi_+\phi_+ - W_\pm\phi_+ W_\pm\phi_+) + \text{h.c}\right). \tag{2.136}$$

Here, the nonrelativistic fields $K, \phi_\pm, \phi_0$ describe kaons, charged and neutral pions, respectively, and $W_K, W_\pm, W_0$ denote the pertinent differential operators. (It is assumed that they act only on the field next to them.) Further, the couplings G_i, H_i describe the kaon decay into the neutral and charged channels, in order. The preceding Lagrangian is used at tree level only, where the calculation gives

$$M_0^{\text{tree}} = G_0 + G_1(p_3^0 - M_{\pi^\pm}) + G_2(p_3^0 - M_{\pi^\pm})^2 + G_3(p_1^0 - p_2^0)^2,$$

$$M_+^{\text{tree}} = H_0 + H_1(p_3^0 - M_{\pi^\pm}) + H_2(p_3^0 - M_{\pi^\pm})^2 + H_3(p_1^0 - p_2^0)^2. \tag{2.137}$$

Here,

$$p_i^0 = \frac{M_K^2 + M_i^2 - s_{jk}^2}{2M_K}, \qquad (ijk) = (123),(312),(231). \tag{2.138}$$

It is seen that at leading order in the low-energy expansion the couplings G_0, H_0 correspond to the constants A_0 and A_+, which we have introduced. The language of the effective Lagrangians, however, allows us to systematically study the expansion of the amplitudes beyond leading order.

The strong part of the Lagrangian describes pion–pion scattering and is similar to the one given in Eqs. (2.105), (2.106) and (2.123). In principle, there are also six-particle vertices present, similar to that in Eq. (2.43). However, the net effect of these vertices is the renormalization of the constants G_i, H_i, which are anyway fitted to data. Hence, it is legitimate to drop the six-particle vertices completely.

Within the effective theory, the tree-level amplitudes, given by Eq. (2.137), get systematically dressed by the loops in the three-pion system, produced by the two-particle Lagrangians. Examples of such loop diagrams up to and including two loops are given in Fig. 2.8. A consistent power counting emerges if one assumes that all three-momenta count as $O(p)$, and the kinetic energies $p_i^0 - M_i$, the mass differences $M_i - M_j$ as well as $M_K - M_i - M_j - M_k$ count as $O(p^2)$. Then, adding a loop is equivalent to adding one power in the p-counting, provided the prescription to calculate Feynman diagrams has been amended. The modification of the prescription at one loop was described in the previous section. The procedure at two loops is slightly more complicated and is described in detail in Ref. [19]. In particular, it can be shown that the diagrams in Fig. 2.8d lead to anomalous thresholds in the decay amplitudes that indicate a richer

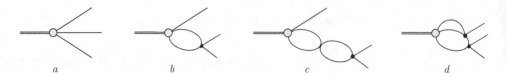

a b c d

Figure 2.8 The kaon decay amplitude up to and including two loops in the nonrelativistic effective theory: (a) the tree-level diagram, (b) the one loop diagram, (c) two elementary bubbles and (d) a diagram with two overlapping pion loops. The large and small circles represent the $K \to 3\pi$ and the four-pion vertices, respectively.

analytic structure of these amplitudes than merely the presence of a unitary cusp. (We recommend, e.g., the textbook [27] for more information on the analytic properties of amplitudes in field theory.)

Next, we would like to discuss the convergence of the approach. In fact, the difference $M_K - 2M_{\pi^0} - M_{\pi^\pm} \simeq 84\,\text{MeV}$ is not that much smaller than the pion masses, casting legitimate doubt on the applicability of the nonrelativistic approach. Fortunately, however, the nonrelativistic approach works better than expected from the preceding argument. First of all, the lowest-lying threshold in the direct channel not taken into account explicitly opens at $5M_\pi$. This is too high to significantly influence the low-energy behavior. Further, if this and all other singularities, which are not explicitly taken into account in this approach, had a large impact on the decay amplitude, this would manifest itself in a non-polynomial behavior, which, as we know, is not the case. Except for the small cusp effect, the amplitude can be parameterized by a polynomial of low degree, and this parameterization fits the experimental data very well. Finally, note that the nonrelativistic expansion is in fact a double expansion, and each power of p is accompanied by an additional power of the $\pi\pi$ scattering lengths, which are small quantities ($a_0 \simeq 0.22$ and $a_2 \simeq -0.04$ in the units of the inverse pion mass). This gives an additional suppression factor, leading to a fast convergence in the loop expansion.

The whole procedure of finding a general parameterization of the decay amplitudes (in both decay channels) that correctly takes into account their singularity structure becomes crystal clear at this stage. All that is needed in order to get such a parameterization at a given order in p is to calculate a full set of Feynman diagrams up to the same order. For illustrative purposes, we display the one-loop expression for the neutral as well as the charged amplitude at $O(p)$:

$$M_0^{1-\text{loop}} = 2C_x H_0 J_{+-}(s_{12}) + C_{00} G_0 J_{00}(s_{12}) + 2C_{+0} G_0 [J_{+0}(s_{23}) + (s_{23} \leftrightarrow s_{31})],$$

$$M_+^{1-\text{loop}} = C_{++} H_0 J_{++}(s_{12}) + [2C_{+-} H_0 J_{+-}(s_{23}) + C_x G_0 J_{00}(s_{23}) + (s_{23} \leftrightarrow s_{31})],$$

$$(2.139)$$

where, at lowest order, the two-body couplings are related to the $\pi\pi$ scattering length according to

$$C_{00} = \frac{16\pi}{3}(a_0 + 2a_2), \quad C_{+0} = 8\pi a_2, \quad C_x = \frac{16\pi}{3}(a_2 - a_0),$$

$$C_{+-} = \frac{8\pi}{3}(2a_0 + a_2), \quad C_{++} = 16\pi a_2, \tag{2.140}$$

and the loop functions $J_{+-}(s), J_{00}(s), J_{+0}(s), J_{++}(s)$, in which different pions run through the loop, are given by Eq. (2.113). As expected, Eq. (2.139) explicitly features the cusp at $s_{12} = 4M_{\pi\pm}^2$. Note also that higher-order results at one loop (from derivative vertices in the one-loop graphs), as well as the two-loop result are available. We, however, refrain from displaying these rather voluminous expressions here, referring the interested reader to the original papers [18, 19]. It is also clear why we use here the covariant formulation of the nonrelativistic theory. In this formulation, all singularities and cuts in the amplitude are at the right places from the beginning. Further, the parameterization is written in terms of the couplings G_i, H_i, as well as the physical scattering lengths, effective ranges and so on. Fitting a general parameterization to the experimental data on the Dalitz plot distributions in both the charged and neutral channels, we could finally determine these parameters to a very high precision.

By using exactly this method, the NA48 collaboration was able to determine the S-wave $\pi\pi$ scattering lengths very accurately [22]. Note that, at the leading order, the cusp depends on $a_0 - a_2$ only, both a_0 and a_2 enter the expression of the amplitudes separately at higher orders and both can be determined from the fit because of the extremely small statistical errors. Working at such a high accuracy, the radiative corrections also start to matter. A systematic inclusion of these corrections has been considered in Ref. [28].

To summarize, the method of the nonrelativistic effective Lagrangians is very convenient for the study of the general low-energy singularity structure of various amplitudes, where the use of the conventional methods from relativistic quantum field theory is complicated. It therefore does not come as a surprise that the method has been successfully used also in the $K_L \rightarrow 3\pi$ decays [29], $\eta \rightarrow 3\pi$ decays [30, 31] and $\eta' \rightarrow \eta\pi\pi$ decays [32]. It also turns out to be very convenient to use the same method for the evaluation of the isospin-breaking corrections in the $K_{\ell 4}$ decays [33], which serve as a very important alternative source of experimental information about the S-wave $\pi\pi$ scattering lengths [34].[7] Last but not least, in recent years nonrelativistic effective theories have been often applied for the calculation of finite-volume effects in the context of lattice QCD calculations. Chapter 6 of this book is dedicated to a detailed discussion of these applications.

2.9 Dimers and the Faddeev Equation

2.9.1 Introducing a Dimer Field at Lowest Order

In the three-particle problem, which was considered in the previous section, the scattering lengths were of natural size (and in fact rather small). Therefore, perturbation

---

[7] For an alternative method using dispersion relations, see Ref. [35].

theory is expected to converge well. In the context of nuclear physics or the physics of ultracold atoms, the situation is often the opposite. As already mentioned, the S-wave NN scattering lengths are very large compared to the typical scale of the problem: the Compton wavelength of the pion, $\lambda_\pi = 1/M_\pi \simeq 1.4\,\text{fm}$. In the case of a large scattering length, a non-perturbative resummation of certain classes of diagrams is required (if one does not want to restrict the EFT to momenta smaller than $1/a$). In the three-particle sector, such a resummation is necessary to obtain the so-called Faddeev equations [36]. (For a nice related textbook, we refer the interested reader to Ref. [37].) In this section, we shall consider this procedure in more detail and introduce the notion of a dimer. Furthermore, in case of zero-range two-body interactions and a vanishing three-body force, the aforementioned equation goes under the name of the Skornyakov–Ter–Martirosian (STM) equation [38], which predates the more general treatment [36].

In the following, we shall restrict ourselves to the case of nonrelativistic kinematics, omitting completely relativistic corrections. Galilean invariance is assumed, that is, the derivative interactions are constructed exclusively by using the Galilean-invariant operator $\overset{\leftrightarrow}{\nabla}$. Furthermore, the case of three identical spinless particles in the CM frame will be considered. Generalizations are possible, but do not add anything to the understanding of the basic principles. Thus, we consider a theory described by the Lagrangian

$$\mathcal{L} = \phi^\dagger \left(i\partial_t - M + \frac{\nabla^2}{2M} \right) \phi + C_0 \phi^\dagger \phi^\dagger \phi \phi + C_1 \left(\phi^\dagger \overset{\leftrightarrow}{\nabla}{}^2 \phi^\dagger \phi \phi + \text{h.c.} \right) + \cdots$$

$$+ D_0 \phi^\dagger \phi^\dagger \phi^\dagger \phi \phi \phi + D_1 \left(\phi^\dagger (\phi^\dagger \overset{\leftrightarrow}{\nabla}{}^2 \phi^\dagger) \phi \phi \phi + \text{h.c.} \right) + \cdots , \tag{2.141}$$

where the ellipsis denotes the higher-order terms with derivative interactions in the two- and three-particle sectors.

In the three-body sector, the derivation of the Faddeev equations is substantially simplified by using the so-called particle-dimer picture [39–41], which is a modern version of the old idea of quasiparticles introduced by Weinberg [42] in the context of this problem. In the particle-dimer picture, the two-particle interactions are represented by an (infinite) sum of s-channel exchanges of fictitious particles (dimers), and the three-particle interactions are given in terms of particle-dimer interactions. In a most transparent way, this can be explained by using the path integral language. The generating functional of a system based on the Lagrangian Eq. (2.141) is given by

$$Z(j, j^\dagger) = \int d\phi \exp\left\{ i \int d^4x \left(\mathcal{L} + j^\dagger \phi + \phi^\dagger j \right) \right\}. \tag{2.142}$$

For simplicity, we shall consider first only nonderivative couplings C_0, D_0 and neglect other possible terms. One can introduce an auxiliary nonrelativistic dimer field $d(x)$ and consider the Lagrangian

$$\mathcal{L}_d = \phi^\dagger \left(i\partial_t - M + \frac{\nabla^2}{2M} \right) \phi + \sigma d^\dagger d + \frac{f_0}{2} \left(d^\dagger \phi \phi + \text{h.c.} \right) + h_0 d^\dagger d \phi^\dagger \phi, \tag{2.143}$$

where $\sigma = \pm 1$ for negative/positive values of the constant C_0; see the following. Note also that, with our choice of normalization, the dimer field has a noncanonical mass dimension. Next, in the generating functional $Z_d(j, j^\dagger)$, which is obtained from the preceding Lagrangian, the dimer field can be integrated out. A straightforward calculation of the Gaussian integral gives

$$Z_d(j, j^\dagger) = \int d\phi \exp\left\{i \int d^4x \left(\phi^\dagger \left(i\partial_t - M + \frac{\nabla^2}{2M}\right)\phi\right.\right.$$
$$\left.\left. - \frac{f_0^2 \phi^\dagger \phi^\dagger \phi \phi}{4(\sigma + h_0 \phi^\dagger \phi)} + j^\dagger \phi + \phi^\dagger j\right)\right\}. \tag{2.144}$$

In order to compare this expression with the original generating functional, one has to expand the integrand in powers of the fields:

$$-\frac{f_0^2 \phi^\dagger \phi^\dagger \phi \phi}{4(\sigma + h_0 \phi^\dagger \phi)} = -\frac{f_0^2 \sigma}{4} \phi^\dagger \phi^\dagger \phi \phi + \frac{f_0^2 h_0}{4} \phi^\dagger \phi^\dagger \phi^\dagger \phi \phi \phi + \cdots, \tag{2.145}$$

where the ellipsis represents the terms that contain eight and more fields. Next, since one is working within a nonrelativistic theory, the sectors containing four and more particles are completely disconnected from the two- and three-body sectors. Thus, if we stay in the sector with at most three particles, the Lagrangians $\mathcal{L}$ and $\mathcal{L}_d$ describe the same theory at tree level, if the following matching conditions are fulfilled:

$$C_0 = -\frac{\sigma f_0^2}{4}, \qquad D_0 = \frac{f_0^2 h_0}{4}. \tag{2.146}$$

From this equation it is clear that the choice of the sign σ is correlated with the sign of the coupling constant C_0, as f_0^2 is always positive.

Further, the full propagator of the dimer field is defined as

$$i\langle 0|T d(x) d^\dagger(y)|0\rangle = \int \frac{d^4P}{(2\pi)^4} e^{-iP(x-y)} D(\mathbf{P}; P^0). \tag{2.147}$$

At tree level, $D(\mathbf{P}; P^0) = -\sigma$ is a constant. The full propagator obeys the Dyson–Schwinger equation:

$$D(\mathbf{P}; P^0) = -\sigma + (-\sigma)\frac{1}{2} f_0^2 J(\mathbf{P}; P^0) D(\mathbf{P}; P^0), \tag{2.148}$$

which yields

$$D(\mathbf{P}; P^0) = \frac{1}{-\sigma - \frac{1}{2} f_0^2 J(\mathbf{P}; P^0)}. \tag{2.149}$$

In the preceding equations, $J(\mathbf{P}; P^0)$ is the two-particle loop:

$$J(\mathbf{P}; P^0) = \int \frac{d^D k}{(2\pi)^D i} \frac{1}{(E_\mathbf{k} - k^0 - i\varepsilon)(E_{\mathbf{P}-\mathbf{k}} - P^0 + k^0 - i\varepsilon)}$$
$$= -\frac{M}{4\pi}\left(2M^2 + \frac{\mathbf{P}^2}{4} - MP^0 - i\varepsilon\right)^{1/2}. \tag{2.150}$$

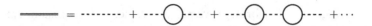

Figure 2.9   The full dimer propagator (double line) given by a sum of all bubble diagrams. The dashed line denotes the free dimer propagator, and a single line the particle propagator, respectively.

Here, dimensional regularization has been used in the expression of the divergent loop integral. Consequently, the full dimer propagator is given by an infinite sum of the bubble diagrams shown in Fig. 2.9.

The two-body scattering amplitude can be algebraically expressed through the full dimer propagator. In Fig. 2.9, this corresponds to equipping each external free dimer propagator with two incoming or outgoing particle lines. At lowest order, the dimer-two particle coupling is given by a constant f_0. Using Eqs. (2.87) and (2.146), we finally get

$$T(\mathbf{P};P^0) = \frac{1}{M}\,\tau(\mathbf{P};P^0) = \frac{f_0^2}{-\sigma - \frac{1}{2}f_0^2 J(\mathbf{P};P^0)} = \frac{8\pi/M}{-1/a - iq_0}, \qquad (2.151)$$

where the quantity q_0, defined as

$$q_0 = i\left(M(2M - P^0 - i\varepsilon) + \frac{\mathbf{P}^2}{4}\right)^{1/2} = \left(M(2P^0 - 2M) - \frac{\mathbf{P}^2}{4}\right)^{1/2}, \qquad (2.152)$$

denotes the magnitude of the relative three-momentum in the nonrelativistic two-particle system, and a is the S-wave two-body scattering length. Hence, at lowest order, we have indeed described the two-particle interactions by an exchange of a (fictitious) dimer in the s-channel.

2.9.2 Higher Orders

Before proceeding further with the derivation of the three-body scattering equations, we would like to briefly comment on the higher-order (derivative) terms. Beyond leading order, we have to deal both with higher partial waves as well as higher-order terms in the ERE in each partial wave. Higher partial waves in the two-body sector can be taken into account by introducing dimers with arbitrary spin. For example, in the model we are considering, P-waves are absent due to Bose symmetry. The contribution from the D-waves in two-body scattering is described by the $O(p^4)$ Lagrangian,

$$\mathcal{L}^D = C_D(\phi^\dagger F_{ij}\phi^\dagger)(\phi F_{ij}\phi), \qquad (2.153)$$

where F_{ij} denotes the second-order differential operator,

$$F_{ij} = \frac{3}{2}\overset{\leftrightarrow}{\nabla}_i\overset{\leftrightarrow}{\nabla}_j - \frac{1}{2}\delta_{ij}\overset{\leftrightarrow}{\nabla}^2. \qquad (2.154)$$

The Lagrangian in Eq. (2.153) can readily be rewritten in the particle-dimer picture. To this end, one has to introduce a spin-2 dimer field $d_{ij}(x)$, which is symmetric in the indices i, j and traceless. Such a field has five linearly independent components that indeed correspond to the number of degrees of freedom of a spin-2 particle. The interaction vertex of a dimer with two elementary fields in the D-wave is given by

$$\mathcal{L}_d^D = \frac{1}{2} f_D \big(d_{ij}^\dagger (\phi F_{ij} \phi) + \text{h.c.} \big). \tag{2.155}$$

Integrating out the auxiliary dimer field d_{ij} in the generating functional, as we did in the S-wave, we obtain the matching condition between couplings C_D and f_D. The generalization to the higher partial waves is straightforward.

Next, we turn to the higher-order derivative terms in a given partial wave, corresponding to the subleading orders in the ERE. There exist different formulations of the particle-dimer framework in the case when such higher-order terms are present in the original theory. It is instructive to consider these formulations separately and to ensure that all of them yield the same result for the on-shell amplitude. Again, let us restrict ourselves to the simplest possible case of the terms with two derivatives in Eq. (2.141), which contains the coupling C_1. Remember that this Lagrangian describes scattering in the S-wave. A straightforward ansatz for the dimer-two particle Lagrangian is, obviously,

$$\mathcal{L}_{d\phi\phi} = \frac{1}{2} d^\dagger \big(f_0 \phi\phi + f_1 \phi \overset{\leftrightarrow}{\nabla}{}^2 \phi \big) + \text{h.c.}. \tag{2.156}$$

Integrating out the dimer field, we arrive at the matching conditions that relate the couplings C_0, C_1 with f_0, f_1. Hence, our goal is fully accomplished. Further, summing up all higher-order terms, the on-shell two-body scattering amplitude in the S-wave takes the form (cf. Eq. (2.151))

$$T = \frac{1}{M} \tau = \frac{8\pi/M}{-1/a + \frac{1}{2} r q_0^2 + \cdots - i q_0} = \frac{8\pi/M}{q_0 \cot \delta(q_0) - i q_0}. \tag{2.157}$$

The procedure just outlined describes one particular scenario of introducing the dimer field. In Ref. [39], an alternative procedure was proposed. Namely, instead of adding derivatives in the dimer-two particle vertex, in this alternative formulation we modify the kinetic part of the dimer field, promoting the free dimer to a *dynamical* degree of freedom:

$$\sigma d^\dagger d \to \sigma d^\dagger \left(i\partial_t - (2M + \Delta) + \frac{\nabla^2}{4M} \right) d, \tag{2.158}$$

where Δ is a coupling of dimension [mass]. As already mentioned, the vertex $\frac{1}{2} y_0 (d^\dagger \phi\phi + \text{h.c.})$ has the same form as the leading-order term in Eq. (2.156). However, here we denote the coupling constant by y_0 to distinguish it from f_0. (These couplings do not even have the same mass dimension, and the dimension of the dimer field differs in the two formalisms as well.)

In order to arrive at the matching condition in the two-particle sector in this second formulation, let us calculate the tree-level two-body scattering amplitude in the CM frame. This is done easily in the dimer picture, where the sole contribution stems from the dimer exchange in the s-channel. The amplitude is given by

$$T^{\text{tree}} = \frac{y_0^2}{\sigma(2M + \Delta - P^0 - i\varepsilon)}. \tag{2.159}$$

The full amplitude is again obtained by summing up all bubbles shown in Fig. 2.9. This results in an additional term, $-\frac{1}{2}y_0^2 J(\mathbf{0}, P^0)$, in the denominator (cf. Eq. (2.151)). On the other hand, the result obtained from the original Lagrangian at tree level is given by

$$T^{\text{tree}} = 4C_0 - 4C_1(\mathbf{p}^2 + \mathbf{q}^2), \tag{2.160}$$

where $\mathbf{p}, \mathbf{q}$ denote the relative three-momenta in the final and initial states, respectively.

The two expressions given in Eqs. (2.159) and (2.160) can be matched on shell, where $P^0 = 2M + \mathbf{p}^2/M$ and $\mathbf{p}^2 = \mathbf{q}^2$. This leads to the following matching conditions:

$$C_0 = \frac{y_0^2}{4\sigma\Delta}, \qquad C_1 = -\frac{y_0^2}{8\sigma M\Delta^2}. \tag{2.161}$$

The generalization to higher orders is straightforward. Note also that the amplitudes in Eqs. (2.159) and (2.160) are different off the energy shell. However, as we already know, this does not upset the equivalence of the two formulations.

At first glance, considering different alternatives for the particle-dimer Lagrangian seems to be a trivial mathematical exercise, where two couplings, f_0, f_1, are traded for another pair, y_0, Δ. In reality, however, the issue is more subtle. In order to understand this, note that the tree-level CM-frame dimer propagator in the second formulation resembles the propagator of a *real* particle with a mass $2M + \Delta$. On the other hand, it is expected that, if a shallow bound state or a resonance in this partial wave indeed exists, the bulk of the interaction at low energies will be provided by an s-channel exchange of this particle. This leads to a nontrivial dependence of the amplitude on the energy that is readily captured by Eq. (2.159). On the contrary, in the first scenario, such an energy dependence emerges non-perturbatively, after the resummation of the subleading effective range term in the dimer propagator. (Recall that $P^0 = 2M + q_0^2/M$ on-shell, and thus Eq. (2.159) has the same energy dependence as Eq. (2.157), where the ERE is truncated after the second term.) Thus, it could be expected that, in the presence of the low-lying excitations or shallow bound states, the formulation with the dynamical dimer field will be more efficient. The remaining corrections from higher orders are expected to be smaller in this scenario.

As a side remark, we would like to also mention that in the context of different effective theories, we often encounter a situation, where a bound state or a resonance gives a large contribution to the coupling constants, once it is integrated out of a theory. For example, the concept of resonance saturation in Chiral Perturbation Theory, which will be considered in Chapter 4, follows the same pattern. In such cases, introducing these bound states/resonances as independent dynamical degrees of freedom (even as auxiliary fields, which do not appear in the in- and out-states), allows us to improve the convergence of the expansion, since the resonance exchange captures a bulk of the low-energy dynamics in the system. However, as we will also see in Chapter 4, the formulation of a consistent power counting in such an approach meets some obstacles.

Finally, note that higher-order three-particle interactions translate into higher-order particle-dimer interactions. For example, the particle-dimer Lagrangian at next-to-leading order can be written as follows:

$$\mathcal{L}_{dd\phi\phi} = h_0 d^\dagger d\phi^\dagger \phi + h_1 \big((d^\dagger \overset{\leftrightarrow}{\nabla}{}^2 \phi^\dagger)(d\phi) + \text{h.c.} \big) + h_P (d^\dagger \overset{\leftrightarrow}{\nabla}_i \phi^\dagger)(d\overset{\leftrightarrow}{\nabla}_i \phi) + \cdots .$$

(2.162)

The tree-level matching conditions for the couplings $h_0, h_1, h_2 \ldots$ can be worked out in a standard manner. We do not consider these conditions explicitly. Note also that the constant h_P describes the P-wave interactions of the particle-dimer pair. The inclusion of higher partial waves as well as further terms in the ERE proceeds straightforwardly.

The main points of our discussion on dimers are briefly summarized as follows:

- Dimers can *always* be introduced in the generating functional to define the Green's functions of a theory. In this sense, the particle-dimer picture is not an approximation. It is as fundamental as the initial effective theory and merely is based on other degrees of freedom.
- The validity of the particle-dimer picture does not rely on the assumption that shallow bound states or the resonances indeed exist in the two-body system. However, if this is the case, then the particle-dimer picture is particularly useful, since it gives a most convenient parameterization of the two-particle scattering amplitude in terms of very few parameters. The remaining higher-order corrections in the channels where the dimers appear are expected to be small at low energies.
- In the particle-dimer picture, there is no restriction on the partial waves that are taken into account. Different partial waves are described by separate dimers with appropriate spin. In the three-particle system, there are two types of angular momenta: the angular momentum of a given pair that can be described by introducing the dimers with arbitrary spin, and the angular momentum between this pair and the spectator. The latter can be taken into account, writing the particle-dimer interaction Lagrangians for all partial waves. An example is given in Eq. (2.162), which features, in particular, the P-wave term. The power-counting rules of the nonrelativistic EFT automatically lead to the truncation in both angular momenta without any further assumptions.
- The dimers can be introduced in different formulations. All these formulations are equivalent on the energy shell but lead to different off-shell Green's functions.
- If dimensional regularization is used in the two-particle sector, the tree-level matching stays intact to all orders. The situation in the three-particle case is more delicate. The integral equations in the three-particle sector cannot be written by using dimensional regularization. Here, cutoff regularization represents a commonly used choice. In this regularization, however, the matching condition for the couplings from the three-body sector gets modified. In what follows, these issues related to the regularization and the cutoff dependence of the couplings are discussed in detail.

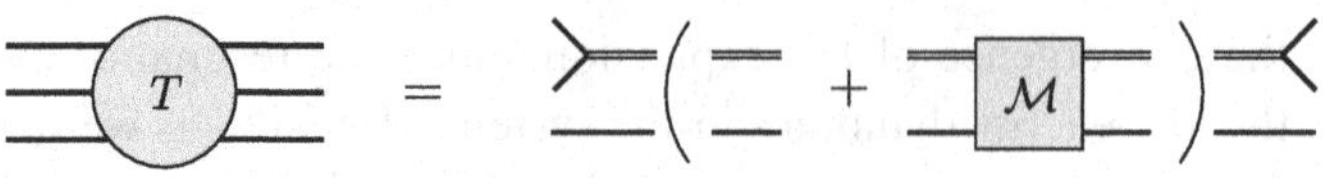

Figure 2.10 Relation between the three-particle scattering amplitude T and the particle-dimer scattering amplitude $\mathcal{M}$. The double and single lines denote a dimer and a particle, respectively, and the $d\phi\phi$ vertex at lowest order is equal to f_0.

2.9.3 Derivation of the Faddeev Equations

In the particle-dimer picture, the three-particle scattering amplitude can be algebraically related to the particle-dimer scattering amplitude. This is schematically shown in Fig. 2.10, and it can be seen that it suffices to equip each external dimer leg with two particle legs, whose coupling to the dimer is described by the pertinent $d\phi\phi$ vertex (at the given order). In order to ease the notation, we shall write down this relation in the CM frame at lowest order (no derivative couplings):

$$T(\{\mathbf{p}\},\{\mathbf{q}\};P^0) = \frac{1}{M} \sum_{\alpha,\beta=1}^{3} \left\{ \tau(-\mathbf{p}_\alpha;P^0)(2\pi)^3 \delta^{(3)}(\mathbf{p}_\alpha - \mathbf{q}_\beta) \right.$$

$$\left. + \tau(-\mathbf{p}_\alpha;P^0)\mathcal{M}(-\mathbf{p}_\alpha,-\mathbf{q}_\beta;P^0)\tau(-\mathbf{q}_\beta;P^0) \right\}. \tag{2.163}$$

Here, $\{\mathbf{p}\}$ denotes the set of the outgoing particle momenta $\mathbf{p}_1,\mathbf{p}_2,\mathbf{p}_3$, and similarly $\{\mathbf{q}\}$ represents the incoming momenta. The dimer momenta in the CM frame are equal to $-\mathbf{p}_1,-\mathbf{p}_2,-\mathbf{p}_3$. Further, τ is proportional to the full dimer propagator, given by Eq. (2.149), and $\mathcal{M}$ is the particle-dimer scattering amplitude, which can be calculated by using the Lagrangian in Eq. (2.143), *treating the dimer as a physical particle*.

In order to derive the integral equation for the particle-dimer scattering amplitude, we have to consider off-shell kinematics, assuming that the outgoing and incoming dimers have four-momenta $p^\mu = (p^0,\mathbf{p})$ and $q^\mu = (q^0,\mathbf{q})$, respectively. Thus, the four-momenta of the outgoing/incoming particles are $(P^0 - p^0,-\mathbf{p})$ and $(P^0 - q^0,-\mathbf{q})$, in order. We denote the off-shell particle-dimer amplitude by $\mathcal{M}(\mathbf{p},p^0,\mathbf{q},q^0;P^0)$. The on-mass-shell amplitude $\mathcal{M}(\mathbf{p},\mathbf{q};P^0)$ is obtained by setting $p^0 = P^0 - E_\mathbf{p}$ and $q^0 = P^0 - E_\mathbf{q}$. Examining the classes of diagrams contributing to M, it is straightforward to see that the off-shell amplitude obeys the equation depicted in Fig. 2.11:

$$\mathcal{M}(\mathbf{p},p^0,\mathbf{q},q^0;P^0) = Z(\mathbf{p},p^0\mathbf{q},q^0;P^0) + \int \frac{dk^0}{2\pi i} \int^\Lambda \frac{d^3\mathbf{k}}{(2\pi)^3} Z(\mathbf{p},p^0,\mathbf{k},k^0;P^0)$$

$$\times \tau(\mathbf{k},k^0) \frac{1}{E_\mathbf{k} - P^0 + k^0 - i\varepsilon} \mathcal{M}(\mathbf{k},k^0,\mathbf{q},q^0;P^0). \tag{2.164}$$

Here, $\int^\Lambda$ represents the cutoff in three-momentum space. Further, Z denotes the kernel of the equation, which is given by a sum of the particle exchange and the local $dd\phi\phi$ vertex. (In the following, we shall refer to this vertex as the three-body force):

$$Z(\mathbf{p},p^0,\mathbf{q},q^0;P^0) = \frac{1}{M(E_{\mathbf{p}+\mathbf{q}} + P^0 - p^0 - q^0 - i\varepsilon)} + \frac{h_0}{Mf_0^2}. \tag{2.165}$$

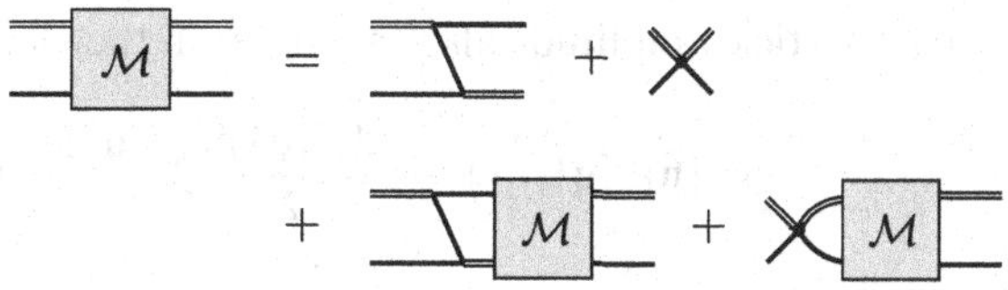

Figure 2.11 Faddeev equation for the particle-dimer scattering amplitude. The double and single lines denote the dimer and the particle, respectively, and the filled dot corresponds to the $dd\phi\phi$ interaction.

Considering the Born series of Eq. (2.164), we could close the Cauchy integration over the variable k^0 in the upper half plane in each term, where only a pole at $k^0 = P^0 - E_\mathbf{k} + i\varepsilon$ survives. Putting the external particle legs on shell, we finally obtains

$$\mathcal{M}(\mathbf{p},\mathbf{q};P^0) = Z(\mathbf{p},\mathbf{q};P^0) + \int^\Lambda \frac{d^3\mathbf{k}}{(2\pi)^3} Z(\mathbf{p},\mathbf{k};P^0)\tau(\mathbf{k},P_0 - E_\mathbf{k})\mathcal{M}(\mathbf{k},\mathbf{q};P^0),$$

(2.166)

where

$$Z(\mathbf{p},\mathbf{q};P^0) = \frac{1}{\mathbf{p}^2 + \mathbf{q}^2 + \mathbf{pq} + M(3M - P^0) - i\varepsilon} + \frac{h_0}{Mf_0^2}.$$

(2.167)

This is the Faddeev equation for the particle-dimer scattering amplitude we are looking for. This equation has a unique solution, owing to the resummation of the pair interactions in all channels. Technically, the resummation is performed by introducing a dimer field and considering a skeleton diagrams with full dimer propagators. It is now crystal clear how the generalization to higher orders proceeds. Additional derivative vertices in the two-body interactions modify the dimer propagator τ, leading to the replacement of $-1/a$ by the on-shell inverse K-matrix $q_0 \cot \delta(q_0)$. The inclusion of higher partial waves in the pair interactions proceeds through the introduction of new dimer fields that carry spin. The Faddeev equations in this case become multichannel matrix equations, which intertwine scattering amplitudes of particles on various dimers. The higher-order terms in the $dd\phi\phi$ vertex will lead to a modification of the kernel Z; that is, the three-body force becomes a polynomial in the momenta $\mathbf{p}, \mathbf{q}$. It can be shown that this momentum dependence can be traded for the energy dependence of the kernel at next-to-leading order, as this does not affect the on-shell three-body scattering amplitude [43]. We shall, however, not consider this option in detail.

In such a system there may exist three-body bound states (called *trimers*). In the vicinity of the bound-state pole $P^0 = E_n$, the particle-dimer scattering amplitude factorizes:

$$\mathcal{M}(\mathbf{p},\mathbf{q};P^0) \to \frac{\psi_n(\mathbf{p})\,\psi_n^\dagger(\mathbf{q})}{E_n - P^0} + \text{regular terms}.$$

(2.168)

Separating this pole contribution on both sides of Eq. (2.166), it is seen that the particle-dimer bound-state wave function ψ_n obeys a homogeneous integral equation:

$$\psi_n(\mathbf{p}) = \int^\Lambda \frac{d^3\mathbf{k}}{(2\pi)^3} Z(\mathbf{p},\mathbf{k};P^0)\tau(\mathbf{k},P_0 - E_\mathbf{k})\psi_n(\mathbf{k}).$$

(2.169)

The three-particle amplitude also factorizes in the vicinity of the pole:

$$T(\{\mathbf{p}\},\{\mathbf{q}\};P^0) \to \frac{\Psi_n(\{\mathbf{p}\})\Psi_n^\dagger(\{\mathbf{q}\})}{E_n - P^0} + \text{regular terms}, \tag{2.170}$$

with, using Eq. (2.163),

$$\sqrt{M}\Psi_n(\{\mathbf{p}\}) = \sum_{\alpha=1}^{3} \tau(-\mathbf{p}_\alpha;P^0)\psi_n(-\mathbf{p}_\alpha). \tag{2.171}$$

The derivation of the normalization condition for the bound-state wave function is a bit tricky, because both Z and τ depend on the energy P^0 in a nontrivial way. (See, e.g., Ref. [44] for a thorough discussion of the normalization condition in this context.) Writing down symbolically the scattering equation as $\mathcal{M} = Z + Z\tau\mathcal{M}$, we get $\mathcal{M}^{-1} = Z^{-1} - \tau$ and, finally,

$$\mathcal{M}(Z^{-1} - \tau)\mathcal{M} = \mathcal{M}. \tag{2.172}$$

In order to arrive at the normalization condition, one has to extract the pole contribution from both sides, letting $P^0 \to E_n$. The potential double pole on the l.h.s. has a zero residue, because the wave function ψ_n obeys Eq. (2.169). Equating the residua at the single poles, one finally gets

$$\int^\Lambda \frac{d^3\mathbf{p}}{(2\pi)^3} \frac{d^3\mathbf{q}}{(2\pi)^3} \, \psi_n^\dagger(\mathbf{p})N(\mathbf{p},\mathbf{q};E_n)\psi_n(\mathbf{q}) = 1, \tag{2.173}$$

where

$$N(\mathbf{p},\mathbf{q};E_n) = \lim_{P^0 \to E_n} \frac{d}{dP^0}\left(Z^{-1}(\mathbf{p},\mathbf{q};P^0) - \tau(\mathbf{p};P^0)(2\pi)^3\delta^{(3)}(\mathbf{p}-\mathbf{q})\right). \tag{2.174}$$

Using the operator identity $\frac{d}{dP^0}Z^{-1} = -Z^{-1}\left[\frac{d}{dP^0}Z\right]Z^{-1}$ and the fact that the wave function obeys the equation $(Z^{-1} - \tau)|\psi_n\rangle = 0$, the preceding expression can finally be rewritten as

$$N(\mathbf{p},\mathbf{q};E_n) = \lim_{P^0 \to E_n}\left(-\tau(\mathbf{p};P^0)\left[\frac{d}{dP^0}Z(\mathbf{p},\mathbf{q};P^0)\right]\tau(\mathbf{q};P^0)\right.$$

$$\left. - \frac{d}{dP^0}\tau(\mathbf{p};P^0)(2\pi)^3\delta^{(3)}(\mathbf{p}-\mathbf{q})\right). \tag{2.175}$$

The normalization condition for the three-particle wave function $\Psi_n(\{\mathbf{p}\})$ can be directly obtained using Eq. (2.171). We do not demonstrate this derivation here.

We conclude this section with few remarks:

- First, note that the approach, which was considered here is non-perturbative, different from the three-body problem considered in Section 2.8. This, in particular, means that the formalism can accommodate the case of a large two-body scattering length. We shall even consider the limit, when this scattering length tends to infinity, the so-called unitary limit.[8] Moreover, bound states that emerge in the two-body subsystems (dimers) and in the three-body sector (trimers) can be readily accommodated in the non-perturbative setting as well.

- Further, since one is considering an EFT approach, a legitimate question about the counting rules, which imply a *perturbative* expansion and a truncation in three-momenta, immediately arises. The answer to this question is that the perturbative expansion up to a given order in p is carried out in the kernel Z, and in the inverse of the dimer propagator τ. (The latter expansion coincides with a well-known ERE for the two-body sector.) After this, the equations are solved exactly. The EFT counting rules hold for Z and τ^{-1} and not for the amplitude $\mathcal{M}$. Note also that Z is not an observable, in contrast to $\mathcal{M}$.

- The Faddeev equations contain an ultraviolet divergence. Since these equations will be solved exactly rather than in perturbation theory, dimensional regularization, which does a very good job in many other calculations, is not an option anymore but it can still be used for the two-particle subsystems. A commonly accepted choice is to use a cutoff on the dimer three-momentum (as already indicated in the pertinent equations). Physical quantities like the particle-dimer amplitude $\mathcal{M}$ should be cutoff-independent. This cutoff independence is provided by the cutoff-dependent three-body couplings $h_0 = h_0(\Lambda), \ldots$. The renormalization should be done non-perturbatively, because the couplings $h_0, h_1, h_2, \ldots$ enter the kernel Z and are thus iterated to all orders. It turns out that the non-perturbative renormalization introduces a RG running of a new type, the so-called log-periodic behavior of the three-body couplings. This will be considered in the next section.

- As already mentioned, taking into account higher-order terms in the two-body sector amounts to using the ERE in the dimer propagator. Here, a problem arises, because the dimer propagator enters the integral equations, with dimer momenta running up to the hard scale given by the cutoff Λ. The ERE is not supposed to work at such a high scale. As a result, the dimer propagator might develop so-called *spurious poles*. These poles emerge below threshold, where $q_0 = i\kappa$. At second order, the denominator of the propagator takes the form

$$-\frac{1}{a} - \frac{1}{2}r\kappa^2 + \kappa = \frac{1}{2}r(\kappa - \kappa_1)(\kappa - \kappa_2), \qquad (2.176)$$

where

$$\kappa_{1,2} = \frac{1 \mp \sqrt{1 - 2r/a}}{r}. \qquad (2.177)$$

The assumption $a \gg r \simeq \Lambda^{-1}$ and $a, r > 0$ gives $\kappa_1 \simeq 1/a$ and $\kappa_2 \simeq 2/r$. In this case, the pole at $\kappa = \kappa_1$ is a low-energy (physical) pole, corresponding to the really existing shallow bound state (stable dimer). The other pole at $\kappa = \kappa_2$ is a spurious one. It emerges because the ERE was applied at energies where it should not. Moreover, it

---

[8] To be precise, the original definition of the unitary limit also entails the vanishing of all other ERE parameters [45]. In what follows, we will be somewhat cavalier and mostly consider an infinite two-body scattering length to define the unitary limit.

can be checked that the residue of this pole has a wrong sign, so it cannot correspond to a physical particle.

The spurious pole by itself does not cause any problems, because it lies outside the range of applicability of the effective theory. However, in the dimer-particle scattering equation, the dimer propagator appears within the integral, and a spurious pole, when integrated over, leads to unwelcome consequences. For example, it starts to contribute to the imaginary part of the particle-dimer amplitude *even at low energies*. Moreover, this contribution has the wrong sign. Hence, a modification of the framework is necessary in order to exclude the violation of the basic principles by the spurious poles.

Obviously, the simplest option to achieve the goal is to choose the cutoff Λ small enough so that the spurious pole cannot be reached kinematically. This option is rather unattractive, since it precludes us from carrying out a full-fledged investigation of the cutoff dependence of the observables and, eventually, from sending the cutoff to infinity. Another option, which was suggested in Ref. [41] – see also Refs. [46, 47] for further implementation – is based on the expansion of the dimer propagator in powers of r and other higher-order ERE parameters prior to the integration (this corresponds to the expansion, shown in Eq. (2.95)). It is clear that such an expansion will modify the high-energy part of the integral over the dimer momenta that can be taken into account by changing the renormalization prescription for the three-body couplings $h_0, h_1, h_2, \ldots$. Other methods to deal with such spurious poles are discussed in Refs. [48–50].

- Obviously, one could work directly in the particle-dimer picture, avoiding the matching to the three-particle sector. (This tree-level matching is, anyway, never used again.) Writing down the most general particle-dimer Lagrangian, one should, however, refrain from using the EOM for the dimer field, in order to reduce the number of terms. The reason for this is that a dimer is a fictitious particle that is merely used to conveniently rewrite the two-particle interactions. In the language of Feynman diagrams, the particle-dimer scattering amplitude that appears in Fig. 2.10 is off-shell. In order to see this, let us consider an outgoing dimer in this figure. It has momentum $-\mathbf{p}_\alpha$ and energy $P^0 - E_{\mathbf{p}_\alpha}$. It is now clear that the energy of a dimer is not determined solely by its three-momentum, as it depends on the total energy as well. Note also that in the case when a dimer really exists (as a physical shallow bound state), the EOM can be safely used for the total energies below the dimer breakup threshold, that is, $P^0 < 3M$.

2.10 Efimov Effect

In this section, we study the cutoff-independence of the observables obtained by solving the scattering equations. We will see that the non-perturbative character of the renormalization leads to a peculiar dependence of the three-body couplings on the cutoff parameter Λ, which is not observed at any finite order in the perturbative expansion.

As already mentioned, a non-perturbative approach is needed when the two-body scattering length has an unnaturally large value. This fact is related to the existence of a shallow bound or virtual state; see Chapter 1. Note that the case of an unnaturally large scattering length is not rare in nature. Apart from the S-wave nucleon–nucleon scattering lengths already mentioned, we would like to draw attention to the scattering of cold atoms. It is particularly fascinating that in this case one can even vary the scattering length by placing atoms in an external magnetic field. The idea behind this effect is very simple (see, e.g., Ref. [51]). The atoms have excited states that contribute to the optical potential in atom–atom scattering. (A nice introduction to the optical potential is given in the classic textbook by Goldberger and Watson [52].) If the total energy in the scattering is close to one of these "shielded" excited levels, then a so-called Feshbach resonance is observed [53, 54]. Since the energy of the excited states depends linearly on the magnetic field, one may tune the field so that a particular excited state is very close to the threshold. Then, the dependence of the atom–atom scattering length on the external magnetic field B is described by a simple formula,

$$a(B) = a' + \frac{a''}{B - B_0}, \qquad (2.178)$$

where the quantities a', a'' in first approximation do not depend on B, and B_0 is the critical value of B, for which the given excited state hits the threshold. Thus, one could tune B in the experiment so that the atom–atom scattering length becomes very large. A particularly interesting aspect of this enterprise is that various physical systems in the limit $a \to \infty$ (the so-called unitary limit) acquire universal properties that allow one to use EFT tools to study these properties. The present section treats the universality exclusively in the context of the three-body problem.

Let us return to the three-body system. In a first approximation, we neglect the three-body coupling and move the cutoff to infinity. Then, in the limit of an infinitely large two-body scattering length, the problem we are considering becomes a no-scale one, because the only available physical scale was set by the inverse of this scattering length. It can be seen that the problem becomes ill-defined in this limit. Indeed, in this limit the quantity $\tau(\mathbf{p}, P^0 - E_{\mathbf{p}})$ from Eq. (2.151) takes the form

$$\tau(\mathbf{p}; P^0 - E_{\mathbf{p}}) = \frac{8\pi}{-\frac{1}{a} + \sqrt{-M(P^0 - 3M) + \frac{3}{4}\mathbf{p}^2}}$$

$$\to \frac{8\pi}{\sqrt{-\kappa^2 + \frac{3}{4}\mathbf{p}^2}}, \qquad (2.179)$$

where $\kappa^2 = M(P^0 - 3M)$. For a three-body bound state $P^0 < 3M$. The homogeneous equation for the bound-state wave function becomes

$$\psi_n(\mathbf{p}) = 8\pi \int \frac{d^3\mathbf{k}}{(2\pi)^3} \frac{1}{\mathbf{p}^2 + \mathbf{k}^2 + \mathbf{p}\mathbf{k} - \kappa^2} \frac{\psi_n(\mathbf{k})}{\sqrt{-\kappa^2 + \frac{3}{4}\mathbf{k}^2}}. \qquad (2.180)$$

Next, we carry out the partial-wave expansion. For the S-wave component of the wave function, denoted by $\psi_n(p)$ (p in the argument stands for $|\mathbf{p}|$), we get the following equation:

$$\psi_n(p) = \frac{2}{\pi} \int_0^\infty \frac{k dk}{p} \ln \frac{p^2 + pk + k^2 + \kappa^2}{p^2 - pk + k^2 + \kappa^2} \frac{\psi_n(k)}{\sqrt{-\kappa^2 + \frac{3}{4}k^2}}. \tag{2.181}$$

This equation coincides with the STM equation in the limit of the infinite two-body scattering length. Exact solutions to this equation have been found by Faddeev and Minlos [55]:

$$\psi_n(p) = \text{const} \cdot \frac{\kappa \sin(s_0 u)}{p}, \qquad u = \ln\left(\frac{\sqrt{3}p}{2\kappa} + \sqrt{1 + \frac{3p^2}{4\kappa^2}}\right). \tag{2.182}$$

Here, s_0 denotes the solution of the following transcendental equation:

$$s_0 \cosh \frac{\pi s_0}{2} = \frac{8}{\sqrt{3}} \sinh \frac{\pi s_0}{6}, \qquad s_0 \simeq 1.00624. \tag{2.183}$$

It is now clearly seen that the equation does not produce a discrete spectrum below the threshold $P^0 = 3M$, as one would expect from the beginning on physical grounds. Rather, the subthreshold spectrum is continuous, as the wave function given in Eq. (2.182) is a solution of Eq. (2.181) for any value of κ. This can be verified by simply substituting Eq. (2.182) into Eq. (2.181) and carrying out the integration explicitly. Rescaling the momenta $p \to \kappa p$, $k \to \kappa k$, one sees that κ disappears from the both sides of the equation; in other words, any real positive κ is an eigenvalue.

In reality, Eq. (2.181) is an idealization, since it implies zero interaction range. Assuming a finite range, r is equivalent to introducing a finite cutoff $\Lambda \sim 1/r$ that renders the integral equation well-defined and leads to a discrete three-body bound-state spectrum κ_n.

The presence of the cutoff will be felt in the wave function only for large momenta, where $p \simeq \Lambda$. For small momenta, the wave function still will be given by the Faddeev–Minlos solution. Next, let us consider shallow bound states with $\kappa_n \sim 1/a \to 0$. In the interval $\kappa_n \ll p \ll \Lambda$, we could use the asymptotic expression

$$u \simeq \ln \frac{p}{\kappa_n} + \frac{\delta}{s_0}, \qquad \psi_n(p) \simeq \text{const} \cdot \frac{\kappa_n}{p} \sin\left(s_0 \ln \frac{p}{\kappa_n} + \delta\right), \tag{2.184}$$

where $\delta = \frac{1}{2} s_0 \ln 3$ is a pure number.

Further, using dimensional arguments, we have $\kappa_n = \bar{\kappa}_n \Lambda$, where the $\bar{\kappa}_n$ are dimensionless numbers. Now, we see that the spectrum has a universal geometric feature independent of the details of the short-range interactions. Indeed, if κ_n is an eigenvalue at a given value of the cutoff Λ, then the asymptotic wave function is invariant under a discrete scale transformation:

$$\Lambda \to \exp\left(\frac{\pi n}{s_0}\right)\Lambda, \qquad n = 0, \pm 1, \pm 2, \cdots. \tag{2.185}$$

Note that the choice of the odd n leads only to an (unobservable) change of sign of the whole wave function $\psi_n(p)$.

One could now reinterpret the transformation in Eq. (2.185) as a shift of the eigenvalue label $\kappa_n \to \kappa_{n+1}$. Hence, the discrete spectrum of the three-body problem is given by

$$\frac{\kappa_{n+1}}{\kappa_n} = \exp\left(\frac{\pi}{s_0}\right), \qquad n = 0, \pm 1, \pm 2, \cdots. \tag{2.186}$$

In other words, fixing a single eigenvalue $\kappa_n = \kappa_*$ at $n = n_*$ is equivalent to fixing the cutoff Λ, and this gives the whole bound-state spectrum. The solution of the recurrence relation in Eq. (2.186) can be written as

$$\kappa_n = S_0^{n-n_*} \kappa_*, \qquad S_0 = \exp\left(-\frac{\pi}{s_0}\right). \tag{2.187}$$

Formally, the sequence runs to infinity on both sides. In reality, however, it is restricted from above and from below by the finite interaction range and the finite scattering length, respectively. For an explicit calculation showing the influence of the effective range, see, for example, Ref. [56].

Next, let us briefly consider the case of a large but finite a. Every physical observable will be a function of a, κ_* and the kinetic energy variables, denoted collectively by E. From the preceding discussion, it is clear that in the case of discrete scale transformations

$$\kappa_* \to \kappa_*, \qquad a \to S_0^m a, \qquad E \to S_0^{-2m} E, \tag{2.188}$$

all physical observables transform according to their mass dimension. Assume, for example, that stable S-wave dimers indeed exist, and consider the particle-dimer scattering cross section below the dimer breakup threshold. The cross section has dimension E^{-2} and, consequently,

$$\sigma_{\phi d}(S_0^{-2m} E, S_0^m a, \kappa_*) = S_0^{2m} \sigma_{\phi d}(E, a, \kappa_*). \tag{2.189}$$

At threshold, $E = 0$, the cross section is expressed through the particle-dimer scattering length:

$$\sigma_{\phi d}(0, a, \kappa_*) = 4\pi a_{\phi d}^2. \tag{2.190}$$

The scattering length on the r.h.s. of this equation depends only on a and κ_*. Using dimensional arguments and discrete scale invariance, we could write

$$a_{\phi d} = f(2s_0 \ln(a\kappa_*)) a, \tag{2.191}$$

where $f(x)$ is a periodic function of x with a period 2π. The periodicity is due to the discrete scale invariance, which is manifest in Eq. (2.189). This latter property is very important, because it translates into a log-periodic dependence of the three-body recombination rate of cold atoms on the quantity $a\kappa_*$ that is measurable in experiment and may provide solid evidence in favor of the validity of the preceding theoretical picture.

The infinite tower of the three-particle bound states, condensing toward the origin, was first discussed in the seminal papers by Efimov [57–59]. A nice example of an (approximate) Efimov state in nuclear physics is the triton, a bound state of three nucleons. This is the only Efimov state in this channel, because of the truncation at high and low energies previously mentioned. Further, in ultracold atom gases, there have been many experimental observations of Efimov states and of the universal log-periodic behavior that is dictated by the discrete scale invariance. For more information, we refer the interested reader to the excellent recent reviews on the subject [60–62].

To briefly summarize the discussion to this point: we have seen that the equation without a cutoff contains mathematical deficiencies. Introducing a cutoff transforms this into a well-defined problem. Further, we can fix the cutoff Λ by considering a single observable, say, a particular energy level, or the particle-dimer scattering length. All other observables can then be expressed in terms of this single one. In this manner, however, we cannot study the limit $\Lambda \to \infty$; the dependence of the observables on Λ has an irregular, log-periodic behavior, and the limit does not exist.

Let us now note that, up to now, we have assumed $h_0 = 0$. According to the commonly accepted wisdom that we have followed up to now, the cutoff dependence in the effective theory should be eaten up by the cutoff dependence of the effective couplings, and thus the observables turn out to be cutoff-independent. This property, however, has been verified so far in perturbation theory, whereas non-perturbative setting leads to the new features, absent in any fixed order in perturbation theory. The main result of Bedaque, Hammer and van Kolck [40, 41] was to show that a single three-body coupling h_0 can absorb the whole Λ-dependence, which results from summing up infinitely many diagrams in the particle-dimer scattering amplitude. In order to verify this statement, let us note that Eq. (2.169) after the partial-wave expansion takes the following form in the S-wave:

$$\psi_n(p) = \frac{2}{\pi} \int_0^\Lambda \frac{kdk}{p} \left(\ln \frac{p^2 + pk + k^2 + \kappa^2}{p^2 - pk + k^2 + \kappa^2} + 2pk \frac{H(\Lambda)}{\Lambda^2} \right) \frac{\psi_n(k)}{\sqrt{-\kappa^2 + \frac{3}{4}k^2}}. \tag{2.192}$$

Here, $2H(\Lambda)/\Lambda^2 = h_0(\Lambda)/(Mf_0^2)$, and one assumes $H(\Lambda) \sim 1$.

Consider now this equation in the intermediate region $1/a \ll p \ll \Lambda$. In this region, one can rewrite the equation in the following form:

$$\psi_n(p) = \frac{2}{\pi} \int_0^\mu \frac{kdk}{p} \left(\ln \frac{p^2 + pk + k^2 + \kappa^2}{p^2 - pk + k^2 + \kappa^2} + 2pk \frac{H(\Lambda)}{\Lambda^2} \right) \frac{\psi_n(k)}{\sqrt{-\kappa^2 + \frac{3}{4}k^2}}$$
$$+ \frac{8}{\sqrt{3}\pi} \int_\mu^\Lambda kdk\, \psi_n(k) \left(\frac{1}{k^2} + \frac{H(\Lambda)}{\Lambda^2} \right). \tag{2.193}$$

Here, μ is an arbitrary scale that obeys $1/a \ll \mu \ll \Lambda$. In the first term, the contribution multiplied by $H(\Lambda)$ is much smaller than that from the logarithm, since both $p/\Lambda \ll 1$ and $k/\Lambda \ll 1$. Hence, the Λ-dependence in the first term disappears completely.

Let us now choose another scale $\Lambda' > \Lambda$ and the new coupling $H(\Lambda')$. The solution of the new equation is denoted by $\psi_n'(p)$. It is seen that $\psi_n'(p)$ and $\psi_n(p)$ obey the same equation for $p \ll \Lambda$, if the following relation holds:

$$\int_\Lambda^{\Lambda'} dk \frac{\psi_n'(k)}{k} + \int_\mu^{\Lambda'} kdk\, \psi_n'(k) \frac{H(\Lambda')}{\Lambda'^2} - \int_\mu^\Lambda kdk\, \psi_n'(k) \frac{H(\Lambda)}{\Lambda^2} = 0. \tag{2.194}$$

Choose now, for instance, $n = n_*$ and use the asymptotic form of the wave function given by Eq. (2.184) up to $p = \Lambda'$. We have

$$\psi_*(p) = \text{const} \cdot \frac{\kappa_*}{p} \sin\left(s_0 \ln \frac{p}{\kappa_*} + \delta\right) = \text{const} \cdot \frac{\kappa_*}{p} \cos\left(s_0 \ln \frac{p}{\Lambda_*}\right), \qquad (2.195)$$

where

$$\Lambda_* = \kappa_* \exp\left(\frac{\pi}{2s_0} - \frac{\delta}{s_0}\right). \qquad (2.196)$$

Using Eq. (2.195) in Eq. (2.194) and performing the integration, We get

$$\frac{1}{\Lambda'}\left[\sin\left(s_0 \ln \frac{\Lambda'}{\Lambda_*} - \arctan \frac{1}{s_0}\right) + H(\Lambda')\sin\left(s_0 \ln \frac{\Lambda'}{\Lambda_*} + \arctan \frac{1}{s_0}\right)\right]$$

$$- (\Lambda' \to \Lambda) + O\left(\frac{\mu}{\Lambda}\right) = 0. \qquad (2.197)$$

The solution of Eq. (2.197) is given by

$$H(\Lambda) = -\frac{\sin\left(s_0 \ln \dfrac{\Lambda}{\Lambda_*} - \arctan \dfrac{1}{s_0}\right)}{\sin\left(s_0 \ln \dfrac{\Lambda}{\Lambda_*} + \arctan \dfrac{1}{s_0}\right)}. \qquad (2.198)$$

It is seen that the three-body force, parameterized by $H(\Lambda)$, exhibits the log-periodic behavior as well, namely, $H(\mathcal{S}_0^{-n}\Lambda) = H(\Lambda)$.

In conclusion, we note that the log-periodic behavior of the coupling constant, which we have just found, corresponds to the so-called limit cycles, which are particular solutions of the renormalization group equations [63]. The relevance of the limit cycle in the context of the three-body problem was first discussed in Ref. [64]. For more information on the subject, see, for example, Refs. [65, 66].

2.11 Including Virtual Photons in Nonrelativistic Theories

2.11.1 Including the Photon Field in the Lagrangian

The application of nonrelativistic effective Lagrangians in particle physics dates back to the seminal work by Caswell and Lepage [67]. The framework laid out in this paper is designed for a systematic treatment of electromagnetic bound states of charged particles like positronium or the hydrogen atom. As one knows from quantum mechanics, the average of the three-momentum squared in the ground state of such an atom is given by $\langle \mathbf{p}^2 \rangle = (\mu\alpha)^2$. Here, μ is the reduced mass of the two-particle system, and $\alpha \simeq 1/137$ denotes the electromagnetic fine-structure constant. Thus, a typical value of the magnitude of the three-momentum in the ground state is smaller than the reduced mass by a factor of order α. Stated differently, the non-relativistic expansion in powers of p/μ translates into an expansion in powers of α in the calculation of the ground state observables. Moreover, the parametric dependence of $\langle \mathbf{p}^2 \rangle$ on α in the excited states is the same, and hence the argument remains valid also for the excited states. It now becomes clear that nonrelativistic QED (NRQED) is tailored to perform bound-state calculations perturbatively. Namely, if one is interested in the result up to a given order in α, one can drop all those higher-dimensional operators in the Lagrangian that do not contribute at this order. Further, the counting rules are also valid for the loop diagrams and hence, at a given order in α, only a finite number of diagrams should be calculated.[9] On the contrary, in a relativistic theory, one does not have counting rules in p, and the calculations involved in using, for example, the Bethe–Salpeter equation or any three-dimensional reduction thereof are much more complicated and less transparent.

Furthermore, a similar approach of nonrelativistic effective Lagrangians has been successfully applied to QCD [68]. The method has been extremely fruitful, for example, for the study of the heavy quarkonia. Moreover, nonrelativistic QCD (NRQCD) is widely used for the heavy quarks on the lattice, because the original relativistic formulation contains disparate scales, related to the QCD scale and to the heavy quark masses, that render the discussion of the continuum limit difficult. We cannot review here this vast and very interesting field of research in any detail and simply provide a few references on the subject for further reading [69, 70].

Last but not least, nonrelativistic effective theories are extremely useful for the study of the finite-volume artifacts on the lattice. The typical three-momenta on a cubic lattice with a box size L are on the order of $2\pi/L$. Then, particles with a mass m, which obey the condition $mL \gg 1$, can be treated within nonrelativistic effective theories. Next, turning on the electromagnetic interactions, we may study the finite-volume effects in the presence of the massless photons. For example, the leading corrections to the charged particle masses in a finite volume have been calculated within this approach. It was shown that in this case the corrections are suppressed only by powers of L, in contrast to massive theories, where the suppression is exponential. Another nice example is the scattering of two particles in the presence of electromagnetic interactions on the lattice, which has also been studied within this approach. All these interesting problems will be discussed in detail in Chapter 5.

In this section, we shall consider the construction of nonrelativistic Lagrangians with virtual photons, the counting rules in the presence of the photons and the matching conditions. In order to simplify the discussion, we shall restrict ourselves to the "standard" formulation, where the energy of a free particle is given by a nonrelativistic expression $p^0 = m + \mathbf{p}^2/(2m)$, with the corrections included perturbatively.

First, following Ref. [67], let us write a nonrelativistic effective Lagrangian of QED, based on the general requirements like rotational and translational invariance, gauge invariance, invariance under the discrete P, C, T symmetries and so on. Equation (2.20),

---

[9] The only exception from this rule, as will be seen in what follows, is given by the exchange of the instantaneous (Coulomb) photons between two charged particles. All ladder diagrams contribute to the same order and should be summed up in order to get the answer at a given order in α. Fortunately, this can be done in closed form.

which describes a particle in the external electromagnetic field, provides the starting point. The full Lagrangian with virtual photons contains more terms:

$$\mathcal{L} = -\frac{1}{4}\mathcal{F}_{\mu\nu}\mathcal{F}^{\mu\nu} + \psi^\dagger\left(iD_t - m + \frac{\mathbf{D}^2}{2m} + c_1\frac{\mathbf{D}^4}{8m^3} + \frac{e}{2m}c_2\boldsymbol{\sigma}\mathbf{B}\right.$$

$$\left. + \frac{ie}{8m^2}c_3\boldsymbol{\sigma}([\mathbf{D}\times\mathbf{E}] - [\mathbf{E}\times\mathbf{D}]) + \frac{e}{8m^2}c_4(\mathbf{DE} - \mathbf{ED}) + \cdots\right)\psi$$

$$- \frac{d_1}{m^2}(\psi^\dagger\psi)^2 - \frac{d_2}{m^2}(\psi^\dagger\boldsymbol{\sigma}\psi)^2 + \cdots + \text{terms with antiparticles}. \qquad (2.199)$$

Here, ψ denotes the electron field, m is its mass, $\mathcal{F}_{\mu\nu} = \partial_\mu\mathcal{A}_\nu - \partial_\nu\mathcal{A}_\mu$, $D_t = \partial_t + ie\mathcal{A}_0$, and $D^i = \nabla^i - ie\mathcal{A}^i$. Further, the c_i and d_i are couplings constants. They will be discussed in more detail in what follows. The added terms are of the two types. First, one has a kinetic term of the photon. At higher orders nonlinear terms like, for example, $(\mathcal{F}_{\mu\nu}\mathcal{F}^{\mu\nu})^2$ also will be present; see later in Section 3.2. Physically, such nonlinear terms emerge, because the electron loop diagram with four external photon lines vanishes in NRQED but not in QED. This mismatch should be accounted for by a local operator in the Lagrangian that produces the diagrams with four external photon lines. Second, there are four-fermion vertices. Again, at the one-loop level (two-photon exchange in the electron–electron scattering) there is a mismatch between the amplitudes calculated in NRQED and QED. The difference is a low-energy polynomial in the three-momenta of the electrons that, at the lowest order, can be removed by the four-fermion terms displayed in the Lagrangian (2.199). Further, note that the higher-order terms in the Lagrangian, which were uniquely determined from the Foldy–Wouthuysen transformation, now come with arbitrary coefficients $c_i = 1 + O(e^2)$. These should be determined from the (perturbative) matching of the relativistic and nonrelativistic amplitudes at a given order in e.

Quantizing the electromagnetic field, one should impose a gauge condition. A crucial simplification arises due to the fact that one is allowed to use different gauge conditions in the relativistic and nonrelativistic theories when performing the matching. This, in turn, is related to the fact that the S-matrix elements, which are used in the matching condition, are gauge invariant. Using this freedom, one could, for instance, use a covariant gauge (say, the Feynman gauge) in the calculations within the relativistic theory. In the nonrelativistic theory, the so-called Coulomb gauge,

$$\nabla_i\mathcal{A}_i = 0, \qquad (2.200)$$

has proven to be the most convenient one. In this gauge, the photon propagator takes the form

$$D^{\mu\nu}(k) = i\int d^4x\, e^{ikx}\langle 0|T\mathcal{A}^\mu(x)\mathcal{A}^\nu(0)|0\rangle, \qquad (2.201)$$

where

$$D^{00}(k) = -\frac{1}{\mathbf{k}^2}\,,$$

$$D^{ij}(k) = -\frac{1}{k^2 + i\varepsilon}\left(\delta^{ij} - \frac{k^i k^j}{\mathbf{k}^2}\right),$$

$$D^{0i}(k) = D^{i0}(k) = 0\,. \tag{2.202}$$

The usefulness of the Coulomb gauge can be traced back to the fact that Coulomb (timelike) and transverse (spacelike) photon vertices have different orders in the power counting. The leading-order Coulomb photon vertex $-e\psi^\dagger \mathscr{A}_0 \psi$ is of order one, whereas the transverse photon vertex $-\frac{ie}{2m}\psi^\dagger(\nabla^i \mathscr{A}^i + \mathscr{A}^i \nabla^i)\psi$ counts as order p. Consequently, the diagrams with Coulomb photons emerge at leading order, whereas the loops with transverse photons give only subleading contributions. On the other hand, the Coulomb exchange is instantaneous, and the pertinent component of the propagator does not depend on k^0. Hence, many diagrams containing the Coulomb photons, except the ladder diagrams corresponding to a multiple exchange of photons in the s-channel, vanish. This circumstance simplifies the perturbative calculations considerably.

Maintaining counting rules in the loops is, however, not straightforward. Here, we again encounter the situation when the hard scale (here, the electron mass m) is explicitly present in the propagator that leads to the violation of the counting rules. In addition, the answer to the question, at which order do the virtual photon momenta count, depends on the topology of a diagram. Apart from the hard photons, whose momenta are of order one, there exist soft photons, whose momenta count as $O(p)$, and also ultrasoft photons, whose momenta should be counted at $O(p^2)$. In order to rectify the power-counting rules, in the calculation of the loop diagrams we have to accurately separate these contributions coming from these different regimes and subtract the part coming from the hard photons. This can be done in many ways. We shall stick to the so-called threshold expansion [71], which is explained in detail in a few explicit examples in the following section.

2.11.2 Electron Self-Energy and Renormalization

We start with the evaluation of the self-energy of the electron at one loop, which may receive contributions from the two diagrams shown in Fig. 2.12. The diagram with the

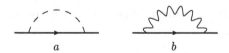

$$a \qquad\qquad\qquad b$$

Figure 2.12   Photon loop corrections to the electron self-energy in NRQED: (a) the (vanishing) Coulomb photon contribution; (b) the transverse photon contribution.

Coulomb photon, shown in Fig. 2.12a, gives the following contribution to the self-energy (in D space-time dimensions):

$$\Sigma_C(\mathbf{p};p^0) = -e^2 \int \frac{d^D k}{(2\pi)^D i} \frac{1}{\mathbf{k}^2} \frac{1}{E_{\mathbf{p}-\mathbf{k}} - p^0 + k^0 - i\varepsilon}. \qquad (2.203)$$

In coordinate space, being proportional to $\theta(x^0 - y^0)$, the electron propagator vanishes for $x_0 < y_0$, and the Coulomb photon propagator contains a factor $\delta(x^0 - y^0)$. It is plausible that their product vanishes. However, one should be careful with this argument, since the product of these two singular distributions is ill-defined at $x^0 = y^0$. A proper definition implies some kind of a regularization and a choice of renormalization prescription, after which the regularization can be removed. Using, for instance, dimensional regularization and performing the integral over k^0 in Eq. (2.203), we get

$$\Sigma_C(\mathbf{p};p^0) = -\frac{e^2}{2} \int \frac{d^d \mathbf{k}}{(2\pi)^d} \frac{1}{\mathbf{k}^2} = 0, \qquad (2.204)$$

where, in the last step, we encounter a no-scale integral that vanishes in dimensional regularization.

On the contrary, the diagram with the transverse photon yields a nonzero contribution:

$$\Sigma_T(\mathbf{p};p^0) = -\frac{e^2}{4m^2} \int \frac{d^D k}{(2\pi)^D i} \frac{1}{k^2 + i\varepsilon} \left(\delta^{ij} - \frac{k^i k^j}{\mathbf{k}^2} \right) \frac{(2p-k)^i (2p-k)^j}{E_{\mathbf{p}-\mathbf{k}} - p^0 + k^0 - i\varepsilon}. \qquad (2.205)$$

Carrying out the integration over k^0, one obtains

$$\Sigma_T(\mathbf{p};p^0) = \frac{e^2}{m^2} \int \frac{d^d \mathbf{k}}{(2\pi)^d} \frac{1}{2|\mathbf{k}|} \left(\mathbf{p}^2 - \frac{(\mathbf{p}\mathbf{k})^2}{\mathbf{k}^2} \right) \frac{1}{E_{\mathbf{p}-\mathbf{k}} - p^0 + |\mathbf{k}| - i\varepsilon}. \qquad (2.206)$$

In this expression, the violation of the counting rules is made explicit, as the denominator of the electron propagator mixes different powers of p:

$$E_{\mathbf{p}-\mathbf{k}} - p^0 + |\mathbf{k}| = m + \frac{\mathbf{p}^2}{2m} - p^0 + \frac{\mathbf{p}\mathbf{k}}{m} + \frac{\mathbf{k}^2}{2m} + |\mathbf{k}|. \qquad (2.207)$$

The counting of various terms in this equation is as follows: the electron three-momentum $\mathbf{p}$ is of order p, and $p^0 - m = O(p^2)$. Thus, the $\mathbf{k}$-independent term in the denominator counts as $O(p^2)$. Assuming now that $|\mathbf{k}|$ is of order one (hard), p (soft) and p^2 (ultrasoft), respectively, we get

$$|\mathbf{k}| = O(1): \quad \underbrace{m + \frac{\mathbf{p}^2}{2m} - p^0}_{O(p^2)} + \underbrace{\frac{\mathbf{p}\mathbf{k}}{m}}_{O(p)} + \underbrace{\frac{\mathbf{k}^2}{2m} + |\mathbf{k}|}_{O(1)},$$

$$|\mathbf{k}| = O(p): \quad \underbrace{m + \frac{\mathbf{p}^2}{2m} - p^0}_{O(p^2)} + \underbrace{\frac{\mathbf{p}\mathbf{k}}{m}}_{O(p^2)} + \underbrace{\frac{\mathbf{k}^2}{2m}}_{O(p^2)} + \underbrace{|\mathbf{k}|}_{O(p)},$$

$$|\mathbf{k}| = O(p^2): \qquad \underbrace{m + \frac{\mathbf{p}^2}{2m} - p^0}_{O(p^2)} + \underbrace{\frac{\mathbf{pk}}{m}}_{O(p^3)} + \underbrace{\frac{\mathbf{k}^2}{2m}}_{O(p^4)} + \underbrace{|\mathbf{k}|}_{O(p^2)} \; . \tag{2.208}$$

This means that, in the case when $|\mathbf{k}| = O(1)$, the electron propagator should be expanded as

$$\frac{1}{E_{\mathbf{p-k}} - p^0 + |\mathbf{k}| - i\varepsilon} = \frac{1}{E_{\mathbf{k}} - m + |\mathbf{k}|} - \frac{1}{(E_{\mathbf{k}} - m + |\mathbf{k}|)^2} \frac{\mathbf{pk}}{m}$$

$$+ \left(-\frac{1}{(E_{\mathbf{k}} - m + |\mathbf{k}|)^2} \left(m + \frac{\mathbf{p}^2}{2m} - p^0 \right) \right.$$

$$\left. + \frac{1}{(E_{\mathbf{k}} - m + |\mathbf{k}|)^3} \frac{(\mathbf{pk})^2}{m^2} \right) + \cdots . \tag{2.209}$$

Integrating term by term, one sees that the contribution from this regime (hard photons) is a low-energy polynomial in momenta that start at order p^2.

Next, it is immediately seen that the whole contribution from the soft photon regime vanishes. Indeed, in this case, $|\mathbf{k}|$ is the leading term in the denominator, and the subsequent terms in the expansion are proportional to $1/|\mathbf{k}|^n$. This leads again to no-scale integrals that vanish in dimensional regularization.

The sole low-energy contribution to the electron self-energy thus comes from the ultrasoft photon regime. The expansion of the electron propagator takes the form

$$\frac{1}{E_{\mathbf{p-k}} - p^0 + |\mathbf{k}| - i\varepsilon} = \frac{1}{E_{\mathbf{p}} - p^0 + |\mathbf{k}| - i\varepsilon} - \frac{1}{(E_{\mathbf{p}} - p^0 + |\mathbf{k}| - i\varepsilon)^2} \frac{\mathbf{pk}}{m} + \cdots . \tag{2.210}$$

At leading order, one has (in d space dimensions)

$$\Sigma_T(\mathbf{p}; p^0) = \frac{e^2}{2m^2} \mathbf{p}^2 \frac{d-1}{d} \int \frac{d^d \mathbf{k}}{(2\pi)^d} \frac{1}{|\mathbf{k}|} \frac{1}{E_{\mathbf{p}} - p^0 + |\mathbf{k}| - i\varepsilon}$$

$$= \frac{e^2}{2m^2} \mathbf{p}^2 \frac{d-1}{d} (E_{\mathbf{p}} - p^0 - i\varepsilon)^{d-2} \int \frac{d^d \mathbf{k}}{(2\pi)^d} \frac{1}{|\mathbf{k}|} \frac{1}{|\mathbf{k}| + 1} . \tag{2.211}$$

Here, we have averaged over the directions of $\mathbf{k}$, replacing $k^i k^j$ by $\delta^{ij} \mathbf{k}^2 / d$. It can now be seen that the contribution from the ulftrasoft region is on the order of p^4, in contrast to the hard regime, where the pertinent contribution to Σ_T in Eq. (2.206) appears at $O(p^2)$, as seen from Eq. (2.209). Further, the value of the initial integral is given by the sum of all contributions from the different regimes. It is easily seen that Eq. (2.208) exhausts all possibilities, which can give a nonzero contribution. Indeed, assuming $\mathbf{k} = O(p^\alpha)$ with an arbitrary α, one arrives at a no-scale integral unless $\alpha = 0, 2$. The same analysis can be applied to a generic Feynman integral, where the energy denominators are polynomials in photon momenta of at most second order.

In order to evaluate the remaining integral, we carry out the integration over angles and change the integration variable, $k = -1 + 1/t$. This gives

$$\int \frac{d^d\mathbf{k}}{(2\pi)^d} \frac{1}{|\mathbf{k}|} \frac{1}{|\mathbf{k}|+1} = \frac{2}{(4\pi)^{d/2}\Gamma(d/2)} \int_0^\infty \frac{dk\, k^{d-2}}{k+1}$$

$$= \frac{2}{(4\pi)^{d/2}\Gamma(d/2)} \int_0^1 dt\, t^{1-d}(1-t)^{d-2}$$

$$= \frac{2\Gamma(2-d)\Gamma(d-1)}{(4\pi)^{d/2}\Gamma(d/2)}, \tag{2.212}$$

in terms of Euler's Γ-function,

$$\Gamma(n+1) = \int_0^\infty dt\, t^n e^{-t}. \tag{2.213}$$

Next, let us consider the renormalization of the self-energy. Dropping the high-energy contribution and expanding Eq. (2.212) in the vicinity of $d = 3$, one gets

$$\Sigma_T(\mathbf{p}; p^0) = -\frac{e^2 \mathbf{p}^2 z(p) \mu^{d-3}}{6\pi^2 m^2} \left(\frac{1}{3-d} + \frac{1}{2}(\Gamma'(1) + \ln 4\pi) + \ln \frac{z(p)}{\mu} + c \right), \tag{2.214}$$

where $z(p) = E_\mathbf{p} - p^0 - i\varepsilon$. Here, μ is the renormalization scale, and c denotes a numerical constant, whose exact value will not be important in the following discussion. As seen, the self-energy is ultraviolet-divergent. In the (modified) $\overline{\text{MS}}$ scheme, renormalization is carried out by dropping the simple pole in $d-3$ together with accompanying factors $\Gamma'(1)$ and $\ln 4\pi$. Defining, *for arbitrary d*, the renormalized self-energy,

$$\Sigma_T^r(\mathbf{p}; p^0) = \frac{e^2}{m^2} \frac{d-1}{d} \frac{\Gamma(2-d)\Gamma(d-1)}{(4\pi)^{d/2}\Gamma(d/2)} \mathbf{p}^2 z(p)^{d-2} - \frac{8e^2 \mathbf{p}^2 z(p)}{3m^2} \lambda, \tag{2.215}$$

where

$$\lambda = \frac{\mu^{d-3}}{16\pi^2} \left(\frac{1}{d-3} - \frac{1}{2}(\Gamma'(1) + \ln 4\pi) \right), \tag{2.216}$$

we verify that it has a well-defined limit at $d \to 3$. Moreover, it is seen that on the mass shell,

$$\lim_{p^0 \to E_\mathbf{p}} \lim_{d \to 3} \Sigma_T^r(\mathbf{p}; p^0) = \lim_{d \to 3} \lim_{p^0 \to E_\mathbf{p}} \Sigma_T^r(\mathbf{p}; p^0) = 0. \tag{2.217}$$

In other words, the electron mass does not get renormalized through this diagram. Note also that the limiting value does not depend on how the order of the limits is performed.

Finding the wave function renormalization factor Z_e is trickier.[10] Note that this factor is not a *constant* anymore because of the lack of Lorentz invariance in the non-relativistic setting, but depends on $\mathbf{p}^2$. As is well known, in the presence of massless particles (photons) Z_e is infrared-divergent. We shall use dimensional regularization

---

[10] Note that this has nothing to do with the fact that we are working with the nonrelativistic framework here. The same line of reasoning applies to relativistic theories (like, e.g., QED) as well.

to tame *both* ultraviolet and infrared divergences. Performing the limit $d \to 3$ first in the unrenormalized part, we arrive at Eq. (2.214), whose derivative with respect to p^0 logarithmically diverges on the mass shell $p^0 \to E_{\mathbf{p}}$. (The counterterm is a polynomial in p^0 and thus its derivative is regular.) In order to define Z_e properly, in dimensional regularization one reverses the order of limits and considers the limit $p^0 \to E_{\mathbf{p}}$ at $d > 3$. Only the contribution of the counterterm survives in this limit, and we get

$$\lim_{p^0 \to E_{\mathbf{p}}} \frac{d}{dp^0} \Sigma_T^r(\mathbf{p}; p^0) = \frac{8e^2 \mathbf{p}^2}{3m^2} \lambda \,. \tag{2.218}$$

The divergence in this expression, however, cannot be of ultraviolet origin, because it emerges only on the mass shell. Namely, the derivative of the self-energy part is finite off-shell as $d \to 3$. Hence, one has to *reinterpret* the divergence as an infrared one. This is made explicit by attaching the subscript "IR" to the quantity λ. Summing up the one-loop self-energy insertions, it is seen that the two-point function in the vicinity of the pole takes the form

$$S_{\alpha\beta}(p) = \delta_{\alpha\beta} \frac{1 - \dfrac{8e^2 \mathbf{p}^2}{3m^2} \lambda_{\text{IR}}}{E_{\mathbf{p}} - p^0 - i\varepsilon} + \text{regular} \,. \tag{2.219}$$

From this, the explicit expression for the factor Z_e can be read off:

$$Z_e = 1 - \frac{8e^2 \mathbf{p}^2}{3m^2} \lambda_{\text{IR}} \,. \tag{2.220}$$

2.11.3 The Electromagnetic Form Factor

The various coupling constants appearing in Eq. (2.199) are related to the electromagnetic properties of the particle under consideration. These properties are given in terms of the corresponding form factors. In this section, we therefore consider in detail how the matching of the form factor is carried out at one loop. In particular, we shall concentrate on the coupling constant c_4 from the Lagrangian (2.199), which is related to the anomalous magnetic moment and the mean square radius of the electron. The latter quantity is infrared-divergent at one loop, and we shall see how this divergence should be handled.

The electromagnetic form factor in the nonrelativistic theory is described by the diagrams displayed in Fig. 2.13. First, let us concentrate on the calculation of the vertex diagram with the transverse photon, which is shown in diagram b of this figure. (The contribution from the Coulomb photons vanishes after integration over k^0, since all poles lay on the same side of the contour.) The matching condition for the coupling

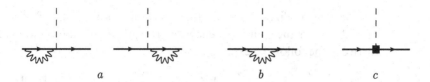

The vertex of a Coulomb photon at one loop in NRQED: (a) insertions into the external lines; (b) vertex corrections; (c) local couplings proportional to c_i. Coulomb and transverse photons are depicted by the dashed and wiggly lines, respectively.

c_4 can be obtained from the matching of the Coulomb photon vertex (coupling to the zeroth component of the photon field $\mathscr{A}^0$), which is given by the following:[11]

$$\Gamma_b^0(p,q) = \frac{e^3}{4m^2} \int \frac{d^D k}{(2\pi)^D i} \frac{(2q-k)^i(2p-k)^j}{(E_{q-k}-q^0+k^0-i\varepsilon)(E_{p-k}-p^0+k^0-i\varepsilon)}$$

$$\times \frac{1}{(k^2+i\varepsilon)} \left(\delta^{ij} - \frac{k^i k^j}{\mathbf{k}^2}\right). \tag{2.221}$$

After integration over k^0, averaging over the directions of $\mathbf{k}$ and expansion of the integrand in the ultrasoft regime, we get

$$\Gamma_b^0(p,q) = -\frac{e^3}{8m^2} \int \frac{d^d \mathbf{k}}{(2\pi)^d} \frac{(2q-k)^i(2p-k)^j}{|\mathbf{k}|(E_{q-k}-q^0+|\mathbf{k}|-i\varepsilon)(E_{p-k}-p^0+|\mathbf{k}|-i\varepsilon)}$$

$$\times \left(\delta^{ij} - \frac{k^i k^j}{\mathbf{k}^2}\right)$$

$$= -\frac{e^3}{2m^2} \mathbf{pq} \left(1-\frac{1}{d}\right) \int \frac{d^d \mathbf{k}}{(2\pi)^d} \frac{1}{|\mathbf{k}|(z(q)+|\mathbf{k}|)(z(p)+|\mathbf{k}|)} + \cdots, \tag{2.222}$$

where $z(p) = E_{\mathbf{p}} - p^0 - i\varepsilon$ and $z(q) = E_{\mathbf{q}} - q^0 - i\varepsilon$.

Like the self-energy diagram, the vertex diagram is ultraviolet-divergent. The renormalized vertex in the $\overline{\text{MS}}$ scheme is given by

$$\Gamma_b^{0,r}(p,q) = \Gamma^0(p,q) - \frac{8e^3}{3m^2} \mathbf{pq}\, \lambda. \tag{2.223}$$

This subtraction renders the vertex function finite everywhere except for the mass shell $z(p) = z(q) = 0$. Going to the mass shell at $d \neq 3$, we discover that the integral vanishes (as a no-scale integral). The sole nonvanishing contribution is provided by a counterterm:

$$\lim_{z(p),z(q)\to 0} \Gamma_b^{0,r}(p,q) = -\frac{8e^3}{3m^2} \mathbf{pq}\, \lambda_{\text{IR}}. \tag{2.224}$$

Here, we have again reinterpreted the divergence at $d \to 3$ as an infrared one.

---

[11] This contribution is spin-independent, i.e., is proportional to $\delta_{s's}$, where s', s denote the spins of the final and initial electrons, respectively. In order to ease the notation, we do not display this δ-symbol explicitly in the following.

The renormalized vertex at this order in e is the sum of all diagrams shown in Fig. 2.13. For the spin-independent part of the Coulomb photon vertex at $O(e^3)$, we arrive at the following expression:

$$\Gamma^{0,r}(p,q) = -eZ_e^{1/2}(\mathbf{p})Z_e^{1/2}(\mathbf{q}) + \Gamma_b^{0,r}(p,q) + \frac{e}{8m^2}c_4^r(\mathbf{p}-\mathbf{q})^2 + \cdots$$

$$= -e\left(1 - \frac{4e^2}{3m^2}(\mathbf{p}-\mathbf{q})^2\lambda_{\mathsf{IR}} - \frac{1}{8m^2}c_4^r(\mathbf{p}-\mathbf{q})^2\right) + \cdots. \qquad (2.225)$$

Here, the ellipses represent the higher-order terms as well as the spin-dependent part proportional to the Pauli matrix $\boldsymbol{\sigma}$. Further, c_4^r denotes the finite part[12] of the UV-divergent coupling c_4:

$$c_4 = \frac{32e^2}{3}\lambda + c_4^r. \qquad (2.226)$$

Let us now consider the relativistic vertex of the electron, given by

$$\langle ps'|j^\mu(0)|qs\rangle = -e\bar{u}(ps')\left(\gamma^\mu F_1(t) + \frac{i}{2m}\sigma^{\mu\nu}(p-q)_\nu F_2(t)\right)u(qs), \qquad (2.227)$$

where $t = (p-q)^2$. Expanding the Dirac (F_1) and Pauli (F_2) form factors in the vicinity of $t = 0$ gives

$$F_1(t) = 1 + \frac{1}{6}\langle r^2\rangle t + O(t^2), \qquad F_2(t) = \kappa + O(t), \qquad (2.228)$$

where $\langle r^2\rangle$ and κ denote the mean squared radius and the anomalous magnetic moment of the electron, respectively. Further,

$$t = (p^0-q^0)^2 - (\mathbf{p}-\mathbf{q})^2 = -(\mathbf{p}-\mathbf{q})^2 + O(p^4). \qquad (2.229)$$

In order to carry out the nonrelativistic reduction in the relativistic expression in Eq. (2.227), we use the known representation of the solutions of the free Dirac equation:

$$u(q,s) = \frac{\slashed{q}+m}{\sqrt{q^0+m}}\chi(s), \qquad \bar{u}(p,s') = \chi^\dagger(s')\frac{\slashed{p}+m}{\sqrt{p^0+m}}. \qquad (2.230)$$

Here, $\chi(s) = u(0,s)$, $\chi^\dagger(s') = \bar{u}(0,s')$ are four-component spinors with vanishing small components. Retaining only the large components and expanding in three-momenta, we gets

$$\frac{1}{\sqrt{4p^0q^0}}\langle ps'|j^0(0)|qs\rangle$$

$$= -e\left\{\delta_{s's}\left(1 - \frac{1}{8m^2}(\mathbf{p}-\mathbf{q})^2\right) + \frac{i}{4m^2}(\boldsymbol{\sigma}_{s's}\cdot[\mathbf{p}\times\mathbf{q}])\right\}F_1(t)$$

$$+ \frac{e}{4m^2}\left\{\delta_{s's}(\mathbf{p}-\mathbf{q})^2 + i(\boldsymbol{\sigma}_{s's}\cdot[\mathbf{p}\times\mathbf{q}])\right\}F_2(t). \qquad (2.231)$$

---

[12] The renormalization is a bit subtle here. Note that in the Lagrangian (2.199) we have already used the EOM. As a result, the counterterms that made the two-point function finite are now lumped together with the couplings c_i, d_i.

Matching the spin-independent part of Eq. (2.231) to Eq. (2.225) and using Eq. (2.228), we obtain

$$c_4^r = 1 + 2\kappa + \frac{4}{3}m^2\langle r^2\rangle - \frac{32}{3}e^2\lambda_{IR}. \tag{2.232}$$

The anomalous magnetic moment at order e^2 is given by the well-known expression $\kappa = e^2/(8\pi^2)$ [72]. It is both ultraviolet- and infrared-finite. The divergence that cancels the term with λ_{IR} comes from the mean squared radius. In order to complete the matching, we have to repeat the calculation of the form factor in QED, again using dimensional regularization to tame both the ultraviolet and infrared divergences. The factor λ now contains a pole in $D-4$ instead of $d-3$. The wave function renormalization constant of the relativistic electron is given by the following expression (the superscript "R" stands for "relativistic"):

$$Z_e^R = 1 + e^2\left\{4\lambda_{IR} + \frac{3}{16\pi^2}\ln\frac{m^2}{\mu^2} - \frac{1}{4\pi^2}\right\}. \tag{2.233}$$

The photon–electron vertex in QED at one loop is given by the diagrams shown in Fig. 2.14. The diagram in Fig. 2.14b corresponds to

$$\Gamma_{R,b}^\mu(p,q) = -e^3\int\frac{d^Dk}{(2\pi)^D i}\frac{1}{k^2}\bar{u}(p,s')\gamma^\nu\frac{1}{m-\not p+\not k}\gamma^\mu\frac{1}{m-\not q+\not k}\gamma_\nu u(q,s). \tag{2.234}$$

Next, one has to use the Feynman trick,

$$\frac{1}{A_1^{\nu_1}\cdots A_n^{\nu_n}} = \frac{\Gamma(\nu_1+\cdots+\nu_n)}{\Gamma(\nu_1)\cdots\Gamma(\nu_n)}\int_0^1 dx_1\cdots dx_n\delta\left(1-\sum_{i=1}^n x_i\right)$$

$$\times\frac{x_1^{\nu_1-1}\cdots x_n^{\nu_n-1}}{(x_1A_1+\cdots+x_nA_n)^{\nu_1+\cdots+\nu_n}}, \tag{2.235}$$

and simplify the numerator, separating the Dirac and Pauli form factors from each other. We do not display the standard, but pretty voluminous calculations here. Adding up all diagrams shown in Fig. 2.14, carrying out the renormalization in the $\overline{MS}$ scheme and evaluating the vertex on the mass shell, we finally obtain the following expression for the Dirac form factor we are looking for:

$$F_1(t) = 1 + 2e^2\left(\lambda_{IR} + \frac{1}{32\pi^2}\ln\frac{m^2}{\mu^2}\right)\left(2 + (t-2m^2)\int_0^1\frac{dx}{G}\right)$$

$$-\frac{e^2}{16\pi^2}\int_0^1 dx\left(-\ln\frac{G}{m^2} + (t-2m^2)\frac{1}{G}\ln\frac{G}{m^2} - 6 + 2(3m^2-t)\frac{1}{G}\right), \tag{2.236}$$

where

$$G = m^2 - x(1-x)t. \tag{2.237}$$

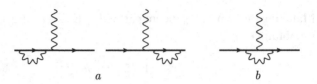

Figure 2.14 The photon-electron vertex in relativistic QED: (a) insertions in the external lines; (b) the vertex correction.

It is easy to verify that $F_1(0) = 1$. Further, the expansion of this expression in t yields the mean charge radius squared:

$$\langle r^2 \rangle = \frac{8e^2}{m^2} \left(\lambda_{\text{IR}} + \frac{1}{32\pi^2} \ln \frac{m^2}{\mu^2} \right) - \frac{3e^2}{16\pi^2 m^2} . \tag{2.238}$$

Finally, using this expression in the matching condition (2.232), we obtain the matching for the coupling c_4^r:

$$c_4^r = 1 + \frac{e^2}{3\pi^2} \ln \frac{m^2}{\mu^2} . \tag{2.239}$$

The same result was obtained in Ref. [73] (cutoff regularization, nonzero photon mass) and Ref. [6] (dimensional regularization). It is clear that the result does not depend on the regularization used. The infrared-divergent terms in any regularization cancel on both sides of the matching condition as they should. (Indeed, matching describes physics at the hard scale, which is insensitive to the low momenta.)

To summarize: we have considered the matching of one particular coupling constant in the nonrelativistic effective Lagrangian in all details. This example was chosen to demonstrate the peculiarities of the matching condition, which emerge due to the presence of infrared divergences. The cancellation of these divergences has been demonstrated explicitly. Now, following Ref. [6], we may even conclude that such a detailed treatment was not needed. In fact, as could be seen, the nonrelativistic loop diagrams always give a vanishing contribution on shell at $d \neq 3$, because they are expressed through no-scale integrals. The only nonvanishing contribution comes from the divergent terms proportional to λ. Since, at the end, all divergent terms should cancel in the matching condition, it suffices to calculate the relativistic diagrams and just drop all such divergent terms. Expanding the result in the external three-momenta gives then the desired matching condition for the nonrelativistic couplings. Using this algorithm, it is possible to perform the one-loop matching in NRQED as well as in NRQCD, where the calculations are much more voluminous [6].

In conclusion, we make two technical remarks. First, in order to simplify the argument we used the same renormalization scale in the relativistic and nonrelativistic theories. In principle, one could relax this requirement at the cost of additional $\sim \ln(\mu/\mu')$ terms in the matching condition (here, μ, μ' denote the scales in two theories). Second, we have used the Feynman gauge in the relativistic theory and the Coulomb gauge in the nonrelativistic one. As we have mentioned, this is justified, because the electromagnetic current (and therefore the form factor) is a gauge-invariant object.

2.11.4 Scattering of Charged Pions: Lagrangian and Matching

As another example, we shall consider a nonrelativistic theory with charged pions interacting with photons. The hard scale in such a theory is given by the pion mass. For pions, the strong and electromagnetic interactions are intertwined, bringing new features into the formalism. The Lagrangian of the system in question can be written in the following form:

$$\mathcal{L} = -\frac{1}{4}\mathscr{F}_{\mu\nu}\mathscr{F}^{\mu\nu} + \phi_0^\dagger\left(i\partial_t - M_0 + \frac{\nabla^2}{2M_0} + \frac{\nabla^4}{8M_0^3} + \cdots\right)\phi_0$$

$$+ \sum_{\pm}\phi_{\pm}^\dagger\left(iD_t - M + \frac{\mathbf{D}^2}{2M} + \frac{\mathbf{D}^4}{8M^3} + \cdots \mp \frac{eh_1}{6M^2}(\mathbf{DE} - \mathbf{ED}) + \cdots\right)\phi_{\pm}$$

$$+ C\phi_+^\dagger\phi_-^\dagger\phi_+\phi_- + C_x(\phi_+^\dagger\phi_-^\dagger\phi_0\phi_0 + \text{h.c.}) + C_0\phi_0^\dagger\phi_0^\dagger\phi_0\phi_0 + \cdots. \qquad (2.240)$$

Here, M and M_0 denote the charged and neutral pion mass, respectively, and the ellipses stand for the omitted higher-order terms.

Let us discuss the matching in the presence of the electromagnetic interactions. The problem we are studying can be formulated as follows. Generally, the couplings h_1, C, C_x, C_0 and so on depend both on the parameters of QCD, the strong coupling constant and the quark masses, as well as on the QED fine-structure constant α. This dependence has the form $h_1 = h_1^0 + h_1^1\alpha + O(\alpha^2)$, and similarly for the other couplings. Here, the quantity h_1^0 denotes the coupling in "pure QCD," obtained at $\alpha = 0$. It can be determined from the matching in the theory with no dynamical photons, as described in the previous sections.[13] So, it is natural to ask whether one can systematically obtain the higher-order coefficients in the expansion in α as well via the matching condition.

We start our discussion by mentioning that the splitting of the coefficients into a "pure QCD" part and the electromagnetic corrections is a rather subtle issue. For example, it is not a priori known how the parameters of "pure QCD" should be chosen. What is, for instance, the pion mass in this theory? Does it coincide with the charged or the neutral mass, or does it take a value that is different from these two? A brief answer to this question is that the definition of such an idealized theory is always a matter of convention, and various prescriptions could be used to fix such a convention. Later, in Section 4.15, we shall discuss such definitions in the context of QCD and provide more details. Here, we merely treat the expansion of the couplings in α in a formal manner, assuming that some (unspecified) convention is chosen. This will suffice for the following discussions.

Next, we consider the coupling h_1, which is related to the mean squared radius of the charged pion. At lowest order in the fine-structure constant α, the matching condition reads

$$h_1 = M^2\langle r_\pi^2\rangle + O(\alpha). \qquad (2.241)$$

[13] Here, we refer to "pure QCD" with the nonequal quark masses. In nature, isospin symmetry is broken both due to the electromagnetic interactions as well as due to $m_u \neq m_d$. Hence, the couplings $h_1^0, \ldots$ still do not describe the isospin-symmetric world. This issue, however, will not affect the discussion here.

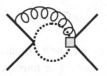

Figure 2.15 An example of a higher-order diagram, where a (transverse) photon is coupled to the four-pion vertex with two π^0 mesons (the shaded square). The solid and dotted lines denote charged and neutral pions, respectively, and the wiggly line represents the transverse photon.

Here, $\langle r_\pi^2 \rangle$ denotes the mean squared radius of the pion in "pure QCD," which is determined by the non-perturbative dynamics of quarks and gluons. If the pion were pointlike, one could calculate the $O(\alpha)$ correction, following the same pattern as in the case of the electron. However, this result will be modified by the strong interactions. The key observation that enables one to carry out the matching is that strong interactions do not change the infrared behavior of the vertex function, and the spin-zero analog of Eq. (2.232) still holds. Hence, calculating the mean squared radius of the pion at $O(\alpha)$ in some (relativistic) theory, we can determine the nonrelativistic coupling h_1 at the same order.

Next, let us turn to the matching of the four-pion couplings. In the isospin limit $m_u = m_d$, the couplings C^0, C_x^0, C_0^0 can be related to the S-wave scattering lengths a_0, a_2:

$$3\bar{M}^2 C^0 = 4\pi(2a_0 + a_2) + \cdots,$$

$$3\bar{M}^2 C_x^0 = 4\pi(a_2 - a_0) + \cdots,$$

$$3\bar{M}^2 C_0^0 = 2\pi(a_0 + 2a_2) + \cdots. \tag{2.242}$$

Here, $\bar{M}$ denotes the pion mass in "pure QCD" (at this stage, we do not specify its value), the subscript "x" refers to the charge exchange channel, $\pi^+\pi^- \to \pi^0\pi^0$, while the subscript "0" denotes neutral particle scattering, $\pi^0\pi^0 \to \pi^0\pi^0$.

Below, we shall concentrate on the matching in the charged sector that allows us to use a nice shortcut that simplifies life a lot and allows one, in addition, to discuss a couple of subtle conceptual issues in an explicit example. As one sees from Eq. (2.240), the neutral pions do not interact with photons at the order we are working. In the one-pion sector, this is true even to all orders due to the C-invariance. In the two-pion sector, one can write down a Lagrangian containing π^+, π^-, two π^0 fields and a photon field. However, the diagrams like the one shown in Fig. 2.15 emerge at a very high order and will therefore be omitted in the following. Hence, one may completely integrate out the π^0-field from the theory, replacing an infinite sum of the neutral pion bubbles, shown in Fig. 2.16, by a single effective charged pion coupling, $C \to \bar{C}$. The price to pay for this is that $\bar{C}$ becomes a function of the total CM energy of the two-particle system and is, in general, complex. Namely, summing up the geometric series in the CM frame of two particles, one obtains

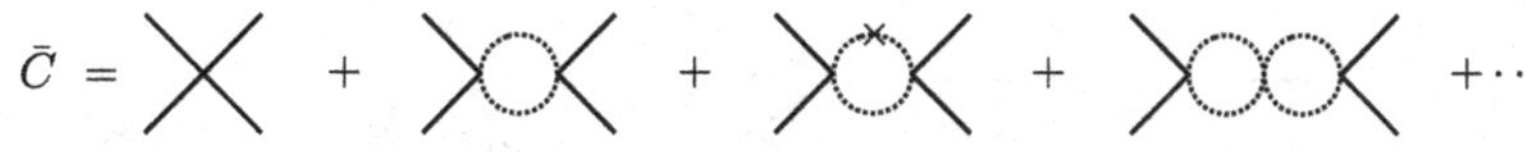

Figure 2.16 The neutral pion bubbles that contribute to the energy-dependent effective coupling $\bar{C}$. The solid and dotted lines denote the charged and neutral pions, respectively. The cross denotes a relativistic insertion into the neutral pion line.

$$\bar{C}(P^0) = C + \frac{2C_x^2 J_0(P^0)}{1 - 2C_0 J_0(P^0)}, \tag{2.243}$$

where

$$\begin{aligned}
J_0(P^0) &= \int \frac{d^3\mathbf{k}}{(2\pi)^3} \left(\frac{1}{2M_0 + \frac{\mathbf{k}^2}{M_0} - P^0} + \frac{\mathbf{k}^4}{4M_0} \frac{1}{\left(2M_0 + \frac{\mathbf{k}^2}{M_0} - P^0\right)^2} + \cdots \right) \\
&= \frac{iM_0}{4\pi} \sqrt{M_0(P^0 - 2M_0)} \left(1 + \frac{5(P_0 - 2M_0)}{8M_0} + \cdots \right) \\
&= \frac{iP^0}{16\pi} \sqrt{(P^0)^2 - 4M_0^2} + \cdots . \tag{2.244}
\end{aligned}$$

Here, the ellipses denote the higher-order relativistic corrections. Thus, $\bar{C}(P^0)$ is complex everywhere in the physical region of charged pion scattering. The imaginary contribution is provided by the neutral pion loops, whose threshold is lower than the charged one.

The issue of the higher-order terms is a bit more subtle. At next-to-leading order, a term of the type $\bar{C}_1(\phi_+^\dagger \overset{\leftrightarrow}{\nabla}{}^2 \phi_-^\dagger \phi_0 \phi_0 + \text{h.c.})$ should be added to the Lagrangian. What is the size of $\bar{C}_1$ as compared to $\bar{C}$? Since the π^0-mesons are integrated out, one would think that the hard scale of the theory is given by $M - M_0$. In QCD, this is a quantity of order α. Since, as seen later, in the bound states we shall be dealing with momenta of order of α, at first glance there will be no suppression at all! Fortunately, as seen from Eq. (2.243), the counting rules of the initial theory do not change. Indeed, let us first consider the higher-order derivative terms in the theory where neutral pions are present. In this case, dimensional analysis for the effective couplings proceeds as usual. Next, let us "hide" the π^0, introducing the effective couplings à la Eq. (2.243), modified to include derivative couplings. Such relations always start with the original coupling (C, in this case) plus bubble corrections that vanish at the charged threshold, if $M - M_0 \to 0$. Thus, the original counting is not modified. It *will be* modified, if one attempts to obtain a local theory by expanding Eq. (2.244) in powers of $P^0 - 2M$ and get rid of the P^0-dependence by using the EOM. As seen, the hard scale corresponding to such an expansion is given by $\sqrt{2M_0(M - M_0)}$. We, however, are not going to perform such an expansion.

Let us now consider the matching of the coupling $\bar{C}$. To this end, consider electromagnetic corrections to the scattering process $\pi^+\pi^- \to \pi^+\pi^-$ within the nonrelativistic framework. Note that in this case we have to keep track of two different countings,

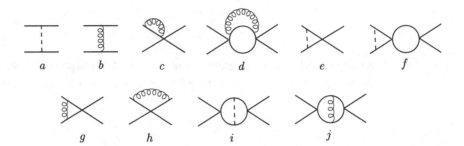

Figure 2.17 A representative set of diagrams for the one-photon corrections to the nonrelativistic $\pi^+\pi^- \to \pi^+\pi^-$ scattering amplitude. Dashed and wiggly lines denote the Coulomb and transverse photons, respectively.

namely the usual power counting scheme in the nonrelativistic theory, and the counting in powers of α, which is explicitly determined by the number of photon vertices. The diagrams at $O(\alpha)$ are shown in Fig. 2.17. Note also that in the scattering amplitude, the contribution from the Coulomb photons does not vanish anymore. In fact, the diagram in Fig. 2.17a provides the most singular part of the scattering amplitude at threshold. In the CM frame, with $\mathbf{p}, \mathbf{q}$ the relative three-momenta of the outgoing and incoming particles, the contribution of this diagram is given by

$$T^{(1C)}(\mathbf{p}, \mathbf{q}) = \frac{e^2}{|\mathbf{p} - \mathbf{q}|^2} . \tag{2.245}$$

This is a quantity of order of $O(p^{-2})$. The diagram with the transverse photon, which has the same topology (Fig. 2.17b), counts as order $O(p^0)$, that is, is suppressed by two powers of p (note that in this diagram the photon is soft and not ultrasoft, i.e., its three-momentum counts as $O(p)$).

It is instructive to consider the exchange of the Coulomb photons at all orders (ladder diagrams). The exchange of two Coulomb photons in the CM frame is given by the expression

$$T^{(2C)}(\mathbf{p}, \mathbf{q}) = \int \frac{d^4k}{(2\pi)^4 i} \frac{e^2}{|\mathbf{p} - \mathbf{k}|^2} \frac{1}{(E_\mathbf{k} - k^0 - i\varepsilon)(E_\mathbf{k} - P^0 + k^0 - i\varepsilon)} \frac{e^2}{|\mathbf{k} - \mathbf{q}|^2}$$

$$= \int \frac{d^3k}{(2\pi)^3} \frac{e^2}{|\mathbf{p} - \mathbf{k}|^2} \frac{1}{2E_\mathbf{k} - P^0 - i\varepsilon} \frac{e^2}{|\mathbf{k} - \mathbf{q}|^2} . \tag{2.246}$$

In (2.246) P^0 denotes the CM energy of the $\pi^+\pi^-$ pair. Since the quantity $P^0 - 2M = O(p^2)$, the above contribution counts as $O(\alpha^2 p^{-3})$. Here, $\alpha = e^2/(4\pi)$ is the fine-structure constant, and we display its power explicitly. The subsequent terms in this expansion,

$$T^{(nC)}(\mathbf{p}, \mathbf{q}) = \int \frac{d^3k_1}{(2\pi)^3} \cdots \frac{d^3k_{n-1}}{(2\pi)^3} \frac{e^2}{|\mathbf{p} - \mathbf{k}_1|^2} \frac{1}{2E_{\mathbf{k}_1} - P^0 - i\varepsilon}$$

$$\cdots \frac{1}{2E_{\mathbf{k}_{n-1}} - P^0 - i\varepsilon} \frac{e^2}{|\mathbf{k}_{n-1} - \mathbf{q}|^2} , \tag{2.247}$$

will be of order $\alpha^3 p^{-4}$, $\alpha^4 p^{-5}$ and so on. Hence, the expansion breaks down for momenta of order $p = O(\alpha M)$. This breakdown is signaled, for example, by the fact that for an attractive potential a tower of Coulomb bound states emerges at such momenta. This clearly is a non-perturbative phenomenon. On the other hand, for momenta $p \gg \alpha M$, the amplitude admits a perturbative expansion in powers of α.

It turns out that the ladder diagrams can be summed up to all orders. In other words, the equation for the exact Green's function in the Coulomb potential,[14]

$$\langle \mathbf{p}|g_C(P^0)|\mathbf{q}\rangle = \frac{(2\pi)^3 \delta^{(3)}(\mathbf{p}-\mathbf{q})}{P^0 - 2E_\mathbf{p}} - \int \frac{d^3 k}{(2\pi)^3} \frac{1}{P^0 - 2E_\mathbf{p}} \frac{4\pi\alpha}{|\mathbf{p}-\mathbf{k}|^2} \langle \mathbf{k}|g_C(P^0)|\mathbf{q}\rangle,$$

(2.248)

can be solved in closed form. The solution is given by [74]:

$$\langle \mathbf{p}|g_C(P^0)|\mathbf{q}\rangle = \frac{(2\pi)^3 \delta^{(3)}(\mathbf{p}-\mathbf{q})}{P^0 - 2E_\mathbf{p}} - \frac{1}{P^0 - 2E_\mathbf{p}} \frac{4\pi\alpha}{|\mathbf{p}-\mathbf{q}|^2} \frac{1}{P^0 - 2E_\mathbf{q}}$$

$$- \frac{1}{P^0 - 2E_\mathbf{p}} 4\pi\alpha v I(P^0; \mathbf{p}, \mathbf{q}) \frac{1}{P^0 - 2E_\mathbf{q}},$$

(2.249)

where $v = (\alpha/2)(-(P^0 - 2M)/M)^{-1/2}$, and

$$I(P^0; \mathbf{p}, \mathbf{q}) = \int_0^1 \frac{x^{-v} dx}{x|\mathbf{p}-\mathbf{q}|^2 + v^2/\alpha^2 (1-x)^2 (2E_\mathbf{p} - P^0)(2E_\mathbf{q} - P^0)}.$$

(2.250)

Expanding the integrand in powers of x, it can be seen that this function develops poles at $v = n = 1, 2, \ldots$, or $P^0 = 2M - M\alpha^2/(4n^2)$. The Coulomb wave functions can be read from the residues of these poles.

Further, using partial integration in the variable x, an alternative representation of the Green's function can be obtained:

$$\langle \mathbf{p}|g_C(P^0)|\mathbf{q}\rangle = \frac{(2\pi)^3 \delta^{(3)}(\mathbf{p}-\mathbf{q})}{P^0 - 2E_\mathbf{p}} - \frac{1}{P^0 - 2E_\mathbf{p}} 4\pi\alpha \bar{I}(P^0; \mathbf{p}, \mathbf{q}) \frac{1}{P^0 - 2E_\mathbf{q}},$$

(2.251)

where

$$\bar{I}(P^0; \mathbf{p}, \mathbf{q}) = \int_0^1 x^{-v} dx \frac{d}{dx} \frac{x}{x|\mathbf{p}-\mathbf{q}|^2 + v^2/\alpha^2 (1-x)^2 (2E_\mathbf{p} - P^0)(2E_\mathbf{q} - P^0)}.$$

(2.252)

Warning: Note that the preceding formulae were obtained in $d = 3$ space dimensions and cannot be used in the dimensionally regularized expressions.

A general remark: The matching is performed by carrying out the expansion of the amplitudes in powers of p/Λ, where Λ denotes the hard scale of the theory. Further, the amplitudes become non-perturbative for momenta of order of αM. Hence, choosing momenta at $\alpha M \ll p \ll \Lambda$, matching can be performed perturbatively, and the use of dimensional regularization can be justified. On the other hand, the study of the behavior of the amplitude in the vicinity of the threshold, as well as the study of

---

[14] In what follows, it is implicitly assumed that P^0 has an infinitesimal positive imaginary part $i\varepsilon$.

the Coulomb bound states (see the next section) cannot proceed perturbatively. Here, Coulomb ladder diagrams should be summed up to all orders.

First, let us consider the perturbative matching. The insertion of the Coulomb exchange in the diagrams with strong vertices leads to a singular behavior at small momenta. We demonstrate this in a few examples. The Coulomb vertex shown in Fig. 2.17e gives the following contribution:

$$T^{(e)}(\mathbf{p},\mathbf{q}) = e^2\bar{C}\int \frac{d^d\mathbf{k}}{(2\pi)^d}\frac{1}{|\mathbf{p}-\mathbf{k}|^2}\frac{1}{2E_\mathbf{k}-P^0-i\varepsilon}\,. \tag{2.253}$$

Here, dimensional regularization is used to tame the infrared divergence. The integration in the above integral can be carried out using the Feynman trick. The on-shell result, where $P^0 = 2M + \mathbf{p}^2/M$, is given by

$$T^{(e)}(\mathbf{p},\mathbf{q}) = e^2\bar{C}M\int_0^1 dx\int \frac{d^d\mathbf{k}}{(2\pi)^d}\frac{1}{(\mathbf{k}^2 - x^2\mathbf{p}^2 - i\varepsilon)^2}$$

$$= \bar{C}\left\{\frac{\pi\alpha M}{4|\mathbf{p}|} + i\alpha\theta_C(|\mathbf{p}|) + O(d-3,|\mathbf{p}|)\right\}. \tag{2.254}$$

Here, θ_C denotes the (infrared-divergent) Coulomb phase:

$$\theta_C = \frac{M}{2|\mathbf{p}|}\mu^{d-3}\left\{\frac{1}{d-3} - \frac{1}{2}(\Gamma'(1) + \ln 4\pi) + \ln\frac{2|\mathbf{p}|}{\mu}\right\}. \tag{2.255}$$

Hence, the correction to the lowest-order strong vertex diagram from the Coulomb photon exchanges between incoming or outgoing pions is given by

$$\bar{C} \to \bar{C} + 2\cdot\bar{C}\left\{\frac{\pi\alpha M}{4|\mathbf{p}|} + i\alpha\theta_C(|\mathbf{p}|)\right\}$$

$$= \exp(2i\theta_C)\left(1 + \frac{\pi\alpha M}{2|\mathbf{p}|}\right)\bar{C} + O(\alpha^2). \tag{2.256}$$

Note that the factor emerging from the Coulomb exchanges in the external lines is *universal* and applies as well for the diagrams with a different topology, like the one shown in Fig. 2.17f. Note also that the correction factor (both real and imaginary parts) is singular as $|\mathbf{p}| \to 0$, albeit this singularity is weaker than the one in Fig. 2.17a.

The diagram in Fig. 2.17i, which contains two strong vertices, leads to an even weaker singularity:

$$T^{(i)}(\mathbf{p},\mathbf{q}) = \bar{C}^2\int \frac{d^d\mathbf{l}}{(2\pi)^d}\frac{d^d\mathbf{k}}{(2\pi)^d}\frac{1}{2E_\mathbf{l}-P^0-i\varepsilon}\frac{e^2}{|\mathbf{l}-\mathbf{k}|^2}\frac{1}{2E_\mathbf{k}-P^0-i\varepsilon}\,, \tag{2.257}$$

where $P^0 = 2E_\mathbf{p} = 2E_\mathbf{q}$. This integral is ultraviolet-divergent and is given by

$$T^{(i)}(\mathbf{p},\mathbf{q}) = -\frac{\alpha M^2}{8\pi}\bar{C}^2\left\{\Lambda(\mu) + 2\ln\frac{2|\mathbf{p}|}{\mu} - 1 - i\pi\right\}, \tag{2.258}$$

where

$$\Lambda(\mu) = \mu^{2(d-3)} \left(\frac{1}{d-3} - \ln 4\pi - \Gamma'(1) \right). \tag{2.259}$$

This ultraviolet divergence can be canceled by the counterterm stemming from the coupling $\bar{C}$. This leads to the replacement of the bare coupling by the renormalized one: $\bar{C} \to \bar{C}^r$ (equivalently, to the renormalization of the couplings C, C_x, C_0 in Eq. (2.244)).

At order α, all other diagrams give a non-singular contribution at $\mathbf{p} \to 0$. Hence, the full on-mass shell amplitude $T(\mathbf{p}, \mathbf{q})$ has a universal behavior in the vicinity of threshold at this order. In order to display this behavior, let us subtract all (both Coulomb and transverse) one-photon exchange diagrams from the full amplitude, defining

$$\bar{T}(\mathbf{p}, \mathbf{q}) = T(\mathbf{p}, \mathbf{q}) - T^{1\gamma}(\mathbf{p}, \mathbf{q}). \tag{2.260}$$

Then, the remainder takes the form

$$e^{-2i\alpha\theta_C} \bar{T}(\mathbf{p}, \mathbf{q}) = \frac{\mathcal{A}_1}{|\mathbf{p}|} + \mathcal{A}_2 \ln \frac{2|\mathbf{p}|}{M} + \mathcal{A} + O(|\mathbf{p}|). \tag{2.261}$$

Here, the coefficients $\mathcal{A}_1, \mathcal{A}_2, \mathcal{A}$ are determined by a single coupling $\bar{C}$:

$$\mathcal{A}_1 = \frac{\pi \alpha M}{2} \bar{C}, \qquad \mathcal{A}_2 = -\frac{\alpha M^2}{4\pi} \bar{C}^2,$$

$$\mathcal{A} = \bar{C} \left\{ 1 - \frac{\alpha M^2}{8\pi} \bar{C} \left(\Lambda(\mu) + \ln \frac{M^2}{\mu^2} - 1 - 2\pi i \right) \right\}. \tag{2.262}$$

Note that, in order to obtain (2.262), one has to include the contribution from the free pion loop and take into account the Coulomb factor from Eq. (2.256). This leads to the change $i\pi \to 2i\pi$, as compared to Eq. (2.258)

Since the relativistic and nonrelativistic amplitudes at low momenta must agree, the behavior of the relativistic amplitude in the vicinity of threshold should also be given by the preceding expression. (In the relativistic theory, however, an infinite string of diagrams contributes to the singular part.) In order to determine the coupling $\bar{C}$ at $O(\alpha)$, one has to consider the matching of the non-singular part of the amplitude $\mathcal{A}$, since the coefficients of the singular terms are already proportional to α. Note also that the scale in the logarithm is chosen arbitrarily. Any change in this scale redefines the constant part $\mathcal{A}$. However, since this scale M should be the same in the relativistic and nonrelativistic theories, it drops out from the matching condition for the constant $\bar{C}$.

The matching of the couplings C_x, C_0 can be carried out similarly. In analogy to Eq. (2.243), one may write

$$\bar{C}_x = \frac{C_x}{1 - 2C_0 J_0}, \qquad \bar{C}_0 = \frac{C_0}{1 - 2C_0 J_0}. \tag{2.263}$$

In the charge exchange and neutral channels, the most singular one-photon exchange diagram is absent, and the diagrams of the type 2.17e emerge only for the charged external lines. Hence, one has

$$e^{-i\alpha\theta_C} T_x(\mathbf{p}, \mathbf{q}) = \frac{\mathcal{A}_{1x}}{|\mathbf{p}|} + \mathcal{A}_{2x} \ln \frac{2|\mathbf{p}|}{M} + \mathcal{A}_x + O(|\mathbf{p}|),$$

$$T_0(\mathbf{p}, \mathbf{q}) = \mathcal{A}_{20} \ln \frac{2|\mathbf{p}|}{M} + \mathcal{A}_0 + O(|\mathbf{p}|), \tag{2.264}$$

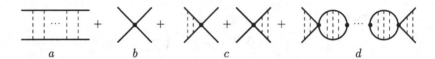

Figure 2.18 Charged pion scattering in the presence of Coulomb interaction (representative diagrams): (a) a ladder with Coulomb photon exchanges between pions; (b) a strong four-pion vertex, no electromagnetic corrections; (c) Coulomb corrections to the strong vertex, either in the initial or the final state; (d) a generic diagram, showing Coulomb photon exchanges in the initial, final and intermediate states. The filled dots stand for the vertex with the effective coupling $\bar{C}$.

where

$$\mathcal{A}_x = 2\bar{C}_x \left\{ 1 - \frac{\alpha M^2}{8\pi} \bar{C} \left(\Lambda(\mu) + \ln \frac{M^2}{\mu^2} - 1 - 2\pi i \right) \right\},$$

$$\mathcal{A}_0 = 4\bar{C}_0 \left\{ 1 - \frac{\alpha M^2}{8\pi} \bar{C} \left(\Lambda(\mu) + \ln \frac{M^2}{\mu^2} - 1 - 2\pi i \right) \right\}. \tag{2.265}$$

2.11.5 Strong Scattering Process in the Presence of the Coulomb Interaction

In the next step, using the nonrelativistic couplings already determined from matching, we study the behavior of the amplitudes for momenta $p \sim \alpha M$ in the nonrelativistic setting. To this end, the resummation of all ladder diagrams of the type shown in Fig. 2.18 should be performed. As for the matching, we shall focus on the scattering in the charged channels; the two other channels can be treated similarly. As already pointed out, one cannot use dimensional regularization here to tame the infrared divergences. Introducing a photon mass is also not an option, because Schwinger's Green's function has been obtained for the exact Coulomb potential. Following Ref. [74], we consider here an off-shell Green's function, with $P^0 = 2M - q_0^2/M$ and $\mathbf{p}^2, \mathbf{q}^2 \neq q_0^2$. Using Eqs. (2.251) and (2.252) is more convenient for our purpose. The integral in Eq. (2.252) will be first evaluated for $P^0 < 2M$ and then analytically continued into the scattering region. Introducing the notation,

$$b = -\frac{M}{4(P^0 - 2M)} (2E_\mathbf{p} - P^0)(2E_\mathbf{q} - P^0) \frac{1}{|\mathbf{p} - \mathbf{q}|^2} > 0, \tag{2.266}$$

one gets

$$\bar{I}(P^0, \mathbf{p}, \mathbf{q}) = \frac{1}{|\mathbf{p} - \mathbf{q}|^2} \int_0^1 dx\, x^{-\nu} \frac{d}{dx} \frac{x}{x + b(1-x)^2}$$

$$= \frac{b^{-\nu}}{|\mathbf{p} - \mathbf{q}|^2} \int_0^{1/b} dx\, x^{-\nu} \frac{d}{dx} \frac{x}{x + (1-xb)^2}. \tag{2.267}$$

In the on-shell limit $b \to 0$ we obtain

$$\bar{I}(P^0, \mathbf{p}, \mathbf{q}) \to -\frac{b^{-\nu}}{|\mathbf{p} - \mathbf{q}|^2} \int_0^\infty \frac{dx\, x^{-\nu}}{(x+1)^2} = -\frac{b^{-\nu}}{|\mathbf{p} - \mathbf{q}|^2} \frac{\nu\pi}{\sin((1-\nu)\pi)}. \tag{2.268}$$

Carrying out the analytic continuation in P^0, the quantity v becomes purely imaginary, $v = i\eta$ with $\eta = \alpha M/(2q_0)$. The Coulomb Green's function in the vicinity of the mass shell takes the following form:

$$\langle \mathbf{p}|g_C(P^0)|\mathbf{q}\rangle \to \frac{(2\pi)^3 \delta^{(3)}(\mathbf{p}-\mathbf{q})}{P^0 - 2E_\mathbf{p}} - \frac{g(\mathbf{p};q_0)}{P^0 - 2E_\mathbf{p}} f(\mathbf{p},\mathbf{q};q_0) \frac{g(\mathbf{p};q_0)}{P^0 - 2E_\mathbf{q}}, \qquad (2.269)$$

where f is the scattering amplitude in the Coulomb potential, which leads to the Rutherford formula for the Coulomb cross section [75]:

$$f(\mathbf{p},\mathbf{q};q_0) = \frac{4\pi\alpha}{|\mathbf{p}-\mathbf{q}|^2} \exp\left(-i\eta \ln \frac{4q_0^2}{|\mathbf{p}-\mathbf{q}|^2}\right), \qquad (2.270)$$

and the quantity g is given by:

$$g(\mathbf{p};q_0) = \mathcal{N}^{1/2} \exp\left(-i\eta \ln \frac{\mathbf{p}^2 - q_0^2}{4q_0^2}\right), \qquad \mathcal{N} = \frac{2\pi\eta}{e^{2\pi\eta} - 1}. \qquad (2.271)$$

Here, $\mathcal{N}$ is the well-known Sommerfeld factor [76–78]; see also [75]. Note that the factor g is singular on shell. This is exactly the infrared singularity, which was regularized by assuming $\mathbf{p}^2 \neq q_0^2$. For instance, performing the Fourier transform, it is seen that for large r,

$$\int \frac{d^3\mathbf{p}}{(2\pi)^3} e^{i\mathbf{p}\mathbf{r}} \frac{g(\mathbf{p};q_0)}{2E_\mathbf{p} - P^0} \sim \frac{M}{4\pi r} \exp\left(ikr + i\eta \ln(2kr) + i\arg\Gamma(1 - i\eta)\right). \qquad (2.272)$$

This is the well-known result for the asymptotic behavior of the scattering wave function in the Coulomb potential [75].

The strong scattering process in the presence of the Coulomb interaction is shown in the diagrams of Fig. 2.18b,c,d. These include resummed Coulomb ladders in both the external pion legs as well as the bubbles. (These bubbles should be also resummed to all orders.) It is straightforward to verify that the sum of all strong plus Coulomb diagrams can be given in the following form:

$$\tilde{T}_{SC}(\mathbf{p},\mathbf{q};q_0) = (1 + V(\mathbf{p};q_0)) \frac{\bar{C}}{1 - \bar{C}B(q_0)} (1 + V(\mathbf{q};q_0)). \qquad (2.273)$$

Here, V and B correspond to the insertions in the external lines and in the bubble, respectively. The former is given by the following expression:

$$V(\mathbf{p};q_0) = 4\pi\alpha \int x^{-i\eta} dx \frac{d}{dx}(xF), \qquad (2.274)$$

where

$$\begin{aligned}
F &= \int \frac{d^3\mathbf{k}}{(2\pi)^3} \frac{1}{x|\mathbf{p}-\mathbf{k}|^2 + v^2/\alpha^2 (1-x)^2 (2E_\mathbf{p} - P^0)(2E_\mathbf{k} - P^0)} \frac{1}{2E_\mathbf{k} - P^0} \\
&= \frac{iM}{8\pi px}\left(\ln \frac{p - q_0 - x(p+q_0) - i\varepsilon}{p + q_0 - x(p - q_0) - i\varepsilon} - \ln \frac{p - q_0 - i\varepsilon}{p + q_0 - i\varepsilon}\right),
\end{aligned} \qquad (2.275)$$

with $p = |\mathbf{p}|$. Using this expression in Eq. (2.274), we get

$$V(\mathbf{p};q_0) = -\frac{iM}{2p}\int_0^1 dx e^{-i\eta}\left(\frac{\delta^{-1}}{1+(-\delta+i\varepsilon)^{-1}x} - \frac{\delta}{1+(-\delta-i\varepsilon)x}\right), \quad (2.276)$$

where $\delta = (p-q_0)/(p+q_0)$. Further, using the formula

$$\int_0^u dx \frac{x^{\mu-1}}{1+\beta x} = \frac{u^\mu}{\mu} {}_2F_1(1,\mu,1+\mu,-\beta u), \quad (2.277)$$

where ${}_2F_1$ denotes the hypergeometric function, and evaluating Eq. (2.276) in the limit $\delta \to 0$, we obtain

$$1 + V(\mathbf{p};q_0) = g(\mathbf{p};q_0)\left(\frac{2\pi\eta}{1-e^{-2\pi\eta}}\right)^{1/2}. \quad (2.278)$$

Removing the factors g in the external legs that lead to the distorted spherical waves, one finally obtains an expression for the strong scattering amplitude in the presence of the Coulomb interactions (cf., for example, with Ref. [79]):

$$T_{SC}(\mathbf{p},\mathbf{q};q_0) = g^{-1}(\mathbf{p};q_0)\tilde{T}_{SC}(\mathbf{p},\mathbf{q};q_0)g^{-1}(\mathbf{q};q_0)$$

$$= \frac{2\pi\eta}{1-e^{-2\pi\eta}}\frac{\bar{C}}{1-\bar{C}B(q_0)}. \quad (2.279)$$

Note that this amplitude can be put on the mass shell, since the infrared singularities contained in the factors g are removed. Note also that, expanding the above expression in powers of $\bar{C}$ and α, one exactly reproduces the $1/p$ factor in Eq. (2.256) in the real part of the amplitude at lowest order.

What remains to be done is to calculate the loop integral $B(q_0)$. The part of this integral that corresponds to the zero and one Coulomb photon exchanges has been evaluated already. The calculation of the multiphoton part is straightforward. There are neither ultraviolet nor infrared divergences here and, hence, Schwinger's formula can be used in combination with dimensional regularization for the zero and one photon parts. Using Eq. (2.258), the final result can be written as

$$B(q_0) = \frac{M}{4\pi}\left\{iq_0 - \frac{\alpha M}{2}\left(\Lambda(\mu) + \ln\frac{-4q_0^2 - i\varepsilon}{\mu^2} - 1\right)\right\} + B_{nC}(q_0), \quad (2.280)$$

where the multiphoton part is given by

$$B_{nC}(q_0) = 4\pi i\alpha\eta\int\frac{d^3\mathbf{p}}{(2\pi)^3}\frac{d^3\mathbf{q}}{(2\pi)^3}\frac{1}{2E_\mathbf{p}-P^0}I(P^0;\mathbf{p},\mathbf{q})\frac{1}{2E_\mathbf{q}-P^0}$$

$$= \frac{\alpha M^2}{4\pi}(\psi(1)-\psi(1-i\eta)). \quad (2.281)$$

Here, $\psi(y) = \Gamma'(y)/\Gamma(y)$ denotes the logarithmic derivative of the Γ-function.

For sure, the exact results that concern the scattering in the Coulomb plus short-range potential were derived within potential scattering theory a long time ago. The re-derivation of the known result, which was presented in this section, serves several purposes. First of all, the present derivation is done in a field theory by using a diagrammatic language. Second, owing to this circumstance, the connection to the relativistic theory can be directly established. Namely, the leading singularities at threshold are the same in both cases and, if needed, the relativistic corrections, as well as the effects due to the transverse photons can be included systematically at any given order in p. Last but not least, it becomes obvious that, even though the amplitudes in the vicinity of a threshold contain non-perturbative contributions, the matching can still be performed perturbatively, and the use of dimensional regularization is justified.

2.12 Spectrum and Decays of Hadronic Atoms

The two-body scattering lengths are the most fundamental characteristics of hadron–hadron interactions at low energy. They can be extracted from the measured scattering phase shifts in the vicinity of the threshold. In practice, however, such experiments involving particles with very small three-momenta represent a major challenge. One factor that contributes to this challenge has already been identified in the previous section. In the case of charged particle scattering, the electromagnetic interactions become sizable at threshold and should be properly included if one aims to extract scattering lengths at a very high precision.

Fortunately, there exist systems that allow one to determine the scattering lengths directly, that is, without analyzing scattering data. These systems are the so-called *hadronic atoms*. These are bound states of two oppositely charged hadrons, held together through the Coulomb interaction. The spectrum of such atoms resembles that of a hydrogen atom (with a different value of the reduced mass μ). The average distance between two hadrons forming the atom is given by the Bohr radius $r_B = (\mu\alpha)^{-1}$ and is very large. For example, in pionic hydrogen (a hydrogen atom, where the electron is replaced by the π^-), the pertinent Bohr radius is approximately equal to 220 fm, much larger than the typical range of the strong interactions (a few fm). The Bohr radius for pionium (a $\pi^+\pi^-$ bound state) is of the order of 400 fm, and so on.

Hadronic atoms differ from ordinary hydrogen atoms in two respects. First of all, their spectrum is almost Coulombic but not exactly. The difference is partly due to the higher-order QED corrections, which arise in the case of ordinary hydrogen as well. Furthermore, a part of the energy level shifts is due to the strong interaction between the involved hadrons. This contribution is very small, because the strong interactions are almost negligible at such large distances. Owing to this fact, perturbation theory can be applied to calculate the strong level shift. The second difference is that such atoms are very short-lived and can decay via multiple mechanisms. In particular, the hadrons composing an atom may decay individually. (For instance, the charged pion in pionic hydrogen may weakly decay, $\pi^- \to \mu\bar{\nu}_\mu$, with a typical decay lifetime $\sim 10^{-8}$ s.) Alternatively, the proton and the pion can strongly annihilate, producing, for example, $\pi^0 n$ or γn in the final state. The total ground-state decay width Γ_1 of the atom is on the order of 1 eV that corresponds to a lifetime on the order of 10^{-15} s. Consequently, the pion in the atom can be safely considered as a stable particle. Further, the ground-state Coulomb binding energy of the pionic hydrogen $E_1 = \frac{1}{2}\mu\alpha^2$ is in the keV-range, and the ratio $\Gamma_1/E_1 \sim 10^{-3}$. Hence, the atom itself can also be considered as a metastable state (to a good approximation).

The crucial feature, which makes the study of hadronic atoms so fascinating, is the existence of two very distinct (electromagnetic and strong) scales. For this reason, the atom "feels" the strong interactions at a very low momenta of order of $\alpha\mu$, that is, practically at threshold. Consequently, the strong energy level shift and the width of an atom should be determined by the scattering lengths. At leading order, these relations, which go under the name of the Deser–Goldberger–Baumann–Thirring (DGBT) formulae, have been known for a long time [80]. For example, in the simplest case of the $\pi^+\pi^-$ bound state, which will be considered here in detail, they can be schematically written in the following form:

$$\Delta E \propto |\tilde{\Psi}_1(0)|^2 (2a_0 + a_2),$$

$$\Gamma \propto |\tilde{\Psi}_1(0)|^2 \sqrt{\Delta M_\pi^2}(a_2 - a_0)^2. \tag{2.282}$$

Here, $\Delta M_\pi^2 = M^2 - M_0^2$ is the pion mass difference, a_0, a_2 denote the S-wave $\pi\pi$ scattering lengths with the isospin $I = 0, 2$, $\tilde{\Psi}_1(\mathbf{r})$ denotes the Coulomb wave function of the atom in coordinate space, and the quantity $|\tilde{\Psi}_1(0)|^2 \propto \alpha^3$ gives the probability of finding the π^+ and the π^- at the same point. Hence, measuring the energy level shift and decay width very accurately, we are able to extract the scattering lengths directly, without analyzing hadron–hadron scattering data.

One may formulate the same idea in a different manner. Roughly speaking, the Coulomb interaction between charged hadrons plays the role of a very extended trap that holds these hadrons together. Further, because the size of this trap is much larger than the typical range of strong interactions, the measured quantities in the experiment will universally depend on the effective-range expansion parameters of the hadron–hadron scattering amplitude, which is on the scattering length in the leading approximation. Note also that there is nothing special about the choice of the Coulomb potential for forming such a trap. For example, the atoms of ultracold gases can be confined in magnetic traps that can be described to a good approximation by an external oscillator potential. By the same token as in the case of hadronic atoms, the displacement of the atom energy levels in such a trap is proportional to the atom–atom scattering length [81]. Another nice example of a trap is provided by a space-time lattice. In fact, the boundary conditions in a finite lattice box amount to trapping the particles inside the box. Trapping two particles on the lattice, one may ensure that the energy level displacements from their free values are again proportional to the two-body scattering length [82]. It is seen that, for all these very different systems, the relation between the discrete spectrum and the two-body scattering length follows the same pattern. Using the nonrelativistic effective Lagrangian, we shall demonstrate below how to arrive at this result in a simple and transparent way. Moreover, our result will not be limited to the leading order, a systematic procedure for calculating corrections will be set up.

We start with briefly mentioning recent experiments with hadronic atoms and their physics case. The DIRAC collaboration at CERN has measured the lifetime of the $\pi^+\pi^-$ and πK atoms that enables one to extract particular linear combinations of the $\pi\pi$ and πK scattering lengths [83–85]. What makes these experiments especially fascinating is the fact that these scattering lengths are very sensitive to the value of the quark condensate in the chiral limit (in the two- and three-flavor cases, respectively) and, thus, provide one with valuable information about the exact scenario of the *spontaneous* chiral symmetry breaking in QCD (see Chapter 3). Further, the Pionic Hydrogen collaboration at PSI has performed very accurate measurements of the ground-state energy and line shape of the pionic hydrogen, as well as of the ground-state energy of the pionic deuterium [86–89]. This resulted in the extraction of the precise values of the S-wave pion–nucleon scattering lengths using suitably tailored NREFTs. In particular, whereas the isospin-odd scattering length was previously known rather well from scattering experiments, the much smaller (chirally suppressed) isospin-even scattering length is more difficult to determine. Knowing this scattering length is important, because it is directly related to the so-called pion–nucleon σ-term, which measures the *explicit* breaking of chiral symmetry in QCD (see Chapter 4 for more discussion). Last but not least, the SIDDHARTA(-2) collaboration at DAΦNE investigated the spectrum and decays of the kaonic hydrogen and kaonic deuterium that will allow one to extend the previous test to the strange quark sector [90, 91]. However, in order to carry out such a beautiful study of the fundamental low-energy parameters of QCD, we need a systematic theory of hadronic atoms that allows us to calculate their characteristics with an accuracy that matches the experimental precision. The accuracy of Eq. (2.282) is not sufficient for this purpose, and the higher-order corrections in α as well as in the strong isospin breaking parameter $m_d - m_u$ should be included. In this section, we shall briefly explain how this goal can be achieved using the nonrelativistic effective Lagrangians just considered.

Below, we shall focus on the case of the bound state of spinless particles, namely pionium, that will enable us to make the discussion more compact. Investing effort in the matching of the nonrelativistic Lagrangian will now pay off. As will be shown, bound-state calculations in the nonrelativistic formalism are done very easily. The aim of the calculations will be to calculate the energy shift and width at next-to-leading order in α. The effective Lagrangian, which is displayed in Eq. (2.240), suffices to achieve this goal, as the terms, which have been dropped there, still do not contribute at this accuracy.

We start with writing down the nonrelativistic Hamiltonian, which can be derived from the effective Lagrangian by using canonical methods. To this end, note first that the EOM for the Coulomb photon field $\mathscr{A}^0(x)$ does not contain the time derivative. Thus, $\mathscr{A}^0(x)$ is not a dynamical field and can be eliminated from the theory with the use of the EOM. The resulting Lagrangian and Hamiltonian are non-local, containing the static Coulomb interaction potential between charged pions explicitly:

$$\mathbf{H} = \mathbf{H}_0 + \mathbf{H}_C + \mathbf{H}_I = \int d^3\mathbf{x}\left(\mathcal{H}_0(x) + \mathcal{H}_I(x)\right) + \int d^3\mathbf{x}\,d^3\mathbf{y}\,\mathcal{H}_C(x,y), \qquad (2.283)$$

where $x^0 = y^0 = 0$. In the calculations, it is more convenient to use the version of the theory, where the neutral pion fields are integrated out. Thus, $\mathcal{H}_0$ denotes the free Hamiltonian of charged pions:

$$\mathcal{H}_0(x) = \sum_{\pm} \phi_{\pm}^{\dagger} \left(i\partial_t - M + \frac{\nabla^2}{2M} \right) \phi_{\pm}, \tag{2.284}$$

$\mathcal{H}_C$ describes the static Coulomb interaction between charged pions

$$\mathcal{H}_C(x,y) = -\frac{e^2}{2} (\phi_+^{\dagger}\phi_+ - \phi_-^{\dagger}\phi_-)(x)\triangle^{-1}(\mathbf{x} - \mathbf{y})(\phi_+^{\dagger}\phi_+ - \phi_-^{\dagger}\phi_-)(y), \tag{2.285}$$

where

$$\triangle^{-1}(\mathbf{x} - \mathbf{y}) = -\int \frac{d^3\mathbf{k}}{(2\pi)^3} \frac{e^{i\mathbf{k}(\mathbf{x}-\mathbf{y})}}{\mathbf{k}^2},$$

and the interaction Hamiltonian $\mathcal{H}_I$ contains the remaining terms that contribute to the (complex) energy level shift at next-to-leading order in α:

$$\mathcal{H}_I(x) = -\bar{C}\phi_+^{\dagger}\phi_-^{\dagger}\phi_+\phi_- - \frac{e^2 h_1}{6M^2} (\phi_+^{\dagger}\phi_+ - \phi_-^{\dagger}\phi_-)^2 - \sum_{\pm} \phi_{\pm}^{\dagger}\frac{\nabla^4}{8M^3}\phi_{\pm}$$

$$\mp \frac{ie}{2M} \sum_{\pm} \mathscr{A}^i(\phi_{\pm}^{\dagger}\nabla^i\phi_{\pm} - \nabla^i\phi_{\pm}^{\dagger}\phi_{\pm})$$

$$= \mathcal{H}_S(x) + e^2\mathcal{H}_F(x) + \mathcal{H}_R(x) + e\mathcal{H}_{\gamma}(x). \tag{2.286}$$

We shall treat the corrections arising from $\mathcal{H}_I$ in perturbation theory. Using the counting rules, it is straightforward to check that adding higher-order terms to the Hamiltonian does not change the expression at next-to-leading order in α. This marks a crucial advantage of the nonrelativistic framework as compared to other approaches, where simple counting rules cannot be defined, and it is not always easy to identify all diagrams that contribute to the bound-state observables at a given order in α.

The scattering state of a free $\pi^+\pi^-$-pair is obtained by acting with the creation operators on the vacuum state:

$$|\mathbf{P}, \mathbf{p}\rangle = a_+^{\dagger}\left(\frac{\mathbf{P}}{2} + \mathbf{p}\right) a_-^{\dagger}\left(\frac{\mathbf{P}}{2} - \mathbf{p}\right)|0\rangle. \tag{2.287}$$

Here, $\mathbf{P}, \mathbf{p}$ denote the total and relative three-momenta of the $\pi^+\pi^-$ pair, respectively.

As we have said, the unperturbed Hamiltonian for our problem is given by the sum $\mathbf{H}_0 + \mathbf{H}_C$. The state vector of the system is given by

$$|\Psi(E; \mathbf{P})\rangle = \int \frac{d^3\mathbf{p}}{(2\pi)^3} \Psi(E; \mathbf{p})|\mathbf{P}, \mathbf{p}\rangle, \tag{2.288}$$

where E is the energy in the CM system and $\Psi(E; \mathbf{p})$ denotes the Schrödinger wave function in momentum space corresponding to the motion in a static Coulomb field.

There are bound states in the spectrum of the unperturbed problem with wave function $\Psi_{n\ell m}(\mathbf{p})$ (here, n, ℓ, m stand for the principal quantum number, orbital angular momentum and its projection on the z-axis, respectively). The bound-state vector obeys the equation

$$\left(\mathbf{H}_0 + \mathbf{H}_C \right)|\Psi_{n\ell m}(\mathbf{P})\rangle = \left(E_n + \frac{\mathbf{P}^2}{4M} \right)|\Psi_{n\ell m}(\mathbf{P})\rangle,$$

$$E_n = 2M - \frac{M\alpha^2}{4n^2}. \tag{2.289}$$

For simplicity, we shall limit ourselves to the ground state with $n = 1$. The indices $\ell, m = 0$ will be suppressed in the following. The wave function of the ground state is given by

$$\Psi_1(\mathbf{p}) = \frac{(64\pi\gamma^5)^{1/2}}{(\mathbf{p}^2 + \gamma^2)^2}, \qquad \gamma = \frac{1}{2}\alpha M. \tag{2.290}$$

The corrections that emerge from the Hamiltonian $\mathbf{H}_I$ can be calculated using Rayleigh–Schrödinger perturbation theory. In order to derive the pertinent expressions, let us consider two resolvents:

$$\mathbf{G}(z) = \frac{1}{z - \mathbf{H}_0 - \mathbf{H}_C - \mathbf{H}_I}, \qquad \mathbf{G}_C(z) = \frac{1}{z - \mathbf{H}_0 - \mathbf{H}_C}. \tag{2.291}$$

The full resolvent obeys the equation

$$\mathbf{G}(z) = \mathbf{G}_C(z) + \mathbf{G}_C(z)\mathbf{H}_I\mathbf{G}(z) = \mathbf{G}_C(z) + \mathbf{G}_C(z)\mathbf{T}(z)\mathbf{G}_C(z). \tag{2.292}$$

Using the spectral decomposition, we may single out the contribution in the Coulomb Green's function, which corresponds to the ground state:

$$\mathbf{G}_C(z) = \int \frac{d^3\mathbf{P}}{(2\pi)^3} \frac{|\Psi_1(\mathbf{P})\rangle\langle\Psi_1(\mathbf{P})|}{z - E_1 - \mathbf{P}^2/(4M)} + \bar{\mathbf{G}}_C(z) \doteq \mathbf{G}_1(z) + \bar{\mathbf{G}}_C(z). \tag{2.293}$$

Since the distance between the energy levels in the Coulomb potential is a quantity of order α^2, the second term in the splitting is analytic and can be expanded in a Taylor series, if $z - E_1 - \mathbf{P}^2/(4M) \ll M\alpha^2$. It will a posteriori turn out that the energy shift of the levels is a quantity of order α^3. Hence, in the Taylor expansion of this term each subsequent term will be suppressed by one power of α.

Further, using translational invariance and choosing the CM system, for any operator $\mathbf{O}$ we define

$$\langle\mathbf{P}, \mathbf{p}|\mathbf{O}|\mathbf{0}, \mathbf{q}\rangle \doteq (2\pi)^3 \delta^{(3)}(\mathbf{P})\langle\mathbf{p}|\mathbf{O}|\mathbf{q}\rangle. \tag{2.294}$$

Note that we are using the same symbol for the operator in different Hilbert spaces, which hopefully does not create any confusion. Thus, in the CM system, $\mathbf{P} = 0$, Eq. (2.293) can be rewritten as

$$\mathbf{G}_C(z) = \frac{|\Psi_1\rangle\langle\Psi_1|}{z - E_1} + \bar{\mathbf{G}}_C(z) = \mathbf{G}_1(z) + \bar{\mathbf{G}}_C(z). \tag{2.295}$$

Now, the operators in the above equation act in the Hilbert state of vectors $|\mathbf{p}\rangle$, where $\mathbf{p}$ is the relative momentum, and $\langle\mathbf{p}|\Psi_1\rangle = \Psi_1(\mathbf{p})$ is the Coulomb wave function of the ground state in the CM frame.

Furthermore, taking the splitting into account, the equation for the full scattering T-matrix can be written as a system of equations in the CM frame:

$$\mathbf{T}(z) = \mathbf{H}_I + \mathbf{H}_I\mathbf{G}_C(z)\mathbf{T}(z) = \boldsymbol{\tau}(z) + \boldsymbol{\tau}(z)\mathbf{G}_1(z)\mathbf{T}(z),$$

$$\boldsymbol{\tau}(z) = \mathbf{H}_I + \mathbf{H}_I \bar{\mathbf{G}}_C(z) \boldsymbol{\tau}(z).$$

(2.296)

The solution of the first equation takes the form

$$\mathbf{T}(z) = \boldsymbol{\tau}(z) + \frac{\boldsymbol{\tau}(z)|\Psi_1\rangle\langle\Psi_1|\boldsymbol{\tau}(z)}{z - E_1 - \bar{\boldsymbol{\tau}}(z)},$$

(2.297)

where

$$\bar{\boldsymbol{\tau}}(z) = \langle\Psi_1|\boldsymbol{\tau}(z)|\Psi_1\rangle = \int \frac{d^3\mathbf{p}}{(2\pi)^3} \frac{d^3\mathbf{q}}{(2\pi)^3} \Psi_1(\mathbf{p})\langle\mathbf{p}|\boldsymbol{\tau}(z)|\mathbf{q}\rangle\Psi_1(\mathbf{q}).$$

(2.298)

The full resolvent takes the form

$$\mathbf{G}(z) = \bar{\mathbf{G}}_C(z) + \bar{\mathbf{G}}_C(z)\boldsymbol{\tau}(z)\bar{\mathbf{G}}_C(z)$$

$$+ \frac{(1 + \bar{\mathbf{G}}_C(z)\boldsymbol{\tau}(z))|\Psi_1\rangle\langle\Psi_1|(1 + \boldsymbol{\tau}(z)\bar{\mathbf{G}}_C(z))}{z - E_1 - \bar{\boldsymbol{\tau}}(z)}.$$

(2.299)

Thus, in the CM frame with $\mathbf{P} = 0$, the position of the pole in the vicinity of ground state is determined from the equation

$$z - E_1 - \bar{\boldsymbol{\tau}}(z) = 0.$$

(2.300)

The real and imaginary parts of the pole position define the energy shift and width of the ground state:

$$z - E_1 = \Delta E_1 - i\frac{\Gamma}{2}.$$

(2.301)

One can now expand the last term in the master equation (2.300) in powers of $(z - E_1)$, yielding

$$z - E_1 = \frac{\bar{\boldsymbol{\tau}}(E_1)}{1 - d\bar{\boldsymbol{\tau}}(E_1)/dE_1} + \cdots.$$

(2.302)

The quantity $\bar{\boldsymbol{\tau}}(E_1)$ should be evaluated perturbatively. Note that the derivative in the denominator of Eq. (2.302) does not contribute at the accuracy we are working and can be safely neglected. In what follows we display all terms in the matrix element of the operator $\boldsymbol{\tau}(z)$, which contribute to the energy shift at next-to-leading order in α:

$$\langle\mathbf{p}|\boldsymbol{\tau}(E_1)|\mathbf{q}\rangle = \langle\mathbf{p}|\mathbf{H}_S + \mathbf{H}_S\bar{\mathbf{G}}_C(E_1)\mathbf{H}_S + e^2\mathbf{H}_F + \mathbf{H}_R + e^2\mathbf{H}_\gamma\bar{\mathbf{G}}_C(E_1)\mathbf{H}_\gamma|\mathbf{q}\rangle.$$

(2.303)

What remains is to evaluate this expression term by term, inserting the free Fock-space states between the operators and using Eq. (2.298). The quantity $\bar{\boldsymbol{\tau}}(z)$ is then split into several pieces:

$$\bar{\boldsymbol{\tau}}(z) = \bar{\boldsymbol{\tau}}_S(z) + \bar{\boldsymbol{\tau}}_F(z) + \bar{\boldsymbol{\tau}}_R(z) + \bar{\boldsymbol{\tau}}_\gamma(z).$$

(2.304)

The "strong" part can be found by inserting the states that contain a $\pi^+\pi^-$ pair between the two strong Hamiltonians $\mathbf{H}_S$. The result is given by

$$\bar{\boldsymbol{\tau}}_S(z) = |\tilde{\Psi}_1(0)|^2\left(-\bar{C} + \bar{C}^2\langle\bar{\mathbf{G}}_C(z)\rangle\right),$$

(2.305)

The value of the integral, containing the Coulomb Green's function, can be read from Eqs. (2.280) and (2.281), performing the analytic continuation to the bound-state energy $q_0 \to i\sqrt{Mz}$, $\eta = \gamma/q_0$, and subtracting the ground-state contribution:

$$\langle \bar{\mathbf{G}}_C(z) \rangle = \int \frac{d^3\mathbf{p}}{(2\pi)^3} \frac{d^3\mathbf{q}}{(2\pi)^3} \langle \mathbf{p}|g_C(z)|\mathbf{q}\rangle - \frac{|\tilde{\Psi}_1(0)|^2}{z - E_1}$$

$$= -B(i\sqrt{Mz}) - \frac{|\tilde{\Psi}_1(0)|^2}{z - E_1}$$

$$= \frac{\alpha M^2}{8\pi}\left(\frac{q_0}{\gamma} + \Lambda(\mu) + \ln\frac{4q_0^2}{\mu^2} + 2(\psi(2-\eta) - \psi(1)) - \frac{3+5\eta}{1+\eta} \right). \quad (2.306)$$

The value of the Coulomb ground-state wave function at the origin is given by

$$\tilde{\Psi}_1(0) = \int \frac{d^3\mathbf{p}}{(2\pi)^3} \frac{(64\pi\gamma^5)^{1/2}}{(\mathbf{p}^2 + \gamma^2)^2} = \frac{\gamma^{3/2}}{\pi^{1/2}}, \quad (2.307)$$

Next, evaluating the contribution in Eq. (2.304), which contains the pion charge radius, one gets

$$\bar{\tau}_F(z) = \frac{e^2 h_1}{3M^2}|\tilde{\Psi}_1(0)|^2 = \frac{1}{6}\alpha^4 M^3 \langle r^2 \rangle. \quad (2.308)$$

The relativistic corrections are equal to

$$\bar{\tau}_R(z) = -\int \frac{d^3\mathbf{p}}{(2\pi)^3} |\Psi_1(\mathbf{p})|^2 \frac{\mathbf{p}^4}{4M^3} = -\frac{5}{64}\alpha^4 M. \quad (2.309)$$

Finally, let us consider the last term in Eq. (2.304), which describes the propagation of one transverse photon in the presence of any number of Coulomb photons. Here, one has to insert the states containing a $\pi^+\pi^-$ pair and one photon between the two operators $\mathbf{H}_\gamma$. The product of the calculated matrix elements should be integrated over the photon momenta,

$$\int \frac{d^3\mathbf{k}}{(2\pi)^3 2|\mathbf{k}|},$$

and summed over photon polarizations

$$\sum_{\text{pol}} \varepsilon^i(\mathbf{k})\varepsilon^{*j}(\mathbf{k}) = \delta^{ij} - \frac{k^i k^j}{\mathbf{k}^2}.$$

Carrying out the calculations, we obtain

$$\langle \mathbf{p}|\tau_\gamma(z)|\mathbf{q}\rangle = \frac{e^2}{M^2}\int \frac{d^3\mathbf{k}}{(2\pi)^3 2|\mathbf{k}|}\left(\delta^{ij} - \frac{k^i k^j}{\mathbf{k}^2} \right)$$

$$\times \left\{ 2p^i q^j \langle \mathbf{p} - \tfrac{1}{2}\mathbf{k}|\bar{g}_C(z - |\mathbf{k}| - \frac{\mathbf{k}^2}{4M})| \mathbf{q} + \tfrac{1}{2}\mathbf{k}\rangle \right.$$

$$\left. - (p^i p^j + q^i q^j)\langle \mathbf{p} - \tfrac{1}{2}\mathbf{k}|\bar{g}_C(z - |\mathbf{k}| - \frac{\mathbf{k}^2}{4M})| \mathbf{q} - \tfrac{1}{2}\mathbf{k}\rangle \right\}. \quad (2.310)$$

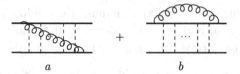

Figure 2.19 Transverse photon contributions to the bound-state energy shift: (a) the transverse photon exchange in the presence of Coulomb photons; (b) the self-energy contribution in the presence of Coulomb photons. The dashed and wiggly lines denote the Coulomb and transverse photons, respectively.

Here,

$$\langle \mathbf{p} | \bar{g}_C(z) | \mathbf{q} \rangle = \langle \mathbf{p} | g_C(z) | \mathbf{q} \rangle - \frac{\Psi_1(\mathbf{p})\Psi_1(\mathbf{q})}{z - E_1}. \tag{2.311}$$

Diagrammatically, the two terms in Eq. (2.310) correspond to the exchange of a transverse photon and the self-energy corrections in the presence of the Coulomb photons; see Fig. 2.19.

In Eq. (2.310) we again encounter a multiscale integral, where the threshold expansion has to be applied. The momenta of the pions, $\mathbf{p}$ and $\mathbf{q}$, always scale as $O(\alpha)$, since $\gamma = O(\alpha)$ is the single scale in the ground-state wave functions. In order to establish the counting of the integration momentum $\mathbf{k}$, let us recall that the argument of the Green's function $z - |\mathbf{k}| - \frac{\mathbf{k}^2}{4M}$ should be of order α^2 and thus $\mathbf{k} = O(\alpha^2)$ (ultrasoft). The threshold expansion then amounts to the *multipole expansion:* The Green's functions should be expanded in photon momentum $\mathbf{k}$, retaining, however, the term $|\mathbf{k}|$ in the denominator. It is easy to see that the integral starts to contribute at order α^5 and thus can be neglected at the accuracy we are working. Indeed, according to Eq. (2.249), the Green's function counts as α^{-5}, and the wave functions count as $\alpha^{-3/2}$. Taking into account the counting rules for momenta, one may easily verify this result.

However, there is one exception to this result. The free term in the first Green's function contains the δ-function $\delta^{(3)}(\mathbf{p} - \mathbf{q} - \mathbf{k})$, which ensures that $\mathbf{k} = O(\alpha)$ (soft) and not $O(\alpha^2)$ (ultrasoft). The integral over $\mathbf{k}$ can be evaluated, leading to

$$\langle \mathbf{p} | \tau_\gamma(z) | \mathbf{q} \rangle = -\frac{e^2}{4M^2} \frac{1}{|\mathbf{p} - \mathbf{q}|} \left((\mathbf{p} + \mathbf{q})^2 - \frac{(\mathbf{p}^2 - \mathbf{q}^2)^2}{(\mathbf{p} - \mathbf{q})^2} \right)$$

$$\times \frac{1}{2M + \dfrac{\mathbf{p}^2}{2M} + \dfrac{\mathbf{q}^2}{2M} + |\mathbf{p} - \mathbf{q}| - z}$$

$$= -\frac{e^2}{4M^2} \frac{1}{(\mathbf{p} - \mathbf{q})^2} \left((\mathbf{p} + \mathbf{q})^2 - \frac{(\mathbf{p}^2 - \mathbf{q}^2)^2}{(\mathbf{p} - \mathbf{q})^2} \right) + \cdots, \tag{2.312}$$

where, in the last line, the threshold expansion has been performed. Folding this expression with the ground-state wave functions and integrating, we immediately obtain

$$\bar{\tau}_\gamma(z) = -\frac{1}{8} \alpha^4 M. \tag{2.313}$$

What remains is to put all the pieces together and complete the calculation of the energy shift of the atom at order α^4. Using the calculated expressions for $\bar{\tau}_S(z)$, $\bar{\tau}_F(z)$, $\bar{\tau}_R(z)$, $\bar{\tau}_\gamma(z)$, it can be checked that the derivative term in the denominator of Eq. (2.302) contributes at $O(\alpha^5)$ and can thus be safely omitted at this order. One finally obtains

$$z - E_1 = \Delta E_{\text{em}} + \Delta E_{\text{str}} - i\frac{\Gamma}{2}, \qquad (2.314)$$

where the electromagnetic shift is given by

$$\Delta E_{\text{em}} = \frac{1}{6}\alpha^4 M^4 \langle r^2 \rangle - \frac{5}{64}\alpha^4 M - \frac{1}{8}\alpha^4 M = \frac{1}{6}\alpha^4 M^4 \langle r^2 \rangle - \frac{13}{64}\alpha^4 M, \qquad (2.315)$$

and the (complex) strong shift takes the following form:

$$\Delta E_{\text{str}} - i\frac{\Gamma}{2} = -\frac{M^3 \alpha^3}{8\pi} \bar{C} \left\{ 1 - \frac{\alpha M^2}{8\pi} \bar{C} \left(\Lambda(\mu) + \ln\frac{4\gamma^2}{\mu^2} - 3 \right) \right\}. \qquad (2.316)$$

In Eq. (2.316) the energy-dependent coupling is evaluated at $z = 2M - \gamma^2/M$. Using now Eqs. (2.243) and (2.263) for the calculation of the imaginary part gives

$$\text{Im}\,\bar{C} = \frac{M}{2\pi} p^* |\bar{C}_x|^2, \qquad p^* = \sqrt{M^2 - M_0^2 - \frac{1}{4}\alpha^2} + \cdots. \qquad (2.317)$$

Next, we evaluate $|\bar{C}_x|^2 = (\text{Re}\,\bar{C}_x)^2 + (\text{Im}\,\bar{C}_x)^2$. Using again unitarity, we have

$$\text{Im}\,\bar{C}_x = \frac{M}{2\pi} p^* \text{Re}\,\bar{C}_x \text{Re}\,\bar{C}_0 + \cdots. \qquad (2.318)$$

In Eqs. (2.317) and (2.318), the ellipses denote the higher-order terms in α. Using these expressions, one can separate the energy shift and the width of the atom:

$$\Delta E_{\text{str}} = -\frac{M^3 \alpha^3}{8\pi} \text{Re}\,\bar{C} \left\{ 1 - \frac{\alpha M^2}{8\pi} \text{Re}\,\bar{C} \left(\Lambda(\mu) + \ln\frac{4\gamma^2}{\mu^2} - 3 \right) \right\} + \cdots,$$

$$\Gamma = \frac{M^4 \alpha^3}{8\pi^2} p^* (\text{Re}\,\bar{C}_x)^2 \left\{ 1 + \frac{M^2}{4\pi^2} (p^*)^2 (\text{Re}\,\bar{C}_0)^2 \right.$$

$$\left. - \frac{\alpha M^2}{4\pi} \text{Re}\,\bar{C} \left(\Lambda(\mu) + \ln\frac{4\gamma^2}{\mu^2} - 3 \right) \right\} + \cdots. \qquad (2.319)$$

In the last step, one should use the matching condition, Eqs. (2.262) and (2.265), in order to express the couplings in terms of the physical amplitudes. As expected, the ultraviolet divergences and the scale-dependence cancel in the bound-state observables, and one obtains

$$\Delta E_{\text{str}} = -\frac{M^3 \alpha^3}{8\pi} \text{Re}\,\mathcal{A} \left\{ 1 - \frac{\alpha M^2}{4\pi} \text{Re}\,\mathcal{A}(\ln\alpha - 1) \right\},$$

$$\Gamma = \frac{M^4 \alpha^3}{32\pi^2} p^* (\text{Re}\,\mathcal{A}_x)^2 \left\{ 1 + \frac{M^2}{64\pi^2} (p^*)^2 (\text{Re}\,\mathcal{A}_0)^2 - \frac{\alpha M^2}{2\pi} \text{Re}\,\mathcal{A}(\ln\alpha - 1) \right\}.$$

$$(2.320)$$

These expressions represent the final result, they express the energy shift and width of the $\pi^+\pi^-$ hadronic atom in terms of threshold amplitudes of $\pi\pi$ scattering at the next-to-leading order in α. At this point, it becomes clear, why the nonrelativistic framework is so efficient. Leaving aside the "pure QED" part of the calculations, the low-momentum behavior of the strong $\pi\pi$ scattering amplitude in a given channel in the nonrelativistic EFT is parameterized by a *single* coupling constant that can be matched to the threshold amplitude. The Lagrangian at the next order contains terms with two derivatives. Consequently, its contribution will be suppressed by two powers of α and already lies beyond the accuracy we are considering. The same is true for the corrections to the strong amplitudes, coming from transverse photons, as none of these diagrams contributes at this order. On the contrary, in the relativistic theory, there exists an infinite number of contributions to the threshold amplitude, and thus, a closed expression for the bound-state observables cannot be obtained.

We may go one step further and ask, whether we can extract the scattering lengths with a given isospin, instead of the threshold amplitudes. This question cannot be answered without addressing the underlying dynamics in QCD plus QED. Chiral Perturbation Theory (ChPT, see later in Chapter 4) provides a proper framework for systematically "purifying" the threshold amplitudes from the isospin-breaking effects coming from both electromagnetic interactions and the quark mass difference. (Here, it is consistent to count $m_d - m_u$ as order α.) At a given order in the chiral expansion, setting, for example, the charged pion mass M as a reference mass $\bar{M}$ (see Eq. (2.242)), one gets

$$\mathcal{A} = \frac{4\pi}{3M^2}(2a_0 + a_2 + \varepsilon),$$

$$\mathcal{A}_x = \frac{8\pi}{3M^2}(a_2 - a_0 + \varepsilon_x), \tag{2.321}$$

where the correction terms $\varepsilon, \varepsilon_x$ vanish when $\alpha, (m_d - m_u) \to 0$. (The third amplitude, $\mathcal{A}_0$, is needed only at the leading order in the above expressions.) Furthermore, a_0, a_2 denote the (dimensionless) S-wave $\pi\pi$ scattering lengths with isospin $I = 0, 2$. Hence, the energy and width can be rewritten in terms of these scattering lengths:

$$\Delta E_{\text{str}} = -\frac{M\alpha^3}{6}(2a_0 + a_2 + \varepsilon)\left\{1 - \frac{\alpha}{3}(2a_0 - a_2)(\ln\alpha - 1)\right\},$$

$$\Gamma = \frac{2}{9}p^\star(a_2 - a_0 + \varepsilon_x)^2\left\{1 + \frac{M^2 - M_0^2}{9M^2}(a_0 + 2a_2)^2\right.$$

$$\left. -\frac{2\alpha}{3}(2a_0 - a_2)(\ln\alpha - 1)\right\}. \tag{2.322}$$

In Refs. [92–94], where this formula for the decay width was first derived, the correction term ε_x was evaluated at the next-to-leading order in ChPT. This has resulted in an estimate of the $\pi^+\pi^-$ atom lifetime, $\tau_{2\pi} = (2.9 \pm 0.1) \cdot 10^{-15}$ s, with an accuracy of 3%. The spectrum and decays of other hadronic atoms have also been studied within this approach. We refer to Ref. [95] for a detailed review on the subject.

Last but not least, note that the electromagnetic mass shift, displayed in Eq. (2.315), contains the charge radius. This term will be there even in the case of ordinary atoms. For example, the mass shift of an ordinary hydrogen atom also contains a (tiny) contribution, depending on the proton charge radius. Such a contribution in the case of muonic hydrogen (an atom formed by a proton and a muon) is much larger, because the Bohr radius of the muonic hydrogen is smaller. Note also that, since the muon does not take part in strong interactions, the contribution from the proton radius is the largest strong contribution to the level shift; the rest can be calculated in pure QCD with point-like particles at a very high precision. Thus, one may extract the proton radius from the precise measurement of the energy levels of the muonic atoms. Exactly this was done in the experiments of the Muonic Hydrogen collaboration at PSI [96, 97]. This beautiful measurement resulted in a small value of the proton radius, $r_p = 0.84184(67)$ fm, 5σ away from the then recommended CODATA value $r_p \simeq 0.88$ fm [98], which was also consistent with the electronic Lamb shift measurements. Such a prominent discrepancy became known as the "proton radius puzzle," leading to a large number of publications, which sometimes offered explanations even beyond the Standard Model. On the other hand, at that time, it was already known that the analysis of the electron–proton scattering data, using dispersion relations for the nucleon form factors, yields a proton radius that is lower than the recommended CODATA value and is consistent with the result of the Muonic Hydrogen experiment [99, 100]. The situation remained unsatisfactory until various experiments on the electronic Lamb shift and ep scattering at very low momentum transfer in 2017–19 beautifully confirmed the small proton radius, consistent with the measurement of the Muonic Hydrogen collaboration (CREMA project). Note that CODATA now recommends the small value. A brief history of the problem and more details can be found in Refs. [101, 102].

2.13 Literature Guide

The particle content of the relativistic field theory and the separation of particles from antiparticles were extensively studied in the early days of quantum field theory. Apart from Foldy and Wouthuysen [1], these issues were addressed in Refs. [103–106]. The textbook of Bjorken and Drell [2] contains a very precise and detailed discussion of the problem.

The reparametrization invariance has been introduced in Ref. [4]; see also [6]. Independently of this, implications of the Lorentz invariance for the kinetic term have been discussed in Ref. [5]. A nice introduction to the issue can be found in the review article [107].

Kaplan, Savage and Wise have pioneered the use of the effective theory with an unnaturally large two-body scattering length for the study of the nucleon–nucleon scattering and nuclear physics phenomena [9–11] (see also Gegelia, Ref. [12]). Inclusion of the Coulomb interactions in this approach is discussed, for example, in the paper by Kong and Ravndal [79]. An alternative power counting, which elevates the one-pion exchange to the leading order, is adopted in the approach proposed by Weinberg [13, 14]. The review article [15] contains a thorough discussion of different approaches.

Using an EFT approach, Bedaque, Hammer and van Kolck have shown that a single nonderivative three-particle coupling suffices for the non-perturbative renormalization of the Faddeev equation in the three-boson system at leading order in the limit of a very large two-body scattering length, or unitary limit [40, 41]. Moreover, it was shown that the non-perturbative nature of the problem leads to a peculiar RG evolution of this coupling (the so-called limit cycle). This is intrinsically related to the emergence of the infinite tower of the three-body bound states, condensing toward threshold in this limit, the celebrated Efimov effect [57, 59]. The absence of a unique solution of the Faddeev equation in the unitary limit was discussed in Refs. [55, 108]. Ref. [60] provides a nice review of the universality in few-body systems with large two-body scattering length.

The bound-states in field theory are described by the Bethe–Salpeter equation, see [109–111] or various quasipotential reductions thereof [112–123]. The perturbation calculations in QED have been done since the early fifties [112, 124]. A nice introduction to these calculations is given in the textbook by Itzykson and Zuber [125]. The nonrelativistic effective theory for the bound-state calculations was invented by Caswell and Lepage [67]. In Ref. [92], it was adapted to study the hadronic atoms, that is, the bound states of two hadrons that experience both strong and electromagnetic interactions. Ref. [95] contains a detailed review of the approach, as well as an extensive list of references on the subject.

References

[1] L. L. Foldy and S. A. Wouthuysen, "On the Dirac theory of spin $1/2$ particles and its nonrelativistic limit," Phys. Rev. **78** (1950) 29.

[2] J. D. Bjorken and S. D. Drell, Relativistic Quantum Mechanics, McGraw-Hill Inc. (1964).

[3] M. Berwein, N. Brambilla, S. Hwang and A. Vairo, "Poincaré invariance in NRQCD and potential NRQCD revisited," Phys. Rev. D **99** (2019) 094008.

[4] M. E. Luke and A. V. Manohar, "Reparametrization invariance constraints on heavy particle effective field theories," Phys. Lett. B **286** (1992) 348.

[5] V. Bernard, N. Kaiser, J. Kambor and U.-G. Meißner, "Chiral structure of the nucleon," Nucl. Phys. B **388** (1992) 315.

[6] A. V. Manohar, "The HQET/NRQCD Lagrangian to order α/m^3," Phys. Rev. D **56** (1997) 230.

[7] J. Heinonen, R. J. Hill and M. P. Solon, "Lorentz invariance in heavy particle effective theories," Phys. Rev. D **86** (2012) 094020.

[8] E. Braaten, "Galilean-invariant effective field theory for the $X(3872)$," Phys. Rev. D **91** (2015) 114007.

[9] D. B. Kaplan, M. J. Savage and M. B. Wise, "Nucleon-nucleon scattering from effective field theory," Nucl. Phys. B **478** (1996) 629.

[10] D. B. Kaplan, M. J. Savage and M. B. Wise, "A new expansion for nucleon-nucleon interactions," Phys. Lett. B **424** (1998) 390.

[11] D. B. Kaplan, M. J. Savage and M. B. Wise, "Two nucleon systems from effective field theory," Nucl. Phys. B **534** (1998) 329.

[12] J. Gegelia, "EFT and NN scattering," Phys. Lett. B **429** (1998) 227.

[13] S. Weinberg, "Nuclear forces from chiral Lagrangians," Phys. Lett. B **251** (1990) 288.

[14] S. Weinberg, "Effective chiral Lagrangians for nucleon-pion interactions and nuclear forces," Nucl. Phys. B **363** (1991) 3.

[15] E. Epelbaum, H. W. Hammer and U.-G. Meißner, "Modern theory of nuclear forces," Rev. Mod. Phys. **81** (2009) 1773.

[16] H. W. Hammer, S. König and U. van Kolck, "Nuclear effective field theory: status and perspectives," Rev. Mod. Phys. **92** (2020) 025004.

[17] D. Lee et al., "Hidden spin-isospin exchange symmetry," Phys. Rev. Lett. **127** (2021) 062501.

[18] G. Colangelo, J. Gasser, B. Kubis and A. Rusetsky, "Cusps in $K \to 3\pi$ decays," Phys. Lett. B **638** (2006) 187.

[19] J. Gasser, B. Kubis and A. Rusetsky, "Cusps in $K \to 3\pi$ decays: a theoretical framework," Nucl. Phys. B **850** (2011) 96.

[20] V. Bernard, M. Lage, U.-G. Meißner and A. Rusetsky, "Resonance properties from the finite-volume energy spectrum," JHEP **08** (2008) 024.

[21] J. R. Batley et al. [NA48/2], "Observation of a cusp-like structure in the $\pi^0\pi^0$ invariant mass distribution from $K^+ \to \pi^+\pi^0\pi^0$ decay and determination of the $\pi\pi$ scattering lengths," Phys. Lett. B **633** (2006) 173.

[22] J. R. Batley et al., "Determination of the S-wave $\pi\pi$ scattering lengths from a study of $K^+ \to \pi^+\pi^0\pi^0$ decays," Eur. Phys. J. C **64** (2009) 589.

[23] P. Budini and L. Fonda, "Pion-pion interaction from threshold anomalies in K^+ decay," Phys. Rev. Lett. **6** (1961) 419.

[24] N. Cabibbo, "Determination of the $a_0 - a_2$ pion scattering length from $K^+ \to \pi^+\pi^0\pi^0$ decay," Phys. Rev. Lett. **93** (2004) 121801.

[25] N. Cabibbo and G. Isidori, "Pion-pion scattering and the $K \to 3\pi$ decay amplitudes," JHEP **03** (2005) 021.

[26] U.-G. Meißner, G. Müller and S. Steininger, "Virtual photons in SU(2) chiral perturbation theory and electromagnetic corrections to $\pi\pi$ scattering," Phys. Lett. B **406** (1997) 154 [erratum: Phys. Lett. B **407** (1997) 454].

[27] R. J. Eden, P. V. Landshoff, D. I. Olive and J. C. Polkinghorne, The Analytic S-Matrix, Cambridge University Press, 1966.

[28] M. Bissegger, A. Fuhrer, J. Gasser, B. Kubis and A. Rusetsky, "Radiative corrections in $K \to 3\pi$ decays," Nucl. Phys. B **806** (2009) 178.

[29] M. Bissegger, A. Fuhrer, J. Gasser, B. Kubis and A. Rusetsky, "Cusps in $K_L \to 3\pi$ decays," Phys. Lett. B **659** (2008) 576.

[30] C. O. Gullstrom, A. Kupsc and A. Rusetsky, "Predictions for the cusp in $\eta \to 3\pi^0$ decay," Phys. Rev. C **79** (2009) 028201.

[31] S. P. Schneider, B. Kubis and C. Ditsche, "Rescattering effects in $\eta \to 3\pi$ decays," JHEP **02** (2011) 028.

[32] B. Kubis and S. P. Schneider, "The cusp effect in $\eta' \to \eta\pi\pi$ decays," Eur. Phys. J. C **62** (2009) 511.

[33] G. Colangelo, J. Gasser and A. Rusetsky, "Isospin breaking in $K_{\ell 4}$ decays," Eur. Phys. J. C **59** (2009) 777.

[34] J. R. Batley et al. [NA48/2], "Precise tests of low energy QCD from K_{e4} decay properties," Eur. Phys. J. C **70** (2010) 635.

[35] V. Bernard, S. Descotes-Genon and M. Knecht, "Isospin breaking in the phases of the K_{e4} form factors," Eur. Phys. J. C **73** (2013) 2478.

[36] L. D. Faddeev, "Scattering theory for a three-particle system," Sov. Phys. JETP **12** (1961) 1014.

[37] W. Glöckle, The Quantum Mechanical Few-Body Problem, Springer-Verlag, 1983.

[38] G. V. Skornyakov and K. A. Ter-Martirosyan, Zh. Eksp. Teor. Fiz. 31 (1956) 775: Sov. Phys. JETP 4 (1956) 648.

[39] D. B. Kaplan, "More effective field theory for nonrelativistic scattering," Nucl. Phys. B **494** (1997) 471.

[40] P. F. Bedaque, H. W. Hammer and U. van Kolck, "Renormalization of the three-body system with short-range interactions," Phys. Rev. Lett. **82** (1999) 463.

[41] P. F. Bedaque, H. W. Hammer and U. van Kolck, "The three-boson system with short-range interactions," Nucl. Phys. A **646** (1999) 444.

[42] S. Weinberg, "Quasiparticles and the Born Series," Phys. Rev. **131** (1963) 440.

[43] P. F. Bedaque, G. Rupak, H. W. Grießhammer and H.-W. Hammer, "Low-energy expansion in the three-body system to all orders and the triton channel," Nucl. Phys. A **714** (2003) 589.

[44] S. Koenig and H.-W. Hammer, "Low-energy p-d scattering and He-3 in pionless EFT," Phys. Rev. C **83** (2011) 064001.

[45] G. A. Baker, "Neutron matter model," Phys. Rev. C **60** (1999) 054311.

[46] C. Ji, D. R. Phillips and L. Platter, "The three-boson system at next-to-leading order in an effective field theory for systems with a large scattering length," Annals Phys. **327** (2012) 1803.

[47] J. Vanasse, "Fully perturbative calculation of nd scattering to next-to-next-to-leading-order," Phys. Rev. C **88** (2013) 044001.

[48] P. Navratil, B. R. Barrett and W. Glöckle, "Spurious states in the Faddeev formalism for few body systems," Phys. Rev. C **59** (1999) 611.

[49] E. Epelbaum et al., "Few nucleon systems with two nucleon forces from chiral effective field theory," Eur. Phys. J. A **15** (2002) 543.

[50] M. Ebert, H.-W. Hammer and A. Rusetsky, "An alternative scheme for effective range corrections in pionless EFT" [arXiv:2109.11982 [hep-ph]].

[51] A. J. Moerdijk, B. J. Verhaar and A. Axelsson, "Resonances in ultracold collisions of $Li - 6$, $Li - 7$, and $Na - 23$," Phys. Rev. A **51** (1995) 4852.

[52] M. L. Goldberger and K. M. Watson, Collision Theory, John Wiley & Sons (1964).

[53] H. Feshbach, "Unified theory of nuclear reactions," Annals Phys. **5** (1958) 357.

[54] H. Feshbach, "A unified theory of nuclear reactions. 2.," Annals Phys. **19** (1962) 287.

[55] L. D. Faddeev and R. A. Minlos, "Comment on the problem of three particles with point interactions," Zh. Eksp. Teor. Fiz. 41 (1961) 1850; Sov. Phys. JETP **14** (1962) 1315,

[56] E. Epelbaum, J. Gegelia, U.-G. Meißner and D. L. Yao, "Renormalization of the three-boson system with short-range interactions revisited," Eur. Phys. J. A **53** (2017) 98.

[57] V. N. Efimov, "Weakly-bound states of three resonantly-interacting particles," Sov. J. Nucl. Phys. **12** (1971) 589.

[58] V. Efimov, "Energy levels of three resonantly interacting particles," Nucl. Phys. A **210** (1973) 157.

[59] V. Efimov, "Low-energyproperties of three resonantly interacting particles," Yad. Fiz. **29** (1979) 1058; Sov. J. Nucl. Phys. **29** (1979) 546.

[60] E. Braaten and H. W. Hammer, "Universality in few-body systems with large scattering length," Phys. Rept. **428** (2006) 259.

[61] H. W. Hammer and L. Platter, "Efimov states in nuclear and particle physics," Ann. Rev. Nucl. Part. Sci. **60** (2010) 207.

[62] P. Naidon and S. Endo, "Efimov physics: a review," Rept. Prog. Phys. **80** (2017) no. 5, 056001.

[63] K. G. Wilson, "The renormalization group and strong interactions," Phys. Rev. D **3** (1971) 1818.

[64] S. Albeverio, R. Hoegh-Krohn and T. T. Wu, "A class of exactly solvable three-body quantum mechanical problems and the universal low-energy behavior," Phys. Lett. A **83** (1981) 105.

[65] E. Braaten and H. W. Hammer, "An infrared renormalization group limit cycle in QCD," Phys. Rev. Lett. **91** (2003) 102002.

[66] R. F. Mohr, R. J. Furnstahl, R. J. Perry, K. G. Wilson and H. W. Hammer, "Precise numerical results for limit cycles in the quantum three-body problem," Annals Phys. **321** (2006) 225.

[67] W. E. Caswell and G. P. Lepage, "Effective Lagrangians for bound state problems in QED, QCD, and other field theories," Phys. Lett. B **167** (1986) 437.

[68] G. P. Lepage, L. Magnea, C. Nakhleh, U. Magnea and K. Hornbostel, "Improved nonrelativistic QCD for heavy quark physics," Phys. Rev. D **46** (1992) 4052.

[69] N. Brambilla, A. Pineda, J. Soto and A. Vairo, "Potential NRQCD: an effective theory for heavy quarkonium," Nucl. Phys. B **566** (2000) 275.

[70] N. Brambilla, A. Pineda, J. Soto and A. Vairo, "Effective field theories for heavy quarkonium," Rev. Mod. Phys. **77** (2005) 1423.

[71] M. Beneke and V. A. Smirnov, "Asymptotic expansion of Feynman integrals near threshold," Nucl. Phys. B **522** (1998) 321.

[72] J. S. Schwinger, "On quantum electrodynamics and the magnetic moment of the electron," Phys. Rev. **73** (1948) 416.

[73] T. Kinoshita and M. Nio, "Radiative corrections to the muonium hyperfine structure. 1. The $\alpha^2(Z\alpha)$ correction," Phys. Rev. D **53** (1996) 4909.

[74] J. Schwinger, "Coulomb Green's function," J. Math. Phys. **5** (1964) 1606.

[75] M. L. Goldberger and K. M. Watson, Collision Theory, Dover Books on Physics, Dover Publications (2004).

[76] A. Sommerfeld, *Atombau und Spektrallinien,* F. Vieweg & Sohn (1921).

[77] G. Gamow, "Zur Quantentheorie des Atomkernes," Z. Phys. **51** (1928) 204.

[78] A. D. Sakharov, "Interaction of an electron and positron in pair production," Zh. Eksp. Teor. Fiz. 18 (1948) 631; Sov. Phys. Usp. **34** (1991) 375.

[79] X. Kong and F. Ravndal, "Coulomb effects in low-energy proton–proton scattering," Nucl. Phys. A **665** (2000) 137.

[80] S. Deser, M. L. Goldberger, K. Baumann and W. E. Thirring, "Energy level displacements in pi mesonic atoms," Phys. Rev. **96** (1954) 774.

[81] Th. Busch, B.-G. Englert, K. Rzazewski and M. Wilkens, "Two cold atoms in a harmonic trap," Found. Phys. **28** (1998) 549.

[82] M. Lüscher, "Volume dependence of the energy spectrum in massive quantum field theories. 2. Scattering states," Commun. Math. Phys. **105** (1986) 153.

[83] B. Adeva et al. "Determination of $\pi\pi$ scattering lengths from measurement of $\pi^+\pi^-$ atom lifetime," Phys. Lett. B **704** (2011) 24.

[84] B. Adeva et al. [DIRAC], "First measurement of a long-lived $\pi^+\pi^-$ atom lifetime," Phys. Rev. Lett. **122** (2019) 082003.

[85] B. Adeva et al. [DIRAC], "Measurement of the πK atom lifetime and the πK scattering length," Phys. Rev. D **96** (2017) 052002.

[86] M. Hennebach, D. F. Anagnostopoulos, A. Dax et al. "Hadronic shift in pionic hydrogen," Eur. Phys. J. A **50** (2014) 190 [erratum: Eur. Phys. J. A **55** (2019) 24].

[87] T. Strauch et al. "Pionic deuterium," Eur. Phys. J. A **47** (2011) 88.

[88] D. Gotta, "Precision spectroscopy of light exotic atoms," Prog. Part. Nucl. Phys. **52** (2004) 133.

[89] A. Hirtl et al. "Redetermination of the strong-interaction width in pionic hydrogen," Eur. Phys. J. A **57** (2021) 70.

[90] A. Scordo et al. "The kaonic atoms research program at DAΦNE: from SID-DHARTA to SIDDHARTA-2," EPJ Web Conf. **181** (2018) 01004.

[91] C. Curceanu et al. "Kaonic atoms to investigate global symmetry breaking," Symmetry **12** (2020) 547.

[92] A. Gall, J. Gasser, V. E. Lyubovitskij and A. Rusetsky, "On the lifetime of the $\pi^+\pi^-$ atom," Phys. Lett. B **462** (1999) 335.

[93] J. Gasser, V. E. Lyubovitskij and A. Rusetsky, "Numerical analysis of the $\pi^+\pi^-$ atom lifetime in ChPT," Phys. Lett. B **471** (1999) 244.

[94] J. Gasser, V. E. Lyubovitskij, A. Rusetsky and A. Gall, "Decays of the $\pi^+\pi^-$ atom," Phys. Rev. D **64** (2001) 016008.

[95] J. Gasser, V. E. Lyubovitskij and A. Rusetsky, "Hadronic atoms in QCD + QED," Phys. Rept. **456** (2008) 167.

[96] R. Pohl et al. "The size of the proton," Nature **466** (2010) 213.

[97] R. Pohl, R. Gilman, G. A. Miller and K. Pachucki, "Muonic hydrogen and the proton radius puzzle," Ann. Rev. Nucl. Part. Sci. **63** (2013) 175.

[98] P. J. Mohr, B. N. Taylor and D. B. Newell, "CODATA recommended values of the fundamental physical constants: 2006," Rev. Mod. Phys. **80** (2008) 633.

[99] M. A. Belushkin, H. W. Hammer and U.-G. Meißner, "Dispersion analysis of the nucleon form-factors including meson continua," Phys. Rev. C **75** (2007) 035202.

[100] I. T. Lorenz, H.-W. Hammer and U.-G. Meißner, "The size of the proton - closing in on the radius puzzle," Eur. Phys. J. A **48** (2012) 151.

[101] H.-W. Hammer and U.-G. Meißner, "The proton radius: from a puzzle to precision," Sci. Bull. **65** (2020) 257.

[102] Y. H. Lin, H. W. Hammer and U.-G. Meißner, "Dispersion-theoretical analysis of the electromagnetic form factors of the nucleon: past, present and future," Eur. Phys. J. A **57** (2021) 255.

[103] H. Feshbach and F. Villars, "Elementary relativistic wave mechanics of spin 0 and spin 1/2 particles," Rev. Mod. Phys. **30** (1958) 24.

[104] N. Kemmer, "The particle aspect of meson theory," Proc. Roy. Soc. Lond. A **173** (1939) 91.

[105] M. Taketani and S. Sakata, "On the wave equation of meson," Proc. Phys. Math. Soc. Japan **22** (1940) 757; reprinted in Prog. Theor. Phys. Suppl. **1** (1955) 84.

[106] W. Heitler and H.-W. Peng, "On the particle equation of the meson," Proc. Roy. Irish Acad. **49** (1943) 1.

[107] M. Neubert, "Heavy quark symmetry," Phys. Rept. **245** (1994) 259.

[108] G. S. Danilov, "On the three-body problem with short-range forces," Sov. Phys. JETP **13** (1961) 349; J. Exptl. Theoret. Phys. **40** (1961) 498.

[109] H. A. Bethe and E. E. Salpeter, Quantum Mechanics of One- and Two-Electron Atoms Springer-Verlag (1957).

[110] M. Gell–Mann and F. Low, "Bound states in quantum field theory," Phys. Rev. **84** (1951) 350.

[111] N. Nakanishi, "A general survey of the theory of the Bethe-Salpeter equation," Prog. Theor. Phys. Suppl. **43** (1969) 1.

[112] E. E. Salpeter, "Mass corrections to the fine structure of hydrogen-like atoms," Phys. Rev. **87** (1952) 328.

[113] R. Blankenbecler and R. Sugar, "Linear integral equations for relativistic multichannel scattering," Phys. Rev. **142** (1966) 1051.

[114] A. A. Logunov and A. N. Tavkhelidze, "Quasioptical approach in quantum field theory," Nuovo Cim. **29** (1963) 380.

[115] F. Gross, "Three-dimensional covariant integral equations for low-energy systems," Phys. Rev. **186** (1969) 1448.

[116] F. Gross, "The relativistic few body problem. 1. Two-body equations," Phys. Rev. C **26** (1982) 2203.

[117] M. H. Partovi and E. L. Lomon, "Field theoretical nucleon-nucleon potential," Phys. Rev. D **2** (1970) 1999.

[118] V. G. Kadyshevsky, "Quasipotential type equation for the relativistic scattering amplitude," Nucl. Phys. B **6** (1968) 125.

[119] I. T. Todorov, "Quasipotential equation corresponding to the relativistic Eikonal approximation," Phys. Rev. D **3** (1971) 2351.

[120] S. J. Wallace and V. B. Mandelzweig, "Covariant two-body equations for scalar and Dirac particles," Nucl. Phys. A **503** (1989) 673.

[121] C. Fronsdal and R. W. Huff, "Two-body problem in quantum field theory," Phys. Rev. D **3** (1971) 933.

[122] G. P. Lepage, "Analytic bound state solutions in a relativistic two-body formalism with applications in muonium and positronium," Phys. Rev. A **16** (1977) 863.

[123] H. Jallouli and H. Sazdjian, "The relativistic two-body potentials of constraint theory from summation of Feynman diagrams," Annals Phys. **253** (1997) 376.

[124] R. Karplus and A. Klein, "Electrodynamics displacement of atomic energy levels. 3. The hyperfine structure of positronium," Phys. Rev. **87** (1952) 848.

[125] C. Itzykson and J. B. Zuber, "Quantum Field Theory," McGraw-Hill (1980) 705 pp. (International Series in Pure and Applied Physics).

3 Symmetries

3.1 Introduction

The construction of any effective field theory (EFT) should obey certain guiding principles. In particular, symmetries provide an extremely powerful tool that enables one to systematically construct the Lagrangian of the EFT. It will on the one hand inherit the symmetry properties of the more fundamental theory it is derived from, but, on the other hand, different patterns of the realization of certain symmetries might also arise. Clearly, similar considerations also apply if one formulates the EFT without recourse to an underlying theory.

Classically, the invariance of a theory under some continuous symmetry group implies the conservation of currents, associated with the generators of this group. This is the celebrated Noether theorem [1]. In quantum field theory, Noether's theorem is translated into Ward identities, which form an infinite tower of equations, relating Green's functions with different numbers of external legs to each other. Further, we may establish Ward identities that are obeyed by the Green's functions corresponding to the effective degrees of freedom only. These identities, as they are a consequence of the underlying symmetry, should survive in the EFT as well. Note that this statement is not limited to the elementary degrees of freedom in the underlying theory. For example, in QCD, where the low-energy spectrum does not contain free quarks and gluons, we consider Ward identities for the Green's functions of colorless composite operators, which describe hadrons that appear in the asymptotic states. An important ingredient to these Ward identities is the lightest particles in QCD, the pions, which are related to the remarkable phenomenon of spontaneous symmetry breaking to be discussed in more detail in what follows.

In the next step, one constructs the Lagrangian of the EFT, requiring that the Green's functions of the effective degrees of freedom obey exactly the same Ward identities. This requirement translates into the invariance of the effective Lagrangian under the symmetry group of the underlying theory, realized on the fields describing effective degrees of freedom. Recalling now that the effective theories respect counting rules, at any given order we may construct a finite set of operators with a definite mass dimension that are invariant under the symmetry transformations. The effective Lagrangian at this order is a sum of all these operators with arbitrary coefficients that are not fixed by symmetry considerations. These parameters are often called *low-energy constants* (LECs).

Finally, note that not all classical symmetries survive at the quantum level. This phenomenon is known under the name of *anomalies,* or, more precisely, anomalous symmetry breaking. Certain Ward identities get modified by anomalies, and the corresponding currents are no longer conserved. In this case, the effective Lagrangian should also contain terms that reproduce the correct anomaly in the underlying theory. An example is given by the Wess–Zumino–Witten (WZW) term in the effective chiral Lagrangian, which yields the anomalous divergence of the singlet axial-vector current in QCD.

As we will see in the following, symmetries can be realized in various ways. Most simply, they can be exact or approximate. However, in many cases symmetries are hidden (spontaneously broken) or anomalous. All these intricate phenomena will appear in QCD and its related effective field theories.

3.2 Euler–Heisenberg Lagrangian

3.2.1 The Role of Symmetry

To elucidate the role of symmetries, we start with a warm-up example, where perturbative calculations can be carried out explicitly in order to verify the result. Consider Quantum Electrodynamics (QED) for momenta/energies much smaller than the electron mass. According to the decoupling theorem, the only relevant degrees of freedom in the EFT for the energies $E \ll m_e$ will be photons (here, E and m_e denote the photon energy and the electron mass, respectively). Consequently, the effective Lagrangian of the theory should be constructed from the photon field $\mathscr{A}_\mu$ only. It is important to realize that this is *not* a theory of free photons: The corresponding Lagrangian contains vertices with 4, 6, ... photons (an odd number is not allowed because of Furry's theorem [2]). These vertices describe interactions that in the original theory are mediated by closed electron loops; see Fig. 3.1.

In order to construct the effective Lagrangian, one writes down all possible terms that can be built using the field $\mathscr{A}_\mu$. In the next step, the couplings in front of these terms should be matched to the underlying theory, in this case QED. Here one arrives at the central question: What is are the criteria for *possible* terms? The corresponding procedure is based on the following rules:

- Use only those fields that correspond to the relevant degrees of freedom at the given energy.
- Respect all symmetries. In our example, Lorentz invariance and the discrete C, P, T symmetries of QED should be maintained. Here, C, P and T refer to charge conjugation, parity transformations and time reversal, in order (see, e.g., Ref. [3]). However, in addition to these general symmetries, QED possesses a $U(1)$ gauge symmetry. In this section we shall demonstrate that the requirement of $U(1)$-invariance of the effective theory severely limits the number of the possible terms. This simplifies the procedure of constructing the effective Lagrangian.

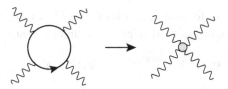

Four-point effective photon vertex emerging from an electron loop. The wiggly and solid lines denote photons and electrons, respectively.

- Respect the counting rules. At a given order in the low-momentum expansion, only the operators with a pertinent mass dimension should be retained in the Lagrangian.

The rest of the present section is dedicated to the study of the implications of the $U(1)$ gauge symmetry in the construction of the effective Lagrangian. To this end, we find it convenient to use the language of the path integral. In an arbitrary covariant gauge (see, e.g., Ref. [3]), the Lagrangian of QED is given by

$$\mathcal{L}_{\text{QED}} = \bar{\psi}(i\gamma^\mu(\partial_\mu + ie\mathscr{A}_\mu) - m_e)\psi - \frac{1}{4}\mathscr{F}_{\mu\nu}\mathscr{F}^{\mu\nu} - \frac{\xi}{2}(\partial^\mu \mathscr{A}_\mu)^2, \qquad (3.1)$$

where ψ and $\mathscr{A}_\mu$ are the electron and the photon fields, respectively, $\mathscr{F}_{\mu\nu} = \partial_\mu\mathscr{A}_\nu - \partial_\nu\mathscr{A}_\mu$ is the electromagnetic field strength tensor, $-e$ is the electron charge and ξ denotes the gauge fixing parameter. Observables (S-matrix elements) do not depend on ξ, but the Green's functions do.

The generating functional of the Green's functions in QED is given by

$$Z(j,\eta,\bar{\eta}) = \int d\psi d\bar{\psi} d\mathscr{A}_\mu \exp\left\{ i\int d^4x(\mathcal{L}_{\text{QED}} + \bar{\eta}\psi + \bar{\psi}\eta + j_\mu \mathscr{A}^\mu) \right\}, \qquad (3.2)$$

where j_μ and η denote external sources for the photon and electron fields, respectively. The Green's functions are obtained in the usual manner, namely by differentiating the generating functional with respect to the sources and, at the end, letting these sources vanish. Since we are interested in the derivation of the effective Lagrangian for the photons only, we may put $\eta = \bar{\eta} = 0$ from the beginning. The generating functional depends then on the argument j_μ only, and we can write

$$Z(j) = Z(j,\eta,\bar{\eta})\big|_{\eta=\bar{\eta}=0}$$

$$= \int d\mathscr{A}_\mu \exp\left\{ i\int d^4x\left(-\frac{1}{4}\mathscr{F}_{\mu\nu}\mathscr{F}^{\mu\nu} - \frac{\xi}{2}(\partial^\mu \mathscr{A}_\mu)^2 + \mathcal{L}_{\text{eff}}(\mathscr{A}_\mu) + j_\mu \mathscr{A}^\mu \right) \right\},$$

$$(3.3)$$

where

$$\exp\left\{ i\int d^4x \mathcal{L}_{\text{eff}}(\mathscr{A}_\mu) \right\} = \int d\psi d\bar{\psi} \exp\left\{ i\int d^4x \bar{\psi}(i\gamma^\mu(\partial_\mu + ie\mathscr{A}_\mu) - m_e)\psi \right\}. \qquad (3.4)$$

Now, let us focus on the role of gauge invariance. It is straightforward to see that the integrand in Eq. (3.4) is invariant under the gauge transformations,

$$\psi(x) \mapsto e^{-i\alpha(x)}\psi(x), \quad \bar{\psi}(x) \mapsto \bar{\psi}(x)e^{i\alpha(x)}, \quad \mathscr{A}_\mu \mapsto \mathscr{A}_\mu + \frac{1}{e}\partial_\mu \alpha(x). \tag{3.5}$$

Here, $\alpha(x)$ denotes the real-valued parameter of the gauge transformation.

Consequently, assuming that the path integral measure is also invariant with respect to the gauge transformations,[1] and performing these transformations in Eq. (3.4), we easily obtain

$$\exp\left\{i\int d^4x \mathcal{L}_{\text{eff}}(\mathscr{A}_\mu)\right\} = \exp\left\{i\int d^4x \mathcal{L}_{\text{eff}}\left(\mathscr{A}_\mu + \frac{1}{e}\partial_\mu \alpha\right)\right\}. \tag{3.6}$$

In Eq. (3.6) $\mathscr{A}_\mu$ is considered as an external classical field. Equation (3.6) holds, if the effective Lagrangian $\mathcal{L}_{\text{eff}}(\mathscr{A}_\mu)$ is gauge-invariant, that is, if it only depends on the gauge-invariant field strength tensor,

$$\mathcal{L}_{\text{eff}}(\mathscr{A}_\mu) = \mathcal{L}_{\text{eff}}(\mathscr{F}_{\mu\nu}). \tag{3.7}$$

Note that the gauge invariance naturally leads to consistent counting rules. Since $\mathscr{F}_{\mu\nu}$ contains field derivatives, insertions of $\mathscr{F}_{\mu\nu}$ into loop diagrams result in the suppression of the loop corrections at low energies.

The nonlinear contributions to the Lagrangian arise first at $O(m_e^{-4})$. To this order, there are only two such terms, consistent with all symmetries:

$$\mathcal{L}_{\text{eff}} = -\frac{\alpha c_0}{4}\mathscr{F}_{\mu\nu}\mathscr{F}^{\mu\nu} + \frac{\alpha^2}{m_e^4}\left\{c_1(\mathscr{F}_{\mu\nu}\mathscr{F}^{\mu\nu})^2 + c_2(\mathscr{F}_{\mu\nu}\tilde{\mathscr{F}}^{\mu\nu})^2\right\} + O(m_e^{-6}), \tag{3.8}$$

where $\tilde{\mathscr{F}}^{\mu\nu} = \frac{1}{2}\varepsilon^{\mu\nu\alpha\beta}\mathscr{F}_{\alpha\beta}$, and $\varepsilon^{\mu\nu\alpha\beta}$ is the totally antisymmetric Levi–Civita tensor. The overall factor m_e^{-4} appears on dimensional grounds, and the factor α^2, where $\alpha = e^2/(4\pi)$ is the electromagnetic fine-structure constant, appears because this term couples with four photons, each carrying a factor e. So, to this order, only two constants c_1, c_2 have to be determined from matching to QED. Note also the constant c_0 is ultraviolet-divergent. Note further that the first term in Eq. (3.8) combines with the free photon kinetic term and leads to a renormalization of the photon field.

Irrespective of the actual values of the constants c_1, c_2, we may investigate, for example, the dependence of the photon-photon scattering cross section on photon energy E at $E \ll m_e$. From the explicit form of the Euler–Heisenberg Lagrangian given in Eq. (3.8), it is straightforward to conclude that the scattering amplitude behaves like

$$A_{2\gamma \to 2\gamma} \sim \frac{\alpha^2 E^4}{m_e^4}, \tag{3.9}$$

---

[1] At first glance, this seems self-evident, since $d\psi d\bar{\psi} = \prod_x d\psi(x)d\bar{\psi}(x) = \prod_x (e^{-i\alpha(x)}d\psi(x))(e^{i\alpha(x)}d\bar{\psi}(x))$. However, a certain care is needed in performing the continuum limit, where the number of integration variables tends to infinity. In particular, one needs to regularize the ultraviolet divergence that emerges in this limit, and one must remove the regularization at the end of the calculations. In the given case, this can be done without a problem, justifying the assumption about the gauge-invariance of the fermionic measure. However, if the gauge transformation contains γ_5, the fermionic measure is, in general, no more gauge-invariant, giving rise to the so-called *anomalies*. In the following, we shall consider this issue in detail.

where the factor E^4 stems from the four derivatives. The cross section behaves as

$$\sigma_{2\gamma \to 2\gamma} \sim \left(\frac{\alpha^2 E^4}{m_e^4}\right)^2 \frac{1}{E^2} = \frac{\alpha^4 E^6}{m_e^8}. \tag{3.10}$$

Note that, in the above expression, the phase-space factor E^{-2} is established on purely dimensional grounds. In the absence of a photon mass, the photon energy E is the only dimensionful parameter on which the phase space factor can depend.

In the subsequent section we shall discuss the matching of the coefficients c_1, c_2 to the underlying theory. The direct method, based on the matching of Feynman integrals, turns out to be very cumbersome. We shall see that using path integral methods allows one to achieve the goal with a substantially smaller effort.

Historical note: While Euler, Heisenberg and Kockel analyzed light-by-light scattering using effective field theory as described (this might in fact be the first use of an EFT) in the mid-1930s, the full calculation of this process based on the finite sum of box diagrams in full QED was performed by Karplus and Neumann only in 1951 [4]. In fact, the low-energy limit of the calculation of Karplus and Neumann exactly recovers the Euler–Heisenberg result. This is a beautiful example that in case of scale separation the EFT approach is much more *effective* than the calculation in the full theory. For a nice discussion on the history of the Euler–Heisenberg approach, see Ref. [5].

3.2.2 Matching of the Couplings in the Effective Lagrangian

In the path integral, Eq. (3.4), we may carry out the Grassmann integration over the variables $\psi, \bar{\psi}$. The answer is given by

$$\exp\left\{i \int d^4x \mathcal{L}_{\text{eff}}(\mathscr{A}_\mu)\right\} = \det(i\gamma^\mu \partial_\mu - e\gamma^\mu \mathscr{A}_\mu - m_e), \tag{3.11}$$

so that

$$\int d^4x \mathcal{L}_{\text{eff}}(\mathscr{A}_\mu) = -i \ln \det(D), \qquad D = i\gamma^\mu \partial_\mu - e\gamma^\mu \mathscr{A}_\mu - m_e. \tag{3.12}$$

In other words, calculating the determinant and expanding in powers of $\mathscr{A}_\mu$, we will reproduce all terms of the effective Lagrangian.

The key observation that simplifies the calculations dramatically [6] consists in the following: In order to establish the coefficients c_1, c_2, it suffices to consider the determinant for *constant* electric and magnetic fields $\mathbf{E}$ and $\mathbf{B}$. Defining the quantities a and b so that

$$a^2 - b^2 = \mathbf{E}^2 - \mathbf{B}^2 = -\frac{1}{2}\mathscr{F}_{\mu\nu}\mathscr{F}^{\mu\nu}, \qquad ab = \mathbf{E} \cdot \mathbf{B} = \frac{1}{4}\mathscr{F}_{\mu\nu}\tilde{\mathscr{F}}^{\mu\nu}, \tag{3.13}$$

it can be shown that (see Section 3.2.3)

$$\mathcal{L}_{\text{eff}}(\mathscr{A}_\mu) = \frac{1}{8\pi^2} \int_0^\infty \frac{ds}{s} e^{-is(m_e^2 - i\varepsilon)} \left(e^2 ab \frac{\cosh(eas)\cos(ebs)}{\sinh(eas)\sin(ebs)} - \frac{1}{s^2}\right), \tag{3.14}$$

where the factor $i\varepsilon$ is needed to make the integral convergent. Expanding in powers of a, b and using Eq. (3.13), we obtain

$$\mathcal{L}_{\mathrm{eff}}(\mathcal{A}_\mu) = \frac{e^2}{24\pi^2}(\mathbf{E}^2 - \mathbf{B}^2) \int_0^\infty \frac{ds}{s} e^{-is(m_e^2 - i\varepsilon)}$$

$$- \left(\frac{e^4}{360\pi^2}(\mathbf{E}^2 - \mathbf{B}^2)^2 + \frac{7e^4}{360\pi^2}(\mathbf{E}\cdot\mathbf{B})^2 \right) \int_0^\infty ds\, s\, e^{-is(m_e^2 - i\varepsilon)} + \cdots .$$

$$(3.15)$$

The ultraviolet divergence at $s = 0$ in the first integral can be removed by the renormalization of the free-photon term $\sim \mathscr{F}_{\mu\nu}\mathscr{F}^{\mu\nu}$ in the Lagrangian. The second term is finite. Performing the integration over s in this term, we finally get

$$\mathcal{L}_{\mathrm{eff}}(\mathcal{A}_\mu) = -\frac{\alpha c_0}{4} \mathscr{F}_{\mu\nu}\mathscr{F}^{\mu\nu} + \frac{\alpha^2}{90m_e^4}(\mathscr{F}_{\mu\nu}\mathscr{F}^{\mu\nu})^2 + \frac{7\alpha^2}{360m_e^4}(\mathscr{F}_{\mu\nu}\tilde{\mathscr{F}}^{\mu\nu})^2 + \cdots ,$$

$$(3.16)$$

where c_0 denotes an ultraviolet-divergent constant, as discussed before. From this equation, we may directly read off the values of c_1, c_2:

$$c_1 = \frac{1}{90}, \qquad c_2 = \frac{7}{360}. \tag{3.17}$$

3.2.3 The Fermion Determinant in a Constant Field

Here, we perform the explicit calculation of the fermion determinant in a constant field. Namely, our final goal will be to derive Eq. (3.14), which was already been used to match the coefficients c_1, c_2 in the Euler–Heisenberg Lagrangian.

Subtracting a constant that does not depend on the field $\mathcal{A}_\mu$, we may define

$$\ln \det(\bar{D}) = \ln \det(D) - \ln \det(i\gamma^\mu \partial_\mu - m_e)$$

$$= \mathrm{Tr} \ln \left((i\slashed{\partial} - e\slashed{A} - m_e)(i\slashed{\partial} - m_e)^{-1} \right), \tag{3.18}$$

where "Tr" denotes the trace both in the coordinate-space and in the space of the Dirac indices. Using $C\gamma_\mu C^{-1} = -\gamma_\mu^T$, where $C = i\gamma^2\gamma^0$, Eq. (3.18) can be rewritten as

$$2\ln \det(\bar{D}) = \mathrm{Tr} \ln \left(((i\slashed{\partial} - e\slashed{A})^2 - m_e^2)((i\slashed{\partial})^2 - m_e^2)^{-1} \right). \tag{3.19}$$

Further, using the relation

$$\ln \frac{\alpha}{\beta} = \int_0^\infty \frac{ds}{s} \left(e^{is(\beta + i\varepsilon)} - e^{is(\alpha + i\varepsilon)} \right), \tag{3.20}$$

Eq. (3.19) can be brought into the form

$$2\ln \det(\bar{D}) = -\int_0^\infty \frac{ds}{s} e^{-is(m_e^2 - i\varepsilon)} \int d^4x \, \mathrm{tr} \left(\langle x| e^{is(i\slashed{\partial} - e\slashed{A})^2} |x\rangle - \langle x| e^{is(i\slashed{\partial})^2} |x\rangle \right), \tag{3.21}$$

where "tr" denotes the trace over the Dirac indices only. In order to further simplify this expression, we use the identity

$$(i\slashed{\partial} - e\slashed{A})^2 = (i\partial_\mu - e\mathscr{A}_\mu)^2 - \frac{e}{2}\sigma_{\mu\nu}\mathscr{F}^{\mu\nu}, \qquad \sigma_{\mu\nu} = \frac{i}{2}[\gamma_\mu, \gamma_\nu]. \qquad (3.22)$$

As we have mentioned, we restrict ourselves to constant electric and magnetic fields:

$$E_3 = \mathscr{F}^{30} = -\mathscr{F}^{03} = a, \qquad B^3 = \mathscr{F}^{12} = -\mathscr{F}^{21} = b. \qquad (3.23)$$

All other entries in the tensor $\mathscr{F}^{\mu\nu}$ are equal to zero. Therefore,

$$-\frac{e}{2}\sigma_{\mu\nu}\mathscr{F}^{\mu\nu} = e\sigma_3 \otimes \begin{pmatrix} b & -ia \\ -ia & b \end{pmatrix}, \qquad (3.24)$$

where σ_3 is a 2×2 Pauli matrix, and

$$\mathrm{tr}\exp\left(-\frac{ise}{2}\sigma_{\mu\nu}\mathscr{F}^{\mu\nu}\right) = 4\cos(ebs)\cosh(eas). \qquad (3.25)$$

Note that for a constant field, the operator $(i\partial_\mu - e\mathscr{A}_\mu)^2$ commutes with $\sigma_{\mu\nu}\mathscr{F}^{\mu\nu}$. Consequently,

$$\ln\det(\bar{D}) = -2\int_0^\infty \frac{ds}{s}\, e^{-is(m_e^2 - i\varepsilon)}\int d^4x \left(\cos(ebs)\cosh(eas)\langle x|e^{is(i\partial_\mu - e\mathscr{A}_\mu)^2}|x\rangle\right.$$

$$\left. - \langle x|e^{is(i\partial_\mu)^2}|x\rangle\right). \qquad (3.26)$$

In order to calculate the matrix elements in Eq. (3.26), we define

$$p_\mu = i\partial_\mu, \qquad [x_\mu, p_\nu] = -ig_{\mu\nu}. \qquad (3.27)$$

Further, the electromagnetic field in our case is

$$\mathscr{A}_0 = \mathscr{A}_2 = 0, \qquad \mathscr{A}_1 = -bx_2, \qquad \mathscr{A}_3 = ax_0. \qquad (3.28)$$

Then,

$$H \doteq (p_\mu - e\mathscr{A}_\mu)^2 = p_0^2 - (p_1 + ebx_2)^2 - p_2^2 - (p_3 - eax_0)^2$$

$$= e^{ip_0p_3/ea}e^{ip_1p_2/eb}(p_0^2 - e^2a^2x_0^2 - p_2^2 - e^2b^2x_2^2)e^{-ip_1p_2/eb}e^{-ip_0p_3/ea}$$

$$= e^{ip_0p_3/ea}(p_0^2 - e^2a^2x_0^2)e^{-ip_0p_3/ea} + e^{ip_1p_2/eb}(-p_2^2 - e^2b^2x_2^2)e^{-ip_1p_2/eb}$$

$$\doteq H_{03} + H_{12}, \qquad (3.29)$$

and

$$\langle x|e^{isH}|x\rangle = \langle x_0x_3|e^{isH_{03}}|x_0x_3\rangle\langle x_1x_2|e^{isH_{12}}|x_1x_2\rangle. \qquad (3.30)$$

In Eq. (3.30) we use the notation $|x\rangle \doteq |x_0x_1x_2x_3\rangle = |x_0\rangle \otimes |x_1\rangle \otimes |x_2\rangle \otimes |x_3\rangle$. Calculating the matrix elements in Eq. (3.30) separately, for the first element we get

$$\langle x_0 x_3 | e^{isH_{03}} | x_0 x_3 \rangle = \int \frac{dp_0 dp_3 dp'_0 dp'_3 dq_0 dq_3 dq'_0 dq'_3}{(2\pi)^8} e^{ix_0(p_0 - p'_0) + ix_3(p_3 - p'_3)}$$

$$\times \langle p_0 p_3 | e^{ip_0 p_3/ea} | q_0 q_3 \rangle \langle q_0 q_3 | e^{is(p_0^2 - e^2 a^2 q_0^2)} | q'_0 q'_3 \rangle$$

$$\times \langle q'_0 q'_3 | e^{-ip_0 p_3/ea} | p'_0 p'_3 \rangle. \tag{3.31}$$

Using the relations

$$\langle p_0 p_3 | e^{\pm ip_0 p_3/ea} | q_0 q_3 \rangle = e^{\pm ip_0 p_3/ea} (2\pi)^2 \delta(p_0 - q_0) \delta(p_3 - q_3),$$

$$\langle q_0 q_3 | e^{is(p_0^2 - e^2 a^2 q_0^2)} | q'_0 q'_3 \rangle = (2\pi) \delta(q_3 - q'_3) \langle q_0 | e^{is(p_0^2 - e^2 a^2 q_0^2)} | q'_0 \rangle, \tag{3.32}$$

we obtain

$$\langle x_0 x_3 | e^{isH_{03}} | x_0 x_3 \rangle = \frac{ea}{4\pi^2} \int dp_0 \langle p_0 | e^{is(p_0^2 - e^2 a^2 x_0^2)} | p_0 \rangle. \tag{3.33}$$

In order to calculate the matrix element in Eq. (3.33), we consider the quantum-mechanical problem of a harmonic oscillator given by the Hamiltonian

$$h_{\text{osc}} = \frac{1}{2} p_0^2 + \frac{\omega_0^2}{2} x_0^2. \tag{3.34}$$

The eigenfunctions of this Hamiltonian are labeled by an integer $n = 0, 1, \cdots$:

$$h_{\text{osc}} |n\rangle = \omega_0 \left(n + \frac{1}{2} \right) |n\rangle. \tag{3.35}$$

Consider now the matrix element

$$\frac{ea}{4\pi^2} \int dp_0 \langle p_0 | e^{2ish_{\text{osc}}} | p_0 \rangle = \frac{ea}{4\pi^2} \sum_{n=0}^{\infty} \int dp_0 |\langle p_0 | n \rangle|^2 \exp \left\{ 2is\omega_0 \left(n + \frac{1}{2} \right) \right\}$$

$$= \frac{ea}{2\pi} \sum_{n=0}^{\infty} \exp \left\{ 2is\omega_0 \left(n + \frac{1}{2} \right) \right\}. \tag{3.36}$$

In order to recover the original matrix element in Eq. (3.33), we have to substitute $\omega_0 \to iea$. Carrying out the summation over n, we finally arrive at the following result:

$$\langle x_0 x_3 | e^{isH_{03}} | x_0 x_3 \rangle = \frac{ea}{4\pi \sinh(eas)}. \tag{3.37}$$

Evaluating the second matrix element in Eq. (3.30) with the same method, we obtain

$$\langle x_1 x_2 | e^{isH_{12}} | x_1 x_2 \rangle = \frac{eb}{4\pi i \sin(ebs)}. \tag{3.38}$$

Finally, substituting Eqs. (3.37) and (3.38) into Eqs. (3.30) and (3.26), we arrive at Eq. (3.14), which was used for matching the couplings of the Euler–Heisenberg Lagrangian.

3.3 Interaction of Long-Wavelength Photons with Atoms

In this section, we shall briefly consider another example with very long-wavelength photons. In particular, we wish to address the scattering of such photons by atoms. This process goes under the name of the Rayleigh scattering. In the following, we mainly follow the reasoning outlined in Ref. [7].

Since the momentum transfer in the process is very small, the atoms can be described nonrelativistically. The pertinent free Lagrangian is given by

$$\mathcal{L}_{\text{atom}} = \Psi^\dagger \left(i\partial_t - M_{\text{atom}} + \frac{\nabla^2}{2M_{\text{atom}}} \right) \Psi, \qquad (3.39)$$

where Ψ denotes the nonrelativistic field

$$\Psi(\mathbf{x},t) = \int \frac{d^3\mathbf{k}}{(2\pi)^3} e^{-ik_0 t + i\mathbf{k}\mathbf{x}} a(\mathbf{k}). \qquad (3.40)$$

Here, $a(\mathbf{k})$ denotes the annihilation operator of the atom.

The interaction Lagrangian of an atom with photons must be gauge-invariant. Since atoms are electrically neutral, the only gauge-invariant objects from which this Lagrangian can be constructed are the atomic fields, along with the electric and magnetic fields $\mathbf{E}$ and $\mathbf{B}$ corresponding to the photon. To the lowest order, the interaction Lagrangian can be written as

$$\mathcal{L}_{\text{int}} = a_0^3 \Psi^\dagger (d_1 \mathbf{E}^2 + d_2 \mathbf{B}^2) \Psi, \qquad (3.41)$$

where, from dimensional counting, a_0 has the dimension of length, and d_1, d_2 are dimensionless. a_0 is the typical size of the atom, which is the only scale in the problem. We further assume that d_1, d_2 are of *natural size*, $d_1, d_2 = O(1)$, so that the operator in Eq. (3.41) is indeed leading. A crucial fact that enables us to establish the counting rules is gauge invariance. Because of gauge invariance, the effective Lagrangian can depend on $\mathbf{E}$ and $\mathbf{B}$, which contain one derivative each, but not on the vector potential $\mathscr{A}_\mu$, which contains no derivative.

According to Eq. (3.41), the scattering amplitude $A_{\gamma-\text{atom}} \sim a_0^3 E^2$, where E is the energy of the photon. The scattering cross section is given by

$$\sigma_{\gamma-\text{atom}} = \text{phase space} \times |A|^2 \sim \text{phase space} \times a_0^6 E^4. \qquad (3.42)$$

From this formula it follows that the phase space factor should be dimensionless and thus energy-independent. Consequently, the cross section for Rayleigh scattering is proportional to the fourth power of the photon energy. This in particular explains why the sky is blue: The blue light is scattered more intensively by the atoms in the atmosphere than is the red light.

3.4 QCD Factsheet

3.4.1 The Lagrangian

After QED, where the perturbative matching to the effective theory could be performed, we turn to a much more challenging endeavor of constructing the effective field theory of QCD.[2] In the beginning, we collect few very well-known facts about QCD, which will be essential for understanding the following material. For more details, the reader is referred to any textbook on QCD such as [9].

QCD is the fundamental theory of the strong interactions. A huge variety of phenomena in particle and nuclear physics are described by the Lagrangian of QCD that fits into a single line,

$$\mathcal{L}_{\text{QCD}} = -\frac{1}{4g^2} F^a_{\mu\nu} F^{\mu\nu a} + \bar{\psi}(i\gamma^\mu D_\mu - \mathcal{M})\psi, \tag{3.43}$$

in terms of the field strength tensor $F^a_{\mu\nu}$ and the gauge-covariant derivative D_μ,

$$F^a_{\mu\nu} = \partial_\mu G^a_\nu - \partial_\nu G^a_\mu + f^{abc} G^b_\mu G^c_\nu, \qquad D_\mu = \partial_\mu \mathbb{1} - \frac{i}{2}\lambda^a G^a_\mu. \tag{3.44}$$

Further, g is the strong coupling constant, and $\mathcal{M}$ denotes the quark mass matrix

$$\mathcal{M} = \text{diag}(m_u, m_d, m_s, m_c, m_b, m_t). \tag{3.45}$$

In the preceding equations, ψ and G^a_μ denote the quark and gluon fields, respectively. Further, the Dirac-spinor ψ is a column in the flavor space, with the components describing u, d, s, c, b, t quarks, and transforms as a vector of the fundamental representation of the color $SU(3)$ group. The quantities λ^a, $a = 1, \cdots, 8$, denote the 3×3 Gell–Mann matrices, which are the generators of the $SU(3)$ color group and obey the following commutation relations:

$$\left[\frac{\lambda^a}{2}, \frac{\lambda^b}{2}\right] = i f^{abc} \frac{\lambda^c}{2}. \tag{3.46}$$

Here, f^{abc} are the structure constants of the $SU(3)$ group; their values are given in the Appendix.

It is useful to define the matrix-valued fields:

$$G_\mu = -\frac{i}{2}\lambda^a G^a_\mu, \quad F_{\mu\nu} = -\frac{i}{2}\lambda^a F^a_{\mu\nu} = \partial_\mu G_\nu - \partial_\nu G_\mu + [G_\mu, G_\nu]. \tag{3.47}$$

The Lagrangian of QCD is invariant under local $SU(3)$ color gauge group transformations

$$\psi'(x) = \Lambda(x)\psi(x), \quad \bar{\psi}'(x) = \bar{\psi}(x)\Lambda(x)^\dagger,$$

---

[2] In fact, there is not only one EFT of QCD. Here, we concentrate on the light quark sector. For systems with heavy quarks, see, e.g., the monograph [8]. More references for the heavy and heavy-light sectors of QCD are given in Chapter 4.

$$G'_\mu(x) = \Lambda(x)G_\mu(x)\Lambda(x)^\dagger - \partial_\mu\Lambda(x)\Lambda(x)^\dagger, \qquad (3.48)$$

where $\Lambda(x) \in SU(3)$. Under these transformations, the field strength tensor transforms covariantly:

$$F'_{\mu\nu}(x) = \Lambda(x)F_{\mu\nu}(x)\Lambda(x)^\dagger. \qquad (3.49)$$

Note that the color transformations are diagonal in the flavor space; color and flavor transformations do not mix. The coupling constant g is flavor-independent. Rescaling the gluon field $G_\mu \to gG_\mu$, it is possible to normalize the kinetic term of G_μ to unity. It is then seen that the universal coupling g describes the quark–gluon interactions, as well as the triple and quartic gluon interactions.

QCD is a gauge theory. In order to carry out the perturbative quantization of such a theory, gauge fixing is necessary. It means that the QCD Lagrangian given in Eq. (3.43) has to be supplemented by a gauge-fixing term and, eventually, by a ghost Lagrangian (in the non-ghost-free gauges, e.g., in the covariant gauge). These (standard) issues, however, will be inessential for the discussions carried out in what follows.

Note that the Lagrangian of QCD contains, in addition, the so-called θ-term,

$$\mathcal{L}_\theta = -\frac{\theta}{32\pi^2} \varepsilon^{\mu\nu\alpha\beta} F^a_{\mu\nu}F^a_{\alpha\beta}, \qquad (3.50)$$

which leads to *CP* violation in QCD. This violation is, however, extremely small (i.e., the effective value of the θ-parameter is very small). The necessity to explain why the violation should be *naturally* small constitutes the essence of the so-called *strong CP problem*. For the moment, we shall not consider the θ-term and return to this problem in Section 4.20.

3.4.2 Renormalization of QCD

The fundamental parameters of QCD are the coupling constant g and the quark masses. All physical observables should be expressed in terms of these parameters. In order to make the observables finite, the bare parameters that enter the Lagrangian should be renormalized. QCD is a renormalizable theory in a conventional sense. This means that it suffices to renormalize a finite number of parameters in the Lagrangian, as well as the fields, in order to make all physical observables finite.

The renormalized coupling constant in QCD obeys the renormalization group (RG) equation, which describes the behavior of the coupling with respect to the variation of the renormalization scale μ (for convenience, we use the $\overline{\text{MS}}$ renormalization scheme throughout this section):

$$\mu\frac{dg_r(\mu)}{d\mu} = \beta(g_r(\mu)), \qquad (3.51)$$

where the perturbative expansion of the β-function starts with a term with a *negative* coefficient:

$$\beta(g) = -\beta_0\frac{g^3}{16\pi^2} - \beta_1\frac{g^5}{(16\pi^2)^2} + O(g^7),$$

$$\beta_0 = \frac{11}{3} N_c - \frac{2}{3} N_f , \quad \beta_1 = \frac{34}{3} N_c^2 - \frac{10}{3} N_c N_f - \frac{(N_c^2 - 1)N_f}{N_c} . \tag{3.52}$$

Here, $N_c = 3$, and N_f are the number of colors and number of flavors, respectively. One sees that, if N_f is sufficiently small, then $\beta_0 > 0$ and the renormalized coupling constant $g_r(\mu)$ *decreases* as $\mu \to \infty$. This leads to the so-called *asymptotic freedom* of the theory and is encountered in *non-Abelian* gauge theories (QED does not possess the property of asymptotic freedom). At small energies, the coupling constant of QCD grows and the theory becomes non-perturbative. The quarks and gluons, which are the fundamental degrees of freedom in QCD, get confined by the strong force, and only colorless bound states are observed in experiments. A typical mass scale of such states is of the order of the nucleon mass, that is, around 1 GeV.

The scale of QCD *to all orders in the strong coupling constant* is defined by

$$\Lambda_{\mathrm{QCD}} = \mu \left(\frac{\beta_0 g_r^2}{16\pi^2} \right)^{-\frac{\beta_1}{2\beta_0^2}} \exp \left(-\frac{16\pi^2}{2\beta_0 g_r^2} \right)$$

$$\times \exp \left(-\int_0^{g_r} dg' \left\{ \frac{1}{\beta(g')} + \frac{16\pi^2}{\beta_0 g'^3} - \frac{\beta_1}{\beta_0^2 g'} \right\} \right) . \tag{3.53}$$

Note that the subtraction in Eq. (3.53) in the integral regularizes the integrand at $g' = 0$, where $\beta(g')$ has a cubic singularity. The other factors, which enter the expression in Eq. (3.53), are chosen so that the quantity Λ_{QCD} is RG-invariant to all orders in perturbation theory:

$$\mu \frac{d\Lambda_{\mathrm{QCD}}}{d\mu} = 0 . \tag{3.54}$$

Typically, the quantity Λ_{QCD} takes values of the order of a couple hundred of MeV, depending on the choice of the renormalization prescription and number of active quark flavors. For a more precise determination of α_S and Λ_{QCD}, see [10].

Note that here we deal with the phenomenon of *dimensional transmutation*: The dimensionless scale-dependent coupling $g_r(\mu)$ is traded for the scale-dependent dimensionful quantity Λ_{QCD}. In order to understand the meaning of this phenomenon, let us for a moment set all quark masses to zero. Since the theory in this case contains a single dimensionless parameter, the coupling constant g, one might come to the conclusion that its spectrum cannot contain massive particles. This conclusion would be, however, premature. As we have seen, the necessity to renormalize the coupling constant g leads to the dynamical generation of the scale Λ_{QCD} and hence a finite mass gap in the theory. In other words, even the theory with massless quarks can accommodate massive asymptotic states, with masses proportional to a single parameter Λ_{QCD} and mass ratios that do not depend on the strong coupling constant. The situation does not change much when (light) quark masses are turned on. The bulk of the hadron masses is still determined by the energy of the gluon field inside hadrons,[3] which is characterized by the parameter Λ_{QCD}.

In nature, the range of the quark mass values varies from a few MeV (u and d quarks) to $\simeq 100$ MeV (s-quark) and higher. (According to the commonly used convention, we fix the scale $\mu = 2$ GeV.) Since the quarks of higher masses (c, b, t) cannot play any role in low-energy physics we are primarily interested in, in what follows we shall concentrate on three light flavors only.

The renormalized quark masses obey RG equations,

$$\mu \frac{dm_r^{(f)}(\mu)}{d\mu} = -\gamma_m(g_r(\mu))m_r^{(f)}(\mu), \tag{3.55}$$

where f is the quark flavor, and the γ-function in the $\overline{MS}$ scheme has the following perturbative expansion:

$$\gamma_m(g) = \gamma_0 \frac{g^2}{16\pi^2} + \gamma_1 \frac{g^4}{(16\pi^2)^2} + O(g^6),$$

$$\gamma_0 = \frac{3(N_c^2 - 1)}{N_c}, \quad \gamma_1 = 3\left(\frac{N_c^2 - 1}{2N_c}\right)^2 + \frac{97}{6}(N_c^2 - 1) - \frac{5N_c(N_c^2 - 1)}{3N_c}. \tag{3.56}$$

Note that the γ_m-function does not depend on the quark flavor that reflects the fact that the strong interactions are flavor-blind. Consequently, the renormalized quark masses run with the scale μ, but the ratios $m_r^{(f)}(\mu)/m_r^{(f')}(\mu)$ are scale-independent in QCD.

To summarize this section, dimensional transmutation in QCD introduces the RG-invariant scale Λ_{QCD} in QCD, which is a substitute of the scale-dependent, dimensionful coupling constant $g_r(\mu)$ and is determined by experimental data. The masses of all hadrons, which do not vanish in the chiral limit (i.e., when quark masses are set to zero), are proportional to Λ_{QCD}, with coefficients that can depend on the number of colors and active flavors, but not on the QCD coupling constant anymore. A typical size of the hadronic scale (around 1 GeV) is given by the mass of the lightest stable hadron, which does not vanish in the chiral limit – the nucleon. Only the Goldstone bosons in the pseudoscalar octet are much lighter due to the (approximate) chiral symmetry of QCD. QCD as a field theory becomes nonlocal below the hadronic scale ~ 1 GeV. Quarks and gluons are not the appropriate degrees of freedom anymore, paving the way for the description in terms of the hadronic fields. Since the pseudoscalar mesons, which are Goldstone bosons of the spontaneously broken chiral symmetry, are the only hadrons whose masses are far below the hadronic scale, the effective theory of QCD at low energies is primarily the theory of these Goldstone bosons.

---

[3] This statement is valid for all hadrons except the light pseudoscalar octet mesons (pions, kaons and η-meson), which are the Goldstone bosons of the spontaneously broken chiral symmetry and whose masses *vanish*, when the quark masses tend to zero. (See more on this issue below.)

3.5 Chiral Symmetry in QCD

If the quark masses are vanishing, the Lagrangian of QCD is invariant under a symmetry group that mixes different quark flavors. Since the Lagrangian of gluons is explicitly invariant under flavor transformations, it suffices to consider the fermionic part of the Lagrangian only:

$$\mathcal{L}_F = \bar{\psi}(i\gamma^\mu D_\mu - \mathcal{M})\psi. \tag{3.57}$$

Further, we shall explicitly consider three light flavors only, $N_f = 3$,

$$\mathcal{M} = \mathrm{diag}\,(m_u, m_d, m_s), \qquad \psi(x) = \begin{pmatrix} u(x) \\ d(x) \\ s(x) \end{pmatrix}. \tag{3.58}$$

At low energies, the heavy quark degrees of freedom decouple from the theory and can be neglected. Furthermore, the masses of three light quarks are much smaller than the typical hadronic scale (given, e.g., by the nucleon mass). For this reason, in a first approximation, one could assume that these quarks are massless. If the masses are turned on, various hadronic characteristics change. It is further assumed that the corrections caused by finite quark masses can be systematically calculated in perturbation theory. In what follows, we shall consider in detail how this approach works in practice.

Let us decompose the quark field $\psi(x)$ in the Lagrangian into its left- and right-handed components,

$$\psi_L(x) = \frac{1}{2}(1 - \gamma_5)\psi(x), \quad \psi_R(x) = \frac{1}{2}(1 + \gamma_5)\psi(x),$$

$$\psi(x) = \psi_L(x) + \psi_R(x). \tag{3.59}$$

We leave it as an exercise to the reader to show that the $P_{L/R} = (1 \mp \gamma_5)/2$ are indeed projectors and that they project out the left- and right-handed parts of the fermion fields. The fermionic part of the Lagrangian in terms of the left- and right-handed fields is rewritten as

$$\mathcal{L}_F = \bar{\psi}_L(i\gamma^\mu D_\mu)\psi_L + \bar{\psi}_R(i\gamma^\mu D_\mu)\psi_R - \bar{\psi}_L\mathcal{M}\psi_R - \bar{\psi}_R\mathcal{M}\psi_L. \tag{3.60}$$

From this expression it is immediately clear that, if the quarks are massless, $\mathcal{M} = 0$, then $\mathcal{L}_F$ is invariant under the *global* group $U(N_f)_L \times U(N_f)_R$:

$$\psi_L(x) \mapsto g_L\psi_L(x), \ \psi_R(x) \mapsto g_R\psi_R(x), \ g_L \in U(N_f)_L, \ g_R \in U(N_f)_R. \tag{3.61}$$

Note again that the transformations of the group $U(N_f)_L \times U(N_f)_R$ do not touch the gluon field, which is flavor blind. Thus, the gluonic (and ghost) parts of the Lagrangian are trivially invariant under the above transformations.

Classically, the symmetry of the Lagrangian results in the conservation of the left- and right-hand currents. The non-singlet Noether currents are defined as

$$L_\mu^i = \bar{\psi}_L\gamma_\mu T^i\psi_L = \bar{\psi}\gamma_\mu\frac{1}{2}(1 - \gamma_5)T^i\psi,$$

$$R^i_\mu = \bar\psi_R \gamma_\mu T^i \psi_R = \bar\psi \gamma_\mu \frac{1}{2}(1+\gamma_5)T^i\psi, \qquad (3.62)$$

where T^i, $i = 1, \ldots N_f^2 - 1$, are the generators of the $SU(N_f)$ group, normalized as $\mathrm{tr}\,(T^i T^j) = \frac{1}{2}\delta^{ij}$. Namely, for $N_f = 3$, the generators are given by $T^i = \frac{1}{2}\lambda^i$, $i = 1, \ldots, 8$, where λ^i denote Gell–Mann matrices. The generators for $N_f = 2$ are given by $T^i = \frac{1}{2}\tau^i$, $i = 1, 2, 3$, where τ^i are Pauli matrices.

The singlet currents take the form

$$L^0_\mu = \bar\psi \gamma_\mu \frac{1}{2}(1-\gamma_5)\psi, \quad R^0_\mu = \bar\psi \gamma_\mu \frac{1}{2}(1+\gamma_5)\psi. \qquad (3.63)$$

If $\mathcal{M} = 0$, then, *at the classical level*, the Noether currents are conserved: $\partial^\mu L^i_\mu = \partial^\mu R^i_\mu = \partial^\mu L^0_\mu = \partial^\mu R^0_\mu = 0$. We shall see below that some of these relations will be modified by anomalies.

The vector- and axial-vector currents are defined by

$$V^i_\mu = R^i_\mu + L^i_\mu, \qquad A^i_\mu = R^i_\mu - L^i_\mu,$$
$$V^0_\mu = R^0_\mu + L^0_\mu, \qquad A^0_\mu = R^0_\mu - L^0_\mu. \qquad (3.64)$$

If $\mathcal{M} = 0$, these currents are conserved along with the left- and right-hand currents.

Assume now that the quark masses are nonzero. Calculating the derivative

$$\partial^\mu V^i_\mu = \bar\psi T^i \gamma_\mu \partial^\mu \psi + (\partial^\mu \bar\psi)\gamma^\mu T^i \psi, \qquad (3.65)$$

and using the equations of motion

$$i\gamma^\mu \left(\partial_\mu - \frac{i}{2}\lambda^a G^a_\mu\right)\psi - \mathcal{M}\psi = 0, \qquad (3.66)$$

we may verify that

$$\partial^\mu V^i_\mu = -i\bar\psi\,[T^i, \mathcal{M}]\,\psi. \qquad (3.67)$$

Analogously,

$$\partial^\mu A^i_\mu = i\bar\psi\,\{T^i, \mathcal{M}\}\gamma_5\psi. \qquad (3.68)$$

This means that the non-singlet vector current is broken by the *differences in the quark masses*, it stays conserved, if $m_u = m_d = m_s \neq 0$. On the contrary, the axial-vector current is violated by the *quark masses themselves*; if the quark masses are nonzero, then it is not conserved.

An important remark is in order. To calculate the divergences of the currents, one has used the equations of motion. Formally, these are exact equations, which are valid beyond perturbation theory. Hence, one would expect that the Noether currents are conserved. However, these currents are composite operators (quark field bilinears taken at the same space-time point). In field theory, such products of field operators are, in general, ultraviolet-singular objects, and calculations should be done with care. For example, one could regularize composite operators by slightly shifting the arguments of different elementary fields apart (this procedure goes under the name of point-split technique), or taming the ultraviolet divergences otherwise. At the end of the day, it turns out that the divergence of the singlet axial-vector current cannot be made vanishing even in the limit $\mathcal{M} \to 0$. The correct result is given by

$$\partial^\mu A_\mu^0 = 2i\bar{\psi}\mathcal{M}\gamma_5\psi + \frac{N_f}{16\pi^2}\varepsilon_{\mu\nu\alpha\beta}\mathrm{tr}_c(F^{\mu\nu}F^{\alpha\beta}),\tag{3.69}$$

where tr_c denotes the trace over color indices. The last term in Eq. (3.69) goes under the name of "axial anomaly." This equation will be derived in section 3.9 in detail.

Finally, note that the singlet vector current V_μ^0 is always conserved. It corresponds to the conservation of the number of quarks minus the number of the antiquarks. Thus, the zeroth component of this current is proportional to the baryon number density.

3.6 Ward Identities

The quarks in nature have masses. The vector and axial-vector symmetries, which we have considered, are *approximate* symmetries broken by the quark masses. In what follows we shall discuss how the formalism of Ward identities, which was designed to treat *exact* symmetries, can be applied in this case.

The Green's functions that we are going to consider in the following are the Green's functions of color-neutral composite operators, namely the quark bilinears of the type $\bar{\psi}(x)\Gamma\psi(x)$, where the matrix Γ collects all Lorentz and flavor indices. For example, the composite operator $P^i(x) = \bar{\psi}(x)i\gamma_5\lambda^i\psi(x)$, $i = 1, 2, 3$, carries the quantum numbers of the π-meson, the operator $V_\mu^i(x) = \bar{\psi}(x)\gamma_\mu\frac{1}{2}\lambda^i\psi(x)$, $i = 1, 2, 3$, those of the ρ-meson and so on.[4] In order to study physical process such as pion–pion scattering, one should evaluate the four-point Green's function of the operators $P^i(x)$ and extract the pion–pion scattering amplitude from this Green's function by using the standard Lehmann–Symanzik–Zimmermann (LSZ) reduction formulae, pretty much the same as in case of the elementary pion fields.[5]

The Green's function of the composite operators can be obtained by equipping the QCD Lagrangian by pertinent external sources. We shall therefore amend the fermionic part of the QCD Lagrangian by external classical scalar (s), pseudoscalar (p), vector (v_μ) and axial-vector (a_μ) sources,

$$\mathcal{L}_F(s, p, v, a) = \bar{\psi}(\gamma^\mu(iD_\mu + v_\mu + \gamma^5 a_\mu) - s + i\gamma^5 p)\psi,\tag{3.70}$$

---

[4] Here, we would like to emphasize that the choice of the bilinear operators does not necessarily have to do with the statement whether the mesons in question indeed *consist* of a quark and an antiquark (whatever the exact meaning of this notion might be within field theory). In fact, we are making a weaker statement that the bilinear operators, written down here, have the same quantum numbers as the pertinent mesons and, therefore, can be used as interpolating meson fields.

[5] For the reformulation of the standard LSZ approach in the case of the composite particles see, e.g., Ref. [11] and references therein. We shall consider this issue in detail later in Chapter 4.

where D^μ denotes the covariant derivative containing the gluon field. The external sources are matrices in the flavor space. The following normalization is chosen for the external fields:[6]

$$s = s^0 \mathbb{1} + s^i \lambda^i, \quad p = p^0 \mathbb{1} + p^i \lambda^i,$$

$$v_\mu = v_\mu^0 \mathbb{1} + \frac{1}{2} v_\mu^i \lambda^i, \quad a_\mu = a_\mu^0 \mathbb{1} + \frac{1}{2} a_\mu^i \lambda^i. \tag{3.71}$$

The generating functional now depends on s, p, v, a:

$$Z(s,p,v,a) = \int d\psi \, d\bar{\psi} \, dG_\mu \, \exp\left\{ i \int d^4x (\mathcal{L}_G + \mathcal{L}_F(s,p,v,a)) \right\}. \tag{3.72}$$

Differentiating this generating functional with respect to the s, p, v, a external sources, and letting them tend to zero at the end, we obtain the Green's functions of the pertinent *quark bilinears* in *massless* QCD. For example, as seen from Eq. (3.70), differentiating with respect to $p^i(x)$ gives the pseudoscalar composite operator $\bar{\psi}(x) i \gamma^5 \lambda^i \psi(x)$, differentiating with respect to $v_\mu^0(x)$ gives the singlet vector current $\bar{\psi}(x) \gamma^\mu \psi(x)$, and so on.

Equation (3.70) does not contain the quark mass term, because the latter would explicitly break the chiral symmetry. In order to obtain the Green's functions with massive quarks, we should expand the generating functional $Z(s,p,v,a)$ not at the point $s = p = v = a = 0$, but at $s = \mathcal{M}$, $p = v = a = 0$. In other words, the trick is that we may work with the generating functional in *massless* QCD, obtain *exact* Ward identities and, at the end, by choosing a different expansion point, obtain the Ward identities for the Green's functions in the *massive* case as well. This is an example of the so-called "spurion technique."

Next, we derive the Ward identities for the generating functional Z. We start by decomposing the fermion field into the left- and right-handed components. Thus, the fermionic part of the Lagrangian in Eq. (3.70) can be rewritten as

$$\mathcal{L}_F = \bar{\psi}_L \gamma^\mu (iD_\mu + v_\mu - a_\mu) \psi_L + \bar{\psi}_R \gamma^\mu (iD_\mu + v_\mu + a_\mu) \psi_R$$

$$- \bar{\psi}_R (s + ip) \psi_L - \bar{\psi}_L (s - ip) \psi_R. \tag{3.73}$$

Defining $r_\mu = v_\mu + a_\mu$ and $l_\mu = v_\mu - a_\mu$, we may check that the fermionic Lagrangian is invariant under the following *local* transformations:

$$\psi_L(x) \mapsto g_L(x) \psi_L(x),$$

$$\psi_R(x) \mapsto g_R(x) \psi_R(x),$$

$$l_\mu(x) \mapsto g_L(x) l_\mu(x) g_L(x)^\dagger + i g_L(x) \partial_\mu g_L(x)^\dagger,$$

$$r_\mu(x) \mapsto g_R(x) r_\mu(x) g_R(x)^\dagger + i g_R(x) \partial_\mu g_R(x)^\dagger,$$

$$(s(x) + ip(x)) \mapsto g_R(x)(s(x) + ip(x)) g_L(x)^\dagger,$$

---

[6] Here we write down the expressions for the $N_f = 3$ case. The two-flavor case can be treated in a similar fashion.

$$(s(x) - ip(x)) \mapsto g_L(x)(s(x) - ip(x))g_R(x)^\dagger, \tag{3.74}$$

where $g_L(x), g_R(x) \in U(N_f)_{L,R}$.

Consider now infinitesimal transformations

$$g_L(x) = \mathbb{1} + i\alpha(x) - i\beta(x), \quad g_R(x) = \mathbb{1} + i\alpha(x) + i\beta(x), \tag{3.75}$$

where

$$\alpha(x) = \alpha^0(x)\mathbb{1} + \alpha^i(x)T^i, \quad \beta(x) = \beta^0(x)\mathbb{1} + \beta^i(x)T^i. \tag{3.76}$$

The external sources transform according to

$$\delta v_\mu = \partial_\mu \alpha + i[\alpha, v_\mu] + i[\beta, a_\mu],$$

$$\delta a_\mu = \partial_\mu \beta + i[\alpha, a_\mu] + i[\beta, v_\mu],$$

$$\delta s = i[\alpha, s] - \{\beta, p\},$$

$$\delta p = i[\alpha, p] + \{\beta, s\}. \tag{3.77}$$

In order to study the invariance of the generating functional, we should know how the path integration measure transforms under the transformations from $U(N_f)_L \times U(N_f)_R$. Since these transformations do not affect the gluon field, we are left with the fermionic measure only. Let us first assume that the fermionic measure $d\psi d\bar{\psi}$ is invariant under all transformations from $U(N_f)_L \times U(N_f)_R$. This statement is not always true in the presence of the vector and axial-vector fields (both the dynamical gluon field as well as external fields v_μ, a_μ). The non-invariance of the measure gives rise to the so-called anomalies, which will be discussed at length in what follows. If we simply neglect the anomalies, the following statement about the invariance of the generating functional can be made:

$$Z(s + \delta s, p + \delta p, v + \delta v, a + \delta a) = Z(s, p, v, a). \tag{3.78}$$

Expanding with respect to the infinitesimal transformation parameters to first order, we obtain

$$\frac{\delta Z}{\delta \alpha^0}\bigg|_{\alpha=\beta=0} = \frac{\delta Z}{\delta \beta^0}\bigg|_{\alpha=\beta=0} = \frac{\delta Z}{\delta \alpha^i}\bigg|_{\alpha=\beta=0} = \frac{\delta Z}{\delta \beta^i}\bigg|_{\alpha=\beta=0} = 0. \tag{3.79}$$

Differentiating with respect to these parameters, we obtain the Ward identities,

$$0 = -\partial_\mu \left(\frac{\delta Z}{\delta v_\mu^0(x)} \right),$$

$$0 = -\partial_\mu \left(\frac{\delta Z}{\delta a_\mu^0(x)} \right) - 2p^0(x)\frac{\delta Z}{\delta s^0(x)} - 2p^i(x)\frac{\delta Z}{\delta s^i(x)} + 2s^0(x)\frac{\delta Z}{\delta p^0(x)}$$

$$+ 2s^i(x)\frac{\delta Z}{\delta p^i(x)},$$

$$0 = -\partial_\mu \left(\frac{\delta Z}{\delta v_\mu^i(x)} \right) - f^{ijk}\left(v_\mu^j(x)\frac{\delta Z}{\delta v_\mu^k(x)} + a_\mu^j(x)\frac{\delta Z}{\delta a_\mu^k(x)} + s^j(x)\frac{\delta Z}{\delta s^k(x)} \right.$$

$$+ p^j(x) \frac{\delta Z}{\delta p^k(x)} \Bigg),$$

$$0 = -\partial_\mu \left(\frac{\delta Z}{\delta a_\mu^i(x)} \right) - f^{ijk} \left(a_\mu^j(x) \frac{\delta Z}{\delta v_\mu^k(x)} + v_\mu^j(x) \frac{\delta Z}{\delta a_\mu^k(x)} \right)$$

$$- 2 \mathrm{tr}\, (T^i \{T^j, T^k\}) \left(p^j(x) \frac{\delta Z}{\delta s^k(x)} - s^j(x) \frac{\delta Z}{\delta p^k(x)} \right)$$

$$- p_i(x) \frac{\delta Z}{\delta s^0(x)} - p^0(x) \frac{\delta Z}{\delta s^i(x)} + s_i(x) \frac{\delta Z}{\delta p^0(x)} + s^0(x) \frac{\delta Z}{\delta p^i(x)}, \tag{3.80}$$

where the f^{ijk} denote the structure constants of the $U(N_f)$ group. Differentiating these functional identities any number of times with respect to the external sources, and setting at the end $s = \mathcal{M}$, $p = v = a = 0$, we get an infinite tower of identities, which relate different Green's functions of the composite quark bilinears in QCD. This infinite number of equations are a direct consequence of the QCD symmetries.

In order to illustrate the physical meaning of the Ward identities, let us consider the first of the identities in Eq. (3.80). Differentiating it with respect to the external sources, we get

$$- \frac{1}{i^{n+1}} \frac{\partial}{\partial x_\mu} \frac{\delta^{n+1} Z(s, p, v, a)}{\delta v_\mu^0(x) \delta j_1(y_1) \cdots \delta j_n(y_n)} \Bigg|_{s = \mathcal{M},\, p = v = a = 0} = 0, \tag{3.81}$$

where $j_m(y_m)$ denotes one of the external sources s, p, v_μ, a_μ. Equivalently,

$$- \frac{\partial}{\partial x_\mu} \langle 0 | T V_\mu^0(x) J_1(y_1) \cdots J_n(y_n) | 0 \rangle = 0, \tag{3.82}$$

where $J_m(y_m)$ denote pertinent quark bilinears. Acting now with the derivative on the T-product, we get

$$0 = - \langle 0 | T (\partial^\mu V_\mu^0(x)) J_1(y_1) \cdots J_n(y_n) | 0 \rangle - \sum_{m=1}^{n} \langle 0 | T \delta(x^0 - y_m^0) [V_0^0(x), J_m(y_m)]$$

$$\times J_1(y_1) \cdots J_{m-1}(y_{m-1}) J_{m+1}(y_{m+1}) \cdots J_n(y_n) | 0 \rangle. \tag{3.83}$$

If it is assumed that we can use canonical commutation relations to calculate the commutators of currents in the above expression (i.e., no Schwinger terms are present [12]), it is straightforward to check that the equal-time commutators of $V_0^0(x)$ with any type of quark bilinears (scalar, pseudo-scalar, vector and axial-vector) vanish. Then, the Ward identity reduces to the statement that an insertion of the divergence of the bilinear quark operator $V_\mu^0(x)$ into all Green's functions of these quark bilinears vanishes. Note that $\partial^\mu V_\mu^0(x) = 0$ corresponds to the conservation of the baryon current.

As the next example, let us consider the T-product of one vector and two axial-vector currents. To this end, we shall differentiate twice the third identity in Eq. (3.80) with respect to the external field a_μ and put all external fields to zero at the end, except for $s \to \mathcal{M}$. The result reads

$$\frac{\partial}{\partial x_\mu} \langle 0|T V_\mu^i(x) A_\nu^j(y) A_\lambda^k(z)|0\rangle = i f^{ijl} \delta^{(4)}(x-y)\langle 0|T A_\nu^l(y) A_\lambda^k(z)|0\rangle$$

$$+ i f^{ikl} \delta^{(4)}(x-z)\langle 0|T A_\nu^j(y) A_\lambda^l(z)|0\rangle + f^{ijk} \bar{s}^j \langle 0|T S^k(x) A_\nu^j(y) A_\lambda^l(z)|0\rangle, \qquad (3.84)$$

where $S^k(x) = \bar{\psi}(x)\lambda^k \psi(x)$ is the non-singlet scalar density and $\bar{s}^j = \mathrm{tr}(\mathcal{M}T^j)$.

Moving now the operator ∂_μ inside the T-product and making use of the canonical commutation relations that imply

$$\delta(x_0 - y_0)[V_0^i(x), A_\nu^j(y)] = i f^{ijl} \delta^{(4)}(x-y) A_\nu^l(y), \qquad (3.85)$$

we arrive at

$$\langle 0|T \partial^\mu V_\mu^i(x) A_\nu^j(y) A_\lambda^k(z)|0\rangle = f^{ijk} \bar{s}^j \langle 0|T S^k(x) A_\nu^j(y) A_\lambda^l(z)|0\rangle$$

$$= -i\langle 0|T \bar{\psi}(x)[T^i, \mathcal{M}]\psi(x) A_\nu^j(y) A_\lambda^l(z)|0\rangle, \qquad (3.86)$$

where we have used the identity

$$-i[T^i, \mathcal{M}] = 2 f^{ijk} \mathrm{tr}(\mathcal{M}T^j) T^k. \qquad (3.87)$$

It is easy to see that this equation reproduces the divergence of the non-singlet vector current, given in Eq. (3.67). The insertion of the operator $\partial^\mu V_\mu^i(x)$ into the Green's functions containing more quark bilinears proceeds in a similar fashion.

3.7 The Triangle Anomaly

3.7.1 Three-Point Function

As we have mentioned, the use of the equations of motion for the calculation of divergences of the vector and axial-vector currents in some cases leads to the wrong result. Broadly spoken, the reason for this is that these currents are products of quark and antiquark fields taken at the same space-time point. In general, such objects are singular in field theory, and a regularization/renormalization is needed to render the Green's functions with the insertion of such composite operators meaningful. If this regularization procedure breaks the symmetry that has led to the conservation of the current through the Noether theorem, it is not a priori guaranteed that the current conservation gets restored after renormalization and removal of the regularization. When it is not the case, we speak of an *anomaly*. In such a case, a given classical symmetry does not survive quantum corrections.

Historically, for the first time the anomalies were encountered in the study of the decay of pseudoscalar mesons (pions) into two photons that proceeds via the fermion loop [13]. Explicitly, for the first time anomalies appear in Ref. [6]. In Ref. [14], a systematic investigation of the anomalies was carried out. The radiative corrections to the

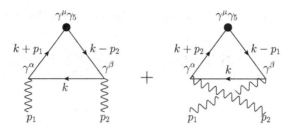

Figure 3.2   Triangle diagram in QED with the insertion of the axial-vector current. Shown are the leading terms. Solid and wiggly lines represent fermions and photons, in order.

axial-vector Ward identities were addressed in Ref. [15]. In this section, we consider the calculation of the triangle diagram in perturbation theory. This is the simplest example that explicitly shows that the anomalies are related to the ultraviolet divergences emerging in Feynman diagrams (albeit the final result is finite). In order to ease the discussion, we again go first back to QED and calculate the three-point Green's function containing the axial-vector current $A^\mu(x) = \bar\psi(x)\gamma^\mu\gamma_5\psi(x)$ and two photon fields, described by the vector-potential $\mathscr{A}^\mu(x)$:

$$\Gamma^{\mu\alpha\beta}(p_1,p_2) = \int d^4x d^4y e^{-ip_1 x - ip_2 y}\langle 0|TA^\mu(0)\mathscr{A}^\alpha(x)\mathscr{A}^\beta(y)|0\rangle$$

$$= T^{\mu\lambda\rho}(p_1,p_2)D_\lambda^\alpha(p_1)D_\rho^\beta(p_2), \qquad (3.88)$$

where $D_\nu^\mu(p)$ is the photon propagator and $T^{\mu\alpha\beta}(p_1,p_2)$ denotes the one-particle irreducible vertex. At lowest order in perturbation theory, this vertex is given by the diagrams shown in Fig. 3.2. A short calculation gives

$$T^{\mu\alpha\beta}(p_1,p_2) = e^2\int\frac{d^4k}{(2\pi)^4 i}\,\mathrm{tr}\left[\gamma^\alpha\frac{1}{m_e-\slashed{k}}\gamma^\beta\frac{1}{m_e-\slashed{k}+\slashed{p}_2}\gamma^\mu\gamma_5\frac{1}{m_e-\slashed{k}-\slashed{p}_1}\right]$$

$$+ [\alpha\leftrightarrow\beta, p_1\leftrightarrow p_2], \qquad (3.89)$$

with m_e the electron mass.

According to the standard power counting (in large momenta), this expression is linearly divergent and should be regularized. However, for a moment we shall not pay attention to this and check the fulfillment of the Ward identities. Performing formal manipulations with the integrand, we have

$$(p_1+p_2)_\mu T^{\mu\alpha\beta}(p_1,p_2) = e^2\int\frac{d^4k}{(2\pi)^4 i}\,\mathrm{tr}\left[\gamma^\alpha\frac{1}{m_e-\slashed{k}}\gamma^\beta\left(\frac{1}{m_e-\slashed{k}+\slashed{p}_2}\gamma_5\right.\right.$$

$$\left.\left.+\gamma_5\frac{1}{m_e-\slashed{k}-\slashed{p}_1}-\frac{1}{m_e-\slashed{k}+\slashed{p}_2}2m_e\gamma_5\frac{1}{m_e-\slashed{k}-\slashed{p}_1}\right)\right]+[\alpha\leftrightarrow\beta, p_1\leftrightarrow p_2].$$

$$(3.90)$$

It can be straightforwardly seen that all terms in the above expression, which are not proportional to m_e, cancel if we are allowed to make the shifts $k\to k-p_1$ or

$k \to k + p_2$ in different parts of the integrand. Then, the whole expression simplifies to

$$(p_1 + p_2)_\mu T^{\mu\alpha\beta}(p_1, p_2) = -e^2 \int \frac{d^4k}{(2\pi)^4 i} \, \text{tr} \left[\gamma^\alpha \frac{1}{m_e - \slashed{k}} \gamma^\beta \frac{1}{m_e - \slashed{k} + \slashed{p}_2} 2m_e \gamma_5 \right.$$

$$\left. \times \frac{1}{m_e - \slashed{k} - \slashed{p}_1} \right] + [\alpha \leftrightarrow \beta, p_1 \leftrightarrow p_2]. \qquad (3.91)$$

The expression on the right-hand side corresponds to the insertion of the pseudoscalar vertex $2m_e \bar\psi \gamma_5 \psi$ into the two-point function of the vector currents. Hence, Eq. (3.91) can be derived from the operator identity *containing no anomaly*:

$$\partial_\mu \left[\bar\psi(x) \gamma^\mu \gamma_5 \psi(x) \right] = 2i m_e \bar\psi(x) \gamma_5 \psi(x). \qquad (3.92)$$

Now we see what can go wrong with the proof. We have taken the freedom to shift the integration variables in different terms, in order to ensure cancellations. This is allowed for convergent integrals but not for the linearly divergent ones we are dealing with. In order to make the formal manipulations with the divergent integrals mathematically sound, one has first to regularize them. It is not immediately clear how dimensional regularization should be performed, since the definition of $\gamma_5 = i\gamma^0 \gamma^1 \gamma^2 \gamma^3$ implies four space-time dimensions (see more in what follows). For this reason, we resort first to the Pauli–Villars regularization [16], which explicitly preserves gauge invariance in the regularized theory. In this regularization, the expression in Eq. (3.88) is modified as

$$T_{\text{reg}}^{\mu\alpha\beta}(p_1, p_2) = T^{\mu\alpha\beta}(p_1, p_2) - T_\Lambda^{\mu\alpha\beta}(p_1, p_2), \qquad (3.93)$$

where the last term is given by the same expression as before but with the mass m_e replaced by Λ (the limit $\Lambda \to \infty$ should be performed at the end of calculations). The shift of the integration momenta can be safely performed in the regularized version and, instead of Eq. (3.91), we have

$$(p_1 + p_2)_\mu T_{\text{reg}}^{\mu\alpha\beta}(p_1, p_2)$$

$$= \left(-e^2 \int \frac{d^4k}{(2\pi)^4 i} \, \text{tr} \left[\gamma^\alpha \frac{1}{m_e - \slashed{k}} \gamma^\beta \frac{1}{m_e - \slashed{k} + \slashed{p}_2} 2m_e \gamma_5 \frac{1}{m_e - \slashed{k} - \slashed{p}_1} \right] \right.$$

$$\left. + e^2 \int \frac{d^4k}{(2\pi)^4 i} \, \text{tr} \left[\gamma^\alpha \frac{1}{\Lambda - \slashed{k}} \gamma^\beta \frac{1}{\Lambda - \slashed{k} + \slashed{p}_2} 2\Lambda \gamma_5 \frac{1}{\Lambda - \slashed{k} - \slashed{p}_1} \right] \right)$$

$$+ [\alpha \leftrightarrow \beta, p_1 \leftrightarrow p_2]. \qquad (3.94)$$

Formally, the last term in the brackets vanishes, if the limit $\Lambda \to \infty$ is performed in the integrand. However, as we know, such cavalier manipulations are not allowed in case of divergent integrands. Using a Feynman parameterization, the Λ-dependent term can be transformed into

$$(p_1 + p_2)_\mu T_\Lambda^{\mu\alpha\beta}(p_1, p_2) = -32 i e^2 \varepsilon^{\alpha\beta\lambda\rho} p_{1\lambda} p_{2\rho} \int_0^1 dx \int_0^{1-x} dy$$

$$\times \int \frac{d^4k}{(2\pi)^4 i} \frac{\Lambda^2}{(\Lambda^2 - k^2 - x(1-x-y)p_1^2 - y(1-x-y)p_2^2 - xy(p_1+p_2)^2)^3}. \qquad (3.95)$$

Note that this integral is ultraviolet (UV)-finite. Performing the integration over four-momentum yields

$$(p_1 + p_2)_\mu T_\Lambda^{\mu\alpha\beta}(p_1, p_2) = -\frac{ie^2}{\pi^2} \varepsilon^{\alpha\beta\lambda\rho} p_{1\lambda} p_{2\rho} \int_0^1 dx \int_0^{1-x} dy$$

$$\times \frac{\Lambda^2}{\Lambda^2 - x(1-x-y)p_1^2 - y(1-x-y)p_2^2 - xy(p_1+p_2)^2}. \tag{3.96}$$

In this expression, the limit $\Lambda \to \infty$ can be readily performed, and we get a finite result in this limit (albeit the original expression before being contracted with $(p_1 + p_2)_\mu$ contained a linear divergence):

$$\lim_{\Lambda \to \infty} (p_1 + p_2)_\mu T_\Lambda^{\mu\alpha\beta}(p_1, p_2) = -\frac{ie^2}{2\pi^2} \varepsilon^{\alpha\beta\lambda\rho} p_{1\lambda} p_{2\rho}. \tag{3.97}$$

It can be seen that at the operator level, the anomalous part of the divergence, given in the Eq. (3.97), can be reproduced through

$$\partial_\mu A^\mu(x) = 2im_e \bar{\psi}(x)\gamma_5 \psi(x) - \frac{e^2}{16\pi^2} \varepsilon_{\mu\nu\alpha\beta} \mathscr{F}^{\mu\nu}(x) \mathscr{F}^{\alpha\beta}(x). \tag{3.98}$$

Indeed, calculating the divergence of the three-point function, given by Eq. (3.88) and substituting Eq. (3.98), to the lowest order in perturbation theory, we get

$$\int d^4x d^4y e^{-ip_1 x - ip_2 y} \langle 0|T \frac{e^2}{16\pi^2} \varepsilon^{\mu\nu\lambda\rho} \mathscr{F}_{\mu\nu}(0) \mathscr{F}_{\lambda\rho}(0) \mathscr{A}^\alpha(x) \mathscr{A}^\beta(y)|0\rangle$$

$$= -\frac{e^2}{2\pi^2} \varepsilon^{\lambda\rho\mu\nu} p_{1\mu} p_{2\nu} D_\lambda^\alpha(p_1) D_\rho^\beta(p_2), \tag{3.99}$$

which exactly matches the anomalous divergence given in Eq. (3.97).

3.7.2 Vector versus Axial-Vector Ward Identities

The result we just obtained deserves quite a few remarks. First of all, the Feynman integrals defining the pertinent Green's functions are UV-divergent (albeit the anomaly itself is UV-finite). In order to make the formal manipulations sensible, the UV divergences should be regularized in one way or another. The crucial point is that the Pauli–Villars regularization, which was used in the preceding section, breaks chiral symmetry,[7] and the symmetry is not restored when the regularization is removed afterwards. Then, a question naturally arises as to whether we may guarantee the absence of anomalous terms by choosing a different regularization, or by changing the renormalization prescription. The answer to this question is *no*, as can be seen from the following argument. The quantity $T^{\mu\alpha\beta}(p_1, p_2)$, which is given by Eq. (3.89), can be renormalized by adding a polynomial in the momenta p_1, p_2. Lorentz invariance and

---

[7] This regularization introduces a fictive "fermion" with a mass Λ and a wrong sign in the residue of the propagator. The chiral symmetry is broken due to the nonzero mass, and the breaking persists even in the limit $\Lambda \to \infty$. In other words, the fictitious particle does not completely decouple in the infinite mass limit. What remains, however, is a polynomial that does not contradict the decoupling theorem.

parity invariance as well as Bose symmetry restrict the form of the polynomial, which is linear in the momenta, to

$$T^{\mu\alpha\beta}(p_1, p_2) \to T^{\mu\alpha\beta}(p_1, p_2) + a\varepsilon^{\mu\alpha\beta\nu}(p_1 - p_2)_\nu, \tag{3.100}$$

where a is some constant. Consequently,

$$(p_1 + p_2)_\mu T^{\mu\alpha\beta}(p_1, p_2) \to (p_1 + p_2)_\mu T^{\mu\alpha\beta}(p_1, p_2) - 2a\varepsilon^{\mu\nu\alpha\beta}p_{1\mu}p_{2\nu}. \tag{3.101}$$

Hence, choosing

$$a = -\frac{ie^2}{4\pi^2}, \tag{3.102}$$

we may ensure that the divergence of the axial-vector current is given by the "naive" expression without the anomalous term; see Eq. (3.92). However, the problem caused by the UV divergences resurfaces at a different place: It is seen that the modified vertex $T^{\mu\alpha\beta}(p_1, p_2)$ does not obey the Ward identities for the vector currents anymore. Indeed, according to these, one should have

$$p_{1\alpha}T^{\mu\alpha\beta}(p_1, p_2) = p_{2\beta}T^{\mu\alpha\beta}(p_1, p_2) = 0. \tag{3.103}$$

Since, for example, $p_{1\alpha}\varepsilon^{\mu\alpha\beta\nu}(p_1 - p_2)_\nu = -\varepsilon^{\mu\alpha\beta\nu}p_{1\alpha}p_{2\nu} \neq 0$, one has to choose $a = 0$ in order to obey the vector Ward identities. Thus there is no choice of the constant a that would guarantee the *simultaneous* fulfillment of the naive Ward identities for the vector and axial-vector currents.

At this point, one has a choice. Usually, the parameter a is chosen so that the divergence of the vector current does not contain anomalies. The reason for this is because in QED the vector current is coupled to the $U(1)$ gauge bosons, and the anomaly would render the theory inconsistent. (For a nice discussion of this issue see, e.g., Ref. [17].) On the contrary, the axial-vector current in QED (and in QCD) is not coupled to the gauge field, and the anomaly is benevolent. Note also that the axial-vector current in the Standard Model *does couple* to the weak gauge fields. What renders the Standard Model a consistent theory is the *cancellation* of anomalies (the anomalies in the lepton and quark sectors cancel against each other), which, in particular, requires that the number of quark and lepton generations be the same.

3.7.3 The Anomaly in Dimensional Regularization

Since the anomaly does not depend on a particular regularization chosen, it is interesting to understand how the same result can be obtained in dimensional regularization, which represents the most convenient tool for the calculation of Feynman integrals in gauge theories. As already mentioned, it is not evident how quantities like γ_5 or $\varepsilon^{\mu\nu\alpha\beta}$ can be generalized to the case of a non-integer dimension, as it is required in dimensional regularization (or any variant thereof). Indeed, the definitions,

$$\{\gamma^\mu, \gamma^\nu\} = 2g^{\mu\nu} \cdot \mathbb{1}, \qquad \{\gamma^\mu, \gamma_5\} = 0, \qquad g^\mu_\mu = D, \tag{3.104}$$

are inconsistent in space-time dimensions other than 4. In order to see this, assume that the above is true. Then, using the cyclic property of trace, we may write

$$D\text{tr}\,(\gamma_5) = \text{tr}\,(\gamma_5\gamma_\mu\gamma^\mu) = \text{tr}\,(\gamma_\mu\gamma_5\gamma^\mu) = -\text{tr}\,(\gamma_5\gamma_\mu\gamma^\mu) = -D\text{tr}\,(\gamma_5), \qquad (3.105)$$

from which it follows that $2D\text{tr}\,(\gamma_5) = 0$. So, the trace of γ_5 should vanish in all dimensions other than $D = 0$. Continuing analytically from $D = 4$, we may thus set $\text{tr}\,(\gamma_5) = 0$ for any D.

Using the same technique, we get

$$D\text{tr}\,(\gamma_5\gamma^\mu\gamma^\nu) = \text{tr}\,(\gamma_5\gamma^\mu\gamma^\nu\gamma_\lambda\gamma^\lambda) = -4g^{\mu\nu}\text{tr}\,(\gamma_5) + (4-D)\text{tr}\,(\gamma_5\gamma^\mu\gamma^\nu). \qquad (3.106)$$

From this, we get $(D-2)\text{tr}\,(\gamma_5\gamma^\mu\gamma^\nu) = 0$ and, after analytic continuation in D, it follows that $\text{tr}\,(\gamma_5\gamma^\mu\gamma^\nu) = 0$ for all D.

Similarly, we obtain

$$(D-4)\text{tr}\,(\gamma_5\gamma^\mu\gamma^\nu\gamma^\alpha\gamma^\beta) = 0. \qquad (3.107)$$

This is where the controversy arises. The trace does not vanish in four dimensions (it is proportional to $\varepsilon^{\mu\nu\alpha\beta}$) and thus cannot be made vanishing for all D after analytic continuation in D.

As shown, for example, in Ref. [18], a consistent definition of the matrix γ_5 in D dimensions is

$$\gamma_5 = i\gamma^0\gamma^1\gamma^2\gamma^3. \qquad (3.108)$$

The matrix γ_5 has the following properties:

$$\{\gamma_5, \gamma^\mu\} = 0 \quad \text{for} \quad \mu = 0, 1, 2, 3,$$

$$[\gamma_5, \gamma^\mu] = 0 \quad \text{for other values of } \mu,$$

$$\gamma_5^2 = 1, \quad \gamma_5^\dagger = \gamma_5. \qquad (3.109)$$

According to this definition,

i) The trace of γ_5 with any odd number of γ-matrices vanishes;
ii) $\text{tr}\,(\gamma_5) = 0$.
iii) $\text{tr}\,(\gamma_5\gamma^\mu\gamma^\nu) = 0$.
iv) $\text{tr}\,(\gamma_5\gamma^\mu\gamma^\nu\gamma^\alpha\gamma^\beta) = -4i\varepsilon^{\mu\nu\alpha\beta}$, where

$$\varepsilon^{\mu\nu\alpha\beta} = \begin{cases} +1, & \text{if } \mu\nu\alpha\beta \text{ is an even permutation of } 0123, \\ -1, & \text{if } \mu\nu\alpha\beta \text{ is an odd permutation of } 0123, \\ 0, & \text{otherwise}. \end{cases} \qquad (3.110)$$

In order to prove these results, we introduce the following notation: $\gamma^\mu = \hat{\gamma}^\mu$ for $\mu = 0, 1, 2, 3$. Then, $\hat{\gamma}_\mu\hat{\gamma}^\mu = 4$ and

$$4\text{tr}\,(\gamma_5) = \text{tr}\,(\gamma_5\hat{\gamma}_\mu\hat{\gamma}^\mu) = -4\text{tr}\,(\gamma_5). \qquad (3.111)$$

This means that, again, $\text{tr}(\gamma_5) = 0$. Using the same method, we see that

$$
8\text{tr}(\gamma_5 \gamma^\mu \gamma^\nu) =
\begin{cases}
4\text{tr}(\gamma_5 \gamma^\mu \gamma^\nu), & \text{if } \mu, \nu = 0,1,2,3, \\
2\text{tr}(\gamma_5 \gamma^\mu \gamma^\nu), & \text{if } \mu = 0,1,2,3, \, \nu > 3 \text{ or vice versa,} \quad (3.112) \\
0, & \text{if } \mu, \nu > 3.
\end{cases}
$$

Thus, $\text{tr}(\gamma_5 \gamma^\mu \gamma^\nu) = 0$ as well. Further, using the same technique, it is straightforward to show that $\text{tr}(\gamma_5 \gamma^\mu \gamma^\nu \gamma^\alpha \gamma^\beta)$ should vanish, if at least one of the μ, ν, α, β is larger than 3. If all four indices belong to the physical subspace $\mu = 0,1,2,3$, then the result is given by the usual formula known from four dimensions. Thus, the proof is complete.

In the following discussion, it will be demonstrated that, using this definition, we recover the correct result for the anomaly. To this end, let us rewrite Eq. (3.89) in dimensional regularization and contract it with $(p_1 + p_2)_\mu$. The numerator in the resulting expression can be transformed by the use of the identity

$$
(\not{p}_1 + \not{p}_2)\gamma_5 = (m_e - \not{k} + \not{p}_2)\gamma_5 + \gamma_5(m_e - \not{k} - \not{p}_1) - 2m_e\gamma_5 + 2(\not{k} - \hat{\not{k}})\gamma_5 .
$$

$$(3.113)$$

In deriving this identity we took into account that both external momenta, p_1, p_2, are defined in four physical dimensions, that is, $p_i - \hat{p}_i = 0$, $i = 1, 2$. The first three terms reproduce the pertinent terms in Eq. (3.91), whereas the last term leads to the anomaly. Introducing Feynman parameters and performing traces while keeping in mind that $\alpha, \beta = 0,1,2,3$, the anomalous part of the divergence transforms into

$$
(p_1 + p_2)_\mu T_{\text{anom}}^{\mu\alpha\beta}(p_1, p_2) = -16ie^2 \varepsilon^{\alpha\beta\mu\nu} p_{1\mu} p_{2\nu} \int_0^1 dx \int_0^{1-x} dy
$$

$$
\times \int \frac{d^D k}{(2\pi)^D i} \frac{(k - \hat{k})^2}{(m_e^2 - k^2 - x(1-x-y)p_1^2 - y(1-x-y)p_2^2 - xy(p_1 + p_2)^2)^3} .
$$

$$(3.114)$$

Following the standard technique for dealing the integrals in dimensional regularization (see, e.g., Ref. [19]), the integration vector can be represented as $k^\mu = (\hat{k}^\mu, s^\mu)$, where s^μ stands for the components in the unphysical dimensions. Rotating to Euclidean space, we get

$$
\int \frac{d^D k_E}{(2\pi)^D} \frac{-s_E^2}{(A^2 + s_E^2 + \hat{k}_E^2)^3} = (A^2)^{D/2-2} \int \frac{d^{D-4} s_E}{(2\pi)^{D-4}} \int \frac{d^4 \hat{k}_E}{(2\pi)^4} \frac{-s_E^2}{(1 + s_E^2 + \hat{k}_E^2)^3}
$$

$$
= -\frac{1}{32\pi^2} + O(D-4) .
$$

$$(3.115)$$

From this, we obtain

$$
(p_1 + p_2)_\mu T_{\text{anom}}^{\mu\alpha\beta}(p_1, p_2) = \frac{ie^2}{2\pi^2} \varepsilon^{\alpha\beta\mu\nu} p_{1\mu} p_{2\nu} ,
$$

$$(3.116)$$

in complete agreement with Eq. (3.97).

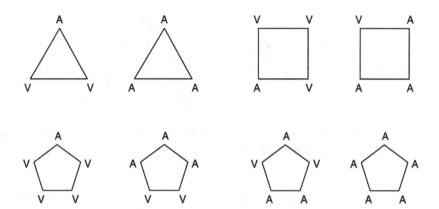

Figure 3.3 Anomalous quark loop diagrams with three, four and five external vertices. V and A denote vector and axial-vector currents, respectively.

3.7.4 Non-Abelian Anomalies

In the Abelian case just considered, only the triangle AVV and AAA vertices are anomalous. However, in case of the non-singlet currents that transform under the non-Abelian group, there exist the so-called box and pentagon anomalies as well, shown in Fig. 3.3. Non-Abelian anomalies will be considered in detail in Section 3.10. Here, we only want to make a brief remark. Whereas the triangle and the box anomalies stem from the divergence of the pertinent quark loop diagrams, the pentagon diagram, containing five quark propagators, is obviously ultraviolet-finite. The anomaly here stems from the contact terms, which should be added to ensure that the Ward identities, obeyed by the vector current, are anomaly free. This argument was first given in Bardeen's seminal paper [14]. The non-Abelian anomalies can be represented in the framework of an EFT in terms of the so-called Wess–Zumino–Witten term, as discussed later.

3.7.5 Non-renormalization of the Axial Anomaly

Finally, note that up to this moment, the calculations were done only at the lowest order in perturbation theory. It remains to be seen how the result will change when radiative corrections are included. To this end, one has to consider the insertion of the current $A_\mu(x)$ into the Green's functions with any number of external electron/positron and photon legs. In particular, we may consider the radiative corrections of the type shown in Fig. 3.4 and ask how the result obtained from the diagrams in Fig. 3.2 change. The same question should be then addressed for a generic Green's function with any number of external legs.

In Ref. [15], by examining Feynman diagrams to all orders in field theory, it was shown that Eq. (3.98), which defines the anomalous divergence of the axial vector current, stays intact when radiative corrections due to dynamical photons are considered.

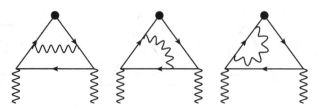

Figure 3.4   Triangle diagram with the insertion of the axial-vector current (filled circle): representative diagrams describing $O(\alpha)$ electromagnetic corrections.

This fundamental statement goes under the name of the *non-renormalization of the axial anomaly*. (To be precise, the proof in Ref. [15] has been carried out within QED and the linear σ-model, but the arguments can be straightforwardly generalized to other Lagrangians as well.) In Ref. [20], this result was checked by explicit calculations at one loop. Further, in Refs. [21, 22], a more compact and elegant proof was provided based on the renormalization group argument. The main idea behind the proof in Ref. [15] is very transparent. One regularizes the photon field by using a method that is explicitly gauge invariant and respects chiral symmetry. To this end, for example, the method of higher covariant derivatives can be used. (See the detailed description of the method, e.g., in Ref. [17].) However, one-loop fermion diagrams with one, two, three, or four vertices cannot be regularized in this way, and the additional regularization should be used on top. We may resort to the Pauli–Villars procedure for the remaining graphs, which also respects gauge invariance but introduces the finite fermion mass and thus breaks chiral symmetry. As we have already seen, when the regulator mass tends toward infinity, the divergence of the axial-vector current acquires an additional finite piece that coincides with the anomaly. The net result of this discussion is that the anomaly can emerge only from one-loop fermion graphs *without* insertions of virtual photons. Those with such insertions, like the ones shown in Fig. 3.4, are automatically anomaly-free.

Note that this argument concerns only the anomaly in the divergence of the axial-vector current. Other types of anomalous symmetry breaking, for example, the trace anomaly, which also will be considered in what follows, get renormalized at higher orders in perturbation theory.

3.8 Anomalies: Point-Split Technique

As noted already, the anomalies stem from the short-distance singularities of the currents, which represent bilinear operators of the quark/antiquark fields with coinciding arguments. In this section we shall consider the example discussed in the previous section from a different viewpoint. Namely, we shall regularize the singular composite operators by using the point-split technique, in which the arguments of the fermion and antifermion fields are taken slightly apart from each other. This alternative derivation enables one to directly observe how the anomaly emerges from the regularization of the singularities.

Naively, the divergence of the axial-vector current $A_\mu(x) = \bar{\psi}(x)\gamma_\mu\gamma_5\psi(x)$ in QED is given by

$$\partial^\mu A_\mu(x) = -i\bar{\psi}(x)i\overleftarrow{\partial}\gamma_5\psi(x) + i\bar{\psi}(x)\gamma^5 i\overrightarrow{\partial}\psi(x) = 2im_e\bar{\psi}(x)\gamma^5\psi(x). \qquad (3.117)$$

Here, $\psi(x)$ denotes the electron field, m_e is the electron mass and the equation of motion $(i\partial\!\!\!/ - e\mathcal{A}\!\!\!/(x) - m_e)\psi(x)$ has been used.

From Eq. (3.117) we may conclude that the axial-vector current should be conserved in massless QED. Such a conclusion would be, however, premature due to the singular character of the product of two fermion field operators. One might try to regularize this product by taking the two operators at two different space-time points $x+\varepsilon/2$ and $x-\varepsilon/2$, separated by an infinitesimal (space-like) vector ε_μ. As $\varepsilon_\mu \to 0$, the original expression for the axial-vector current is restored. The gauge-invariant expression for the axial-vector field takes the form

$$A_\mu(x;\varepsilon) = \bar{\psi}(x+\varepsilon/2)\gamma_5\gamma_\mu\psi(x-\varepsilon/2)\exp\left(-ie\int_{x-\varepsilon/2}^{x+\varepsilon/2}dy_\nu\mathcal{A}^\nu(y)\right). \qquad (3.118)$$

Here, we treat $\mathcal{A}_\mu$ as an external source. No loops with virtual photons are considered at this stage. The line integral in Eq. (3.118) ensures gauge-invariance (which we leave to be checked by the reader). Calculating the divergence of both sides, we get

$$\partial^\mu A_\mu(x;\varepsilon) = 2im_e\bar{\psi}(x+\varepsilon/2)\gamma_5\psi(x-\varepsilon/2)\exp\left(-ie\int_{x-\varepsilon/2}^{x+\varepsilon/2}dy_\nu\mathcal{A}^\nu(y)\right)$$
$$+ ieA_\mu(x;\varepsilon)\left(-\frac{\partial}{\partial x_\mu}\int_{x-\varepsilon/2}^{x+\varepsilon/2}dy_\nu\mathcal{A}^\nu(y) + \mathcal{A}^\mu(x+\varepsilon/2) - \mathcal{A}^\mu(x-\varepsilon/2)\right).$$
$$(3.119)$$

Expanding now the electromagnetic field in powers of ε leads to

$$\partial^\mu A_\mu(x;\varepsilon) = 2im_e\bar{\psi}(x+\varepsilon/2)\gamma_5\psi(x-\varepsilon/2)\exp\left(-ie\int_{x-\varepsilon/2}^{x+\varepsilon/2}dy_\nu\mathcal{A}^\nu(y)\right)$$
$$+ ieA_\mu(x;\varepsilon)\varepsilon_\nu\mathcal{F}^{\nu\mu}(x) + O(\varepsilon^2), \qquad (3.120)$$

where $\mathcal{F}^{\nu\mu}(x) = \partial^\nu\mathcal{A}^\mu(x) - \partial^\mu\mathcal{A}^\nu(x)$ is the electromagnetic field strength tensor. When $\varepsilon_\mu \to 0$, the first term in Eq. (3.120) reproduces the result obtained by the use of Noether's theorem. Naively, one would think that the second term vanishes in this limit, but this is not true since the current is singular. In order to single out the singularity, we consider the vacuum expectation value of the current. It is convenient to perturbatively expand this quantity in powers of e, since the subsequent terms are less and less singular and, hence, only a finite number of terms should be taken into account. (We remind the reader that, for the time being, the electromagnetic field is considered as a classical external source.)

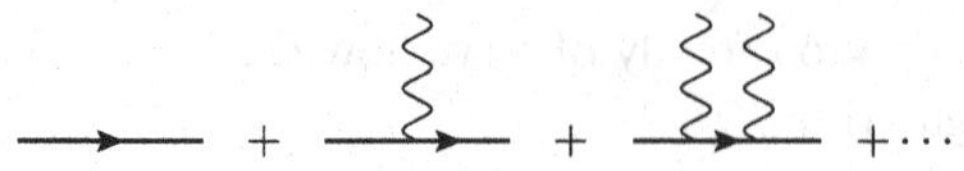

Figure 3.5 The propagator of an electron in an external electromagnetic field.

In order to carry out the perturbative expansion, note that

$$\langle 0|A_\mu(x;\varepsilon)|0\rangle = i\mathrm{tr}\left[\gamma_\mu\gamma_5 S_A(x-\varepsilon/2,x+\varepsilon/2)\right], \tag{3.121}$$

where $S_A(x,y)$ is the electron propagator in the external photon field. The perturbative expansion of this quantity in powers of e, which is shown in Fig. 3.5, takes the form

$$S_A(x-\varepsilon/2,x+\varepsilon/2) - S_0(-\varepsilon)$$

$$-e\int d^4y\, S_0(x-\varepsilon/2-y)\gamma_\lambda \mathcal{A}^\lambda(y)S_0(y-x-\varepsilon/2) + \cdots. \tag{3.122}$$

Here, $S_0(x)$ denotes the free electron propagator and the ellipses denote the non-singular terms in ε.

The first term of Eq. (3.122), inserted into Eq. (3.121), gives a vanishing result, because the trace over the Dirac matrices is zero. The only nonvanishing contribution comes from the second term. Rewriting the propagators in momentum space, we get

$$i\mathrm{tr}\left[\gamma_\mu\gamma_5 S_A(x-\varepsilon/2,x+\varepsilon/2)\right]$$

$$= -ie\int d^4y\int \frac{d^4p}{(2\pi)^4}\frac{d^4q}{(2\pi)^4} e^{-ip(x-\varepsilon/2-y)-iq(y-x-\varepsilon/2)}\mathcal{A}^\lambda(y)$$

$$\times \frac{\mathrm{tr}\left[\gamma_\mu\gamma_5(m_e+\not p)\gamma_\lambda(m_e+\not q)\right]}{(m_e^2-p^2)(m_e^2-q^2)}. \tag{3.123}$$

Calculating the trace and making the change of variables $p-q=k$, $(p+q)/2=l$ gives

$$i\mathrm{tr}\left[\gamma_\mu\gamma_5 S_A(x-\varepsilon/2,x+\varepsilon/2)\right] = 4e\varepsilon_{\mu\lambda\alpha\beta}\int d^4y\int \frac{d^4k}{(2\pi)^4}\frac{d^4l}{(2\pi)^4} e^{-ik(x-y)+il\varepsilon}$$

$$\times l^\alpha k^\beta \mathcal{A}^\lambda(y)\frac{1}{(m_e^2-(l+k/2)^2)(m_e^2-(l-k/2)^2)}. \tag{3.124}$$

Consequently,

$$\varepsilon_\nu\langle 0|A_\mu(x;\varepsilon)|0\rangle = 4ie\varepsilon_{\mu\lambda\alpha\beta}\int d^4y\int \frac{d^4k}{(2\pi)^4}\frac{d^4l}{(2\pi)^4} e^{-ik(x-y)+il\varepsilon}\mathcal{A}^\lambda(y)k^\beta$$

$$\times \frac{\partial}{\partial l^\nu}\frac{l^\alpha}{(m_e^2-(l+k/2)^2)(m_e^2-(l-k/2)^2)}. \tag{3.125}$$

Let us now consider the integral

$$I_\nu^\alpha(k) = \int \frac{d^4l}{(2\pi)^4 i} e^{il\varepsilon} \frac{\partial}{\partial l^\nu} \frac{l^\alpha}{(m_e^2 - (l+k/2)^2)(m_e^2 - (l-k/2)^2)}. \tag{3.126}$$

First, note that, in the limit $\varepsilon_\mu \to 0$, the value of this integral does not depend on k. In order to prove this, consider the quantity $I_\nu^\alpha(k) - I_\nu^\alpha(0)$. In the limit $\varepsilon_\mu \to 0$, it becomes an integral over a total derivative, with an integrand that vanishes quickly at the infinity, so that the surface term disappears. Consequently, $I_\nu^\alpha(k) - I_\nu^\alpha(0) \to 0$ in this limit. In the expression of $I_\nu^\alpha(0)$ itself, however, the limit $\varepsilon_\mu \to 0$ cannot be immediately performed, because the integrand does not fall off sufficiently fast (we remind the reader that ε_μ plays the role of an ultraviolet regulator). In order to calculate this quantity, we may use the following trick. First, using Lorentz invariance allows one to write

$$I_\nu^\alpha(0) = a g_\nu^\alpha + b \varepsilon^\alpha \varepsilon_\nu, \tag{3.127}$$

where a and b are just numbers. Next, note that, in order to perform the limit $\varepsilon_\mu \to 0$ in this expression, we should specify from which particular direction this limit is approached. We choose to perform the limit symmetrically, averaging over all dimensions of the (spacelike) vector ε_μ. This results in $\langle \varepsilon^\alpha \varepsilon_\nu \rangle = -\frac{1}{4} g_\nu^\alpha$ and, hence,

$$\langle I_\nu^\alpha(0) \rangle = \left(a - \frac{b}{4} \right) g_\nu^\alpha. \tag{3.128}$$

On the other hand, summing up over the Lorentz indices, we get

$$I_\nu^\nu(0) = 4a - b = \int \frac{d^4l}{(2\pi)^4 i} e^{il\varepsilon} \frac{\partial}{\partial l^\nu} \frac{l^\nu}{(m_e^2 - l^2)^2}$$

$$= 4m_e^2 \int \frac{d^4l}{(2\pi)^4 i} e^{il\varepsilon} \frac{1}{(m_e^2 - l^2)^3} = \frac{1}{8\pi^2} + O(\varepsilon). \tag{3.129}$$

This finally gives

$$I_\nu^\alpha(k) = \frac{g_\nu^\alpha}{32\pi^2} + O(\varepsilon), \tag{3.130}$$

and

$$\lim_{\varepsilon_\nu \to 0} \varepsilon_\nu \langle 0|A_\mu(x;\varepsilon)|0 \rangle = -\frac{ie\varepsilon_{\mu\nu\beta\lambda}}{8\pi^2} \partial^\beta \mathscr{A}^\lambda(x). \tag{3.131}$$

The divergence of the axial-vector current in the limit $\varepsilon_\nu \to 0$ is given by

$$\partial^\mu A_\mu(x) = 2im_e \bar\psi(x) \gamma_5 \psi(x) - \frac{e^2}{16\pi^2} \varepsilon_{\mu\nu\alpha\beta} \mathscr{F}^{\mu\nu}(x) \mathscr{F}^{\alpha\beta}(x). \tag{3.132}$$

From this expression, which agrees with Eq. (3.98), it is immediately seen that the naive divergence of the axial-vector current (given by the first term) is modified through the interaction with the electromagnetic field. Moreover, the reason for this modification is the singular behavior of the product of two field operators, taken at the same space-time point. Using the point-split technique, we have seen that this product behaves like $\varepsilon^\nu / \varepsilon^2$, where ε^ν is the shift of the argument of one operator with respect to the other.

The following remarks are in order. First, to separate the singular term in the current operator, we have considered the vacuum expectation value of this operator. It can be seen that this approach suffices to find *all* singularities. Indeed, an Operator Product Expansion (OPE) can be applied to the product of two operators in $A_\mu(x;\varepsilon)$:

$$A_\mu(x;\varepsilon) = \sum_n C_n(\varepsilon)\mathcal{O}_n(x). \qquad (3.133)$$

Here $\mathcal{O}_n(x)$ denote renormalized operators, and $C_n(\varepsilon)$ are the c-number Wilson coefficients (see, for example, a nice introduction to OPE in the textbook [19]). The operators $\mathcal{O}_n(x)$ and the coefficients $C_n(\varepsilon)$ carry Lorentz indices (not shown explicitly), ensuring that the l.h.s. of Eq. (3.133) is a Lorentz vector. Furthermore, the singularities at $\varepsilon \to 0$ are contained in the c-number coefficients only, and their strength depends on the engineering dimension of the operator $\mathcal{O}_n(x)$ that multiplies the given $C_n(\varepsilon)$. Among all operators, the unit operator $\mathbb{1}$ (dimension 0) has the lowest dimension. All other local operators, which can be constructed by using the fermion fields $\psi, \bar{\psi}$ (dimension $\frac{3}{2}$) and the photon field $\mathscr{A}_\mu$ (dimension 1), will have positive dimension. Hence, one can write

$$A_\mu(x;\varepsilon) = C_\mu(\varepsilon)\mathbb{1} + \cdots, \qquad C_\mu(\varepsilon) = \langle 0|A_\mu(x;\varepsilon)|0\rangle, \qquad (3.134)$$

where ellipses stand for the terms, power suppressed in $|\varepsilon|$. Finally, from this we may infer that, since the most singular term gives a finite contribution to the anomaly, the contribution from the suppressed term should vanish and we have indeed calculated all contributions to the anomaly.

Further, note that, from the beginning, we have relied on the algebra of γ-matrices in four space-time dimensions. For this reason, one is not allowed to utilize dimensional regularization at a later stage, after evaluating the traces. At first glance, this might be regarded as counterintuitive since, as already mentioned, the emerging integrals are finite. However, these are only conditionally finite. The power counting indicates a divergence but, after invariant integration over directions (equivalently – averaging over all directions of ε_μ), the divergent term exactly cancels. Let us recall now that the invariant integration leads to different overall coefficient in the space-time with D dimensions: $1/D$ instead of $1/4$. Then it can be seen why dimensional regularization fails to restore the result obtained in four dimensions (in fact, the anomalous term vanishes, if dimensional regularization is used at this stage). Thus, in order to reproduce the correct result in dimensional regularization, one has to use it from the beginning.

3.9 Fujikawa Determinant

In the path-integral derivation of the Ward identities, both fields and currents are classical c-number objects and not operators. So, the formal manipulations with the Lagrangian, which are needed to derive Ward identities, are justified. As shown by Fujikawa in his seminal papers [23, 24] (see also the textbook [25]), in this formulation the anomalies stem from the non-invariance of the fermionic measure in the path integral. We consider the derivation for the example of QCD with N_f flavors and set, for the moment, all external fields, s, p, v, a, to zero. This means, in particular, that the quark masses are vanishing and the system has an exact chiral symmetry at the classical level.

Furthermore, the path integral, which defines the generating functional, implies an integration over quark and gluon fields (as well as the ghost fields, if one is using gauges that are not ghost-free). We may agree to perform the integration over quarks first (this would correspond to considering other fields as external classical sources) and integrate over gluons and ghosts later (corresponding to higher-order corrections, which were briefly discussed in the context of the non-renormalization of the axial anomaly). Thus, we start with the path integral over massless fermion fields in the presence of the external gluon field G_μ:

$$Z = \int d\psi d\bar{\psi} \exp\left\{ i \int d^4x \bar{\psi}(x) i \slashed{D} \psi(x) \right\}, \tag{3.135}$$

where $D_\mu = \partial_\mu + G_\mu$ is the covariant derivative, the so-called *Dirac operator*.

The path integral is a well-defined object in Euclidean space, where the oscillating behavior of the exponential is replaced by an exponential damping factor. The expression of the Euclidean generating functional is given by:[8]

$$Z = \int d\psi d\bar{\psi} \exp\left\{ - \int d^4x \bar{\psi}(x) \gamma_\mu D_\mu \psi(x) \right\}. \tag{3.136}$$

Switching to the Euclidean γ-matrices from Minkowski space, we have

$$\gamma^0 \to \gamma_0, \qquad \gamma^k \to i\gamma_k, \tag{3.137}$$

and the Euclidean versions of γ_5 and $\sigma_{\mu\nu}$ are defined through Euclidean γ-matrices as

$$\gamma_5 = \gamma_0 \gamma_1 \gamma_2 \gamma_3, \qquad \sigma_{\mu\nu} = \frac{i}{2} [\gamma_\mu, \gamma_\nu]. \tag{3.138}$$

In Euclidean space, the following relations hold:

$$\{\gamma_\mu, \gamma_\nu\} = 2\delta_{\mu\nu}, \; \{\gamma_\mu, \gamma_5\} = 0, \; \gamma_5^2 = 1, \; \mathrm{tr}\,(\gamma_5 \gamma_\mu \gamma_\nu \gamma_\alpha \gamma_\beta) = 4\varepsilon_{\mu\nu\alpha\beta}, \tag{3.139}$$

where $\varepsilon_{0123} = 1$.

Anticipating the fact that the anomaly will emerge in the flavor-singlet current only, here we restrict ourselves to *local* singlet axial transformations:

$$\psi(x) \mapsto (1 + i\beta^0(x)\gamma^5)\psi(x), \qquad \bar{\psi}(x) \mapsto \bar{\psi}(x)(1 + i\beta^0(x)\gamma^5). \tag{3.140}$$

Using partial integration, one sees that the action functional under this transformation takes the form

$$\int d^4x \, \bar{\psi}(x) \gamma_\mu D_\mu \psi(x) \mapsto \int d^4x \left[\bar{\psi}(x) \gamma_\mu D_\mu \psi(x) - i\beta^0(x) \partial_\mu (\bar{\psi}(x)\gamma^\mu \gamma^5 \psi(x)) \right]. \tag{3.141}$$

---

[8] In order to avoid clutter in the notation, we refrain from attaching subscripts and use the same symbols for the objects defined in Minkowski and Euclidean spaces.

If the fermionic measure $d\psi d\bar{\psi}$ were invariant, then the conservation of the singlet axial-vector current would follow:

$$\partial^\mu A_\mu^0(x) = \partial^\mu(\bar{\psi}(x)\gamma^\mu\gamma^5\psi(x)) = 0. \tag{3.142}$$

However, according to Refs. [23–25], this is not the case: The fermionic measure is not invariant, and the divergence of the singlet axial-vector current does not vanish even in the chiral limit.

To be more specific, note that the fermionic measure is not a well-defined object, due to the presence of ultraviolet divergences. In order to introduce the regularization, let us define the eigenvectors of the massless Euclidean Dirac operator in the external field G_μ, corresponding to real eigenvalues λ:

$$\slashed{D}\psi_\lambda = i\lambda\psi_\lambda, \qquad \bar{\psi}_\lambda\,(\overleftarrow{\slashed{D}}) = i\lambda\,\bar{\psi}_\lambda, \qquad \int d^4x\,\bar{\psi}_\lambda(x)\psi_{\lambda'}(x) = \delta_{\lambda\lambda'}, \tag{3.143}$$

which fulfill the completeness relation

$$\sum_\lambda \psi_\lambda(x)\bar{\psi}_\lambda(y) = \delta^4(x-y) \tag{3.144}$$

and expand the fields in the basis of the eigenvectors

$$\psi(x) = \sum_\lambda a_\lambda\psi_\lambda(x), \qquad \bar{\psi}(x) = \sum_\lambda \bar{\psi}_\lambda(x)\bar{a}_\lambda. \tag{3.145}$$

The fermionic measure is then given by

$$d\psi d\bar{\psi} = \prod_\lambda da_\lambda d\bar{a}_\lambda. \tag{3.146}$$

Let us now consider the Jacobian of the infinitesimal transformation given in Eq. (3.140). The variables a_λ transform as

$$a_\lambda \mapsto \int d^4x\,\bar{\psi}_\lambda(x)(1+i\beta^0(x)\gamma^5)\psi(x) = \sum_{\lambda'}\int d^4x\,\bar{\psi}_\lambda(x)(1+i\beta^0(x)\gamma^5)\psi_{\lambda'}(x)a_{\lambda'}$$

$$\doteq \sum_{\lambda'}(\delta_{\lambda\lambda'}+C_{\lambda\lambda'})a_{\lambda'}, \tag{3.147}$$

where

$$C_{\lambda\lambda'} = i\int d^4x\,\beta^0(x)\,\bar{\psi}_\lambda(x)\gamma^5\psi_{\lambda'}(x). \tag{3.148}$$

The Jacobian is given by

$$d\psi d\bar{\psi} \mapsto J^{-2}d\psi d\bar{\psi}, \tag{3.149}$$

where

$$J = \det(\mathbb{1}+C) = \exp(\mathrm{Tr}(\ln(\mathbb{1}+C))) = \exp\left\{\sum_\lambda C_{\lambda\lambda} + O(\beta^2)\right\}. \tag{3.150}$$

This gives

$$\ln J = i\int d^4x\,\beta^0(x)\sum_\lambda\bar{\psi}_\lambda(x)\gamma^5\psi_\lambda(x) \doteq i\int d^4x\,\beta^0(x)S(x). \tag{3.151}$$

Formally, we may think that the quantity $S(x)$ in Eq. (3.151) is equal to zero. Indeed, using the closure relation for the eigenvectors of the Dirac operators, one would get $\sum_\lambda \bar{\psi}_\lambda(x)\gamma_5\psi_\lambda(y) = -\mathrm{tr}\,(\gamma_5)\delta^{(4)}(x-y) = 0$. However, in the limit $x \to y$ the δ-function and thus the whole expression becomes undefined. Stated differently, the fermionic measure contains ultraviolet divergences and needs to be regularized in order to justify the formal manipulations with the path integral. This divergence is evident from Eq. (3.151), as the sum over the eigenvalues is divergent at large values of $|\lambda|$. A regularization that preserves gauge invariance (and thus is similar in spirit to the Pauli–Villars regularization already used before) is given by

$$\sum_\lambda \bar{\psi}_\lambda(x)\gamma^5\psi_\lambda(x) \to \lim_{M\to\infty}\sum_\lambda e^{-\lambda^2/M^2}\bar{\psi}_\lambda(x)\gamma^5\psi_\lambda(x) = \lim_{M\to\infty} S_M(x). \quad (3.152)$$

In order to ensure that Eq. (3.152) indeed provides a cutoff at large Euclidean momenta, note that for large momenta we may neglect the presence of the gluon field in the operator $\slashed{D}$, which now becomes a free Dirac operator. Recall that the eigenfunctions of the free Euclidean Dirac operator are plane waves $\sim e^{\pm ikx}$ with eigenvalues $\lambda^2 = k^2 > 0$. Thus, the cutoff at large λ corresponds to a cutoff of large Euclidean momenta.

Equation (3.152) can be rewritten as

$$S_M(x) = \sum_\lambda \bar{\psi}_\lambda(x)\gamma^5 e^{\slashed{D}^2/M^2}\psi_\lambda(x) = \left\langle x\left|\mathrm{tr}\left(\gamma^5 e^{\slashed{D}^2/M^2}\right)\right|x\right\rangle, \quad (3.153)$$

where "tr" stands for the trace in Dirac, color and flavor indices. Since

$$\slashed{D}^2 = D^2 - \frac{i}{2}\sigma_{\mu\nu}F_{\mu\nu}, \quad (3.154)$$

then

$$S_M(x) = \mathrm{tr}_D(\gamma^5 H_M(x)), \quad H_M(x) = \left\langle x\left|e^{(D^2+\Sigma)/M^2}\right|x\right\rangle, \quad \Sigma = -\frac{i}{2}\sigma_{\mu\nu}F_{\mu\nu}, \quad (3.155)$$

where "tr_D" denotes the trace over the Dirac indices only. We note that the method is closely related to heat kernel regularization, which will be discussed in what follows. The quantity $H_M(x)$ is most easily calculated in momentum space:

$$H_M(x) = \int \frac{d^4p}{(2\pi)^4} e^{-ipx} e^{(D^2+\Sigma)/M^2} e^{ipx}. \quad (3.156)$$

Using the relation

$$D_\mu e^{ipx} = e^{ipx}(ip_\mu + D_\mu) \quad (3.157)$$

and rescaling $p_\mu \to Mp_\mu$ leads to

$$H_M(x) = M^4 \int \frac{d^4p}{(2\pi)^4} e^{-p^2} e^{2ip_\mu D_\mu/M + (D^2+\Sigma)/M^2}. \quad (3.158)$$

Expanding the second exponent up to and including terms of order M^{-4} and performing the momentum integrations by using the following relations:

$$\int \frac{d^4 p}{(2\pi)^4} e^{-p^2} = \frac{1}{16\pi^2},$$

$$\int \frac{d^4 p}{(2\pi)^4} p_\mu p_\nu e^{-p^2} = \frac{1}{32\pi^2} \delta_{\mu\nu},$$

$$\int \frac{d^4 p}{(2\pi)^4} p_\mu p_\nu p_\alpha p_\beta e^{-p^2} = \frac{1}{64\pi^2} (\delta_{\mu\nu}\delta_{\alpha\beta} + \delta_{\mu\alpha}\delta_{\nu\beta} + \delta_{\mu\beta}\delta_{\nu\alpha})$$

$$\int \frac{d^4 p}{(2\pi)^4} p_\mu e^{-p^2} = \int \frac{d^4 p}{(2\pi)^4} p_\mu p_\nu p_\alpha e^{-p^2} = 0, \tag{3.159}$$

we get

$$H_M(x) = \frac{M^4}{16\pi^2} \left\{ 1 + \frac{1}{M^2} \Sigma + \frac{1}{M^4} \left(\frac{1}{2} \Sigma^2 + \frac{1}{12} [D_\mu, D_\nu][D_\mu, D_\nu] \right. \right.$$

$$\left. \left. + \frac{1}{6} [D_\mu, [D^\mu, \Sigma]] \right) + O(M^{-6}) \right\}. \tag{3.160}$$

Multiplying this expression by γ^5 and calculating the trace over Dirac and flavor indices, we obtain

$$\lim_{M\to\infty} S_M(x) = \frac{N_f}{32\pi^2} \varepsilon_{\mu\nu\alpha\beta} \text{tr}_c(F_{\mu\nu}F_{\alpha\beta}), \tag{3.161}$$

where the remaining trace is performed over the color indices only.

Finally, putting all pieces together and performing the limit $M \to \infty$, the Jacobian is given by

$$J = \exp\left\{ i \int d^4x \beta^0(x) \frac{N_f}{32\pi^2} \varepsilon_{\mu\nu\alpha\beta} \text{tr}_c(F_{\mu\nu}F_{\alpha\beta}) + O(\beta^2) \right\}. \tag{3.162}$$

So, the Jacobian is a *phase*. We note in passing that one sees a posteriori – only the singlet axial-vector transformation is anomalous in this case, as the flavor trace for any non-singlet transformation would vanish identically.

Taking into account the Jacobian, the transformed generating functional takes the form

$$Z = \int d\psi d\bar\psi \exp\left\{ - \int d^4x \left(\bar\psi \not{D} \psi - i\beta^0(x) \left(\partial^\mu (\bar\psi \gamma_\mu \gamma^5 \psi) \right. \right. \right.$$

$$\left. \left. \left. - \frac{N_f}{16\pi^2} \varepsilon_{\mu\nu\alpha\beta} \text{tr}_c(F_{\mu\nu}F_{\alpha\beta}) \right) \right) \right\}, \tag{3.163}$$

which gives the divergence of the axial-vector current in massless QCD:

$$\partial_\mu A_\mu^0(x) = \frac{N_f}{16\pi^2} \varepsilon_{\mu\nu\alpha\beta} \text{tr}_c(F_{\mu\nu}F_{\alpha\beta}). \tag{3.164}$$

Note that this expression is similar to the one in QED.

If the quarks are massive, Eq. (3.164) is modified. Going back to Minkowski space at the end of calculations, we get

$$\partial^\mu A_\mu^0(x) = 2i\bar{\psi}\gamma^5 \mathcal{M}\psi + \frac{N_f}{16\pi^2}\,\varepsilon^{\mu\nu\alpha\beta}\,\mathrm{tr}_c(F_{\mu\nu}F_{\alpha\beta}). \tag{3.165}$$

Here, one should again note that, at the end of the day, we have to carry out the integration over the gluon (and ghost) fields as well. As in the Abelian case, it can be shown that the form of Eq. (3.165) stays intact. This property is intuitively plausible because the gluon field is not affected by the flavor transformations. More rigorously, this can be proven by using the regularization with higher covariant derivatives (which is rather cumbersome in the non-Abelian case).

3.10 Non-Abelian Anomalies

In the previous sections, we have set the external scalar, pseudoscalar, vector and axial-vector fields to zero. In this section, we extend the preceding discussion to the general case, where all these external fields are present. The integral over the quark fields in Euclidean space now takes the form

$$Z(s,p,v,a;G) = \int d\psi d\bar{\psi}\exp\left(-\int d^4x\,\bar{\psi}(x)D\psi(x)\right) = \det D, \tag{3.166}$$

where

$$D = \gamma_\mu(\partial_\mu - i\hat{v}_\mu(x) - ia_\mu(x)\gamma_5) + s(x) - i\gamma_5 p(x), \quad \hat{v}_\mu = v_\mu + G_\mu. \tag{3.167}$$

In general, it is not possible to obtain a closed expression for the determinant $\det D$. However, it is possible to write down how this determinant transforms under the transformations belonging to the $U(N_f) \times U(N_f)$ chiral group (an analog to Eq. (3.162)). In what follows, we shall derive this expression, using a novel regularization method, which is different from the one used in the previous section. Namely, we shall closely follow Ref. [26] and tame the divergences of the determinant with the help of the ζ-function technique.

Here we briefly explain the essentials of the method. Consider first a finite-dimensional matrix A, whose eigenvalues a_n, $n = 1,\ldots N$, all have a positive real part. The quantity

$$F(s) = \frac{\mu^{2s}}{\Gamma(s)}\int_0^\infty d\lambda\,\lambda^{s-1}\mathrm{tr}(e^{-\lambda A}) = \frac{\mu^{2s}}{\Gamma(s)}\int_0^\infty d\lambda\,\lambda^{s-1}\sum_{n=1}^N e^{-\lambda a_n}, \quad s > 0,$$

$$\tag{3.168}$$

where μ is an arbitrary positive real parameter, is well defined, since the integral converges at both limits. Note that large (small) values of λ correspond to the infrared (ultraviolet) region. Evaluating the integral, we get

$$F(s) = \mu^{2s} \sum_{n=1}^{N} (a_n)^{-s}. \tag{3.169}$$

From this, we straightforwardly obtain

$$-\frac{d}{ds} F(s)\bigg|_{s=0} = -N \ln \mu^2 + \sum_N \ln a_n = -N \ln \mu^2 + \mathrm{tr}\,(\ln A)$$

$$= \mathrm{const} + \ln \det A, \tag{3.170}$$

that is, up to an irrelevant constant, $\ln \det A$ is given by the expression on the l.h.s. of Eq. (3.170).

We shall generalize the preceding formula to calculate the quantity $\det D$. It should be pointed out first that the method is not applicable immediately in this case, because the operator iD is not Hermitean (and, consequently, the eigenvalues of $-D^2$ are not positively definite). However, here one is interested only in the perturbative expansion of the determinant in the external fields, and hence it suffices if the pertinent differential operator is Hermitean at the origin of the expansion. Now, it is seen that the operator $\bar{D} = \gamma_5 D$ is Hermitean, if $p = a_\mu = 0$, $s = \mathcal{M}$, whereas v_μ and the gluon field G_μ can be arbitrary. One can thus define the determinant according to

$$\ln \det D = -\frac{d}{ds} \frac{\mu^{2s}}{2\Gamma(s)} \int_0^\infty d\lambda\, \lambda^{s-1} \mathrm{Tr}\,(e^{-\lambda \bar{D}^2})\bigg|_{s=0}. \tag{3.171}$$

Here the symbol "Tr" apart from usual flavor, color and Dirac indices, implies the integration over the four-dimensional Euclidean space as well. Note that, unlike in the finite-dimensional case, the determinant depends on the scale μ due to the presence of the UV divergences (see what follows).

Under the local chiral transformations given in Eq. (3.77), the operator $\bar{D}$ transforms as

$$\delta \bar{D} = -i[\bar{D}, \alpha] - i\{\bar{D}, \beta \gamma_5\}. \tag{3.172}$$

From the preceding relation, using the cyclic property of the trace, we obtain

$$\delta \mathrm{Tr}\,\exp(-\lambda \bar{D}^2) = -2\lambda \mathrm{Tr}\,[\delta \bar{D} \bar{D} \exp(-\lambda \bar{D}^2)] = 4i\lambda \frac{d}{d\lambda}[\beta \gamma_5 \exp(-\lambda \bar{D}^2)]. \tag{3.173}$$

Substituting this expression in Eq. (3.171), we obtain the transformation rule for the determinant under local chiral transformations. By the way, from the preceding expression it is already seen that the vector Ward identities are automatically preserved by this regularization, because the parameter α, which describes vector transformations, never appears there.

Our next aim is to derive an explicit expression for $\delta \ln \det D$. To this end, the heat kernel method will be used. In a first step, the expression for $\bar{D}^2$ is simplified by using the Dirac algebra:

$$\bar{D}^2 = (-\gamma_\mu \partial_\mu - r_+)(\gamma_\nu \partial_\nu + r_-), \tag{3.174}$$

where

$$r_{\pm} = -i\gamma_{\mu}(v_{\mu} + a_{\mu}\gamma_5) \mp (s - i\gamma_5 p). \tag{3.175}$$

Further, we introduce the notations

$$\Gamma_{\mu} = \frac{1}{2}(r_+\gamma_{\mu} + \gamma_{\mu}r_-) = -i(\hat{v}_{\mu} - i\sigma_{\mu\nu}\gamma_5 a_{\nu} + \gamma_{\mu}\gamma_5 p),$$

$$\xi_{\mu} = \frac{1}{2}(r_+\gamma_{\mu} - \gamma_{\mu}r_-) = \sigma_{\mu\nu}\hat{v}_{\nu} + i\gamma_5 a_{\mu} - \gamma_{\mu}s. \tag{3.176}$$

Equation (3.174) can be then rewritten as

$$\bar{D}^2 = -\partial_{\mu}\partial_{\mu} - \partial_{\mu}(\Gamma_{\mu} - \xi_{\mu}) - (\Gamma_{\mu} + \xi_{\mu})\partial_{\mu} - r_+r_- = -D_{\mu}D_{\mu} + \sigma, \tag{3.177}$$

where $D_{\mu} = \partial_{\mu} + \Gamma_{\mu}$ and the explicit expression for the quantity σ is [26]:

$$\sigma = \Gamma_{\mu}\Gamma_{\mu} + (\partial_{\mu}\xi_{\mu}) - r_+r_-$$

$$= (2a_{\mu}a_{\mu} + s^2 + 3p^2) - \gamma_{\mu}(\partial_{\mu}s - i[\hat{v}_{\mu}, s] + 2\{a_{\mu}, p\})$$

$$- i\sigma_{\mu\nu}(i\partial_{\mu}\hat{v}_{\nu} + \hat{v}_{\mu}\hat{v}_{\nu} + a_{\mu}a_{\nu}) + i\gamma_{\mu}\gamma_5[a_{\mu}, s]$$

$$+ i\gamma_5(\partial_{\mu}a_{\mu} - i[\hat{v}_{\mu}, a_{\mu}] - \{s, p\}). \tag{3.178}$$

Consider now the operator $(x|e^{-\lambda\bar{D}^2}|y)$ in coordinate space,[9] and single out the most singular behavior at short distances, which is given by the differential operator $\Box = \partial_{\mu}\partial_{\mu}$ that contains no external fields at all. Using the relation

$$(x|e^{\lambda\Box}|y) = \int \frac{d^4p}{(2\pi)^4} e^{-ip(x-y)-\lambda p^2} = \frac{1}{(4\pi\lambda)^2} \exp\left(-\frac{z^2}{4\lambda}\right), \tag{3.179}$$

where $z = x - y$, we define

$$(x|e^{-\lambda\bar{D}^2}|y) = \frac{1}{(4\pi\lambda)^2} \exp\left(-\frac{z^2}{4\lambda}\right) H(x|\lambda|y), \tag{3.180}$$

where the kernel H obeys the differential equation:

$$\frac{\partial}{\partial\lambda} H + \frac{1}{\lambda} z_{\mu}D_{\mu}H - D_{\mu}D_{\mu}H + \sigma H = 0. \tag{3.181}$$

Further, H obeys the boundary condition:

$$H(x|0|x) = 1. \tag{3.182}$$

Now, one can expand the function H as a Taylor series in λ:

$$H(x|\lambda|y) = H_0(x|y) + \lambda H_1(x|y) + \lambda^2 H_2(x|y) + \cdots. \tag{3.183}$$

The coefficients of this expansion are known under the name of the Seeley–DeWitt coefficients, see, for example, Ref. [27]. They obey the following recursion relations:

$$(n + 1 + z_{\mu}D_{\mu})H_{n+1} + (-D_{\mu}D_{\mu} + \sigma)H_n = 0, \qquad z_{\mu}D_{\mu}H_0 = 0. \tag{3.184}$$

---

[9] Note that the round bracket notation employed here and in what follows defines the heat kernel in coordinate space.

Substituting this expansion into the integral, one can calculate both $\ln \det D$ and its variation $\delta \ln \det D$ under local chiral transformation. The expression for $\ln \det D$ takes the form

$$\ln \det D = -\frac{d}{ds} \frac{\mu^{2s}}{32\pi^2 \Gamma(s)} \int_0^\infty d\lambda \, \lambda^{s-3} \text{Tr}\left(\sum_{n=0}^\infty \lambda^n H_n(x|x)\right)\bigg|_{s=0}. \tag{3.185}$$

Here, we have to introduce some kind of an infrared (IR) cutoff in the integral to cope with the divergences at large values of λ. These divergences arise from the Taylor expansion of the kernel H. One could, for example, cut the integral at some value $\lambda = \lambda_{max}$. Thus, Eq. (3.185) can be used to get information about ultraviolet divergences that arise at small values of λ (and to which only a finite number of terms contribute), but cannot provide a closed expression of the determinant itself. Note also that the quantity defined by Eq. (3.185) is finite. The ultraviolet divergence manifests itself in the dependence of the determinant on the scale μ. It is easy to see that, differentiating with respect to s and letting $s \to 0$, only the term with $n - 2$ gives the μ-dependent contribution with

$$\mu \frac{\partial}{\partial \mu} \ln \det D = -\frac{1}{(4\pi)^2} \int d^4x \, \text{tr} \, H_2(x|x). \tag{3.186}$$

Here, the symbol "tr" implies traces over color, flavor and Dirac indices only. This scale dependence reproduces the one obtained by using the conventional dimensional regularization.

In contrast to $\ln \det D$, an explicit expression for $\delta \ln \det D$ *can* be obtained in a closed form. Indeed, performing a partial integration in λ and dropping surface terms, we get

$$\delta \ln \det D = \frac{d}{ds} \frac{2is\mu^{2s}}{\Gamma(s)} \int_0^\infty d\lambda \, \lambda^{s-1} \text{Tr}(\beta \gamma_5 e^{-\lambda \bar{D}^2})\bigg|_{s=0}$$

$$= \frac{d}{ds} \frac{is\mu^{2s}}{8\pi^2 \Gamma(s)} \int_0^\infty d\lambda \, \lambda^{s-3} \text{Tr}\left(\beta \gamma_5 \sum_{n=0}^\infty \lambda^n H_n(x|x)\right)\bigg|_{s=0}. \tag{3.187}$$

It is easy to check that, because of the presence of the additional power of s in the numerator, all terms in this sum vanish in the limit $s \to 0$, except for the term with $n = 2$. Thus, the expression for $\delta \ln \det D$ can be written in a closed form in terms of the second Seeley–DeWitt coefficient:

$$\delta \ln \det D = -\frac{i}{8\pi^2} \int d^4x \, \text{tr}\left[\beta(x) \gamma_5 H_2(x|x)\right]. \tag{3.188}$$

What remains now is to find an explicit expression for $H_2(x|x)$ from the recursion relations given in Eq. (3.184). First, as a consequence of the boundary condition, we may set $H_0(x|x) = 1$. Further, acting on the relation $z_\mu D_\mu H_0 = 0$ with the operator $D_\nu D_\nu$ and letting $z \to 0$ at the end, we get

$$\lim_{x \to y} D_\nu D_\nu H_0(x|y) = 0, \tag{3.189}$$

and, consequently,

$$H_1(x|x) = -\sigma(x). \tag{3.190}$$

At the next step we obtain from the recursion relations at $z = 0$:

$$H_2(x|x) = \frac{1}{2} D_\mu D_\mu H_1(x|x) - \frac{1}{2} H_1(x|x). \tag{3.191}$$

The first term in this expression can be obtained by acting on the recursion relation

$$(1 + z_\mu D_\mu)H_1 + (-D_\mu D_\mu + \sigma)H_0 = 0, \tag{3.192}$$

with the operator $D_\nu D_\nu$ and letting $z \to 0$. Consequently, one obtains

$$3D_\mu D_\mu H_1 = (D_\mu D_\mu)^2 H_0 - [D_\mu, [D_\mu, \sigma]]H_0 - 2[D_\mu, \sigma]D_\mu H_0 - \sigma D_\mu D_\mu H_0. \tag{3.193}$$

Next, acting repeatedly on the equation $z_\mu D_\mu H_0 = 0$ with the operator D_ν, we get the following relations:

$$D_\mu H_0 + z_\lambda D_\mu D_\lambda H_0 = 0,$$

$$\{D_\mu, D_\nu\}H_0 + z_\lambda D_\mu D_\nu D_\lambda H_0 = \{D_\mu, D_\nu\}H_0 + z_\lambda D_\nu D_\mu D_\lambda H_0 = 0,$$

$$2D_\mu D_\mu D_\nu D_\nu H_0 + 2D_\mu D_\nu D_\nu D_\mu H_0 + z_\lambda D_\mu D_\mu D_\nu D_\nu D_\lambda H_0 = 0. \tag{3.194}$$

Recalling now that the commutator of two covariant derivatives gives the gauge field tensor, which is the function of the fields only and does not contain derivatives,

$$[D_\mu, D_\nu] = \partial_\mu \Gamma_\nu - \partial_\nu \Gamma_\mu + [\Gamma_\mu, \Gamma_\nu] \doteq \Gamma_{\mu\nu}, \tag{3.195}$$

from the last two relations of Eq. (3.194) we obtain

$$-2D_\mu D_\nu D_\nu D_\mu H_0 = D_\mu D_\nu \Gamma_{\mu\nu} H_0 + D_\mu D_\mu D_\nu D_\nu H_0 + z_\lambda D_\mu D_\nu D_\nu D_\mu D_\lambda H_0. \tag{3.196}$$

Letting $z \to 0$ gives

$$D_\mu H_0(x|x) = 0,$$

$$D_\mu D_\nu H_0(x|x) = \frac{1}{2} \Gamma_{\mu\nu},$$

$$D_\mu D_\mu D_\nu D_\nu H_0(x|x) = \frac{1}{2} \Gamma_{\mu\nu} \Gamma_{\mu\nu}. \tag{3.197}$$

Finally, using these relations, the expression for the second Seeley–DeWitt coefficient is obtained:

$$H_2(x|x) = \frac{1}{12} \Gamma_{\mu\nu} \Gamma_{\mu\nu} + \frac{1}{2} \sigma^2 - \frac{1}{6} [D_\mu, [D_\mu, \sigma]]. \tag{3.198}$$

The calculation of the higher coefficients proceeds along the same path. However, they will not be needed here.

Substituting Eq. (3.198) into Eq. (3.188) and calculating the traces over Dirac indices, we arrive at a rather lengthy expression, which consists of two parts. The first part contains the tensor $\varepsilon_{\mu\nu\alpha\beta}$ and is complex (i.e., contributes to the *phase* of the effective action). The second part is real and does not contain $\varepsilon_{\mu\nu\alpha\beta}$. Thus, the quantity $\delta \ln \det D$ can be written in the form

$$\delta \ln \det D = \frac{i}{16\pi^2} \int d^4x \left\{ \varepsilon_{\mu\nu\alpha\beta} \mathrm{tr}_c (F_{\mu\nu} F_{\alpha\beta}) \mathrm{tr}_f \beta + N_c \mathrm{tr} (\beta\Omega) \right\}$$
$$+ R(s, p, v, a), \tag{3.199}$$

where tr_c, tr_f stand for the color and flavor traces, respectively, and $N_c = 3$ is the number of colors. The quantity Ω is given by

$$\Omega = \varepsilon_{\alpha\beta\mu\nu} \left\{ v_{\alpha\beta} v_{\mu\nu} + \frac{4}{3} \nabla_\alpha a_\beta \nabla_\mu a_\nu + \frac{2i}{3} \{ v_{\alpha\beta}, a_\mu a_\nu \} + \frac{8i}{3} a_\mu v_{\alpha\beta} a_\nu \right.$$
$$\left. + \frac{4}{3} a_\alpha a_\beta a_\mu a_\nu \right\}, \tag{3.200}$$

where $v_{\alpha\beta}$ and $\nabla_\alpha a_\beta$ are defined as

$$v_{\alpha\beta} = \partial_\alpha v_\beta - \partial_\beta v_\alpha - i[v_\alpha, v_\beta],$$
$$\nabla_\alpha a_\beta = \partial_\alpha a_\beta - i[v_\alpha, a_\beta], \tag{3.201}$$

and R denotes the real part of $\delta \ln \det D$ (a rather lengthy expression, which is not displayed here). It is important to stress that $R = R(s, p, v, a)$ does not depend on the gluon fields. Further, it turns out that R is irrelevant, for the following reason: It is always possible to add a polynomial $P(s, p, v, a)$ containing *only* external fields s, p, v, a to the effective action $\ln \det D$. Such an additional polynomial changes the Green's functions of the external currents only by local terms and thus amounts to a choice of the renormalization prescription. In Ref. [26], an explicit expression for the polynomial P is given, with the property

$$\delta P(s, p, v, a) = R(s, p, v, a). \tag{3.202}$$

Consequently, the transformation rule of the modified effective action,

$$\ln \det D \to \ln \det D - \int d^4x P(s, p, v, a), \tag{3.203}$$

is given by Eq. (3.199) with the real part $R(s, p, v, a)$ set to zero.[10] Note also that the first term in Eq. (3.199), containing the gluon field, was already obtained in previous sections. What is new here is the second part, containing external fields. The full expression for this part was first obtained in Ref. [14].

Note also that by changing the renormalization prescription

$$\ln \det D \to \ln \det D$$
$$+ \frac{1}{24\pi^2} \varepsilon_{\alpha\beta\mu\nu} \int d^4x \mathrm{tr} \left(i \hat{v}_{\mu\nu} \{ \hat{v}_\alpha, a_\beta \} - a_\alpha a_\beta a_\mu \hat{v}_\nu - \hat{v}_\alpha \hat{v}_\beta \hat{v}_\mu a_\nu \right), \tag{3.204}$$

---

[10] As we have seen in previous sections, we cannot do the same with the imaginary part without violating the vector Ward identities.

the contributions from the left- and right-handed sources in the expression of the anomaly are separated:

$$\delta \ln \det D = \frac{1}{16\pi^2} \int d^4x \left(\mathrm{tr}((\alpha+\beta)A(\hat{F}^R)) + \mathrm{tr}((\alpha-\beta)A(\hat{F}^L)) \right), \quad (3.205)$$

where $\hat{F}^R_\mu = \hat{v}_\mu + a_\mu$ and $\hat{F}^L_\mu = \hat{v}_\mu - a_\mu$. The functional $A(F)$ is a total derivative:

$$A(F) = \frac{2}{3} \varepsilon_{\alpha\beta\mu\nu} \partial_\alpha \left(F_\beta \partial_\mu F_\nu - \frac{i}{2} F_\beta F_\mu F_\nu \right). \quad (3.206)$$

However, it should be also noted that in this renormalization prescription the vector current is not anomaly-free.

Now, we are in a position to update the discussion of the Ward identities, which were carried out in Section 3.6, and take into account the anomalies. Switching back from Euclidean to Minkowski space, we define the Lagrangian, equipped with the external fields:

$$\mathcal{L} = \mathcal{L}_G - \frac{\theta(x)}{16\pi^2} \varepsilon^{\mu\nu\alpha\beta} \mathrm{tr}_c(F_{\mu\nu}F_{\alpha\beta}) + \bar{\psi}(\gamma^\mu(iD_\mu + v_\mu + \gamma^5 a_\mu) - s + i\gamma^5 p)\psi. \quad (3.207)$$

Note that here an additional (flavor singlet) source $\theta(x)$ has been introduced, in order to account for the anomaly in the singlet axial current. The transformations, given in Eq. (3.77), are amended by the transformation of θ:

$$\delta v_\mu = \partial_\mu \alpha + i[\alpha, v_\mu] + i[\beta, a_\mu],$$

$$\delta a_\mu = \partial_\mu \beta + i[\alpha, a_\mu] + i[\beta, v_\mu],$$

$$\delta s = i[\alpha, s] - \{\beta, p\},$$

$$\delta p = i[\alpha, p] + \{\beta, s\},$$

$$\delta \theta = -2\mathrm{tr}_f \beta. \quad (3.208)$$

Thus, Eq. (3.78) in the presence of anomalies is modified to:

$$-\ln Z(s + \delta s, p + \delta p, v + \delta v, a + \delta a, \theta + \delta\theta)$$

$$= -\ln Z(s, p, v, a, \theta) + \frac{iN_c}{16\pi^2} \int d^4x \, \mathrm{tr}_f[\beta(x)\Omega(x)], \quad (3.209)$$

where Ω is a polynomial in the external fields, which is given by Eq. (3.200). The Ward identities are obtained from this relation in a standard manner, namely by Taylor-expanding it with respect to the external fields s, p, v, a, θ. In particular, from the structure of Ω, it is seen that only the Green's functions of the vector and axial-vector currents with up to and including five external legs are anomalous, as anticipated.

Up to now, we have treated $\theta(x)$ as a purely formal object that is present in the generating functional, namely, as an external field that couples to the winding number density. In order to obtain the Green's functions, we have to expand the generating functional in all external sources. The question arises now, whether this expansion should be carried out in the vicinity of $\theta(x) = 0$, or some other nonzero value $\theta(x) = \theta_0 \neq 0$ that would signal CP violation in QCD. (The pertinent term in the Lagrangian is CP-odd, as the reader might easily verify.) It has been shown that such a term naturally emerges in QCD due to the nontrivial topological structure of vacuum gluon field configurations, see, for example, Refs. [28, 29], and a very clear introduction to the subject is contained in the textbook [30]. The question why the strong CP violation is not observed in nature, which, in the present context, boils down to a fine-tuning of the parameter θ_0, is known under the name of the strong CP problem. We shall address this problem in more detail in Section 4.20. Note further that we will encounter the last term in Eq. (3.209) in the discussion of the WZW term in Section 4.16.

Finally, note that we have considered only infinitesimal transformations in this section. Finite transformations can be built as a sequence of infinitesimal ones. It is easy to see that the generating functional possesses certain periodicity properties with respect to the parameter θ. Let us, for simplicity, set the vector and axial-vector sources to zero (which leads to the vanishing of the non-Abelian anomaly, encoded in the functional Ω), and consider a one-parameter set of transformations $g_R = \exp(-i\gamma)$, $g_L = \mathbb{1}$. The parameter γ does not depend on x (otherwise, setting $v_\mu = a_\mu = 0$ is inconsistent). Changing γ continuously from 0 to 2π, we see that s and p return to their original values, whereas θ changes by 2π. This means that the functional Z with $v_\mu = a_\mu = 0$ is periodic in $\theta(x)$ with the period 2π:

$$Z(s, p, 0, 0, \theta + 2\pi) = Z(s, p, 0, 0, \theta). \tag{3.210}$$

In particular, the vacuum angle $\bar{\theta}$ is defined modulo 2π.

3.11 Atiyah–Singer Index Theorem

In this section, we shall demonstrate that the singlet anomaly (i.e., the anomaly that remains after setting $s = p = v = a = 0$ in Eq. (3.199)) is directly related to the so-called zero modes of the massless Dirac operator in a given external gauge field G_μ. To this end, consider the Euclidean Dirac operator $\gamma_\mu D_\mu = \gamma_\mu(\partial_\mu + G_\mu)$. The eigenvectors of this operator obey the following equation:

$$\gamma_\mu D_\mu \psi_\lambda(x) = i\lambda \psi_\lambda(x). \tag{3.211}$$

Further, if $\psi_\lambda(x)$ is an eigenvector, corresponding to the eigenvalue λ, then, obviously, $\gamma_5 \psi_\lambda(x)$ will be an eigenvector, too, corresponding to the eigenvalue $-\lambda$. Since $i\gamma_\mu D_\mu$ is a Hermitean operator, these two eigenvectors are orthogonal, if $\lambda \neq 0$:

$$\int d^4x \, \bar{\psi}_\lambda(x) \gamma_5 \psi_\lambda(x) = 0. \tag{3.212}$$

However, if $\lambda = 0$, the argument is not applicable anymore, because λ and $-\lambda$ are the same. Let $\psi_0^i(x)$, $i = 1, \ldots, n$, be the zero modes of the Dirac operator (we assume that the number of zero modes is finite):

$$\gamma_\mu D_\mu \psi_0^i(x) = 0. \tag{3.213}$$

Further, if $\psi_0^i(x)$ is a zero mode, so is $\gamma_5 \psi_0^i(x)$. Thus, the eigenfunctions can be chosen so that they have either positive or negative chirality:

$$\gamma^5 \psi_0^i(x) = \pm \psi_0^i(x). \tag{3.214}$$

Denote by n_+ and n_- the number positive- and negative-chirality eigenmodes (with $n_+ + n_- = n$). Then, it can be straightforwardly checked that

$$\int d^4x \sum_\lambda \bar{\psi}_\lambda(x) \gamma_5 \psi_\lambda(x) = n_+ - n_-. \tag{3.215}$$

On the other hand, we have derived that (see Eq. (3.161))

$$\sum_\lambda \bar{\psi}_\lambda(x) \gamma_5 \psi_\lambda(x) = \frac{N_f}{32\pi^2} \varepsilon_{\mu\nu\alpha\beta} \mathrm{tr}_c \left[F_{\mu\nu}(x) F_{\alpha\beta}(x) \right]. \tag{3.216}$$

Comparing Eqs. (3.214) and (3.215), we arrive at the Atiyah–Singer index theorem [31]:

$$n_+ - n_- = \frac{N_f}{32\pi^2} \varepsilon^{\mu\nu\alpha\beta} \int d^4x \, \mathrm{tr}_c \left[F_{\mu\nu}(x) F_{\alpha\beta}(x) \right]. \tag{3.217}$$

This is a remarkable result, as it links the topology of the QCD vacuum to the zero modes of the Dirac operator.

3.12 Chiral Symmetry and Anomalies on the Lattice

3.12.1 Lattice QCD Factsheet

From the discussion in the previous sections you might be tempted to conclude that the anomalies emerge because it is not possible to introduce an ultraviolet regularization, which would respect all symmetries of the classical Lagrangian simultaneously. This might give the impression that the issue is purely technical and perhaps related to the mathematical deficiencies in the regularization procedure, whereas in reality the situation is opposite. In order to better understand the problem, here we consider axial anomalies in lattice QCD, which, in contrast to regularized continuum QCD, is a well-defined quantum-mechanical system even at a finite lattice spacing. In this formulation, the ultraviolet divergences are tamed by defining the theory on a discrete space-time grid, so lattice QCD can be considered as a particular ultraviolet regularization of continuum QCD. Further, the path integral measure on the lattice contains a *finite* number of integration variables and is therefore finite, thus there is no need for an additional regularization. Consequently, the Jacobian of the fermion measure transformation is equal to unity and no anomalies are present there. However, for consistency reasons, the anomalies *must* emerge in the lattice formulation somewhere. It is both very interesting and extremely instructive to trace back the origin of the anomalies in this case. The lesson learned from the discussion will be that anomalies are a fundamental property of the theory, not related to the choice of a particular regularization procedure. If they are eliminated at one place, they have to appear at another one.

Presenting a detailed description of lattice QCD is beyond the scope of the present book. A nice introduction to the subject can be found, for example, in the textbooks [32–35]. As in the case of continuum QCD, we shall restrict ourselves to a brief factsheet including all necessary information that is needed for the discussion of anomalies.

Let us start with fermions, which play a central role in the emergence of anomalies. The continuum Euclidean action of free fermions with mass m takes the form

$$S_F = \int d^4x \, \bar{\psi}(x)(\gamma_\mu \partial_\mu + m)\psi(x) \,. \tag{3.218}$$

Note that in case of N_f flavors, m is replaced by the mass matrix $\mathcal{M}$. The partition function in the continuum is given by

$$Z = \int d\psi d\bar{\psi} \, e^{-S_F} \,, \tag{3.219}$$

and Green's functions are obtained in a standard manner by introducing external sources, differentiating with respect to them and letting them vanish at the end.

On the lattice, the integral in Eq. (3.218) is replaced by a sum over the discrete points of a (hypercubic) lattice,[11]

$$x_\mu = a n \hat{\mu} \,, \qquad n = 0, 1, \ldots N - 1 \,, \tag{3.220}$$

where $\hat{\mu}$ denotes a unit vector in the direction of $\mu = 0, 1, 2, 3$, a is the lattice spacing, $L = Na$ is the side length of a hypercube and N^4 is the total number of lattice points. A straightforward discretization of the continuum Euclidean action is given by

$$S_F^{\text{naive}} = a^4 \sum_x \sum_\mu \bar{\psi}(x)\gamma_\mu \frac{1}{2}(\partial_\mu + \partial_\mu^*)\psi(x) + a^4 \sum_x \bar{\psi}(x)m\psi(x) \,, \tag{3.221}$$

where the fermion field $\psi(x)$ lives on lattice grid points, the so-called sites (see Fig. 3.6), and ∂_μ and ∂_μ^* denote the lattice derivatives in their most simple form:

$$\partial_\mu \psi(x) = \frac{1}{a}\left[\psi(x + a\hat{\mu}) - \psi(x)\right] \,,$$

$$\partial_\mu^* \psi(x) = \frac{1}{a}\left[\psi(x) - \psi(x - a\hat{\mu})\right] \,. \tag{3.222}$$

Further, the fermions obey periodic/antiperiodic boundary conditions in the spatial/temporal directions (for a more detailed discussion, see, e.g., [35]):

$$\psi(x + N\hat{\mu}) = +\psi(x) \,, \qquad \mu = 1, 2, 3 \,,$$

---

[11] Here, we adopt the simplest choice of the lattice geometry, with the same number of intervals as well as lattice spacing in the spatial and temporal directions.

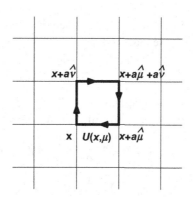

Figure 3.6 Definition of the fermion and gauge fields on the lattice (for demonstrative purposes, the picture is drawn in two dimensions $\hat{\mu}, \hat{\nu}$). The fermion fields live on the grid points $x, x + a\hat{\mu}, x + a\hat{\nu}, x + a\hat{\mu} + a\hat{\nu}, \dots$, whereas gauge fields live on the links connecting each two grid points. The bold square in the middle of the figure denotes the elementary plaquette on the lattice.

$$\psi(x + N\hat{\mu}) = -\psi(x), \qquad \mu = 0. \tag{3.223}$$

The Green's functions in the discretized Euclidean theory are given by

$$\langle \psi(x_1) \cdots \psi(x_n) \bar{\psi}(y_1) \cdots \bar{\psi}(y_n) \rangle$$

$$= \frac{1}{Z} \int \prod_x (d\psi_x d\bar{\psi}_x) \, e^{-S_F} \, \psi(x_1) \cdots \psi(x_n) \bar{\psi}(y_1) \cdots \bar{\psi}(y_n), \tag{3.224}$$

where

$$Z = \int \prod_x (d\psi_x d\bar{\psi}_x) \, e^{-S_F}. \tag{3.225}$$

The gauge fields are introduced in order to preserve the exact gauge symmetry on the lattice. These fields are described by the so-called link variables $U(x, \mu)$, which live on a link connecting the lattice points x and $x + a\hat{\mu}$. In the continuum, the link variables correspond to the parallel transporter along any path $\mathscr{C}$, connecting the points y and x:

$$U(\mathscr{C}) = P \exp\left(-\int_y^x dz_\mu G_\mu(z) \right), \tag{3.226}$$

where G_μ is the gauge field and the symbol "P" denotes path ordering. Applying *continuum* gauge transformations of fermion and gauge fields,

$$\psi(x) \mapsto \Lambda(x)\psi(x), \qquad \bar{\psi}(x) \mapsto \bar{\psi}(x)\Lambda(x)^\dagger,$$

$$G_\mu(x) \mapsto \Lambda(x) G_\mu(x) \Lambda(x)^\dagger - \partial_\mu \Lambda(x) \Lambda(x)^\dagger, \tag{3.227}$$

it is seen that the parallel transporter, defined by Eq. (3.226), transforms as

$$U(\mathscr{C}) \mapsto \Lambda(x) U(\mathscr{C}) \Lambda(y)^\dagger. \tag{3.228}$$

On the lattice, $U(x, \mu) = U(\mathscr{C})$, where $\mathscr{C}$ is a straight line, connecting $x + a\hat{\mu}$ and x. In the limit $a \to 0$, we have

$$U(x, \mu) \simeq \exp\left[-aG_\mu(x)\right] \simeq 1 + \frac{i\lambda^b}{2} aG_\mu^b(x), \tag{3.229}$$

with b the index corresponding to the $SU(N_c)$ gauge symmetry. It is straightforward to check that the lattice covariant derivatives, defined as

$$D_\mu \psi(x) = \frac{1}{a}\left[U(x, \mu)\psi(x + a\hat{\mu}) - \psi(x)\right],$$

$$D_\mu^* \psi(x) = \frac{1}{a}\left[\psi(x) - U(x - a\hat{\mu}, \mu)^{-1}\psi(x - a\hat{\mu})\right], \tag{3.230}$$

transform like $\psi(x)$ under local gauge transformations, if the link transforms according to Eq. (3.228). Hence, the lattice action, which is a straightforward generalization of Eq. (3.221), can be written as

$$S_F^{\text{naive}} = \frac{1}{2} a^3 \sum_x \sum_\mu \bar{\psi}(x)\gamma_\mu\left[U(x, \mu)\psi(x + a\hat{\mu}) - U(x - a\hat{\mu}, \mu)^{-1}\psi(x - a\hat{\mu})\right]$$

$$+ ma^4 \sum_x \bar{\psi}(x)\psi(x). \tag{3.231}$$

It is invariant under local gauge transformations. Moreover, expanding in powers of a, it is readily seen that, at leading order, the continuum action functional for the fermions that interact with the gauge field G_μ is reproduced.

In order to write the kinetic term for the gauge field, consider the closed loop containing a single lattice cell (the so-called *plaquette*); see Fig. 3.6. The plaquette field

$$P_{\mu\nu}(x) = U(x, \mu)U(x + a\hat{\mu}, \nu)U(x + a\hat{\nu}, \mu)^{-1}U(x, \nu)^{-1} \tag{3.232}$$

under the local gauge transformations transforms as

$$P_{\mu\nu}(x) \mapsto \Lambda(x)P_{\mu\nu}(x)\Lambda(x)^\dagger. \tag{3.233}$$

Consequently, $\operatorname{tr}P_{\mu\nu}(x)$ is a gauge-invariant field. Expanding in the continuum limit $a \to 0$, we get

$$\operatorname{tr}P_{\mu\nu}(x) = N_c - \frac{1}{2} a^4 \operatorname{tr}\left[F_{\mu\nu}(x)F_{\mu\nu}(x)\right] + \cdots. \tag{3.234}$$

Consequently, the gluonic part of the lattice action can be defined as

$$S_G = \frac{1}{g_0^2} \sum_x \sum_{\mu\nu} \operatorname{Re}\operatorname{tr}\left[1 - P_{\mu\nu}(x)\right], \tag{3.235}$$

where g_0 denotes the bare coupling constant on the lattice.

In lattice QCD, the Green's functions of the gauge-singlet composite fields (meson and baryon fields) are calculated in a standard manner, expressing them through path integrals over the quark and gluon fields and using the discretized action. Since the fermionic part of the action is bilinear with respect to the fermion fields, the integral over these can be done analytically. At the end, the (finite-dimensional) integral over the gluon variables $U(x, \mu)$ can be evaluated by using Monte Carlo methods.

3.12.2 Absence of Anomalies with Naive Fermions

Now, we are in a position to formulate our problem. As already mentioned, the lattice represents a particular regularization scheme, where all ultraviolet divergences are cut off at the momenta of order of the inverse lattice spacing, $1/a$. However, this does not boil down to merely placing a cutoff in all Feynman diagrams that leads to an ill-defined (non-unitary) theory for a finite cutoff. In contrast to this, the lattice formulation represents a consistent field-theoretical framework even for nonzero values of the lattice spacing a; in particular, it preserves the exact gauge symmetry for any finite a. Further, the (finite-dimensional) fermionic measure needs no regularization and is thus explicitly gauge-invariant. In addition, the theory is formulated in four space-time dimensions from the beginning, so the definition of the matrix γ_5 is free of any problem. At first glance, it seems that we have exhausted all loopholes, which would allow the anomaly to sneak in. Such a conclusion would, however, be premature. Indeed, consider the fermionic part of the action and, for simplicity, neglect the gauge fields at all, setting $U(x,\mu)=1$. The action takes the form

$$
S_F^{\text{naive}} = \frac{1}{2} a^3 \sum_x \sum_\mu \bar{\psi}(x)\gamma_\mu\left[\psi(x+a\hat{\mu})-\psi(x-a\hat{\mu})\right] + ma^4 \sum_x \bar{\psi}(x)\psi(x)
$$

$$
= a^4 \sum_{xy} \bar{\psi}(x)K(x,y)\psi(y), \tag{3.236}
$$

where

$$
K(x,y) = \frac{1}{2a}\sum_\mu \gamma_\mu\left(\delta_{x+a\hat{\mu},y}-\delta_{x-a\hat{\mu},y}\right) + m\delta_{x,y}. \tag{3.237}
$$

Defining the Fourier transformation,

$$
\tilde{\psi}(p) = a^4 \sum_x e^{-ipx}\psi(x), \tag{3.238}
$$

it is possible to rewrite the fermion action in momentum space:

$$
S_F^{\text{naive}} = (Na)^{-4}\sum_p \tilde{\bar{\psi}}(p)\tilde{K}(p)\tilde{\psi}(p), \tag{3.239}
$$

where

$$
\tilde{K}(p) = \frac{1}{2a}\sum_\mu \gamma_\mu(e^{iap_\mu}-e^{-iap_\mu})+m = \frac{i}{a}\sum_\mu \gamma_\mu \sin(ap_\mu)+m. \tag{3.240}
$$

The sum over momenta runs over the discrete values of the momentum,

$$
p_\mu = \frac{2\pi}{L}n_\mu, \qquad n_\mu \in \mathbb{Z}^4, \tag{3.241}
$$

up to the highest value in the first Brillouin zone:

$$
|p_\mu| \leq \frac{\pi}{a}. \tag{3.242}
$$

When $L \to \infty$, the sum over momenta in the action transforms into the integral over the first Brillouin zone, $-\pi/a \leq p_\mu \leq \pi/a$.

The fermion propagator in momentum space is the inverse of $\tilde{K}(p)$:

$$S(p) = \tilde{K}^{-1}(p) = \frac{m - i\gamma_\mu s_\mu}{m^2 + s^2}, \qquad s_\mu = \frac{1}{a}\sin(ap_\mu). \tag{3.243}$$

In the limit $a \to 0$, the function $S(p)$ reduces to the well-known continuum propagator:

$$s_\mu = p_\mu + O(a^2), \qquad S(p) = \frac{m - i\gamma_\mu p_\mu}{m^2 + p^2} + O(a^2). \tag{3.244}$$

In particular, if the parameter $m = 0$, the propagator has a pole at $p_\mu = 0$, corresponding to a massless particle. However, other (unwanted) poles also emerge in the discretized theory. Namely, it can easily be seen that s_μ vanishes not only at $p_\mu = 0$, but also at the edges of the Brillouin zone, which are 16 points in total:

$$p_\mu^{(A)} = \frac{\pi}{a} n_\mu^{(A)}, \qquad A = \varnothing, 0, 1, 2, \ldots, 01, 02, \ldots, 0123.$$

$$n_\mu^{(\varnothing)} = (0,0,0,0), \quad \cdots$$

$$n_\mu^{(3)} = (0,0,0,1), \quad \cdots$$

$$n_\mu^{(23)} = (0,0,1,1), \quad \cdots$$

$$n_\mu^{(0123)} = (1,1,1,1). \tag{3.245}$$

Writing $p_\mu = p_\mu^{(A)} + k_\mu$, where k_μ denotes a "small" momentum, in the continuum limit, we get

$$S(p) \to \frac{m - i\gamma_\mu^{(A)} k_\mu}{m^2 + k^2}, \qquad \gamma_\mu^{(A)} = \pm\gamma_\mu. \tag{3.246}$$

More precisely, the sign of $\gamma_\mu^{(A)}$ is equal to $(-1)^{n_A}$, where n_A is the number of the nonzero components in $n_\mu^{(A)}$. Introducing the notation

$$\{S_A\} = \{\mathbb{1}, S_\rho, S_\rho S_\sigma, S_\rho S_\sigma S_\tau, \gamma_5\}, \qquad S_\rho = i\gamma_\rho \gamma_5, \qquad \rho \neq \sigma \neq \tau, \tag{3.247}$$

we may rewrite Eq. (3.246) in the form

$$S(p^{(A)} + k) \to S_A^\dagger \frac{m - i\gamma_\mu k_\mu}{m^2 + k^2} S_A. \tag{3.248}$$

At this point, we realize that in the continuum limit the discretized theory does not describe a single fermion with a mass m as was naively hoped for. (This explains the superscript "naive" attached to the discretized version of the fermion action functional.) Rather, 15 additional species of fermions with the same mass m, called *doublers*, emerge from the different corners of the Brillouin zone. This is exactly the price that one pays for the absence of anomalies.

3.12.3 Anomalies and Doublers

Following the argument given in Ref. [36], it can be explicitly shown that the anomalies are canceled by the doubler contributions. Indeed, continuing analytically the lattice propagator to Minkowski space, in the vicinity of the poles we get

$$S(p^{(A)} + k) \to S_A^\dagger \frac{m + \gamma_0 k_0 - \boldsymbol{\gamma}\mathbf{k}}{m^2 - k_0^2 + \mathbf{k}^2 - i\varepsilon} S_A$$

$$= \sum_s \left\{ S_A^\dagger \frac{u(\mathbf{k},s)\bar{u}(\mathbf{k},s)}{2w(\mathbf{k})(w(\mathbf{k}) - k_0 - i\varepsilon)} S_A - S_A^\dagger \frac{v(-\mathbf{k},s)\bar{v}(-\mathbf{k},s)}{2w(\mathbf{k})(w(\mathbf{k}) + k_0 - i\varepsilon)} S_A \right\}. \tag{3.249}$$

Hence, the wave function of the free fermion (antifermion) with the label "A" is given by $S_A^\dagger u(\mathbf{k},s)$ $(S_A^\dagger v(-\mathbf{k},s))$.

Next, we derive an expression of the axial-vector vertex (three-point function) in lattice regularization. Using Noether's theorem, we start with the action functional, given by Eq. (3.231), and perform an axial transformation,

$$\psi(x) \to [1 + i\beta(x)\gamma_5]\psi(x),$$

$$\bar{\psi}(x) \to \bar{\psi}(x)[1 + i\beta(x)\gamma_5]. \tag{3.250}$$

For an infinitesimal variation, we get

$$\delta S_F^{\text{naive}} = ia^3 \sum_x \sum_\mu (\beta(x + a\hat{\mu}) - \beta(x))A_\mu(x) + ia^4 \sum_x 2m\beta(x)\bar{\psi}(x)\gamma_5\psi(x)$$

$$= -ia^4 \sum_x \beta(x) \left[\sum_\mu (\partial_\mu^* A_\mu) - 2m\bar{\psi}(x)\psi(x) \right], \tag{3.251}$$

where the lattice version of the axial current is given by the following (for simplicity, we assumed everywhere a single quark flavor, but the generalization to any number of flavors is straightforward):

$$A_\mu(x) = \frac{1}{2} \left[\bar{\psi}(x)\gamma_\mu\gamma_5 U(x,\mu)\psi(x + a\hat{\mu}) + \bar{\psi}(x + a\hat{\mu})\gamma_\mu\gamma_5 U^{-1}(x,\mu)\psi(x) \right]. \tag{3.252}$$

At lowest order in the coupling constant, $U(x,\mu) = 1$ and

$$A_\mu(x) \simeq \frac{1}{2} \left[\bar{\psi}(x)\gamma_\mu\gamma_5\psi(x + a\hat{\mu}) + \bar{\psi}(x + a\hat{\mu})\gamma_\mu\gamma_5\psi(x) \right]. \tag{3.253}$$

Note that two fermion operators in the preceding expressions, by analogy to the point-split technique, are defined at different space-time points.

Performing again the Fourier transform, (3.238), we may read off the axial-vector vertex in momentum space:

$$\Gamma_\mu^5(p,q) = \exp\left(-\frac{i}{2}a(p - q)_\mu \right) \cos\left(\frac{1}{2}a(p + q)_\mu \right) \gamma_\mu\gamma_5. \tag{3.254}$$

In the vicinity of the fermion labeled "A", we may write $p = p^{(A)} + r$, $q = p^{(A)} + l$ and expand in the small momenta r, l. At leading order, the coupling of this fermion to the axial-vector current is then described by the vertex

$$\Gamma^{5,A}_{\mu} = \cos\left(ap^{(A)}_{\mu}\right)\gamma_{\mu}\gamma_{5}, \tag{3.255}$$

and the axial charge is obtained by sandwiching this vertex between the free fermion wave functions. If all components of p^{A}_{μ} are zero, then $S_A = 1$ and we get the usual expression:

$$\bar{u}(\mathbf{k},s)S^{\dagger}_{A}\gamma_{\mu}\gamma_{5}S_{A}u(\mathbf{k},s) = +\bar{u}(\mathbf{k},s)\gamma_{\mu}\gamma_{5}u(\mathbf{k},s). \tag{3.256}$$

Exploring all 16 possible doublers, it is straightforward to check that the sign $+(-)$ on the r.h.s. of the preceding equation emerges, if the vector $p^{(A)}_{\mu}$ has an even (odd) number of the nonzero components. In other words, in the continuum limit we have eight fermions with positive and eight fermions with negative axial charge. The anomalies produced by these fermions cancel, and thus the net anomaly vanishes.

3.12.4 Wilson Fermions

To summarize, the naive fermion formulation in the continuum limit yields 16 fermion species instead of one. In order to arrive at the correct limit, one should modify the discretized Lagrangian, aiming at the elimination of the doublers. In the literature, many alternative formulations can be found. Space does not allow to discuss all these various formulations. Instead, for simplicity, we choose to work with Wilson fermions [37] and write down the Ward identities for the vector and axial-vector currents in this formulation.

In its simplest form, the Lagrangian of QCD with Wilson fermions is given by a sum of gluonic and fermionic parts. The gluonic part is identical to Eq. (3.235), whereas the fermionic part contains an additional term, the so-called Wilson term:

$$S^{\text{Wilson}}_{F} = a^4 \sum_{x} \bar{\psi}(x)(D_W + m)\psi(x), \tag{3.257}$$

where

$$D_W = \frac{1}{2}\left[\gamma_{\mu}(D^*_{\mu} + D_{\mu}) - arD^*_{\mu}D_{\mu}\right]. \tag{3.258}$$

The parameter r takes values from 0 to 1. If $r = 0$, the naive fermion action is restored. It should be emphasized that the Wilson term is of order a and disappears in the continuum limit $a \to 0$. Hence, the Eq. (3.257) represents a valid discretization prescription for the continuum theory.

The propagator of free Wilson fermions can be found analogously to the case of naive fermions. Performing the Fourier-transform, it is seen that the quantity $\tilde{K}(p)$ from Eq. (3.240) gets modified:

$$\tilde{K}_W(p) = \frac{i}{a}\sum_{\mu}\gamma_{\mu}\sin(ap_{\mu}) + m + \frac{r}{a}\sum_{\mu}[1 - \cos(ap_{\mu})]. \tag{3.259}$$

Hence, the free fermion propagator is given by

$$S_W(p) = \frac{m(p) - i\gamma_\mu s_\mu}{m(p)^2 + s^2}, \qquad m(p) = m + \frac{r}{a}\sum_\mu [1 - \cos(ap_\mu)]. \qquad (3.260)$$

At the edges of the Brillouin zone, $p_\mu = p_\mu^{(A)} + k_\mu$ and

$$S_W(p^{(A)} + k) = \frac{m_A - i\gamma_\mu^{(A)} k_\mu}{m_A^2 + k^2}, \qquad m_A = m + \frac{2r}{a} n_A, \qquad (3.261)$$

where n_A is the number of the nonzero components in the vector $p_\mu^{(A)}$. It is readily seen that, if $r \neq 0$, the doublers with $n_A \neq 0$ get masses of order $1/a$ and thus are eliminated from the theory in the continuum limit. In other words, the continuum limit of the theory with Wilson fermions contains a single fermion species.

However, one also sees that the elimination of the doublers comes at a cost. The theory with Wilson fermions does not possess chiral symmetry anymore, that is, $\gamma_5 D_W \gamma_5 \neq -D_W$. Thus, the lattice regularization has broken chiral symmetry, and there exists no guarantee that the symmetry will be restored in the continuum limit. The situation is already familiar to us: Those symmetries, which do not survive the regularization, contain anomalous terms in the divergence of the pertinent currents.

It is instructive to see how such terms emerge in the Wilson formulation. To this end, we first write down the fermion action explicitly in terms of the variable $U(x, \mu)$:

$$S_F^{\text{Wilson}} = -\frac{1}{2}a^3 \sum_x \sum_\mu [\bar{\psi}(x)(r - \gamma_\mu)U(x, \mu)\psi(x + a\hat{\mu})$$

$$+ \bar{\psi}(x + a\hat{\mu})(r + \gamma_\mu)U(x, \mu)^{-1}\psi(x)] + ma^4 \sum_x \bar{\psi}(x)\psi(x). \qquad (3.262)$$

We shall derive the expression for the vector and axial-vector currents from the Wilson action using again Noether's theorem. Performing first the vector transformations,

$$\psi(x) \to [1 + i\alpha(x)]\psi(x),$$

$$\bar{\psi}(x) \to \bar{\psi}(x)[1 - i\alpha(x)], \qquad (3.263)$$

we get

$$\delta S_F^{\text{Wilson}} = ia^3 \sum_x \sum_\mu (\alpha(x + a\hat{\mu}) - \alpha(x))V_\mu(x)$$

$$= -ia^4 \sum_x \alpha(x) \sum_\mu (\partial_\mu^* V_\mu), \qquad (3.264)$$

where $V_\mu(x)$ is the vector current:

$$V_\mu(x) = -\frac{1}{2}[\bar{\psi}(x)(r - \gamma_\mu)U(x, \mu)^{-1}\psi(x + a\hat{\mu}) - \bar{\psi}(x + a\hat{\mu})(r + \gamma_\mu)U(x, \mu)\psi(x)].$$

$$(3.265)$$

From Eq. (3.264), it is seen that the vector current is conserved for Wilson fermions.

The axial transformations, Eq. (3.250), are more tricky. The action functional is changed by

$$\delta S_F^{\text{Wilson}} = ia^3 \sum_x \sum_\mu (\beta(x+a\hat{\mu}) - \beta(x))A_\mu(x) + ia^4 \sum_x 2m\beta(x)\bar{\psi}(x)\psi(x)$$

$$- \frac{ir}{2}a^3 \sum_x \sum_\mu (\beta(x+a\hat{\mu}) + \beta(x))$$

$$\times \left[\bar{\psi}(x+a\hat{\mu})\gamma_5 U(x,\mu)\psi(x) + \bar{\psi}(x)\gamma_5 U(x,\mu)^{-1}\psi(x+a\hat{\mu})\right], \quad (3.266)$$

where the lattice version of the axial-vector current is defined as

$$A_\mu(x) = \frac{1}{2}\left[\bar{\psi}(x)\gamma_\mu\gamma_5 U(x,\mu)^{-1}\psi(x+a\hat{\mu}) + \bar{\psi}(x+a\hat{\mu})\gamma_\mu\gamma_5 U(x,\mu)\psi(x)\right].$$

$$(3.267)$$

Note that the axial-vector current is the coefficient appearing in front of $\partial_\mu\beta(x) = (\beta(x+a\hat{\mu}) - \beta(x))/a$, whereas the rest, which is proportional to $\beta(x+a\hat{\mu}) + \beta(x)$, defines the divergence of the axial-vector current. From Eq. (3.266), the divergence takes the form

$$\partial_\mu^* A_\mu(x) = 2m\psi(x)\gamma_5\psi(x) + A(x), \quad (3.268)$$

where the first term vanishes in the chiral limit and the last term reproduces the anomaly. In order to show this, similarly to the continuum case, one can integrate out the quark fields in the preceding relation [36].[12] As a result of this procedure, the anomalous term turns into $\langle A(x)\rangle$, where

$$\langle A(x)\rangle = \frac{r}{2a}\sum_\mu\left(\text{tr}\left[\gamma_5 S(x-a\hat{\mu},x)\right]U(x-a\hat{\mu},\mu)\right.$$

$$+ \text{tr}\left[\gamma_5 S(x,x-a\hat{\mu})\right]U(x-a\hat{\mu},\mu)^{-1} + \text{tr}\left[\gamma_5 S(x,x+a\hat{\mu})\right]U(x,\mu)$$

$$+ \left.\text{tr}\left[\gamma_5 S(x+a\hat{\mu},x)\right]U(x,\mu)^{-1}\right) - 4\text{tr}\left[\gamma_5 S(x,x)\right], \quad (3.269)$$

and $S(x,y)$ denotes the fermion propagator in the external field. Further, $S(x,y)$ can be expanded in powers of the (external) gluon field. In analogy with the calculations by using the point-split technique, it can be shown that, as $a \to 0$, the anomalous term is given by [36]

$$\langle A(x)\rangle = \frac{1}{16\pi^2}\varepsilon_{\mu\nu\alpha\beta}\text{tr}_c(F_{\mu\nu}F_{\alpha\beta}), \quad (3.270)$$

in accordance with our previous calculations in the continuum.

To summarize, the Wilson formulation eliminates the doublers on the lattice. At the same time, it breaks chiral symmetry explicitly, and the divergence of the axial-vector

[12] This is an analog of using the point-split method and finding the most singular contribution by evaluating the vacuum expectation value of the divergence of the axial-vector current; see concluding remarks in Section 3.8.

current, which emerges from the Wilson term, exactly reproduces the anomaly that we have obtained previously in the continuum.

3.12.5 Nielsen–Ninomiya Theorem

The result, obtained in the previous section for one particular formulation of lattice fermions, strongly supports a general statement, known under the name of the *Nielsen–Ninomiya no-go theorem* [38]. This theorem states that any lattice formulation of chiral fermions with a *local* action, which has the property of reflection positivity and obeys cubic symmetry, *must* produce doublers. The argument behind this statement is rather transparent. Indeed, consider the free fermion action in momentum space, which, by analogy with the naive fermions, Eq. (3.239), is written as

$$S_F = (Na)^{-4} \sum_p \tilde{\bar{\psi}}(p) \tilde{K}(p) \tilde{\psi}(p). \tag{3.271}$$

Here, $K(p)$ is an arbitrary 4×4 matrix function that reduces to $K(p) \simeq i\gamma_\mu p_\mu$ for small values of p. (The choice of $K(p)$ is equivalent to a choice of a particular formulation.) Let us now see what the restrictions are that are imposed on $K(p)$ from the general requirements of the theory. First, reflection positivity (which is the counterpart of unitarity in the Euclidean theory) implies the invariance of the action with respect to the following antiunitary transformations:

$$\tilde{\psi}(\mathbf{p}, p_0) \to \tilde{\bar{\psi}}(\mathbf{p}, -p_0)\gamma_0, \qquad \tilde{\bar{\psi}}(\mathbf{p}, p_0) \to \gamma_0 \tilde{\psi}(\mathbf{p}, -p_0). \tag{3.272}$$

From this, it follows that

$$K(\mathbf{p}, p_0) = \gamma_0 K^\dagger(\mathbf{p}, -p_0)\gamma_0. \tag{3.273}$$

Further, using cubic symmetry on the lattice, we conclude that this property holds for each $\mu = 0, 1, 2, 3$:

$$K(p_0, \ldots, p_\mu, \ldots p_3) = \gamma_\mu K^\dagger(p_0, \ldots, -p_\mu, \ldots p_3)\gamma_\mu. \tag{3.274}$$

Next, demanding invariance of the action under chiral transformations,

$$\tilde{\psi}(p) \to \exp(-i\beta\gamma_5)\tilde{\psi}(p), \qquad \tilde{\bar{\psi}}(p) \to \tilde{\bar{\psi}}(p)\exp(-i\beta\gamma_5), \tag{3.275}$$

we can derive

$$K(p) = -\gamma_5 K(p)\gamma_5. \tag{3.276}$$

Combining the preceding two properties, we get

$$K(p) = -K(-p). \tag{3.277}$$

These requirements are fulfilled, if and only if $K(p)$ has the following form:

$$K(p) = \gamma_\mu K_\mu(p), \qquad K_\mu(-p) = -K_\mu(p), \tag{3.278}$$

where $K_\mu(p)$ are real functions of the argument p.

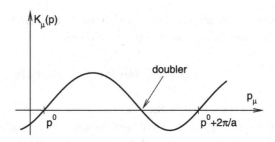

The zeros of the function $K_\mu(p)$. Since this function is both continuous and periodic with period $2\pi/a$, a zero corresponding to a doubler should necessarily exist in the interval $\left[p^0, p^0 + 2\pi/a\right]$.

Finally, we mention that locality requires the functions $K_\mu(p)$ be smooth functions of p, so that the Fourier transform thereof vanishes sufficiently fast as $|x - y| \to \infty$. On the other hand, the functions $K_\mu(p)$ are periodic on the lattice, that is, shifting the momentum $p_\mu \to p_\mu + (2\pi/a)n_\mu$, where n_μ is a vector with integer components, does not change either the value of the function or its derivative. Consequently, if the functions $K_\mu(p)$ have a zero at some $p = p^0$, from continuity it follows that it will have another zero in the interval $\left[p^0, p^0 + 2\pi/a\right]$, which corresponds to a doubler, as depicted in Fig. 3.7. In other words, the doublers necessarily emerge in any formulation if exact chiral symmetry is imposed on the lattice.

3.12.6 Chiral Fermions

As seen in the previous section, the fermion differential operator on the lattice can obey exact chiral symmetry,

$$\gamma_5 K[U] + K[U]\gamma_5 = 0, \tag{3.279}$$

only at the cost of fermion doubling (here, $K[U]$ denotes a differential operator, including the gauge fields). However, we may argue that chiral symmetry may be realized differently on the lattice. For example, using block spin transformations in a theory, which is exactly chirally invariant in the continuum, Ginsparg and Wilson [39] have suggested that, if $a \neq 0$, the differential operator may instead obey a weaker relation:

$$\gamma_5 K[U] + K[U]\gamma_5 = aK[U]2R\gamma_5 K[U], \tag{3.280}$$

where R denotes some *local* operator, which in position space vanishes exponentially as $|x - y| \to \infty$. In the following, we shall consider a simplified version with $2R = 1$. In the continuum limit $a \to 0$, the right-hand side of the preceding equation formally disappears, and chiral symmetry is restored. Moreover, in the literature, there exist formulations, in which the fermion differential operator obeys the Ginsparg–Wilson relation, Eq. (3.280). These are, for example, the domain wall fermions [40–42], overlap fermions [43–45], and the fermions described by the so-called perfect action [46].

Using the Ginsparg–Wilson relation, Lüscher has shown that the chiral transformations on the lattice can be modified so that the fermion action possesses exact chiral symmetry like in the continuum [47]. The modified infinitesimal transformations take the form

$$\psi \mapsto \left\{ 1 + i\beta\gamma_5 \left(1 - \frac{a}{2}K[U] \right) \right\} \psi,$$

$$\bar{\psi} \mapsto \bar{\psi} \left\{ 1 + i\beta \left(1 - \frac{a}{2}K[U] \right) \gamma_5 \right\}. \tag{3.281}$$

It can be straightforwardly checked that the Lagrangian is invariant under these transformations:

$$\bar{\psi}K[U]\psi \mapsto \bar{\psi} \left\{ 1 + i\beta \left(1 - \frac{a}{2}K[U] \right) \gamma_5 \right\} K[U] \left\{ 1 + i\beta\gamma_5 \left(1 - \frac{a}{2}K[U] \right) \right\} \psi$$

$$= \bar{\psi}K[U]\psi + \bar{\psi}i\beta \left(\gamma_5 K[U] + K[U]\gamma_5 - aK[U]\gamma_5 K[U] \right) \psi + O(\beta^2)$$

$$= \bar{\psi}K[U]\psi + O(\beta^2). \tag{3.282}$$

On the other hand, the fermion determinant is no more invariant under the modified transformations. The Jacobian for the transformation of the fermion measure $d\psi d\bar{\psi}$ is given by

$$J = 1 - i\mathrm{Tr}\left(\beta\gamma_5(2 - aK[U]) \right) = 1 - ia\beta^0 \mathrm{Tr}\left(\gamma_5 K[U] \right). \tag{3.283}$$

Thus, similarly to the continuum, an anomalous divergence emerges from the measure and not from the non-invariance of the Lagrangian. This circumstance enables one to find a lattice formulation for the gauge theories, where gauge bosons interact with chiral currents (e.g., the Standard Model) [48, 49].

3.12.7 Index Theorem on the Lattice

Next, we wish to find out how the Atiyah–Singer index theorem, see Section 3.11, can be represented on the lattice. We mainly follow Ref. [50]. If z is a complex number, not contained in the spectrum of the operator $K[U]$, using the Ginsparg–Wilson relation, it can be seen that

$$a(z - K[U])\gamma_5(z - K[U]) = z(2 - az)\gamma_5$$

$$- (1 - az)\left[(z - K[U])\gamma_5 + \gamma_5(z - K[U]) \right]. \tag{3.284}$$

Multiplying both sides of this equation by $(z - K[U])^{-1}$ from the right and taking the trace, we get

$$-a\mathrm{Tr}\left[\gamma_5 K[U] \right] = z(2 - az)\mathrm{Tr}\left[\gamma_5(z - K[U])^{-1} \right]. \tag{3.285}$$

Finally, dividing both sides by $z(2 - az)$ and integrating over a small closed contour in the complex z-plane including the origin $z = 0$, which contains no other eigenvalues of $K[U]$ except zero, we obtain

$$-\frac{a}{2} \mathrm{Tr}\left[\gamma_5 K[U]\right] = \oint \frac{dz}{2\pi i} \mathrm{Tr}\left[\gamma_5 \frac{1}{z - K[U]}\right]$$

$$= \sum_{i=\text{zero modes}} \oint \frac{dz}{2\pi i} \frac{\langle \psi_0^i | \gamma_5 | \psi_0^i \rangle}{z} = n_+ - n_-, \qquad (3.286)$$

where the $|\psi_0^i\rangle$ are eigenvectors of the operator $K[U]$ with eigenvalue zero and with positive/negative chirality:

$$K[U]|\psi_0^i\rangle = 0, \qquad \gamma_5 |\psi_0^i\rangle = +|\psi_0^i\rangle. \qquad (3.287)$$

Further, n_+/n_- denotes the number of zero modes with positive/negative chirality. Note also here that, if $K[U]|\psi_0^i\rangle = 0$, then

$$K[U]\gamma_5|\psi_0^i\rangle = \left(aK[U]\gamma_5 K[U] - \gamma_5 K[U]\right)|\psi_0^i\rangle = 0. \qquad (3.288)$$

In other words, if $|\psi_0^i\rangle$ is a zero mode, then $\gamma_5|\psi_0^i\rangle$ is a zero mode as well and thus the basis in the subspace of the zero modes can be constructed from the vectors with definite chirality.

To summarize, Eq. (3.286) represents a lattice analog of the Atiyah–Singer index theorem, Eq. (3.217). The generalization of the operator of the topological charge (that appears on the r.h.s. of Eq. (3.217)) on the lattice can be defined as

$$Q = \frac{a}{2} \mathrm{Tr}\left[\gamma_5 K[U]\right]. \qquad (3.289)$$

Further, the expectation value of any operator $\mathcal{O}$ on the lattice is given by Eq. (3.289). (Here, the gluon is considered as an external source, and the integration is performed over the fermion fields only.)

$$\langle \mathcal{O} \rangle = \int d\psi d\bar{\psi} \mathcal{O}(\psi, \bar{\psi}) \exp\left(-a^4 \sum xy \bar{\psi}(x) K[U; x, y] \psi(y)\right). \qquad (3.290)$$

Let us, for simplicity, assume that $\mathcal{O}$ is chirally invariant. Performing (global) chiral transformation, given by Eq. (3.281), we get

$$\langle \delta \mathcal{O} \rangle = a\beta^0 \mathrm{Tr}\left[\gamma_5 K[U]\right]\langle \mathcal{O} \rangle. \qquad (3.291)$$

This is nothing but the anomalous Ward identity known from the continuum field theory.

3.13 Trace Anomaly, Dimensional Transmutation and the Proton Mass

The classically conserved currents that correspond to the chiral transformations are not the only currents that acquire an anomalous divergence through quantum corrections. The other example is the scale invariance in QCD, which is also broken by an anomaly. The physical importance of this effect can be realized immediately. To this end, consider the chiral limit in QCD, where the light quark masses $m_u, m_d, m_s \to 0$. (The heavy quarks are already integrated out of the theory.) In this limit, QCD is a theory containing only a single dimensionless constant g and hence should be invariant with respect to the scale transformations $x_\mu \to \lambda x_\mu$, where λ is a nonzero real number. By the same token, massive particles such as the nucleons could never emerge in a theory that contains no scale. Of course, we know that the solution of this puzzle in QCD is given by dimensional transmutation; the nucleon mass is proportional to the scale Λ_{QCD} in the chiral limit. It remains to be seen what the mechanism of breaking the scale invariance is, and how these two phenomena – the breakdown of the scale invariance and dimensional transmutation – are related to each other.

Let us start from the Lagrangian of classical QCD, which, for convenience, is rewritten in a symmetrized form:

$$\mathcal{L}_{QCD} = \frac{1}{2g^2} \mathrm{tr}_c(F_{\mu\nu}F^{\mu\nu}) + \bar{\psi}\left(\frac{i}{2}\gamma^\mu \overset{\leftrightarrow}{D}_\mu - \mathcal{M}\right)\psi. \tag{3.292}$$

Note that this Lagrangian differs from the Lagrangian in Eq. (3.43) by a total derivative and hence describes the same theory.

Applying Noether's theorem, we may calculate the *classical* energy-momentum tensor:

$$\theta_{\mu\nu} = \frac{\partial \mathcal{L}_{QCD}}{\partial \partial_\mu G^a_\lambda} \partial_\nu G^a_\lambda + \frac{\partial \mathcal{L}_{QCD}}{\partial \partial_\mu \psi} \partial_\nu \psi + \partial_\nu \bar{\psi} \frac{\partial \mathcal{L}_{QCD}}{\partial \partial_\mu \bar{\psi}} - g_{\mu\nu}\mathcal{L}_{QCD}$$

$$= \frac{i}{2} \bar{\psi}\gamma_\mu \overset{\leftrightarrow}{\partial}_\nu \psi + 2\mathrm{tr}_c\left(F_{\mu\lambda}\partial_\nu G^\lambda - \frac{1}{4}g_{\mu\nu}F_{\lambda\rho}F^{\lambda\rho}\right). \tag{3.293}$$

The so defined energy-momentum tensor is conserved, $\partial^\mu \theta_{\mu\nu} = 0$, but not gauge-invariant. The latter deficiency can be cured by subtracting the total derivative of the antisymmetric tensor:

$$\Delta\theta_{\mu\nu} = 2\mathrm{tr}_c(F_{\mu\lambda}(\partial^\lambda G_\nu - [G_\nu, G^\lambda])) - i\bar{\psi}\gamma_\mu G_\nu \psi = 2\mathrm{tr}_c(\partial^\lambda(F_{\mu\lambda}G_\nu)), \tag{3.294}$$

where the last equality is obtained by using the EOM. Now, taking into account the obvious identity $\partial^\mu(\Delta\theta_{\mu\nu}) = 0$ that stems from the antisymmetry of $F_{\mu\lambda}$, we may define a modified energy-momentum tensor, which is conserved and gauge-invariant at the same time:

$$\bar{\theta}_{\mu\nu} = \theta_{\mu\nu} - \Delta\theta_{\mu\nu} = \frac{i}{2}\bar{\psi}\gamma_\mu \overset{\leftrightarrow}{D}_\nu \psi + 2\mathrm{tr}_c\left(F_{\mu\lambda}F^\lambda_\nu - \frac{1}{4}g_{\mu\nu}F_{\lambda\rho}F^{\lambda\rho}\right). \tag{3.295}$$

With the use of the EOM, for the trace of this improved energy-momentum tensor we obtain:

$$\bar{\theta}_\mu^\mu = \bar{\psi}\mathcal{M}\psi. \tag{3.296}$$

That is, the trace of the modified energy-momentum tensor vanishes in the chiral limit.

The total four-momentum operator is given by

$$P_\nu = \int d^3\mathbf{x}\, \theta_{0\nu}(0,\mathbf{x}) = \int d^3\mathbf{x}\, \bar{\theta}_{0\nu}(0,\mathbf{x}). \tag{3.297}$$

Note that $\Delta\theta_{\mu\nu}$ does not contribute to the four-momentum.

Let us assume for a moment that we may rely on all these formal manipulations for calculating the derivatives and using the EOM for the Fock space operators as well (this is analogous to assuming the naive Ward identities). Consider the matrix element of the operator $\bar{\theta}_{\mu\nu}$ between one-hadron states with the four-momentum k. Due to Lorentz invariance, this matrix element can be written as

$$\langle k|\bar{\theta}_{\mu\nu}(0)|k\rangle = ak_\mu k_\nu + bg_{\mu\nu}, \tag{3.298}$$

where a and b are unknown constants. In order to determine these, let us use Eq. (3.297):

$$\langle k|P_\nu|k'\rangle = (2\pi)^3\delta^{(3)}(\mathbf{k}-\mathbf{k}')2k_0 k_\nu$$

$$= \int d^3\mathbf{x}\,\langle k|\bar{\theta}_{0\nu}(\mathbf{x},0)|k'\rangle = (2\pi)^3\delta^{(3)}(\mathbf{k}-\mathbf{k}')(ak_0 k_\nu + bg_{0\nu}). \tag{3.299}$$

From this, one immediately obtains $a = 2$ and $b = 0$. Further, calculating the trace of Eq. (3.298), we get

$$\langle k|\bar{\theta}_\mu^\mu(0)|k\rangle = 2M^2, \tag{3.300}$$

where M denotes the mass of the hadron under consideration. Recalling finally Eq. (3.296), one immediately comes to the conclusion that, in the chiral limit $M \to 0$, all hadrons should be massless. The physical meaning of this result has already been discussed: It merely states that, if there is no dimensionful parameter in the theory, all hadron masses should be zero, in order to avoid any inconsistency. However, intuitive and appealing the preceding argument might seem, it is in flat contradiction with the experimentally observed fact that the hadrons in nature have (quite large) masses, even those that are composed from the almost massless u and d quarks, such as the nucleon.

Further, as already mentioned, in the chiral limit QCD does not possess dimensionful quantities and is thus scale-invariant. The scale transformations are given by:[13]

$$\psi(x) \mapsto \lambda^{3/2}\psi(\lambda x), \quad G_\mu(x) \mapsto \lambda G_\mu(\lambda x), \qquad \lambda \in \mathbb{R}\backslash\{0\}. \tag{3.301}$$

The massless Lagrangian, $\mathcal{M} = 0$, under scale transformations turns into

$$\mathcal{L}_{\text{QCD}}(x) \mapsto \lambda^4 \mathcal{L}_{\text{QCD}}(\lambda x). \tag{3.302}$$

---

[13] Recall that the canonical engineering dimensions of the fermion and vector fields are 3/2 and 1, respectively.

The mass term $\mathcal{M} \neq 0$ breaks the scale invariance. Assume now that $\lambda = 1 - \varepsilon/2$. The infinitesimal scale transformations in the massless case up to the terms of order ε^2 take the form

$$\delta \psi(x) = -\frac{\varepsilon}{2}\left(x_\nu \frac{\partial}{\partial x_\nu} + \frac{3}{2}\right)\psi(x),$$

$$\delta \bar{\psi}(x) = -\frac{\varepsilon}{2}\left(x_\nu \frac{\partial}{\partial x_\nu} + \frac{3}{2}\right)\bar{\psi}(x),$$

$$\delta G_\mu(x) = -\frac{\varepsilon}{2}\left(x_\nu \frac{\partial}{\partial x_\nu} + 1\right)G_\mu(x),$$

$$\delta \mathcal{L}_{QCD}(x) = -\frac{\varepsilon}{2}\left(x_\nu \frac{\partial}{\partial x_\nu} + 4\right)\mathcal{L}_{QCD}(x) = -\frac{\varepsilon}{2}\partial_\mu(x_\nu g^{\mu\nu}\mathcal{L}_{QCD}(x)). \quad (3.303)$$

On the other hand, the variation of the Lagrangian under the preceding transformations is given by

$$\delta \mathcal{L}_{QCD}(x) = \partial_\mu\left\{\frac{\partial \mathcal{L}_{QCD}}{\partial \partial_\mu \psi}\delta\psi + \delta\bar{\psi}\frac{\partial \mathcal{L}_{QCD}}{\partial \partial_\mu \bar{\psi}} + \frac{\partial \mathcal{L}_{QCD}}{\partial \partial_\mu G_\lambda^a}\delta G_\lambda^a\right\}$$

$$= -\frac{\varepsilon}{2}\partial_\mu\left\{\frac{\partial \mathcal{L}_{QCD}}{\partial \partial_\mu \psi}\left(x^\nu \partial_\nu + \frac{3}{2}\right)\psi + \bar{\psi}\left(\overleftarrow{\partial_\nu}x^\nu + \frac{3}{2}\right)\frac{\partial \mathcal{L}_{QCD}}{\partial \partial_\mu \bar{\psi}}\right.$$

$$\left. + \frac{\partial \mathcal{L}_{QCD}}{\partial \partial_\mu G_\lambda^a}(x^\nu \partial_\nu + 1)G_\lambda^a\right\}. \quad (3.304)$$

Note that, in the first line of this equation, we have already used the equations of motion. Equating now the quantity $\delta\mathcal{L}_{QCD}(x)$ that appears in Eqs. (3.303) and (3.304), we conclude that the dilatation current should be conserved:

$$\partial_\mu J_D^\mu(x) = 0, \quad (3.305)$$

where

$$J_D^\mu(x) = \frac{\partial \mathcal{L}_{QCD}}{\partial \partial_\mu \psi}\left(x^\nu \partial_\nu + \frac{3}{2}\right)\psi + \bar{\psi}\left(\overleftarrow{\partial_\nu}x^\nu + \frac{3}{2}\right)\frac{\partial \mathcal{L}_{QCD}}{\partial \partial_\mu \bar{\psi}}$$

$$+ \frac{\partial \mathcal{L}_{QCD}}{\partial \partial_\mu G_\lambda^a}(x^\nu \partial_\nu + 1)G_\lambda^a - x_\nu g^{\mu\nu}\mathcal{L}_{QCD}. \quad (3.306)$$

After straightforward calculations, we finds that the dilatation current is given by

$$J_D^\mu(x) = x_\nu \bar{\theta}^{\mu\nu}(x) + 2\partial_\lambda(x_\nu \text{tr}_c(F^{\mu\lambda}G^\nu)), \quad (3.307)$$

where the first term contains the improved energy-momentum tensor $\bar{\theta}^{\mu\nu}$ given in Eq. (3.295). The second term is trivially conserved (i.e., conserved without use of the equations of motion, just because of antisymmetry in indices). We can omit this term and define the improved dilatation current as

$$\bar{J}_D^\mu(x) = x_\nu \bar{\theta}^{\mu\nu}(x). \quad (3.308)$$

This current is conserved in the massless theory. If the quark masses are turned on, we have

$$\partial_\mu \bar{J}_D^\mu(x) = \bar{\theta}_\mu^\mu(x) + x_\nu \partial_\mu \bar{\theta}^{\mu\nu}(x) = \bar{\theta}_\mu^\mu(x) = \bar{\psi}\mathcal{M}\psi. \tag{3.309}$$

To summarize, according to the preceding equation, *classically* the divergence of the dilatation current is equal to the trace of the improved energy-momentum tensor and vanishes in the chiral limit. As a consequence, massive particles cannot be present in the spectrum of QCD. However, as we shall see in the following discussion, Eq. (3.309) gets modified by quantum corrections, and the result in our notation is given by [51], see also [52–54]:

$$\bar{\theta}_\mu^\mu(x) = (1 + \gamma_m(g_r))[\psi(x)\mathcal{M}_r\psi(x)]_r - \frac{\beta(g_r)}{g_r^3}[\mathrm{tr}_c(F_{\mu\nu}(x)F^{\mu\nu}(x))]_r. \tag{3.310}$$

Here, g_r and $\mathcal{M}_r$ denote the renormalized strong coupling constant and the quark mass matrix, respectively. Further, the RG functions $\beta(g)$ and $\gamma_m(g)$ are defined in Section 3.4.2. And finally, $[O(x)]_r$ denotes the renormalized operator $O(x)$ in a particular renormalization scheme (normal product in dimensional regularization, defined in Ref. [55]).

Equation (3.310) closely resembles the expression for the anomalous divergence of the axial-vector current. However, there are important differences as well. Namely, there is no counterpart to the non-renormalization theorem in this case, and the coefficients in front of the renormalized operators $\bar{\psi}\mathcal{M}\psi$ and $\mathrm{tr}_c(F_{\mu\nu}(x)F^{\mu\nu}(x))$ *are* modified in perturbation theory. It should be also noted that the left-hand side of this equation (the trace of the energy-momentum tensor) is UV finite and scale-independent. So is the right-hand side: albeit the renormalized operators and the parameters $g_r, \mathcal{M}_r$ depend on the scale, the particular combination given there is scale-independent; see, for example, Ref. [56].

The second term in Eq. (3.310) represents the anomaly and does not vanish in the chiral limit. In this limit the nucleon mass is determined by the matrix element of the renormalized operator, $\mathrm{tr}_c(F_{\mu\nu}(x)F^{\mu\nu}(x))$, that describes the energy of the gluon field inside the nucleon:

$$2m_N^2 = \langle N|\bar{\theta}_\mu^\mu|N\rangle = \left\langle N\left|(1 + \gamma_m(g_r))[\bar{\psi}\mathcal{M}_r\psi]^r - \frac{\beta(g_r)}{g_r^3}[\mathrm{tr}_c(F_{\mu\nu}F^{\mu\nu})]^r\right|N\right\rangle$$

$$\rightarrow \left\langle N\left|-\frac{\beta(g_r)}{g_r^3}[\mathrm{tr}_c(F_{\mu\nu}F^{\mu\nu})]^r\right|N\right\rangle \quad \text{as } \mathcal{M}_r \rightarrow 0. \tag{3.311}$$

Moreover, this term gives the bulk contribution to the nucleon mass even at physical values of the quark masses, whereas the term containing the quark mass matrix can be treated as a perturbation.

A complete derivation of Eq. (3.310) is beyond the scope of the present book. However, using the method of Fujikawa [57], we may check rather straightforwardly that Eq. (3.310) holds at the lowest order, where one neglects the contribution from the gluon loops. In the following, we shall closely follow the discussion in Refs. [57, 58], which replicates the derivation of the axial anomaly in the path integral formalism. In the approximation we are working, gluons can be described by an external field G_μ. This amounts to considering the generating functional:

$$Z(G_\mu) = \int d\psi d\bar\psi \exp\left\{i \int d^4x \mathcal{L}_{QCD}\right\}. \tag{3.312}$$

In order to get the Green's functions, one should equip the preceding generating functional with external sources that are coupled to the elementary/composite operators. We skip this step here, since this has already been discussed in the context of chiral Ward identities. Further, in order to get the full generating functional in QCD, we have to integrate $Z(G_\mu)$ over the gluon and ghost fields, introducing the gauge fixing term and the ghost Lagrangian explicitly. This, however, goes beyond the approximation used here. In order to derive the Ward identities for the generating functional $Z(G_\mu)$, let us perform the following infinitesimal transformation of the variables in the integral:

$$\psi(x) \mapsto \left(1 - \frac{\varepsilon(x)}{2}\right)\psi(x), \qquad \bar\psi(x) \mapsto \left(1 - \frac{\varepsilon(x)}{2}\right)\bar\psi(x). \tag{3.313}$$

Under this transformation, the Lagrangian transforms as

$$\mathcal{L}_{QCD} \mapsto \mathcal{L}_{QCD} - \varepsilon(x)\bar\psi\left(\frac{i}{2}\gamma^\mu \overleftrightarrow{D}_\mu - \mathcal{M}\right)\psi + O(\varepsilon^2)$$

$$= \mathcal{L}_{QCD} - \varepsilon(x)(\bar\theta_\mu^\mu(x) - \bar\psi(x)\mathcal{M}\psi(x)) + O(\varepsilon^2). \tag{3.314}$$

Consequently,

$$Z(G_\mu) = \int d\psi d\bar\psi\, J^{-2} \exp\left\{i \int d^4x \left(\mathcal{L}_{QCD} - \varepsilon(\bar\theta_\mu^\mu(x) - \bar\psi(x)\mathcal{M}\psi(x))\right)\right\}, \tag{3.315}$$

where J denotes the pertinent Jacobian, emerging from the fermionic measure. In order to derive the preceding expression, we have used Eq. (3.295). Finally, to first order in $\varepsilon(x)$ we obtain the following Ward identity:

$$\int d\psi d\bar\psi \left(i \int d^4x\, \varepsilon(x)(\bar\theta_\mu^\mu(x) - \bar\psi(x)\mathcal{M}\psi(x)) + 2\ln J\right)\exp\left\{i \int d^4x \mathcal{L}_{QCD}\right\} = 0. \tag{3.316}$$

It is immediately seen that, if the Jacobian is equal to unity, the trace of the energy-momentum tensor obeys Eq. (3.296). The anomaly, as for the singlet axial-vector current, emerges from the fermionic measure.

The calculation of J proceeds in a close analogy to Section 3.9.[14] Namely, similar to Eq. (3.153),

$$\ln J = \ln(\det(e^{-\varepsilon/2})) = -\lim_{M\to\infty}\left\{\int d^4x \frac{\varepsilon(x)}{2}\langle x|\mathrm{tr}\left(e^{(i\slashed{D})^2/M^2}\right)|x\rangle\right\}. \tag{3.317}$$

---

[14] We display all final results in Minkowski space. It is understood that, similar to Section 3.9, the calculations are done in Euclidean space, and the analytic continuation back to the Minkowski space is performed at the end.

Using Eq. (3.160) and the identity $[D_\mu, D_\nu] = F_{\mu\nu}$, this equation can be rewritten in the following form:

$$
\ln J = -\lim_{M \to \infty} \int d^4x \, \frac{\varepsilon(x)}{2} \frac{iM^4}{16\pi^2} \, \mathrm{tr} \left\{ 1 - \frac{1}{2M^2} \sigma_{\mu\nu} F^{\mu\nu} \right.
$$

$$
\left. + \frac{1}{M^4} \left(\frac{1}{8} \sigma_{\mu\nu} \sigma_{\alpha\beta} F^{\mu\nu} F^{\alpha\beta} + \frac{1}{12} F_{\mu\nu} F^{\mu\nu} + \frac{1}{12} \sigma_{\alpha\beta} [D_\mu, [D^\mu, F^{\alpha\beta}]] \right) \right.
$$

$$
\left. + O(M^{-6}) \right\}. \tag{3.318}
$$

The first term in this expression is an unessential constant, which can be included in the normalization. Performing the traces and taking the limit $M \to \infty$ leads to

$$
\ln J = i \int d^4x \, \frac{\varepsilon(x)}{2} \frac{N_f}{24\pi^2} \, \mathrm{tr}_c(F_{\mu\nu}(x) F^{\mu\nu}(x)). \tag{3.319}
$$

Consequently, the trace of the energy-momentum tensor is given by

$$
\bar{\theta}^\mu_\mu = \bar\psi \mathcal{M} \psi - \frac{N_f}{24\pi^2} \, \mathrm{tr}_c(F_{\mu\nu}(x) F^{\mu\nu}(x)). \tag{3.320}
$$

The second term is the anomaly. It survives in the chiral limit and contributes to the divergence of the dilatation current as well.

We can now check that the preceding equation agrees with the general result, displayed in Eq. (3.310). For this, note that, within the approximation used, $\gamma_m(g_r)$ vanishes, since the quark mass is renormalized only through loops containing gluons. Further, there is a single contribution to $\beta(g_r)$ that does not contain gluon loops, namely the lowest-order vacuum polarization diagram with the quark loop, which is proportional to N_f. Hence, neglecting the loops with gluons altogether, Eqs. (3.52) and (3.56) reduce to

$$
\beta(g_r) = \frac{g_r^3 N_f}{24\pi^2}, \qquad \gamma_m(g_r) = 0. \tag{3.321}
$$

Using these values, it can be immediately seen that Eqs. (3.310) and (3.320) lead to the same result.

3.14 Low-Energy Spectrum of QCD: Goldstone Theorem

3.14.1 Commutation Relations for Q_V^i, Q_A^i

After discussing in detail the chiral Ward identities and their breaking through anomalies, we shall next consider the low-energy spectrum of QCD, which, as will be shown in what follows, is determined by the phenomenon of the *spontaneous chiral symmetry breaking*. To this end, we recall that, owing to the anomaly, the chiral symmetry in the massless case in the absence of the external sources is broken down $U(N_f)_L \times U(N_f)_R \to$

$SU(N_f)_L \times SU(N_f)_R \times U(1)_V$ (the other diagonal subgroup, $U(1)_A$, is violated by the anomaly). This is not the whole story, however. As it is well known, in nature this symmetry can be realized in different ways, depending whether the transformations leave the vacuum of the theory invariant or not. Let us now see how this works in the case of chiral symmetry.

To start with, let us restrict ourselves to the subgroup $SU(N_f)_L \times SU(N_f)_R$ and consider the conserved charges:

$$Q_V^i(t) = \int d^3\mathbf{x}\, V_0^i(\mathbf{x},t), \quad Q_A^i(t) = \int d^3\mathbf{x}\, A_0^i(\mathbf{x},t), \quad i = 1,\cdots,N_f^2 - 1. \quad (3.322)$$

In the massless case, we have

$$\dot{Q}_{V,A}^i(t) = i[H, Q_{V,A}^i(t)] = 0. \quad (3.323)$$

Here, H denotes the Hamiltonian of the system. Since the $Q_{V,A}^i(t)$ do not depend on the time, we further drop the argument t in this expression.

Integrating the Ward identities for the vector and axial-vector currents obtained in Section 3.6 (see, e.g., Eq. (3.85)) over the pertinent space volume $d^3\mathbf{x}\, d^3\mathbf{y}$, it can be shown that the charges $Q_{V,A}^i$ obey commutation relations that replicate the Lie algebra of the group $SU(N_f)_L \times SU(N_f)_R$. Namely,

$$[Q_V^i, Q_V^j] = if^{ijk} Q_V^k, \quad [Q_V^i, Q_A^j] = if^{ijk} Q_A^k, \quad [Q_A^i, Q_A^j] = if^{ijk} Q_V^k, \quad (3.324)$$

where f^{ijk} denotes the structure constants of the $SU(N_f)$ group. Moreover, defining

$$Q_L^i = \frac{1}{2}(Q_V^i - Q_A^i), \qquad Q_R^i = \frac{1}{2}(Q_V^i + Q_A^i), \quad (3.325)$$

it is seen that the commutators of the left and right charges decouple,

$$[Q_R^i, Q_R^j] = if^{ijk} Q_R^k, \quad [Q_L^i, Q_L^j] = if^{ijk} Q_L^k, \quad [Q_R^i, Q_L^j] = 0, \quad (3.326)$$

showing that Q_R^i, Q_L^i are, in fact, the generators of the independent groups $SU(N_f)_R$, $SU(N_f)_L$, respectively.

3.14.2 The Symmetry of the Vacuum State

Let Q^i denote a generator of a symmetry group G of a system (for the time being, we keep the discussion as general as possible). This means that Q^i commutes with the Hamiltonian H. It is possible to have two different *modes of realization* of the symmetry, depending on whether the corresponding charges annihilate the vacuum state or not:

- Wigner–Weyl mode: $Q^i|0\rangle = 0$.
- Nambu–Goldstone mode: $Q^i|0\rangle \neq 0$.

Consider first the symmetry realization in the Wigner–Weyl mode. Assume further that there is a mass gap in the theory, and consider the one-particle state with mass m in the rest frame. Since Q^i commutes with the Hamiltonian,

$$HQ^i|\mathbf{p} = \mathbf{0}\rangle = Q^iH|\mathbf{p} = \mathbf{0}\rangle = mQ^i|\mathbf{p} = \mathbf{0}\rangle \,. \tag{3.327}$$

This means that $Q^i|\mathbf{p} = \mathbf{0}\rangle$ gives either zero or is also a one-particle state with the same mass as the original state. In other words, the massive particles form *symmetry multiplets*, where the individual states are labeled by some index α. Introducing creation operators, corresponding to the zero three-momentum, we may write

$$Q^i a_\alpha^\dagger|0\rangle = [Q^i, a_\alpha^\dagger]|0\rangle + a_\alpha^\dagger Q^i|0\rangle = (T^i)_{\alpha\beta} a_\beta^\dagger|0\rangle \,. \tag{3.328}$$

Here, we have used the condition $Q^i|0\rangle = 0$ and the fact that, acting with an operator Q^i on any member of a multiplet, we can only get a linear combination (with some coefficients $(T^i)_{\alpha\beta}$) of states in this multiplet, but nothing else. Moreover, acting with the commutator $[Q^i, Q^j]$ on a one-particle state and using the commutation relations from Eq. (3.326), we get

$$[Q^i, Q^j]a_\alpha^\dagger|0\rangle = [T^i, T^j]_{\alpha\beta} a_\beta^\dagger|0\rangle = if^{ijk}(T^k)_{\alpha\beta} a_\beta^\dagger|0\rangle \,. \tag{3.329}$$

Hence, the Hermitean matrices $(T^i)_{\alpha\beta}$ obey the same commutation relations as the generators of the group G:

$$[T^i, T^j]_{\alpha\beta} = if^{ijk}(T^k)_{\alpha\beta} \,. \tag{3.330}$$

Furthermore, taking the trace from both sides of the preceding equation, it is seen that T^i should be traceless. Now, choosing the appropriate basis in the space spanned by the vectors $a_\beta^\dagger|0\rangle$, it is possible to bring the matrices T^i to the canonical form of the generators of the group G. Then, the symmetry group G has a linear representation in this space. The matrices of this linear representation are given by

$$U(\boldsymbol{\omega})_{\alpha\beta} = \exp(i\boldsymbol{\omega}\mathbf{T})_{\alpha\beta} = \exp(i\omega^i T^i)_{\alpha\beta} \,, \tag{3.331}$$

where the components of the vector ω denote the parameters of the symmetry group G.

It is now easily seen that the symmetry is linearly realized in the basis of the fields, describing the particles in the multiplet. Indeed, defining the free fields,[15]

$$\phi_\alpha(x) = \int \frac{d^3\mathbf{k}}{(2\pi)^3 2w(\mathbf{k})} \left(e^{-ikx}a_\alpha(\mathbf{k}) + e^{ikx}a_\alpha^\dagger(\mathbf{k})\right), \tag{3.332}$$

we get

$$[Q^i, \phi_\alpha(x)] = (T^i)_{\alpha\beta}\phi_\beta(x) \,. \tag{3.333}$$

The invariance of the theory then implies that the Lagrangian can contain only invariant operators that are built out of the fields $\phi_\alpha(x)$, belonging to one of the (irreducible) representation of the symmetry group G.

In the case of chiral symmetry, the implications would be far-reaching. In this case, Q^i should be identified either with Q_V^i or with Q_A^i. Assume now that chiral symmetry in QCD is realized in the Wigner–Weyl mode. Since under parity $Q_A^i \mapsto -Q_A^i$, one

---

[15] For simplicity of notation, we consider the case of Hermitean scalar fields here and do not attach additional indices.

should have *multiplets with opposite parity and with the same mass* in QCD.[16] If the quark masses are turned on, the masses in these multiplets would become slightly different but, for small quark masses, the pattern should be easily recognizable. This, however, is not what one observes in experiment. For example, the lightest pseudoscalar bosons (the pions) have mass around 140 MeV, whereas the lightest scalar boson has mass around 450 MeV and width exceeding 500 MeV. A similar picture emerges for other multiplets. For example, the lowest excited state of the nucleon with negative parity, the $S_{11}(1535)$ with the quantum numbers $1/2^-$, lies approximately 600 MeV above the nucleon ground state. It is evident that the symmetry pattern of the QCD spectrum at low energies cannot be consistently described within this picture.

3.14.3 Goldstone Theorem

Let us now consider the alternative picture, where some charges Q^i *do not* annihilate the vacuum (we shall see in what follows that, in nature, this is the case with the axial-vector charges $Q^i = Q^i_A$). This is equivalent to the statement that there exists a set of (elementary or composite) operators B^i, whose vacuum expectation value is not invariant under the group transformations. For infinitesimal transformations, the preceding condition is equivalent to

$$
\begin{aligned}
\mathcal{B}^{ki} &= \langle 0|[B^k, Q^i_A]|0\rangle \\
&= \sum_n \int d^3\mathbf{x} \left(\langle 0|B^k|n\rangle\langle n|A^i_0(\mathbf{x},t)|0\rangle - \langle 0|A^i_0(\mathbf{x},t)|n\rangle\langle n|B^k|0\rangle \right) \\
&= \sum_n (2\pi)^3 \delta^{(3)}(\mathbf{P}_n) \\
&\quad \times \left(e^{iE_n t}\langle 0|B^k|n\rangle\langle n|A^i_0(0)|0\rangle - e^{-iE_n t}\langle 0|A^i_0(0)|n\rangle\langle n|B^k|0\rangle \right) \neq 0,
\end{aligned}
$$

$$(3.334)$$

where $A^i_\mu(x)$ denotes the axial-vector current, and $P^\mu_n = (E_n, \mathbf{P}_n)$ is the four-momentum of the state $|n\rangle$. The matrix $\mathcal{B}^{ki}$ should be non-singular, that is, $\det(\mathcal{B}) \neq 0$. In this case, one speaks about the *spontaneous breaking of chiral symmetry*.

---

[16] We do not claim here that the positive- and negative-parity multiplets in this case will contain the *same number* of particles. On the contrary, it is very easy to construct a model (e.g., the linear σ-model in the unbroken phase, see Chapter 4), which violates this requirement. We just claim that the multiplets should contain particles with either parity. Theoretically, there is also a possibility that particles of only definite parity (even or odd) are present. Since the generators Q^i_A are parity-odd, they should annihilate all state vectors in this multiplet. Furthermore, since the commutator of two axial-vector charges gives the vector charge, the generators Q^i_V should also annihilate all state vectors in this multiplet. Hence, this *hypothetical* multiplet can appear in the trivial representation of the $SU(N_f) \times SU(N_f)$ only.

On the other hand, since Q_A^i is conserved, we may write

$$0 = \frac{d}{dt} \langle 0|[B^k, Q_A^i]|0 \rangle$$

$$= \sum_n (2\pi)^{(3)} \delta^3(\mathbf{P}_n) i E_n$$

$$\times \left(e^{iE_n t} \langle 0|B^k|n \rangle \langle n|A_0^i(0)|0 \rangle + e^{-iE_n t} \langle 0|A_0^i(0)|n \rangle \langle n|B^k|0 \rangle \right). \qquad (3.335)$$

Since the different terms in the preceding sum are multiplied by the exponents corresponding to different energies E_n, and the equalities must hold at all times t, each term in Eq. (3.335) should vanish individually. This is possible, if and only if the states $|n\rangle$, which are produced by acting of Q^i on the vacuum state, are massless, that is, $E_n = 0$ at $\mathbf{P}_n = \mathbf{0}$. This leads to $\langle n|A_0^i(0)|0 \rangle \neq 0$ but $E_n(\mathbf{P} = \mathbf{0})\langle n|A_0^i(0)|0 \rangle = 0$. Consequently, in this case massless *Nambu–Goldstone bosons* emerge in the spectrum.

In the relativistic theory that is considered here, each symmetry generator Q_A^i, which does not annihilate the vacuum, gives rise to one massless boson.[17] In fact, it is straightforward to see that the number of the Goldstone boson states is equal to the number of the generators that do not annihilate the vacuum. Indeed, label the states as $|n_\alpha\rangle$, $\alpha = 1, \ldots n_G$. Since Q_V^i annihilates the vacuum, $|n_\alpha\rangle$ must form a basis of an irreducible representation of the group $SU(N_f)_V$. On the other hand, from the commutation relations in Eq. (3.324) it follows that Q_A^i and hence $A_0^i(x)$ transform as irreducible tensor operators in the adjoint representation of the $SU(N_f)_V$ group. Then, according to the Wigner–Eckart theorem, $|n_\alpha\rangle$ should belong to the adjoint representation as well. Otherwise, the overlap $\langle 0|A_0^i(0)|n_\alpha\rangle$ would vanish. From this, we immediately obtain that the number of the Goldstone bosons is equal to the number of the generators Q_A^i, which do not annihilate the vacuum, that is, $n_G = N_f^2 - 1$.

Goldstone theorem: If the conserved charges do not annihilate the vacuum, there are massless particles present in the spectrum, one for each generator of the spontaneously broken symmetry $Q^i|0\rangle \neq 0$.

If in QCD the chiral symmetry is spontaneously broken with $Q_A^i|0\rangle \neq 0$, then, instead of multiplets with opposite parity, eight pseudoscalar Goldstone bosons will arise. These (pseudoscalar) bosons would have exactly zero mass in the chiral limit $m_u, m_d, m_s \to 0$. The measured masses then emerge as a result of the explicit chiral symmetry breaking by the quark masses. This is exactly what is observed in nature. The masses of the eight pseudoscalar mesons, the pions, the kaons and the η-meson, are much smaller than the masses of all other hadrons. For example, the charged and neutral pion masses (~ 140 MeV) are almost seven times smaller than the nucleon

---

[17] Note that, if the symmetry is spontaneously broken, special care should be taken in working out the infinite-volume limit of the theory, which is defined in a finite volume. For example, it can be shown that the state $Q_A^i|0\rangle$ is not normalizable in this limit [59]. Further, since the massless Goldstone bosons have infinite correlation length, they cannot be generated in a finite volume. Consequently, a correct treatment of the spontaneous breaking implies considering *first* the infinite-volume limit and *then* the limit where the explicit symmetry breaking parameters (e.g., the quark masses) tend to zero. These limits cannot be interchanged.

mass (~ 940 MeV). The masses of the kaons and the η are larger owing to the presence of the valence strange quark(s), but still smaller than the nucleon mass (by about a factor of two). Consequently, the low-energy spectrum of QCD provides strong evidence in favor of the conjecture that the chiral symmetry in nature is realized in the Nambu–Goldstone mode.

Finally, note that, if the singlet axial-vector current were also conserved in the chiral limit, there would be an associated ninth Goldstone boson in the spectrum. However, the conservation of the singlet current is broken by an anomaly, and the would-be Goldstone boson η' has a mass around 1 GeV (as discussed later).

3.14.4 Order Parameter for the Chiral Symmetry Breaking

If the members of the pseudoscalar meson octet are Nambu–Goldstone bosons, then, according to Eq. (3.334), the matrix element of the axial-vector current between the vacuum and one-meson states does not vanish in the chiral limit. Using Lorentz invariance and taking into account the fact that $SU(N_f)_V$ is an exact symmetry in the chiral limit, this matrix element can be expressed through a single constant F_0:

$$\langle 0 | A_\mu^k(x) | \pi^i(q) \rangle = iq_\mu \delta^{ik} F_0 e^{-iqx}. \tag{3.336}$$

Here, $|\pi^i(q)\rangle$, $i = 1, \dots, 8$ denote the states in the pseudoscalar octet in the chiral limit. The quantity F_0 represents the true *order parameter* in QCD, with $F_0 \neq 0$ corresponding to the spontaneously broken phase. Namely, the chiral symmetry is spontaneously broken, if and only if $F_0 \neq 0$.

On the other hand, the spontaneous symmetry breaking is signaled, for example, by the fact that

$$\langle 0 | \bar{\psi}(0)\psi(0) | 0 \rangle = \langle 0 | \bar{\psi}_L(0)\psi_R(0) + \bar{\psi}_R(0)\psi_L(0) | 0 \rangle \neq 0. \tag{3.337}$$

Note that the chiral limit $\mathcal{M} \to 0$ is implicit in the preceding expression. This quantity is invariant under the vector subgroup $g_L = g_R$, but not under general chiral transformations. Consequently, having this quantity nonvanishing is *a sufficient but not a necessary condition* of the spontaneous chiral symmetry breaking. Thus, the *quark condensate* defined by Eq. (3.337), if nonvanishing, also serves as an order parameter. However, the vanishing of the quark condensate does not necessarily imply the absence of spontaneous breaking of chiral symmetry. In fact, there is a large number of possible order parameters, but the quark condensate enjoys a special role between these, as discussed in what follows.

3.14.5 Banks–Casher Relation

The quark condensate is exactly zero to all orders in perturbation theory in QCD at $\mathcal{M} \to 0$. This is easy to understand. In dimensional regularization, the condensate in the chiral limit is defined as a sum of no-scale Feynman integrals to all orders. Consequently, a nonzero condensate is a direct indication of non-perturbative effects. It is instructive to see how these effects can emerge in QCD.

The Banks–Casher theorem [60] establishes the relation between the quark condensate and the spectrum of the Dirac operator in an external gluon field. Consider, for example, the vacuum expectation value of the operator $\bar{u}u$, which is given by the following (properly regularized)[18] path integral over the quark and gluon fields in Euclidean space:

$$\langle 0|\bar{u}u|0\rangle = \int d\psi \, d\bar{\psi} \, dG_\mu \, \bar{u}(x)u(x)$$

$$\times \exp\left\{-\int d^4x \left(\bar{\psi}(\slashed{D}+\mathcal{M})\psi + \frac{1}{2g^2}\mathrm{tr}(F_{\mu\nu}F_{\mu\nu})\right)\right\}. \tag{3.338}$$

Integrating first over the fermion fields, one obtains

$$\langle 0|\bar{u}u|0\rangle = \int dG_\mu \, \mathrm{Tr}\, S_u(G)\det(\slashed{D}+\mathcal{M})\exp\left\{-\frac{1}{2g^2}\int d^4x\,\mathrm{tr}(F_{\mu\nu}F_{\mu\nu})\right\}, \tag{3.339}$$

where $S_u(G)$ is the quark propagator with the flavor $f = u$ in the external gauge field G_μ,

$$\mathrm{Tr}\, S_f(G) = \int d^4x \left\langle x \left| \frac{1}{\slashed{D}+m_f} \right| x \right\rangle, \qquad f = u, d, s, \tag{3.340}$$

and $\slashed{D}$ denotes the massless Dirac operator in the external gauge field.

The Euclidean operator $i\slashed{D}$ is Hermitean. Its real nonzero eigenvalues are paired; if λ is such an eigenvalue, $\slashed{D}\psi_\lambda = i\lambda\,\psi_\lambda$, then $-\lambda$ is also an eigenvalue, $\slashed{D}\gamma^5\psi_\lambda = -i\lambda\gamma^5\psi_\lambda$. In addition, there can be $N \neq 0$ zero eigenvalues that are not necessarily paired.

It is easy to see that the Fermion determinant is positive definite:

$$\det(\slashed{D}+\mathcal{M}) = \prod_f m_f^N \prod_{\lambda \neq 0}(m_f - i\lambda) = \prod_f m_f^N \prod_{\lambda > 0}(m_f^2 + \lambda^2) > 0. \tag{3.341}$$

This means that the measure in the path integral, which is given by a product of this determinant and the exponential with the Yang–Mills action functional, is positive definite.[19]

Next, for the quantity $\mathrm{Tr}\, S_f(G)$ we get the following spectral representation:

$$\mathrm{Tr}\, S_f(G) = \int_{-\infty}^{+\infty} d\lambda \, \frac{\rho(\lambda)}{m_f - i\lambda}, \tag{3.342}$$

where

$$\rho(\lambda) = \int d^4x |\langle \psi_\lambda | x \rangle|^2 \tag{3.343}$$

is the density of eigenvalues of the massless Dirac operator $-i\slashed{D}$.

---

[18] Here, following Ref. [61], we present a slightly cavalier approach to the problem in order to highlight the essentials. It is possible to carefully reformulate the proof, using explicitly the lattice regularization of the path integral. The astute reader is referred to the original paper [60].

[19] The determinant contains the factor $\prod_f m_f^N$, which vanishes in the chiral limit. However, this factor does not depend on the external field G_μ and hence can be included in the normalization of the path integral.

Consider now the limit $m_f \to 0$ in this expression. Using

$$\lim_{m_f \to 0} \frac{1}{\lambda - im_f} = \text{P.V.} \frac{1}{\lambda} + i\pi\delta(\lambda), \tag{3.344}$$

and taking into account the fact that $\rho(\lambda)$ in an even function of λ, we get

$$\lim_{m_f \to 0} \text{Tr} S_f(G) = \langle 0 | \bar{\psi}_f \psi_f | 0 \rangle = i\pi\rho(0). \tag{3.345}$$

The following remark is needed here. In fact, all the preceding manipulations were formal and did not take into account the fact that the path integral has been regularized. The regularization implies, in particular, putting the system in a finite-sized box and considering the infinite-volume limit at the end. Recall now that, in a finite volume, the spectrum of the Dirac operator is always discrete and the integrals are replaced by sums. Equivalently, the spectral density $\rho(\lambda)$ is not a smooth function but a sum of the δ-function-like terms, corresponding to the discrete spectrum $\lambda_k(G)$ in the external gluon field G_μ:

$$\rho(\lambda) = \sum_k \delta\left(\lambda - \lambda_k(G)\right). \tag{3.346}$$

Consequently, the justification of Eq. (3.345) becomes questionable and should be put under renewed scrutiny.

Eq. (3.345) in fact implies two limiting procedures:

i) The infinite-volume limit, when the size of the box tends to infinity. In this limit, the eigenvalues of the Dirac operator condense and the discrete energy levels turn into the continuum description with the smooth spectral function $\rho(\lambda)$.

ii) The chiral limit $\mathcal{M} \to 0$, where the smallest mass m_u sets the "resolution scale" for the energy levels. If the distance between the neighboring levels is much smaller than m_u for small values of λ, individual levels cannot be distinguished and the continuum description is appropriate.

To summarize, the infinite volume limit and the limit $\mathcal{M} \to 0$ cannot be interchanged. Otherwise, the lowest-mass excitations of QCD (the pseudoscalar Goldstone bosons) will have a Compton wavelength that exceeds the size of the box, and hence the low-lying spectrum of QCD will receive large finite-volume artifacts. This Compton wavelength, on the other hand, should be much larger than the intrinsic scale implicitly present in the function $\rho(\lambda)$, in order to ensure that one is already in the chiral limit. On dimensional grounds, this scale can only be given by Λ_{QCD}.

At the final stage, Eq. (3.345) should be integrated over all possible gluon configurations G_μ (the spectral density $\rho(\lambda)$ is a functional of G_μ). If the absolute value of the spectral density of the Dirac operator at the origin is bound from below by a constant, independent from the gluon field G_μ, then the vacuum expectation value of the operator $\bar{u}u$ is different from zero. This follows from the fact that the integration measure over the gluon fields is positive definite. Thus, the emergence of a nonzero quark condensate is directly related to the properties of the spectrum of the massless Dirac operator in an external gluon field.

3.14.6 The Goldstone Theorem in the Nonrelativistic Case

As already discussed, *in the relativistic theory,* the number of the Goldstone bosons that emerge as a result of the spontaneous symmetry breaking of a group to its sub-group is equal to the number of the generators, which do not annihilate the vacuum state. Remarkably, this statement should be modified in the nonrelativistic theory, see, for example, Refs. [62–67]. Let us consider a completely general case, where the symmetry, described by a compact, connected, semi-simple Lie group G, is spontaneously broken down to the subgroup H. The unbroken/broken generators, which are equal to the volume integral over the zeroth component of the pertinent Noether currents, are denoted, as before, by Q_V^i, Q_A^l. Let the number of these generators be $n - d$ and d, respectively. Further, consider the matrix

$$\rho^{kl} = \lim_{V \to \infty} \frac{-i}{V} \langle 0 | [Q_A^k, Q_A^l] | 0 \rangle , \qquad (3.347)$$

where V denotes the volume of the system. Quite generally, the commutator in the preceding expression will be a linear combination of all generators. In the relativistic theory, the vacuum expectation value of all Q_V^i, Q_A^l vanishes, since a nonzero vacuum expectation value of the Noether currents $V_\mu^i(x), A_\mu^l(x)$ would lead to the spontaneous breaking of Lorentz symmetry. This restriction is, however, lifted in the nonrelativistic theory. The charges can have nonvanishing vacuum expectation values and hence the matrix ρ^{kl} is different from zero. In this case, the number of the Goldstone bosons is equal to [64, 65]

$$n_G = d - \frac{1}{2} \operatorname{rank} \rho , \qquad (3.348)$$

where the rank of a matrix is the maximum number of linearly independent column (or row) vectors. In order to justify this statement, consider the states $|n_i\rangle$ that contain exactly one Goldstone boson. These states have nonzero overlaps with the states that are obtained by acting with the charges Q_A^l, $l = 1, \ldots, d$, on the vacuum state:

$$\text{for all } i, \quad \langle n_i | Q_A^l | 0 \rangle \neq 0 \quad \text{at least for some } l. \qquad (3.349)$$

The states $|n_i\rangle$ form a basis of some linear subspace of the Fock space. First, let us show that the number of such linearly independent states cannot be larger than d. Indeed, assume the opposite, namely, that there is one extra state. Let it form a one-particle state:

$$|\psi\rangle = c_1 |n_1\rangle + \cdots + c_{d+1} |n_{d+1}\rangle . \qquad (3.350)$$

Using the methods of linear algebra, it is straightforward to show that it is always possible to choose the coefficients c_i such that the condition $\langle 0 | Q_A^l | \psi \rangle = 0$, $l = 1, \ldots, d$, is fulfilled. Indeed, these coefficients must obey the following equation:

$$\sum_{i=1}^{d+1} x_{li} c_i = 0 , \qquad x_{li} = \langle 0 | Q_A^l | n_i \rangle . \qquad (3.351)$$

This equation always has a nontrivial solution, since the number of unknown coefficients is larger than the number of equations. Hence, our initial assumption was not correct, and such a state cannot be regarded as a Goldstone boson. It follows then that the number of the Goldstone bosons does not exceed d.

Further, as we have seen in Section 3.14.3, in the relativistic case the upper bound on the number of Goldstone bosons is, in fact, saturated. This statement is, however, based on the fact that the charges Q_V^i annihilate the vacuum, and does not hold in the nonrelativistic theory anymore. In the latter case, some of the states $|n_i\rangle$ turn out to be linearly dependent.

In what follows, we give heuristic arguments in favor of Eq. (3.348). (For a rigorous proof, the reader is referred to Refs. [64, 65].) We start by noting that applying orthogonal transformations, the real antisymmetric matrix ρ^{kl} can be brought into the form

$$
\rho^{kl} = \left(
\begin{array}{cc|c}
\begin{matrix} 0 & \lambda_{12} \\ -\lambda_{12} & 0 \end{matrix} & & \\
 & \begin{matrix} 0 & \lambda_{34} \\ -\lambda_{34} & 0 \end{matrix} & 0 \\
 & \quad\cdots & \\
\hline
 & 0 & 0
\end{array}
\right),
\tag{3.352}
$$

where λ_{ij} are all real numbers. In the new basis, the generators Q_A^i are split into pairs (Q_A^1, Q_A^2), (Q_A^3, Q_A^4), ..., whose commutators taken between the vacuum states are nonzero, and the "unpaired" generators that correspond to the vanishing entries on the diagonal in the bottom-right corner of the matrix ρ^{kl}. We want first to show that each pair of generators leads, in fact, to a single Goldstone boson state (and not two). Consider, for simplicity, a single pair of generators Q_A^1, Q_A^2 and the states $|n_1\rangle, |n_2\rangle$. If these states were linearly independent, one could always form linear combinations $|n_1'\rangle, |n_2'\rangle$ thereof, so that

$$
\langle 0|Q_A^l|n_i'\rangle = x_l \delta_{li}, \qquad i = 1, 2.
\tag{3.353}
$$

In this case, saturating the commutator $\langle 0|[Q_A^1, Q_A^2]|0\rangle$ by one-particle states gives zero, which contradicts our initial assumption about the form of the matrix ρ^{kl}; see Eq. (3.352). For this reason, $|n_1\rangle$ and $|n_2\rangle$ cannot be linearly independent and $|n_2\rangle = c|n_1\rangle$.

On the other hand, the "unpaired" generators correspond to a single state each. In what follows we shall complete the proof by verifying the preceding statement. To this end, let us choose one generator in each pair, say, $Q_A^1, Q_A^3, \ldots$, and add all "unpaired" generators. In the resulting set of $n_G = d - \frac{1}{2}\operatorname{rank}\rho$ generators, which we denote by $\bar{Q}_A^i$, all commutators have vanishing vacuum expectation value, and all space vectors, corresponding to the generators from this set, are linearly independent. Indeed, assume that this is not the case. Then, it is possible to find m (complex) coefficients C_{pi}, $i = 1 \ldots, m$, for which

$$
C_{pi}\bar{Q}_A^i|0\rangle = 0, \qquad p = 1, \cdots, m.
\tag{3.354}
$$

Further, it is seen that the vectors $C^*_{pi}\bar{Q}^i_A|0\rangle$ are all linearly independent. Indeed, assuming the opposite, a set of the of the coefficients D_p should exist, such that $D_p C^*_{pi}\bar{Q}^i_A|0\rangle = 0$. Introduce the notation $k_i = D_p C^*_{pi}$ and $|v^i\rangle = \bar{Q}^i_A|0\rangle$. Multiplying Eq. (3.354) by D^*_p and summing over p, in this case we finally get

$$k_i|v^i\rangle = 0, \qquad k_i^*|v^i\rangle = 0, \qquad \langle v^i|k_i = 0, \qquad \langle v^i|k_i^* = 0. \tag{3.355}$$

Using Eq. (3.334), we could then write

$$k_i\mathcal{B}^{ki} = -k_i\langle 0|[\bar{Q}^i_A, B^k]|0\rangle = \langle 0|B^k k_i|v^i\rangle - \langle v^i|k_i B^k|0\rangle = 0. \tag{3.356}$$

This would mean that the matrix $\mathcal{B}^{ki}$ has a zero mode and is thus singular, contradicting the initial assumption.

On the basis of the preceding arguments, one can finally conclude that

$$\text{Re}\,(C_{pi})|v^i\rangle = -i\text{Im}\,(C_{pi})|v^i\rangle \doteq |\psi_p\rangle \neq 0. \tag{3.357}$$

Then,

$$\langle 0|[\text{Re}\,(C_{pi})\bar{Q}^i_A, \text{Im}\,(C_{pj})\bar{Q}^j_A]|0\rangle = 2i\langle\psi_p|\psi_p\rangle \neq 0, \tag{3.358}$$

which contradicts the assumption $\langle 0|[\bar{Q}^i_A, \bar{Q}^j_A]|0\rangle = 0$. Hence, the number of the linearly independent states is indeed given by Eq. (3.348). Finally, since the states $\bar{Q}^i_A|0\rangle$ are linearly independent, the corresponding set of the one-particle states $|\bar{n}_i\rangle$ is also linearly independent.

It is very interesting to note that the dispersion law for the Goldstone bosons that correspond to the "paired" and "unpaired" generators is different. As a rule,[20] the energy of the boson, corresponding to an "unpaired" generator, vanishes as the first order of the magnitude of the three-momentum (similar to the zero-mass relativistic particle). On the other hand, the energy of a "paired" mode is proportional to the square of the three-momentum that is reminiscent of the nonrelativistic dispersion law. We do not give a proof of this statement here. Instead, for a simple model with a doublet complex scalar field at nonzero chemical potential, it will be explicitly demonstrated that the breaking of Lorentz invariance leads to the modification of the dispersion law of the Goldstone bosons. Our argumentation will closely follow the discussion of kaon condensation in the medium [68, 69]. The Lagrangian of the model is given by

$$\mathcal{L} = (\partial_0 + i\mu)\Phi^\dagger(\partial_0 - i\mu)\Phi - \boldsymbol{\nabla}\Phi^\dagger\boldsymbol{\nabla}\Phi - m^2\Phi^\dagger\Phi - \lambda(\Phi^\dagger\Phi)^2$$

$$= \partial_\mu\Phi^\dagger\partial^\mu\Phi - (m^2 - \mu^2)\Phi^\dagger\Phi - \lambda(\Phi^\dagger\Phi)^2 + i\mu(\Phi^\dagger\partial_0\Phi - \partial_0\Phi^\dagger\Phi), \tag{3.359}$$

where $\Phi = (\phi, \tilde{\phi})^T$ is a doublet of scalar complex fields $\phi = (\phi_1 + i\phi_2)/\sqrt{2}$ and $\tilde{\phi} = (\tilde{\phi}_1 + i\tilde{\phi}_2)/\sqrt{2}$, and μ denotes the chemical potential. Mathematically, introducing the chemical potential is equivalent to placing the system in a constant background electromagnetic field, $A_0 = \mu$, $\mathbf{A} = 0$. It is clear that at nonzero chemical potential, Lorentz invariance is explicitly broken, as well as charge symmetry. As a consequence of the latter, the masses of the charge-conjugated particles split.

---

[20] Excluding exceptional cases that involve some fine-tuning of the parameters; see, e.g., Ref. [66].

Suppose first that $m^2 > 0$ and $\mu^2 < m^2$. Then, the symmetry $SU(2) \times U(1)$ of the original Lagrangian is not broken. In the basis of real fields, the quadratic part of the Lagrangian can be written as

$$\mathcal{L}_0 = \frac{1}{2}(\phi_1, \phi_2)M\begin{pmatrix}\phi_1\\\phi_2\end{pmatrix} + \frac{1}{2}(\tilde{\phi}_1, \tilde{\phi}_2)\tilde{M}\begin{pmatrix}\tilde{\phi}_1\\\tilde{\phi}_2\end{pmatrix}, \tag{3.360}$$

where, in momentum space,

$$M = \tilde{M} = \begin{pmatrix} p^2 + \mu^2 - m^2 & 2i\mu p_0 \\ -2i\mu p_0 & p^2 + \mu^2 - m^2 \end{pmatrix}. \tag{3.361}$$

The particle spectrum is determined by the equations $\det M = 0$ and $\det \tilde{M} = 0$. This gives

$$((p_0 - \mu)^2 - m^2 - \mathbf{p}^2)((p_0 + \mu)^2 - m^2 - \mathbf{p}^2) = 0. \tag{3.362}$$

The preceding equation has four roots: $p_0 = \omega + \mu$, $p_0 = \omega - \mu$, $p_0 = -(\omega + \mu)$ and $p_0 = -(\omega - \mu)$, with $\omega = \sqrt{m^2 + \mathbf{p}^2}$. Increasing the parameter μ continuously, it is seen that a massless mode at $\mu = m$ emerges. Further, as seen from Eq. (3.360), exactly at this value of the parameter μ, the configuration $\Phi = 0$ becomes a local maximum of energy, and the field Φ starts to develop a vacuum expectation value:

$$|\langle 0|\Phi|0\rangle| = v, \qquad v^2 = \frac{\mu^2 - m^2}{2\lambda}. \tag{3.363}$$

Defining the shifted field

$$\Phi = \begin{pmatrix} 0 \\ v \end{pmatrix} + \frac{1}{\sqrt{2}}\begin{pmatrix} \phi_1 + i\phi_2 \\ \tilde{\phi}_1 + i\tilde{\phi}_2 \end{pmatrix}, \tag{3.364}$$

we may read off the matrices M and $\tilde{M}$ in this case:

$$M = \begin{pmatrix} p_0^2 - \mathbf{p}^2 & 2i\mu p_0 \\ -2i\mu p_0 & p_0^2 - \mathbf{p}^2 \end{pmatrix},$$

$$\tilde{M} = \begin{pmatrix} p_0^2 - \mathbf{p}^2 - 2(\mu^2 - m^2) & 2i\mu\omega \\ -2i\mu\omega & p_0^2 - \mathbf{p}^2 \end{pmatrix}. \tag{3.365}$$

The spectrum of the system is given by

$$p_0 = \sqrt{\mu^2 + \mathbf{p}^2} \pm \mu,$$

$$p_0 = \sqrt{3\mu^2 - m^2 + \mathbf{p}^2 \pm \sqrt{(3\mu^2 - m^2)^2 + 4\mu^2\mathbf{p}^2}}. \tag{3.366}$$

It can be directly seen that there are two massless modes, with *quadratic* and *linear* dispersion relations near $|\mathbf{p}| \simeq 0$, respectively:

$$p_0 = \frac{\mathbf{p}^2}{2\mu} + \cdots \quad \text{and} \quad p_0 = \sqrt{\frac{\mu^2 - m^2}{3\mu^2 - m^2}}|\mathbf{p}| + \cdots. \tag{3.367}$$

At first glance, this result comes at a surprise, since a nonzero vacuum expectation value of the field, introduced in Eq. (3.364), breaks the symmetry along the pattern $SU(2) \times U(1) \to U(1)$. Since the broken group $SU(2)$ has three generators, one would expect three Goldstone bosons instead of two. Recall, however, that according to the Noether theorem, the generators of the broken $SU(2)$ group are given by

$$Q_A^i = \int d^3 \mathbf{x} \left(\dot{\Phi}^\dagger \frac{i}{2} \tau^i \Phi - \Phi^\dagger \frac{i}{2} \tau^i \dot{\Phi} - \mu \Phi^\dagger \tau^i \Phi \right), \qquad (3.368)$$

where the τ^i are the Pauli matrices. These broken generators obey the commutation relations:

$$[Q_A^i, Q_A^j] = -i\varepsilon^{ijk} Q_A^k. \qquad (3.369)$$

Now, from the explicit expression for the generators, Eq. (3.368), it is seen that the vacuum expectation value of Q_A^3 at tree level is equal to

$$\langle 0|Q_A^3|0\rangle = \mu v^2 V, \qquad (3.370)$$

where V denotes the volume of the box. Consequently, Q_A^1, Q_A^2 are "paired" generators and give rise to one Goldstone boson only.

Physically, this result is easy to interpret. Namely, if the dispersion law is linear, the corresponding EOM is of second order in time derivatives (reminiscent of the Klein–Gordon equation). The solutions contain both positive- and negative-energy components, and a *real* field suffices to describe a particle. On the contrary, if the dispersion law is quadratic, then the EOM resembles the Schrödinger equation, and a *complex* field (*two* real fields) is needed to describe a particle; the number of particles in the spectrum is halved [63]. Alternatively, as seen from Eq. (3.359), if the dispersion law is quadratic, then the term $\partial_0\Phi^\dagger \partial_0\Phi$ is much smaller at low energy than $\Phi^\dagger \partial_0\Phi - \partial_0\Phi^\dagger \Phi$, and can be dropped. Then, it is seen that the fields ϕ_1 and ϕ_2 ($\tilde{\phi}_1$ and $\tilde{\phi}_2$) are canonically conjugated to each other and, in fact, describe a single particle [69].

3.15 Vafa–Witten Theorem

In previous sections, it was argued that the $SU(N_f)_L \times SU(N_f)_R$ symmetry in broken down to $SU(N_f)_V$ in the QCD vacuum. As a consequence, a multiplet of pseudoscalar Goldstone bosons emerges in the low-energy spectrum. It would be very interesting to find out whether such a pattern of spontaneous symmetry breaking is completely accidental or follows from the theory.

As a part of the solution of the preceding problem, Vafa and Witten have shown that the vector-like global symmetries (like isospin and baryon number) are not spontaneously broken in QCD, if the vacuum angle θ is equal to zero [61]. In this section we shall give intuitive arguments that support this claim.

Following Ref. [61], it can be stated that the spontaneous symmetry breaking of, say, isospin is signaled by a nonzero vacuum expectation value of any operator, which does not transform under the trivial representation of the isospin group (exact isospin symmetry in the Lagrangian is assumed). For example, we may consider the quark bilinear $\bar{u}u - \bar{d}d$, which transforms like the third component of isospin. Acting analogously as in the case of the quark condensate (see Section 3.14.5), we get

$$\langle 0|\bar{u}u - \bar{d}d|0\rangle = \int dG_\mu \, (\mathrm{Tr}\, S_u(G) - \mathrm{Tr}\, S_d(G))\det(\slashed{D} + \mathcal{M})$$

$$\times \exp\left\{-\frac{1}{2g^2}\int d^4x \, \mathrm{tr}(F_{\mu\nu}F_{\mu\nu})\right\}. \tag{3.371}$$

The difference $\mathrm{Tr}\, S_u(G) - \mathrm{Tr}\, S_d(G)$ takes the form

$$\mathrm{Tr}\, S_u(G) - \mathrm{Tr}\, S_d(G) = (m_d - m_u)\int_{-\infty}^{+\infty} d\lambda \, \frac{\rho(\lambda)}{(m_u - i\lambda)(m_d - i\lambda)}. \tag{3.372}$$

Consider now the limit of this expression, when $m_u - m_d \to 0$ but the quark masses stay nonzero. The denominator of the integrand in Eq. (3.372) stays non-singular in this limit, because the values of λ are limited to the real axis. Consequently, the preceding difference tends to zero in the isospin limit $m_u - m_d \to 0$.

This is not the whole story, however, since the preceding expression has still to be integrated over all gluon field configurations. As shown in Section 3.14.5, the gluon measure is positive definite at $\theta = 0$. Hence, it suffices to show that the absolute value of the integral over λ in Eq. (3.372) is uniformly bounded from above (i.e., bounded by a constant that does not depend on the gluon field G_μ). In order to demonstrate this, let us consider lattice regularization of the path integral. Let a denote the lattice spacing and V be its Euclidean volume. The total number of the lattice sites is V/a^4. If the fermion field has n components, the lattice Dirac operator is a $Vn/a^4 \times Vn/a^4$ matrix with Vn/a^4 eigenvalues, that is, n/a^4 eigenvalues in the unit volume. Consequently the function $\rho(\lambda)$ on the lattice is normalized by

$$\int_{-\infty}^{+\infty} d\lambda \, \rho(\lambda) = \frac{n}{a^4}. \tag{3.373}$$

Further, since $|(m + i\lambda)^{-1}| \leq m^{-1}$, from Eq. (3.372) we get a rigorous upper bound:

$$\mathrm{Tr}\, S_u(G) - \mathrm{Tr}\, S_d(G) \leq \frac{|m_d - m_u|}{m_u m_d}\int_{-\infty}^{+\infty} d\lambda \, \rho(\lambda) = \frac{n}{a^4}\frac{|m_d - m_u|}{m_u m_d}. \tag{3.374}$$

It is seen that the latter expression indeed does not depend on the gluon field G_μ. Thus, the integration over all gluon field configurations with a positive definite measure cannot change the conclusion that this quantity vanishes as $(m_d - m_u) \to 0$ for any fixed value of a. We finally get

$$\langle 0|\bar{u}u - \bar{d}d|0\rangle \to 0 \quad \text{as} \quad m_u - m_d \to 0 \quad \text{and} \quad m_u, m_d \neq 0. \tag{3.375}$$

The proof just given has loopholes. Namely, it assumes that the quantity $\langle 0|\bar{u}u - \bar{d}d|0\rangle$ is the order parameter for the isospin symmetry breaking. Vafa and Witten have further refined the proof, making it devoid of such loopholes. They consider *any operator* $J(x)$ with a nonzero isospin or baryon number and show that the two-point function $\langle 0|TJ(x)J(0)|0\rangle$ decays at large Euclidean distances as $\exp(-\beta|x|)$, $\beta \neq 0$. This means that there are no massless particles in the theory carrying the quantum numbers of $J(x)$. According to the Goldstone theorem, the presence of such particles would signal the spontaneous breaking of the symmetry. For a complete proof of the theorem, the reader is referred to the original paper [61].

The main lesson that one learns from the Vafa–Witten theorem is that there exists a clear difference between vector and nonvector symmetries. While the vector symmetries, under very plausible assumptions, cannot be spontaneously broken in QCD, other symmetries can be and actually are spontaneously broken.

3.16 Quark Mass Dependence of Hadron Masses

3.16.1 Quark Mass Dependence at Leading Order

The pseudoscalar Goldstone bosons, which are generated by the spontaneous chiral symmetry breaking, are massless, if the quark masses in the Lagrangian are exactly zero. In addition, the low-energy spectrum of QCD contains particles, whose mass is nonvanishing in the chiral limit, like, for example, the nucleons. However, as we know, the u, d, s quarks have small but nonzero masses. Owing to this fact, the Goldstone bosons acquire masses, and the masses of other particles are also shifted from their values in the chiral limit. In this section, we consider these shifts at the first order in perturbation theory in the (small) quark masses. Our discussion here mainly follows Ref. [70].

The QCD Hamiltonian can be written in the following form (we take $N_f = 3$):

$$H_{\text{QCD}} = H_0 + H_m,$$

$$H_m = \int d^3\mathbf{x}\, (m_u \bar{u}(x)u(x) + m_d \bar{d}(x)d(x) + m_s \bar{s}(x)s(x)), \tag{3.376}$$

where H_0 denotes the Hamiltonian in the chiral limit. Let now $|p,n\rangle$ be the covariantly normalized 1-particle eigenstates of H_0 with a mass M_n,

$$H_0|p,n\rangle = E_n(\mathbf{p})|p,n\rangle, \qquad \langle p',n|p,n\rangle = (2\pi)^3\delta^{(3)}(\mathbf{p}' - \mathbf{p})2E_n(\mathbf{p}), \tag{3.377}$$

where $E_n(\mathbf{p}) = (M_n^2 + \mathbf{p}^2)^{1/2}$ and the label n collects all quantum numbers of the state.

Calculating the energy shift to first order, we have

$$\langle p',n|H_m|p,n\rangle = (2\pi)^3\delta^{(3)}(\mathbf{p}' - \mathbf{p})\langle p',n|m_u\bar{u}u + m_d\bar{d}d + m_s\bar{s}s|p,n\rangle$$

$$= \delta E_n(\mathbf{p})(2\pi)^3\delta^{(3)}(\mathbf{p}' - \mathbf{p})2E_n(\mathbf{p}). \tag{3.378}$$

From the preceding relation, one immediately obtains

$$2E_n(\mathbf{p})\delta E_n(\mathbf{p}) = \delta M_n^2 = \langle p,n|m_u\bar{u}u + m_d\bar{d}d + m_s\bar{s}s|p,n\rangle. \tag{3.379}$$

If the mass M_n does not vanish in the chiral limit, we may write

$$M_n^2 = \overset{\circ}{M}_n^2 + m_u B_n^u + m_d B_n^d + m_s B_n^s + \cdots,$$

$$B_n^f = \lim_{\mathcal{M} \to 0} \langle p, n | \bar{\psi}_f \psi_f | p, n \rangle, \quad f = u, d, s, \tag{3.380}$$

where $\overset{\circ}{Q}$ denotes the chiral limit value of the quantity Q. Note that $\overset{\circ}{M}_n$ and B_n^f depend only on a single parameter, Λ_{QCD}. Obviously, for Goldstone bosons the parameter $\overset{\circ}{M}_n = 0$.

Further, for the hadrons whose masses do not vanish in the chiral limit, Eq. (3.380) can be rewritten in a form that is linear in the hadron masses:

$$M_n = \overset{\circ}{M}_n + m_u \tilde{B}_n^u + m_d \tilde{B}_n^d + m_s \tilde{B}_n^s + \cdots. \tag{3.381}$$

The relation between the parameters B_n^f and $\tilde{B}_n^f$ can be easily established by applying a Taylor expansion. This procedure, however, fails, if we are dealing with Goldstone bosons, for which the quantity $\overset{\circ}{M}_n$ vanishes. In this case, we have to stick to the quadratic form given in Eq. (3.380).

There is, however, a subtlety in the derivation just presented. It namely assumes that the matrix elements of the Hamiltonian H_m between the unperturbed states form a diagonal matrix. This statement is not true, for example, for the $SU(3)_V$ octet. In order to see this, we may use the fact that, in the chiral limit, QCD has an exact $SU(3)_V$ symmetry. To this end, we write first

$$\bar{u}u = u^0 + \frac{1}{2} u^3 + \frac{1}{2\sqrt{3}} u^8,$$

$$\bar{d}d = u^0 - \frac{1}{2} u^3 + \frac{1}{2\sqrt{3}} u^8,$$

$$\bar{s}s = u^0 - \frac{1}{\sqrt{3}} u^8, \tag{3.382}$$

where $u^i = \bar{\psi} \lambda^i \psi$, $i = 1, \ldots, 8$ and $u^0 = (\bar{u}u + \bar{d}d + \bar{s}s)/3$. It is straightforward to check that u^i transforms like a pertinent component of the $SU(3)_V$-octet, whereas u^0 is a scalar. The Hamiltonian H_m can then be rewritten as a linear combination of u^0, u^3 and u^8. Furthermore, since the states $|p, n\rangle$, which are the eigenstates of the QCD Hamiltonian *in the chiral limit*, form $SU(3)_V$-multiplets, we may use the Wigner–Eckart theorem to relate the matrix elements B_n^f with a different n and f. This gives

$$\langle p, n | u^0 | p, n' \rangle = a \delta_{nn'}, \tag{3.383}$$

$$\langle p, n | u^i | p, n' \rangle = x \langle p, n | T^i | p, n' \rangle + y \langle p, n | D^i | p, n' \rangle, \tag{3.384}$$

where T^i denote the generators of the $SU(3)$ group and $D^i = \frac{2}{3} d^{ijk} T^j T^k$. (The symbol d^{ijk} is defined in the Appendix.) Using now the explicit expressions for the operators T^i and D^i, it can be straightforwardly shown that the third and the eighth components of the octet mix, if $m_u \neq m_d$. The physical states can be defined by

$$|\pi^0\rangle = \cos\theta |\pi^3\rangle - \sin\theta |\pi^8\rangle,$$

$$|\eta\rangle = \sin\theta |\pi^3\rangle + \cos\theta |\pi^8\rangle. \tag{3.385}$$

where θ is chosen in the following way:

$$\tan 2\theta = \frac{\sqrt{3}}{2}\frac{m_u - m_d}{\hat{m} - m_s}, \qquad \hat{m} = \frac{1}{2}(m_u + m_d). \tag{3.386}$$

(Here, as an example, we consider the pseudoscalar meson octet.) The Hamiltonian is diagonalized in the new basis, and the masses can be easily read off.

In order to keep things as simple as possible, below we display the mass formulae for the pseudoscalar octet in the case $M_{\pi^+} = M_{\pi^-}$, all octet masses can be expressed in terms of two linearly independent matrix elements:

$$M_\pi^2 = 2\hat{m}B_0 + m_s\delta\cdots,$$

$$M_K^2 = (\hat{m} + m_s)B_0 + \hat{m}\delta\cdots,$$

$$M_\eta^2 = \frac{2}{3}(\hat{m} + 2m_s)B_0 + \frac{1}{3}(4\hat{m} - m_s)\delta\cdots, \tag{3.387}$$

where the ellipses stand for the higher-order terms in the quark masses, and B_0, δ are given by

$$B_0 = \langle\pi^0|\bar{u}u|\pi^0\rangle = \langle\pi^0|\bar{d}d|\pi^0\rangle,$$

$$\delta = \langle\pi^0|\bar{s}s|\pi^0\rangle. \tag{3.388}$$

Both these matrix elements should be evaluated in the chiral limit, where all quark masses are set to zero. We shall demonstrate in what follows that chiral symmetry imposes further restriction on the independent matrix elements, resulting in $\delta = 0$. Thus, at the first order in quark masses, all masses in the Goldstone boson octet are determined by a single matrix element B_0.

As expected, in the isospin symmetry limit, the Gell–Mann–Okubo formula holds:

$$4M_K^2 = 3M_\eta^2 + M_\pi^2. \tag{3.389}$$

This relation is fulfilled in nature within a few percent accuracy.

Finally, note that higher-order corrections to hadron masses can be also calculated using perturbation theory. This is not a trivial enterprise, however, and the result is not merely a Taylor expansion of hadron masses in quark masses, with fixed coefficients. The reason for this is that the energy denominators that appear beyond the first order lead to infrared singularities in the presence of the (massless) Goldstone particles. Hence, apart from the conventional polynomial terms, logarithms of the quark masses (the so-called chiral logarithms) as well as fractional powers of the quark masses appear in higher orders. Most straightforwardly, the expansion of the hadron masses in the quark masses is performed within Chiral Perturbation Theory as will be explicitly demonstrated in the next chapter.

3.16.2 Feynman–Hellmann Theorem

In the previous section, the expansion of the hadron masses in the vicinity of the chiral limit was considered. It is also interesting to perform such an expansion in the vicinity of the physical quark masses. The coefficient of this expansion in the first order is given by the so-called Feynman–Hellmann theorem [71, 72] and characterizes the *explicit* chiral symmetry breaking effect in a given hadron at the physical point (in contrast to the *spontaneous* breaking, which arises even at zero quark masses).

Consider a quantum-mechanical Hamiltonian $H(\lambda)$ depending on some external parameter λ. The eigenvalues and the eigenvectors of this Hamiltonian are also functions of λ:

$$H(\lambda)|\psi(\lambda)\rangle = E(\lambda)|\psi(\lambda)\rangle, \qquad \langle\psi(\lambda)|\psi(\lambda)\rangle = 1. \tag{3.390}$$

Differentiating $E(\lambda)$ with respect to λ, we get

$$\frac{dE(\lambda)}{d\lambda} = \frac{d}{d\lambda}\langle\psi(\lambda)|H(\lambda)|\psi(\lambda)\rangle$$

$$= \left\langle\psi(\lambda)\left|\frac{dH(\lambda)}{d\lambda}\right|\psi(\lambda)\right\rangle + \left\langle\frac{\psi(\lambda)}{d\lambda}\left|H(\lambda)\right|\psi(\lambda)\right\rangle + \left\langle\psi(\lambda)\left|H(\lambda)\right|\frac{d\psi(\lambda)}{d\lambda}\right\rangle$$

$$= \left\langle\psi(\lambda)\left|\frac{dH(\lambda)}{d\lambda}\right|\psi(\lambda)\right\rangle + E(\lambda)\frac{d}{d\lambda}\langle\psi(\lambda)|\psi(\lambda)\rangle$$

$$= \left\langle\psi(\lambda)\left|\frac{dH(\lambda)}{d\lambda}\right|\psi(\lambda)\right\rangle. \tag{3.391}$$

In the case of QCD, we may identify the external parameter λ with the quark masses m_f. From the Feynman–Hellmann theorem it then follows that for any hadron state H,

$$\frac{dM_H^2}{dm_f} = \langle H|\bar{\psi}_f\psi_f|H\rangle. \tag{3.392}$$

Note that, as we have shown, due to the relativistic normalization used, the mass squared M_H^2 instead of M_H emerges. We emphasize here once more that Eq. (3.392) is valid to all orders in the quark masses and does not imply the chiral limit $m_f \to 0$.

The matrix element on the r.h.s. of Eq. (3.392) is scale-dependent in QCD. It is useful to define a scale-independent quantity:

$$\sigma_H^f = \frac{m_f}{2M_H}\langle H|\bar{\psi}_f\psi_f|H\rangle. \tag{3.393}$$

This quantity is referred to as the sigma-term of a given hadron H and plays a very important role in low-energy hadron phenomenology. As mentioned, it serves as a measure of the *explicit* chiral symmetry breaking in QCD.

Using the σ-terms, we may define a quantity called the strangeness content of a hadron y_H,

$$y_H = \frac{2\langle H|\bar{s}s|H\rangle}{\langle H|\bar{u}u + \bar{d}d|H\rangle}, \tag{3.394}$$

which obviously measures the "admixture" of strange quarks in a hadron. It is expected that in conventional non-strange hadrons like pions, ρ-mesons or the nucleon, the strangeness content should be small albeit not vanishing (due to the presence of virtual $\bar{s}s$ pairs). Note also that the quantity y_H is scale-independent, because the scale-dependence of the operators $\bar{\psi}_f\psi_f$ in QCD is universal (it does not depend on the flavor).

Finally, we mention that in the preceding quantum-mechanical derivation we did not pay attention to issues like renormalization (due to which, for example, m_f is to be identified with the renormalized quark mass). The treatment can be done more rigorously, see, for example, Ref. [73]. In the same paper, we may find the generalization of the Feynman–Hellmann theorem for *resonances,* where on the l.h.s. of Eq. (3.392) the *complex resonance pole position* appears instead of real-valued M_H^2. Note also that, as proposed in Ref. [73], the relation of the type Eq. (3.392) can be used to determine the nature of the states and resonances, measured on the lattice. Indeed, for example, a large measured strangeness content of the particles, which, according to the simple quark model, consist of u- and d-quarks only, would strongly testify in favor of the exotic nature of these particles. To summarize, the Feynman–Hellmann theorem provides a powerful tool for the calculation of the σ-terms and the strangeness content in lattice QCD (because in that case we would be able to freely change the values of the quark masses, which is not possible in nature) and, in addition, provides a tool to study the nature of stable particles and resonances on the lattice.

3.17 Soft-Pion Technique

The soft-pion technique discussed next was been a premier tool of hadron and particle physics in the 1960s and 1970s; see, for example, the textbook [12]. By now we know that this technique, also called current algebra, amounts to nothing but a tree-level calculation of the corresponding effective Lagrangian. Still, some features of low-energy interactions can be nicely worked out using this by now old-fashioned approach and, furthermore, it will also be interesting to see how the modern use of effective Lagrangians gives a very different tool to arrive at these results and the corrections thereof.

3.17.1 Gell–Mann–Oakes–Renner Relation

As shown in previous sections, all masses in the pseudoscalar multiplet are given to first order in terms of a single parameter, B_0, defined in Eq. (3.383). It is possible to relate this quantity to the quark condensate, considered in Section 3.14.5, which is one of the fundamental parameters of QCD characterizing the spontaneous chiral symmetry breaking. To this end, we use here the so-called soft-pion technique (an ancestor to Chiral Perturbation Theory), which enables one to relate matrix elements containing different number of pions in the in- and out-states in the chiral limit. We shall briefly consider this method in what follows, since, apart from calculating B_0, the method enables one to get important information about the interactions of Goldstone bosons at small momenta.

Let us start from the matrix element of the axial-vector current between the vacuum and the one pseudoscalar Goldstone boson state (called pion for brevity); see Eq. (3.336). Differentiating both sides of this equation, we get

$$\langle 0|\partial^\mu A_\mu^k(x)|\pi^i(q)\rangle = M_\pi^2 \delta_{ik} F_\pi e^{-iqx}. \tag{3.395}$$

This means that the operator

$$\phi^k(x) = \frac{1}{F_\pi M_\pi^2} \partial^\mu A_\mu^k(x), \tag{3.396}$$

which carries the quantum numbers of a pseudoscalar boson, is properly normalized and can be used as interpolating field operator in the calculation of the matrix elements. The equation (3.396) is usually referred to as the hypothesis of the partial conservation of the axial-vector current (PCAC).

Let us explain the technique on the example of the following matrix element, which contains one pseudoscalar boson in the out state,

$$\mathcal{M}(q) = \langle \pi^i(q), n; out|\mathcal{O}(0)|m; in\rangle, \tag{3.397}$$

where $\mathcal{O}(0)$ is some (local) operator and n, m are arbitrary states. Using the LSZ reduction formula, we may contract the boson from the out-state. To this end, let us consider the (off-shell) quantity

$$\mathcal{T}(q) = i\frac{M_\pi^2 - q^2}{F_\pi M_\pi^2} \int d^4x\, e^{iqx} \langle n; out|T\partial^\mu A_\mu^i(x)\mathcal{O}(0)|m; in\rangle. \tag{3.398}$$

Here, $q^2 \neq M_\pi^2$. The matrix element $\mathcal{M}(q)$ is related to $\mathcal{T}(q)$ as

$$\mathcal{M}(q) = \lim_{q^2 \to M_\pi^2} \mathcal{T}(q). \tag{3.399}$$

Next, let us consider the limit $q^\mu = 0$ in the quantity $\mathcal{T}(q)$. Performing a partial integration, we can write

$$\mathcal{T}(q) = i\frac{M_\pi^2 - q^2}{F_\pi M_\pi^2} \int d^4x\, e^{iqx} \langle n; out|-\delta(x_0)[A_0^i(x), \mathcal{O}(0)]|m; in\rangle + q^\mu R_\mu^i(q)$$

$$= -i\frac{M_\pi^2 - q^2}{F_\pi M_\pi^2} \langle n; out|[Q_A^i, \mathcal{O}(0)]|m; in\rangle + q^\mu R_\mu^i(q), \tag{3.400}$$

where

$$R_\mu^i(q) = \frac{M_\pi^2 - q^2}{F_\pi M_\pi^2} \int d^4x\, e^{iqx} \langle n; out|TA_\mu^i(x)\mathcal{O}(0)|m; in\rangle. \tag{3.401}$$

Since we assume that $M_\pi \neq 0$, the quantity $R_\mu^i(q)$ is regular as $q^\mu \to 0$, and the last term in Eq. (3.400) vanishes in this limit. We thus obtain

$$\mathcal{T}(0) = \lim_{q_\mu \to 0} \mathcal{T}(q) = -\frac{i}{F_\pi} \langle n; out|[Q_A^i, \mathcal{O}(0)]|m; in\rangle. \tag{3.402}$$

The essence of the approach, based on the PCAC hypothesis, consists in the assumption that the amplitude $\mathcal{T}(q)$ is a smooth function of q^μ and does not change much during the extrapolation from $q^2 = 0$ to $q^2 = M_\pi^2$, namely, $\mathcal{M}(q) \simeq \mathcal{T}(0)$. Note also that $\mathcal{T}(q)$ at $q^2 \to M_\pi^2$ equals the physical amplitude $\mathcal{M}(q)$ and is thus independent of the choice of the interpolating pion field. On the contrary, $\mathcal{T}(0)$ depends on this choice, and the statement concerning the smoothness is made for a particular choice of the pion interpolating field displayed in Eq. (3.396).

In addition, note that the PCAC approach builds on the assumption that the most singular behavior of the matrix element $\langle n; out | TA_\mu^i(x)\mathcal{O}(0) | m; in \rangle$ is given by the one-pion exchange contribution. The factor $(M_\pi^2 - q^2)$ in front of the Fourier transform of this matrix element removes just the one-pion singularity, and the rest is assumed to be smoothly behaved.

At the next step, we may repeat the same manipulation with any number of pions in the in- and out-states. We shall do this for the matrix element given in Eq. (3.383) with $\mathcal{O}(0) = \bar{u}(0)u(0)$, and perform the chiral limit at the end. As a result, we get

$$B_0 = \lim_{\mathcal{M} \to 0} \langle \pi^0 | \bar{u}u | \pi^0 \rangle = - \lim_{\mathcal{M} \to 0} \frac{1}{F_\pi^2} \langle 0 | [Q_A^3, [Q_A^3, \bar{u}u]] | 0 \rangle. \tag{3.403}$$

Using the canonical commutation relations for the quark fields, one obtains

$$B_0 = - \lim_{\mathcal{M} \to 0} \frac{1}{F_\pi^2} \langle 0 | \bar{u}u | 0 \rangle. \tag{3.404}$$

By the same token, we get

$$\delta = \lim_{\mathcal{M} \to 0} \langle \pi^0 | \bar{s}s | \pi^0 \rangle = - \lim_{\mathcal{M} \to 0} \frac{1}{F_\pi^2} \langle 0 | [Q_A^3, [Q_A^3, \bar{s}s]] | 0 \rangle = 0. \tag{3.405}$$

Finally, using Eq. (3.404) and Eqs. (3.404) in Eq. (3.387), we arrive at the famous Gell–Mann–Oakes–Renner (GOR) relation, which is exact at lowest order in the quark masses [74]. Note that the quantity B_0 is properly defined in QCD only in the chiral limit $\mathcal{M} \to 0$. If $\mathcal{M} \neq 0$, this quantity is UV-divergent and acquires a scale dependence after renormalization.

3.17.2 Interactions of the Goldstone Bosons at Low Energies

Using the soft-pion technique, we may prove that the Goldstone boson interactions must be vanishing at zero momenta in the chiral limit, both with other Goldstone bosons and with the particles whose masses do not vanish in the chiral limit. Such particles are often called "matter fields." This property of Goldstone bosons is extremely important and, in fact, is quintessential for the construction of a systematic low-energy perturbation theory. Indeed, the interactions of Goldstone bosons are always weak at low energies, irrespective of the value(s) of the coupling constant(s)!

The simple arguments that prove the preceding statement can be found, for example, in Ref. [75]. Here we shall follow a similar line of reasoning. Consider the case of the exact symmetry $\mathcal{M} = 0$, where the axial-vector current is conserved and the pion

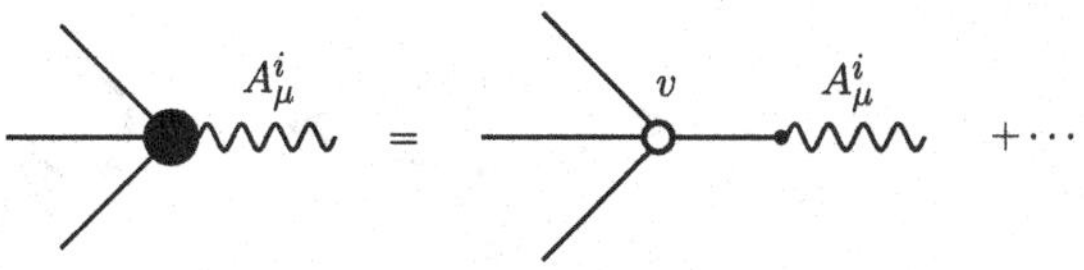

Figure 3.8 The matrix element of the axial-vector current between the vacuum and the three-pion state. At low momenta, this matrix element is dominated by a pole diagram that contains the massless pion, where v denotes the four-pion vertex.

mass is equal to zero.[21] Consider further, for example, the amplitude of the three-pion creation from the vacuum by the axial-vector current

$$M_\mu^{i,i_1,i_2,i_3}(p_1,p_2,p_3) = \langle \pi^{i_1}(p_1)\pi^{i_2}(p_2)\pi^{i_3}(p_3); out|A_\mu^i(0)|0\rangle, \qquad (3.406)$$

with isospin indices i,i_1,i_2,i_3. This quantity at small momenta is dominated by the contribution of the pion pole, as depicted in Fig. 3.8. The dominant part is given by

$$M_\mu^{i,i_1,i_2,i_3}(p_1,p_2,p_3) = v^{j,i_1,i_2,i_3}(p;p_1,p_2,p_3)\frac{-i}{p^2}\langle \pi^j(p)|A_\mu^i(0)|0\rangle + \cdots$$

$$= v^{i,i_1,i_2,i_3}(p;p_1,p_2,p_3)\frac{F_\pi p_\mu}{p^2} + \cdots, \qquad (3.407)$$

where $p = p_1 + p_2 + p_3$ and $v^{j,i_1,i_2,i_3}(p;p_1,p_2,p_3)$ denotes the four-pion interaction vertex, also shown in the figure. Further, the conservation of the axial-vector current implies that

$$p^\mu M_\mu^{i,i_1,i_2,i_3}(p_1,p_2,p_3) = F_\pi v^{i,i_1,i_2,i_3}(p;p_1,p_2,p_3) = 0, \qquad (3.408)$$

when $p_1,p_2,p_3 \to 0$. Since $F_\pi \neq 0$ in the chiral limit, this means that the four-pion interaction vertex vanishes when the external momenta vanish. This is exactly the property that we wanted to demonstrate. Note that later we will show how corrections to this leading order result can be obtained systematically. This is not possible within the soft-pion approach.

3.18 Witten–Veneziano Formula

Up to now, we have restricted our discussion to the eight Goldstone bosons, which correspond to the symmetry breaking pattern $SU(N_f)_L \times SU(N_f)_R \to SU(N_f)_V$. There exists a ninth candidate for the Goldstone bosons, the η', corresponding to the broken symmetry $U(1)_A$. However, due to the singlet anomaly, its mass is nonzero even in the chiral limit [76]. In nature, the mass of the η' is around 1 GeV, which is much larger than the mass of any member of the pseudoscalar octet.

---

[21] Note that here we have interchanged the zero-momentum and the chiral limit. According to the PCAC hypothesis, however, we have to arrive at the same result at the end.

Figure 3.9 The connected (left panel) and disconnected (right panel) diagrams contributing to the two-point function of the axial-vector currents, denoted by filled circles. The disconnected diagram arises only in the case of the singlet current and is suppressed in the large-N_c limit.

In order to relate the mass of the η' meson to the anomaly in the singlet current, we define the composite operator:

$$2N_f \omega(x) = \frac{N_f}{16\pi^2} \varepsilon^{\mu\nu\alpha\beta} \text{tr}_c \left(F_{\mu\nu} F_{\alpha\beta} \right). \tag{3.409}$$

For simplicity, we work in the chiral limit $\mathcal{M} = 0$. In this case, the divergence of the singlet axial-vector current is equal to

$$\partial^\mu A_\mu^0(x) = -2N_f \omega(x). \tag{3.410}$$

Let us consider the two-point function of the ω-field. Inserting a complete set of the intermediate states, we get

$$(2N_f)^2 \int d^4x e^{ipx} \langle 0|T \omega(x)\omega(0)|0\rangle = \langle 0|\omega(0)|\eta'(p)\rangle \frac{i(2N_f)^2}{p^2 - M_{\eta'}^2} \langle \eta'(p)|\omega(0)|0\rangle + \cdots, \tag{3.411}$$

where we have singled out the contribution from the η'-meson. The matrix elements entering Eq. (3.411) can be related to the divergence of the axial-vector current

$$(2N_f)\langle 0|\omega(0)|\eta'(p)\rangle = -\langle 0|\partial^\mu A_\mu^0(0)|\eta'(p)\rangle = ip^\mu \langle 0|A_\mu^0(0)|\eta'(p)\rangle. \tag{3.412}$$

At this stage, one can apply large-N_c arguments, see, for example, Refs. [77, 78]. As it is well known, in this limit the strong coupling constant g scales like $N_c^{-1/2}$. If now, in order to get a standard normalization of the gluon field, we rescale $G_\mu \to gG_\mu$, then the anomaly $\omega(x)$ will acquire a prefactor, g^2. Consequently, in the large-N_c limit, the anomaly $\omega(x)$ vanishes like N_c^{-1} and the singlet and non-singlet pseudoscalar states obey a larger $U(N_f)$ symmetry.

From this fact we may conclude the following:

i) In the limit $N_c \to \infty$, the masses of the singlet and non-singlet pseudoscalar mesons are degenerate, that is, the mass of the η' meson vanishes (in the chiral limit). *Hint:* Consider the two-point function of axial-vector currents, shown in Fig. 3.9. The masses are given by the poles in the external momentum squared. The only difference between the nonsinglet and singlet cases is the disconnected diagram shown in Fig. 3.9, right panel. However, the contribution of this diagram is suppressed in the large-N_c limit (this corresponds to the well-known OZI rule [79–81]), and thus the $U(N_f)$ symmetry is restored.

ii) Using the normalization

$$\langle 0|A_\mu^0(0)|\eta'(p)\rangle = ip_\mu\sqrt{2N_f}F_{\eta'}\,, \tag{3.413}$$

it can be shown that $F_{\eta'} = F_\pi$ to the leading order in N_c. *Hint:* Not only the pole position of the two-point function, but also the residue of the pole can be determined in the large-N_c limit by using the $U(N_f)$ symmetry.

Since, except for the η' meson, there are no more particles in the singlet channel whose masses vanish in the large-N_c limit, it is immediately seen that, to the leading order in this limit, only a one-particle intermediate state with the η' meson contributes. Further, substituting $p = 0$ in Eq. (3.411), one straightforwardly obtains

$$M_{\eta'}^2 = \frac{2N_f}{F_\pi^2}\int d^4x\,\langle 0|T\omega(x)\omega(0)|0\rangle\,, \tag{3.414}$$

which, at the leading order in N_c^{-1} and in the chiral limit, relates the mass of the η'-meson with the anomaly. The integral on the r.h.s. of Eq. (3.414) is referred to as the *topological susceptibility*. Equation (3.414) is known under the name of the "Witten–Veneziano formula" [82, 83]; see also Ref. [76].

Finally, note that the two-point function in Eq. (3.411) is ill-defined due to the ultraviolet divergences; thus, the pertinent dispersion relation requires two subtractions. This renders the notion of the topological susceptibility ambiguous. On the other hand, defining the (RG-invariant) winding number operator:

$$\nu = \int d^4x\,\omega(x)\,, \tag{3.415}$$

one may show that the susceptibility can be unambiguously defined as the mean square value of the winding number per unit volume $V^{-1}\langle\nu^2\rangle$. A detailed discussion of this issue can be found, for example, in Ref. [84].

3.19 Anomaly Matching

Next, we discuss the so-called anomaly matching that goes back to the paper by 't Hooft [85]. The basic idea behind it is very simple. Suppose one considers a theory with some global symmetry and assumes that some of the Noether currents corresponding to this symmetry are anomalous (the situation is reminiscent of QCD, with global flavor symmetries and vector/axial-vector currents). At some lower momentum scale, new degrees of freedom come into play, and the Green's functions of the currents are given in terms of Feynman diagrams build from the effective degrees of freedom (Goldstone bosons, nucleons, ..., in our case). The anomaly matching condition states that the form of the anomalous Ward identities remains unchanged in the effective theory.

In order to prove this statement, the following line of reasoning is suggested in Ref. [85]. Let us promote the abovementioned global symmetry to a local one. Thus, one obtains a gauge theory with anomalies, which is inconsistent. In order to arrive at a consistent theory, one could cancel these anomalies, for example, by adding massless fermions, which are coupled to the newly introduced gauge bosons only. At low energies, we have an effective theory, which is also consistent. In other words, the anomalies that are produced by the additional fermions are canceled by the anomalies that emerge in the effective theory. Since the additional fermion sector in the original and effective theories is identical, we may finally conclude that the anomalies of the original theories should be reproduced by the anomalies in the effective theory.

The anomaly matching condition looks very plausible, almost self-evident. At the same time, when combined with the assumption of confinement, this condition has powerful implications on the possible spectrum of QCD at low energies. To demonstrate this, we start by mentioning that, as first noted in Ref. [86], the anomalies are responsible for the emergence of the peculiar pole-like behavior in the Green's functions of the currents. For illustration of this statement, let us in the beginning consider the diagrams shown in Fig. 3.2. For simplicity, assume further that $p_1^2 = p_2^2 = 0$. An explicit calculation for the amplitude from Eq. (3.89) in the Pauli–Villars regularization gives

$$
T^{\mu\alpha\beta}(p_1, p_2) = -8ie^2 \left[p_2^\beta \varepsilon^{\alpha\mu\sigma\lambda} p_{1\sigma} p_{2\lambda} - p_1^\alpha \varepsilon^{\beta\mu\sigma\lambda} p_{1\sigma} p_{2\lambda} \right] (J_1(m^2, q^2) - J_1(\Lambda^2, q^2))
$$

$$
+ 16ie^2 \left[p_2^\alpha \varepsilon^{\beta\mu\sigma\lambda} p_{1\sigma} p_{2\lambda} - p_1^\beta \varepsilon^{\alpha\mu\sigma\lambda} p_{1\sigma} p_{2\lambda} - \frac{1}{2} q^2 \varepsilon^{\alpha\beta\mu\sigma} (p_{2\sigma} - p_{1\sigma}) \right]
$$

$$
\times (J_2(m^2, q^2) - J_2(\Lambda^2, q^2)), \tag{3.416}
$$

where $q^2 = (p_1 + p_2)^2$, and

$$
J_1(m^2, q^2) = \frac{1}{16\pi^2} \int_0^1 \rho\, d\rho \int_0^1 d\tau \frac{\rho(1-\rho) + 2\rho^2 \tau(1-\tau)}{m^2 - \rho^2 \tau(1-\tau)q^2},
$$

$$
J_2(m^2, q^2) = \frac{1}{16\pi^2} \int_0^1 \rho\, d\rho \int_0^1 d\tau \frac{\rho^2 \tau(1-\tau)}{m^2 - \rho^2 \tau(1-\tau)q^2}, \tag{3.417}
$$

where ρ and τ are Feynman parameters. Consider, for instance, the function $J_2(m^2, q^2)$. After performing the integration over the Feynman parameter ρ, partial integration and change of variables, this integral can be rewritten as

$$
J_2(m^2, q^2) = -\frac{1}{32\pi^2 q^2} \left[1 + \int_0^1 \frac{m^2 d\tau}{\tau(1-\tau)q^2} \ln \frac{m^2 - \tau(1-\tau)q^2}{m^2} \right]
$$

$$
= \frac{1}{32\pi^2} \int_{4m^2}^\infty ds \frac{2m^2}{s^2} \ln \frac{1 + \sqrt{1 - 4m^2/s}}{1 - \sqrt{1 - 4m^2/s}} \frac{1}{s - q^2}. \tag{3.418}
$$

This equation has the form of a dispersion integral

$$J_2(m^2, q^2) = \frac{1}{\pi} \int ds \, \frac{\operatorname{Im} J_2(m^2, s)}{s - q^2}, \tag{3.419}$$

and the discontinuity can be read off straightforwardly:

$$\operatorname{Im} J_2(m^2, s) = \frac{1}{32\pi} \frac{2m^2}{s^2} \ln \frac{1 + \sqrt{1 - 4m^2/s}}{1 - \sqrt{1 - 4m^2/s}}. \tag{3.420}$$

It can be further seen that in the sense of distributions the preceding expression tends to const $\times \delta(s)$ as $m \to 0$. Indeed, consider a smooth test function $\varphi(s)$. Then,

$$\int_{-\infty}^{\infty} ds \, \varphi(s) \operatorname{Im} J_2(m^2, s) = \frac{1}{32\pi} \int_{4m^2}^{\infty} ds \, \varphi(s) \frac{2m^2}{s^2} \ln \frac{1 + \sqrt{1 - 4m^2/s}}{1 - \sqrt{1 - 4m^2/s}}$$

$$= \frac{1}{64\pi} \int_1^{\infty} \frac{ds}{s^2} \varphi(4m^2 s) \ln \frac{1 + \sqrt{1 - 1/s}}{1 - \sqrt{1 - 1/s}}$$

$$\to \frac{1}{32\pi} \varphi(0). \tag{3.421}$$

Consequently, the function $J_2(m^2, q^2)$ itself, which is given by the dispersion representation (3.420), contains a massless pole $\sim 1/q^2$ in the variable q^2 in this limit.

In the case of QCD, which is a confining theory, the preceding result that was obtained at one loop cannot be trusted, of course. However, using Ward identities, we may show that a pole still emerges. Here we shall generally follow the method outlined in Ref. [87]. Consider the currents that are free of the *gluon anomaly* (in other words, all currents except the singlet axial current). Further, let us concentrate on the three-point function of one axial-vector and two vector currents, which is given by

$$T_{\alpha\beta\mu}^{ijk} = \int d^4x \, d^4y \, e^{i(p_1 x + p_2 y)} \langle 0 | T V_\alpha^i(x) V_\beta^j(y) A_\mu^k(0) | 0 \rangle. \tag{3.422}$$

The flavor indices in the preceding equations can be singled out, yielding a totally symmetric tensor $C^{ikj} = \operatorname{tr}(T^i \{T^j, T^k\})$, where T^i denote the generators of the flavor group. Turning to the Lorentz structure, for simplicity, we choose the kinematics $p_1^2 = p_2^2 = p^2$. Using Bose and Lorentz symmetry, the amplitude can be expanded in invariant scalar amplitudes as

$$T_{\alpha\beta\mu}^{ijk} = C^{ijk} \left\{ \left[p_2^\beta \varepsilon^{\alpha\mu\sigma\lambda} p_{1\sigma} p_{2\lambda} - p_1^\alpha \varepsilon^{\beta\mu\sigma\lambda} p_{1\sigma} p_{2\lambda} \right] A_1(p^2, q^2) \right.$$

$$\left. + \left[p_2^\alpha \varepsilon^{\beta\mu\sigma\lambda} p_{1\sigma} p_{2\lambda} - p_1^\beta \varepsilon^{\alpha\mu\sigma\lambda} p_{1\sigma} p_{2\lambda} \right] A_2(p^2, q^2) + \varepsilon^{\alpha\beta\mu\sigma} (p_{2\sigma} - p_{1\sigma}) A_3(p^2, q^2) \right\}. \tag{3.423}$$

The vector Ward identity, $p_1^\alpha T_{\alpha\beta\gamma}^{ijk} = p_2^\beta T_{\alpha\beta\gamma}^{ijk} = 0$, gives

$$-p^2 A_1(p^2, q^2) - \left(p^2 - \frac{q^2}{2} \right) A_2(p^2, q^2) + A_3(p^2, q^2) = 0. \tag{3.424}$$

Further, assuming a naive axial-vector Ward identity in the chiral limit ($p_1 + p_2)^\mu T^{ijk}_{\alpha\beta\gamma} = 0$, we get

$$A_3(p^2, q^2) = 0. \tag{3.425}$$

From Eq. (3.424), assuming first $p^2 = 0$ and then $q^2 = 0$, we get

$$A_2(0, q^2) = 0, \qquad A_1(p^2, 0) = -A_2(p^2, 0). \tag{3.426}$$

Finally, taking both arguments to zero, we obtain

$$A_1(0,0) = A_2(0,0) = A_3(0,0) = 0. \tag{3.427}$$

However, we have already checked that this statement does not hold even at one loop order in perturbation theory due to the presence of the anomalies. Consequently, we have to find a way to modify the preceding statement, sacrificing the naive axial-vector identity. A possible way forward is the following. As we have seen, the anomaly is a polynomial in the external momenta with a vanishing imaginary part. Hence, one might loosen the requirements on the amplitude, assuming that the Ward identities Eq. (3.424) and (3.425) hold only for the imaginary parts of the amplitudes, not for the amplitudes themselves. Furthermore, one assumes that the asymptotic behavior of the amplitudes at large momenta is given by naive dimensional analysis. (This statement is plausible because of the asymptotic freedom of QCD, it holds up to logarithmic corrections.) Thus, we may assume that, at large values of q^2,

$$\operatorname{Im}A_1(p^2, q^2) \sim \frac{1}{(q^2)^2}, \quad \operatorname{Im}A_2(p^2, q^2) \sim \frac{1}{(q^2)^2}, \quad \operatorname{Im}A_3(p^2, q^2) \sim \frac{1}{q^2}. \tag{3.428}$$

Consequently, we may write down an unsubtracted dispersion relation for the amplitudes

$$A_i(p^2, q^2) = \frac{1}{\pi} \int ds \, \frac{\operatorname{Im}A_i(p^2, s)}{s - q^2}, \qquad i = 1, 2, 3. \tag{3.429}$$

Using now the Ward identities *for the imaginary parts*, we get

$$-p^2 A_1(p^2, q^2) - \left(p^2 - \frac{q^2}{2}\right) A_2(p^2, q^2) + A_3(p^2, q^2) + \frac{1}{2\pi} \int ds \operatorname{Im}A_2(p^2, s) = 0,$$

$$A_3(p^2, q^2) = 0. \tag{3.430}$$

The integral, which appears in the preceding equation, is dimensionless. Since it is finite in the chiral limit, it can only be a constant (independent of p^2):

$$C = \frac{1}{2\pi} \int ds \operatorname{Im}A_2(p^2, s). \tag{3.431}$$

Thus, a *simultaneous* fulfillment of the vector and axial-vector Ward identities requires that the superconvergence relation $C = 0$ holds. However, as we know, it does not hold, and we have a choice about which Ward identity we want to violate. For the preceding choice of the invariant amplitudes, the axial-vector identity holds and the vector identity does not. Making the subtraction in the dispersion integral for A_3,

$$A_3'(p^2, q^2) = A_3(p^2, q^2) + C, \tag{3.432}$$

we may ensure the fulfillment of the vector Ward identity at the cost of the axial-vector one:

$$-p^2 A_1(p^2, q^2) - \left(p^2 - \frac{q^2}{2}\right) A_2(p^2, q^2) + A_3'(p^2, q^2) = 0,$$

$$A_3'(p^2, q^2) = -C. \tag{3.433}$$

Finally, taking $p^2 = 0$ in this equation, we get

$$q^2 A_2(0, q^2) = -2A_3'(0, q^2) = 2C, \tag{3.434}$$

or

$$q^2 \operatorname{Im} A_2(0, q^2) = 0, \qquad C = \frac{1}{2\pi} \int ds \operatorname{Im} A_2(0, s). \tag{3.435}$$

This equation has the solution

$$\operatorname{Im} A_2(0, q^2) = C\delta(q^2). \tag{3.436}$$

Consequently, the amplitude itself has a massless pole. Moreover, the value of the constant C, which defines the anomalous divergence of the axial vector current, is fully fixed by the underlying one-loop quark diagram due to the non-renormalization of the axial anomaly.

At this stage, we may invoke the 't Hooft anomaly matching condition and ask how these poles in the external lines, corresponding to the insertion of a non-singlet axial-vector currents, emerge in the *effective* theory. We remind the reader that, owing to confinement, there are no massless quarks present. In fact, there exist only two possible mechanisms:

i) The Goldstone bosons, emerging from the spontaneous symmetry breaking, couple to the axial-vector current and produce poles in the external momenta.
ii) As we have just seen, *massless* fermions may also produce poles in the external momenta. These fermions cannot be quarks, due to the confinement. They can be, however, the color-singlet bound states of quarks, say, massless baryons.

If existing, the bound massless fermions would form an irreducible representation of the group $SU(N_f)_L \times SU(N_f)_R \times U(1)_V$. Using group-theoretical methods and invoking the Appelquist–Carrazzone decoupling theorem, it can be seen that the anomaly matching condition cannot be satisfied for all values of N_f and N_c. For example, there exists no solution for $N_f = N_c = 3$. Consequently, the anomaly matching condition leads to an important statement about the low-energy spectrum of QCD: In the case of QCD with three flavors, the chiral symmetry *must* be broken spontaneously to its vector subgroup, and the nucleons *must* be massive. Note also that, in case of only two flavors, there is a solution that allows for massless nucleons and massive pions in the unbroken phase of chiral symmetry. This solution can be described by the linear σ-model; see the next chapter.

3.20 Symmetries and Spectrum of (Partially) Quenched QCD

At present, lattice QCD is the only available tool that, from first principles, enables one to calculate hadronic observables in terms of the input parameters of the QCD Lagrangian in a systematic and controlled fashion. In this approach, the Euclidean Green's functions of composite meson and baryon fields are evaluated from the discretized path integral by using stochastic methods. The calculations become extensively time-expensive as the number of lattice points grows (considering finer lattices and/or larger volumes). For this reason, it is important to optimize these calculations and to use analytical methods to estimate the lattice artifacts.

The Euclidean partition function in lattice QCD is given by

$$Z = \int d\psi \, d\bar{\psi} dG_\mu \exp\left\{ -\int d^4x \big(\mathcal{L}_G + \bar{\psi}(\slashed{D} + \mathcal{M})\psi\big) \right\}$$

$$= \int dG_\mu \det(\slashed{D} + \mathcal{M}) \exp\left\{ -\int d^4x \mathcal{L}_G \right\}. \tag{3.437}$$

The calculation of the fermion determinant is a particularly costly enterprise, and the costs explode when the quark masses become small. For this reason one sometimes uses two masses for a given quark flavor, namely those of the valence and sea quarks. The sea quarks are those used to evaluate the determinant, and the valence quarks appear in the propagators that connect different vertices in the diagrams describing the Green's functions one is calculating. Taking sea quark masses much larger than those of the valence quarks, the calculational costs can be substantially reduced. In the extreme case, when sea quark masses are set to infinity (the determinant is dropped altogether), one speaks about the *quenched* approximation.[22] The case when the sea quarks have large but finite masses is referred to as *partial quenching*. And, finally, when the masses are equal, the original theory is restored. In the rest of this section, we shall mainly be concerned with the partial quenching.

The (partial) quenching might make the calculational costs smaller, but itself does not come at zero cost. The structure of the theory is different from QCD and, a priori, we do not even know what the symmetries and low-energy spectrum of this new theory are. In this section, we shall address these issues that will enable us to construct a consistent effective field theory framework that smoothly interpolates between the partially quenched case and full QCD with equal valence and sea quark masses.

For the study of the (partially) quenched QCD, two methods are available at present: First, the graded-symmetry method [88–90], which is based on the ideas of the earlier work in Ref. [91], and the so-called replica method [92, 93]. Both methods give the same results at one loop in the effective theory, and it is conceivable that the agreement persists to all orders. In what follows we shall concentrate on the first method only, following mainly the discussion in Ref. [89]. The partition function in Eq. (3.437) with

---

[22] As will be discussed in Section 4.19, this approximation does not lead to a well-defined quantum field theory.

different values of the sea and valence quarks can be rewritten in the following form:

$$Z = \int d\psi_v d\bar\psi_v \, d\psi_s d\bar\psi_s \, d\psi_g d\bar\psi_g \, dG_\mu$$

$$\times \exp\left\{ -\int d^4x \left(\mathcal{L}_G + \bar\psi_v(\slashed{D} + \mathcal{M}_v)\psi_v + \bar\psi_s(\slashed{D} + \mathcal{M}_s)\psi_s + \bar\psi_g(\slashed{D} + \mathcal{M}_v)\psi_g \right) \right\}$$

$$= \int dG_\mu \frac{\det(\slashed{D} + \mathcal{M}_v)}{\det(\slashed{D} + \mathcal{M}_v)} \det(\slashed{D} + \mathcal{M}_s) \exp\left\{ -\int d^4x \mathcal{L}_G \right\}$$

$$= \int dG_\mu \det(\slashed{D} + \mathcal{M}_s) \exp\left\{ -\int d^4x \mathcal{L}_G \right\}. \tag{3.438}$$

Here, ψ_v, ψ_s, ψ_g are the valence, sea and so-called ghost quarks, respectively. In general, the number of the valence and sea quarks need not be the same (the number of the ghost and valence quarks, as well as their masses, are always the same). We assume that there are N_v valence and N_s sea quarks (the fully quenched QCD then corresponds to $N_s = 0$). Note also that the ghost quarks are unphysical particles described by *commuting* spinor fields (similar to the Faddeev–Popov ghosts, which are described by the anticommuting scalar fields). The trick with introducing ghosts allows one to cancel the valence quark determinant, so only the sea quark determinant is left in the final expression. Hence, writing the partition function with *different* sea and valence quark masses as a path integral has been achieved through enlarging the quark sector of the theory. The pertinent Lagrangian is also given in Eq. (3.438).

The following questions need to be addressed in this context:

- What is the symmetry group of the enlarged Lagrangian?
- What is the symmetry breaking pattern?
- What is the spectrum of the Goldstone bosons?

The answers to these questions enable one to construct the low-energy effective theory of (partially) quenched QCD that will allow us to understand better the results that are obtained in (partially) quenched lattice calculations.

As can be seen from Eq. (3.438), naively,[23] the fermionic sector of the theory is invariant under the transformations belonging to the graded group $SU(N_v + N_s | N_v)_L \times SU(N_v + N_s | N_v)_R \times U(1)_V$. The singlet axial-vector current, corresponding to the $U(1)_A$ as in conventional QCD, must be anomalous. (We remind the reader that the anomalous part of the divergence does not depend on the quark masses, and, if the valence and sea quark masses are set equal, and $N_v = N_s$, partially quenched QCD reduces to the conventional QCD.) All information about the graded groups, which will be

---

[23] As discussed, e.g., in Ref. [89], the true symmetry of the generating functional is different from the one given here, and is different in the quark and ghost sectors, due to the requirement of the convergence of the path integral. However, it was also shown that the Ward identities that emerge from the "true" symmetry are identical to those obtained on the basis of the "fake" symmetry displayed here (the terminology used in Ref. [89]). For simplicity, we proceed further with the "fake" symmetry, referring the interested reader to the original publication.

needed, is given in what follows. The graded group matrices contain both commuting and anticommuting elements. If g is a matrix belonging to $SU(N_v + N_s | N_v)$, then it can be written in the form

$$g = \begin{pmatrix} A & B \\ C & D \end{pmatrix} . \tag{3.439}$$

$$\underbrace{}_{N_v + N_s} \underbrace{}_{N_v}$$

Here, the matrix elements of the square matrices A and D are usual numbers, whereas the matrix elements of the matrices B and C represent anticommuting Grassmann variables. The complex conjugation for the Grassmann variables is defined as

$$(\eta_1 \eta_2)^* = \eta_2^* \eta_1^* = -\eta_1^* \eta_2^* . \tag{3.440}$$

To proceed, we require the "supertrace" (str) and "superdeterminant" (sdet). These are the generalization of the ordinary trace and determinant for the matrices that contain both commuting and anticommuting elements. They are defined in the following manner:

$$\mathrm{str}\, g = \mathrm{tr}\, A - \mathrm{tr}\, D , \qquad \mathrm{str}(g_1 g_2) = \mathrm{str}(g_2 g_1) , \tag{3.441}$$

and

$$\mathrm{sdet}(g) = \exp[\mathrm{str}(\ln(g))] = \det(A - BD^{-1}C)/\det(D) ,$$

$$\mathrm{sdet}(g_1 g_2) = \mathrm{sdet}(g_1)\mathrm{sdet}(g_2) . \tag{3.442}$$

Further, note that the conditions under which the Vafa–Witten theorem holds are valid in case of the graded symmetry as well [89]. Consider now the vacuum expectation value:

$$\Omega_{ab} = \langle 0 | \bar{\psi}_a \psi_b | 0 \rangle , \qquad a, b = v, s, g . \tag{3.443}$$

If nonvanishing, the quantity Ω_{ab} can serve as an order parameter for the chiral symmetry breaking.

Let us (anti)commute the fermion fields in Eq. (3.443) and consider the quantity

$$\tilde{\Omega}_{ba} = \langle 0 | \psi_{b\tau} \bar{\psi}_{a\tau} | 0 \rangle , \tag{3.444}$$

where τ subsumes other indices (Dirac, color), and the sum over all these indices is implicit. Since, according to the Vafa–Witten theorem, the vector symmetry is unbroken, the matrix $\tilde{\Omega}_{ba}$ should obey the following relation:

$$\tilde{\Omega} = g_V \tilde{\Omega} g_V^\dagger , \qquad g_V \in SU(N_v + N_s | N_v) . \tag{3.445}$$

Hence,

$$\tilde{\Omega}_{ab} = \omega \delta_{ab} , \tag{3.446}$$

where ω is a number independent of a, b. From this, we get

$$\Omega_{ab} = -\omega \delta_{ab} \varepsilon_a , \qquad \varepsilon_a = \begin{cases} 1, & a = v, s \\ -1, & a = g \end{cases} . \tag{3.447}$$

Now, setting the sea and valence quark masses equal at $N_v = N_s$, one sees that $\omega \neq 0$ in the chiral limit (because this is so in conventional QCD). Consequently, the chiral symmetry in partially quenched QCD for $N_v = N_s$ should be broken down to the diagonal subgroup. Note that this argument does not apply in fully quenched QCD with $N_s = 0$. In that case spontaneous symmetry breaking is an additional assumption.

Next, let us count the number of Goldstone particles. To this end, introduce the generators T^α, $\alpha = 1, \ldots, (2N_v + N_s)^2$ of the graded group $U(N_v + N_s | N_v)$, which are $(2N_v + N_s) \times (2N_v + N_s)$ matrices. Further, consider the axial-vector current and the pseudoscalar density:

$$A_\mu^\alpha(x) = \bar{Q}(x) \gamma_\mu \gamma_5 T^\alpha Q(x), \quad p^\alpha(x) = \bar{Q}(x) \gamma_5 T^\alpha Q(x), \quad Q = \begin{pmatrix} \psi_v \\ \psi_s \\ \psi_g \end{pmatrix}. \quad (3.448)$$

Using Eq. (3.447), it can be shown that the Fourier transform of the vacuum matrix element of the T-product,

$$C_\mu^{\alpha\beta}(p) = ip_\mu C^{\alpha\beta}(p^2) = i \int d^4 x \, e^{ipx} \langle 0 | T j_\mu^\alpha(x) p^\beta(0) | 0 \rangle, \quad (3.449)$$

obeys the Ward identity:

$$p^2 C^{\alpha\beta}(p^2) = -2i\omega \, \mathrm{str}(T^\alpha T^\beta). \quad (3.450)$$

Here, it was assumed that the current is non-anomalous, that is, $\mathrm{str}\, T^\alpha = 0$. If the current is anomalous, an additional contribution proportional to $\mathrm{str}\, T^\alpha \cdot \mathrm{str}\, T^\beta$ emerges on the r.h.s. of the preceding equation. Consequently, the Goldstone theorem may be formulated as follows: A massless particle corresponding to a pole $C^{\alpha\beta}(p^2)$ ar $p^2 = 0$ arises each time when $\mathrm{str}\, T^\alpha \cdot \mathrm{str}\, T^\beta = 0$ and $\mathrm{str}(T^\alpha T^\beta) \neq 0$. On the other hand, if the r.h.s. of Eq. (3.450) contains an additional term coming from anomaly, the two terms can cancel each other and the left-hand side vanishes in the limit $p^2 \to 0$. The existence of a massless pole then does not follow from the Ward identity.

It can be shown that the off-diagonal generators can be always chosen so that they are orthogonal to all diagonal generators and obey the relations $\mathrm{str}\, T^\alpha = 0$ and $\mathrm{str}(T^\alpha T^\beta) = \pm \delta^{\alpha\beta}$. That is, each off-diagonal generator gives one Goldstone particle. The treatment of the diagonal generators is more subtle. Let λ^α be the $N_v + N_s - 1$ generators of $SU(N_v + N_s)$, normalized by $\mathrm{tr}(\lambda^\alpha \lambda^\beta) = \delta^{\alpha\beta}$. The set of $N_v - 1$ generators for the group $SU(N_v)$ is denoted by $\tilde{\lambda}^\alpha$. In case of partially quenched QCD, a full set of $2N_v + N_s$ diagonal generators of the group $U(N_v + N_s | N_v)$ can be chosen as

$$T^\alpha = \begin{pmatrix} \lambda^\alpha & 0 \\ 0 & 0 \end{pmatrix}, \quad \begin{pmatrix} 0 & 0 \\ 0 & \tilde{\lambda}^\alpha \end{pmatrix}, \quad \frac{-1}{\sqrt{N_v N_s (N_v + N_s)}} \begin{pmatrix} N_v & 0 \\ 0 & N_v + N_s \end{pmatrix},$$

$$\frac{1}{\sqrt{N_s}} \begin{pmatrix} 1 & 0 \\ 0 & -1 \end{pmatrix}. \quad (3.451)$$

The normalization is again given by $\mathrm{str}(T^\alpha T^\beta) = \pm \delta^{\alpha\beta}$. Note that only the last generator has a nonzero supertrace (anomaly) and, hence, does not correspond to a massless particle. Consequently, there are in total $(2N_v + N_s)^2 - 1$ massless Goldstone particles in the chiral limit. From these, there are $2(N_v + N_s)N_v$ Goldstone particles that consist of one ghost and one "normal" quark (valence or sea). These particles are fermions. Further, $(N_v + N_s)^2 - 1$ particles are bosons consisting of sea and valence quarks, and N_v^2 are bosons consisting of two ghosts. The latter have negative residue in the propagator because of the negative sign in the normalization condition of the pertinent generators.

The situation is slightly more subtle in the fully quenched theory. The equation (3.451) does not make sense at $N_s = 0$. Instead, the last two generators can be defined as follows:

$$T^{2N_v-1} = \begin{pmatrix} 1 & 0 \\ 0 & 1 \end{pmatrix}, \qquad T^{2N_v} = \frac{1}{2N_V} \begin{pmatrix} 1 & 0 \\ 0 & -1 \end{pmatrix}. \tag{3.452}$$

Again, the last matrix has the nonzero supertrace (anomaly). However, the normalization condition changes:

$$\mathrm{str}(T^{2N_v-1}T^{2N_v-1}) = \mathrm{str}(T^{2N_v}T^{2N_v}) = 0,$$

$$\mathrm{str}(T^{2N_v-1}T^{2N_v}) = \mathrm{str}(T^{2N_v}T^{2N_v-1}) = 1. \tag{3.453}$$

Using these relations, one can straightforwardly check that the r.h.s. of the Ward identity always contains only one term. Hence, each generator in the fully quenched theory corresponds to a Goldstone particle, and the total number of Goldstone particles is equal to $(N_v + N_s)^2$. The additional Goldstone boson is nothing but a counterpart of the familiar η' meson, which also becomes light in the fully quenched case.

3.21 Literature Guide

A comprehensive bibliography on the Euler–Heisenberg Lagrangian is contained in the recent review article by Dunne [5], and we refer the interested reader to this article. A quite detailed discussion of this problem is contained also in the textbook by Itzykson and Zuber [3], as well as that of Dittrich and Reuter [94]. The scattering of light off atoms is described, for example, in the lectures of Holstein [95] and Kaplan [96].

There exists a vast amount of literature on the Ward identities and anomalies. We just mention our favorite textbooks here: [3, 17, 25, 97, 98], see also [99]. To the papers that were already cited in the text, we would like to add Ref. [100], where the decay of the π^0-meson into 2γ has been studied within the linear σ-model. Ref. [101] was one of the first papers devoted to a systematic study of the anomalous axial-vector vertex in spinor QED. In Ref. [102] the consistency conditions, which are obeyed by the anomalous part of the fermion determinant in an external fields, have been derived. A very clear and concise review on anomalies is given in Ref. [103]. For the definition of the renormalized composite operators, apart from Ref. [55], already cited in the text, see also Ref. [104].

Again, there exists a vast amount of literature on lattice QCD that forces us to restrict ourselves to a few textbooks: [32–35]. A very concise discussion of the chiral symmetry in the context of lattice QCD is given in Ref. [50] and Ref. [105]. For the use of the heat kernel method to calculate the fermion determinant in an external field, see Ref. [26], and also Ref. [106] as well as the review by Ball [27].

Spontaneous symmetry breaking is routinely considered in modern textbooks. To the literature, cited in the text, we wish to add a few time-honored reviews on the subject [107–110]. The nonrelativistic Goldstone theorem is considered in the recent review by Watanabe [111], who gives a comprehensive list of references. For the anomaly matching, we note the recent article [112], which discusses in detail the treatment of the anomalies at different momentum scales. Finally, nice reviews of the applications of Chiral Perturbation Theory to lattice QCD (including the partially quenched case) are given in the lectures by Sharpe [113] and Golterman [114].

References

[1] E. Noether, "Invariant variation problems," Gött. Nachr. **1918** (1918) 235 [Transp. Theory Statist. Phys. **1** (1971) 186].

[2] W. H. Furry, "A symmetry theorem in the positron theory," Phys. Rev. **51** (1937) 125.

[3] C. Itzykson and J. B. Zuber, "Quantum Field Theory," McGraw-Hill (1980) 705 pp. (International Series In Pure and Applied Physics).

[4] R. Karplus and M. Neuman, "The scattering of light by light," Phys. Rev. **83** (1951) 776.

[5] G. V. Dunne, "The Heisenberg-Euler effective action: 75 years on," Int. J. Mod. Phys. A **27** (2012) 1260004.

[6] J. S. Schwinger, "On gauge invariance and vacuum polarization," Phys. Rev. **82** (1951) 664.

[7] A. V. Manohar, "Effective field theories," arXiv:hep-ph/9606222.

[8] A. V. Manohar and M. B. Wise, "Heavy quark physics," Camb. Monogr. Part. Phys. Nucl. Phys. Cosmol. **10** (2000) 1.

[9] F. Yndurain, "Quantum Chromodynamics: An Introduction to the Theory of Quarks and Gluons," Texts and Monographs in Physics, Springer-Verlag, 1983.

[10] P. A. Zyla et al. [Particle Data Group], "Review of particle physics," PTEP **2020** (2020) 083C01.

[11] K. Huang and H. A. Weldon, "Bound state wave functions and bound state scattering in relativistic field theory," Phys. Rev. D **11** (1975) 257.

[12] V. de Alfaro, S. Fubini, G. Furlan and C. Rossetti, "Currents in Hadron Physics," North-Holland Publishing, 1973.

[13] J. Steinberger, "On the use of subtraction fields and the lifetimes of some types of meson decay," Phys. Rev. **76** (1949) 1180.

[14] W. A. Bardeen, "Anomalous Ward identities in spinor field theories," Phys. Rev. **184** (1969) 1848.

[15] S. L. Adler and W. A. Bardeen, "Absence of higher order corrections in the anomalous axial vector divergence equation," Phys. Rev. **182** (1969) 1517.

[16] W. Pauli and F. Villars, "On the invariant regularization in relativistic quantum theory," Rev. Mod. Phys. **21** (1949) 434.

[17] L. D. Faddeev and A. A. Slavnov, "Gauge fields. Introduction to quantum theory," Front. Phys. **50** (1980) 1 [Front. Phys. (1991) 1].

[18] G. 't Hooft and M. J. G. Veltman, "Regularization and renormalization of gauge fields," Nucl. Phys. B **44** (1972) 189.

[19] J. C. Collins, Renormalization: An Introduction to Renormalization, The Renormalization Group, and the Operator Product Expansion, Cambridge University Press (2010).

[20] S. L. Adler, R. W. Brown, T. F. Wong and B. L. Young, "Vanishing of the second order correction to the triangle anomaly in landau-gauge, zero-fermion-mass quantum electrodynamics," Phys. Rev. D **4** (1971) 1787.

[21] A. Zee, "Axial vector anomalies and the scaling property of field theory," Phys. Rev. Lett. **29** (1972) 1198.

[22] J. H. Lowenstein and B. Schroer, "Comment on the absence of radiative corrections to the anomaly of the axial-vector current," Phys. Rev. D **7** (1973) 1929.

[23] K. Fujikawa, "Path integral measure for gauge invariant fermion theories," Phys. Rev. Lett. **42** (1979) 1195.

[24] K. Fujikawa, "Path integral for gauge theories with fermions," Phys. Rev. D **21** (1980) 2848 [Erratum-ibid. D **22** (1980) 1499].

[25] K. Fujikawa and H. Suzuki, Path Integrals and Quantum Anomalies, Clarendon (2004) 284 p.

[26] J. Gasser and H. Leutwyler, "Chiral perturbation theory to one loop," Annals Phys. **158** (1984) 142.

[27] R. D. Ball, "Chiral gauge theory," Phys. Rept. **182** (1989) 1.

[28] C. G. Callan, Jr., R. F. Dashen and D. J. Gross, "The structure of the gauge theory vacuum," Phys. Lett. B **63** (1976) 334 [Phys. Lett. **63B** (1976) 334].

[29] R. Jackiw and C. Rebbi, "Vacuum periodicity in a Yang-Mills quantum theory," Phys. Rev. Lett. **37** (1976) 172.

[30] R. Rajaraman, Solitons and Instantons. An Introduction to Solitons and Instantons in Quantum Field Theory, North-Holland Publishing (1982) 409 pp.

[31] M. F. Atiyah and I. M. Singer, "Dirac operators coupled to vector potentials," Proc. Nat. Acad. Sci. **81** (1984) 2597.

[32] I. Montvay and G. Münster, Quantum Fields on a Lattice, Cambridge Monographs on Mathematical Physics, Cambridge University Press (1994).

[33] H. J. Rothe, "Lattice gauge theories: An introduction," World Sci. Lect. Notes Phys. **43** (1992) 1 [World Sci. Lect. Notes Phys. **59** (1997) 1] [World Sci. Lect. Notes Phys. **74** (2005) 1] [World Sci. Lect. Notes Phys. **82** (2012) 1].

[34] J. Smit, "Introduction to quantum fields on a lattice: A robust mate," Cambridge Lect. Notes Phys. **15** (2002) 1.

[35] C. Gattringer and C. B. Lang, "Quantum chromodynamics on the lattice," Lect. Notes Phys. **788** (2010) 1.

[36] L. H. Karsten and J. Smit, "Lattice fermions: Species doubling, chiral invariance, and the triangle anomaly," Nucl. Phys. B **183** (1981) 103.

[37] K. G. Wilson, "Quarks: From paradox to myth," Subnucl. Ser. **13** (1977) 13.

[38] H. B. Nielsen and M. Ninomiya, "No go theorem for regularizing chiral fermions," Phys. Lett. **105B** (1981) 219.

[39] P. H. Ginsparg and K. G. Wilson, "A remnant of chiral symmetry on the lattice," Phys. Rev. D **25** (1982) 2649.

[40] C. G. Callan, Jr. and J. A. Harvey, "Anomalies and fermion zero modes on strings and domain walls," Nucl. Phys. B **250** (1985) 427.

[41] D. B. Kaplan, "A method for simulating chiral fermions on the lattice," Phys. Lett. B **288** (1992) 342.

[42] Y. Shamir, "Chiral fermions from lattice boundaries," Nucl. Phys. B **406** (1993) 90.

[43] R. Narayanan and H. Neuberger, "Chiral fermions on the lattice," Phys. Rev. Lett. **71** (1993) 3251.

[44] R. Narayanan and H. Neuberger, "Chiral determinant as an overlap of two vacua," Nucl. Phys. B **412** (1994) 574.

[45] H. Neuberger, "A practical implementation of the overlap Dirac operator," Phys. Rev. Lett. **81** (1998) 4060.

[46] P. Hasenfratz and F. Niedermayer, "Perfect lattice action for asymptotically free theories," Nucl. Phys. B **414** (1994) 785.

[47] M. Lüscher, "Exact chiral symmetry on the lattice and the Ginsparg-Wilson relation," Phys. Lett. B **428** (1998) 342.

[48] M. Lüscher, "Abelian chiral gauge theories on the lattice with exact gauge invariance," Nucl. Phys. B **549** (1999) 295.

[49] M. Lüscher, "Weyl fermions on the lattice and the non-Abelian gauge anomaly," Nucl. Phys. B **568** (2000) 162.

[50] S. Chandrasekharan and U. J. Wiese, "An introduction to chiral symmetry on the lattice," Prog. Part. Nucl. Phys. **53** (2004) 373.

[51] J. C. Collins, A. Duncan and S. D. Joglekar, "Trace and dilatation anomalies in gauge theories," Phys. Rev. D **16** (1977) 438.

[52] R. J. Crewther, "Nonperturbative evaluation of the anomalies in low-energy theorems," Phys. Rev. Lett. **28** (1972) 1421.

[53] M. S. Chanowitz and J. R. Ellis, "Canonical anomalies and broken scale invariance," Phys. Lett. **40B** (1972) 397.

[54] N. K. Nielsen, "The energy momentum tensor in a nonabelian quark gluon theory," Nucl. Phys. B **120** (1977) 212.

[55] J. C. Collins, "Normal products in dimensional regularization," Nucl. Phys. B **92** (1975) 477.

[56] R. Tarrach, "The renormalization of FF," Nucl. Phys. B **196** (1982) 45.

[57] K. Fujikawa, "Energy momentum tensor in quantum field theory," Phys. Rev. D **23** (1981) 2262.

[58] J. F. Donoghue, E. Golowich and B. R. Holstein, "Dynamics of the standard model," Camb. Monogr. Part. Phys. Nucl. Phys. Cosmol. **2** (1992) 1 [Camb. Monogr. Part. Phys. Nucl. Phys. Cosmol. **35** (2014)].

[59] E. Fabri and L. E. Picasso, "Quantum field theory and approximate symmetries," Phys. Rev. Lett. **16** (1966) 408.

[60] T. Banks and A. Casher, "Chiral symmetry breaking in confining theories," Nucl. Phys. B **169** (1980) 103.

[61] C. Vafa and E. Witten, "Restrictions on symmetry breaking in vector-like gauge theories," Nucl. Phys. B **234** (1984) 173.

[62] H. B. Nielsen and S. Chadha, "On how to count Goldstone bosons," Nucl. Phys. B **105** (1976) 445.

[63] H. Leutwyler, "Nonrelativistic effective Lagrangians," Phys. Rev. D **49** (1994) 3033.

[64] Y. Hidaka, "Counting rule for Nambu-Goldstone modes in nonrelativistic systems," Phys. Rev. Lett. **110** (2013) 091601.

[65] H. Watanabe and H. Murayama, "Effective Lagrangian for nonrelativistic systems," Phys. Rev. X **4** (2014) 031057.

[66] H. Watanabe and T. Brauner, "On the number of Nambu-Goldstone bosons and its relation to charge densities," Phys. Rev. D **84** (2011) 125013.

[67] Y. Nambu, "Spontaneous breaking of Lie and current algebras," J. Statist. Phys. **115** (2004) 7.

[68] V. A. Miransky and I. A. Shovkovy, "Spontaneous symmetry breaking with abnormal number of Nambu-Goldstone bosons and kaon condensate," Phys. Rev. Lett. **88** (2002) 111601.

[69] T. Schäfer, D. T. Son, M. A. Stephanov, D. Toublan and J. J. M. Verbaarschot, "Kaon condensation and Goldstone's theorem," Phys. Lett. B **522** (2001) 67.

[70] J. Gasser and H. Leutwyler, "Quark Masses," Phys. Rept. **87** (1982) 77.

[71] H. Hellmann, *Einführung in die Quantenchemie*, Deuticke, (1937).

[72] R. P. Feynman, "Forces in molecules," Phys. Rev. **56** (1939) 340.

[73] J. Ruiz de Elvira, U.-G. Meißner, A. Rusetsky and G. Schierholz, "Feynman-Hellmann theorem for resonances and the quest for QCD exotica," Eur. Phys. J. C **77** (2017) 659.

[74] M. Gell–Mann, R. J. Oakes and B. Renner, "Behavior of current divergences under $SU(3) \times SU(3)$," Phys. Rev. **175** (1968) 2195.

[75] H. Leutwyler, "On the foundations of chiral perturbation theory," Annals Phys. **235** (1994) 165.

[76] S. Weinberg, "The $U(1)$ problem," Phys. Rev. D **11** (1975) 3583.

[77] G. 't Hooft, "A planar diagram theory for strong interactions," Nucl. Phys. B **72** (1974) 461.

[78] E. Witten, "Baryons in the $1/n$ expansion," Nucl. Phys. B **160** (1979) 57.

[79] S. Okubo, "Phi meson and unitary symmetry model," Phys. Lett. **5** (1963) 165.

[80] G. Zweig, "An SU(3) model for strong interaction symmetry and its breaking II," in Development in the Quark Theory of Hadrons, vol. 1, edited by D. B. Lichtenberg and S. P. Rosen (Hadronic Press, 1980).

[81] J. Iizuka, "Systematics and phenomenology of meson family," Prog. Theor. Phys. Suppl. **37** (1966) 21.

[82] E. Witten, "Current algebra theorems for the U(1) Goldstone boson," Nucl. Phys. B **156** (1979) 269.

[83] G. Veneziano, "U(1) without instantons," Nucl. Phys. B **159** (1979) 213.

[84] R. Kaiser and H. Leutwyler, "Large N_c in chiral perturbation theory," Eur. Phys. J. C **17** (2000) 623.

[85] G. 't Hooft, "Naturalness, chiral symmetry, and spontaneous chiral symmetry breaking," NATO Sci. Ser. B **59** (1980) 135.

[86] A. D. Dolgov and V. I. Zakharov, "On conservation of the axial current in massless electrodynamics," Nucl. Phys. B **27** (1971) 525.

[87] Y. Frishman, A. Schwimmer, T. Banks and S. Yankielowicz, "The axial anomaly and the bound state spectrum in confining theories," Nucl. Phys. B **177** (1981) 157.

[88] C. W. Bernard and M. F. L. Golterman, "Partially quenched gauge theories and an application to staggered fermions," Phys. Rev. D **49** (1994) 486.

[89] S. R. Sharpe and N. Shoresh, "Partially quenched chiral perturbation theory without Φ_0," Phys. Rev. D **64** (2001) 114510.

[90] S. R. Sharpe and N. Shoresh, "Physical results from unphysical simulations," Phys. Rev. D **62** (2000) 094503.

[91] A. Morel, "Chiral logarithms in quenched QCD," J. Phys. (France) **48** (1987) 1111.

[92] P. H. Damgaard and K. Splittorff, "Partially quenched chiral perturbation theory and the replica method," Phys. Rev. D **62** (2000) 054509.

[93] P. H. Damgaard, "Partially quenched chiral condensates from the replica method," Phys. Lett. B **476** (2000) 465.

[94] W. Dittrich and M. Reuter, "Effective Lagrangians in quantum electrodynamics," Lect. Notes Phys. **220** (1985) 1.

[95] B. R. Holstein, "Effective field theory and χpt," AIP Conf. Proc. **520** (2000) 271.

[96] D. B. Kaplan, "Five lectures on effective field theory," nucl-th/0510023.

[97] R. A. Bertlmann, "Anomalies in quantum field theory," Clarendon (1996) 566 pp. (International Series of Monographs on Physics: 91).

[98] S. Treiman, R. Jackiw and D. J. Gross, "Lectures on current algebra and its applications," Princeton University Press (2016).

[99] W. Dittrich and M. Reuter, "Selected topics in gauge theories," Lect. Notes Phys. **244** (1986) 1.

[100] J. S. Bell and R. Jackiw, "A PCAC puzzle: $\pi^0 \to \gamma\gamma$ in the σ model," Nuovo Cim. A **60** (1969) 47.

[101] S. L. Adler, "Axial vector vertex in spinor electrodynamics," Phys. Rev. **177** (1969) 2426.

[102] J. Wess and B. Zumino, "Consequences of anomalous Ward identities," Phys. Lett. **37B** (1971) 95.

[103] J. L. Petersen, "Nonabelian chiral anomalies and Wess-Zumino effective actions," Acta Phys. Polon. B **16** (1985) 271.

[104] J. H. Lowenstein, "Differential vertex operations in Lagrangian field theory," Commun. Math. Phys. **24** (1971) 1.

[105] F. Niedermayer, "Exact chiral symmetry, topological charge and related topics," Nucl. Phys. Proc. Suppl. **73** (1999) 105 [hep-lat/9810026].

[106] A. P. Balachandran, G. Marmo, V. P. Nair and C. G. Trahern, "A nonperturbative proof of the nonabelian anomalies," Phys. Rev. D **25** (1982) 2713.

[107] G. S. Guralnik, C. R. Hagen and T. W. B. Kibble, "Broken symmetries and the Goldstone theorem," Adv. Part. Phys. **2** (1968) 567.

[108] E. S. Abers and B. W. Lee, "Gauge theories," Phys. Rept. **9** (1973) 1.

[109] S. R. Coleman, "Secret symmetry: An introduction to spontaneous symmetry breakdown and gauge fields," Subnucl. Ser. **11** (1975) 139.

[110] J. Bernstein, "Spontaneous symmetry breaking, gauge theories, the Higgs mechanism and all that," Rev. Mod. Phys. **46** (1974) 7 Erratum: [Rev. Mod. Phys. **47** (1975) 259] Erratum: [Rev. Mod. Phys. **46** (1974) 855].

[111] H. Watanabe, "Formula for the number of Nambu-Goldstone modes," Ann. Rev. Condensed Matter Phys. **11** (2020) 169.

[112] G. Dvali, "Topological origin of chiral symmetry breaking in QCD and in gravity," arXiv:1705.06317 [hep-th].

[113] S. R. Sharpe, "Applications of chiral perturbation theory to lattice QCD," hep-lat/0607016.

[114] M. Golterman, "Applications of chiral perturbation theory to lattice QCD," arXiv:0912.4042 [hep-lat].

Low-Energy Effective Field Theory of QCD

4.1 Introduction

The low-energy effective field theory of QCD rests on two pillars.[1] The first is the knowledge of the low-energy spectrum of QCD, that is, the identification of the relevant effective degrees of freedom. The EFT is constructed by using the physical fields corresponding to the observed low-mass hadrons, namely the Goldstone bosons and stable baryon octet states, rather than the quark and gluon fields. These are simply not resolved in low-energy processes, as it requires probes with high momentum transfer (much larger than the nucleon mass) to investigate the quark-gluon substructure of nucleons or pions. Seen from another angle, a crucial role is played by color confinement, which implies that the singularities (the poles and cuts) of the Green's functions of the (composite) meson and baryon fields in QCD emerge from the hadronic intermediate states only. Hence, in order to produce such Green's functions *perturbatively* in some field theory, we have to start from a Lagrangian that contains hadronic fields only. In contradistinction, the masses of the fundamental fields (quarks and gluons) determine the location of singularities of the Green's functions in QCD in perturbation theory, and this has nothing in common with the real structure of the singularities, because perturbation theory in QCD cannot describe the on-shell singularities at all energies.[2]

From the preceding discussion it can be concluded that the Lagrangian of the low-energy effective theory of QCD should be built from meson and baryon fields. So the question is: Up to which energies could one describe the Green's functions in QCD using a *local* effective Lagrangians? We remind the reader that in perturbation theory the heavy scale, after which the theory becomes nonlocal, is defined by the mass of the heavy particle that is integrated out; see Chapter 1. In the case at hand, the elementary constituents are not observed at any energy, and the heavy scale is implicitly defined by

---

[1] Note that, in fact, there are different EFTs of QCD for light and/or heavy quarks. In what follows, we concentrate entirely on the light quark sector. A nice introduction to the heavy quark EFT can be found, e.g., in Refs. [1–3]. Extensions to heavy-light systems also exist; see, e.g., Refs. [4–6].

[2] Perturbation theory in QCD is applicable, when the typical momentum transfer in Feynman diagrams is large, at least of the order of a few GeV. The study of the singularity structure of various Green's functions, however, cannot be considered separately from the study of the hadronization phenomenon, when the quarks and gluons are combined in colorless clusters that can move far away from the interaction region, giving rise to the singularities in the Green's functions. The relative momenta in such clusters are low, rendering the use of perturbation theory meaningless.

a multiple of the QCD parameter Λ_{QCD}, which is the scale relevant for confinement. We shall see here that in QCD, the heavy scale (the hadronic scale) turns out to be of the order of 1 GeV. Beyond this scale, sometimes also called the hard scale, the perturbative expansion on the basis of the low-energy effective Lagrangian ceases to makes sense, rendering the theory nonlocal.

The second pillar is provided by the symmetries of QCD and the pattern of the symmetry breaking, which poses severe restrictions both on the low-energy spectrum and the interactions. Namely, as seen from Chapter 3, the $SU(3)_L \times SU(3)_R$ chiral symmetry of QCD is spontaneously broken down to the diagonal vector subgroup $SU(3)_V$, giving rise to the multiplet of the pseudo-Goldstone bosons. If the quark masses were exactly zero, the masses of the Goldstone bosons also would vanish. In nature, however, the masses of u, d, s quarks are small but not exactly zero. In this context, small refers to the scale of QCD just mentioned.[3] As a result, the Goldstone bosons acquire masses, which are, however, small compared to the masses of other hadrons. Consequently, the Goldstone bosons determine the structure of the lowest singularities. Further, the presence of anomalies has two important implications on the spectrum. First, in the absence of the massless spin-1/2 baryons, the chiral symmetry in QCD *should indeed be* spontaneously broken according to the anomaly matching condition. Second, the mass of the ninth Goldstone boson is nonvanishing even for massless quarks and pushed up to the hard scale Λ due to the anomaly, according to the Witten–Veneziano relation.

The chiral symmetry of QCD also has implications for the interactions of the Goldstone bosons, both for self-interactions and interactions with other matter fields (e.g., spin-1/2 baryons). Namely, as discussed in Chapter 3, these interactions are of derivative nature and vanish at threshold in the massless case. This means that the interactions are weak at small momenta, justifying the power counting in the effective Lagrangian and the use of the perturbation theory at low energies, even if generic dimensionless couplings are of order unity.

Finally, the chiral symmetry of the underlying theory sets guiding principles, which are used to derive the Lagrangian of the effective theory. Namely, the fields from which the Lagrangian is constructed should form a basis for a representation of the symmetry group (linear or nonlinear), whereas the Lagrangian should be composed from the chirally invariant building blocks. Together with the counting rules, this defines a systematic recipe for the construction of the effective Lagrangian up to a given chiral order.

In this chapter, based on the general principles just outlined, we shall systematically construct Chiral Perturbation Theory (ChPT), which is the low-energy effective field

---

[3] Under the notion of quark masses, we always refer to the so-called "current" quark masses, or the running quark masses that are perturbatively defined in QCD at some running scale μ using, say, the $\overline{MS}$ renormalization scheme. In the literature, the notion of the "constituent" quark masses can be also found, mainly in the context of the constituent quark models and other related approaches. In our opinion, introducing such a notion can be useful only within the preceding models and does not allow for a meaningful model-independent generalization.

theory of QCD. First, the purely mesonic sector will be considered, followed by the one-nucleon sector. The structure of the effective Lagrangians, power counting and renormalization will be discussed in detail, and few examples of explicit calculations of observables will be given. We will not discuss the extension to systems with two and more nucleons, but rather refer to the original papers by Weinberg [7, 8] and the recent book [9].

4.2 The *S*-Matrix of Composite Particles

The mesons and baryons are bound states of the elementary constituents, namely quarks and gluons. The Lagrangian of QCD is written in terms of these elementary fields and does not contain composite fields at all. Consequently, in order to describe hadron interactions in QCD, we have to deal with the calculation of the *S*-matrix elements between states that are composite objects. The problem is not new, and the solution within quantum field theory has been known since the 1950s [10–16]. We have already anticipated this solution in Chapter 3, where we used quark bilinears as interpolating pion fields in the proof of the soft-pion theorems. Here, we shall address the issue in a more systematic fashion. It should also be noted that confinement is not critical for the discussion given here. The chain of arguments can be equally applied to the cases without confinement (say, the scattering of the deuteron, which is a bound state of a proton and a neutron, in an effective theory containing elementary proton and neutron fields). In this case, we have to define a rule for the calculation of the *S*-matrix elements between the states that contain both elementary and composite particles.

In order to simplify the discussion, let us restrict ourselves to meson states in what follows. The one-particle states of stable mesons with the mass M and momentum p^μ will be denoted as $|p\rangle$ (spin and internal symmetry indices are suppressed for simplicity). The completeness condition in the Fock space runs over the states with an arbitrary number of mesons. Separating the one-particle contribution explicitly, we get

$$\mathbb{1} = \int \frac{d^3\mathbf{p}}{(2\pi)^3 2p^0} |p\rangle\langle p| + \cdots. \tag{4.1}$$

Again, the sum over all possible spin and internal symmetry indices is not explicitly shown.

From the assumption that the mesons are bound states of a quark and an antiquark, it can be concluded that the following matrix element does not vanish:

$$\langle p|\Phi(x)|0\rangle \neq 0, \qquad \Phi(x) = \bar\psi(x)\Gamma\psi(x). \tag{4.2}$$

Here, $\Phi(x)$ denotes the composite field that describes the meson under consideration, and Γ contains all factors (Dirac and internal symmetry matrices, derivatives ∂_μ) that are needed to correctly describe the quantum numbers of the meson.

A few important remarks are needed:

- The choice of the composite field $\Phi(x)$ is not unique. Any field that has the appropriate quantum numbers and a nonvanishing matrix element between the one-particle state and the vacuum represents a legitimate choice. For example, in our case, the tetraquark field $\Phi'(x) = \bar{\psi}(x)\Gamma_1\psi(x)\,\bar{\psi}(x)\Gamma_2\psi(x)$ could be used as well. This, in particular, shows that in field theory the notion of a "composite object consisting of a fixed number of elementary constituents" cannot be rigorously defined without further ado. Indeed, using the field $\Phi'(x)$ instead of $\Phi(x)$, one could conclude that a meson is a bound state of four quarks, and so on. Refining the argument, one could, for instance, argue that if the numerical value of the matrix element $\langle p|\Phi(x)|0\rangle$ (termed "overlap" in lattice QCD) for a particular choice of the interpolating field $\Phi(x)$ is much larger as compared to other choices, then the state is *predominantly* a bound state of elementary constituents that are explicitly displayed in the operator $\Phi(x)$ (this argument is often used to identify the content of various states in lattice QCD). While intuitively appealing, the argument nevertheless comes with a grain of salt: as seen from the following remark, the overlaps are not observable quantities, depending, for instance, on the renormalization scale. Consequently, the identification of states turns scale-dependent as well.
- Bare composite fields like $\Phi(x)$ are, in general, not well-defined objects in quantum field theory, meaning that the Green's functions containing $\Phi(x)$ are ultraviolet-divergent, even if all Green's functions with elementary quark and gluon fields are made finite by renormalization. Defining, for example, $\Phi(x) = \lim_{y\to x} T(\bar{\psi}(y)\Gamma\psi(x))$, it is seen that an additional ultraviolet divergence arises in the limit $y \to x$, where the arguments of two field operators coincide. Subtracting the divergence, one arrives at the renormalized field operator $\Phi_r(x;\mu)$, which depends on the renormalization scale μ. In addition, mixing with operators with engineering (canonical) dimension less or equal to that of $\Phi(x)$ should be taken into account. In what follows, we implicitly assume that the renormalization is performed and we are dealing with the renormalized composite fields. In order to ease the notation, we drop the subscript "r" and the explicit dependence on the scale μ.

Now, we are in the position to repeat the derivation of the Lehmann–Symanzik–Zimmermann (LSZ) reduction formula for the case of composite particles. The intuitive derivation, which we present here, focuses on the singularity structure of the Green's functions of composite operators. Consider, for example, the Green's function of $n+m$ composite fields:

$$G(x_1,\cdots,x_n,y_1,\cdots,y_m) = \langle 0|T(\Phi(x_1)\cdots\Phi(x_n)\Phi^\dagger(y_1)\cdots\Phi^\dagger(y_m))|0\rangle. \tag{4.3}$$

These can be obtained by equipping the Lagrangian with the external sources coupled to the composite operators $\Phi(x)$:

$$\mathcal{L}(x) \to \mathcal{L}(x) + j(x)\Phi(x), \tag{4.4}$$

and expanding the generating functional in powers of $j(x)$. The coefficients of this expansion yield Green's functions with different number of external legs. Furthermore, the Fourier transform of the preceding expression describes the process m mesons $\to$ n mesons:

$$\int \prod_i^n dx_i \prod_j^m dy_j \exp\left(i\sum_i p_i x_i - i\sum_j q_j y_j\right) G(x_1,\cdots,x_n,y_1,\cdots,y_m)$$

$$= (2\pi)^4 \delta^{(4)}\left(\sum_i p_i - \sum_j q_j\right) G(p_1,\cdots,p_n,q_1,\cdots,q_m). \tag{4.5}$$

The δ-function corresponds to energy-momentum conservation.

Our aim is to single out the on-mass-shell singularities of the Green's function G and relate the residue to the amplitude for the process $m \to n$. To this end, let us first multiply $G(x_1,\cdots,x_n,y_1,\cdots,y_m)$ from Eq. (4.3) by $\exp\{ip_1 x_1\}$ and integrate over $d^4 x_1$. From this expression, we shall pick up the singular term when the first particle is on the mass shell, $p_1^2 \to M^2$ and $p_1^0 > 0$. For this purpose, we rewrite the T-product in the form

$$T(\Phi(x_1)\cdots\Phi(x_n)\Phi^\dagger(y_1)\cdots\Phi^\dagger(y_m))$$

$$= \Phi(x_1)\theta\left(x_1^0 - \max(x_2^0,\cdots,y_m^0)\right) T(\Phi(x_2)\cdots\Phi(x_n)\Phi^\dagger(y_1)\cdots\Phi^\dagger(y_m)) + \cdots, \tag{4.6}$$

where the ellipses stand for the terms where x_1^0 is smaller than all other arguments, or x_1^0 is bound both from above and below. We shall see that all such terms can be neglected in the calculation of the singular piece. Further, we shall use the completeness condition, inserting a full set of states between the operator $\Phi(x_1)$ and the rest, and single out the contribution of the one-particle state $|k\rangle$. Using translational invariance, we get

$$\int d^4 x_1 e^{ip_1 x_1} G(x_1,\cdots,x_n,y_1,\cdots,y_m)$$

$$= \int \frac{d^3\mathbf{k}}{(2\pi)^3 2k^0} \int d^4 x_1\, e^{i(p_1-k)x_1}\, \theta\left(x_1^0 - \max(x_2^0,\cdots,y_m^0)\right)$$

$$\times \langle 0|\Phi(0)|k\rangle\langle k|T(\Phi(x_2)\cdots\Phi(x_n)\Phi^\dagger(y_1)\cdots\Phi^\dagger(y_m))|0\rangle + \cdots, \tag{4.7}$$

where the ellipses stand for the neglected terms in Eq. (4.6), as well as the contribution from other states. The integration over $d^3\mathbf{x}_1$ yields $(2\pi)^3 \delta^{(3)}(\mathbf{k}-\mathbf{p}_1)$. Integrating over x_1^0, we get

$$\int d^4 x_1 e^{ip_1 x_1} G(x_1,\cdots,x_n,y_1,\cdots,y_m) = \frac{e^{i(p_1^0-k^0+i\varepsilon)x_1^0}}{i(p_1^0-k^0+i\varepsilon)2k^0}\Bigg|_{x_1^0=\max(x_2^0,\cdots,y_m^0)}^{x_1^0 \to +\infty}$$

$$\times \langle 0|\Phi(0)|p_1\rangle\langle p_1|T(\Phi(x_2)\cdots\Phi(x_n)\Phi^\dagger(y_1)\cdots\Phi^\dagger(y_m))|0\rangle + \cdots, \tag{4.8}$$

where $k^0 = \sqrt{M^2 + \mathbf{p}_1^2}$ and the $+i\varepsilon$ prescription ensures that the integral converges on the upper limit. Assuming now that $p_1^0 \to k^0$ and using

$$\frac{e^{i(p_1^0 - k^0 + i\varepsilon)\max(x_2^0, \cdots, y_m^0)}}{(p_1^0 - k^0 + i\varepsilon)2k^0} = \frac{1}{(p_1^0)^2 - k_0^2 + i\varepsilon} + \cdots = \frac{1}{p_1^2 - M^2 + i\varepsilon} + \cdots, \tag{4.9}$$

where the ellipses stand for the terms that remain regular at $p_1^0 \to k^0$ (or, at $p_1^2 \to M^2$), Eq. (4.7) becomes

$$\int d^4 x_1 e^{ip_1 x_1} G(x_1, \cdots, x_n, y_1, \cdots, y_m)$$

$$= \frac{iZ^{1/2}}{p_1^2 - M^2 + i\varepsilon} \langle p_1 | T(\Phi(x_2) \cdots \Phi(x_n) \Phi^\dagger(y_1) \cdots \Phi^\dagger(y_m)) | 0 \rangle + \cdots, \tag{4.10}$$

where the wave function renormalization constant $Z^{1/2} = \langle 0 | \Phi(0) | p_1 \rangle$ does not depend on the momentum p_1 because of Lorentz invariance. (Here, we assume that the mesons have spin zero, albeit the treatment can be straightforwardly generalized for particles with any spin.)

From the preceding expressions, it is straightforward to verify that the contributions from the intermediate states different from one-particle states $|p\rangle$ are always regular in the limit $p_1^2 \to M^2$. Indeed, the invariant mass k_n^2 of a generic intermediate state $|n\rangle$ with the quantum number of a given meson obeys the inequality $k_n^2 > M^2$ (otherwise, the bound state would be unstable). Consequently, a denominator similar to the one in Eq. (4.10) cannot become singular at $p_1^2 \to M^2$. Further, as is clear from the above derivation, the singularity emerges because the integral over x_1^0 diverges on the upper limit of the integration, if $p_1^0 = k^0$ (the oscillatory integrand, which emerges at $p_1^0 \neq k^0$, leads to a convergent integral). It is now seen that the terms that are not displayed explicitly in Eq. (4.6) lead to regular contributions. Indeed, the integrals where x_1^0 is bound both from above and below cannot diverge, as well as the integral where x_1^0 is smaller than all other arguments. In the latter case, the exponential factor is given by $\exp\{i(p_1 + k_n)x\}$, whose argument never vanishes identically.

Next, we could repeat the procedure for the second particle and single out the contribution containing two bound states $|k_1, k_2; out\rangle$ (in the case of a positive/negative argument in the exponent, the outgoing/ingoing states should be used). We obtain

$$\int d^4 x_1 d^4 x_2 e^{ip_1 x_1 + ip_2 x_2} G(x_1, \cdots, x_n, y_1, \cdots, y_m)$$

$$= \int \frac{d^3 k_1}{(2\pi)^3 2k_1^0} \int \frac{d^3 k_2}{(2\pi)^3 2k_2^0} \langle p_1 | \Phi(0) | k_1, k_2; out \rangle$$

$$\times \frac{iZ^{1/2}}{p_1^2 - M^2 + i\varepsilon} \frac{1}{2!} \frac{i}{p_1^0 + p_2^0 - k_1^0 - k_2^0 + i\varepsilon} (2\pi)^3 \delta^{(3)}(\mathbf{k}_1 + \mathbf{k}_2 - \mathbf{p}_1 - \mathbf{p}_2)$$

$$\times \langle k_1, k_2; out | T(\Phi(x_3) \cdots \Phi(x_n) \Phi^\dagger(y_1) \cdots \Phi^\dagger(y_m)) | 0 \rangle + \cdots. \tag{4.11}$$

The first matrix element in this equation contains both disconnected and connected pieces:

$$\langle p_1|\Phi(0)|k_1,k_2;out\rangle = \langle 0|[a(\mathbf{p}_1),\Phi(0)]|k_1,k_2;out\rangle + \langle 0|\Phi(0)a(\mathbf{p}_1)|k_1,k_2;out\rangle$$

$$= \langle p_1|\Phi(0)|k_1,k_2;out\rangle_c + Z^{1/2}(2\pi)^3 2k_1^0\delta^{(3)}(\mathbf{k}_1-\mathbf{p}_1) + Z^{1/2}(2\pi)^3 2k_2^0\delta^{(3)}(\mathbf{k}_2-\mathbf{p}_1).$$

$$(4.12)$$

Here, $a(\mathbf{p})$ is an annihilation operator. The disconnected and connected pieces are schematically shown in Fig. 4.1. In Eq. (4.12), the disconnected piece contains two terms, corresponding to the situation where the particle with the momentum $\mathbf{p}_1 = \mathbf{k}_1$ ($\mathbf{p}_1 = \mathbf{k}_2$) propagates freely and is not attached to the rest of the diagram. Further, it can be easily seen that the connected piece (denoted by the subscript "*c*") produces a non-singular result after integration and thus can be dropped. Retaining only the disconnected terms and performing the momentum integrations, we get

$$\int d^4x_1 d^4x_2 e^{ip_1x_1+ip_2x_2}G(x_1,\cdots,x_n,y_1,\cdots,y_m)$$

$$= \frac{iZ^{1/2}}{p_1^2-M^2+i\varepsilon}\frac{iZ^{1/2}}{p_2^2-M^2+i\varepsilon}$$

$$\times \langle p_1,p_2;out|T(\Phi(x_3)\cdots\Phi(x_n)\Phi^\dagger(y_1)\cdots\Phi^\dagger(y_m))|0\rangle + \cdots.\qquad(4.13)$$

We can continue the procedure further until all particles, except the last one, are in the initial- or final-state vectors. (To ease notation, we assume that none of the initial-state momenta coincides with none of the final-state momenta, so the disconnected contributions are absent in the process $m \to n$. We also drop the subscript "*c*" in all matrix elements.) This results in

$$G(p_1,\cdots,p_n,q_1,\cdots,q_m) = \prod_{i=1}^{n}\frac{iZ^{1/2}}{p_i^2-M^2+i\varepsilon}\prod_{j=1}^{m-1}\frac{iZ^{1/2}}{q_i^2-M^2+i\varepsilon}$$

$$\times \langle p_1,\cdots,p_n;out|\Phi^\dagger(0)|q_1,\cdots,q_{m-1};in\rangle + \cdots.\qquad(4.14)$$

The last remaining field cannot, however, be eliminated in a similar way, and the asymptotic condition should be used, in addition. First, we note that the matrix element on the right-hand side of this equation is singular on the mass shell as well. One could single out this singularity explicitly by writing

Figure 4.1 A schematic representation of connected and disconnected contributions in the matrix element from Eq. (4.12). The disconnected diagrams, by definition, consist of two separate pieces, which are not connected with each other by any line corresponding to a particle propagator. The filled black circles correspond to the Z-factor.

$$\langle p_1, \cdots, p_n; out | \Phi^\dagger(0) | q_1, \cdots, q_{m-1}; in \rangle$$

$$= \frac{iZ^{1/2}}{q_m^2 - M^2 + i\varepsilon} \langle p_1, \cdots, p_n; out | - iJ^\dagger(0) | q_1, \cdots, q_{m-1}; in \rangle, \qquad (4.15)$$

where

$$J^\dagger(x) = Z^{-1/2}(\Box + M^2)\Phi^\dagger(x), \qquad (4.16)$$

and

$$q_m = \sum_{i=1}^{n} p_i - \sum_{j=1}^{m-1} q_j. \qquad (4.17)$$

Next, the matrix element of the operator J will be calculated with the use of the asymptotic condition. The following relation plays the central role (here, $q^0 = \sqrt{M^2 + \mathbf{q}^2}$):

$$\int d^4 y \, e^{-iqy} (\Box + M^2)\Phi^\dagger(y)$$

$$= \int d^4 y \frac{\partial}{\partial y^0} \left[e^{-iqy} (\dot{\Phi}^\dagger(y) + iq^0 \Phi^\dagger(y)) \right]$$

$$= \lim_{y^0 \to +\infty} \int d^3\mathbf{y} \, e^{-iqy} (\dot{\Phi}^\dagger(y) + iq^0 \Phi^\dagger(y))$$

$$- \lim_{y^0 \to -\infty} \int d^3\mathbf{y} \, e^{-iqy} (\dot{\Phi}^\dagger(y) + iq^0 \Phi^\dagger(y)). \qquad (4.18)$$

Thus,

$$iZ^{1/2}(2\pi)^4 \delta^{(4)} \left(\sum_{i=1}^{n} p_i - \sum_{j=1}^{m} q_j \right) \langle p_1, \cdots, p_n; out | - iJ^\dagger(0) | q_1, \cdots, q_{m-1}; in \rangle$$

$$= \lim_{y^0 \to +\infty} \int d^3\mathbf{y} \, e^{-iq_m y} \langle p_1, \cdots, p_n; out | \dot{\Phi}^\dagger(y) + iq_m^0 \Phi^\dagger(y) | q_1, \cdots, q_{m-1}; in \rangle$$

$$- \lim_{y^0 \to -\infty} \int d^3\mathbf{y} \, e^{-iq_m y} \langle p_1, \cdots, p_n; out | \dot{\Phi}^\dagger(y) + iq_m^0 \Phi^\dagger(y) | q_1, \cdots, q_{m-1}; in \rangle. $$

$$(4.19)$$

In order to calculate the first term, we insert the sum over all *out*-states,

$$\langle p_1, \cdots, p_n; out | \dot{\Phi}^\dagger(y) + iq_m^0 \Phi^\dagger(y) | q_1, \cdots, q_{m-1}; in \rangle$$

$$= \sum_{l=0}^{\infty} \frac{1}{l!} \prod_{i=1}^{l} \frac{d^3\mathbf{k}_i}{(2\pi)^3 2k_i^0} \langle p_1, \cdots, p_n; out | \dot{\Phi}^\dagger(y) + iq_m^0 \Phi^\dagger(y) | k_1, \cdots, k_l; out \rangle$$

$$\times \langle k_1, \cdots, k_l; out | q_1, \cdots, q_{m-1}; in \rangle, \qquad (4.20)$$

and use the asymptotic condition, according to which the matrix element between the *out*-states factorizes[4] in the limit $y^0 \to +\infty$, that is, only the disconnected contributions survive:

[4] Loosely speaking, in case of an *elementary* field, the asymptotic condition implies that, in the limit $y^0 \to \infty$, the field $Z^{-1/2}\Phi(y)$ can be replaced by a field $\Phi_{out}(y)$, which obeys the free Klein–Gordon equation. Now,

$$\lim_{y^0 \to +\infty} \int d^3 \mathbf{y} \, e^{-i q_m y} \langle p_1, \cdots, p_n; out | \dot{\Phi}^\dagger(y) + i q_m^0 \Phi^\dagger(y) | k_1, \cdots, k_{n-1}; out \rangle$$

$$= \sum_{i=1}^n \lim_{y^0 \to +\infty} \int d^3 \mathbf{y} \, e^{-i q_m y} \langle p_i | \dot{\Phi}^\dagger(y) + i q_m^0 \Phi^\dagger(y) | 0 \rangle$$

$$\times \langle p_1, \cdots, p_{i-1}, p_{i+1}, \cdots, p_n; out | k_1, \cdots k_{n-1}; out \rangle$$

$$= i Z^{1/2} \sum_{\mathrm{perm}(i_1, \cdots, i_n)} (2\pi)^3 2 p_{i_1}^0 \delta^{(3)}(\mathbf{p}_{i_1} - \mathbf{k}_1) \cdots (2\pi)^3 2 p_{i_{n-1}}^0 \delta^{(3)}(\mathbf{p}_{i_{n-1}} - \mathbf{k}_{n-1})$$

$$\times (2\pi)^3 2 p_{i_n}^0 \delta^{(3)}(\mathbf{p}_{i_n} - \mathbf{q}_m). \tag{4.21}$$

Here, $l = n - 1$; otherwise, the matrix element vanishes in the limit $y^0 \to +\infty$, and "perm$(i_1, \cdots, i_n)$" denotes permutations over the indices. It is seen that this expression gives a contribution only to the disconnected pieces for the process $m \to n$ and vanishes, since $\mathbf{q}_m$ is not equal to any of the $\mathbf{p}_i$.

The nonzero contribution emerges in the limit $y^0 \to -\infty$. Inserting a complete set of *in*-states, we get

$$\langle p_1, \cdots, p_n; out | \dot{\Phi}^\dagger(y) + i q_m^0 \Phi^\dagger(y) | q_1, \cdots, q_{m-1}; in \rangle$$

$$= \sum_{l=0}^\infty \frac{1}{l!} \prod_{i=1}^l \frac{d^3 \mathbf{k}_i}{(2\pi)^3 2 k_i^0} \langle p_1, \cdots, p_n; out | k_1, \cdots, k_l; in \rangle$$

$$\times \langle k_1, \cdots, k_l; in | \dot{\Phi}(y) + i q_m^0 \Phi(y) | q_1, \cdots, q_{m-1}; in \rangle. \tag{4.22}$$

Using the asymptotic condition for $y_0 \to -\infty$, we arrive at

$$\lim_{y^0 \to -\infty} \int d^3 \mathbf{y} \, e^{-i q_m y} \langle k_1, \cdots, k_m; in | \dot{\Phi}^\dagger(y) + i q_m^0 \Phi^\dagger(y) | q_1, \cdots, q_{m-1}; in \rangle$$

$$= \sum_{i=1}^m \lim_{y^0 \to -\infty} \int d^3 \mathbf{y} \, e^{-i q_m y} \langle k_i | \dot{\Phi}^\dagger(y) + i q_m^0 \Phi^\dagger(y) | 0 \rangle$$

$$\times \langle k_1, \cdots, k_{i-1}, k_{i+1}, \cdots, k_m; in | q_1, \cdots q_{m-1}; in \rangle$$

$$= i Z^{1/2} \sum_{\mathrm{perm}(i_1, \cdots, i_m)} (2\pi)^3 2 q_{i_1}^0 \delta^{(3)}(\mathbf{q}_{i_1} - \mathbf{k}_1) \cdots (2\pi)^3 2 q_{i_m}^0 \delta^{(3)}(\mathbf{q}_{i_m} - \mathbf{k}_m). \tag{4.23}$$

Here, $l = m$. Using now Eqs. (4.19, 4.22), we finally obtain

$$(2\pi)^4 \delta^{(4)}\left(\sum_{i=1}^n p_i - \sum_{j=1}^m q_j \right) \langle p_1, \cdots, p_n; out | -i J^\dagger(0) | q_1, \cdots, q_{m-1}; in \rangle$$

$$= \langle p_1, \cdots, p_n; out | q_1, \cdots, q_m; in \rangle. \tag{4.24}$$

$\Phi(y)$ is *not* an elementary field. We, however, *postulate* that the asymptotic condition remains the same – namely, that the asymptotic states consist of stable composite objects that behave exactly in the same way as the elementary particles. In particular, it is assumed that the matrix element of any *local* operator $\Phi(y)$ factorizes in the limit $y^0 \to \infty$. More details about the asymptotic condition can be found in any standard field theory textbook; see, e.g., Ref. [17].

Consequently, the matrix element on the left-hand side is equal to the amplitude for the process $m \to n$.

Let us now summarize our findings:

- We describe composite mesons by local fields $\Phi(x)$, which carry the quantum numbers of the meson under consideration. The explicit form of the field operator does not play any role. Only the normalization matters. A properly normalized field operator is given by $Z^{-1/2}\Phi(x)$, where $Z^{1/2} = \langle 0|\Phi(0)|p\rangle$.

- In order to calculate the scattering amplitude, corresponding to the process m mesons $\to n$ mesons, we have to first calculate the Green's function of $n+m$ composite fields, defined in Eqs. (4.3) and (4.5). These Green's functions contain poles in the momenta of all external particles:

$$G(p_1,\cdots,p_n,q_1,\cdots,q_m) = \prod_{i=1}^{n} \frac{iZ^{1/2}}{p_i^2 - M^2 + i\varepsilon} A(\{p_i\},\{q_j\}) \prod_{j=1}^{m} \frac{iZ^{1/2}}{q_i^2 - M^2 + i\varepsilon},$$

(4.25)

where $A(\{p_i\},\{q_j\})$ on mass shell $p_i^2 = q_j^2 = M^2$ is the scattering amplitude we are looking for:

$$\langle p_1,\cdots,p_n; out|q_1,\cdots,q_m; in\rangle = (2\pi)^4 \delta^{(4)}\left(\sum_{i=1}^{n} p_i - \sum_{j=1}^{m} q_j\right) A(\{p_i\},\{q_j\}).$$

(4.26)

- In the derivation, we never used the fact that the meson is a composite object. For this reason, the same formula is applicable to the scattering of elementary particles, in cases without confinement. This also shows that the distinction between elementary and composite particles in field theory is a subtle issue and, in many cases, cannot be uniquely defined. S-matrix elements can thus be described in terms of elementary as well as composite fields.

- In order to avoid clutter of notation, most of the preceding formulae are written for spinless particles, and the internal symmetry indices are ignored. However, the approach can be straightforwardly generalized for particles with any spin (mesons as well as baryons) and any internal symmetry group.

Our last remark concerns the use of local composite fields. Up to now, only such fields were considered. This limitation is unnecessary. In fact, instead of $\Phi(x)$ one could consider a more general construction, for example:

$$\Phi(x) = \int d^4\xi\, \bar{f}(\xi)\, T\left[\bar{\psi}\left(x+\frac{\xi}{2}\right)\Gamma\psi\left(x-\frac{\xi}{2}\right)\right],$$

$$\Phi^\dagger(y) = \int d^4\eta\, f(\eta)\, T\left[\bar{\psi}\left(y-\frac{\eta}{2}\right)\gamma^0\Gamma^\dagger\gamma^0\psi\left(y+\frac{\eta}{2}\right)\right].$$

(4.27)

Note that "smeared" operators of this kind are often used in lattice QCD. In principle, one could consider also more complicated operators (say, nonlocal tetraquark operators, etc.), but we restrict ourselves to the simplest choice.

The "vertex function" $\bar{f}(\xi)$ should obey a normalization condition, but is otherwise arbitrary. Namely, instead of the normalization condition $Z^{-1/2}\langle 0|\Phi(0)|p\rangle = 1$, one requires

$$\int d^4\xi\, \bar{f}(\xi)\chi(\xi;p) = 1\,,$$

$$\int d^4\eta\, \bar{\chi}(\eta;p)f(\eta) = 1\,, \tag{4.28}$$

where $\chi,\bar{\chi}$ are the so-called Bethe–Salpeter amplitude and its conjugate:

$$\chi(\xi;p) = \left\langle 0\left|T\bar{\psi}\left(\frac{\xi}{2}\right)\Gamma\psi\left(-\frac{\xi}{2}\right)\right|p\right\rangle$$

$$\bar{\chi}(\eta;p) = \left\langle p\left|T\bar{\psi}\left(-\frac{\eta}{2}\right)\gamma^0\Gamma^\dagger\gamma^0\psi\left(\frac{\eta}{2}\right)\right|0\right\rangle. \tag{4.29}$$

The Fourier transform reads

$$\chi(k;p) = \int d^4\xi\, e^{-ik\xi}\chi(\xi;p)\,, \qquad \bar{\chi}(l;p) = \int d^4\eta\, e^{il\eta}\bar{\chi}(\eta;p)\,. \tag{4.30}$$

In order to derive the LSZ reduction formula in this case, one starts from the Green's function of $n+m$ composite fields and performs the Fourier transform in the meson center-of-mass momenta p_i, q_j (conjugate to each of the x_i, y_j). Using exactly the same way of reasoning, it can be shown that this Green's function has poles in each external momentum p_i, q_j. Further, using the normalization condition, it can be demonstrated that the residue coincides with the amplitude for the process $m \to n$, as in the case of a local field.

As seen, apart from fulfilling the normalization condition, the form of the vertex $f(\xi)$ function can be chosen arbitrarily. For instance, choosing $f(\xi) = c\delta^{(4)}(\xi)$, we arrive at the case of a local field considered before. A particular choice for the vertex function $f(\xi)$, which is making use of the normalization condition for the Bethe–Salpeter wave functions, leads to a transparent expression for the amplitude of the mesonic process we are considering. To describe this choice, consider the four-point Green's function G_2, which, taking into account translational invariance, can be written in the following form:

$$G_2(x-y;\xi,\eta) = \left\langle 0\left|T\bar{\psi}\left(x+\frac{\xi}{2}\right)\Gamma\psi\left(x-\frac{\xi}{2}\right)\bar{\psi}\left(y-\frac{\eta}{2}\right)\gamma^0\Gamma^\dagger\gamma^0\psi\left(y+\frac{\eta}{2}\right)\right|0\right\rangle. \tag{4.31}$$

Performing now the Fourier transform, we get:

$$G_2(p;k,l) = \int d^4z\, d^4\xi\, d^4\eta\, e^{ipz+ik\xi-il\eta}G_2(z;\xi,\eta)\,. \tag{4.32}$$

In the vicinity of a single meson on its mass shell, $p^2 \to M^2$, this function exhibits a pole singularity:

$$G_2(p;k,l) \to i\frac{\chi(k;p)\bar{\chi}(l;p)}{p^2 - M^2} + \cdots. \tag{4.33}$$

The inverse of the four-point Green's function is defined as

$$\int \frac{d^4q}{(2\pi)^4} G_2(p;k,q) G_2^{-1}(p;q,l) = (2\pi)^4 \delta^{(4)}(k-l).$$ (4.34)

The Bethe–Salpeter equation for the wave function can be written in terms of the inverse Green's function (here, $p^2 = M^2$):

$$\int \frac{d^4k}{(2\pi)^4} G_2^{-1}(p;l,k) \chi(k;p) = 0,$$

$$\int \frac{d^4l}{(2\pi)^4} \bar{\chi}(l;p) G_2^{-1}(p;l,k) = 0.$$ (4.35)

In order to derive the normalization condition for the Bethe–Salpeter wave function, we write the identity $G_2 = G_2 G_2^{-1} G_2$ and extract the pole at $p^2 \to M^2$ from both sides of this identity. This leads to

$$i \int \frac{d^4k}{(2\pi)^4} \int \frac{d^4l}{(2\pi)^4} \bar{\chi}(k;p) \left[G_2^{-1}(p;k,l) \right]' \chi(l;p) = 1,$$ (4.36)

where[5]

$$\left[G_2^{-1}(p;k,l) \right]' = \lim_{p^2 \to M^2} \frac{d}{dp^2} G_2^{-1}(p;k,l).$$ (4.37)

From the normalization condition, it is clear that the Fourier transform of the conjugate vertex function, $\bar{f}(k) = \int d^4\xi\, e^{ik\xi}\, \bar{f}(\xi)$, can be chosen as follows:

$$\bar{f}(k) = i \int \frac{d^4l}{(2\pi)^4} \bar{\chi}(l;p) \left[G_2^{-1}(p;l,k) \right]'.$$ (4.38)

A similar expression is obtained for the Fourier transform of the vertex function $f(\eta)$.

Using this choice of the vertex function, the amplitude for the process $m \to n$ can be written in a simple and intuitive form. Note that all factors $(p_i^2 - M^2)$ and $(q_j^2 - M^2)$ in the external legs cancel against those arising in the normalization condition. Defining the truncated Green's function (here, the integration over the relative momenta k_i, l_j is implicit),

$$G_{\mathrm{tr}}(p_1, \cdots, p_n, q_1, \cdots, q_m)$$

$$= G_2^{-1}(p_1) \cdots G_2^{-1}(p_n) G(p_1, \cdots, p_n, q_1, \cdots, q_m) G_2^{-1}(q_1) \cdots G_2^{-1}(q_m),$$ (4.39)

we finally arrive at the following expression:

$$A(\{p_i\}, \{q_j\}) = \bar{\chi}(p_1) \cdots \bar{\chi}(p_n) G_{\mathrm{tr}}(p_1, \cdots, p_n, q_1, \cdots, q_m) \chi(q_1) \cdots \chi(q_m).$$ (4.40)

The physical interpretation of the preceding result is transparent. All interactions of the quark–antiquark pair, forming a meson with a momentum p, are summed up in the

---

[5] The derivative with respect to p^2 can be defined as $\lim\limits_{p^2 \to M^2} \frac{d}{dp^2}(\cdots) = \lim\limits_{p^0 \to w(\mathbf{p})} \frac{1}{2p^0} \frac{d}{dp^0}(\cdots).$

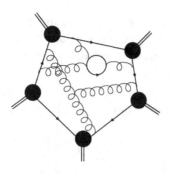

Figure 4.2 A schematic representation of a mesonic scattering amplitude as a convolution of the interior block (as explained in the text), and the Bethe–Salpeter wave functions, attached to each external meson leg (filled circles). The double lines correspond to the external mesons.

four-point function $G_2(p)$. Thus, in the truncated Green's functions, all Feynman diagrams that describe interactions within the external legs are removed (they are already included in the Bethe–Salpeter wave function).

The preceding particular choice of the (nonlocal) interpolating composite field, which allows us to express the transition amplitudes in QCD as a convolution of an "interior block," identified with the truncated Green's function, and the Bethe–Salpeter wave functions for each external hadronic leg, has been extensively used in phenomenological applications. Schematically, the hadron–hadron amplitude in this representation is given in Fig. 4.2. From the preceding discussion it should be, however, clear that all representations for the interpolating hadron fields in terms of elementary constituents are exactly equivalent. We shall use this freedom and stick to local fields in the following.

4.3 Weinberg's Theorem

The following section plays a central role for understanding of the material contained in this book. An effective theory of QCD at low energy should enable one to systematically calculate the S-matrix elements describing transitions between the pertinent mesons and baryons. These can be obtained from the Green's functions of composite operators; see Section 4.2. Further, in QCD the Green's functions are obtained by expanding the generating functional,

$$Z_{\mathsf{QCD}}(j) = \int d\psi \, d\bar{\psi} \, dG_\mu \exp\left\{ iS_{\mathsf{QCD}}(\psi, \bar{\psi}, G_\mu; j) \right\}, \tag{4.41}$$

with respect to the external sources near the point, where the scalar source is equal to the quark mass matrix, and all other sources are set to zero; see Chapter 3:

$$Z_{\mathsf{QCD}}(j) = \sum_{n=0}^{\infty} \frac{1}{n!} \int d^4x_1 \cdots d^4x_n \, G_{\mathsf{QCD},n}(x_1, \cdots, x_n) \, j(x_1) \cdots j(x_n). \tag{4.42}$$

In the preceding expressions, $j(x)$ stands for the set of external sources, coupling both to the meson and baryon composite fields, and S_{QCD} is the QCD action equipped with the external sources. An example of such an action, containing meson sources only, is given in Chapter 3.

On the other hand, the generating functional in the effective theory is given by the path integral over the meson and baryon fields,

$$Z_{\text{eff}}(j) = \int d\Psi d\bar{\Psi} d\phi \exp\left\{ iS_{\text{eff}}(\Psi, \bar{\Psi}, \phi; j) \right\}, \tag{4.43}$$

and the Green's functions are defined through the expansion,

$$Z_{\text{eff}}(j) = \sum_{n=0}^{\infty} \frac{1}{n!} \int d^4 x_1 \cdots d^4 x_n \, G_{\text{eff},n}(x_1, \cdots, x_n) j(x_1) \cdots j(x_n). \tag{4.44}$$

The equivalence of the two theories implies that the Green's functions in both representations coincide. To be more specific, let us consider the Fourier transform of these Green's functions, which depend on the external momenta, quark masses m_q and the QCD scale Λ_{QCD}. Instead of the last two parameters, we may consider the Goldstone boson masses and the hadronic scale $\Lambda \sim 1\,\text{GeV}$, which is proportional to Λ_{QCD}. As we shall see, effective field theory enables one to calculate the momentum-space Green's functions $G_{\text{eff},n}(p_1, \cdots, p_n)$ in a systematic expansion in small parameters, p_i/Λ and m_q/Λ (in general, this expansion contains logarithms of these small parameters, which, however, do not affect the power of the terms in the expansion). If $G_{\text{QCD},n}(p_1, \cdots, p_n)$ is expanded in a series in these small parameters as well, the equivalence of the two theories then states that

$$G_{\text{QCD},n}(p_1, \cdots, p_n) = G_{\text{eff},n}(p_1, \cdots, p_n), \tag{4.45}$$

order by order in this expansion. It is also seen that this statement makes no sense when the momenta become of the order of the hadronic scale Λ, since the expansion starts to diverge.[6]

The following questions naturally arise:

- Is it possible to construct the effective action functional $S_{\text{eff}}(\Psi, \bar{\Psi}, \phi; j)$ out of the meson and baryon fields only, which leads to the theory that is equivalent to QCD at low energies, as just described?
- What are the guiding principles for the construction of such an action functional?
- What are the rules for the construction of the effective Lagrangian?

The answer to the first two questions was given in Weinberg's seminal paper [18]; see also Ref. [19]. Quoting Ref. [18],

> "Quantum field theory itself has no content besides analyticity, unitarity, cluster decomposition and symmetry. This can be put more precisely in the context of perturbation theory: If one writes down the most general possible Lagrangian, including all

---

[6] Here we consider the situation, when the heavy quarks are already integrated out. Their effect is implicitly taken into account in the parameters of the QCD Lagrangian that contains light quark fields only.

terms consistent with assumed symmetry principles, and then calculates matrix elements with this Lagrangian to any given order of perturbation theory, the result will simply be the most general possible S-matrix consistent with analyticity, perturbative unitarity, cluster decomposition and the assumed symmetry principles."

The keyword here is "most general," we have to make sure that one includes all terms in the Lagrangian (with arbitrary coefficients), which are allowed by symmetry. On the other hand, it is clear that, in general, the number of such terms will be infinite. So, in order to possess predictive power, certain counting rules should be applicable, along the lines described in Chapter 1, which enable one to calculate the matrix elements order by order in the expansion in a certain (some) small parameter(s). In the following, we shall consider the construction of the effective Lagrangian and the counting rules in detail.

The last question, concerning the construction of the effective Lagrangian, is a subtle one. The symmetry, imposed on the theory, requires the invariance of the action functional and not of the Lagrangian, which is allowed to change by a total derivative. It can, however, be shown that in the case when there are no anomalies present, QCD symmetries lead to the invariance of the effective Lagrangian [20]; see also Ref. [21]. Moreover, in the presence of anomalies, only a single non-invariant term (the so-called Wess–Zumino–Witten term) is contained in the effective action, which takes care of the anomalies in the underlying theory and is not renormalized in higher orders. We shall return to the proof of this theorem later, once the rules for the construction of the effective Lagrangian are set. At this moment, we merely assume that these statements hold.

4.4 The σ-Model

4.4.1 The Linear σ-Model

Before addressing the problem in full generality, in this section we construct a simple model, containing meson fields only, in which the chiral symmetry breaking pattern of QCD is realized. Notwithstanding this simplicity, the model, which goes under the name of the linear σ-model [22], allows one to gain deep insight into many peculiar properties of QCD at low energies and to reproduce the results of current algebra with surprising ease (that, frankly, does not come at a surprise at all, since these results are a consequence of chiral symmetry alone. This property is featured by the linear σ-model from the beginning). What is also important, one could consider the linear σ-model as the first (and very instructive) step toward the construction of a general effective field theory of QCD.

In the following, for simplicity, we restrict ourselves to the two lightest flavors $N_f = 2$ and consider first the meson sector of the theory only, which is formed by the pion triplet. The simplest choice for the composite pion field is the (unnormalized) quark bilinear operator $\bar{\psi}(x)\gamma^5 \boldsymbol{\tau}\psi(x)$, where the τ^i ($i = 1, 2, 3$) are the Pauli matrices in isospin space. It can be straightforwardly checked that under infinitesimal $SU(2)_L \times SU(2)_R$ chiral transformations with:

$$g_L = \mathbb{1} + i\alpha - i\beta, \quad g_R = \mathbb{1} + i\alpha + i\beta, \qquad \alpha = \boldsymbol{\alpha}\frac{\boldsymbol{\tau}}{2}, \quad \beta = \boldsymbol{\beta}\frac{\boldsymbol{\tau}}{2}, \qquad (4.46)$$

the pertinent quark bilinears transform as:

$$(\bar{\psi}\gamma^5\tau^i\psi) \mapsto (\bar{\psi}\gamma^5\tau^i\psi) - \varepsilon^{ijk}\alpha^j(\bar{\psi}\gamma^5\tau^k\psi) + \beta^i(\bar{\psi}\psi),$$

$$(\bar{\psi}\psi) \mapsto (\bar{\psi}\psi) - \beta^i(\bar{\psi}\gamma^5\tau^i\psi). \qquad (4.47)$$

This means that the triplet $(\bar{\psi}\gamma^5\tau^i\psi)$ is not closed with respect to all transformations of $SU(2)_L \times SU(2)_R$. Consequently, if one wishes to build a *linear* realization of the group transformations in the basis of elementary field operators, the triplet pion field $\boldsymbol{\pi}(x)$ does not suffice and we have to introduce a field with the quantum numbers of the composite operator $(\bar{\psi}\psi)$.

The model therefore contains four fields: The pion triplet $\boldsymbol{\pi}(x)$ and the scalar-isoscalar field $\sigma(x)$ (the latter gave the name to the model). The transformation laws of these fields are identical to those of the quark bilinears, given in Eq. (4.47):

$$\pi^i \mapsto \pi^i - \varepsilon^{ijk}\alpha^j\pi^k + \beta^i\sigma,$$

$$\sigma \mapsto \sigma - \beta^i\pi^i. \qquad (4.48)$$

Taking into account the fact that the group $SO(4)$ is isomorphic to $SU(2) \times SU(2)$, one can combine σ and $\boldsymbol{\pi}$ into an $SO(4)$-vector $\phi^A = (\sigma, \boldsymbol{\pi})$. It can be straightforwardly checked that the square of the vector $\phi^A\phi^A = \sigma^2 + \boldsymbol{\pi}^2$ is invariant under the transformations given in Eq. (4.48).

Another useful representation of the fields is given in terms of an 2×2 matrix:

$$\Sigma = \sigma\mathbb{1} + i\boldsymbol{\pi}\cdot\boldsymbol{\tau}. \qquad (4.49)$$

Under the $SU(2)_L \times SU(2)_R$ transformations, the matrix Σ transforms as

$$\Sigma \mapsto g_R\Sigma g_L^\dagger. \qquad (4.50)$$

The infinitesimal form of these transformations coincides with Eq. (4.48).

The Lagrangian of the linear σ-model is given by

$$\mathcal{L} = \frac{1}{4}\langle\partial^\mu\Sigma\partial_\mu\Sigma^\dagger\rangle - \frac{m_0^2}{4}\langle\Sigma\Sigma^\dagger\rangle - \frac{\lambda}{16}\langle\Sigma\Sigma^\dagger\rangle^2 + \frac{c}{4}\langle\Sigma + \Sigma^\dagger\rangle = \mathcal{L}_0 + c\mathcal{L}_1, \qquad (4.51)$$

where $\langle\cdots\rangle$ denotes the trace of 2×2 matrices in the space of isospin indices. It is clear that $\mathcal{L}_0$ is invariant under the $SU(2)_L \times SU(2)_R$ group, whereas the term $c\mathcal{L}_1$ breaks this symmetry down to the diagonal $SU(2)_V$ subgroup. This term serves as an analog for the explicit chiral symmetry breaking term in QCD, which is proportional to the quark mass (in the isospin limit) and also transforms as a fourth component of an $SO(4)$ vector.

In terms of the component fields the Lagrangian takes the form

$$\mathcal{L} = \frac{1}{2}(\partial \boldsymbol{\pi})^2 + \frac{1}{2}(\partial \sigma)^2 - \frac{m_0^2}{2}(\boldsymbol{\pi}^2 + \sigma^2) - \frac{\lambda}{4}(\boldsymbol{\pi}^2 + \sigma^2)^2 + c\sigma. \tag{4.52}$$

The vector and axial-vector currents, which can be straightforwardly obtained by using the Noether theorem, are given by

$$V_\mu^i = \varepsilon^{ijk}\pi^j \partial_\mu \pi^k, \qquad A_\mu^i = \sigma \partial_\mu \pi^i - \pi^i \partial_\mu \sigma. \tag{4.53}$$

If $c \neq 0$, the axial-vector current is not conserved (an analog to PCAC; see Eq. (3.396)):

$$\partial^\mu A_\mu^i = -c\pi^i. \tag{4.54}$$

At the classical level, the configuration of the ground state of the system, described by the Lagrangian in Eq. (4.52), depends on the sign of the parameter m_0^2 (the other parameter λ is always positive; otherwise, the vacuum state would not be stable). If the parameter $m_0^2 < 0$, the vacuum expectation value of the field σ must be nonzero in the state with a minimum energy. (A nonzero vacuum expectation value of the pion field would amount to the spontaneous breaking of parity, which is not observed in nature, and which we wish to avoid in the model.) For a moment, let us set the explicit symmetry-breaking parameter c to zero. Furthermore, we require that the parity is not spontaneously broken (like in QCD). This means that the pseudoscalar field $\boldsymbol{\pi}$ cannot develop a vacuum expectation value in the broken phase, and we may thus take $\boldsymbol{\pi} = 0$ while minimizing the effective potential. Hence, in the translationally invariant vacuum, the vacuum expectation value of the field σ is obtained from the minimization of the potential:

$$V(\sigma^2 + \pi^2) \rightarrow V(\sigma^2) = \frac{m_0^2}{2}\sigma^2 + \frac{\lambda}{4}\sigma^4 + \cdots \tag{4.55}$$

where the ellipsis denotes terms not relevant for the further discussion. As we have mentioned, here we have to distinguish between two cases: $m_0^2 > 0$ and $m_0^2 < 0$. In the first case, the minimum of the potential occurs at $\sigma = 0$; see Fig. 4.3, left panel. The masses of the pion and σ are both equal to m_0, and the symmetry is realized in the Wigner–Weyl mode. The second case is more interesting. The equation for a local extremum is given by $dV(\sigma^2)/d\sigma = m_0^2\sigma + \lambda\sigma^3 = 0$. As seen from Fig. 4.3, right panel, now the solution $\sigma = 0$ corresponds to a local maximum of the potential. The minima occur at $\sigma = \pm v$, where $v^2 = -m_0^2/\lambda > 0$. Introducing the shift $\sigma \rightarrow \sigma + v$, so that the new field σ has a vanishing vacuum expectation value $\langle 0|\sigma|0 \rangle = 0$, and expanding the Lagrangian in the vicinity of the new minimum, up to an inessential constant, we get

$$\mathcal{L}_0 = \frac{1}{2}(\partial \boldsymbol{\pi})^2 + \frac{1}{2}(\partial \sigma)^2 - \frac{(m_0^2 + 3\lambda v^2)}{2}\sigma^2 - \frac{(m_0^2 + \lambda v^2)}{2}\boldsymbol{\pi}^2$$
$$- \lambda v\sigma(\boldsymbol{\pi}^2 + \sigma^2) - \frac{\lambda}{4}(\boldsymbol{\pi}^2 + \sigma^2)^2. \tag{4.56}$$

From this expression, we can immediately read off the masses

$$M_\pi^2 = m_0^2 + \lambda v^2 = 0, \qquad M_\sigma^2 = m_0^2 + 3\lambda v^2 = -2m_0^2 > 0. \tag{4.57}$$

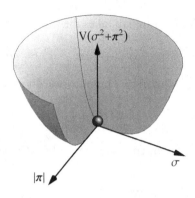 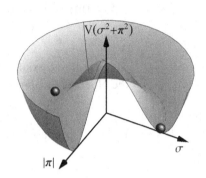

Tree-level effective potential for the pion and σ-fields. Left panel: Wigner–Weyl representation with the trivial vacuum at $\sigma = 0, |\boldsymbol{\pi}| = 0$. Right panel: Nambu–Goldstone representation with the minimum at $\sigma = \pm v, |\boldsymbol{\pi}| = 0$, corresponding to the spontaneously broken chiral symmetry (the so-called "Mexican hat" potential). Figure courtesy of Feng-Kun Guo.

Consequently, when the symmetry is spontaneously broken, $v \neq 0$, the pion becomes a Goldstone boson with a vanishing mass. The σ particle remains massive. This is nothing but a consequence of the Goldstone theorem, and the pions are the Goldstone bosons. Stated differently, along the rim of the potential, one finds massless excitations (the pions), whereas in the orthogonal direction, the second derivative of the potential is nonvanishing, leading to the mass of the corresponding excitation (the sigma).

In terms of the shifted fields, the axial-vector current takes the form

$$A_\mu^i = v\partial_\mu \pi^i + (\sigma\partial_\mu \pi^i - \pi^i\partial_\mu \sigma). \tag{4.58}$$

In order to establish the relation between the parameters of the model and the pion decay constant F_π, introduced in Chapter 3, we have to calculate the matrix element of the axial-vector current between the vacuum and the one-pion state. At tree level (that means neglecting all diagrams that contain meson loops), this matrix element is equal to

$$\langle 0|A_\mu^i(x)|\pi^j(q)\rangle = \langle 0|v\partial_\mu \pi^i(x) + \cdots|\pi^j(q)\rangle$$
$$= -iq_\mu v\delta^{ij}e^{-iqx} + \cdots = iq_\mu F_\pi \delta^{ij}e^{-iqx}. \tag{4.59}$$

From this expression, we may immediately identify $F_\pi = -v + \cdots$, where the ellipsis stands for the higher-order corrections in perturbation theory.

Next, we consider the case when the explicit symmetry breaking parameter $c \neq 0$. Since the effect of the explicit symmetry breaking is considered to be small, all expressions will be expanded in powers of c. The minimum of the potential is at

$$\left.\frac{dV(\sigma^2)}{d\sigma}\right|_{\sigma=v} = m_0^2 v + \lambda v^3 - c = 0. \tag{4.60}$$

Figure 4.4   Diagrams contributing to the $\pi\pi$ scattering amplitude in the linear σ-model at tree level. Dashed and solid lines correspond to the pions and the σ-meson, respectively.

This can be visualized as a small tilt of the Mexican hat potential away from the z-axis defined in Fig. 4.3. The solution of this equation is given by

$$v = \pm\sqrt{\frac{-m_0^2}{\lambda} - \frac{c}{2m_0^2}} + O(c^2).$$ (4.61)

The pion mass $M_\pi^2 = m_0^2 + \lambda v^2$ does not vanish, when $c \neq 0$. Instead,

$$vM_\pi^2 = -F_\pi M_\pi^2 = m_0^2 v + \lambda v^3 = c.$$ (4.62)

From this, we finally obtain, in the tree approximation,

$$M_\pi^2 = -\frac{c}{F_\pi}.$$ (4.63)

Together with Eq. (4.54), this leads to the PCAC condition $\pi^i = (F_\pi M_\pi^2)^{-1}\partial^\mu A_\mu^i$.

4.4.2 Pion–Pion Scattering at Tree Level

It is very instructive to consider $\pi\pi$ scattering within the linear σ-model. The scattering amplitude for the process $\pi^i(p_1) + \pi^j(p_2) \to \pi^k(p_3) + \pi^l(p_4)$ has the following general representation because of Lorentz invariance and Bose symmetry:

$$T^{ij,kl}(p_1, p_2; p_3, p_4) = \delta^{ij}\delta^{kl}A(s,t,u) + \delta^{ik}\delta^{jl}A(t,u,s) + \delta^{il}\delta^{jk}A(u,s,t),$$ (4.64)

where s, t, u are the usual Mandelstam variables

$$s = (p_1 + p_2)^2 = (p_3 + p_4)^2,$$

$$t = (p_1 - p_3)^2 = (p_2 - p_4)^2,$$

$$u = (p_1 - p_4)^2 = (p_2 - p_3)^2,$$ (4.65)

subject to the constraint $s + t + u = 4M_\pi^2$.

In the tree approximation, the amplitude is described by the diagrams shown in Fig. 4.4. After a straightforward calculation using the Lagrangian displayed in Eq. (4.56), we obtain

$$A(s,t,u) = A(s) = \frac{4\lambda^2 v^2}{M_\sigma^2 - s} - 2\lambda.$$ (4.66)

If the symmetry breaking term $c = 0$, the pions are massless and the elastic threshold is located at $s = 4M_\pi^2 = 0$. It can be straightforwardly verified that in this case the amplitude $A(s)$ vanishes at threshold:

$$A(s) = \frac{(4\lambda^2 v^2 - 2\lambda M_\sigma^2) + 2\lambda s}{M_\sigma^2 - s} = \frac{2\lambda s}{M_\sigma^2 - s}. \tag{4.67}$$

In other words, the diagrams with the σ-exchange and the local four-pion interaction exactly cancel at threshold. This cancellation is a direct consequence of chiral symmetry, because the Goldstone bosons have *derivative interactions,* which vanish as four-momenta go to zero. This leads to the so-called *Adler zeros* in the amplitudes with Goldstone bosons.

If $c \neq 0$, the amplitude takes the form

$$A(s) = \frac{2\lambda(s - M_\pi^2)}{M_\sigma^2 - s}. \tag{4.68}$$

An important remark is needed. We have seen that the pions are Goldstone bosons, whose mass is protected by chiral symmetry. The mass of the σ is not protected, and there exists no particular reason why its mass should be small. For this reason, it is reasonable to investigate the limit $M_\pi \ll M_\sigma$, where the σ-meson decouples from physics at low energies. This makes sense from the phenomenological point of view as well, since experimentally a narrow low-lying chiral partner of pions is not observed.[7] Performing the limit $M_\sigma \to \infty$ in Eq. (4.68), we immediately obtain

$$A(s) = \frac{s - M_\pi^2}{F_\pi^2} + O(M_\sigma^{-4}), \tag{4.69}$$

and the Adler zero is at $s = M_\pi^2$, which is below the threshold located at $s = 4M_\pi^2$. Note that the result is *universal:* In the large-M_σ limit, it does not contain the specific parameters of the linear σ-model anymore. Rather, the final result is expressed in terms of the physical observables M_π and F_π. As we shall see later, Eq. (4.69) is a consequence of chiral symmetry alone and holds in QCD as well.

4.4.3 Chiral Logarithms

All results in the previous section have been obtained at tree level. It is interesting to carry out calculations at one loop. Among other things, this calculation will explicitly demonstrate the emergence of *chiral logarithms*. The presence of chiral logarithms, which arise from the loops containing Goldstone bosons, shows that the physical quantities cannot be expanded in a Taylor series in the symmetry-breaking parameter c, starting from the symmetric world with $c = 0$. The expansion contains, apart from polynomials in c, nonanalytic contributions, say, of the type $c^2 \ln c$. (Later, we shall see that, in the baryon sector, non-integer powers of c, like $c^{3/2}$, also arise.) In terms of conventional Rayleigh–Schrödinger perturbation theory, such nonanalytic terms emerge because the energy denominators in the perturbation series, which correspond to the intermediate states with Goldstone bosons, become infrared-singular as $c \to 0$ (see, e.g., Ref. [24]).

[7] On the basis of the solution of Roy equations for pion–pion scattering, a pole at $M_\sigma = (441 - i272)$ MeV is found in the S-matrix [23]. This result, however, can by no means be considered as a justification of the phenomenological relevance of the linear σ-model.

Figure 4.5   Diagrams contributing to the pion self-energy at one loop in the linear σ-model. The last two graphs depict the so-called tadpoles. As in Fig. 4.4, dashed and solid lines denote the pions and the σ-meson, respectively.

In this section, we shall carry out the calculation of the pion mass at one loop and explicitly demonstrate the nonanalytic contributions to the mass. This example strongly resembles the calculation of the light particle mass in the toy model with two particles, which was considered in Chapter 1. More information about one-loop calculations in the linear σ-model can be found in Ref. [25].

The self-energy of the pion at one loop in the linear σ-model is described by the diagrams shown in Fig. 4.5, and the physical pion mass squared, M_P^2, is given by the pole position in the dressed pion propagator. Using the results of Section 1.6, the equation for the pion pole position can be written as

$$M_\pi^2 - M_P^2 + \Sigma(M_P^2) = 0, \tag{4.70}$$

where the self-energy is the sum of five terms, corresponding to the five diagrams shown in Fig. 4.5:

$$\Sigma(p^2) = 5\lambda T(M_\pi^2) + \lambda T(M_\sigma^2) - (2\lambda v)^2 J(p^2, M_\pi^2, M_\sigma^2)$$

$$- \frac{3(2\lambda v)^2}{2M_\sigma^2} \left(T(M_\pi^2) + T(M_\sigma^2)\right). \tag{4.71}$$

In the preceding expressions, M_π and M_σ denote the masses of the pion and σ, calculated at the lowest order:

$$M_\pi^2 = m_0^2 + \lambda v^2 = -\frac{c}{F_\pi} + O(c^2),$$

$$M_\sigma^2 = m_0^2 + 3\lambda v^2 = -2m_0^2 - \frac{3c}{F_\pi} + O(c^2). \tag{4.72}$$

Further, the tadpole graph in D space-time dimensions is given by

$$T(m^2) = \int \frac{d^D l}{(2\pi)^D i} \frac{1}{m^2 - l^2} = m^2 \left\{ 2L - \frac{1}{16\pi^2} \left(1 - \ln \frac{m^2}{\mu^2}\right) \right\}, \tag{4.73}$$

and the loop integral, $J(p^2, M_\pi^2, M_\sigma^2)$, is given in Eq. (1.90),

$$J(p^2, M_\pi^2, M_\sigma^2) = \int \frac{d^D l}{(2\pi)^D i} \frac{1}{(M_\sigma^2 - l^2)(M_\pi^2 - (p-l)^2)}$$

$$= -2L - \frac{1}{16\pi^2} \left\{ \frac{1}{2} \left(1 - \frac{M_\sigma^2 - M_\pi^2}{p^2}\right) \ln \frac{M_\pi^2}{\mu^2} + \frac{1}{2} \left(1 + \frac{M_\sigma^2 - M_\pi^2}{p^2}\right) \ln \frac{M_\sigma^2}{\mu^2} \right.$$

$$- \frac{\lambda^{1/2}}{2p^2} \left(\ln \frac{\frac{1}{2}\left(1 - \frac{M_\sigma^2 - M_\pi^2}{p^2}\right) - \frac{\lambda^{1/2}}{2p^2}}{\frac{1}{2}\left(1 - \frac{M_\sigma^2 - M_\pi^2}{p^2}\right) + \frac{\lambda^{1/2}}{2p^2}} - \ln \frac{-\frac{1}{2}\left(1 + \frac{M_\sigma^2 - M_\pi^2}{p^2}\right) - \frac{\lambda^{1/2}}{2p^2}}{-\frac{1}{2}\left(1 + \frac{M_\sigma^2 - M_\pi^2}{p^2}\right) + \frac{\lambda^{1/2}}{2p^2}} \right) - 2 \right\},$$

$$\tag{4.74}$$

where $\lambda = \lambda(p^2, M_\sigma^2, M_\pi^2)$ denotes the Källén triangle function, $\lambda(x,y,z) = x^2 + y^2 + z^2 - 2xy - 2xz - 2yz$. Here, μ is the scale of dimensional regularization. The divergent constant L is defined as in Eq. (1.89).

Next, expanding simultaneously in $(p^2 - M_\pi^2)$ and in inverse powers of M_σ^2, we get

$$J(p^2, M_\pi^2, M_\sigma^2) = -2L - \frac{1}{16\pi^2}\left\{ b_0 + b_1(p^2 - M_\pi^2) + O((p^2 - M_\pi^2)^2) \right\},$$

$$b_0 = \left(-1 + \ln \frac{M_\sigma^2}{\mu^2}\right) - \frac{M_\pi^2}{2M_\sigma^2}\left(1 - 2\ln \frac{M_\sigma^2}{M_\pi^2}\right)$$

$$- \frac{M_\pi^4}{3M_\sigma^4}\left(5 - 6\ln \frac{M_\sigma^2}{M_\pi^2}\right) + O(M_\sigma^{-6}),$$

$$b_1 = -\frac{1}{2M_\sigma^2} - \frac{M_\pi^2}{6M_\sigma^4}\left(11 - 6\ln \frac{M_\sigma^2}{M_\pi^2}\right) + O(M_\sigma^{-6}). \tag{4.75}$$

The physical mass (the position of the pole in the inverse propagator of the π-meson) at this order is given by the solution of Eq. (4.71) and is equal to

$$M_P^2 = M_\pi^2 + 5\lambda T(M_\pi^2) + \lambda T(M_\sigma^2) - (2\lambda v)^2\left(-2L - \frac{b_0}{16\pi^2}\right)$$

$$- \frac{3(2\lambda v)^2}{2M_\sigma^2}\left(T(M_\pi^2) + T(M_\sigma^2)\right). \tag{4.76}$$

Expanding in inverse powers of M_σ^2, we finally obtain

$$M_P^2 = M_\pi^2\left\{1 + 3\lambda\left(2L + \frac{1}{16\pi^2}\ln \frac{M_\sigma^2}{\mu^2} - \frac{1}{12\pi^2}\right)\right\}$$

$$+ \frac{\lambda M_\pi^4}{48\pi^2 M_\sigma^2}\left\{288\pi^2 L + 6\ln \frac{M_\sigma^2}{\mu^2} + 3\ln \frac{M_\pi^2}{\mu^2} - 16\right\} + O(M_\sigma^6). \tag{4.77}$$

As we can see, this expression still contains the divergent quantity L, which should disappear, if the bare couplings are replaced by the renormalized ones. The renormalization of the σ-model at one loop is considered, for example, in Ref. [25]. Here we need the results of that paper only in the isospin symmetry limit. In the $\overline{\text{MS}}$ scheme, the renormalized and bare couplings are related by

$$m_0^2 = m_r^2(1 - 12\lambda_r L) + O(\lambda_r^2),$$

$$\lambda = \lambda_r(1 - 24\lambda_r L) + O(\lambda_r^3),$$

$$c = c_r + O(\lambda_r^2). \tag{4.78}$$

Using the preceding equations, the factor M_π^2 in the first term in the r.h.s. of Eq. (4.77) can be rewritten as

$$M_\pi^2 = \frac{c}{v} = c\left(\frac{\lambda}{-m_0^2}\right)^{1/2} + \frac{c^2\lambda}{2m_0^4}$$

$$= c_r\left(\frac{\lambda_r}{-m_r^2}\right)^{1/2}(1-6\lambda_r L) + \frac{c_r^2\lambda_r}{2m_r^4} + \cdots, \qquad (4.79)$$

where the ellipsis denotes the higher-order terms is λ_r and c_r. Defining the renormalized masses *at leading order in* c_r via

$$\bar{M}^2 = c_r\left(\frac{\lambda_r}{-m_r^2}\right)^{1/2}, \qquad \bar{M}_\sigma^2 = -2m_r^2, \qquad (4.80)$$

we immediately see that all divergent terms cancel (as they should):

$$M_P^2 = \bar{M}^2\left\{1 + 3\lambda_r\left(\frac{1}{16\pi^2}\ln\frac{\bar{M}_\sigma^2}{\mu^2} - \frac{1}{12\pi^2}\right)\right\}$$

$$- \frac{\bar{M}^4}{\bar{M}_\sigma^2}\left\{1 - \frac{\lambda_r}{16\pi^2}\left(\ln\frac{\bar{M}^2}{\mu^2} - \ln\frac{\bar{M}_\sigma^2}{\mu^2} - \frac{25}{3}\right)\right\} + \cdots. \qquad (4.81)$$

Further, it is seen that the contributions at zeroth order in M_π cancel. This means that the pion stays massless in the limit $c \to 0$, also when loop corrections are taken into account. Moreover, one sees that, up to $O(c)$, the pion mass is a *polynomial*. The nonanalytic term has a form $c^2\ln c$ and thus emerges at *higher order in* c (we count all expressions $c^n\ln^m c$ as $O(c^n)$). Using the tree-level relation $M_{\sigma,r}^2 = 2\lambda_r F_\pi^2$, one can rewrite this *chiral logarithm* in a form that contains only physical observables

$$M_P^2 = A\bar{M}^2 + \frac{\bar{M}^4}{32\pi^2 F_\pi^2}\ln\frac{\bar{M}^2}{\mu^2} + B\bar{M}^4 + O(\bar{M}^6). \qquad (4.82)$$

This equation tells us that the form of the chiral logarithm generated from the pion loops is universal and is determined by chiral symmetry alone. The same result will be obtained in what follows in ChPT. In contrast to this, the coefficients A, B in the polynomial are not universal, depending on the parameters of the model. The parameter A describes the radiative corrections to the vacuum expectation value of the σ-field; see Ref. [25]. Note further that the scale dependence in the chiral logarithm is balanced by a similar one in the coefficient B, rendering the physical pion mass scale-independent.

4.4.4 The Nonlinear σ-Model

The linear σ-model, which was considered in previous sections, exhibits the same symmetry-breaking pattern as QCD. However, the linear representation of chiral symmetry, which was adopted there, precludes one from a systematic inclusion of the higher-order terms that is needed for the construction of the effective theory. Indeed, adding operators $(\sigma^2 + \boldsymbol{\pi}^2)^n$, $n = 3, 4, \ldots$ to the Lagrangian with arbitrary couplings would lead to a theory with the same symmetry. The individual contributions from such terms, say, to the pion–pion scattering amplitude at low energies, will not be suppressed by a small parameter, and the cancellations, which would turn more and more complicated at higher orders, will only ensure that the net contribution to the amplitude vanishes at threshold linearly with s (in the massless case). Therefore, this means that we are not allowed to truncate the Lagrangian, even if we want to calculate the amplitude at a given accuracy. Since the total number of the chirally symmetric operators (and, hence, the number of the independent couplings associated with them) is infinite, the linear σ-model cannot be elevated to the effective theory of QCD just by adding all possible terms allowed by the symmetry. It stays a model.

The property of the linear σ-model, which we have mentioned, stems from two reasons, which are closely related to each other. First, in nature, the pions are Goldstone bosons, and their interactions contain derivatives (i.e., these are weak for small momenta). In the linear σ-model, the pions and σ couple nonderivatively with each other, and the weakness of the interactions at small momenta results from an elaborate cancellation of a number of Feynman diagrams. Next, the chiral transformations necessarily link the pions with the σ-particle. Whereas the pion mass is small (as it is protected by chiral symmetry), the mass of the σ can be arbitrary and on the order of a heavy scale. Thus, multiple scales, including the heavy scale, are *explicitly* present in the theory, hindering us from establishing a consistent power-counting scheme.

Both of the preceding problems can be removed in a single move along the path already discussed at the end of Section 4.4.2. Namely, since nothing precludes M_σ to be on the order of the heavy scale, one could integrate it out and write down a theory that contains pion fields only. We have to sacrifice, however, the linear realization of chiral symmetry, if one decides to follow this path – there exists no chiral partner of the pion in the theory anymore.

Here, it will be demonstrated how the σ-particle can be eliminated from the theory, and what the chiral transformation for the fields looks like. To this end, we choose a new set of fields: Instead of σ and $\boldsymbol{\pi}$, we choose the radial field ρ and a 2×2 unitary matrix U. The matrix field Σ, introduced in Eq. (4.49), can be rewritten in the new parameterization as

$$\Sigma = v\rho U, \qquad UU^\dagger = 1. \tag{4.83}$$

Here, the vacuum expectation value v at the lowest order is equal to $-F_\pi$. Substituting the preceding equation in the Lagrangian, we get

$$\mathcal{L} = \frac{v^2}{2} \partial_\mu \rho \partial^\mu \rho + \frac{v^2 \rho^2}{4} \langle \partial_\mu U \partial^\mu U^\dagger \rangle - \frac{m_0^2 v^2}{2} \rho^2 - \frac{\lambda v^4}{4} \rho^4 + \frac{cv\rho}{4} \langle U + U^\dagger \rangle. \tag{4.84}$$

Let us consider the limit of this expression for $M_\sigma^2 = -2m_0^2 \to \infty$ and F_π fixed. In this limit the kinetic term for the ρ-field can be neglected, and the equation of motion for this field simplifies to

$$-m_0^2 v^2 \rho - \lambda v^4 \rho^3 = 0. \tag{4.85}$$

Expressing the vacuum expectation value v in terms of m_0^2 and λ, the relevant solution of the equation of motion is

$$\rho = 1, \tag{4.86}$$

and, consequently, in this limit

$$\mathcal{L} = \frac{F_\pi^2}{4} \langle \partial_\mu U \partial^\mu U^\dagger \rangle - \frac{cF_\pi}{4} \langle U + U^\dagger \rangle. \tag{4.87}$$

As expected, the Higgs-like field ρ disappears from the Lagrangian, which now contains the triplet of the pion fields only.

The matrix U can be parameterized in terms of the pion field $\boldsymbol{\pi}$ in an arbitrary way, provided unitarity is observed. For example, the so-called σ-model parameterization has proven to be convenient in many calculations:

$$U = \sqrt{1 - \frac{\boldsymbol{\pi}^2}{F_\pi^2}} + \frac{i}{F_\pi} \boldsymbol{\tau} \cdot \boldsymbol{\pi}. \tag{4.88}$$

Another parameterization, which also is often used in the calculations, is the so-called exponential parameterization:[8]

$$U = \exp\left(\frac{i}{F_\pi} \boldsymbol{\tau} \cdot \boldsymbol{\pi} \right). \tag{4.89}$$

We may also use a different parameterization, if one wishes. For example, we may use a general parameterization for U. Up to and including the fourth order in the expansion in $\boldsymbol{\pi}$, any parameterization can be characterized by a single real parameter α:

$$U = 1 + \frac{1}{F_\pi} \boldsymbol{\tau} \cdot \boldsymbol{\pi} - \frac{\boldsymbol{\pi}^2}{2F_\pi^2} - \frac{i\alpha}{F_\pi^3} \boldsymbol{\pi}^2 \boldsymbol{\tau} \cdot \boldsymbol{\pi} + \frac{8\alpha - 1}{8F_\pi^4} (\boldsymbol{\pi}^2)^2 + \dots. \tag{4.90}$$

Here, α takes the values $\alpha = 1/6$ for the exponential and $\alpha = 0$ for the σ-model parameterization. All these parameterizations differ from each other by a local transformation of the field variable; the theorem discussed in Chapter 1 ensures that all physical observables are parameterization-independent.

Next, it is immediately seen that the symmetry-conserving part of the Lagrangian in Eq. (4.87) is invariant under the $SU(2)_L \times SU(2)_R$ chiral transformation:

$$U \to g_R U g_L^\dagger. \tag{4.91}$$

This transformation, however, turns nonlinear on the pion fields. Using, for instance, the σ-model parameterization, it is seen that the action of the infinitesimal chiral transformations on the pion field is again given by Eq. (4.48). The field σ, however, is given by a nonlinear function of the pion field; namely, it is proportional to the square root that is present in Eq. (4.88). This is the price that one pays for integrating out the σ-field.

---

[8] Albeit very convenient, the σ-model parameterization can be defined for $N_f = 2$ flavors only. In the three-flavor case, the exponential parameterization is used.

In the following, we shall stick to the σ-model parameterization and expand the Lagrangian from Eq. (4.87) in powers of the pion field. Expressing the coupling c in terms of the pion mass, up to the terms that include six and more fields, the result is given by

$$\mathcal{L} = \frac{1}{2}\partial_\mu\boldsymbol{\pi}\partial^\mu\boldsymbol{\pi} - \frac{M_\pi^2}{2}\boldsymbol{\pi}^2 + \frac{(\boldsymbol{\pi}\cdot\partial_\mu\boldsymbol{\pi})(\boldsymbol{\pi}\cdot\partial^\mu\boldsymbol{\pi})}{2F_\pi^2} - \frac{M_\pi^2(\boldsymbol{\pi}^2)^2}{8F_\pi^2} + O((\boldsymbol{\pi}^2)^3). \quad (4.92)$$

As can be seen from this, the pion–pion interactions are weak: They are suppressed either by the momenta that appear in the (derivative) vertex, or by the factor M_π^2 that is proportional to the quark mass. Further, it can be straightforwardly checked that, at tree level, this Lagrangian yields the pion–pion scattering amplitude that is given in Eq. (4.69). No mysterious cancellations are needed.

Finally, we shall briefly discuss the counting rules within the nonlinear σ-model (we shall dwell on this issue in much more detail later). For simplicity, consider the symmetric case $c = 0$. Any term that will be added to the Lagrangian in Eq. (4.87) must be invariant under global chiral transformations displayed in Eq. (4.91). Only the matrix field U and derivatives thereof can enter. All such terms are given by the trace of a matrix, which is a product of the type

$$\partial U\partial U^\dagger\partial U\partial U^\dagger\ldots\partial U\partial U^\dagger, \quad (4.93)$$

with any number of derivatives, denoted by the symbol ∂, acting on the U's. In the end, all Lorentz-indices in the derivatives are summed up, so that the Lagrangian is a Lorentz-scalar. The key observation is that a string of U's and $U^\dagger$'s without derivatives is not allowed, since this string can be reduced to 1 due to the unitarity condition $UU^\dagger = 1$. Thus, adding terms with more fields automatically leads to more derivatives and to a stronger suppression at low momenta. In other words, a consistent power-counting scheme exists. This is different in the linear realization, where strings with no derivatives $\Sigma\Sigma^\dagger\Sigma\Sigma^\dagger\ldots\Sigma\Sigma^\dagger$ are allowed, thus upsetting a consistent power counting.

Up to now, all results in the nonlinear σ-model have been obtained at tree level. It is legitimate to ask, what happens when the loop contributions are included. We shall return to this question later in this chapter.

4.4.5 Higher-Order Terms, External Sources

In the previous section, the lowest-order Lagrangian of the nonlinear σ-model has been derived; see Eq. (4.87). The question naturally arises as to which operators emerge in the Lagrangian at higher orders. Moreover, one eventually intends to equip the Lagrangian with external sources, and write down the terms that are invariant with respect to transformations of both the pion fields and the external sources. A study of the limit $M_\sigma \to \infty$ in the linear σ-model with external sources, as done in Ref. [26], enables us to gain the necessary intuition, since in this limit the sub-leading operators are systematically produced. Here we shall consider this limit in the classical theory only, and the reader who is interested in the one-loop results is referred to the original paper [26].

It will be convenient to work in the vector notation. As before, we introduce the vector $\phi^A = (\sigma, \pi)$ with $A = 0, 1, 2, 3$. The Lagrangian $\mathcal{L}_0$, given in Eq. (4.51), in terms of the field ϕ^A is rewritten in a form that is manifestly invariant with respect to the $O(4) = SU(2) \times SU(2)$ group:

$$\mathcal{L}_0 = \frac{1}{2} \partial_\mu \phi^T \partial^\mu \phi - \frac{m_0^2}{2} \phi^T \phi - \frac{\lambda}{4} (\phi^T \phi)^2, \tag{4.94}$$

where the superscript "T" denotes "transposed." We proceed further by equipping this Lagrangian with external sources. Namely, we define the covariant derivative

$$\nabla_\mu \phi^A = \partial_\mu \phi^A + F_\mu^{AB} \phi^B, \tag{4.95}$$

where the external vector and axial-vector sources are related to the antisymmetric tensor F_μ^{AB} is the following manner:

$$F_\mu^{0i} = -F_\mu^{i0} = a_\mu^i, \quad F_\mu^{ij} = -\varepsilon^{ijk} v_\mu^k, \quad A, B = 0, 1, 2, 3, \quad i, j = 1, 2, 3. \tag{4.96}$$

Note that the preceding procedure is consistent with the definition of the vector- and axial-vector currents in Eq. (4.53).

In addition, we introduce the four-vector $f = (s, \mathbf{p})$, consisting of the scalar and the pseudoscalar sources. The modified Lagrangian is given by

$$\mathcal{L} = \frac{1}{2} \nabla^\mu \phi^T \nabla_\mu \phi - \frac{m_0^2}{2} \phi^T \phi - \frac{\lambda}{4} (\phi^T \phi)^2 + f^T \phi + h \, \text{tr} \, F_{\mu\nu} F^{\mu\nu}, \tag{4.97}$$

where $F_{\mu\nu} = [\nabla_\mu, \nabla_\nu]$. The last term is added to the Lagrangian in order to ensure the finiteness of all Green's functions in the theory. Namely, if such a term is not added, some Green's functions with 2,3,4 external vector and axial-vector legs would be divergent. Note that the constant h multiplies an operator that contains external sources only and therefore contributes a polynomial to all Green's functions. Such constants are hereafter referred to as "high-energy constants." As we will see in more detail later, such high-energy constants accompany operators that are solely used for renormalization; thus, no observable in the EFT can ever depend on such a parameter.

This Lagrangian is invariant under the transformations

$$\phi^0 \mapsto \phi^0 - \beta^i \phi^i,$$

$$\phi^i \mapsto \phi^i - \varepsilon^{ijk} \alpha^j \phi^k + \beta^i \phi^0,$$

$$s \mapsto s - \beta^i p^i,$$

$$p^i \mapsto p^i - \varepsilon^{ijk} \alpha^j p^k + \beta^j s,$$

$$a_\mu^i \mapsto a_\mu^i + \partial_\mu \beta^i - \varepsilon^{ijk} \beta^j v_\mu^k - \varepsilon^{ijk} \alpha^j a_\mu^k,$$

$$v_\mu^i \mapsto v_\mu^i + \partial_\mu \alpha^i + \varepsilon^{ijk} \alpha^j v_\mu^k + \varepsilon^{ijk} \beta^j a_\mu^k. \tag{4.98}$$

The transformations for the external sources are identical to those given in Eq. (3.77), for $\alpha^0 = \beta^0 = 0$, $s^i = p^0 = v^0 = a^0 = 0$. Consequently, for such a choice of the external sources, the generating functional of the linear σ-model obeys *exactly the same Ward identities* under $SU(2)_L \times SU(2)_R$ transformations as the generating functional in QCD,

$$Z_\sigma(s,p,v,a) = Z_\sigma(s+\delta s, p+\delta p, v+\delta v, a+\delta a). \tag{4.99}$$

Further, in analogy with the previous section, we find it useful to investigate the limit of large M_σ, where only the pion degrees of freedom survive in the theory. Here, we consider the tree approximation, where it suffices to substitute the solutions of the classical equations of motion ϕ_c into the Lagrangian. These equations take the following form:

$$\nabla_\mu \nabla^\mu \phi_c^A = f^A - m_0^2 \phi_c^A - \lambda (\phi_c^T \phi_c) \phi_c^A. \tag{4.100}$$

In order to solve this equation by iterations, it is convenient to use the following parameterization of the classical field, similar to Eq. (4.83),

$$\phi_c^A = -v\rho U^A, \qquad U^T U = 1, \tag{4.101}$$

where $v = -(-m_0^2/\lambda)^{1/2}$ and ρ is the "radial" field. In this parameterization, the equations of motion can be rewritten as

$$\Box \rho + \rho(U^T \nabla_\mu \nabla^\mu U) = \chi_0^T U - m_0^2 \rho(1-\rho^2),$$

$$\rho(\nabla_\mu \nabla^\mu U^A - U^A(U^T \nabla_\mu \nabla^\mu U)) = \chi_0^A - U^A(\chi^T U) - 2\partial_\mu \rho \nabla^\mu U^A, \tag{4.102}$$

where

$$\chi_0^A = -\frac{1}{v} f^A. \tag{4.103}$$

Let us find a solution of this equation by expanding in inverse powers of m_0^2 (equivalently, of M_σ^2). At lowest order, the solution is trivial: $\rho = 1$. At higher orders,

$$\rho = 1 + \delta\rho_1 + \delta\rho_2 + \cdots, \qquad \delta\rho_n = O(m_0^{-2n}). \tag{4.104}$$

The solutions at sub-leading order can be found easily by solving the equations of motion iteratively:

$$\delta\rho_1 = -\frac{1}{2m_0^2}(\chi_0^T U + \nabla_\mu U^T \nabla^\mu U)$$

$$\delta\rho_2 = -\frac{3}{2}\delta\rho_1^2 - \frac{1}{2m_0^2}\delta\rho_1(\nabla_\mu U^T \nabla^\mu U) + \text{total derivative}, \tag{4.105}$$

and so on. Note that the iterative procedure imposes *counting rules* on the *external sources*. Namely, the expansion is carried out in the dimensionless parameter p/m_0, where p denotes the (small) momentum of the pion. The fields ρ and U count both as $O(p^0)$ and the derivative $\partial_\mu = O(p)$. Then, a consistent counting is achieved, if we assign

$$v_\mu^i = O(p), \quad a_\mu^i = O(p), \quad \chi_0^A = O(p^2). \tag{4.106}$$

On the other hand, the classical action functional in terms of the fields ρ, U can be rewritten in the following manner:

$$S(\phi_c) = \int d^4x \mathcal{L}(\phi_c) = -\frac{m_0^2}{2\lambda} \int d^4x \left(\rho(\chi_0^T U) - \frac{m_0^2}{2}\rho^4 + h \operatorname{tr} F_{\mu\nu}F^{\mu\nu} \right). \qquad (4.107)$$

Substituting the expansion $\rho = 1 + \delta\rho_1 + \delta\rho_2 + \cdots$, we get

$$S(\phi_c) = -\frac{m_0^2}{\lambda} \int d^4x \left\{ \left(\frac{1}{2}\nabla_\mu U^T \nabla^\mu U + \chi_0^T U \right) \right.$$
$$\left. - \frac{1}{4m_0^2}(\chi_0^T U + \nabla_\mu U^T \nabla^\mu U)^2 + \frac{h}{F_\pi^2}\operatorname{tr} F_{\mu\nu}F^{\mu\nu} + O(p^6) \right\}. \qquad (4.108)$$

Finally, expressing m_0^2 through F_π, we obtain the desired Lagrangian up to and including order p^4,

$$S(\phi_c) = F_\pi^2 \int d^4x \left\{ \left(\frac{1}{2}\nabla_\mu U^T \nabla^\mu U + \chi_0^T U \right) \right.$$
$$\left. + \frac{1}{4\lambda F_\pi^2}(\chi_0^T U + \nabla_\mu U^T \nabla^\mu U)^2 + \frac{h}{F_\pi^2}\operatorname{tr} F_{\mu\nu}F^{\mu\nu} + O(p^6) \right\}. \qquad (4.109)$$

As anticipated, the final result contains only the pion fields, encoded in U, and the external sources s, p, v, a. The σ-meson has disappeared. The effective Lagrangian contains a tower of operators at $O(p^2)$, $O(p^4)$, etc. The coefficients of these operators are all determined through the tree-level matching in terms of the parameters m_0^2, λ, h of the original model. In general, it can be shown that there are eight independent constants at this order. The coefficients of the operators, which contain the field U, are called *low-energy constants*, in contradistinction to the high-energy constant h, which multiplies the operator depending only on the external sources.

To end this section, note here once more that, due to the condition $U^0 U^0 + U^i U^i = 1$, the $SU(2)_L \times SU(2)_R$ symmetry on the fields U is realized in a *nonlinear manner* (cf. Eq. (4.98)):

$$U^i \mapsto U^i - \varepsilon^{ijk}\alpha^j U^k + \beta^i \sqrt{1 - U^i U^i}. \qquad (4.110)$$

As noted before, this nonlinear transformation law is the price to pay for having eliminated the σ-meson from the theory.

4.5 The σ-Model with Nucleons

4.5.1 Nucleon Field Transformations

In the previous section we have considered a simple model, where the chiral symmetry breaking pattern in QCD was linearly realized on the $SU(2)_L \times SU(2)_R$ multiplet, consisting of three elementary pion fields and a scalar-isoscalar field, σ. Here, we want to extend this model to include the nucleon field. To this end, we have first to choose under which representation of the group $SU(2)_L \times SU(2)_R$ should the nucleon field transform. As in the case of the $(\sigma, \boldsymbol{\pi})$ multiplet, in order to gain some insight, it is instructive (albeit not necessary) to consider quark composite fields, which are used for the description of baryons. It has been shown that, for $N_f = 2$ flavors, there exists only two linearly independent three-quark fields with positive parity, which do not contain derivatives [27]:

$$\eta_1(x) = (\psi_a^T C \psi_b)_{I=0} \gamma_5 \psi_c \varepsilon^{abc},$$

$$\eta_2(x) = (\psi_a^T \gamma^5 C \psi_b)_{I=0} \psi_c \varepsilon^{abc}. \qquad (4.111)$$

Here, $C = i\gamma^2\gamma^0$ is the charge conjugation matrix and a, b, c denote the color indices, whereas Lorentz and flavor indices are implicit. It can be further shown that, using Fierz transformations, the preceding expressions can be expressed as linear combinations of the so-called vector and tensor nucleon currents, which are more familiar in the context of the so-called baryon sum rules. The combination of two quark fields with zero isospin ($I = 0$) is given by

$$(\psi_a^T C \psi_b)_{I=0} = \psi_{a}iC\tau^2\psi_b = u_a C d_b - d_a C u_b,$$

$$(\psi_a^T C \gamma^5 \psi_b)_{I=0} = \psi_{a}iC\gamma^5\tau^2\psi_b = u_a C\gamma^5 d_b - d_a C\gamma^5 u_b. \qquad (4.112)$$

Here, τ^2 is a Pauli matrix acting in the isospin space.

It is straightforward to check that the quark bilinears, defined in Eq. (4.112), are invariant under $SU(2)_L \times SU(2)_R$ chiral transformations. Indeed, rewriting, for example, the first of these bilinears in terms of the left- and right-handed quark components, we get

$$(\psi_a^T C \psi_b)_{I=0} = \psi_{L,a}iC\tau^2\psi_{L,b} + \psi_{R,a}iC\tau^2\psi_{R,b}, \qquad (4.113)$$

and, similarly, for the other quark bilinear. Performing now the chiral transformations $\psi_{L,R} \mapsto g_{L,R}\psi_{L,R}$ and using the identity $\tau^2\tau^i\tau^2 = -(\tau^i)^T$, it can be shown that the quark bilinears stay invariant. Consequently, the transformation properties of the nucleon composite field operator in Eq. (4.111) under chiral transformations coincide with those of a quark field:

$$(\eta_{1,2})_L \mapsto g_L(\eta_{1,2})_L, \qquad (\eta_{1,2})_R \mapsto g_R(\eta_{1,2})_R. \qquad (4.114)$$

Consequently, in the case of the nucleon field, one *could* construct a linear realization of the chiral symmetry group, considering the left- and right-handed components of the nucleon field N and imposing the transformation:

$$(1): \qquad N_L \mapsto g_L N_L, \qquad N_R \mapsto g_R N_R. \qquad (4.115)$$

However, this is not the only option. Adapting the analysis from [28] to the case of $N_f = 2$ flavors, we may consider different alternatives, namely N_L and N_R transforming under the fundamental representations of either $SU(2)_L$ or $SU(2)_R$. This means that, apart from the transformation (4.115), the transformations (2): $N_L \mapsto g_R N_L$, $N_R \mapsto g_L N_R$, as well as (3): $N_L \mapsto g_R N_L$, $N_R \mapsto g_R N_R$ and (4): $N_L \mapsto g_L N_L$, $N_R \mapsto g_L N_R$ should be included. All this, however, does not lead to different physics, since these various alternatives can be related to each other by field transformations. Namely, recalling that the meson field Σ, introduced in Eq. (4.49), transforms like $\Sigma \mapsto g_R \Sigma g_L^\dagger$, the alternative (2) reduces to (1) by the nucleon field redefinition $N_L \to \Sigma^\dagger N_L$ and $N_R \to \Sigma N_R$ (similarly for other alternatives). As the physics cannot depend on the choice of the interpolating field, we restrict ourselves to the alternative (1), written in Eq. (4.115).

4.5.2 The Linear σ-Model with Nucleons

The model consists of the pion and the σ-fields coupled to the doublet nucleon field in the isospin space. As discussed in the previous section, the realization of the chiral $SU(2)_L \times SU(2)_R$ symmetry in the basis of the preceding fields is given by

$$\Sigma \mapsto g_R \Sigma g_L^\dagger, \qquad N_R \mapsto g_R N_R, \quad N_L \mapsto g_L N_L. \tag{4.116}$$

It is immediately seen that chiral symmetry forbids a nucleon mass term: the expression $m_N \bar{N} N = m_N (\bar{N}_L N_R + \bar{N}_R N_L)$ is not invariant under chiral transformations. As we shall see, in this model the nucleon acquires its mass through spontaneous chiral symmetry breaking.

The Lagrangian of the linear σ-model with nucleons is given by $\mathcal{L} = \mathcal{L}_0 + c\sigma$, where the symmetric part of the Lagrangian takes the form

$$\mathcal{L}_0 = \bar{N}\big[i\,\slashed{\partial} + g(\sigma - i\gamma_5 \boldsymbol{\tau} \cdot \boldsymbol{\pi})\big]N + \frac{1}{2}(\partial_\mu \boldsymbol{\pi})^2 + \frac{1}{2}(\partial_\mu \sigma)^2 - \frac{m_0^2}{2}(\sigma^2 + \boldsymbol{\pi}^2)$$

$$- \frac{\lambda}{4}(\sigma^2 + \boldsymbol{\pi}^2)^2. \tag{4.117}$$

As seen from the preceding Lagrangian, the constant g at tree level coincides with the pseudoscalar pion–nucleon coupling constant $g_{\pi NN} \simeq 13$. Because of chiral symmetry, the coupling of the (hypothetical) σ-meson to nucleons is described by the same constant g.

It is straightforward to verify that the Lagrangian is invariant under the $SU(2)_L \times SU(2)_R$ symmetry group, since

$$\bar{N} i \,\slashed{\partial} N = \bar{N}_L i \,\slashed{\partial} N_L + \bar{N}_R i \,\slashed{\partial} N_R,$$

$$\bar{N}\big[\sigma - i\gamma_5 \boldsymbol{\tau}\boldsymbol{\pi}\big]N = \bar{N}_L\big[\sigma - i\boldsymbol{\tau}\cdot\boldsymbol{\pi}\big]N_R + \bar{N}_R\big[\sigma + i\boldsymbol{\tau}\cdot\boldsymbol{\pi}\big]N_L$$

$$= \bar{N}_L \Sigma^\dagger N_R + \bar{N}_R \Sigma N_L. \tag{4.118}$$

Hence, both expressions are explicitly invariant under $SU(2)_L \times SU(2)_R$ transformations, given in Eq. (4.116).

The vector and axial-vector currents are given by

$$V_\mu^i = \bar{N}\gamma_\mu \frac{\tau^i}{2} N + \varepsilon^{ijk}\pi^j \partial_\mu \pi^k,$$

$$A_\mu^i = \bar{N}\gamma_\mu \gamma_5 \frac{\tau^i}{2} N + (\sigma \partial_\mu \pi^i - \pi^i \partial_\mu \sigma). \tag{4.119}$$

Next, let us consider the axial-vector form factor of the nucleon, which is given by the matrix element of the axial-vector current between the one-nucleon states. Using only Lorentz invariance, invariance under charge conjugation and isospin symmetry, this matrix element can be expressed in terms of two scalar form factors, which are functions of a single variable $q^2 = (p' - p)^2$ (for a review, see Ref. [29]):

$$\langle p', s' | A_\mu^i(0) | p, s \rangle = \bar{u}(p', s') \frac{\tau^i}{2} \left[\gamma_\mu \gamma_5 G_A(q^2) + q_\mu \gamma_5 G_P(q^2) \right] u(p, s). \quad (4.120)$$

Here, $G_A(q^2)$ and $G_P(q^2)$ are the axial and the induced pseudoscalar form factor, respectively. The axial-vector coupling is defined as $g_A = G_A(q^2 = 0)$. Its experimental value is $g_A \simeq 1.28$. In the linear σ-model, at tree level, we obtain:

$$G_A(q^2) = 1, \qquad G_P(q^2) = 0, \quad (4.121)$$

which corresponds to $g_A = 1$ at this order.

Next, consider the spontaneous symmetry breaking and shift the σ-field:

$$\sigma \to \sigma + v + O(c). \quad (4.122)$$

After this substitution, the Lagrangian of the linear σ-model with nucleons takes the form

$$\mathcal{L} = \bar{N}(i\slashed{\partial} + gv)N + g\bar{N}(\sigma - i\gamma_5 \boldsymbol{\tau}\boldsymbol{\pi})N + \frac{1}{2}(\partial_\mu \boldsymbol{\pi})^2 + \frac{1}{2}(\partial_\mu \sigma)^2$$
$$- \frac{M_\sigma^2}{2}\sigma^2 - \frac{M_\pi^2}{2}\boldsymbol{\pi}^2 - \lambda v \sigma(\sigma^2 + \boldsymbol{\pi}^2) - \frac{\lambda}{4}(\sigma^2 + \boldsymbol{\pi}^2)^2. \quad (4.123)$$

As can be seen from this, the shift of the σ-field gives rise to the nucleon mass term. The nucleon mass at tree level can be directly read from the preceding Lagrangian:

$$m_N = -gv = gF_\pi. \quad (4.124)$$

Note that the preceding expression is a particular case of the more general Goldberger–Treiman relation $g_A m_N = g_{\pi NN} F_\pi$, which is a consequence of chiral symmetry and holds exactly in the chiral limit [30]. We remind the reader that at tree level in the linear σ-model, g_A is equal to 1.

Finally, it should be emphasized that the linear σ-model provides a beautiful illustration of the anomaly matching condition, discussed in detail in Section 3. The anomaly matching condition in QCD, in case of $N_f = 2$ flavors, has two alternative solutions: either an effective theory with massless bound colorless fermions, or a theory with massive fermions and massless Goldstone bosons, one for each broken generator. It is seen that, in the case of the linear σ-model, these two alternatives are realized by the choice of a different sign in the boson mass term. It should also be noted that no such alternative exists for QCD with $N_f > 2$ flavors. Chiral symmetry *must* be spontaneously broken in this case.

4.5.3 The Nonlinear σ-Model with Nucleons

The nonlinear σ-model with nucleons can be obtained along the same lines as the one without nucleons. We again introduce the radial and angular meson fields

$$\Sigma = v\rho U\,, \qquad U^\dagger U = UU^\dagger = 1\,, \tag{4.125}$$

where $\rho \to 1$ when $M_\sigma^2 \to \infty$. In addition, we define the 2×2 matrix field u through

$$u^2 = U\,, \tag{4.126}$$

and redefine the nucleon field

$$N_R = u\Psi_R\,, \qquad N_L = u^\dagger \Psi_L\,, \qquad \Psi = \Psi_L + \Psi_R\,. \tag{4.127}$$

It is straightforward to verify that

$$g\bar{N}(\sigma - i\gamma_5 \boldsymbol{\tau}\boldsymbol{\pi})N = -m_N \bar{\Psi}\Psi\,, \tag{4.128}$$

where the tree-level relation $m_N = -gv$ has been used. In other words, we may directly write down the chirally invariant mass term through the transformed fields Ψ. The kinetic term, however, gets more complicated in terms of the new fields:

$$\begin{aligned}
\bar{N}i\partial\!\!\!/N &= \bar{N}_L i\partial\!\!\!/N_L + \bar{N}_R i\partial\!\!\!/N_R = \bar{\Psi}_L u i\partial\!\!\!/(u^\dagger \Psi_L) + \bar{\Psi}_R u^\dagger i\partial\!\!\!/(u\Psi_R) \\
&= \bar{\Psi}i\partial\!\!\!/\Psi + \bar{\Psi}\frac{1}{2}(u^\dagger i\partial\!\!\!/u + u i\partial\!\!\!/u^\dagger)\Psi - \bar{\Psi}\gamma_5 \frac{1}{2}(u^\dagger i\partial\!\!\!/u - u i\partial\!\!\!/u^\dagger)\Psi \\
&= \bar{\Psi}i\gamma_\mu D^\mu \Psi - \frac{1}{2}\bar{\Psi}\gamma_5 \gamma_\mu u^\mu \Psi\,,
\end{aligned} \tag{4.129}$$

where

$$D^\mu = \partial^\mu + \Gamma^\mu\,,$$

$$\Gamma^\mu = \frac{1}{2}(u^\dagger \partial^\mu u + u\partial^\mu u^\dagger)\,,$$

$$u^\mu = i(u^\dagger \partial^\mu u - u\partial^\mu u^\dagger)\,. \tag{4.130}$$

The first term in the last line of Eq. (4.129) describes the "covariant derivative" of the nucleon field, and the second term describes the pion–nucleon coupling. As expected, we have a derivative coupling, that is, the Goldstone boson interactions are weak at small momentum.

It is interesting to note that the transformed nucleon field Ψ transforms only under the unbroken vector subgroup $SU(2)_V$. In order to see this, observe that the field U under the transformations from the group $SU(2)_L \times SU(2)_R$ transforms as

$$U \mapsto g_R U g_L^\dagger\,. \tag{4.131}$$

The preceding transformation induces the following transformation on the field $u = \sqrt{U}$:

$$u \mapsto g_R u h^\dagger = u' \quad \text{or} \quad u \mapsto h u g_L^\dagger = u'\,. \tag{4.132}$$

(Both these transformations are equivalent.) Here, g_L, g_R are the elements of the global $SU(2)$, which *do not depend on the field u.* In contrast, the matrix

$$h \in SU(2), \quad h = h(u, g_L, g_R) \tag{4.133}$$

does depend on u. For this reason, one speaks of a nonlinear realization of chiral symmetry.

Next, we have to verify that both of the preceding definitions of h are consistent (i.e., one should be able to find such a single $h \in SU(2)$ that both relations in Eq.(4.132) are fulfilled). Indeed, from both alternative definitions we get

$$h = u' g_L u^\dagger \quad \text{and} \quad h^\dagger = u^\dagger g_R^\dagger u'. \tag{4.134}$$

It is seen that these definitions are consistent, that is,

$$h h^\dagger = u' g_L u^\dagger u^\dagger g_R^\dagger u' = u' U'^\dagger u' = 1. \tag{4.135}$$

It can also be easily seen that on the unbroken vector subgroup $g_L = g_R$, the matrix h does not depend on the field u. Indeed, Eq. (4.132) has a trivial solution for the transformations from the vector subgroup: $g_L = g_R = h$. Parameterizing the matrix u in terms of the elementary pion fields, say as $u = \exp(i\boldsymbol{\tau}\boldsymbol{\pi}/(2F_\pi))$, it is seen that the pion field transforms linearly with respect to the vector subgroup.

Next, we consider the transformation of the redefined nucleon field Ψ:

$$\Psi_R = u^\dagger N_R \mapsto h u^\dagger g_R^\dagger g_R N_R = h\Psi_R,$$

$$\Psi_L = u N_L \mapsto h u g_L^\dagger g_L N_L = h\Psi_L. \tag{4.136}$$

Hence, the transformation of the field Ψ simplifies to

$$\Psi = \Psi_L + \Psi_R \mapsto h\Psi. \tag{4.137}$$

It is seen that the field Ψ also forms a linear representation of the vector subgroup.

The quantities Γ^μ and u^μ transform in the following manner:

$$\Gamma^\mu = \frac{1}{2}(u^\dagger \partial^\mu u + u \partial^\mu u^\dagger) \mapsto \frac{1}{2}\left[h u^\dagger g_R^\dagger \partial^\mu (g_R u h^\dagger) + h u g_L^\dagger \partial^\mu (g_L u^\dagger h^\dagger)\right]$$

$$= \frac{1}{2} h(u^\dagger \partial^\mu u + u \partial^\mu u^\dagger)h^\dagger + h \partial^\mu h^\dagger$$

$$= h\Gamma^\mu h^\dagger + h \partial^\mu h^\dagger, \tag{4.138}$$

and

$$u^\mu = i(u^\dagger \partial^\mu u - u \partial^\mu u^\dagger) \mapsto i\left[h u^\dagger g_R^\dagger \partial^\mu (g_R u h^\dagger) - h u g_L^\dagger \partial^\mu (g_L u^\dagger h^\dagger)\right]$$

$$= ih(u^\dagger \partial^\mu u - u \partial^\mu u^\dagger)h^\dagger = h u^\mu h^\dagger. \tag{4.139}$$

The covariant derivative of the field Ψ transforms as it should:

$$D^\mu \Psi \mapsto (\partial^\mu + h\Gamma^\mu h^\dagger + h \partial^\mu h^\dagger)h\Psi = h(\partial^\mu + \Gamma^\mu)\Psi = h D^\mu \Psi. \tag{4.140}$$

Here, we have used the identity

$$0 = \partial^\mu(h^\dagger h) = \partial^\mu h^\dagger h + h^\dagger \partial^\mu h. \qquad (4.141)$$

At this moment, we could temporarily abandon the large-M_σ limit of the linear σ-model and ask a more general question, namely, how would one construct a generic Lagrangian, which is invariant under the nonlinear transformations (4.132) and (4.137). The answer to this question is simple – the Lagrangian should be made from the invariant building blocks that could be constructed out of the pion and nucleon fields. Using the transformation properties of these fields, it is straightforward to write such building blocks:

$$\bar\Psi\Psi, \quad \bar\Psi i\gamma_\mu D^\mu\Psi, \quad \bar\Psi\gamma_\mu\gamma_5 u^\mu\Psi, \quad \cdots. \qquad (4.142)$$

The power counting in a theory with nucleons is more complicated than for the case of mesons only, since the zeroth component of the nucleon momenta is of order one. However, the power-counting rules can still be established, since the three-momentum components, as well as the kinetic energy $p^0 - m$ are small quantities of order p. We shall address this issue in more detail later. Here, we merely write down the lowest-order chiral pion–nucleon Lagrangian without further discussion:

$$\mathcal{L} = \frac{F_\pi^2}{4}\langle\partial_\mu U\partial^\mu U^\dagger\rangle - \frac{cF_\pi}{4}\langle U + U^\dagger\rangle + \bar\Psi(i\slashed{D} - m_N)\Psi - \frac{g_A}{2}\bar\Psi\gamma_\mu\gamma_5 u^\mu\Psi + \cdots. \qquad (4.143)$$

When $c \neq 0$, the chiral symmetry is explicitly broken by the pion mass term.

As was already pointed out before, a nucleon mass term is not forbidden a priori in the nonlinear realization. Moreover, as seen from the preceding expression, the lowest-order pion–nucleon pseudovector coupling comes with an arbitrary constant, g_A. If one starts from the linear σ-model, then $g_A = 1$. However, the value of this coupling is not fixed by chiral symmetry, because the pertinent operator is explicitly invariant. Consequently, one is allowed to use any value of g_A here, generalizing the result obtained from the matching to the linear σ-model. (This already shows that this model is not a faithful representation of QCD.)

In order to be used in calculations, the preceding Lagrangian should be expanded in the pion fields. Using, for example, the σ-model parameterization, we get

$$\mathcal{L} = \frac{1}{2}(\partial_\mu\boldsymbol{\pi})^2 - \frac{M_\pi^2}{2}\boldsymbol{\pi}^2 + \frac{(\boldsymbol{\pi}\cdot\partial_\mu\boldsymbol{\pi})^2}{2F_\pi^2} - \frac{M_\pi^2\boldsymbol{\pi}^4}{8F_\pi^2}$$

$$+ \bar\Psi(i\slashed{\partial} - m_N)\Psi - \frac{i}{4F_\pi^2}\varepsilon^{ijk}\bar\Psi\partial_\mu\pi^j\pi^k\gamma^\mu\Psi + \frac{g_A}{2F_\pi}\bar\Psi\gamma_\mu\gamma_5\boldsymbol{\tau}\cdot\partial^\mu\boldsymbol{\pi}\Psi + \cdots. \qquad (4.144)$$

As explicitly seen, the coupling of the pions to the nucleons is also of derivative nature and vanishes for small momenta.

For illustration, we shall use the preceding Lagrangian to calculate the tree-level scattering amplitude for the process $N(p,s) + \pi^i(q) \to N(p',s') + \pi^j(q')$, where (p,s), (p',s')

Figure 4.6   Pion–nucleon scattering at tree level in the nonlinear σ-model. Solid (dashed) lines denote nucleons (pions).

are the momenta and spins of the initial and final nucleons, whereas (q,i), (q',j) are the momenta and isospin labels of the initial and final pions, respectively. The Feynman diagrams describing this process at tree level are shown in Fig. 4.6. These consist of the direct and crossed nucleon pole diagrams plus the direct four-particle interaction (the so-called Weinberg–Tomozawa term), which emerges from the covariant derivative of the nucleon field. The scattering amplitude takes the form

$$
T^{ji} = \frac{g_A^2}{4F_\pi^2} \bar{u}(p',s') \gamma_\mu \gamma_5 q'^\mu \tau^j \frac{1}{m_N - \not{p}' - \not{q}'} \gamma_\nu \gamma_5 q^\nu \tau^i u(p,s)
$$

$$
+ \frac{g_A^2}{4F_\pi^2} \bar{u}(p',s') \gamma_\nu \gamma_5 q^\nu \tau^i \frac{1}{m_N - \not{p}' + \not{q}} \gamma_\mu \gamma_5 q'^\mu \tau^j u(p,s)
$$

$$
- \frac{i}{4F_\pi^2} \bar{u}(p',s') \varepsilon^{ijk} \tau^k (q'+q)_\mu \gamma^\mu u(p,s). \tag{4.145}
$$

Let us perform the isospin algebra first. The amplitude can be written as the sum of symmetric and antisymmetric parts in the indices j,i:

$$
T^{ji} = \delta^{ji} T^+ + i\varepsilon^{jik} \tau^k T^-. \tag{4.146}
$$

It can be shown that the projectors onto the states with total isospin $I = \frac{1}{2}, \frac{3}{2}$ are given by

$$
\Pi^{1/2} = \frac{1}{3} \tau^j \tau^i, \qquad \Pi^{3/2} = \delta^{ji} - \Pi^{1/2}, \tag{4.147}
$$

or

$$
\delta^{ji} = \Pi^{3/2} + \Pi^{1/2}, \qquad i\varepsilon^{ijk} \tau^k = 2\Pi^{1/2} - \Pi^{3/2}. \tag{4.148}
$$

From this we get

$$
T^{1/2} = T^+ + 2T^-, \qquad T^{3/2} = T^+ - T^-. \tag{4.149}
$$

Let us evaluate the amplitude at threshold. Accidentally, due to the algebra of the γ-matrices, the nucleon pole terms are suppressed at threshold. Indeed, anticommuting the γ_5 matrices until we can use the identity $\gamma_5^2 = 1$, the expression for the pole terms turns into

$$
T_{\mathrm{pole}}^{ji} = -\frac{1}{4F_\pi^2} \bar{u}(p',s') \not{q}' \frac{1}{m_N + \not{p}' + \not{q}'} \not{q} \tau^j \tau^i u(p,s)
$$

$$
- \frac{1}{4F_\pi^2} \bar{u}(p',s') \not{q} \frac{1}{m_N + \not{p}' - \not{q}} \not{q}' \tau^i \tau^j u(p,s). \tag{4.150}
$$

At threshold, $\not{q}' = \not{q} = M_\pi \gamma_0$, and the denominators are of order $2m_N + O(M_\pi)$. Thus, the whole expression is of order M_π^2 and we obtain

$$T^+ = O(M_\pi^2).$$

$$T^- = \frac{1}{4F_\pi^2} \bar{u}(p',s') \gamma_\mu (q'+q)^\mu u(p,s) \to \frac{m_N M_\pi}{F_\pi^2}. \tag{4.151}$$

The S-wave scattering lengths are then given by

$$a^\pm = \frac{T^\pm}{8\pi(m_N + M_\pi)}. \tag{4.152}$$

From this we get the following:[9]

$$a^+ = O(M_\pi^2) \simeq 0,$$

$$a^- = \frac{1}{8\pi(m_N + M_\pi)} \frac{m_N M_\pi}{F_\pi^2} + O(M_\pi^3) \simeq 80 \cdot 10^{-3}/M_\pi. \tag{4.153}$$

Experimentally, $a^+ = (7.6 \pm 3.1) \cdot 10^{-3}/M_\pi$ and $a^- = (86.1 \pm 0.9) \cdot 10^{-3}/M_\pi$ [31], and the tree-level theoretical prediction is fulfilled very well. This simple calculation serves as a beautiful demonstration of the predictive power of the approach based on chiral effective Lagrangians [32, 33]. Clearly, it remains to work out systematically the corrections to the leading-order results, as will be done later.

4.6 Nonlinear Realization of Chiral Symmetry

In the previous sections, the limit of the large M_σ has been considered in detail in the linear σ-model. It was shown that a nonlinear realization of chiral symmetry naturally emerges in this limit, leading to the derivative pion couplings and consistent counting rules. The reason why such an effort was invested in a study of a particular model is simple. It helps one considerably to develop the necessary intuition to easily understand the rather abstract mathematical construction contained in the seminal papers by Callan, Coleman, Wess and Zumino [34, 35], where a prescription for constructing the generic chirally invariant Lagrangians is given; see also the work by Weinberg for the two-flavor case [36]. In the following we shall refer to this as to the CCWZ formalism.

In Refs. [34, 35], the following mathematical problem is posed. Let G be a generic compact, connected, semisimple Lie group with n parameters and H a continuous subgroup of G, to which the symmetry is spontaneously broken. We wish to describe the most general nonlinear realization of the group G on some set of elementary fields (both Goldstone boson fields and matter fields), which linearizes on the subgroup H (i.e., both the Goldstone boson and matter fields form a basis of some *linear* representation of the subgroup H). Furthermore, constructing the most general Lagrangian

---

[9] Note that, for historical reasons, the sign convention for the scattering lengths is different from the definition in Eq. (1.10).

that is invariant under the symmetry group G is considered. The strategy, adopted in Refs. [34, 35], consists in first constructing a single realization of the symmetry with the required properties, called the "standard realization," and then showing that any realization can be obtained from the standard realization by field redefinitions. Since, as we already know, field redefinitions leave the S-matrix of the theory invariant, it thus suffices to restrict ourselves to the standard realization.

Let us begin by constructing a standard realization. Let V^i, $i = 1, \ldots, n - d$ be the generators of H and A^l, $l = 1, \ldots, d$ be the remaining generators, chosen so that Hermitean matrices V^i and A^l together form a complete set of orthonormal generators of G. For example, in case of the group $SU(N_f)_L \times SU(N_f)_R$, which has been considered in previous sections, one could choose $V^i = T^i$ and $A^l = \gamma^5 T^l$, where $i, l = 1, \cdots (N_f^2 - 1)$ and T^i represent the generators of the $SU(N_f)$ group.[10]

Next, let (ξ, η) denote the (real) coordinates on the group manifold, associated with the generators A and V, respectively. Introducing the shorthand notations $\xi A = \sum_l^d \xi_l A^l$ and $\eta V = \sum_i^{n-d} \eta_i V^i$, it is seen that any group element $g \in G$ can be written down in a form $g = e^{i\xi A} e^{i\eta V}$. Now, for any element $g_0 \in G$, we may write:

$$g_0 e^{i\xi A} = e^{i\xi' A} e^{i\eta' V}, \qquad \xi' = \xi'(\xi, g_0), \quad \eta' = \eta'(\xi, g_0). \qquad (4.154)$$

Further, let Ψ denote a matter field (say, a nucleon field), and

$$h: \quad \Psi \mapsto D(h)\Psi, \qquad (4.155)$$

with $h \in H$, be a linear unitary representation of the subgroup H. Then the following statement is true: The transformations

$$g_0: \quad \xi \mapsto \xi', \quad \Psi \mapsto D(e^{i\eta' V})\Psi \qquad (4.156)$$

give a nonlinear realization of G. It is important to emphasize that the Goldstone boson fields in this framework are identified with the functions ξ^l from the parameter space of the symmetry group G, associated with the generators A^l, which are orthogonal to the generators of the unbroken subgroup V^i.

In order to prove this statement, we have to show that the preceding structure of the transformations reproduces the composition law of two group elements. To this end, let us consider the combination of transformations g_0, g_1. Since

$$g_1 e^{i\xi' A} = e^{i\xi'' A} e^{i\eta'' V}, \qquad (4.157)$$

then

$$g_1 g_0 e^{i\xi A} = e^{i\xi'' A} e^{i\eta''' V}, \quad \text{where} \quad e^{i\eta''' V} = e^{i\eta'' V} e^{i\eta' V}. \qquad (4.158)$$

---

[10] In the context of purely mesonic effective theories, introducing γ^5 looks a bit unnatural. However, writing the generators this way just serves to ensure that the Lie algebra takes the form $[V^i, V^j] = [A^i, A^j] = i f^{ijk} V^k$ and $[V^i, A^j] = i f^{ijk} A^k$. Alternatively, one could introduce any matrix $E \neq \mathbb{1}$ with $E^2 = \mathbb{1}$ and define $V^i = \mathbb{1} \otimes T^i$, $A^l = E \otimes T^l$ that leads to the same result.

Moreover, since D is a linear representation of the invariant subgroup H, then

$$D(e^{i\eta'''V}) = D(e^{i\eta''V})D(e^{i\eta'V}). \tag{4.159}$$

This means that in the transformations corresponding to the product, $g_1 g_0$ is given by

$$\xi \mapsto \xi'' = \xi''(\xi'(\xi, g_0), g_1),$$

$$\Psi \mapsto D(e^{i\eta'''V})\Psi = D(e^{i\eta''V})D(e^{i\eta'V})\Psi. \tag{4.160}$$

Thus, the group composition property is obeyed, that is, these transformations provide a nonlinear realization of the symmetry group G. We shall call this realization the "standard realization."

Next, we shall demonstrate that the preceding transformation linearizes on the invariant subgroup H. Indeed, take $g_0 = h \in H$ and write

$$g_0 e^{i\xi A} = h e^{i\xi A} = h e^{i\xi A} h^{-1} h. \tag{4.161}$$

Let us, for simplicity, consider an infinitesimal transformation,

$$h = 1 + i\sum_i \eta_i V^i + O(\eta^2). \tag{4.162}$$

Then,

$$h\xi A h^{-1} = \xi A + i\sum_{il} \eta_i \xi_l [V^i, A^l] + O(\eta^2). \tag{4.163}$$

Since both V^i and A^l are elements of Lie algebra, we have

$$[V^i, A^l] = \sum_j b_{jli} V^j + \sum_k c_{kli} A^k, \tag{4.164}$$

where b_{jli} and c_{kli} are numerical coefficients. Projecting out the coefficients b_{jli}, we obtain

$$b_{jli} = \text{const tr}([V^i, A^l]V^j) = \text{const tr}([V^j, V^i]A^l)$$

$$= (if_{ijm}) \text{const tr}(V^m A^l) = 0, \tag{4.165}$$

where f_{ijm} are the structure constants of the subgroup H and we used the fact that all generators are orthogonal to each other. This finally means that

$$h\xi A h^{-1} = \sum_l \xi_l A^l + \sum_{ilk} \xi_l \eta_i c_{kli} A^k + O(\eta^2) = \sum_l \xi_l' A^l, \tag{4.166}$$

where

$$\xi_l' = \xi_l + \sum_{ik} \eta_i c_{lki} \xi_k + O(\eta^2). \tag{4.167}$$

Now, using the Baker–Hausdorff formula:

$$e^{i\eta V} e^{i\xi A} e^{-i\eta V} = e^{i\xi A + \frac{1}{1!}[i\eta V, i\xi A] + \frac{1}{2!}[i\eta V, [i\eta V, i\xi A]] + \cdots} e^{i\eta V} e^{-i\eta V}$$

$$= e^{i\xi A + [i\eta V, i\xi A] + O(\eta^2)}, \tag{4.168}$$

it is seen that

$$he^{i\xi A}h^{-1} = e^{i\xi' A} \tag{4.169}$$

where, for an infinitesimal h, the quantity ξ' is again given by Eq. (4.167). Note that the mapping $\xi \mapsto \xi'$ is linear, that is, contains the field ξ only linearly. Considering higher-order terms in the Baker–Hausdorff formula and using Eq. (4.165), it is straightforward to verify that this property holds for finite transformations as well. In other words, for finite transformations we may conclude that

$$\xi' = D^{(b)}(h)\xi, \tag{4.170}$$

where the matrix $D^{(b)}$ is uniquely determined by the structure of G and its symmetry breaking pattern:

$$D^{(b)}(h)_{lk} = \left[e^{i\eta c} \right]_{lk}, \qquad \left[\eta c \right]_{lk} = \eta_i c_{lki} . \tag{4.171}$$

Furthermore, using the identity

$$[V^j,[V^i,A^l]] - [V^i,[V^j,A^l]] = [[V^j,V^i],A^l] = if^{jik}[V^k,A^l], \tag{4.172}$$

where the f^{jik} are the structure constants of the subgroup H, we straightforwardly obtain

$$c_{nmj}c_{mli} - c_{nmi}c_{mlj} = if^{jik}c_{nlk}. \tag{4.173}$$

This means that the matrices $(c^i)_{nl} = c_{nli}$ obey the same commutation relations as the generators $(V^i)_{nl}$. In other words, $D^{(b)}$ is a linear representation of the subgroup H, realized on the fields ξ. Furthermore, it is evident that for $g = h = e^{i\eta'V}$, the quantity η' does not depend on ξ, and the transformation

$$\Psi \mapsto D(e^{i\eta'V})\Psi = D(h)\Psi \tag{4.174}$$

also linearizes on the unbroken subgroup H.

Back to the case of the spontaneously broken $SU(N_f)_L \times SU(N_f)_R$ symmetry in QCD, note that generators V^i and A^l transform differently under parity operation P,

$$V^i \mapsto P(V^i) = V^i, \qquad A^l \mapsto P(A^l) = -A^l . \tag{4.175}$$

Thus, from the transformation

$$g_0 e^{i\xi A} = e^{i\xi' A} e^{i\eta' V} \tag{4.176}$$

it follows that

$$P(g_0)e^{-i\xi A} = e^{-i\xi' A} e^{i\eta' V} . \tag{4.177}$$

We may combine these two expressions, eliminating the quantity η' altogether. The transformation law for the Goldstone boson field ξ then simplifies to

$$g_0 e^{2i\xi A} P(g_0^{-1}) = e^{2i\xi' A} . \tag{4.178}$$

This is nothing but the nonlinear transformation law for the Goldstone boson matrix $U = e^{2i\xi A} \mapsto g_R U g_L^\dagger$, with $g_0 = g_R = e^{i\beta A} e^{i\alpha V}$ and $P(g_0^{-1}) = g_L^\dagger = e^{-i\alpha V} e^{i\beta A}$ (here, (α, β) are group parameters that define the element g_0).

Next, let us consider the construction of the chirally invariant Lagrangians, using the Goldstone boson (ξ) and matter (Ψ) fields. This Lagrangian should be constructed from the fields and their derivatives. The derivatives of the fields do not transform covariantly, since the (nonlinear) transformation parameters in their turn depend on the fields. It is, however, straightforward to construct covariant derivatives with the required transformation properties. To this end, let us consider the Maurer–Cartan form which can be represented as a linear combination of all generators of the group (here, we use the same notations and normalization as in Section 4.5.3):

$$ie^{-i\xi A} \partial_\mu e^{i\xi A} = i\Gamma_\mu V + \frac{1}{2} u_\mu A. \tag{4.179}$$

From the transformation law for the field ξ,

$$g \partial_\mu e^{i\xi A} = (\partial_\mu e^{i\xi' A}) e^{i\eta' V} + e^{i\xi' A} (\partial_\mu e^{i\eta' V}), \tag{4.180}$$

one obtains

$$e^{-i\xi' A} \partial_\mu e^{i\xi' A} = e^{i\eta' V} e^{-i\xi A} \partial_\mu e^{i\xi A} e^{-i\eta' V} - (\partial_\mu e^{i\eta' V}) e^{-i\eta' V}. \tag{4.181}$$

Using Eq. (4.179), one sees that Γ_μ and u_μ obey the following relations:

$$u'_\mu A = e^{i\eta' V} u_\mu e^{-i\eta' V},$$

$$\Gamma'_\mu V = e^{i\eta' V} \Gamma_\mu V e^{-i\eta' V} - (\partial_\mu e^{i\eta' V}) e^{-i\eta' V}. \tag{4.182}$$

Further, defining the covariant derivative of a matter field Ψ as

$$D_\mu \Psi = \partial_\mu \Psi + \Gamma_\mu \bar{T} \Psi, \tag{4.183}$$

where $\bar{T}$ denotes the pertinent generator of the subgroup H in the representation D, in which the field Ψ transforms, we finally get the transformation rule:

$$(D_\mu \Psi)' = e^{i\eta' \bar{T}} D_\mu \Psi. \tag{4.184}$$

Clearly, this is the transformation rule that should be obeyed by the covariant derivative. Now, taking into account Eqs. (4.182) and (4.184), which describe the blocks that transform covariantly under the nonlinear transformations, the construction of the invariant Lagrangians becomes a straightforward task.

Finally, note that, in case of the chiral $SU(N_f)_L \times SU(N_f)_R$ group, due to a special structure of the Lie algebra, namely, the property of the generators displayed in Eq. (4.175) it is possible to express the quantities u_μ and Γ_μ in terms of the Maurer–Cartan form separately:

$$\Gamma_\mu V = \frac{1}{2} (e^{-i\xi A} \partial_\mu e^{i\xi A} + e^{i\xi A} \partial_\mu e^{-i\xi A}),$$

$$u_\mu A = i(e^{-i\xi A} \partial_\mu e^{i\xi A} - e^{i\xi A} \partial_\mu e^{-i\xi A}). \tag{4.185}$$

The analogy with the large-M_σ limit in the linear σ-model is perfect (cf. Eq. (4.130)). We see that the redefined fields U and Ψ in the nonlinear σ-model, which is obtained in this limit, provide a basis for a standard nonlinear realization of chiral symmetry.

It remains to be shown that every nonlinear realization of G, which becomes linear when the group is restricted to its subgroup H, is equivalent to the standard realization up to a change of field variables. We do not present this (very elegant) proof here, referring the interested reader instead to the original paper [34].

4.7 Chiral Perturbation Theory in the Meson Sector: Lagrangian and Generating Functional

In this and the following sections, we finally put all the pieces together and construct the Lagrangian, which contains the external sources s, p, v, a, and is invariant under the chiral $SU(N_f)_L \times SU(N_f)_R$ group. Correspondingly, we do not consider singlet vector and axial-vector transformations, restricting ourselves to the traceless sources $\langle v_\mu \rangle = \langle a_\mu \rangle = 0$. The anomalies are put aside for the moment. It should be stressed that one could have started here, without reference to the σ-model discussed before, but we consider it easier to understand the concepts using this model first. As before, we start with the purely mesonic case and consider the inclusion of the baryons later.

One crucial point here is the inclusion of the explicit chiral symmetry breaking due to the quark mass term. As discussed in Chapter 3, one would systematically treat these effects by first introducing the scalar source s instead of quark masses. This renders the action functional explicitly chirally symmetric, because the sources also transform under the $SU(N_f)_L \times SU(N_f)_R$ group. At a later stage, the Green's functions in the theory with massive quarks are obtained by expanding the generating functional around $s = \mathcal{M}$, the quark mass matrix, instead of $s = 0$. This trick, which goes under the name of the spurion technique, allows one to deal with an explicitly chirally symmetric Lagrangian at all stages, and the construction described in the previous section can be readily used. Another crucial point is that the QCD generating functional with the external sources is invariant under *local $SU(N_f)_L \times SU(N_f)_R$* transformations – the external fields v_μ and a_μ play the role of gauge fields; see also Section 3.6. This severely limits the choice of the invariant building blocks for the Lagrangian.

In the mesonic sector, we can use the unitary $N_f \times N_f$ matrix U to describe the Goldstone bosons. As already mentioned, this is a counterpart of the quantity $e^{2i\xi A}$ from the previous section and transforms as

$$U(x) \mapsto g_R(x)\, U(x)\, g_L^\dagger(x). \tag{4.186}$$

Note that here we have promoted chiral symmetry to a local symmetry using space-time dependent transformations. The transformation of the sources $s(x)$, $p(x)$, $v_\mu(x)$, $a_\mu(x)$ are written down in Eq. (3.74).

In the exponential parameterization, the matrix U is expressed through the pion fields

$$U(x) = \exp\left(\frac{i}{F_0}\,\Phi(x)\right),\tag{4.187}$$

where F_0 is a constant with the dimension of mass, and $\Phi(x) = \lambda^i \phi^i(x)$ or $\Phi(x) = \tau^i \phi^i(x)$ in the three- and two-flavor cases, respectively. Later, we shall identify F_0 with the pion decay constant in the chiral limit. It is seen that Eq. (4.186) corresponds to the nonlinear transformations of the pion fields. From here on, we will not display the x-dependence of U and the Goldstone boson fields any more.

It is straightforward to check that the following quantities transform covariantly under local $SU(3)_L \times SU(3)_R$ transformations:

$$D_\mu U = \partial_\mu U - i r_\mu U + i U l_\mu\,, \qquad D_\mu U \mapsto g_R(x) D_\mu U g_L^\dagger(x)\,,$$

$$D_\mu \chi = \partial_\mu \chi - i r_\mu \chi + i \chi l_\mu\,, \qquad D_\mu \chi \mapsto g_R(x) D_\mu \chi g_L^\dagger(x)\,,$$

$$r_{\mu\nu} = \partial_\mu r_\nu - \partial_\nu r_\mu - i[r_\mu, r_\nu]\,, \qquad r_{\mu\nu} \mapsto g_R(x) r_{\mu\nu} g_R^\dagger(x)\,,$$

$$l_{\mu\nu} = \partial_\mu l_\nu - \partial_\nu l_\mu - i[l_\mu, l_\nu]\,, \qquad l_{\mu\nu} \mapsto g_L(x) l_{\mu\nu} g_L^\dagger(x)\,,\tag{4.188}$$

where we have defined

$$\chi = 2B_0(s + ip)\,.\tag{4.189}$$

Here, B_0 is another constant with the dimension of mass.[11] Later, we shall relate this constant to the quark condensate.

As already briefly mentioned, in order to outline a systematic procedure for the construction of the Lagrangian, the preceding symmetry considerations should be supplemented by certain counting rules.[12] Stated differently, it is possible to construct infinitely many operators, which are invariant under the symmetry transformations. The Lagrangian is the sum of all these operators. It is clear that, if all operators had to be considered, the effective field theory would be useless. The EFT makes sense, if and only if there exists a principle that tells us that some operators are more important than the others. In other words, some hierarchy between these operators is introduced, making the infinite series of the operators tractable. We have already encountered examples of such *counting rules*. In particular, we have seen that in low-energy effective theories, the expansion parameter is the (small) momentum of the light particles, or mass thereof, divided by some hard scale in the theory. We may use the same philosophy here and assume all momenta p of the Goldstone bosons to be small (compared to the hard scale $\Lambda \simeq 1\,\mathrm{GeV}$). In other words, we shall count

[11] The symbols F_0, B_0 are reserved for the case of $N_f = 3$ flavors. For $N_f = 2$ flavors, we use the symbols F, B instead. These quantities are related to each other, since the two-flavor effective theory can be obtained from the three-flavor one by integrating out the strange quark. In what follows, this procedure will be considered in detail.

[12] We would like to emphasize from the beginning that here we exclusively discuss the counting rules for various operators in the Lagrangian. The counting rules for the S-matrix elements are more involved, since they also imply assigning chiral orders to loop diagrams, which are obtained by using this Lagrangian (cf. Section 1.8). We shall discuss these counting rules in detail in what follows.

$$U = O(p^0), \qquad D_\mu U = O(p). \tag{4.190}$$

Further, since the fields r_μ, l_μ enter in the covariant derivative along with the derivative ∂_μ, it is natural to count them as

$$r_\mu = O(p), \qquad l_\mu = O(p). \tag{4.191}$$

Finally, the quantity χ contains the scalar field $s = \mathcal{M} + \cdots$. According to the Gell–Mann–Oakes–Renner relation (see Eqs. (3.387) and (3.404)), the square of the Gold-stone boson masses is proportional to the quark masses; thus, a consistent counting emerges if we assume

$$\chi = O(p^2). \tag{4.192}$$

Note that the preceding counting rules are the same as in case of the σ-model; see Eq. (4.106).

Using the counting rules and symmetry considerations, the effective Lagrangian can be written as an infinite tower of operators at order $p^2, p^4, \ldots$ (note that, due to the Lorentz invariance, only even powers are allowed):

$$\mathcal{L} = \mathcal{L}^{(2)} + \mathcal{L}^{(4)} + \cdots. \tag{4.193}$$

It can be seen that, at a given order in p, only a finite number of operators emerge in the Lagrangian. For example, at leading order, there are two such operators,

$$\mathcal{L}^{(2)} = \frac{F_0^2}{4} \langle D^\mu U^\dagger D_\mu U + \chi^\dagger U + U^\dagger \chi \rangle, \tag{4.194}$$

where the brackets denote the trace in flavor space. Note that the overall constant, multiplying the Lagrangian, was chosen to normalize the kinetic term of the Goldstone boson terms to its canonical form. In the two-flavor case, the lowest-order Lagrangian has exactly the same form, with F_0, B_0 replaced by F, B. Often, $\mathcal{L}^{(2)}$ is called the LO (leading-order) Lagrangian.

The number of independent terms in the case of $N_f = 3$ and $N_f = 2$ flavors, however, differs already at order p^4. In the case of three flavors, the pertinent NLO (next-to-leading-order) Lagrangian is given by [37]:

$$\mathcal{L}^{(4)} = \sum_{i=1}^{10} L_i O_i + \sum_{i=1}^{2} H_i \tilde{O}_i, \tag{4.195}$$

with

$$O_1 = \langle D^\mu U^\dagger D_\mu U \rangle^2, \quad O_2 = \langle D_\mu U^\dagger D_\nu U \rangle \langle D^\mu U^\dagger D^\nu U \rangle,$$

$$O_3 = \langle D^\mu U^\dagger D_\mu U D^\nu U^\dagger D_\nu U \rangle, \quad O_4 = \langle D^\mu U^\dagger D_\mu U \rangle \langle \chi^\dagger U + \chi U^\dagger \rangle,$$

$$O_5 = \langle D^\mu U^\dagger D_\mu U (\chi^\dagger U + U^\dagger \chi) \rangle, \quad O_6 = \langle \chi^\dagger U + \chi U^\dagger \rangle^2,$$

$$O_7 = \langle \chi^\dagger U - \chi U^\dagger \rangle^2, \quad O_8 = \langle \chi^\dagger U \chi^\dagger U + \chi U^\dagger \chi U^\dagger \rangle,$$

$$O_9 = -i \langle r_{\mu\nu} D^\mu U D^\nu U^\dagger + l_{\mu\nu} D^\mu U^\dagger D^\nu U \rangle, \quad O_{10} = \langle U^\dagger r_{\mu\nu} U l^{\mu\nu} \rangle, \tag{4.196}$$

and

$$\tilde{O}_1 = \langle r_{\mu\nu}l^{\mu\nu} + l_{\mu\nu}l^{\mu\nu}\rangle, \quad \tilde{O}_2 = \langle \chi^\dagger \chi\rangle. \tag{4.197}$$

Here, L_i and H_i denote the low-energy and the high-energy constants, respectively. The high-energy constants multiply the operators that contain external sources only. The contribution from these operators to the S-matrix elements is a polynomial and thus is physically irrelevant. However, these terms are still needed to render all Green's functions finite, since they provide counterterms for the one-loop graphs.

In the case of $N_f = 2$, the standard choice of the operators is given in Ref. [19] (see also, e.g., Ref. [38], where matrix notations have been used instead of $O(4)$ vector notations):

$$\mathcal{L}^{(4)} = \sum_{i=1}^{7} l_i o_i + \sum_{i=1}^{3} h_i \tilde{o}_i, \tag{4.198}$$

with

$$o_1 = \frac{1}{4}\langle D^\mu U^\dagger D_\mu U\rangle^2, \quad o_2 = \frac{1}{4}\langle D_\mu U^\dagger D_\nu U\rangle\langle D^\mu U^\dagger D^\nu U\rangle,$$

$$o_3 = \frac{1}{16}\langle \chi^\dagger U + \chi U^\dagger\rangle^2, \quad o_4 = \frac{1}{4}\langle D^\mu U^\dagger D_\mu \chi + D^\mu \chi^\dagger D_\mu U\rangle,$$

$$o_5 = \langle U^\dagger r_{\mu\nu} U l^{\mu\nu}\rangle, \quad o_6 = \frac{i}{2}\langle r_{\mu\nu} D^\mu U D^\nu U^\dagger + l_{\mu\nu} D^\mu U^\dagger D^\nu U\rangle,$$

$$o_7 = \langle \chi^\dagger U - \chi U^\dagger\rangle^2, \tag{4.199}$$

and

$$\tilde{o}_1 = \frac{1}{4}\langle \chi^\dagger \chi\rangle + \frac{1}{2}\mathrm{Re}\,(\det \chi), \quad \tilde{o}_2 = \langle r_{\mu\nu} r^{\mu\nu} + l_{\mu\nu} l^{\mu\nu}\rangle,$$

$$\tilde{o}_3 = \frac{1}{4}\langle \chi^\dagger \chi\rangle - \frac{1}{2}\mathrm{Re}\,(\det \chi). \tag{4.200}$$

The two-flavor Lagrangian contains fewer terms, because the use of Cayley–Hamilton identities allows one to reduce the number of the independent operators. In order to demonstrate the idea behind this reduction on a simple example without further ado, we check that, for the 2×2 matrices, the operator O_3 from Eq. (4.196) is a linear combination of O_1 and O_2. Indeed, the operator O_3 can be rewritten as

$$O_3 = \langle X_\mu X^\mu X_\nu X^\nu\rangle, \quad X_\mu = U D_\mu U^\dagger. \tag{4.201}$$

Using now the relation

$$e^{-i\phi T}\partial_\mu e^{i\phi T} = i\partial_\mu \phi T - \frac{i^2}{2!}[\phi T, \partial_\mu \phi T] + \frac{i^3}{3!}[\phi T, [\phi T, \partial_\mu \phi T]] + \cdots, \tag{4.202}$$

we can straightforwardly verify that the Maurer–Cartan form X_μ must be a linear combination of the generators of the $SU(N_f)$ group T^i:

$$X_\mu = \sum_i T^i X_\mu^i. \tag{4.203}$$

Now, note that, in the case of the $SU(2)$ group, the generators are proportional to the Pauli matrices. Recalling the expression for the trace of the four Pauli matrices,

$$\langle \tau^i \tau^j \tau^k \tau^l \rangle = 2\delta^{ij}\delta^{kl} - 2\delta^{ik}\delta^{jl} + 2\delta^{il}\delta^{jk}, \tag{4.204}$$

one can easily check that O_3 indeed reduces to a linear combination of O_1 and O_2. On the other hand, in case of three flavors, the trace of four generators has a more complicated form, and O_1, O_2, O_3 are all linearly independent.

In contrast, it should be mentioned that $\det \chi$ is chirally invariant in the two-flavor case, but not in the three-flavor case. Hence, the $SU(2)$ effective Lagrangian contains three high-energy constants at $O(p^4)$ instead of two (cf. Eqs. (4.197) and (4.200)).

Finally, note that the generalization of the preceding procedure for the construction of the effective Lagrangian to higher orders in p is straightforward (but tedious). The construction of the pertinent meson Lagrangian at $O(p^6)$ is discussed, for example, in Ref. [39].

Once the Lagrangian is known, one can write the generating functional of the effective theory:

$$Z(s, p, v, a) = \int dU \, \exp\left\{ i \int d^4x \, \mathcal{L}(U, s, p, v, a) \right\}. \tag{4.205}$$

The generating functional is invariant under *local $SU(N_f)_L \times SU(N_f)_R$* transformations and thus obeys the same Ward identities as the generating functional of QCD (apart from the anomalies, as mentioned in the beginning of this section). By construction, it is the most general generating functional that has this property and, in addition, generates correct one- and many-particle singularities *in perturbation theory*, produced by the intermediate states containing Goldstone bosons. Consequently, according to Weinberg's theorem, the theory constructed in this manner is the effective field theory of QCD at low energy, where only the lightest degrees of freedom (the Goldstone bosons) matter. Note also that, as seen from Eq. (4.205), the value of the generating functional in the effective theory does not depend on the parameterization of the field U in terms of the Goldstone fields, cf. Eq. (4.187), because U is merely an integration variable in the path integral. Further, the generating functional of the effective theory is calculated perturbatively. The justification of this procedure is based on the fact that the interactions of Goldstone bosons become weak at small momenta.

All low-energy constants $F_0, B_0, L_i, \ldots$ and $F, B, l_i, \ldots$, whose values are not restricted by symmetry considerations, can, in principle, be expressed through the sole fundamental parameter Λ_{QCD} (say, by carrying out lattice simulations). We remind the reader of the special role played by the pion decay constant (cf. Section 2.14.4), as its nonvanishing is a necessary and sufficient condition for spontaneous symmetry breaking. There is, however, still a long way to go for pinning down all these LECs from lattice QCD. One the other hand, we may also treat these quantities as phenomenological parameters and try to extract their values from data directly. At present, the latter method is an important source of information about the numerical values of the low-energy constants. In the description of particular processes, ChPT will be useful, if the perturbation series converges fast, before a large number of the unknown low-energy constants at higher orders start to contribute, and the uncertainty in the results of the calculations blows up.

4.8 Tree-Level Calculations

From here on, we start to implement the program we have outlined for the calculation of various physical quantities. First, we consider the lowest order (LO), where only the Lagrangian $\mathcal{L}^{(2)}$ given in Eq. (4.194) contributes. All calculations will be carried out at tree level. As we shall see, loop diagrams build from the LO Lagrangian $\mathcal{L}^{(2)}$ start to contribute at order p^4.

4.8.1 The Quark Condensate

The quark condensate in the chiral limit is given by

$$\lim_{\mathcal{M} \to 0} \langle 0 | \bar{\psi}_f(0) \psi_f(0) | 0 \rangle = -N_f^{-1} \frac{1}{i} \frac{\delta Z(s, p, v, a)}{\delta s^0(0)} \bigg|_{s=p=v=a=0}, \qquad f = u, d, s.$$
(4.206)

On the other hand, the term with $s^0(x)$ in the lowest-order effective Lagrangian from Eq. (4.194) is given by

$$\mathcal{L}^{(2)} = 2B_0 s^0(x) \frac{F_0^2}{4} \langle U + U^\dagger \rangle + \cdots$$

$$= \left(B_0 F_0^2 N_f s^0(x) - 2B_0 s^0(x) \phi^i(x) \phi^i(x) + O(\phi^4) \right) + \cdots,$$
(4.207)

where the ellipses denote the terms that do not depend on $s^0(x)$. Note that, in order to obtain Eq. (4.207), we have used the exponential parameterization given in Eq. (4.187) and expanded the matrix U in powers of the field Φ.

It is easy to see that the powers of the $\phi^i(x)$ in the Lagrangian produce loop corrections to the quark condensate. For example, the contribution of the second term is proportional to $\sum_i \langle 0 | T \phi^i(x) \phi^i(x) | 0 \rangle = -i(N_f^2 - 1)D(0)$, where $D(0)$ is given by a closed pion loop (in the chiral limit, where pions are massless):

$$D(0) = -i \int \frac{d^D p}{(2\pi)^D} \frac{1}{-p^2}.$$
(4.208)

Note that the quantity $D(0)$ vanishes in dimensional regularization, because this is a no-scale integral. The same statement is true for two, three, or more loops, if dimensional regularization is used.

Using now Eqs. (4.206) and (4.207), we may straightforwardly read off

$$\lim_{\mathcal{M} \to 0} \langle 0 | \bar{\psi}_f(0) \psi_f(0) | 0 \rangle = -F_0^2 B_0 \qquad \text{(no sum over } f\text{)}.$$
(4.209)

Moreover, it is easy to see that this result will not change at higher orders. First of all, the loops will not contribute, because we always arrive at no-scale integrals. The higher-order terms in the Lagrangian could contribute powers of $\mathcal{M}$. However, all such contributions vanish in the chiral limit, $\mathcal{M} \to 0$.

Contrary to the condensate in the chiral limit, the condensate at $\mathcal{M} \neq 0$ cannot be related to observable quantities. This is easy to see, for example, from Eqs. (4.195) and (4.196). The contribution to the condensate from the operator $\tilde{O}_2$ is equal to $-2H_2 N_f^{-1} B_0^2 \operatorname{tr} \mathcal{M}$. Note that now that H_2 is a high-energy constant, it appears in front of a polynomial in Green's functions and therefore does not appear in the S-matrix elements (which are the residues at the *poles* of Green's functions). From this it follows that H_2 cannot be expressed through physical observables.

4.8.2 Goldstone Boson Masses

The Goldstone boson masses can be read off from the two-point function of two pseudoscalar sources:

$$
\int \frac{d^4 p}{(2\pi)^4 i} e^{-ip(x-y)} D^{ij}(p^2) = \langle 0|T(\bar{\psi}(x)i\gamma^5\lambda^i\psi(x))(\bar{\psi}(y)i\gamma^5\lambda^j\psi(y))|0\rangle
$$

$$
= \frac{\delta}{i\delta p^i(x)} \frac{\delta}{i\delta p^j(y)} Z(s,p,v,a)\Big|_{s=\mathcal{M},p=v=a=0}, \tag{4.210}
$$

where $\mathcal{M}$ is the quark mass matrix. Expanding the LO effective Lagrangian $\mathcal{L}^{(2)}$ up to the second order in the fields, we get

$$
\mathcal{L}^{(2)} = \frac{1}{2}\partial_\mu\phi^i\partial^\mu\phi^i
$$

$$
- \frac{B_0}{2}\Bigg\{ m_u\left(\phi^1\phi^1 + \phi^2\phi^2 + \phi^4\phi^4 + \phi^5\phi^5 + \phi^3\phi^3 + \frac{2\sqrt{3}}{3}\phi^3\phi^8 + \frac{1}{3}\phi^8\phi^8\right)
$$

$$
+ m_d\left(\phi^1\phi^1 + \phi^2\phi^2 + \phi^6\phi^6 + \phi^7\phi^7 + \phi^3\phi^3 - \frac{2\sqrt{3}}{3}\phi^3\phi^8 + \frac{1}{3}\phi^8\phi^8\right)
$$

$$
+ m_s\left(\phi^4\phi^4 + \phi^5\phi^5 + \phi^6\phi^6 + \phi^7\phi^7 + \frac{4}{3}\phi^8\phi^8\right)\Bigg\} + 2F_0 B_0 p^i\phi^i + \cdots. \tag{4.211}
$$

Here, we already assumed that $s = \mathcal{M}$ and $v_\mu = a_\mu = 0$.

The position of the poles in the two-point function is determined by the form of the second-order differential operator that acts on the Goldstone boson fields ϕ_i. However, if $m_u \neq m_d$, the third and the eighth components of the octet mix, and the resulting operator does not have a diagonal form. Defining the physical fields for the pions, the kaons and the η-meson via the following equations:[13]

---

[13] In case of no isospin breaking, this choice is consistent with the de Swart phase convention for the $SU(3)$ group; see, e.g., the textbook [40], Eq. (I.66). If the fields are chosen in this way, the one-particle states in the octet are normalized as $\langle 0|\phi^a|b\rangle = \delta^{ab}$, where a, b denote the physical states.

$$\pi^\pm = \frac{\mp\phi^1 + i\phi^2}{\sqrt{2}}, \qquad K^\pm = \frac{\mp\phi^4 + i\phi^5}{\sqrt{2}},$$

$$K^0 = \frac{-\phi^6 + i\phi^7}{\sqrt{2}}, \qquad \bar{K}^0 = \frac{-\phi^6 - i\phi^7}{\sqrt{2}},$$

$$\pi^0 = \cos\varepsilon\,\phi^3 + \sin\varepsilon\,\phi^8, \qquad \eta = -\sin\varepsilon\,\phi^3 + \cos\varepsilon\,\phi^8, \qquad (4.212)$$

the quadratic form in Eq. (4.211) can be diagonalized, if the mixing angle is chosen in the following way:

$$\tan 2\varepsilon = \frac{\sqrt{3}}{2}\frac{m_d - m_u}{m_s - \hat{m}}, \qquad \hat{m} = \frac{1}{2}(m_u + m_d). \qquad (4.213)$$

Rewriting Eq. (4.211) in terms of physical fields, we may easily read off the masses at the lowest order:

$$M_{\pi^\pm}^2 = (m_u + m_d)B_0,$$

$$M_{K^\pm}^2 = (m_u + m_s)B_0,$$

$$M_{K^0}^2 = M_{\bar{K}^0}^2 = (m_d + m_s)B_0,$$

$$M_{\pi^0}^2 = (m_u + m_d)B_0 - \frac{4}{3}(m_s - \hat{m})B_0\frac{\sin^2\varepsilon}{\cos 2\varepsilon},$$

$$M_\eta^2 = \frac{2}{3}(\hat{m} + 2m_s)B_0 + \frac{4}{3}(m_s - \hat{m})B_0\frac{\sin^2\varepsilon}{\cos 2\varepsilon}. \qquad (4.214)$$

From these equations it becomes evident that within Chiral Perturbation Theory one can make precise statements about quark mass ratios only; see, for example, Refs. [41, 42]. In the isospin symmetric case, $m_u = m_d$, this reproduces the result given in Eq. (3.387). Of course, this result is valid at the lowest order in the quark mass expansion and receives corrections at higher orders. From these relations, it is obvious that $m_s \gg m_u, m_d$, as

$$\frac{m_s}{m_u + m_d} = \frac{m_s}{2\hat{m}} = \frac{M_{K^\pm}^2 + M_{K^0}^2 - M_{\pi^\pm}^2}{M_{\pi^\pm}^2} \simeq 24. \qquad (4.215)$$

Note also that the quantity $\sin^2\varepsilon/\cos 2\varepsilon$ contributes at the second order in the isospin breaking; so, for example, the charged and neutral pion mass splitting in QCD is of order $(m_d - m_u)^2$. The reason for this is the absence of the d-symbol in the $SU(2)$ Lie algebra; see Exercise 18.

4.8.3 Pion Decay Constant

For simplicity, let us assume that quark masses are equal and work in the basis with the Hermitean fields ϕ^i. If needed, the case with different quark masses can be addressed along the same pattern. The two-point function with one axial-vector and one pseudoscalar external leg is given by

$$\int \frac{d^4 p}{(2\pi)^4 i} e^{-ip(x-y)} A_\mu^{ij}(p) = \langle 0|T(\bar{\psi}(x)i\gamma_5\lambda^i\psi(x))(\bar{\psi}(y)\gamma_\mu\gamma^5 T^j\psi(y))|0\rangle$$

$$= \frac{\delta}{i\delta p^i(x)} \frac{\delta}{i\delta a_\mu^j(y)} Z(s,p,v,a)\bigg|_{s=\mathcal{M}, p=v=a=0}. \tag{4.216}$$

Let us insert a one-particle state between two quark bilinears and consider the limit $p^2 \to M^2$, with M the Goldstone boson mass. Recalling the definition of the pion decay constant,

$$-iF_\pi q_\mu \delta^{ij} e^{iqy} = \langle \pi^i(q)|\bar{\psi}(y)\gamma_\mu\gamma^5 T^j\psi(y)|0\rangle, \tag{4.217}$$

we get

$$A_\mu^{ij}(p) = p_\mu \delta^{ij} A(p^2)$$

$$= \int d^4x e^{ipx} \int \frac{d^3\mathbf{q}}{(2\pi)^3 2q^0} (\theta(x^0)e^{-iqx} + \theta(-x^0)e^{iqx})(-iq_\mu)\delta^{ij}Z^{1/2}F_\pi + \cdots, \tag{4.218}$$

where

$$Z^{1/2}\delta^{ij}e^{iqy} = \langle \pi^i(q)|\bar{\psi}(y)i\gamma^5\lambda^j\psi(y)|0\rangle, \tag{4.219}$$

and the ellipses denote the contributions of multiparticle states.

Performing the integral in Eq. (4.218), we get

$$A(p^2) = \frac{-Z^{1/2}F_\pi}{M^2 - p^2} + \cdots. \tag{4.220}$$

The terms, which are not explicitly shown, are regular at $p^2 \to M^2$.

On the other hand, calculating this two-point function at lowest order in the effective theory, we have to expand the Lagrangian in the presence of the sources p^i and a_μ^i:

$$\mathcal{L}^{(2)} = \frac{1}{2}\partial_\mu\phi^i\partial^\mu\phi^i - \frac{M^2}{2}\phi^i\phi^i + 2F_0 B_0 p^i\phi^i - F_0\partial^\mu\phi^i a_\mu^i + \cdots. \tag{4.221}$$

Then, using Eq. (4.216), it is seen that at LO, the quantity $A(p^2)$ takes the form

$$A(p^2) = \frac{-(2F_0 B_0)F_0}{M^2 - p^2}. \tag{4.222}$$

Further, calculating the quantity $D^{ij}(p^2)$ from Eq. (4.210) again at LO, we get:

$$D^{ij}(p^2) = \delta^{ij}\frac{(2F_0 B_0)^2}{M^2 - p^2}. \tag{4.223}$$

This means that the wave function renormalization constant in Eq. (4.219) is equal to

$$Z^{1/2} = 2B_0 F_0. \tag{4.224}$$

From this, we finally conclude that

$$F_\pi = F_0 \tag{4.225}$$

at leading order in the chiral expansion. Note that at higher orders we have to differentiate between the various meson decay constants F_π, F_K and F_η, because the loop corrections to the physical matrix elements break $SU(3)$ unitary symmetry, if the particles running in the loops have different masses.

4.8.4 Pion–Pion Scattering Lengths

According to the LSZ formalism, which was discussed at length in Section 4.2, in order to calculate the $\pi\pi$ scattering S-matrix element we have to start from the Green's function of quark bilinears. This quantity can be readily evaluated from the generating functional by differentiating it with respect to the external sources $p^i(x)$. After this, we have to single out the poles in the external legs and extract the residues. Finally, we have to multiply each external leg with $Z^{-1/2}$, where Z is the wave function renormalization constant, given by Eq. (4.219). In practice, in calculations within the effective theory framework, we use a shortcut: we set the external sources to zero and calculate an S-matrix element that corresponds to the transition between particles described by the elementary fields ϕ^i. We wish to show here that this leads to the correct answer. This statement, naturally, is not restricted to $\pi\pi$ scattering and applies to the calculation of a generic S-matrix element with any number of external legs.

For simplicity here, we restrict ourselves to $N_f = 2$ flavors and speak about pions (of course, the argument does not depend on the number of flavors). One can read off from the Lagrangian that the single pseudoscalar external source couples to $1, 3, 5, \ldots$ pion fields. At higher orders, there are terms, where two and more external sources couple at the same point. Correspondingly, we could group the Feynman diagrams, describing the Green's function with external sources, into distinct classes; see Fig. 4.7. Namely, a given source can couple to the rest of a diagram either through an exchange of a single dressed pion (the so-called one-particle reducible diagrams a,b), or could be connected to the rest of the diagram with multiple pion propagators (the one-particle irreducible diagram c). Several sources can also meet at the same vertex, which is connected to the rest of the diagram with one or several pion lines (diagram d). These are all possible Feynman diagram topologies. It should be also noted that the vertex, which describes the coupling of a pion to an external source, receives corrections from pion loops, as well as tree-level corrections from the higher-order Lagrangians.

The complicated structure of the Green's functions with external currents considerably simplifies near the mass shell of the external particles. Namely, only one-particle irreducible diagrams produce a pole, whereas the diagrams with multiparticle states, or with more than one external source, are regular in the vicinity of the mass shell. Consequently, the diagrams of types c and d do not contribute on the mass shell.

Next, let us consider the behavior of the remaining contributions in the vicinity of a pole. The dressed propagator of a pion *field*, shown in diagrams a and b as a solid line with a shaded blob, close to the mass shell takes the form

$$D_\pi^{ij}(p^2) \to \frac{Z_\pi \delta^{ij}}{M_\pi^2 - p^2} + \text{regular terms}, \tag{4.226}$$

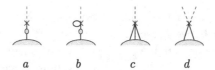

Figure 4.7 Attaching the external pseudoscalar source, depicted by a dashed line, to a generic diagram. Distinct topologies are possible. For example, the source can couple via the exchange of a single dressed pion (diagram a), or through one-particle irreducible diagrams (diagram c). Possible corrections in the vertex also should be included (diagram b). Finally, two or several external sources can meet at one point in the diagram (diagram d). The solid lines correspond to pions.

Figure 4.8 The same as in Fig. 4.7 for the particular case of a two-point function. Both one-particle reducible (a) and irreducible (b) diagrams are present. All vertex corrections should be explicitly included.

where Z_π is the wave function renormalization constant of the pion *field,* which is expressed in a standard way through the derivative of the self-energy (the shaded blob in the diagrams of Fig. 4.7a,b), evaluated on the mass shell. However, the external leg, corresponding to the external source, carries another factor, which is described by the sum of all corrections to the vertex of the external source coupling to a pion field. For diagram b, this is symbolically depicted by a single pion loop, attached to the vertex. The sum of all such diagrams, evaluated on the mass shell, gives a constant, which we denote as Z_χ. Finally, one concludes that each external leg in a diagram gives a factor $\delta^{ij} Z_\pi Z_\chi / (M_\pi^2 - p^2)$ plus terms that are regular in the vicinity of the mass shell.

It remains to calculate the wave function renormalization factor. To this end, we consider the two-point function of the pseudoscalar currents, shown in Fig. 4.8. Again, there are one-particle reducible (a) and one-particle irreducible (b) diagrams. Only the former can lead to a pole. Repeating the arguments just given, it is seen that the two-point function of the pseudoscalar currents has the following behavior in the vicinity of the pole (note that there are one self-energy blob and two vertices):

$$D^{ij}(p^2) \to \frac{\delta^{ij} Z_\pi Z_\chi^2}{M_\pi^2 - p^2} + \text{regular terms.} \tag{4.227}$$

Further, according to the LSZ formula discussed in Section 4.2, in order to arrive at the S-matrix elements, in the Green's function one should multiply each external leg with $(M_\pi^2 - p^2)$ together with the normalization factor, which, according to Eq. (4.227), is equal to $(Z_\pi Z_\chi^2)^{-1/2} = Z_\pi^{-1/2} Z_\chi^{-1}$. Now, we see that this prescription boils down to simply removing all external legs in the diagrams (i.e., considering proper diagrams with pion *fields*) and multiplying the result by $Z_\pi Z_\chi Z_\pi^{-1/2} Z_\chi^{-1} = Z_\pi^{1/2}$. This is exactly the prescription used to calculate the S-matrix in case of elementary pion fields. Note also that Z_χ completely cancels from the final expressions.

At this stage, we may ask why we made such a detour, introducing first the external sources and letting them tend to zero at the end. It could be easier to write the invariant Lagrangian in terms of the pion fields from the beginning. The reason why this is not done in this way is simple. The CCWZ formalism allows us to construct Lagrangians in the case of the exact symmetry, but it cannot account for an explicit symmetry breaking. In order to illustrate this point, it is instructive to look back to the early days of chiral effective field theories, when mutually contradicting predictions about the $\pi\pi$ scattering lengths have been made, based on a different form of the symmetry-breaking term [32, 43, 44]. For example, in Ref. [43], three models were introduced, differing by a choice of the parameterization of the interpolating pion field. The Lagrangian in these models took the form:

$$\mathcal{L} = \frac{1}{2}\partial_\mu \boldsymbol{\pi}\partial^\mu \boldsymbol{\pi} - \frac{M_\pi^2}{2}\boldsymbol{\pi}^2 + \frac{(\boldsymbol{\pi}\cdot\partial_\mu\boldsymbol{\pi})(\boldsymbol{\pi}\cdot\partial^\mu\boldsymbol{\pi})}{2F_\pi^2} - \beta\frac{M_\pi^2(\boldsymbol{\pi}^2)^2}{4F_\pi^2} + \cdots, \qquad (4.228)$$

where the parameter β that took different values in different models, leading to different values of the $\pi\pi$ scattering lengths at tree level. It is also obvious that the parameter β cannot be fixed, if we start from the symmetric Lagrangian containing elementary pion fields, since the symmetry considerations alone do not fix the explicit symmetry breaking pattern.

Introducing the external sources solves the problem, because this turns it into a symmetric problem again. The symmetric Lagrangian can be constructed at any order, using the CCWZ formalism. A crucial point is that the explicit symmetry breaking pattern in QCD is known, it proceeds through quarks acquiring small masses. This corresponds to setting the scalar source equal to the quark mass matrix. Thus, the spurion formalism allows one to translate the knowledge about the explicit symmetry breaking pattern from QCD into the effective field theory, and to calculate the explicit symmetry breaking effect not only at leading order, but in higher orders as well. In particular, this formalism fixes the correct value of the parameter, $\beta = 1/2$, which reproduces the lowest-order Lagrangian in Eq. (4.92). The lowest-order $\pi\pi$ scattering amplitude, obtained from this Lagrangian, is given by Eqs. (4.64) and (4.69).

If u and d quark masses are equal, isospin is conserved. The general pion–pion scattering amplitudes in the channels with total isospin $I = 0, 1, 2$ can be expressed through one single amplitude A in Eq. (4.64):

$$T^0(s,t) = 3A(s,t,u) + A(t,u,s) + A(u,s,t),$$

$$T^1(s,t) = A(t,u,s) - A(u,s,t),$$

$$T^2(s,t) = A(t,u,s) + A(u,s,t), \qquad (4.229)$$

since the total isospin of the $\pi\pi$-system can be $I = 0, 1, 2$. On the mass shell, these amplitudes depend on two kinematical variables, say, s, t, whereas $u = 4M_\pi^2 - s - t$. The corresponding partial-wave expansion takes the form

$$T^I(s,t) = 32\pi \sum_{\ell=0}^{\infty} (2\ell+1) P_\ell(\cos\theta) t_\ell^I(s),$$

$$s = 4(M_\pi^2 + p^2), \qquad t = -2p^2(1 - \cos\theta), \tag{4.230}$$

and ℓ labels the angular momentum of the $\pi\pi$-system, $\ell = 0, 1, 2, \ldots$. Further, unitarity implies that

$$t_\ell^I = \frac{\sqrt{s}}{4ip} \left(e^{2i\delta_\ell^I(s)} - 1 \right), \tag{4.231}$$

where $\delta_\ell^I(s)$ denotes the (real) $\pi\pi$ scattering phase shifts.

In the S-wave ($\ell = 0$), the expansion of the phase shift near threshold takes the form (the well-known effective range expansion)

$$p \cot \delta_0^I(s) = -\frac{1}{A^I} + \frac{1}{2} r^I p^2 + O(p^4), \tag{4.232}$$

where A^I and r^I stand for the S-wave scattering length and the effective range, respectively.

In case of the $\pi\pi$ scattering, it is customary to define the dimensionless scattering length $a_I = -M_\pi A^I$. At LO in chiral expansion, using Eq. (4.69), one obtains [32]:

$$a_0 = \frac{7M_\pi^2}{32\pi F_\pi^2}, \qquad a_2 = -\frac{M_\pi^2}{16\pi F_\pi^2}. \tag{4.233}$$

Numerically, this leads to $a_0 = 0.16$ and $a_2 = -0.046$.

4.9 Chiral Perturbation Theory at One Loop

4.9.1 The Role of Unitarity

Up to now, we have exclusively restricted ourselves to the tree approximation in the effective theory. The story obviously cannot end here, because, for example, tree-level amplitudes are real and thus break unitarity. So, in order to arrive at a consistent theory, loops should be considered and renormalization should be carried out (as such loop diagrams are, in general, UV divergent). This is a nontrivial issue, because non-linear Lagrangians describe local interactions that were termed "non-renormalizable" in the old terminology. That means that S-matrix elements cannot be made finite by adding a *finite* number of operators to the Lagrangian. Since the total number of such terms is necessarily infinite, the issue of the renormalization is closely tied to the validity of the counting rules, which state that, *at a given order in the chiral expansion*, only a finite number of terms in the Lagrangian is sufficient for the calculation of all S-matrix elements. This, indeed, is the modern view of renormalization, which states that all QFTs are indeed EFTs, as we do not know (and do not need to know) what happens at very high energies; see, for example, Ref. [45].

In the following sections we shall consider a faithful perturbative expansion of the S-matrix elements with the use of chiral Lagrangians, as well as the systematic removal of the UV divergences by counterterms, which are contained in these Lagrangians. As an appetizer, in this section we shall derive the next-to-leading-order $\pi\pi$ scattering amplitude, based on unitarity and analyticity alone, without resorting to the Feynman diagram technique. This has been done first in Ref. [46] (for a further development in this direction, namely, the so-called reconstruction theorem, we refer the reader to Ref. [47]; see also Ref. [48]). In particular, it will be seen that these general arguments together with chiral symmetry lead to the conclusion that chiral logarithms should be present in the amplitude. Also, in the derivation, we shall become familiar with the chiral power counting for the amplitudes at an intuitive level, which will be useful for understanding the more formal definitions that will be introduced in the next sections. For further discussion of the role of unitarity in Chiral Perturbation Theory, see, for example, Ref. [49] (and references therein).

To make the formulae more compact, we take $N_f = 2$ and assume the chiral limit $M_\pi = 0$. Let us consider the scattering process from the initial state $|i\rangle = |\pi^i(p_1)\pi^j(p_2); in\rangle$ into the final state $|f\rangle = |\pi^k(p_3)\pi^l(p_4); out\rangle$ (here, i, j, k, l denote the isospin indices of the pions). The unitarity condition for the scattering amplitude is

$$\frac{1}{2i}\langle f|T - T^\dagger|i\rangle = \frac{1}{2}\sum_n \langle f|T|n\rangle (2\pi)^4\delta^{(4)}(p_1 + p_2 - p_n)\langle n|T^\dagger|i\rangle. \tag{4.234}$$

The sum over the intermediate states is given by

$$\sum_n(\cdots) = \sum_{n=2}^\infty \sum_{m_1,\cdots,m_n} \frac{1}{n!} \int \frac{d^3\mathbf{q}_1}{(2\pi)^3 2q_1^0} \cdots \frac{d^3\mathbf{q}_n}{(2\pi)^3 2q_n^0} (\cdots), \tag{4.235}$$

where $q_1,\ldots,q_n$ and $m_1,\ldots,m_n$ denote the four-momenta and isospin indices of the intermediate pions, respectively.

Further, it will be convenient to rewrite the unitarity condition in terms of the connected amplitudes, repeating the arguments from Section 4.2. The reason for this is that the chiral counting rules, which we shall derive in the following, apply only to the connected matrix elements. Consider, for instance, the set of all Feynman diagrams, contributing to the amplitude $\langle f|T|n\rangle$. The disconnected piece will consist of a subset that describes a free propagation of a particle with a momentum $p_1 = q_k$, $k = 1,\ldots,n$, which is not attached to the rest of the diagram. Symmetrizing with respect to $p_1 \leftrightarrow p_2$ yields a factor $1/2!$ and, finally, we may write

$$\langle f|T|n\rangle = \frac{1}{2!}\sum_{a=1,2}\sum_{b=1,\cdots,n} (2\pi)^3 2q_b^0\delta^{(3)}(\mathbf{p}_a - \mathbf{q}_b)\delta^{i_a m_b}\langle f_a|T|n_b\rangle_c + \langle f|T|n\rangle_c. \tag{4.236}$$

Here, the subscript c denotes the connected amplitude, and the states $|f_a\rangle$, $|n_k\rangle$ are obtained from the states $|f\rangle$, $|n\rangle$ by extracting the particles $\pi^{i_a}(p_a)$ and π^{m_b}/q_b), respectively. Using the preceding expression in the unitarity condition, we may carry out the integration over the δ-functions and rewrite this condition in terms of the connected amplitudes only. We do not display the final (rather voluminous) result here, which will not be needed. Note only that the integration over the δ-functions effectively reduces the number of particles in the intermediate state.

Next, let us turn to the power counting. As we know, the tree-level amplitudes are of order p^2 in chiral counting. Now, let us demonstrate that analyticity and unitarity allow one to evaluate the amplitude at order p^4 (the meson–meson scattering amplitudes contain even powers of p only, so that any $O(p^3)$ term is absent). To this end, let us first calculate the imaginary part at this order. It is straightforward to see that all amplitudes $2 \to n$ are of order p^2. Indeed, since each pion field in the Lagrangian is accompanied by $1/F_\pi$, see, for example, Eqs. (4.88) and (4.89), the *connected* amplitudes $\langle f|T|n \rangle_c$ and $\langle n|T^\dagger|i \rangle_c$ can be expanded in a power series in this parameter, with coefficients that depend only on pion momenta and masses, that is, they do not contain a heavy scale anymore.[14] It can be checked that the first term in this expansion is proportional to $1/F_\pi^n$. Further, let us consider the unitarity relation in Eq. (4.235). The mass dimension of the n-particle phase space and the δ-function are $2n$ and -4, respectively. From this, we may conclude that these count as p^{2n} and p^{-4} in chiral counting, because there is no heavy scale present there. Now, since the left-hand side of Eq. (4.234) is dimensionless, it is immediately seen that the expansion of the n-particle amplitudes should start at M^2/F_π^n, where M denotes some small scale. Consequently, $\langle f|T|n \rangle_c = O(p^2)$ and $\langle n|T^\dagger|i \rangle_c = O(p^2)$. Putting everything together, one finally sees that, working at $O(p^4)$ in the chiral expansion, it suffices to truncate the unitarity condition, retaining only the terms with $n = 2$ (after rewriting this condition in terms of the connected diagrams). The result has the following form:

$$\frac{1}{2i} \left(T^{ij,kl}(p_1, p_2; p_3, p_4) - (T^{kl,ij}(p_3, p_4; p_1, p_2))^* \right)$$

$$= \frac{1}{4} \int \sum_{rs} \frac{d^3\mathbf{q}_1}{(2\pi)^3 2q_1^0} \frac{d^3\mathbf{q}_2}{(2\pi)^3 2q_2^0} (2\pi)^4 \delta^{(4)}(p_1 + p_2 - q_1 - q_2)$$

$$\times T^{rs,kl}(q_1, q_2; p_3, p_4)(T^{rs,ij}(q_1, q_2; p_1, p_2))^*$$

$$+ \frac{1}{4} \int \sum_{rs} \frac{d^3\mathbf{q}_1}{(2\pi)^3 2q_1^0} \frac{d^3\mathbf{q}_2}{(2\pi)^3 2q_2^0} (2\pi)^4 \delta^{(4)}(p_1 - p_3 - q_1 - q_2)$$

$$\times T^{rs,jl}(q_1, q_2; -p_2, p_4)(T^{rs,ik}(q_1, q_2; p_1, -p_3))^*$$

$$+ \frac{1}{4} \int \sum_{rs} \frac{d^3\mathbf{q}_1}{(2\pi)^3 2q_1^0} \frac{d^3\mathbf{q}_2}{(2\pi)^3 2q_2^0} (2\pi)^4 \delta^{(4)}(p_1 - p_4 - q_1 - q_2)$$

$$\times T^{rs,kj}(q_1, q_2; p_3, -p_2)(T^{rs,il}(q_1, q_2; p_1, -p_4))^*. \tag{4.237}$$

---

[14] As we shall see in the next sections, this argument is valid in dimensional regularization (or any mass-independent regularization scheme). Using, e.g., cutoff regularization would destroy this power counting, because the cutoff plays the role of the hard scale in the coefficients. In the following, let us assume that dimensional regularization is used.

Here, the first, second and third terms correspond to the s-, t- and u-channel contributions, respectively. It can be straightforwardly checked that the t- and u-channel contributions emerge from the disconnected parts.

What remains is to substitute the $O(p^2)$ expression for the scattering amplitude, given by Eq. (4.69),

$$A(s,t,u) = \frac{s}{F_\pi^2},\tag{4.238}$$

into the right-hand side of Eq. (4.237) and calculate the left-hand side at $O(p^4)$. In order to simplify the calculation, note that the amplitude $T^{ij,kl}(p_1,p_2;p_3,p_4)$ can be decomposed into parts with total isospin $I = 0,1,2$:

$$T^{ij,kl}(p_1,p_2;p_3,p_4) = P_0^{ij,kl}T^0(s,t,u) + P_1^{ij,kl}T^1(s,t,u) + P_2^{ij,kl}T^2(s,t,u),\tag{4.239}$$

where the projectors onto the states with definite isospin subject to the constraint $P_I^{ij,rs}P_{I'}^{rs,kl} = \delta_{II'}P_I^{ij,kl}$ are given by

$$P_0^{ij,kl} = \frac{1}{3}\delta^{ij}\delta^{kl},$$

$$P_1^{ij,kl} = \frac{1}{2}(\delta^{ik}\delta^{jl} - \delta^{il}\delta^{jk}),$$

$$P_2^{ij,kl} = \frac{1}{2}(\delta^{ik}\delta^{jl} + \delta^{il}\delta^{jk}) - \frac{1}{3}\delta^{ij}\delta^{kl}.\tag{4.240}$$

Therefore, the scattering amplitudes with a definite isospin at $O(p^2)$ are

$$T^0(s,t) = \frac{1}{F_\pi^2}(3s+t+u) = \frac{2s}{F_\pi^2},$$

$$T^1(s,t) = \frac{1}{F_\pi^2}(t-u),$$

$$T^2(s,t) = \frac{1}{F_\pi^2}(t+u) = -\frac{s}{F_\pi^2}.\tag{4.241}$$

These expressions should be substituted into Eq. (4.237), and the integrations and summations over isospin indices should be carried out. To this end, it is convenient to expand the amplitudes in the partial waves with the use of Eq. (4.230) and the relations $s = 4p^2$, $t = -2p^2(1-\cos\theta)$, $u = -2p^2(1+\cos\theta)$. The result is

$$t_0^0(s) = \frac{s}{16\pi F_\pi^2}, \qquad t_1^1(s) = \frac{s}{96\pi F_\pi^2}, \qquad t_0^2(s) = -\frac{s}{32\pi F_\pi^2},\tag{4.242}$$

and all other partial-wave amplitudes vanish at this order.

Using now the definition

$$\frac{1}{2i}\left(T^{ij,kl}(p_1,p_2;p_3,p_4) - (T^{kl,ij}(p_3,p_4;p_1,p_2))^*\right)$$

$$= \delta^{ij}\delta^{kl}\mathrm{Im}A(s,t,u) + \delta^{ik}\delta^{jl}\mathrm{Im}A(t,u,s) + \delta^{il}\delta^{jk}\mathrm{Im}A(u,s,t),\tag{4.243}$$

a straightforward calculation finally yields

$$\text{Im}A(s,t,u) = \frac{1}{32\pi F_\pi^4}\left(s^2\theta(s) + \frac{1}{3}t(t-u)\theta(t) + \frac{1}{3}u(u-t)\theta(u)\right),\qquad(4.244)$$

where $\theta(x)$ denotes the Heaviside function.

To summarize, knowing the amplitude at $O(p^2)$ and using only unitarity, we are able to derive the imaginary part of the amplitude at next-to-leading order. Now, we may use the known analytic properties of the amplitude and restore the real part thereof at $O(p^4)$ as well. To this end, it suffices to ask, which analytic function has the imaginary part given by Eq. (4.244)? Adding the leading-order term, which is real, the answer at $O(p^4)$ can be written down immediately [46]:

$$A(s,t,u) = \frac{s}{F_\pi^2} - \frac{1}{32\pi^2 F_\pi^4}\left(s^2\ln\frac{-s-i\varepsilon}{\mu_1^2}\right.$$

$$\left. + \frac{1}{3}t(t-u)\ln\frac{-t-i\varepsilon}{\mu_2^2} + \frac{1}{3}u(u-t)\ln\frac{-u-i\varepsilon}{\mu_2^2}\right),\qquad(4.245)$$

where μ_1, μ_2 are two arbitrary parameters with dimension of mass. We remind the reader that, due to Bose–symmetry, $A(s,t,u)$ is symmetric with respect to $t \leftrightarrow u$.

We finally note that dispersion relations can restore the real part of the amplitude up to a polynomial with arbitrary coefficients. The most general second-order polynomial of s,t,u, symmetric under $t \leftrightarrow u$, is given by $as^2 + b(t-u)^2$, with a and b arbitrary constants. Stated differently, the amplitude at $O(p^4)$ depends on two arbitrary constants, which are not fixed by analyticity, unitarity and symmetries. As we will see, the role of these constants in the effective field theory framework is played by the $O(p^4)$ couplings in the Lagrangian. This freedom is, however, already taken into account in the expressions of the amplitude in Eq. (4.245), which depends on two arbitrary scales denoted by μ_1, μ_2. For this reason, no additional polynomial is added.

As will become clear in the following sections, the expression of the $\pi\pi$ scattering amplitude at $O(p^4)$ can be obtained within perturbation theory at one loop in a straightforward manner and with less effort. The reason why the preceding detailed derivation has been presented is to provide the reader with an alternative view, which could be useful for studying more complicated problems in the EFT framework. For example, it is seen that the emergence of the chiral logarithms at the next-to-leading order in not an accident, but a direct consequence of unitarity, analyticity, crossing symmetry and chiral symmetry. Moreover, at $O(p^4)$, the coefficient of this chiral logarithm can be predicted in a model-independent manner. On the contrary, the polynomial part of the amplitude contains two arbitrary parameters, which cannot be fixed through general arguments.

Finally, we would like to note that the higher-order corrections, which are required by unitarity, indeed improve the agreement of the theory with experiment (here, we refer to the full theory with nonzero quark masses). Indeed, as already mentioned, at order p^2 the scattering lengths are $a_0 = 0.16$ and $a_2 = -0.046$, which agrees reasonably with the experiment. However, higher-order corrections to a_0 (but not a_2) are sizable; for example, calculations at order p^4 give $a_0 = 0.20$ [19]. A combined analysis, which

Figure 4.9 Self-energy of the pion at order p^4. There are two types of diagrams: (a) the one-loop tadpole diagram, generated from the leading-order Lagrangian $\mathcal{L}^{(2)}$, and (b) the tree diagram generated by the next-to-leading-order Lagrangian $\mathcal{L}^{(4)}$.

uses the results of calculations at order p^6, together with the constraints imposed by analyticity and unitarity, gives a prediction that is accurate to better than 3 %, namely $a_0 = 0.220 \pm 0.005$ [50]. This prediction agrees very well with the present experimental results, obtained, for example, from the study of the $K \to 3\pi$ decays [51]:

$$(a_0 - a_2) = 0.2571 \pm 0.0048 \text{ (stat.)} \pm 0.0025 \text{ (syst.)} \pm 0.0014 \text{ (ext.)},$$

$$a_2 = -0.024 \pm 0.013 \text{ (stat.)} \pm 0.009 \text{ (syst.)} \pm 0.002 \text{ (ext.)}, \tag{4.246}$$

as well as from the study of the K_{e4} decays [52]:

$$a_0 = 0.2220 \pm 0.0128 \text{ (stat.)} \pm 0.0050 \text{ (syst.)} \pm 0.0037 \text{ (th.)},$$

$$a_2 = -0.0432 \pm 0.0086 \text{ (stat.)} \pm 0.0034 \text{ (syst.)} \pm 0.0028 \text{ (th.)}. \tag{4.247}$$

In particular, it is immediately seen from these results that the present theoretical uncertainty is smaller than the experimental one.

4.9.2 Loops and Power Counting

In this section, we discuss a systematic perturbative expansion of the S-matrix elements by using the effective chiral Lagrangian. Let us again start from the simplest example and calculate the pion mass at order p^4, which is given by the pole position in the two-point function of pseudoscalar sources at this order. (For simplicity, we consider here the case of $N_f = 2$ flavors and assume isospin symmetry $m_u = m_d = \hat{m}$.) According to the discussion in Section 4.8.4, we may instead calculate the two-point function of the elementary pion fields, as the pole position is the same. This two-point function is given by

$$\delta^{ij} D_\pi(p^2) = i \int d^4x e^{ipx} \langle 0|T\phi^i(x)\phi^j(0)|0\rangle = \frac{\delta^{ij}}{M^2 - p^2 - \Sigma_\pi(p^2)}, \tag{4.248}$$

where $M^2 = 2B\hat{m}$ is the pion mass at $O(p^2)$ (the exact Gell–Mann–Oakes–Renner relation; see Eq. (4.214)), and $\Sigma_\pi(p^2)$ denotes the self-energy of the pion. At order p^4, there are two types of contributions to the self-energy, depicted in Fig. 4.9. These are the pion loop, generated by the $O(p^2)$ Lagrangian (see Eq. (4.194)), and the tree contribution from the $O(p^4)$ Lagrangian given in Eqs. (4.198) and (4.199):

$$\Sigma_\pi(p^2) = \Sigma_a(p^2) + \Sigma_b(p^2), \tag{4.249}$$

$$\Sigma_a(p^2) = \frac{p^2}{F^2} I_0 + \frac{1}{F^2} I_2 - \frac{5M^2}{2F^2} I_0 \,, \tag{4.250}$$

$$\Sigma_b(p^2) = -\frac{2M^4}{F^2} l_3 \,. \tag{4.251}$$

Here, I_0 and I_2 correspond to the pion tadpoles:

$$I_0 = \int \frac{d^D k}{(2\pi)^D i} \frac{1}{M^2 - k^2} \,, \tag{4.252}$$

$$I_2 = \int \frac{d^D k}{(2\pi)^D i} \frac{(ik_\mu)(-ik^\mu)}{M^2 - k^2} = M^2 I_0 \,. \tag{4.253}$$

The physical mass at this order is given by

$$M_P^2 = M^2 - \Sigma_\pi(M^2) = M^2 + \frac{M^2}{2F^2} I_0 + \frac{2M^4}{F^2} l_3 \,. \tag{4.254}$$

Further, carrying out the calculation of the tadpole I_0 in dimensional regularization, we get

$$I_0 = 2M^2 \lambda + \frac{M^2}{16\pi^2} \ln \frac{M^2}{\mu^2} \,, \tag{4.255}$$

where λ diverges as $D \to 4$:

$$\lambda = \frac{\mu^{D-4}}{16\pi^2} \left(\frac{1}{D-4} - \frac{1}{2} (\Gamma'(1) + \ln 4\pi + 1) \right) \,. \tag{4.256}$$

Note that the divergent quantity L, which was defined in Eq. (1.89), is similar but not identical to λ. For historical reasons, we shall use λ in ChPT calculations.

Now, from Eqs. (4.254) and (4.255) it is clear that l_3 should contain the UV divergence that absorbs the divergence from the loops. Namely,

$$l_3 = -\frac{1}{2} \lambda + l_3^r(\mu) \,, \tag{4.257}$$

where $l_3^r(\mu)$ is the finite (renormalized) part of the low-energy coupling. Using the preceding equation, the expression for the pion mass at order p^4 can be rewritten in the following fashion:

$$M_P^2 = M^2 + \frac{M^4}{F^2} \left(\frac{1}{32\pi^2} \ln \frac{M^2}{\mu^2} + 2l_3^r(\mu) \right) \,. \tag{4.258}$$

It is straightforward to check that the right-hand side of the preceding equation does not depend on the renormalization scale μ: the explicit dependence on μ in the logarithm is exactly compensated by the μ-dependence of the low-energy constant $l_3^r(\mu)$.

This short calculation serves as an example for more complicated cases. We do not consider these cases in detail, restricting ourselves to a few general remarks:

• Closing a loop with vertices from the lower-order Lagrangian, one arrives at the amplitude at a higher order. An example is provided by the quantity Σ_a, which contains a vertex from $\mathcal{L}^{(2)}$ and contributes to the mass at $O(p^4)$. Note that the chiral powers can be counted without actually performing the integration. Namely, the diagram in Fig. 4.9a contains a vertex of order p^2, a single pion propagator $1/(M^2 - k^2)$ that counts as order p^{-2}, and the loop integral d^4k (in $D = 4$ dimensions) that counts as order p^4. In total, the diagram counts as order p^{2-2+4}, that is, at $O(p^4)$, as confirmed by the explicit calculation. Here this counting will be carried out for generic amplitudes. It is also important to reiterate that this counting holds in the case of dimensional regularization. On the contrary, the sharp cutoff regularization, which introduces a hard scale explicitly, violates the counting rules considered here.

- As already mentioned, the number of diagrams, which are ultraviolet-divergent in this theory, is infinite. It is important, however, that the divergences, which emerge by using the lower-order vertices in the loops, are canceled by the counterterms that appear at higher orders.

- Each term in the Lagrangian at a given order describes an infinite number of amplitudes with a different number of external legs. Namely, expanding the matrix U in pion fields, it is seen that the $O(p^2)$ Lagrangian produces terms that describe Green's functions with $2, 4, 6, \ldots$ external legs at tree level. Calculating one-loop graphs with the vertices from $\mathcal{L}^{(2)}$, one arrives at the Green's functions with $2, 4, 6, \ldots$ external legs at $O(p^4)$, all of them divergent. These divergences should be canceled by the counterterms appearing in $\mathcal{L}^{(4)}$. The latter at tree level also produces Green's functions with an arbitrary high (even) number of external pion legs, but the number of *independent* low-energy couplings is finite. So, in order to make all one-loop Green's functions finite, the UV divergences, which emerge in the Green's functions at one loop, should be correlated with each other. It is straightforward to see that such correlations are guaranteed by chiral symmetry. Indeed, since dimensional regularization does not break chiral symmetry, the regularized Green's functions obey Ward identities for arbitrary D. This means that the divergent parts of these Green's functions at one loop, which are polynomials in momenta, obey the Ward identities separately. On the other hand, it is seen that all these polynomials can be obtained from the most general $O(p^4)$ Lagrangian at tree level. This means that a finite number of low-energy constants (LECs) in the $O(p^4)$ Lagrangian suffices to remove the UV divergences in all Green's functions at this order. The argument can be straightforwardly generalized to higher chiral orders.

- Contrary to this, chiral symmetry does not fix the finite parts of the LECs, like $l_3^r(\mu)$ in Eq. (4.258). In principle, these all could be calculated in terms of the single parameter Λ_{QCD} (note that these quantities do not depend, by definition, on the quark masses). This objective cannot, however, be achieved by using the analytic tools that are presently at our disposal. One might, however, consider the LECs as the free parameters of the theory and try to determine them from observables measured in experiment. Then, one can make predictions for other processes in which these LECs appear. Another method, which is becoming increasingly popular, consists in matching these couplings to observables determined from lattice calculations. The latter method is particularly useful for finding the LECs, which describe the dependence of observables on the quark masses (the constants in front of those terms in the Lagrangian that contain the quantity χ). The reason for this is that on the lattice, unlike in experiments, one could freely tune the quark masses. In fact, we can group the LECs into two groups. On the one hand, we have the so-called *dynamical LECs* that parameterize operators with derivatives only and are present even in the limit of vanishing quark masses. On the other hand, the so-called *symmetry-breaking LECs* parameterize the operators that have at least one quark mass insertion, as just discussed.

- As already pointed out, the number of the LECs at any given order is finite. However, their number quickly inflates, as one moves to higher orders (this holds for the effective Lagrangian but not necessarily for specific processes, where the growth in the number of LECs can be moderate). Consequently, the predictive power of the theory rests on the fact that the chiral expansion converges, before the large number of poorly known couplings makes the uncertainty in the predictions too large. This is the case for most observables in $N_f = 2$ flavor ChPT in the meson sector. For the example just considered, a thorough study, carried out in Ref. [53], showed that more than 94% of the pion mass comes from the $O(p^2)$ term, whereas the uncertainty, introduced by the low-energy constant l_3^r, is very small. Another example with good convergence is provided by the S-wave pion–pion scattering lengths, which were quoted in the previous section. In case of $N_f = 3$ flavor ChPT, as well as ChPT in the baryon sector, convergence is slower.

- Eq. (4.258) features the chiral logarithm that emerges in the one-loop diagram and is exactly the same as found in the linear σ-model; see Eq. (4.82). This is by no means surprising, since the coefficient of the logarithm at this order is fixed by chiral symmetry, that is, depends on the pion decay constant F only. This property holds for a generic Green's function at $O(p^4)$. These particular terms are a signal of the long-range nature of the "pion cloud," which extends to infinity in the chiral limit. (For a more detailed discussion on this issue, see [54] and references therein.)

- As seen from Eq. (4.258), the small expansion parameter in ChPT can be identified with $M^2/(4\pi F)^2 \simeq 0.015$. The convergence is, however, not as fast as suggested by the smallness of this parameter. The reason for this is mainly the presence of chiral logarithms, which, for some observables (e.g., for the $\pi\pi$ scattering length a_0), have rather large coefficients. The physics behind this enhancement is the strong rescattering (unitarity corrections) in the $\pi\pi$ S-wave with isospin zero.

- In the preceding calculation, the hard scale turned out to be $\Lambda = 4\pi F_\pi \simeq 1.2\,\text{GeV}$; see also Ref. [55]. More generally, the appearance of resonances limits the range of applicability of the chiral expansion, such as the $\rho(770)$ meson in P-wave pion–pion scattering. In that case, one would set $\Lambda = M_\rho \simeq 775\,\text{MeV}$. We will come back to this issue when discussing the inclusion of baryons in ChPT.

The calculations of other observables, such as the pion–pion scattering amplitude, which was considered in the previous section, proceed in a completely similar fashion. We do not aim here at considering the details of such calculations. Rather, we shall merely describe the universal counting rules to be applied in the calculation of any of these observables. The arguments are quite similar to the ones used in the case of the pion mass calculation.

Consider the calculation of a generic amplitude that is given by a sum of Feynman diagrams. Further, consider the contribution of a particular diagram with I internal lines and L loops. Such a diagram will, in general, contain vertices of different orders. Let N_d be number of vertices of order d. For example, the diagram in Fig. 4.9a contains one vertex with $d = 2$, that is, $N_2 = 1$ and $N_d = 0$ for $d \neq 2$. Counting the number of integration momenta, pion propagators and momenta flowing into and out of the vertices, the contribution of this diagram to the amplitude can be schematically written as follows:[15]

$$I \propto \int (d^4 k)^L \left(\frac{1}{k^2} \right)^I \prod_d k^{dN_d} . \tag{4.259}$$

Consequently, the chiral order of this diagram is

$$N_\chi = 4L - 2I + \sum_d dN_d . \tag{4.260}$$

On the other hand, it is easy to check that the number of internal lines, vertices and loops obey the following relation:[16]

$$L = I - \sum_d N_d + 1 . \tag{4.261}$$

This finally gives

$$N_\chi = 2L + 2 + \sum_d (d - 2)N_d . \tag{4.262}$$

This expression is interesting for a number of reasons. First, because of chiral symmetry, the Lagrangian starts at $O(p^2)$. (We remind the reader that the Goldstone bosons do not interact at zero momenta.) Since $d - 2$ is always nonnegative, only a finite number of diagrams contribute to a given amplitude at a given chiral order. (In particular, one can easily check that $N_\chi = 4$ for both diagrams in Fig. 4.9a,b.) Further, adding a loop increases the chiral power by two. To summarize, the perturbative expansion of the amplitudes on the basis of the chiral Lagrangian can be unambiguously mapped onto the chiral expansion of these amplitudes in external momenta and quark masses. In practice, these two expansion parameters are treated as one, which is legitimate as long as we keep the ratio m_{quark}/p^2 fixed [37]. In order to obtain an expression to a given chiral order, it suffices to write down the Lagrangian at the same order and calculate a finite number of loop diagrams. According to Weinberg's theorem, this yields the chiral expansion of a generic amplitude in QCD up to this order.

---

[15] The calculations are done in dimensional regularization. In a somewhat cavalier notation, we set $D = 4$ in the derivation of the counting rules, aiming at a simplification of the notation.

[16] The easiest way to verify this relation is to calculate how many independent loop momenta exist for a Feynman graph with a given topology. At the beginning, we have I momenta, one for each internal line. Each vertex contains the δ-function, which can be used to express one momentum in terms of others. This decreases the number of independent momenta by $\sum_d N_d$. Finally, one of these δ-functions corresponds to the total energy-momentum conservation and thus cannot be used to reduce the number of independent momenta.

4.9.3 Generating Functional at One Loop

In the previous section, we fixed the divergent part of the low-energy constant l_3 from the requirement that the physical pion mass is ultraviolet-finite. Choosing an appropriate set of Green's functions, this method can be applied to find the divergences related to the other low- and high-energy constants as well. Chiral symmetry guarantees that the divergences in all Green's functions at $O(p^4)$ are canceled. In this section, we shall consider another method for achieving this. It is based on the evaluation of the path integral with the use of the saddle-point technique. This method allows one to directly evaluate the divergent part of the one-loop generating functional and thus to establish the coefficients of the divergent parts of all constants simultaneously.

We shall first explain the essence of the method for the simple case of a single self-interacting scalar field, described by the Lagrangian

$$\mathcal{L} = \frac{1}{2}(\partial\phi)^2 - \frac{m^2}{2}\phi^2 - V(\phi).$$

(4.263)

Let us further assume that $V(\phi)$ depends on the field ϕ but not on its derivatives. In order to simplify the bookkeeping of different orders in perturbation theory, it is convenient to temporarily restore the reduced Planck constant $\hbar$ in the formulae. Then, the generating functional is given by

$$Z(j) = e^{\frac{i}{\hbar}W(j)} = \int d\phi \exp\left\{\frac{i}{\hbar}\int d^4x\left(\frac{1}{2}(\partial\phi)^2 - \frac{m^2}{2}\phi^2 - V(\phi) + j\phi\right)\right\}.$$

(4.264)

Here, $W(j)$ denotes the generating functional of the connected Green's functions.

The saddle-point technique for the calculation of the preceding path integral utilizes the expansion of the integrand in the vicinity of the classical solution, which is given by the solution of the EOM:

$$(\Box + m^2)\phi_c + V'(\phi_c) = j.$$

(4.265)

In other words, ϕ_c is a functional of the external source j. Writing now

$$\phi = \phi_c + \xi,$$

(4.266)

and expanding the action functional up to second order in the small fluctuation ξ, we get

$$S(\phi) = \int d^4x(\mathcal{L}(\phi) + j\phi) = S(\phi_c) + \frac{1}{2}\int d^4x\,\xi\left(\Box + m^2 - V''(\phi_c)\right)\xi + O(\xi^3).$$

(4.267)

Note that the first-order term vanishes due to the EOM. Inserting this expansion into Eq. (4.264), neglecting the $O(\xi^3)$ term and evaluating the remaining Gaussian integral, we arrive at

$$e^{\frac{i}{\hbar}W(j)} = e^{\frac{i}{\hbar}S(\phi_c)}\left(\det(\Box + m^2 - V''(\phi_c))\right)^{-1/2},$$

(4.268)

which leads to

$$W(j) = S(\phi_c) + \frac{i\hbar}{2} \ln\left(\det\left(\Box + m^2 - V''(\phi_c)\right)\right) + O(\hbar^2), \qquad (4.269)$$

where the second term represents the one-loop correction to the classical effective action functional. The preceding expression demonstrates that the loop expansion can be mapped into the expansion of the effective action in powers of $\hbar$.

Next, we expand the logarithm as

$$\ln\left(\det\left(\Box + m^2 - V''(\phi_c)\right)\right)$$

$$= \ln\left(\det\left(\Box + m^2\right)\right) - \int d^4x D(0) V''(\phi_c(x))$$

$$- \frac{1}{2} \int d^4x d^4y D(x-y) D(y-x) V''(\phi_c(x)) V''(\phi_c(y)) + \cdots, \qquad (4.270)$$

with

$$D(x-y) = \langle x|(\Box + m^2)^{-1}|y\rangle = \int \frac{d^4p}{(2\pi)^4} \frac{e^{-ip(x-y)}}{m^2 - p^2}. \qquad (4.271)$$

The first term in Eq. (4.270) is an inessential constant and can be dropped. Further, using dimensional regularization, it is straightforward to show that

$$D(0) = 2im^2\lambda + \frac{im^2}{16\pi^2} \ln \frac{m^2}{\mu^2},$$

$$D(x-y) D(y-x) = -2i\lambda \delta^{(4)}(x-y) - i \int \frac{d^4p}{(2\pi)^4} e^{-ip(x-y)} J(p^2), \qquad (4.272)$$

where

$$J(p^2) = \frac{1}{16\pi^2} \int_0^1 dx \left(\ln \frac{m^2 - x(1-x)p^2}{\mu^2} + 1 \right). \qquad (4.273)$$

The one-loop contributions that contain three and more propagators are UV-finite. For this reason, they are not shown in Eq. (4.270). Using the preceding results, the Eq. (4.269) can be rewritten in the following manner:

$$W(j) = S(\phi_c) + \hbar\lambda m^2 \int d^4x V''(\phi_c) - \frac{1}{2} \hbar\lambda \int d^4x \left(V''(\phi_c)\right)^2 + \text{finite} + O(\hbar^2).$$

$$(4.274)$$

It is now seen that, in order to cancel the ultraviolet divergences arising from the quantum one-loop corrections at $O(\hbar)$, the potential $V(\phi)$ should be equipped with a bunch of divergent counterterms. Further, the divergences in the couplings can be directly read from Eq. (4.274). For example, choosing

$$V(\phi) = \frac{g + \delta g}{4} \phi^4 + \frac{\delta m^2}{2} \phi^2 + O(\hbar^2), \qquad (4.275)$$

where $g, m^2 = O(1)$ and $\delta g, \delta m^2 = O(\hbar)$, we get

$$g + \delta g = -18\hbar\lambda g^2 + g_r(\mu),$$

$$m^2 + \delta m^2 = 6\hbar\lambda g m^2 + m_r^2(\mu). \qquad (4.276)$$

These formulae determine the running of the strong coupling constant and the mass at one loop in the φ^4-theory; see Chapter 1.

We would like to apply the same method to calculate all divergences in ChPT at one loop. (We set the Planck constant $\hbar = 1$ from now on.) The classical EOM can be obtained from the $O(p^2)$ Lagrangian in Eq. (4.194) using the Euler–Lagrange method as follows: In order to impose the constraint $\det U = 1$, we may use the method of Lagrange multipliers. The lowest-order Lagrangian takes the form

$$\mathcal{L}^{(2)} = \frac{F_0^2}{4} \langle D^\mu U^\dagger D_\mu U + \chi^\dagger U + U^\dagger \chi \rangle + r(\det U - 1). \qquad (4.277)$$

Treating the Lagrange multiplier r as an independent variable, one immediately sees that its EOM reproduces the constraint $\det U = 1$.

In order to derive the EOM for the matrix U, we have to calculate the change of the action functional with respect to small variations of U. To this end, we have to take into account that the variations of U and $U^\dagger$ are not independent. From unitarity it follows that

$$\delta U U^\dagger + U \delta U^\dagger = 0, \qquad (4.278)$$

or

$$\delta U^\dagger = -U^\dagger \delta U U^\dagger. \qquad (4.279)$$

Further, the variation of the determinant can be found in the following manner:

$$\delta \det U = \delta \exp\{\operatorname{tr} \ln U\} = \exp\{\operatorname{tr} \ln U\} \operatorname{tr}(U^\dagger \delta U) = \operatorname{tr}(U^\dagger \delta U). \qquad (4.280)$$

Here, we have used the constraint $\exp\{\operatorname{tr} \ln U\} = \det U = 1$.

Using partial integration and the cyclic property of the trace, the variation of the lowest-order Lagrangian can be rewritten in the following form:

$$\delta \mathcal{L}^{(2)} = \frac{F_0^2}{4} \langle (U^\dagger (D_\mu D^\mu U) U^\dagger - D_\mu D^\mu U^\dagger + \chi^\dagger - U^\dagger \chi U^\dagger) \delta U \rangle + r \langle U^\dagger \delta U \rangle. \qquad (4.281)$$

Since δU is arbitrary, the Euler–Lagrange equation takes the form

$$\frac{F_0^2}{4} ((D_\mu D^\mu U) U^\dagger - U D_\mu D^\mu U^\dagger + U \chi^\dagger - \chi U^\dagger) + r\mathbb{1} = 0, \qquad (4.282)$$

where $\mathbb{1}$ denotes the unit matrix in flavor space. Taking the trace from both sides of this equation and using the fact that the matrices $U D_\mu U^\dagger$, $U^\dagger D_\mu U$ are traceless, one can determine r:

$$N_f r = -\frac{F_0^2}{4} \langle U \chi^\dagger - \chi U^\dagger \rangle. \qquad (4.283)$$

Substituting this relation into Eq. (4.282), we finally obtain the EOM for the matrix U at leading order:

$$(D_\mu D^\mu U)U^\dagger - U D_\mu D^\mu U^\dagger + U\chi^\dagger - \chi U^\dagger - \frac{1}{N_f}\langle U\chi^\dagger - \chi U^\dagger\rangle \mathbb{1} = 0. \qquad (4.284)$$

Next, we turn to the calculation of the one-loop generating functional in the saddle-point approximation, following Ref. [37]. Let U_c denote the classical solution of the preceding EOM (the analog of ϕ_c from the example just considered). Define $U_c = u^2$ and introduce the matrix of small fluctuations ξ according to

$$U = u\exp\left(i\frac{\xi}{F_0}\right)u = U_c + \frac{i}{F_0}u\xi u - \frac{1}{2F_0^2}u\xi^2 u + O(\xi^3). \qquad (4.285)$$

Here, $\xi = \tau^a\xi^a$ and $\xi = \lambda^a\xi^a$ for $N_f = 2$ and $N_f = 3$ flavors, respectively, and a is the flavor index, $a = 1,\ldots N_f^2 - 1$. Note that here we use the symbol F_0 for the pion decay constant in the two- and the three-flavor chiral limit to ease notation (being aware that these are not the same quantities). Expanding the action functional up to the second order in ξ yields

$$S(U) = S(U_c) + \frac{1}{4}\int d^4x\Big\langle D_\mu(u^\dagger\xi u^\dagger)D^\mu(u\xi u) - \frac{1}{2}D_\mu U_c^\dagger D^\mu(u\xi^2 u)$$

$$-\frac{1}{2}D_\mu U_c D^\mu(u^\dagger\xi^2 u^\dagger) - \frac{1}{2}\xi^2(u\chi^\dagger u + u^\dagger\chi u^\dagger)\Big\rangle. \qquad (4.286)$$

Note that the linear term in ξ is absent due to the EOM. We further introduce the notations

$$\Gamma_\mu = \frac{1}{2}[u^\dagger, \partial_\mu u] - \frac{i}{2}u^\dagger r_\mu u - \frac{i}{2}u l_\mu u^\dagger,$$

$$\Delta_\mu = \frac{1}{2}u^\dagger D_\mu U_c u^\dagger = -\frac{1}{2}u D_\mu U_c^\dagger u, \qquad (4.287)$$

and define the covariant derivative:

$$d_\mu\xi = \partial_\mu\xi + [\Gamma_\mu, \xi],$$

$$(d_\mu\xi)^a = (\partial_\mu\delta^{ab} + \Gamma_\mu^{ab})\xi^b = (d_\mu)^{ab}\xi^b. \qquad (4.288)$$

Here,

$$\Gamma_\mu^{ab} = -2\langle[T^a, T^b]\Gamma_\mu\rangle, \qquad (4.289)$$

and the T^a denote the generators of the group $SU(N_f)$.

In this notation, the expansion of the action functional given in Eq. (4.286) can be rewritten in a compact form:

$$S(U) = S(U_c) + \frac{1}{4}\int d^4x\langle d_\mu\xi d^\mu\xi - [\Delta_\mu, \xi][\Delta^\mu, \xi] - \xi^2\sigma_0\rangle$$

$$= S(U_c) - \frac{1}{2}\int d^4x\,\xi^a G^{ab}\xi^b, \qquad (4.290)$$

where

$$G^{ab} = (d_\mu)^{ac}(d^\mu)^{cb} + \sigma^{ab},$$

$$\sigma^{ab} = 2\langle[T^a, \Delta_\mu][T^b, \Delta^\mu]\rangle + \langle\{T^a, T^b\}\sigma_0\rangle,$$

$$\sigma_0 = \frac{1}{2}(u^\dagger\chi u^\dagger + u\chi^\dagger u). \tag{4.291}$$

Similar to the ϕ^4-model, the one-loop effective action is given by

$$W(j) = S(U_c) + \frac{i}{2}\ln\det G. \tag{4.292}$$

The divergent part of the one-loop action can be readily evaluated with the use of the technique given in Section 3.10 (for more details, see Ref. [19]). In D-dimensional Minkowski space we have

$$\ln\det G = -\int_0^{i\infty}\frac{dt}{t}\,\mathrm{Tr}\,e^{-tG}$$

$$= -i(4\pi)^{D/2}\int_0^{i\infty}dt\,t^{-1-D/2}\int d^Dx\,\mathrm{tr}H(x|t|x), \tag{4.293}$$

where

$$(x|e^{-tG}|y) = i(4\pi t)^{-D/2}\exp\left(\frac{z^2}{4t}\right)H(x|t|y). \tag{4.294}$$

The heat kernel $H(x|t|y)$ can be expanded in a Taylor series in t; see Eq. (3.183). Substituting this expansion in Eq. (4.293) and evaluating the integral over t, it is easily seen that the divergence for $D = 4$ occurs in a single term:

$$\frac{i}{2}\ln\det G = -\frac{1}{16\pi^2}\frac{1}{D-4}\int d^4x\,\mathrm{tr}H_2(x,x) + \text{finite terms}. \tag{4.295}$$

Using Eq. (3.198), we finally obtain

$$\frac{i}{2}\ln\det G = -\frac{1}{16\pi^2}\frac{1}{D-4}\int d^4x\,\mathrm{tr}\left(\frac{1}{12}\Gamma_{\mu\nu}\Gamma^{\mu\nu} + \frac{1}{2}\sigma^2\right) + \text{finite terms}, \tag{4.296}$$

where

$$\Gamma_{\mu\nu} = \partial_\mu\Gamma_\nu - \partial_\nu\Gamma_\mu + [\Gamma_\mu, \Gamma_\nu], \quad \Gamma_{\mu\nu}^{ab} = -2\langle[T^a, T^b]\Gamma_{\mu\nu}\rangle. \tag{4.297}$$

Note that, in Eqs. (4.293), (4.295) and (4.296), the trace is calculated over the indices of the adjoint representation, for the $(N_f^2 - 1) \times (N_f^2 - 1)$ matrices G^{ab}, $\Gamma_{\mu\nu}^{ab}$ and σ^{ab}. What remains to be done is to use the explicit expressions for Γ_μ and σ, given in Eqs. (4.287) and (4.291), and evaluate the expression of the divergent term. This rather straightforward but cumbersome procedure is considered in detail in Ref. [37]. Here, we just quote the final result:

$$\text{tr}\left(\Gamma_{\mu\nu}\Gamma^{\mu\nu}\right) = \frac{N_f}{4}\langle D^\mu U_c^\dagger D^\nu U_c D_\mu U_c^\dagger D_\nu U_c\rangle - \frac{N_f}{4}\langle D^\mu U_c^\dagger D_\mu U_c D^\nu U_c^\dagger D_\nu U_c\rangle$$

$$- iN_f\langle r_{\mu\nu}D^\mu U_c D^\nu U_c^\dagger\rangle - iN_f\langle l_{\mu\nu}D^\mu U_c^\dagger D^\nu U_c\rangle$$

$$- N_f\langle r_{\mu\nu}U_c l^{\mu\nu}U_c^\dagger\rangle - \frac{N_f}{2}\langle r_{\mu\nu}r^{\mu\nu} + l_{\mu\nu}l^{\mu\nu}\rangle, \tag{4.298}$$

and

$$\text{tr}\left(\sigma^2\right) = \frac{1}{8}\langle D^\mu U_c^\dagger D_\mu U_c\rangle^2 + \frac{1}{4}\langle D^\mu U_c^\dagger D^\nu U_c\rangle\langle D_\mu U_c^\dagger D_\nu U_c\rangle$$

$$+ \frac{N_f}{8}\langle D^\mu U_c^\dagger D_\mu U_c D^\nu U_c^\dagger D_\nu U_c\rangle + \frac{1}{4}\langle D^\mu U_c^\dagger D_\mu U_c\rangle\langle \chi^\dagger U_c + U^\dagger\chi\rangle$$

$$+ \frac{N_f}{4}\langle D^\mu U_c^\dagger D_\mu U_c(\chi^\dagger U_c + U_c^\dagger\chi)\rangle + \frac{N_f^2+2}{8N_f^2}\langle \chi^\dagger U_c + U_c^\dagger\chi\rangle^2$$

$$+ \frac{N_f^2-4}{8N_f}\langle \chi^\dagger U_c\chi^\dagger U_c + U_c^\dagger\chi U_c^\dagger\chi + 2\chi^\dagger\chi\rangle. \tag{4.299}$$

Finally, we must reduce the preceding expressions to the operator basis, displayed in Eqs. (4.195) and (4.196), and Eqs. (4.198) and (4.199), for $N_f = 2$ and $N_f = 3$ flavors, respectively. This allows one to straightforwardly read off the divergent parts of the effective couplings, which cancel the divergences in the one-loop generating functional. In case of two flavors, $N_f = 2$, the renormalized couplings are given by

$$l_i = \gamma_i\lambda + l_i^r(\mu), \qquad i = 1,\ldots,7,$$

$$h_i = \delta_i\lambda + h_i^r(\mu), \qquad i = 1,3, \tag{4.300}$$

where

$$\gamma_1 = \frac{1}{3}, \quad \gamma_2 = \frac{2}{3}, \quad \gamma_3 = -\frac{1}{2}, \quad \gamma_4 = 2 \quad \gamma_5 = -\frac{1}{6}, \quad \gamma_6 = -\frac{1}{3}, \quad \gamma_7 = 0,$$

$$\delta_1 = 2, \quad \delta_2 = \frac{1}{12}, \quad \delta_3 = 0. \tag{4.301}$$

Note also that we can define

$$l_i^r(\mu) = \frac{\gamma_i}{32\pi^2}\left(\bar{l}_i + \ln\frac{M^2}{\mu^2}\right), \qquad M^2 = B(m_u + m_d). \tag{4.302}$$

The new couplings $\bar{l}_i$ are scale-independent but contain chiral logarithms.

For $N_f = 3$ flavors, the renormalized couplings follow as

$$L_i = \Gamma_i\lambda + L_i^r(\mu), \qquad i = 1,\ldots,10,$$

$$H_i = \Delta_i\lambda + H_i^r(\mu), \qquad i = 1,2, \tag{4.303}$$

where

$$\Gamma_1 = \frac{3}{32}, \quad \Gamma_2 = \frac{3}{16}, \quad \Gamma_3 = 0, \quad \Gamma_4 = \frac{1}{8}, \quad \Gamma_5 = \frac{3}{8},$$

$$\Gamma_6 = \frac{11}{144}, \quad \Gamma_7 = 0, \quad \Gamma_8 = \frac{5}{48}, \quad \Gamma_9 = \frac{1}{4}, \quad \Gamma_{10} = -\frac{1}{4},$$

$$\Delta_1 = -\frac{1}{8}, \quad \Delta_2 = \frac{5}{24}. \tag{4.304}$$

These counterterms ensure that *any* Green's function in the theory, calculated up to and including $O(p^4)$, is free of UV divergences. This solves the problem that was posed in the beginning: the divergent parts of all couplings are determined simultaneously, without resorting to the calculation of individual observables. Note once more that, in dimensional regularization, all divergences emerging after closing a loop with vertices from $\mathcal{L}^{(2)}$ can be removed by the counterterms that are contained in $\mathcal{L}^{(4)}$. As already mentioned before, this property does not hold for an arbitrary regularization, which implies the introduction of a heavy scale.

Finally, we wish to note that the saddle-point technique, considered in the present section, can be applied in case of two or more loops. A very instructive and transparent introduction to the two-loop calculations is given in Ref. [56].

4.9.4 Matching of the $SU(2)$ and $SU(3)$ LECs

Formally, ChPT with $N_f = 2,3$ flavors, which are EFTs of QCD containing two or three light quarks, are distinct theories. For this reason, we have reserved different notations $F, B, l_i, \ldots$ and $F_0, B_0, L_i, \ldots$ for the effective couplings in these theories. It is, however, fully legitimate to ask, what happens to the three-flavor effective theory when one of the quark masses (the strange quark mass) approaches the heavy scale, whereas the strong coupling constant at a given renormalization scale stays fixed? The answer is that such a theory can be described by a two-flavor ChPT (the strange particles are integrated out), and certain relations emerge between the effective couplings in the two theories. Moreover, the matching can be carried out perturbatively, since the calculations are restricted to the effective theories only. We have already seen examples of such calculations in Chapter 1.

A full-fledged matching of the effective couplings in $SU(2)$ and $SU(3)$ theories at $O(p^4)$ is done in Ref. [37]. This implies the evaluation of the one-loop effective action at this order (and not only the divergent part thereof). Here, we do not repeat these calculations, restricting ourselves to a simple example, where the relations between the certain low-energy couplings can be easily established.

The pion mass at $O(p^4)$ in the three-flavor case can be easily derived along the same path as in Eq. (4.258) for the two-flavor case. The result has the following form (for simplicity, we again assume $m_u = m_d = \hat{m}$):

$$M_\pi^2 = 2\hat{m}B_0 \left\{ 1 + \mu_\pi - \frac{1}{3}\mu_\eta + 2\hat{m}\frac{8B_0}{F_0^2}(2L_8^r - L_5^r) \right.$$

$$\left. + (2\hat{m} + m_s)\frac{16B_0}{F_0^2}(2L_6^r - L_4^r) \right\}, \qquad (4.305)$$

where the following notations were introduced:

$$\mu_\pi = \frac{2\hat{m}B_0}{32\pi^2 F_0^2} \ln \frac{2\hat{m}B_0}{\mu^2},$$

$$\mu_K = \frac{(\hat{m} + m_s)B_0}{32\pi^2 F_0^2} \ln \frac{(\hat{m} + m_s)B_0}{\mu^2},$$

$$\mu_\eta = \frac{\frac{2}{3}(\hat{m} + 2m_s)B_0}{32\pi^2 F_0^2} \ln \frac{\frac{2}{3}(\hat{m} + 2m_s)B_0}{\mu^2}. \qquad (4.306)$$

We easily recognize the π, K, η-meson masses at $O(p^2)$. (Cf. Eq.(4.214), entering the preceding expressions.)

We further introduce a special notation for the leading-order K, η-meson masses in the $SU(2)$ chiral limit with $m_u, m_d \to 0$ and m_s fixed:

$$\bar{M}_K^2 = m_s B_0, \qquad \bar{M}_\eta^2 = \frac{4}{3}m_s B_0, \qquad (4.307)$$

so that, up to and including the terms of order p^2 in $SU(2)$ ChPT,

$$\mu_K = \bar\mu_K + \frac{M^2}{64\pi^2 F_0^2}\left(\ln\frac{\bar{M}_K^2}{\mu^2} + 1\right) + \cdots, \qquad \bar\mu_K = \frac{\bar{M}_K^2}{32\pi^2 F_0^2}\ln\frac{\bar{M}_K^2}{\mu^2},$$

$$\mu_\eta = \bar\mu_\eta + \frac{M^2}{96\pi^2 F_0^2}\left(\ln\frac{\bar{M}_\eta^2}{\mu^2} + 1\right) + \cdots, \qquad \bar\mu_\eta = \frac{\bar{M}_\eta^2}{32\pi^2 F_0^2}\ln\frac{\bar{M}_\eta^2}{\mu^2}, \qquad (4.308)$$

and $M^2 = 2\hat{m}B_0$. Substituting the preceding expansion into the expression of the pion mass at order p^2, given by Eq. (4.305), we get

$$M_\pi^2 = M^2 \left\{ 1 - \frac{1}{3}\bar\mu_\eta + \frac{16\bar{M}_K^2}{F_0^2}(2L_6^r - L_4^r) + \mu_\pi \right.$$

$$\left. + \frac{2M^2}{F_0^2}\left(4(2L_8^r - L_5^r) + 8(2L_6^r - L_4^r) - \frac{1}{576\pi^2}\left(\ln\frac{\bar{M}_\eta^2}{\mu^2} + 1\right)\right)\right\} + \cdots. \qquad (4.309)$$

Comparing this expression to the same quantity calculated in the two-flavor ChPT, Eq. (4.258), the matching condition for two combinations of the effective couplings can be read off:

$$B = B_0\left\{ 1 - \frac{1}{3}\bar\mu_\eta + \frac{16\bar{M}_K^2}{F_0^2}(L_4^r - 2L_6^r)\right\}$$

$$l_3^r = 4(2L_8^r - L_5^r) + 8(2L_6^r - L_4^r) - \frac{1}{576\pi^2}\left(\ln\frac{\bar{M}_\eta^2}{\mu^2} + 1\right). \qquad (4.310)$$

The same pattern is followed by the other effective couplings. Here we give the full result for the matching without derivation; the details can be found in Ref. [37]. Introducing the notation,

$$v_P = \frac{1}{32\pi^2 F_0^2} \left(\ln \frac{\bar{M}_P^2}{\mu^2} + 1 \right), \qquad P = K, \eta, \tag{4.311}$$

the relations between the $SU(2)$ and $SU(3)$ couplings take the form

$$B = B_0 \left\{ 1 - \frac{1}{3}\bar{\mu}_\eta + \frac{16\bar{M}_K^2}{F_0^2}(L_4^r - 2L_6^r) \right\}, \qquad F = F_0 \left\{ 1 - \bar{\mu}_K + \frac{8\bar{M}_K^2}{F_0^2} L_4^r \right\},$$

$$l_1^r = 4L_1^r + 2L_3^r - \frac{1}{24} v_K, \qquad l_2^r = 4L_2^r - \frac{1}{12} v_K,$$

$$l_3^r = 8L_8^r - 4L_5^r + 16L_6^r - 8L_4^r - \frac{1}{18} v_K, \qquad l_4^r = 8L_4^r + 4L_5^r - \frac{1}{2} v_K,$$

$$l_5^r = L_{10}^r + \frac{1}{12} v_K, \qquad l_6^r = -2L_9^r + \frac{1}{6} v_K,$$

$$l_7 = \frac{F_0^2}{8\bar{M}_K^2} \left(1 + \frac{10}{3}\bar{\mu}_\eta \right) + 4(L_4^r - L_6^r - 9L_7 - 3L_8^r) + \frac{1}{2} v_K,$$

$$h_1^r = 8L_4^r + 4L_5^r - 4L_8^r + 2H_2^r - \frac{1}{2} v_K, \qquad h_2^r = -\frac{1}{4}L_{10}^r - \frac{1}{2}H_1^r - \frac{1}{24} v_K,$$

$$h_3^r = 4L_8^r + 2H_2^r - \frac{1}{2} v_K - \frac{1}{3} v_\eta + \frac{1}{96\pi^2}. \tag{4.312}$$

This shows that if the strange quark is considered heavy, its effects are encoded in the LECs of the two-flavor theory.

4.9.5 Physics of the LECs

Here, we want to discuss in more detail the physics underlying the LECs (we work in $SU(3)$ for definiteness). First, we remind the reader that these quantities are scale-dependent. Indeed, since the bare quantities L_i do not depend on the renormalization scale μ, Eq. (4.303) gives

$$\mu \frac{d}{d\mu} L_i^r(\mu) = -\frac{\Gamma_i}{(4\pi)^2}. \tag{4.313}$$

Thus, the running of the LECs is logarithmic:

$$L_i^r(\mu_2) = L_i^r(\mu_1) + \frac{\Gamma_i}{(4\pi)^2} \ln \left(\frac{\mu_1}{\mu_2} \right). \tag{4.314}$$

In actual ChPT calculations, one uses $\mu = 0.5 \ldots 1.0\,\text{GeV}$, so that the LECs show only a slow variation under such scale changes. It is also important to stress that observables are, of course, scale-independent, so that any variation of μ in fact amounts to shuffling strength between the loops and the counterterm contributions. In the equations just displayed, we notice the prefactor $1/(4\pi)^2$, which is also accompanying any one-loop integral. Since the logarithms are of order unity, the natural scale of the LECs can therefore be estimated as $c/(4\pi)^2 \simeq 6 \cdot 10^{-3}$, with c a number of order one.

The preceding argument about the natural size of the LECs requires further scrutiny in some cases. As already mentioned, a natural hard scale of ChPT is of order of $4\pi F_\pi \sim 1\,\text{GeV}$. At this energy, the perturbative loop expansion should cease to converge. However, in certain channels, this may occur at lower energies, signaled by the emergence of low-lying resonances in these channels. The most prominent examples of such resonances in the interactions of pions are the very broad $f_0(500)$-resonance in the scalar-isoscalar channel or the $\rho(770)$ in the P-wave with isospin one. If such low-lying resonances emerge, one could consider an option to increase the radius of convergence by including these resonances as explicit degrees of freedom in the effective Lagrangian or using some resummation technique like, for example, the inverse amplitude method. We will not further discuss such schemes here in detail but show how to estimate the effect of such resonances at low energies, where they do not appear as dynamical degrees of freedom anymore.

In the previous section, we had seen that integrating out the strange quark leaves imprints in the values of the LECs of the two-flavor EFT. More generally, one can show that integrating out heavy mesons indeed allows one to understand the numerical values of the LECs to good accuracy, as shown in Refs. [57, 58]. This is termed "resonance saturation."

We start with an example to clarify the physics underlying this. Consider the pion vector form factor, that is, the matrix element of the vector current V_μ^i between pion states,

$$\langle \pi^i(p')|V_\mu^k|\pi^j(p)\rangle = \varepsilon^{ijk}(p'+p)_\mu F_\pi^V(t)\,, \tag{4.315}$$

where $t = (p'-p)^2$ is the invariant momentum transfer squared and $F_\pi^V(0) = 1$ (with the charge of the pion measured in units of the elementary charge e). The chiral expansion of the pion vector form factor to one loop reads

$$F_\pi^V(t) = 1 + t\left\{\frac{2L_9^r}{F_\pi^2} + \underbrace{\Phi_\pi(t) + \Phi_K(t)}_{\text{loops}}\right\}\,, \tag{4.316}$$

where we have not displayed the μ-dependence. The loop contribution $\sim \Phi_{\pi,K}$ starts at order p^2 but is rather small in this case (for a typical choice of the scale μ).

On the other hand, a time-honored phenomenological approach to the pion vector form factor is *vector meson dominance* (VMD), see, for example, Ref. [59]. In that scheme, the photon couples only through vector mesons to all hadrons with a universal coupling for each vector meson, as depicted for $F_\pi^V(t)$ in Fig. 4.10. Taking into account the exchange of the ρ-meson only and expanding the propagator, in VMD the form factor for small t is given by

$$F_\pi^V(t) = 1 + \frac{t}{M_\rho^2} + O(t^2)\,, \tag{4.317}$$

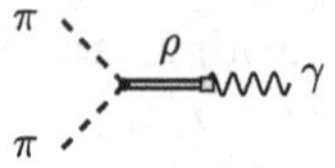

Figure 4.10   Vector meson dominance representation of the pion vector form factor.

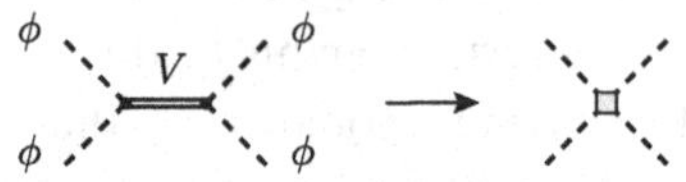

Figure 4.11   Vector meson contribution to elastic Goldstone boson scattering with an explicit vector meson (left) and the vector meson integrated out (right).

with $M_\rho \simeq 775\,\text{MeV}$ the ρ-meson mass. Comparing Eqs. (4.316) and (4.317) immediately leads to

$$L_9^{\text{VMD}} = \frac{F_\pi^2}{2M_\rho^2} \simeq 7.2 \cdot 10^{-3} ,\qquad (4.318)$$

which is quite consistent with the empirical determination, $L_9^r(M_\rho) = 6.9(7) \cdot 10^{-3}$. This can be considered as the QCD version of VMD. Note, however, that the LEC, estimated from ρ-exchange, is scale-independent, so we have tacitly assumed that we are working at a scale set by the vector meson mass.

We only outline the general analysis of resonance saturation here and refer to Ref. [57] for details. Consider meson resonances R, coupled to the Goldstone bosons in a chiral-invariant manner, and integrate out the heavy degrees of freedom. This can be symbolically written as

$$\int dR dU \exp\left\{ i \int \tilde{\mathcal{L}}_{\text{eff}}[U,R] \right\} = \int dU \exp\left\{ i \int \mathcal{L}_{\text{eff}}[U] \right\} ,\qquad (4.319)$$

where we consider only lowest-order (p^2) coupling between the heavy fields and the Goldstone bosons, such as $G_V \langle W_{\mu\nu} u_\mu u_\nu \rangle$ in terms of the tensor-field notation for the vector mesons. (Here, $W_{\mu\nu}$ is an antisymmetric tensor field that transforms like any matter field under the nonlinearly realized chiral symmetry; see Section 4.6.) Integrating out the heavy fields R is best understood by an example (see also Section 1). Consider vector exchange between Goldstone bosons, as depicted in Fig. 4.11. In the limit $M_V \to \infty$ but keeping G_V/M_V fixed, the explicit vector meson exchange reduces to a series of contact interactions, the leading one being of order p^4. In the symbolic notation, this can be written as

$$T(\phi\phi \to \phi\phi) \sim \frac{G_V^2}{M_V^2} \langle u_\mu u_\nu \rangle^2 + O(p^6) .\qquad (4.320)$$

The term on the r.h.s. corresponds to the contribution to one of the operators from $\mathcal{L}^{(4)}$, with the LEC fixed in terms of the vector meson coupling and mass.

More generally, performing resonance saturation, the LECs can be expressed as

$$L_i^r(\mu) = \sum_{R=A,V,S,P} L_i^{\text{RES}} + \hat{L}_i(\mu) .\qquad (4.321)$$

L_1^r	L_2^r	L_3^r	L_4^r	L_5^r	L_6^r	L_7^r	L_8^r	L_9^r	L_{10}^r
1.0(1)	1.6(2)	$-3.8(3)$	0.0(3)	1.2(1)	0.0(4)	$-0.3(2)$	0.5(2)	6.9(7)	$-5.2(1)$
0.9	1.8	-4.8	0	1.1	0	-0.2	0.5	7.2	-5.4

Here, $\hat{L}_i(\mu)$ is the remainder with $\hat{L}_i(\mu) \simeq 0$ at $\mu = M_R$; otherwise, the concept would not make sense. In fact, the dominant contribution comes from the vector (V) and axial-vector (A) mesons, whenever they can contribute; but one also needs the scalars (S) and the pseudoscalars (P), notably the η', at the mass scale of about 1 GeV, to complete the picture. In Table 4.1 we show the present status of the determination of the LECs at one loop and the corresponding resonance saturation values (adopted from [60]). Thus, the spectrum of QCD leaves its imprint in the low-energy EFT through the numerical values of the LECs. Remarkably, this "chiral VMD" (or chiral duality) works extremely well, as the LECs are really saturated by the resonance contributions.

4.10 Chiral Lagrangian with Nucleons

4.10.1 Construction of the Lagrangian

In Section 4.6 we showed how to couple matter fields to the Goldstone bosons of the spontaneously broken chiral symmetry of QCD. The prime example for such matter fields are the low-lying baryons (the ground state octet) in the three-flavor case or the nucleons in the case of chiral $SU(2)$. Already in the days of current algebra, which is the lowest-order approximation of Chiral Perturbation Theory, much has been learned about the chiral dynamics of baryons; see, for example, Ref. [61]. In particular, it was shown that the expansion around the chiral limit cannot be described by simple perturbation theory but leads to terms like $\varepsilon \ln \varepsilon$ or $\varepsilon^{1/2}$, where ε parameterizes the explicit chiral symmetry breaking. Such effects are due to the long-range component of the strong interaction generated from the Goldstone bosons (often called the "pion cloud").

A systematic study of baryon ChPT has started from the seminal article [62]. In what follows, we shall briefly review the main findings of that paper. However, before proceeding with the construction of the effective chiral Lagrangian in the one-baryon sector, we make a comment about decoupling of sectors with different numbers of baryons. (Here, for simplicity, we shall consider the $N_f = 2$ case.) It is important that, due to baryon number conservation and the fact that the nucleon mass counts as order unity in the chiral counting (i.e., it does not disappear in the chiral limit), the sectors that contain a different number of nucleons can be decoupled from each other.

Consider, for example, the purely mesonic sector. If the nucleons are present in the theory, they contribute to the mesonic Green's functions via closed nucleon loops only. Considering all external momenta small, we may expand nucleon loops in a Taylor series over these momenta (the typical hard scale associated with this expansion is the nucleon mass). This amounts to shrinking the nucleon loops and replacing them by a bunch of local counterterms. Since such (purely mesonic) terms are present in the chiral Lagrangian anyway, one could get rid of all nucleon loops in the meson sector merely by a redefinition of the renormalization prescription. The same argument holds in a sector with any number of nucleons; nucleon loops can thus be excluded from the theory completely.[17]

Further, in a process where we have only nucleons (or only antinucleons), all meson momenta can be simultaneously soft, and it is therefore (in principle) possible to introduce a consistent power counting. On the contrary; for example, the nucleon–antinucleon case is different, because these can annihilate into pions, some of which will necessarily have momenta of order of the heavy scale.

From the preceding discussion it is clear that, considering the one-nucleon sector, we need to include bilinear operators of the type $\bar{\Psi}\mathcal{O}\Psi$ in the Lagrangian, because the operators that contain higher powers of nucleon fields can couple to the one-nucleon sector only through fermion loops. The lowest-order chiral pion–nucleon Lagrangian is given by (cf. with Eq. (4.143))

$$\mathcal{L}^{(1)} = \bar{\Psi}(i\slashed{D} - \mathring{m})\Psi + \frac{\mathring{g}_A}{2}\bar{\Psi}\slashed{u}\gamma_5\Psi. \tag{4.322}$$

Here, $\mathring{m}$ and $\mathring{g}_A$ denote the nucleon mass and the axial-vector coupling constant in the chiral limit $m_u, m_d \to 0$ (and m_s fixed at its physical value). The covariant derivative is given by

$$D_\mu = \partial_\mu + \Gamma_\mu, \qquad \Gamma_\mu = \frac{1}{2}[u^\dagger, \partial_\mu u] - \frac{i}{2}u^\dagger r_\mu u - \frac{i}{2}u l_\mu u^\dagger, \tag{4.323}$$

and

$$u_\mu = i(u^\dagger(\partial_\mu - ir_\mu)u - u(\partial_\mu - il_\mu)u^\dagger). \tag{4.324}$$

From the preceding discussion, it is clear that $\Gamma_\mu = O(p)$. Then, in order to arrive at a consistent power counting, it is necessary to count $i\slashed{D} - \mathring{m}$ as a quantity of order p. With this assignment, the Lagrangian in Eq. (4.322) indeed counts as $O(p)$. The higher-order Lagrangians can be readily constructed by using the chirally invariant building blocks, and accounting also for the other symmetries of QCD. A very detailed description of this procedure is given in Ref. [63]. The result is written in the following form:

$$\mathcal{L}^{(2)} = \sum_{i=1}^{7} c_i\bar{\Psi}\mathcal{O}_i^{(2)}\Psi, \qquad \mathcal{L}^{(3)} = \sum_{i=1}^{23} d_i\bar{\Psi}\mathcal{O}_i^{(3)}\Psi, \qquad \mathcal{L}^{(4)} = \sum_{i=1}^{118} e_i\bar{\Psi}\mathcal{O}_i^{(4)}\Psi, \tag{4.325}$$

[17] One should, however, make sure that such a redefinition does not affect the validity of the Ward identities. We shall discuss this subtle issue in more detail in Section 4.11 that deals with the restoration of the power counting rules.

Table 4.2 Operator basis of the $O(p^2)$ pion–nucleon Lagrangian.

i	$\mathcal{O}_i^{(2)}$
1	$\langle \chi_+ \rangle$
2	$-\dfrac{1}{4\overset{\circ}{m}^2} \langle u_\mu u_\nu \rangle (D^\mu D^\nu + \text{h.c.})$
3	$\dfrac{1}{2} \langle u_\mu u^\mu \rangle$
4	$\dfrac{i}{4} \sigma^{\mu\nu}[u_\mu, u_\nu]$
5	$\hat{\chi}_+$
6	$\dfrac{1}{8\overset{\circ}{m}} \sigma^{\mu\nu} F_{\mu\nu}^+$
7	$\dfrac{1}{8\overset{\circ}{m}} \sigma^{\mu\nu}\langle F_{\mu\nu}^+ \rangle$

which is the Lagrangian required for complete one-loop calculations up to and including order $O(p^4)$. It is important to note that the nucleon Lagrangian, unlike the purely mesonic Lagrangian, also contains terms with odd powers of p. This happens because odd powers of p in the meson Lagrangian can emerge only from factors $\partial_\mu, v_\mu, a_\mu$, each of them of order p. However, due to Lorentz invariance all these factors should come in pairs. In contrast, the Lorentz indices in the nucleon sector can be contracted with γ_μ or $\sigma_{\mu\nu}$, which count as $O(1)$. Hence, all powers of p (even and odd) are allowed.

The operator basis at $O(p^2)$ is given in Table 4.2. The following notations are used:

$$\chi_\pm = u^\dagger \chi u^\dagger \pm u\chi^\dagger u, \qquad \hat{\chi}_+ = \chi_+ - \frac{1}{2}\langle \chi_+ \rangle,$$

$$F_{\mu\nu}^\pm = u^\dagger r_{\mu\nu} u \pm u l_{\mu\nu} u^\dagger, \qquad \hat{F}_{\mu\nu}^+ = F_{\mu\nu}^+ - \frac{1}{2}\langle F_{\mu\nu}^+ \rangle. \qquad (4.326)$$

We note that the explicit chiral symmetry breaking appears only at this order. This means that, unlike the meson sector, at leading order all baryon masses are equal (we have briefly returned to the three flavor case). Further, the operator basis for the $O(p^3)$ pion–nucleon Lagrangian, which is shown in Table 4.3, is taken from Ref. [64] (see also Ref. [65]), and its extension to three flavors can be found in Ref. [66]; see also Ref. [67].

The natural size of the LECs c_i of the dimension-two chiral pion–nucleon Lagrangian is $g_A/\Lambda \sim 1\,\text{GeV}^{-1}$ [68]. (Here Λ stands for the characteristic hard scale of ChPT.) The empirical determination of these LECs shows that some of them are sizably larger than this naive dimensional estimate; see, for example, Refs. [69, 70]. This enhancement can be traced back to the closeness of the $\Delta(1232)$ resonance. Using again resonance saturation, the Δ enhancement is easily understood in terms of factors like $\Lambda/(m_\Delta - m_N) \sim 3$ [69].

Table 4.3 Operator basis at $O(p^3)$. The abbreviation "sym" denotes the symmetrization of indices.

i	$\mathcal{O}_i^{(3)}$
1	$-\dfrac{1}{2\overset{\circ}{m}}\left([u_\mu,[D_\nu,u^\mu]]D^\nu + \text{h.c.}\right)$
2	$-\dfrac{1}{2\overset{\circ}{m}}\left([u_\mu,[D^\mu,u_\nu]]D^\nu + \text{h.c.}\right)$
3	$\dfrac{1}{12\overset{\circ}{m}^3}\left([u_\mu,[D_\nu,u_\lambda]](D^\mu D^\nu D^\lambda + \text{sym}) + \text{h.c.}\right)$
4	$-\dfrac{1}{2\overset{\circ}{m}}\left(\varepsilon^{\mu\nu\alpha\beta}\langle u_\mu u_\nu u_\alpha\rangle D_\beta + \text{h.c.}\right)$
5	$\dfrac{i}{2\overset{\circ}{m}}\left([\chi_-,u_\mu]D^\mu - \text{h.c.}\right)$
6	$\dfrac{i}{2\overset{\circ}{m}}\left([D^\mu,\hat{F}^+_{\mu\nu}]D^\nu - \text{h.c.}\right)$
7	$\dfrac{i}{2\overset{\circ}{m}}\left([D^\mu,\langle F^+_{\mu\nu}\rangle]D^\nu - \text{h.c.}\right)$
8	$\dfrac{i}{2\overset{\circ}{m}}\left(\varepsilon^{\mu\nu\alpha\beta}\langle\hat{F}^+_{\mu\nu}u_\alpha\rangle D_\beta - \text{h.c.}\right)$
9	$\dfrac{i}{2\overset{\circ}{m}}\left(\varepsilon^{\mu\nu\alpha\beta}\langle F^+_{\mu\nu}\rangle u_\alpha D_\beta - \text{h.c.}\right)$
10	$\dfrac{1}{2}\gamma^\mu\gamma_5 u_\mu\langle u^\nu u_\nu\rangle$
11	$\dfrac{1}{2}\gamma^\mu\gamma_5 u^\nu\langle u_\mu u_\nu\rangle$
12	$-\dfrac{1}{8\overset{\circ}{m}^2}\left(\gamma^\mu\gamma_5 u_\mu\langle u_\nu u_\lambda\rangle\{D^\nu,D^\lambda\} + \text{h.c.}\right)$
13	$-\dfrac{1}{8\overset{\circ}{m}^2}\left(\gamma^\mu\gamma_5 u_\lambda\langle u_\mu u_\nu\rangle\{D^\nu,D^\lambda\} + \text{h.c.}\right)$
14	$\dfrac{i}{4\overset{\circ}{m}}\left(\sigma^{\mu\nu}\langle u_\nu[D_\lambda,u_\mu]\rangle D^\lambda - \text{h.c.}\right)$
15	$\dfrac{i}{4\overset{\circ}{m}}\left(\sigma^{\mu\nu}\langle u_\mu[D_\nu,u_\lambda]\rangle D^\lambda - \text{h.c.}\right)$
16	$\dfrac{1}{2}\gamma^\mu\gamma_5\langle\chi_+\rangle u_\mu$
17	$\dfrac{1}{2}\gamma^\mu\gamma_5\langle\chi_+ u_\mu\rangle$
18	$\dfrac{i}{2}\gamma^\mu\gamma_5[D_\mu,\chi_-]$
19	$\dfrac{i}{2}\gamma^\mu\gamma_5[D_\mu,\langle\chi_-\rangle]$
20	$-\dfrac{i}{8\overset{\circ}{m}^2}\left(\gamma^\mu\gamma_5[\hat{F}^+_{\mu\nu},u_\lambda]\{D^\nu,D^\lambda\} - \text{h.c.}\right)$
21	$\dfrac{i}{2}\gamma^\mu\gamma_5[\hat{F}^+_{\mu\nu},u^\nu]$
22	$\dfrac{i}{2}\gamma^\mu\gamma_5[D^\nu,F^-_{\mu\nu}]$
23	$\dfrac{1}{2}\varepsilon^{\mu\nu\alpha\beta}\gamma_\mu\gamma_5\langle u_\nu F^-_{\alpha\beta}\rangle$

Next, since the free nucleon propagator counts at $O(p^{-1})$, the lowest-order one-loop diagram, which contains two pion–nucleon vertices at $O(p)$, one pion and one nucleon propagator, would *naively* count at $O(p^3)$. An example of such a diagram

One-nucleon sector: Examples of diagrams that (naively) contribute at $O(p^3)$. The self-energy of the nucleon is depicted in diagram a, whereas in diagram b, a diagram contributing to pion–nucleon scattering is displayed. Solid and dashed lines denote nucleons and pions, respectively.

is given by the nucleon self-energy graph, shown in Fig. 4.12a. Another example of this type is provided by a typical one-loop diagram that contributes to πN scattering at one loop; see Fig. 4.12b. For this reason, one might expect that the second-order LECs c_i in Eq. (4.325) are UV-finite, whereas the UV divergences arise in the third-order LECs d_i, which will cancel the divergences from the leading one-loop diagrams and the LECs. Likewise, the LECs e_i should cancel the divergences that appear in the one-loop diagrams at fourth order. However, all the preceding arguments are upset by the presence of hard scale $\overset{\circ}{m}$ in the propagators, as discussed next.

4.10.2 Breakdown of the Counting Rules

As discussed in previous sections, an important property of meson ChPT in dimensional regularization is that it preserves chiral power counting. Contrary to this, as we shall see, loops in ChPT in the one-baryon sector break the power counting, when using the same regularization and renormalization as in the meson case. A detailed discussion of this issue was originally provided in Ref. [62].

We shall explain the arguments of Ref. [62] for a simple example. Consider the nucleon self-energy (Fig. 4.12a). Using the Lagrangians at order p, p^2, p^3, which were given in the preceding section, a straightforward calculation of the physical nucleon mass m_N in the isospin limit $m_u = m_d = \hat{m}$ to third order gives:

$$m_N - \overset{\circ}{m} + 4M^2 c_1 - \Sigma(p)\big|_{\not{p}=m_N} = 0, \qquad M^2 = 2B_0\hat{m}. \tag{4.327}$$

We note in particular the appearance of the nucleon mass $\overset{\circ}{m}$ in the chiral limit, which is the distinguishing feature between Goldstone bosons and matter fields like the nucleons. The effect of the explicit chiral symmetry breaking is only a perturbation to the nucleon mass, as already discussed in the context of the trace anomaly in Section 3.13. Further, the self-energy $\Sigma(p)$ is given by

$$\Sigma(p) = \frac{3\overset{\circ}{g}_A^2}{4F^2} \int \frac{d^D k}{(2\pi)^D i} \frac{\not{k}(\overset{\circ}{m} - \not{p} + \not{k})\not{k}}{(\overset{\circ}{m}^2 - (p-k)^2)(M^2 - k^2)}. \tag{4.328}$$

At this order, we may replace $\overset{\circ}{m}, \overset{\circ}{g}_A, M^2, F$ by the physical values m_N, g_A, M_π^2, F_π, respectively. Simplifying the numerator in the preceding expression and using Lorentz invariance, one can write

$$\Sigma(p) = \frac{3g_A^2}{4F_\pi^2} \left(-(\slashed{p}+m_N)I_N + M_\pi^2(\slashed{p}+m_N)I_{\pi N} \right.$$

$$\left. + \frac{\slashed{p}(p^2 - m_N^2)}{2p^2} \left(I_\pi - I_N + (M_\pi^2 + p^2 - m_N^2)I_{\pi N} \right) \right), \tag{4.329}$$

where

$$I_N = \int \frac{d^D k}{(2\pi)^D i} \frac{1}{m_N^2 - (p-k)^2} = 2m_N^2 \lambda + \frac{m_N^2}{16\pi^2} \ln \frac{m_N^2}{\mu^2},$$

$$I_\pi = \int \frac{d^D k}{(2\pi)^D i} \frac{1}{M_\pi^2 - k^2} = 2M_\pi^2 \lambda + \frac{M_\pi^2}{16\pi^2} \ln \frac{M_\pi^2}{\mu^2},$$

$$I_{\pi N} = \int \frac{d^D k}{(2\pi)^D i} \frac{1}{(M_\pi^2 - k^2)(m_N^2 - (p-k)^2)} = -2\lambda$$

$$- \frac{1}{16\pi^2} \left(\ln \frac{m_N^2}{\mu^2} - 1 + \frac{p^2 - m_N^2 - M_\pi^2}{2p^2} \ln \frac{M_\pi^2}{m_N^2} + \frac{2M_\pi m_N}{p^2} F(\Omega) \right). \tag{4.330}$$

Here,

$$F(\Omega) = \begin{cases} \sqrt{\Omega^2 - 1} \ln(-\Omega - \sqrt{\Omega^2 - 1}), & \Omega \le -1, \\ \sqrt{1 - \Omega^2} \arccos(-\Omega), & -1 < \Omega < 1, \\ \sqrt{\Omega^2 - 1} \ln(\Omega + \sqrt{\Omega^2 - 1}), & \Omega \ge 1, \end{cases} \tag{4.331}$$

and Ω is a variable of chiral order unity:

$$\Omega = \frac{p^2 - m_N^2 + M_\pi^2}{2m_N M_\pi}. \tag{4.332}$$

From the preceding explicit expressions, it is immediately seen that the quantities $I_N, I_\pi, I_{\pi N}$ are of order p^0, p^2, p^0, respectively,[18] whereas a naive power counting would yield p^3, p^2, p^1. The chiral order of the quantity $\Sigma(p) = O(1)$ also does not correspond to the naive power-counting rules. In other words, loop corrections in baryon ChPT break the naive power counting. The reason for this can be seen easily in this example; the parameter m_N and the external momentum p_μ are not "soft," and the argument with the rescaling of momenta in the Feynman integral does not apply. Hence, the contributions, which break the power counting, emerge from the region where the integration momentum k is on the order of the nucleon mass.

Examining Eq. (4.330) carefully, it is seen that the quantities $I_N, I_\pi, I_{\pi N}$ contain both polynomial and nonanalytic parts. The nonanalytic terms are of two types and depend either on kinematical variables (such as the function $F(\Omega)$ in the preceding equations) or on the quark masses (the factor $\ln(M_\pi^2/\mu^2)$). The crucial observation is that all non-analytic contributions respect the power counting. This can be checked explicitly from the expressions just given. Intuitively, this fact is also very easy to understand. Indeed,

---

[18]   In contrast to the meson case, this statement refers, in general, only to the leading order in the chiral expansion. For example, $I_{\pi N}$ contains all chiral powers, starting from $O(p^0)$.

let us split the momentum integration into two parts, corresponding to small k (on the order of the pion mass) and large k (on the order of the nucleon mass). In the first part, one can rescale the momenta and ensure that power counting is obeyed. In the second part, one is allowed to Taylor-expand the integrand in the small kinematical variables and quark masses that leads to a low-energy polynomial after integration. Thus, the conjecture that power counting is broken only by the low-energy polynomials is justified.

From the preceding discussion we may conclude that the terms that break the power counting can be removed by an appropriate choice of the renormalization prescription. Indeed, putting $\not{p} = m_N$ in our example, we get:

$$m_N = \overset{\circ}{m} - 4M_\pi^2 c_1 + \frac{3g_A^2 m_N}{2F_\pi^2} \left(-I_N + M_\pi^2 I_{\pi N} \right). \tag{4.333}$$

It is seen that the "wrong" terms at this order disappear after renormalization:

$$\overset{\circ}{m} \rightarrow \overset{\circ}{m} + \frac{3g_A^2 m_N^3}{2F_\pi^2} \left(2\lambda + \frac{1}{16\pi^2} \ln \frac{m_N^2}{\mu^2} \right),$$

$$c_1 \rightarrow c_1 - \frac{3g_A^2 m_N}{8F_\pi^2} \left(2\lambda + \frac{1}{16\pi^2} \left(\ln \frac{m_N^2}{\mu^2} - 1 \right) \right). \tag{4.334}$$

One could cancel all "wrong" terms in other diagrams, following a similar path. The situation resembles power counting in the cutoff regularization, discussed in Section 1.2.4. At each order, we will have to renormalize the parameters of the Lagrangian at all orders, up to and including the order at which the calculations are carried out.

Besides being not transparent and technically very demanding, the procedure just described has one important drawback. Namely, it is not clear a priori whether this renormalization prescription, when applied to the individual diagrams, preserves all Ward identities; albeit it should, of course, if done consistently. For this reason, alternative approaches have been developed, which boil down to the modification of the rules for the calculation of Feynman diagrams. The idea for such a modification is best explained in a simple example of a one-dimensional integral:

$$H = \int_{-\infty}^{+\infty} \frac{dk}{2\pi} \frac{k^2}{(a^2 + k^2)(b^2 + k^2)} = \frac{1}{2(a+b)}. \tag{4.335}$$

In analogy to baryon ChPT, let us assume that a and b denote the soft and hard scales of the theory (analogous to the pion and nucleon masses). Therefore, we have $a \ll b$, and the result can be expanded in a Taylor series in a/b:

$$\frac{1}{2(a+b)} = \frac{1}{2b} \left(1 - \frac{a}{b} + \frac{a^2}{b^2} + \cdots \right). \tag{4.336}$$

This expression contains both nonanalytic (odd powers of a) as well as analytic (even powers of a) terms in a^2. Further, a naive power counting would assume that k is of order of a, and thus $I = O(a)$. It is seen that the leading-order analytic term in the preceding expression breaks the power counting.

Imagine now that we had a method of separating the analytic and nonanalytic terms in the integral without explicitly calculating it. (The latter might be problematic, if we have to deal with the integrals emerging from the actual Feynman diagrams.) Since one knows that the analytic terms can be removed by changing the renormalization prescription, one could alternatively change the rules for calculating the integral, *declaring* that the result of the integral is equal to its nonanalytic part. This method has the advantage that it automatically preserves the Ward identities since these identities obviously hold for the analytic and nonanalytic parts separately. On the other hand, in order to respect the power counting, it suffices to drop only the first few terms in the expansion of the analytic part and not the whole analytic part. In this case, however, one should ensure that this is done consistently in all Green's functions, so that the Ward identities stay intact.

Next, we shall demonstrate in our simple example, how one can separate the analytic and nonanalytic parts without calculating the integrals explicitly. In order to render all integrals finite after Taylor expansion of the integrands, we shall use dimensional regularization with $D \to 1$ at the end. Further, let us introduce a cutoff Λ with $a \ll \Lambda \ll b$ and split the integral in two parts:

$$H = \int \frac{d^D k}{(2\pi)^D} \frac{k^2 \left(\theta(|k| - \Lambda) + \theta(\Lambda - |k|) \right)}{(a^2 + k^2)(b^2 + k^2)} . \tag{4.337}$$

For $|k| < \Lambda$, one can expand the propagator, containing the heavy mass b, in the integrand. For $|k| > \Lambda$, the other propagator can be expanded. The result is given by

$$H = \int_{|k|<\Lambda} \frac{d^D k}{(2\pi)^D} \frac{k^2}{b^2(a^2 + k^2)} \left(1 - \frac{k^2}{b^2} + \cdots \right)$$
$$+ \int_{|k|>\Lambda} \frac{d^D k}{(2\pi)^D} \frac{1}{(b^2 + k^2)} \left(1 - \frac{a^2}{k^2} + \cdots \right) . \tag{4.338}$$

Note that both integrals are finite even for $D = 1$.

Let us now extend the integration in the first integral up to $\Lambda \to \infty$ (the integral is still finite in dimensional regularization) by rewriting the preceding expression as:

$$H = \int \frac{d^D k}{(2\pi)^D} \frac{k^2}{b^2(a^2 + k^2)} \left(1 - \frac{k^2}{b^2} + \cdots \right)$$
$$- \left\{ \int_{|k|>\Lambda} \frac{d^D k}{(2\pi)^D} \frac{k^2}{b^2(a^2 + k^2)} \left(1 - \frac{k^2}{b^2} + \cdots \right) \right.$$
$$+ \left. \int_{|k|>\Lambda} \frac{d^D k}{(2\pi)^D} \frac{1}{(b^2 + k^2)} \left(1 - \frac{a^2}{k^2} + \cdots \right) \right\} . \tag{4.339}$$

The propagator in the second integral can now be expanded in powers of a^2/k^2. Further, the expression in the curly brackets does not depend on Λ because the original integral did not. One could use this property and perform analytic continuation of the result to $\Lambda \to 0$. The second integral vanishes in this limit, because it contains only no-scale integrals of the type $\int d^D k\, k^\alpha$, which are zero in dimensional regularization. We finally get

$$H = \int \frac{d^D k}{(2\pi)^D} \frac{k^2}{b^2(a^2+k^2)} \left(1 - \frac{k^2}{b^2} + \cdots\right)$$

$$+ \int \frac{d^D k}{(2\pi)^D} \frac{1}{(b^2+k^2)} \left(1 - \frac{a^2}{k^2} + \cdots\right) = I + R. \tag{4.340}$$

Performing the integral in dimensional regularization and setting $D \to 1$ at the end, one observes that the first and the second integral reproduce the nonanalytic and the analytic terms in the original expression, respectively. This can be verified even without calculating the integrals explicitly. Indeed, rescaling the momenta in the integrals for a non-integer D, one sees that the first and second integrals scale like $(a^2)^{D/2+n}$ and $(a^2)^n$, respectively, with n a positive integer. Consequently, they reproduce the non-polynomial and polynomial parts, respectively, in space-time with D dimensions. Note that the technique, used here, is similar to the so-called threshold expansion [71], or expansion by regions [72], already considered in Chapter 2. Namely, the characteristic scales of the problem are a and b. The integration momentum can scale either as $k \sim a$ or as $k \sim b$. In either region, we may Taylor-expand the pertinent propagators in the integrand. Carrying out the integration in dimensional regularization and adding the contributions from all regions together, we arrive at the correct result for the whole integral, which was just displayed. The analytic (R) and nonanalytic (I) terms within this approach emerge from different regions of the integration momenta and are clearly separated from each other.

Next, we briefly consider another approach that allows one to arrive at the same result without expanding the integrand. Using again dimensional regularization and applying the Feynman trick to combine the two propagators, we get

$$H = \frac{D\Gamma(1-D/2)}{2(4\pi)^{D/2}} \int_0^1 dx \left(xb^2 + (1-x)a^2\right)^{D/2-1}. \tag{4.341}$$

Performing the limit $D \to 1$ and carrying out the integration in x, we again arrive at the exact result, given in Eq. (4.335).

Note now that the nonanalytic behavior of the preceding integral in a can be traced back to the endpoint singularity at $x = 0$. Indeed, setting $D = 1$, one sees that the integral is finite at $a^2 = 0$, but its derivatives with respect to a^2 diverge at $a^2 = 0$. For this reason, we may extend the integration to $x \to \infty$, writing

$$H = \frac{D\Gamma(1-D/2)}{2(4\pi)^{D/2}} \left\{\int_0^\infty - \int_1^\infty\right\} dx \left(xb^2 + (1-x)a^2\right)^{D/2-1} = I + R. \tag{4.342}$$

In the first integral, we may rescale $x \to a^2 x$. The result is

$$I = \frac{D\Gamma(1-D/2)}{2(4\pi)^{D/2}} a^D \int_0^\infty dx \left(1 + (b^2-a^2)x\right)^{D/2-1}. \tag{4.343}$$

Unlike the original integral, the expansion of the integrand in powers of a^2 in the preceding expression does not generate divergences and is thus allowed. Taking into account the factor a^D in front, one concludes that this expansion produces only non-integer powers of a in a space with D dimensions and thus corresponds to the non-analytic term. On the other hand, it is easy to ensure that the expansion of R produces only integer powers of a^2.

Let us briefly summarize our findings. We have seen that, due to the presence of a hard scale $\mathring{m}$ in the propagators, the loops in baryon ChPT break the power counting. The breaking emerges only in the polynomial terms and can be taken care of with a modified renormalization prescription. Here, it is important to verify that such a modification preserves the Ward identities, which lie at the foundation of our approach, which is based on chiral symmetry. The most convenient way is to alter the rules for the calculation of the Feynman diagrams, in order to ensure that the unwanted contributions disappear. There exist different strategies to achieve this:

- One can Taylor-expand the propagators, explicitly containing the heavy scale, in order to ensure that the contribution from the region, where the integration momenta are of order of the heavy scale, disappear. A most convenient way to do this is to perform such an expansion at the Lagrangian level and produce Feynman diagrams by using the expanded Lagrangian. This approach is known under the name of Heavy Baryon ChPT. Note that the nonanalytic part in this approach also comes expanded in chiral powers. Also, the fermion loops, which were discussed in the previous section, are automatically set to zero here, because their contribution is purely analytic.

- We can use the trick with the extension of the integration limits in Feynman parameters. This approach comes under the name of the Infrared Regularization. The nonanalytic part in this approach is resummed to all orders.

- As mentioned already, it is possible to subtract only the polynomial terms that break the power counting. This procedure is known under the name of the Extended On-Mass-Shell approach.

In the next section, we briefly consider all these three approaches.

4.11 Consistent Formulations of Baryon Chiral Perturbation Theory

4.11.1 Heavy Baryon Chiral Perturbation Theory

As we saw, the fully relativistic treatment of the baryons leads to severe complications in the low-energy structure of the EFT. So the question arises, how can we restore the one-to-one correspondence between the loop and the small momentum expansion? The first and maybe most intuitive solution is the so-called heavy baryon approach. If one considers the baryons as extremely heavy, only baryon momenta relative to the rest mass will count, and these can be small. The emerging picture is that of a very heavy source surrounded by a cloud of light (almost massless) particles. This is exactly the same idea that is used in heavy quark EFTs; see, for example, Ref. [73]. Therefore, it appears natural to apply the insight gained from heavy quark EFTs to the pion–nucleon sector. A formulation of baryon ChPT based on these ideas was given in Ref. [74]; it is called Heavy Baryon ChPT (HBChPT). In this approach, one takes the extreme nonrelativistic limit of the fully relativistic theory and expands in powers of the inverse baryon mass,[19] similar to the time-honored Foldy–Wouthuysen [78] transformation of the Dirac equation for a very heavy fermion; see Chapter 2.

Let us first spell out the underlying ideas using a very simple model before we come back to the pion–nucleon (πN) system. Our starting point is a free Dirac field with mass m:

$$\mathcal{L} = \bar{\Psi}(i\slashed{\partial} - m)\Psi. \tag{4.344}$$

Consider the spin-1/2 particle as very heavy. This allows us to write its four-momentum as

$$p_\mu = mv_\mu + l_\mu, \tag{4.345}$$

with v_μ the four-velocity satisfying $v^2 = 1$, and l_μ a small off-shell momentum, $v \cdot l \ll m$. We can now construct eigenstates of the velocity projection operator $P_v = \frac{1}{2}(1 + \slashed{v})$ via

$$\Psi = e^{-imv \cdot x}(H + h), \qquad \slashed{v}H = H, \qquad \slashed{v}h = -h. \tag{4.346}$$

In the nucleon rest-frame, $v_\mu = (1, 0, 0, 0)$, this leads to the standard nonrelativistic reduction of a spinor into upper and lower components. Substituting Eq. (4.346) into Eq. (4.344), we find

$$\mathcal{L} = \bar{H}(iv \cdot \partial)H - \bar{h}(iv \cdot \partial + 2m)h + \bar{H}i\slashed{\partial}^{\perp}h + \bar{h}i\slashed{\partial}^{\perp}H, \tag{4.347}$$

with $\slashed{\partial}^{\perp}$ the transverse part of the Dirac operator, $\slashed{\partial} = \slashed{v}(v \cdot \partial) + \slashed{\partial}^{\perp}$. From this it follows that the large component field H obeys a free Dirac equation (making use of the EOM for h):

$$v \cdot \partial H = 0, \tag{4.348}$$

modulo corrections that are suppressed by powers of $1/m$. The corresponding propagator of H reads

$$S(\omega) = \frac{-1}{v \cdot k + i\varepsilon}, \qquad \varepsilon > 0, \tag{4.349}$$

with $\omega = v \cdot k$.[20] The Fourier transform of Eq. (4.349) gives the space-time representation of the heavy baryon propagator. Its explicit form, $\tilde{S}(t, \mathbf{r}) = \theta(t)\delta^{(3)}(\mathbf{r})$, illustrates

---

[19]   Notice that earlier, the authors of Refs. [75, 76] considered a static source model for the baryons in their determination of quark mass ratios from the baryon spectrum. Also, the HBChPT approach can be formulated without recourse to the relativistic approach, but this requires the use of reparameterization invariance for a correct implementation [77]. We consider the matching to the relativistic approach more natural.

[20]   In principle, such a propagator can lead to an unwanted IR singular behavior. However, in most calculations it is outside the physical region and of no relevance. In the calculation of box diagrams for the nucleon–nucleon potential, we encounter this difficulty, which requires a special treatment, as discussed in Ref. [8].

very clearly that the field H represents an (infinitely heavy) static source. The mass-dependence now resides entirely in new vertices, which can be ordered according to their power in $1/m$. A more elegant path integral formulation is given in Ref. [79]. There is one more point worth noticing. In principle, the field H should carry a label 'v' since it has a definite velocity. For the purposes of our discussion, we do not need to worry about this label and will therefore drop it.

Let us now return to the πN system. The reasoning is completely analogous to that just discussed. Consider the nucleon four-momentum,

$$p_\mu = \mathring{m} v_\mu + l_\mu \,, \tag{4.350}$$

with $v^2 = 1$ and $v \cdot l \ll \mathring{m}$, with $\mathring{m}$ the nucleon mass in the chiral two-flavor limit. Velocity projection of the nucleon spinor Ψ amounts to

$$\Psi = \exp(-i\mathring{m} v \cdot x)(H + h), \qquad \slashed{v} H = H, \qquad \slashed{v} h = -h. \tag{4.351}$$

The EOM for the small component h takes the form

$$h = \frac{1}{2}(1 - \slashed{v})\frac{1}{2\mathring{m}}\left(i\slashed{D} + \frac{1}{2}\mathring{g}_A \slashed{u}\gamma_5\right)H + O(1/\mathring{m}^2)\,, \tag{4.352}$$

which allows us to eliminate it, and the leading-order Lagrangian takes the form

$$\mathcal{L}_{\pi N}^{(1)} = \bar{H}\left(iv \cdot D + \mathring{g}_A S \cdot u\right)H + O(1/\mathring{m})\,. \tag{4.353}$$

Here, S_μ is the Pauli–Lubanski spin-operator, $S_\mu = i\gamma_5 \sigma_{\mu\lambda} v^\lambda/2$. The latter obeys the relations (in D space–time dimensions)

$$S \cdot v = 0, \quad S^2 = \frac{1 - D}{4}, \quad \{S_\mu, S_\nu\} = \frac{1}{2}\left(v_\mu v_\nu - g_{\mu\nu}\right), \quad [S_\mu, S_\nu] = i\varepsilon_{\mu\nu\alpha\beta} v^\alpha S^\beta. \tag{4.354}$$

In the rest-frame, $\mathbf{S} = \boldsymbol{\sigma}/2$. We notice again that the mass term has disappeared from the effective Lagrangian and that the mass dependence resides in a tower of higher-order operators that are suppressed by powers of $1/\mathring{m}$.

To construct these higher-order terms, we follow here the systematic analysis of quark currents in flavor $SU(2)$ of Ref. [80]. We will derive the effective Lagrangian for heavy nucleons in terms of path integrals. In this formulation, the $1/m_N$ corrections are easily constructed. Consider the generating functional for the chiral Lagrangian of the πN system:

$$Z[\eta, \bar{\eta}, s, p, v, a] = \int d\Psi d\bar{\Psi} dU \exp\left(i\left\{S_{\pi N} + S_{\pi\pi} + \int d^4 x(\bar{\eta}\psi + \bar{\psi}\eta)\right\}\right), \tag{4.355}$$

where

$$S_{\pi N} = \int d^4 x \bar{\Psi}\left(i\slashed{D} - \mathring{m} + \frac{\mathring{g}_A}{2}\slashed{u}\gamma_5\right)\Psi + \int d^4 x\left\{\mathcal{L}_{\pi N}^{(2)} + \mathcal{L}_{\pi N}^{(3)} + \ldots\right\},$$

$$S_{\pi\pi} = \int d^4 x\left\{\mathcal{L}_{\pi\pi}^{(2)} + \mathcal{L}_{\pi\pi}^{(4)} + \ldots\right\}. \tag{4.356}$$

The aim is to integrate out the heavy degrees of freedom. To this end, the nucleon field Ψ is split into upper and lower components with fixed four-velocity v, according to Eq. (4.351). In terms of these components, the action $S_{\pi N}$ may be rewritten as

$$S_{\pi N} = \int d^4x \left\{ \bar{H} \mathcal{A} H + \bar{h} \mathcal{B} H + \bar{H} \gamma_0 \mathcal{B}^\dagger \gamma_0 h - \bar{h} \mathcal{C} h \right\} . \tag{4.357}$$

The operators $\mathcal{A}$, $\mathcal{B}$ and $\mathcal{C}$ have the low-energy expansions

$$\mathcal{X} = \mathcal{X}^{(1)} + \mathcal{X}^{(2)} + \dots , \quad \mathcal{X} = (\mathcal{A}, \mathcal{B}, \mathcal{C}) , \tag{4.358}$$

where $\mathcal{X}^{(i)}$ is a quantity of order p^i. The explicit expressions read

$$\mathcal{A}^{(1)} = i(v \cdot D) + \overset{\circ}{g}_A(u \cdot S),$$

$$\mathcal{A}^{(2)} = \frac{\overset{\circ}{m}}{F^2} \Big(c_1 \langle \chi_+ \rangle + c_2 (v \cdot u)^2 + c_3 u \cdot u + c_4 [S^\mu, S^\nu] u_\mu u_\nu$$

$$+ c_5 \left(\chi_+ - \frac{1}{2} \langle \chi_+ \rangle \right) - \frac{i}{4\overset{\circ}{m}} [S^\mu, S^\nu] ((1 + c_6) F_{\mu\nu}^+ + c_7 \langle F_{\mu\nu}^+ \rangle) \Big),$$

$$\mathcal{B}^{(1)} = i \not{D}^\perp - \frac{1}{2} \overset{\circ}{g}_A (v \cdot u) \gamma_5,$$

$$\mathcal{B}^{(2)} = \frac{\overset{\circ}{m}}{F^2} \Big(\left(-\frac{c_2}{4} i[u_\mu, u_\nu] - \frac{c_6}{4} F_{\mu\nu}^+ \right) \sigma^{\mu\nu} - \frac{c_4}{4\overset{\circ}{m}} \langle u^\mu u^\nu v_\mu v_\nu \rangle i(v \cdot \overset{\leftrightarrow}{D}) \Big),$$

$$\mathcal{C}^{(1)} = i(v \cdot D) + 2\overset{\circ}{m} + \overset{\circ}{g}_A(u \cdot S),$$

$$\mathcal{C}^{(2)} = -\mathcal{A}^{(2)} . \tag{4.359}$$

Here, $\not{D}^\perp = \gamma^\mu (g_{\mu\nu} - v_\mu v_\nu) D^\nu$ is the transverse part of the covariant derivative, which satisfies $\{\not{D}^\perp, \not{v}\} = 0$. We have taken advantage of the simplifications for the Dirac algebra in the heavy mass formulation. It allows us to express any Dirac bilinear $\bar{\Psi} \Gamma_\mu \Psi$ ($\Gamma_\mu = 1, \gamma_\mu, \gamma_5, \dots$) in terms of the four-velocity velocity v_μ and the spin-vector S_μ. In terms of these quantities, the standard Dirac bilinears can be rewritten as

$$\bar{H} \gamma_\mu H = v_\mu \bar{H} H, \ \bar{H} \gamma_5 H = 0, \ \bar{H} \gamma_\mu \gamma_5 H = 2\bar{H} S_\mu H,$$

$$\bar{H} \sigma^{\mu\nu} H = 2\varepsilon^{\mu\nu\alpha\beta} v_\alpha \bar{H} S_\beta H, \ \bar{H} \gamma_5 \sigma^{\mu\nu} H = 2i(v^\mu \bar{H} S^\nu H - v^\nu \bar{H} S^\mu H). \tag{4.360}$$

Therefore, the Dirac algebra is very simple in the extreme nonrelativistic limit.

Let us return to the discussion of the generating functional. The source term in Eq. (4.355) is also rewritten in terms of the fields H and h,

$$\int d^4x (\bar{\eta} \Psi + \bar{\Psi} \eta) = \int d^4x (\bar{R} H + \bar{H} R + \bar{\rho} h + \bar{h} \rho) \tag{4.361}$$

with

$$R = \frac{1}{2}(1 + \not{v}) e^{i m v \cdot x} \eta , \quad \rho = \frac{1}{2}(1 - \not{v}) e^{i m v \cdot x} \eta . \tag{4.362}$$

Differentiating with respect to the source R yields the Green's functions of the projected fields H. The heavy degrees of freedom, h, may now be integrated out. Shifting variables $h' = h - C^{-1}(\mathcal{B}H + \rho)$ and completing the square, the generating functional becomes

$$Z[R, \bar{R}, \rho, \bar{\rho}, s, p, v, a]$$

$$= \int dH d\bar{H} dU \Delta_h \exp i \left\{ S'_{\pi N} + S_{\pi\pi} + \int d^4 x (\bar{R}H + \bar{H}R) + \dots \right\}, \qquad (4.363)$$

where

$$S'_{\pi N} = \int d^4 x \bar{H} \left(\mathcal{A} + (\gamma_0 \mathcal{B}^\dagger \gamma_0) C^{-1} \mathcal{B} \right) H, \qquad (4.364)$$

and the ellipsis stands for terms with the sources ρ and $\bar{\rho}$ [81]. In Eq. (4.364), Δ_h denotes the determinant coming from the Gaussian integration over the small component field, that is,

$$\Delta_h = \exp\left\{ \frac{1}{2} \operatorname{tr} \ln C \right\}$$

$$= \mathcal{N} \exp\left\{ \frac{1}{2} \operatorname{tr} \ln \left(1 + C^{(1)-1} (i(v \cdot D) + \mathring{g}_A (S \cdot u) + C^{(2)} + \dots) \right) \right\}, \qquad (4.365)$$

with $\mathcal{N}$ an irrelevant constant. As noted in Ref. [79], the space-time representation of the h propagator, $C^{(1)-1}$, implies that Δ_h is just a constant. Indeed, in coordinate space, each propagator is proportional to $\exp(-2i\mathring{m}(x_0 - y_0))\theta(x_0 - y_0)$, where x, y are the endpoints of the fermion line. The loop contains a product of several operators with endpoints in the vertices of the loop diagram. Since in the loop the starting and the end points are the same, the product of the theta-functions necessarily vanishes and, thus, the determinant Δ_h becomes an (inessential) constant that can be included in the normalization of the path integral.[21]

The next step consists in expanding the nonlocal functional (4.363) in a series of operators of increasing dimension. This corresponds to an expansion of the matrix C^{-1} in a power series in $1/\mathring{m}$

$$C^{-1} = \frac{1}{2\mathring{m}} - \frac{i(v \cdot D) + \mathring{g}_A (u \cdot S)}{(2\mathring{m})^2} + O(p^2). \qquad (4.366)$$

Thus the effective heavy nucleon Lagrangian up to $O(p^3)$ is given as

$$S'_{\pi N} = \int d^4 x \bar{H} \left\{ \mathcal{A}^{(1)} + \mathcal{A}^{(2)} + \mathcal{A}^{(3)} + (\gamma_0 \mathcal{B}^{(1)\dagger} \gamma_0) \frac{1}{2\mathring{m}} \mathcal{B}^{(1)} \right.$$

$$+ \frac{(\gamma_0 \mathcal{B}^{(1)\dagger} \gamma_0) \mathcal{B}^{(2)} + (\gamma_0 \mathcal{B}^{(2)\dagger} \gamma_0) \mathcal{B}^{(1)}}{2\mathring{m}}$$

$$\left. - (\gamma_0 \mathcal{B}^{(1)\dagger} \gamma_0) \frac{i(v \cdot D) + \mathring{g}_A (u \cdot S)}{(2\mathring{m})^2} \mathcal{B}^{(1)} \right\} H + O(p^4). \qquad (4.367)$$

---

[21] This discussion contains a small loophole. Namely, the θ function at a vanishing argument is not uniquely defined. We have already encountered a similar problem before and have seen that it can be easily handled within the dimensional regularization.

Note that the neglected terms of $O(p^4)$ are suppressed by inverse powers of either $\overset{\circ}{m}$ or $\Lambda = 4\pi F_\pi$. These two scales are treated on the same footing, the only thing that counts is the power of the low momentum p. It is important to note that this expansion of the nonlocal action makes the closed fermion loops disappear from the theory because at any finite order in $1/\overset{\circ}{m}$, $S'_{\pi N}$ is local. To complete the expansion of the generating functional up to order p^3, we have to add the one-loop corrections with vertices from $\mathcal{A}^{(1)}$ only. Working to order p^4 (which still includes only one-loop diagrams), one also has to include vertices from $\mathcal{A}^{(2)}$ and from $(\gamma_0 \mathcal{B}^{(1)\dagger}\gamma_0)\mathcal{B}^{(1)}/(2\overset{\circ}{m})$.

The disappearance of the nucleon mass term to leading order in $1/\overset{\circ}{m}$ in the Lagrangian for the field H allows for a consistent chiral power counting. The nucleon propagator is of the form Eq. (4.349), that is, has chiral power p^{-1}. Consequently, the dimension N_χ of any Feynman diagram is given by the following:[22]

$$N_\chi = 4L - 2I_M - I_B + \sum_d d(N_d^M + N_d^{MB}) \tag{4.368}$$

with L the number of loops, I_M (I_B) the number of internal meson (baryon) lines and N_d^M, N_d^{MB} the number of vertices of dimension d from the meson and the meson-baryon Lagrangian, in order. Consider now the case of a single baryon line running through the diagram [82]. In that case, we have

$$\sum_d N_d^{MB} = I_B + 1 . \tag{4.369}$$

Together with the general topological relation

$$L = I_M + I_B - \sum_D (N_d^M + N_d^{MB}) + 1 , \tag{4.370}$$

we obtain

$$N_\chi = 2L + 1 + \sum_d (d-2)N_d^M + \sum_d (d-1)N_d^{MB} . \tag{4.371}$$

Clearly, $N_\chi \geq 2L + 1$ so that we have a consistent power-counting scheme in analogy to the one in the meson sector (cf. Eq. (4.262)). In particular, the coefficients appearing in $\mathcal{L}_{\pi N}^{(1)}$ and $\mathcal{L}_{\pi N}^{(2)}$ are not renormalized at any loop order since $N_\chi \geq 3$ for $L \geq 1$ (if one uses, e.g., dimensional regularization). This is in marked contrast to the infinite renormalization of $\overset{\circ}{g}_A$ and $\overset{\circ}{m}$ in the relativistic approach; see Eq. (4.334). As stated before, all mass dependence now resides in the vertices of the local pion–nucleon Lagrangian, that is, all vertices now consist of a string of operators with increasing powers in $1/\overset{\circ}{m}$. For example, the pion–nucleon vertex reads

$$\frac{\overset{\circ}{g}_A}{F} \tau^i S \cdot q + O(1/\overset{\circ}{m}) . \tag{4.372}$$

For a list of all Feynman rules in the heavy baryon approach, we refer to the reviews [68, 83].

---

[22] Since this power-counting argument is general, we consider mesons (M) and baryons (B) for a while (instead of pions and nucleons).

The renormalization of the HB Lagrangian proceeds along the same lines as discussed for the mesonic Lagrangian in Section 4.9.3. The one-loop renormalization at $O(p^3)$ can be found in Refs. [64, 81] and the extension to fourth order in Ref. [84]. An elegant approach, which allows us to treat the bosonic and fermionic degrees of freedom in a setting that resembles the supersymmetric theories, can be found in Refs. [85–87]. Here we list (without derivation) the divergent parts of the $O(p^3)$ LECs from Ref. [65]:[23]

$$d_i = \frac{\beta_i}{F^2}\lambda + d_i^r(\mu).$$

(4.373)

Here,

$$\beta_1 = -\frac{1}{6}\overset{\circ}{g}_A^4, \quad \beta_2 = -\frac{1}{12} - \frac{5}{12}\overset{\circ}{g}_A^2, \quad \beta_3 = \frac{1}{2} + \frac{1}{6}\overset{\circ}{g}_A^4, \quad \beta_5 = -\frac{5}{24} + \frac{5}{24}\overset{\circ}{g}_A^2,$$

$$\beta_6 = -\frac{1}{6} - \frac{5}{6}\overset{\circ}{g}_A^2, \quad \beta_{10} = \frac{1}{2}\overset{\circ}{g}_A + \frac{5}{2}\overset{\circ}{g}_A^3 + 2\overset{\circ}{g}_A^5, \quad \beta_{11} = \frac{1}{2}\overset{\circ}{g}_A - \frac{3}{2}\overset{\circ}{g}_A^3 - \frac{2}{3}\overset{\circ}{g}_A^5,$$

$$\beta_{12} = -2\overset{\circ}{g}_A - \overset{\circ}{g}_A^3, \quad \beta_{13} = \overset{\circ}{g}_A^3 + \frac{2}{3}\overset{\circ}{g}_A^5, \quad \beta_{14} = \frac{1}{3}\overset{\circ}{g}_A^4, \quad \beta_{16} = \frac{1}{2}\overset{\circ}{g}_A + \overset{\circ}{g}_A^3,$$

$$\beta_{18} = \overset{\circ}{g}_A, \quad \beta_{19} = -\frac{1}{2}\overset{\circ}{g}_A, \quad \beta_{20} = \overset{\circ}{g}_A + \overset{\circ}{g}_A^3 \quad \beta_{21} = -\overset{\circ}{g}_A^3,$$

(4.374)

and $\beta_i = 0$ for $i = 4, 7, 8, 9, 15, 17, 22, 23$.

As an example of how the heavy baryon approach works, consider the leading one-loop correction to the nucleon mass, which should scale as M_π^3 on dimensional grounds. An explicit HBChPT calculation gives

$$\delta m_N^{(3)} = -\frac{3g_A^2 M_\pi^3}{32\pi F_\pi^2}.$$

(4.375)

Here, we have set the appearing couplings and masses on their physical values, as the differences to the leading-order expressions will show up only in higher orders. Contrary to the relativistic ChPT result, the loop correction in the heavy baryon approach is finite and vanishes in the chiral limit. Notice further that Eq. (4.375) represents the leading contribution to the nucleon mass, which is nonanalytic in quark masses. It can be obtained from Eqs. (4.330), (4.331) and (4.333), if the function $F(\Omega)$ on the mass shell $p^2 = m_N^2$ is expanded in powers of $\Omega = M_\pi/(2m_N)$ and the leading nonanalytic term, proportional to M_π^3, is singled out in the nucleon mass. It also agrees with the time-honored result of Pagels, Langacker and others [88]. It is therefore instructive to perform the calculations in HBChPT that lead to Eq. (4.375). Consider the Feynman diagram in Fig. 4.12a, where the nucleon emits a pion of momentum k and absorbs the

---

[23] There is one subtlety here. After deriving the divergent part of the one-loop generating functional, we have to use the EOM in order to reduce the mesonic part of the effective action at $O(p^3)$ to the standard form given in Eqs. (4.198), (4.199) and (4.200). As a result of using the EOM in the meson sector, some coefficients in the divergent parts in the nucleon sector may change. In this manner, we ensure that the divergences in the meson and nucleon sectors are consistent with each other, when the operator basis in the meson sector is chosen in the standard form.

same pion (which is incoming with momentum $-k$). Using Eqs. (4.349), (4.372) and the relativistic propagator for the pion, the mass shift $\delta m_N^{(3)}$ in D dimensions is given by

$$\delta m_N^{(3)} = \frac{3g_A^2}{F_\pi^2} \int \frac{d^D k}{(2\pi)^D i} \frac{1}{M_\pi^2 - k^2 - i\varepsilon} \frac{1}{-v \cdot k + i\varepsilon} S \cdot (-k) S \cdot k, \qquad (4.376)$$

making use of $\tau^i \tau^i = 3$. From the anti-commutation relation of two spin matrices, Eq. (4.354), and by completing the square we have

$$S_\mu S_\nu k^\mu k^\nu = \frac{1}{4} \left(v \cdot k \, v \cdot k + M_\pi^2 - k^2 - M_\pi^2 \right), \qquad (4.377)$$

so that

$$\delta m_N^{(3)} = -\frac{3g_A^2}{4F_\pi^2} \int \frac{d^D k}{(2\pi)^D i} \left[\frac{1}{v \cdot k - i\varepsilon} + \frac{v \cdot k}{M_\pi^2 - k^2 - i\varepsilon} \right.$$

$$\left. - \frac{M_\pi^2}{(M_\pi^2 - k^2 - i\varepsilon)(v \cdot k - i\varepsilon)} \right]. \qquad (4.378)$$

It is easy to show that the contribution of the first two terms in the dimensional regularization vanishes. Indeed, consider the rest-frame of the nucleon with $v_\mu = (1,0,0,0)$. Then, the sum of the first two terms gives

$$\frac{1}{k_0 - i\varepsilon} + \frac{k_0}{w_\pi^2 - k_0^2 - i\varepsilon} = \frac{w_\pi^2}{(k_0 - i\varepsilon)(w_\pi^2 - k_0^2 - i\varepsilon)}, \qquad (4.379)$$

where $w_\pi = \sqrt{M_\pi^2 + \mathbf{k}^2}$. The Cauchy integration over k_0 can be performed and yields a constant independent of $\mathbf{k}$. Integration over $\mathbf{k}$ in dimensional regularization then leads to a vanishing result. So we are left with

$$\delta m_N^{(3)} = \frac{3g_A^2}{4F_\pi^2} J(0) M_\pi^2,$$

$$J(0) = \int \frac{d^D k}{(2\pi)^D i} \frac{1}{(M_\pi^2 - k^2 - i\varepsilon)(v \cdot k - i\varepsilon)}. \qquad (4.380)$$

The remaining task is to evaluate $J(0)$. For that, we use the identity

$$\frac{1}{AB} = \int_0^\infty dy \frac{2}{[A + 2yB]^2}, \qquad (4.381)$$

define $k' = k - yv$, complete the square and use $v^2 = 1$,

$$J(0) = \int_0^\infty dy \int \frac{d^D k'}{(2\pi)^D i} \frac{1}{[M_\pi^2 + y^2 + k'^2 - i\varepsilon]^2}$$

$$= \frac{2}{(2\pi)^D} \int_0^\infty dy \int_0^\infty dk' \frac{(k')^{D-1}}{[M_\pi^2 + y^2 + k'^2]^2} \frac{2\pi^{D/2}}{\Gamma(D/2)}, \qquad (4.382)$$

where we have performed a Wick rotation, $k_0 \to ik_0$, and dropped the $i\varepsilon$. The last factor in Eq. (4.382) is the surface of the sphere in D dimensions. Introducing polar coordinates, $y = r\cos\phi$ and $k' = r\sin\phi$, and noting that the Jacobian of this transformation is r, we have

$$J(0) = \frac{4(4\pi)^{-D/2}}{\Gamma(D/2)} \int_0^\infty dr \, \frac{r^D}{(r^2 + M_\pi^2)^2} \int_0^{\pi/2} d\phi \, (\sin\phi)^{D-1} . \tag{4.383}$$

Both integrals appearing in Eq. (4.383) can be expressed in terms of products of Euler Gamma-functions with the result

$$J(0) = M_\pi^{D-3} (4\pi)^{-D/2} \Gamma\left(\frac{1}{2}\right) \Gamma\left(\frac{3-D}{2}\right) = -\frac{M_\pi}{8\pi} , \tag{4.384}$$

where in the last step we have set $D = 4$. This leads us to the final result given in Eq. (4.375) and shows the relative simplicity of the heavy baryon approach. Many processes have been analyzed in this scheme. Arguably the most intriguing findings were the modification of the low-energy theorem for neutral pion photoproduction off the nucleon [89] and the observation of the $1/M_\pi$ divergence of the nucleons' electric and magnetic polarizabilities in the chiral limit [90].

We end this section with a few remarks on HBChPT:

- It is clear that the Ward identities in the HBChPT stay intact at every order in a simultaneous expansion in $1/\Lambda$ and $1/\mathring{m}$. This is simply because the Lagrangian of HBChPT is chirally symmetric by construction. Closed nucleon loops vanish because all poles in such a loop lie on one side of the integration contour in k_0 (in position space, this corresponds to the product of the theta-functions, which vanishes for any closed loop; see the preceding discussion). Moreover, the antinucleons, which appear in the loops in the relativistic case, here are integrated out and are consistently included in the LECs. Consequently, the decoupling of the sectors with different number of nucleons, which was mentioned, indeed takes place – the fermion determinant in HBChPT is equal to one.

- The extreme nonrelativistic limit may have an impact on the analytic structure of certain amplitudes. Prime examples are the pion–nucleon partial wave amplitudes $f_\pm^1(t)$ that have a logarithmic singularity on the second Riemann sheet [91] (originating from the projection of the nucleon pole terms in the invariant πN scattering amplitudes) located at

$$t_c = 4M_\pi^2 - M_\pi^4/m_N^2 = 3.98 M_\pi^2 , \tag{4.385}$$

very close to the physical threshold $t_0 = 4M_\pi^2$ of the spectral functions of the nucleon isovector form factors (for details, see Ref. [92]). In the limit $m_N \to \infty$, these two thresholds coalesce, and thus the analytic structure is distorted, leading to an infinite slope of the nucleon isovector spectral functions at threshold. Similar effects are also observed in the scalar nucleon form factor or the nucleon Compton scattering amplitude at the opening of the pion threshold. This, however, is a problem that is overcome at higher orders; for example, the two-loop calculation of nucleon spectral functions indeed leads to the correct behavior at threshold [93]. Another possibility is the use of non-expanded kinematical variables; see, for example, Refs. [94].

- We had already discussed that some of the LECs c_i are unnaturally large due to the presence of the close-by Δ-resonance. Instead of working with such enhanced LECs, we can also entertain the possibility of including the low-lying decuplet states in the EFT [95]. A systematic approach based on counting the mass difference $m_\Delta - m_N$ as another small parameter was developed in Ref. [96]. While this is a viable approach, it, of course, introduces additional LECs, and further, care has to be taken when discussing the chiral limit, as nucleon-delta mass difference remains finite [97].

4.11.2 Infrared Regularization

We have seen that the heavy baryon approach has certain deficiencies that are related to the extreme nonrelativistic limit (the strict $1/m_N$-expansion). In a manifestly covariant formulation of baryon ChPT that, however, must also allow for a consistent power counting, these problems should be overcome. A first step in this direction was the observation in Refs. [98, 99] that relativistic one-loop integrals can be separated into "soft" and "hard" parts. While for the former a similar power counting as in HBChPT applies, the contributions from the latter can be absorbed in certain LECs. In this way, one can combine the advantages of both methods.

A more formal and rigorous implementation of such a program was worked out in Ref. [100], and the resulting method was termed "infrared regularization." The essence of the method is as follows: any one-loop integral H is split into an infrared-singular and a regular part by a particular choice of Feynman parameterization. Consider first the regular part, called R. If one chirally expands these terms, we generate polynomials in momenta and quark masses. Consequently, to any order, R can be absorbed in the LECs of the effective Lagrangian. On the other hand, the infrared-singular part I has the same analytic properties as the full integral H in the low-energy region, and its chiral expansion leads to the nontrivial momentum and quark-mass dependencies of ChPT, like, for example, the chiral logarithms or fractional powers of the quark masses. It is this infrared-singular part I that is closely related to the heavy baryon expansion. There, the relativistic nucleon propagator is replaced by a heavy baryon propagator plus a series of $1/m_N$-suppressed insertions. Summing up all heavy baryon diagrams with all internal-line insertions yields the infrared-singular part of the corresponding relativistic diagram.

To be specific, let us consider the again the nucleon self-energy diagram, Fig. 4.12a. In D dimensions, the corresponding scalar loop integral is

$$H(p^2) = \frac{1}{i} \int \frac{d^D k}{(2\pi)^D} \frac{1}{[M_\pi^2 - k^2 - i\varepsilon][m_N^2 - (p-k)^2 - i\varepsilon]}, \quad \varepsilon > 0, \quad (4.386)$$

with p the four-momentum of the nucleon. At threshold, $p^2 = s_0 = (M_\pi + m_N)^2$, this results in (for more details, see [100])

$$H(s_0) = c(D) \frac{M_\pi^{D-3} + m_N^{D-3}}{M + m} = I + R , \qquad (4.387)$$

where $c(D)$ is some constant depending on the dimensionality of space-time. The infrared-singular piece I is characterized by fractional powers in the pion mass and is generated by loop momenta of order M_π. For these soft contributions, the power counting is fine and proceeds along the lines discussed in the previous section. On the other hand, the infrared regular part R is characterized by integer powers in the pion mass and is generated by internal momenta of the order of the nucleon mass (the large mass scale). These are the terms that lead to the violation of the power counting in the standard dimensional regularization discussed. For the self-energy integral, this splitting can be achieved in the following way:

$$
\begin{aligned}
H &= \int \frac{d^D k}{(2\pi)^D} \frac{1}{AB} = \int_0^1 dx \int \frac{d^D k}{(2\pi)^D} \frac{1}{[(1-x)A + xB]^2} \\
&= \left\{ \int_0^\infty - \int_1^\infty \right\} dx \int \frac{d^D k}{(2\pi)^D} \frac{1}{[(1-x)A + xB]^2} = I + R ,
\end{aligned} \qquad (4.388)
$$

with $A = M_\pi^2 - k^2 - i\varepsilon$, $B = m_N^2 - (P-k)^2 - i\varepsilon$, and the Feynman parameter integral is most easily solved by rescaling $x = \alpha u$ in terms of the mass ratio,

$$\alpha = M_\pi / m_N . \qquad (4.389)$$

The integral Eq. (4.388) develops an infrared singularity as M_π tends to zero, coming from the low-momentum region of the integration (which is the name-giving property for this scheme). The high-momentum part is free of infrared singularities and thus leads to a contribution that can be expanded in an ordinary power series. These correspond to the soft and hard momentum parts discussed before.

It can also be shown that infrared regularization leads to a unique, that is, process-independent result, in accordance with the chiral Ward identities of QCD [100]. This is essentially based on the fact that terms with fractional versus integer powers in the pion mass must be separately chirally symmetric. Consequently, the transition from any one-loop graph H to its infrared-singular piece I defines a symmetry-preserving regularization. Note also that fermion loops do not contain the infrared-singular pieces at all and, therefore, they are set to zero from the beginning in infrared regularization.

Some remarks concerning renormalization within this scheme are appropriate here. To leading order, the infrared-singular parts coincide with the heavy baryon expansion, in particular the infinite parts of loop integrals are the same. Therefore, the β-functions for the LECs, which absorb these infinities, are identical. However, the infrared-singular parts of relativistic loop integrals also contain infinite parts that are suppressed by powers of α, which hence cannot be absorbed as long as one only introduces counter terms to a finite order: exact renormalization only works up to the order at which one works; higher-order divergences have to be removed by hand.

Closely related to this problem is that of the mass scale μ, which we have to introduce in the process of regularization and renormalization. In dimensional regularization and related schemes, loop diagrams depend logarithmically on μ. This $\log\mu$ dependence is compensated for by running coupling constants, the running behavior being determined by the corresponding β-functions. In the same way as the contact terms cannot consistently absorb higher-order divergences, their β-functions cannot compensate for the scale dependence, which is suppressed by powers of α. In order to avoid this unphysical scale dependence in physical results, in Ref. [100] it was argued that the nucleon mass m_N serves as a "natural" scale in a relativistic baryon ChPT loop calculation and that therefore one should set $\mu = m_N$ everywhere when using the infrared regularization scheme. This was already suggested in Ref. [80] for the framework of a relativistic theory with ordinary dimensional regularization. Consequently, the chiral logarithms, $\log(M_\pi/\mu)$, will always appear as $\log \alpha$ in the IR scheme.

It is useful to be able to compare the results of the IR scheme to the equivalent heavy baryon ones. Note that one can always regain the heavy baryon result from the infrared regularized relativistic one by performing a strict chiral expansion of all involved loop functions. As a simple example, we again calculate the leading one-loop correction to the nucleon mass at third order, but this time using infrared regularization. The result is

$$\delta m_N^{(3)} = \frac{3g_A^2}{2F_\pi^2} m_N M_\pi^2 I(m_N^2) , \tag{4.390}$$

where the loop integral $I(m_N^2)$ is given by

$$I(m_N^2) = -\alpha^2 \left(L + \frac{1}{16\pi^2} \log \alpha \right) + \frac{\alpha}{16\pi^2} \left\{ \frac{\alpha}{2} - \sqrt{4 - \alpha^2} \arccos\left(-\frac{\alpha}{2} \right) \right\} . \tag{4.391}$$

Expanding this to leading order, we find

$$I(m_N^2) = -\frac{\alpha}{16\pi} + O(\alpha^2) , \tag{4.392}$$

which leads to the same (and well-known) result as the HB approach (cf. Eq. (4.375)), yielding the time-honored contribution nonanalytic in the quark masses. As explained before, the loop function $I(m_N^2)$ contains a non-leading divergence, which cannot be absorbed to this order, but will be canceled by an appropriate contact term at fourth order. The calculation of the nucleon mass to fourth order and the reduction to the heavy baryon limit is done in detail in Ref. [100].

An essential ingredient to the treatment of loop integrals as just described is the fact that higher-order effects are included as compared to the "strict" chiral expansion in the heavy baryon formalism. This was justified in Ref. [100] by improved convergence properties in the low-energy region. It introduces, however, a certain amount of arbitrariness as to which of these higher-order terms to keep and which to dismiss. It is therefore mandatory to exactly describe the treatment of these terms. The philosophy in Ref. [100] was, above all, to preserve the correct relativistic analyticity properties. This was achieved by keeping the full denominators of loop integrals (and evaluating them by the infrared regularization prescription), while expanding the numerators to the desired chiral order only. In addition, for example, crossing symmetry is to be conserved. A different approach was explored in Ref. [101]. As studies in HBChPT reveal, certain observables like the neutron electric form factor are very sensitive to recoil effects. Thus, it can be advantageous to keep all terms that occur according to the infrared regularization prescription and, in particular, not expand the numerators of loop integrals. Some care has also been taken in the treatment of certain LECs that appear in loop diagrams; see, for example, Ref. [101] for a detailed discussion. For yet another way of dealing with the higher-order terms, namely the promotion of some of such operators to a lower order, see Ref. [102].

We end this section with a few remarks:

- The analysis of the self-energy can be generalized to arbitrary one-loop graphs, which carry factors of the loop momentum in the numerator. These can always be reduced to combinations of the scalar ones,

$$H_{mn} = \frac{1}{i} \int \frac{d^D k}{(2\pi)^D} \frac{1}{a_1 \cdots a_m b_1 \cdots b_n} , \qquad (4.393)$$

corresponding to m meson and n baryon propagators. The representation of the corresponding infrared parts in terms of Feynman parameters coincides with the one for H_{mn} except that the integration over the parameter that combines the meson propagators with the nucleon ones runs from 0 to ∞ instead of 0 to 1. Special cases are $I_{m0} = H_{m0}$, $R_{m0} = 0$ and $I_{0n} = 0$, $R_{0n} = H_{0n}$. For more details on these calculations; see Ref. [100].

- One can show that the regular part R has a cut along the negative real axis. While this is not felt for most processes in the low-energy region, like, for example, pion–nucleon scattering in the threshold and subthreshold regime [103], it can lead to distortions in observables even at low energies that require differentiation with respect to some external parameter, like, for example, the photon virtuality in virtual Compton scattering [104].

- Extensions of this approach to the case of two heavy particles [105], to the spin-3/2 Δ baryon [106], to spin-1 resonances (vector mesons) [107], and vector mesons coupled to baryons [108] also exist. Note, however, that the application of IR to the case of vector mesons is a very subtle issue, since the number of vector mesons (unlike the number of baryons) is not a conserved quantity. Therefore, the ρ-meson can decay into two pions with hard momenta. This complication is discussed in more detail in Ref. [107].

- An extension of the approach to two loop diagrams is discussed in Ref. [109].

4.11.3 Extended On-Mass-Shell Renormalization

The extended on-mass-shell (EOMS) scheme is another covariant approach to baryon ChPT, that obeys a consistent power counting and has become popular lately. It was formulated in Ref. [110], based on earlier works [111, 112]. Here, we discuss only the underlying ideas. The basic idea consists in performing additional subtractions of dimensionally regularized diagrams beyond the modified minimal subtraction scheme. More precisely, the EOMS scheme consists of providing a rule determining which terms of a given diagram should be subtracted in order to satisfy the "naive" power counting by which one associates a well-defined power with the diagram in question. The terms to be subtracted are polynomials in the small parameters (external momenta and quark masses) and can thus be realized by a suitable adjustment of the counter terms of the most general effective Lagrangian.

This framework can be most easily understood by considering again the nucleon self-energy diagram, Fig. 4.12a, obtained by first expanding the integrand in small quantities and then performing the integration for each term. To be specific, for massless pions, the corresponding scalar one-loop integral H is expressed as

$$
H = \int \frac{d^D k}{(2\pi)^D i} \sum_{n=0}^{\infty} \frac{(p^2 - m_N^2)^n}{n!}
$$

$$
\times \left[\left(\frac{1}{2p^2} p_\mu \frac{\partial}{\partial p_\mu} \right)^n \frac{1}{(k^2 + i\varepsilon)[k^2 - 2k \cdot p + (p^2 - m_N^2) + i\varepsilon]} \right]_{p^2 = m_N^2}
$$

$$
= \int \frac{d^D k}{(2\pi)^D i} \left[\frac{1}{(k^2 + i\varepsilon)(k^2 - 2k \cdot p + i\varepsilon)} \right]_{p^2 = m_N^2}
$$

$$
- (p^2 - m_N^2) \left[\frac{1}{2m_N^2} \frac{1}{(k^2 - 2k \cdot p + i\varepsilon)^2} - \frac{1}{2m_N^2} \frac{1}{(k^2 + i\varepsilon)(k^2 - 2k \cdot p + i\varepsilon)} \right.
$$

$$
\left. - \frac{1}{(k^2 + i\varepsilon)(k^2 - 2k \cdot p + i\varepsilon)^2} \right]_{p^2 = m_N^2} + \dots \tag{4.394}
$$

The use of the subtraction point $p^2 = m_N^2$ in this equation gave rise to the name EOMS for the renormalization condition by analogy with the on-mass-shell renormalization scheme in renormalizable theories. A formal definition of the renormalization scheme follows. Subtract from the integrand of H those terms of the series of Eq. (4.394) that violate the power counting. These terms are always analytic in the small parameters and do not contain infrared singularities. In the preceding example we only need to subtract the first term. All the higher-order terms contain infrared singularities. For example, the last term of the second coefficient in Eq. (4.394) would generate a behavior $\sim k^3/k^4$ of the integrand for $D = 4$. Thus, in the EOMS approach we have $H^r = H - H^{\text{sub}}$ with

$$
H^{\text{sub}} = -i \int \frac{d^D k}{(2\pi)^D} \left[\frac{1}{(k^2 + i\varepsilon)(k^2 - 2k \cdot p + i\varepsilon)} \right]_{p^2 = m_N^2}, \tag{4.395}
$$

which is indeed the first term in the expansion. This so-called power-counting breaking (PCB) term can be absorbed in a LEC (or combinations of LECs) of the chiral effective pion–nucleon Lagrangian. This method can easily be generalized to the case of massive pions [110] and also to two loops.

The main advantage of EOMS is that it automatically leads to relativistic amplitudes with the correct analytic structure, in contrast to the HBChPT, where the kinematical variables are expanded in powers of $1/m_N$ that lead to the displacement of the kinematical singularities (poles and cuts), as well as to the baryon ChPT in the infrared regularization, where the extension of the integration range in the Feynman parameter leads to the emergence of spurious singularities. In some cases, these spurious singularities can come rather close to the physical region and affect the observables. On the

Figure 4.13 Two types of nucleon loops in the relativistic theory in the one-nucleon sector: (a) nucleon loops that are attached to the nucleon line only via pion lines, and (b) nucleon loops that are directly attached to the nucleon line via a four-Fermi vertex with the coupling constant C.

other hand, the EOMS approach is not completely free of deficiencies either. In particular, while in the HBChPT and the IR schemes the Ward identities are observed by construction, special care has to be taken in the EOMS approach not to break chiral symmetry. In what follows, we shall briefly discuss this issue.

First, we would like to discuss the role of closed nucleon loops. In the EOMS approach, one starts from the relativistic theory and faithfully evaluates Feynman diagrams in dimensional regularization. The Ward identities hold, of course. Let us restrict ourselves to the one-nucleon sector. In this case, two types of nucleon loops arise; see Fig. 4.13. These are (a) nucleon loops that are attached to the single nucleon line, which goes through the diagram, only via meson propagators, and (b) nucleon loops that are directly attached to the nucleon line via, for example, the vertex arising from the Fermi-type Lagrangian $\mathcal{L}_F = C(\bar{\Psi}\Psi)^2$. There are also vertices that contain a larger number of fermion fields as well as vertices with more derivatives.

A Ward identity is a relation among different Green's functions. Expanding all Green's functions in a perturbation series, it is straightforward to single out the contributions coming from the nucleon loops. Let us first prove that, after dropping all loops, the Ward identities remain the same. Indeed, the loops of type a can be included in the couplings of the effective meson Lagrangian, which remains chirally invariant. Further, the loops of type b come with different couplings (C, in our example). Since these couplings are allowed to take any value, the Ward identities should be fulfilled for the C-independent and C-dependent parts separately. Consequently, after dropping all closed nucleon loops, the Ward identities will still hold.[24]

What remains to consider are the PCB terms that emerge in the integrals of the type displayed in Eq. (4.394). Naively, these terms can be canceled by the first few terms in the expansion of the regular parts of the same diagrams in the IR scheme. Since these regular parts respect the Ward identities separately, so will the first few terms of the expansion. We might therefore conclude that the set of Green's functions in the EOMS scheme obeys the Ward identities.

The issue is, however, more subtle, because the definition of the kinematic variables for the expansion is not unique. More precisely, by redefining the external momenta,

---

[24] In order to have an intuitive picture, one might consider the calculation of the electron propagator in QED. One could quench all electron loops and use the Euler–Heisenberg Lagrangian instead of the usual photon kinetic term. Define further the electron residual momentum l_μ as $p_\mu = m_e v_\mu + l_\mu$, with $v^2 = 1$. The result for the self-energy will be the same as in QED, if the components of the residual momentum are much smaller than the electron mass m_e.

one could change the subleading powers in the chiral expansion. Therefore, it is not a priori clear whether the Ward identities hold at the order one is working, albeit it definitely works for the PCB terms. We remind the reader that we want to stay in the one-nucleon sector and do not want to extend the proof to the sectors with any number of nucleons. In order to refine the argument, we first define the kinematics. The four-momenta of the incoming and outgoing nucleons will be denoted by p and $p' = p + r$, respectively, and the momenta of the external (incoming and outgoing) pions are $q_1, \ldots, q_n$ (r is a linear combination of q_i with the coefficients ± 1). The Ward identities are *linear* relations between the Green's functions $G(p; q_1, \ldots, q_n)$ with different n. Moreover, these identities do not contain the nucleon momenta p, p' explicitly. An easy way to check this is to differentiate the generating functional with respect to the external sources, coupled to the nucleon fields, and set these sources to zero. There emerge no derivatives on the nucleon fields in this operation. Furthermore, performing the (nonlinear) chiral transformation leads to the derivatives acting on the pion fields rather than the nucleon fields. Thus, the preceding conjecture is verified.

At the next stage, we introduce a (uniquely defined) power counting:

$$p = O(1), \qquad p^2 - m^2 = O(\varepsilon), \qquad \not{p} - m = O(\varepsilon),$$

$$q_1, \cdots, q_n = O(\varepsilon). \tag{4.396}$$

In the Green's functions, the momenta are rescaled according to

$$p \to p, \qquad q_i \to \varepsilon q_i. \tag{4.397}$$

We simultaneously expand all Green's functions in a given Ward identity in a Taylor series in ε. The PCB terms are those with a wrong power in ε, and they are consequently dropped. Now, since the Ward identity is linear with coefficients that are *homogeneous in ε*, it follows that these identities hold separately for the PCB terms as well, justifying the use of the EOM scheme in calculations in the one-nucleon sector.

We end this section with a few remarks on EOMS:

- By construction, the EOMS approach leads to the heavy baryon results when the properly renormalized amplitudes, that is, the amplitudes after subtraction of all PCB terms, are expanded in $1/m_N$.

- The IR approach can also be obtained from the EOMS scheme, as discussed in Ref. [113].

- A systematic removal of the PCB terms can be obtained using the background field method and heat kernel expansion techniques for the generating functional. This is worked out in detail for spinless matter fields coupled to Goldstone bosons in the three-flavor case in Refs. [114, 115]. Using the path integral approach, one can explicitly verify the fulfillment of the whole tower of Ward identities in one go.

- Extensions of the EOMS approach to include vector-mesons [116], the spin-3/2 Δ-resonance [117] and the Roper resonance [118] also exist.

4.11.4 Baryon Masses and σ-Terms

In this section, we want to discuss the chiral expansion of the baryon masses in the case of two and three flavors, respectively, and the related σ-terms. This is an extension of the discussion of the lowest quark mass dependence in QCD (cf. Section 3.16.1), and it serves a number of purposes: First, the masses and σ-terms contain information about the explicit chiral symmetry breaking in QCD. Second, it is important to understand the relation of the chiral expansion of the baryon masses and the trace anomaly, already discussed in Section 3.13. Third, these observables nicely display the differences in the chiral expansion that can arise when the strange quarks that are sizably more heavy than the up and down quarks are involved.

Consider first the complete one-loop expression of the nucleon mass to fourth order in the isospin limit. In HBChPT, it takes the form [119, 120]

$$m_N = \overset{\circ}{m} - 4c_1 M_\pi^2 - \frac{3g_A^2 M_\pi^3}{32\pi F_\pi^2} + k_1 M_\pi^4 \ln \frac{M_\pi}{m_N} + k_2 M_\pi^4 + O(M_\pi^5) \,,$$

$$k_1 = -\frac{3}{32\pi^2 F_\pi^2 m_N} \left(g_A^2 + m_N(-8c_1 + c_2 + 4c_3) \right) \,,$$

$$k_2 = \bar{e}_1(m_N) - \frac{3}{128\pi^2 F_\pi^2 m_N} \left(2g_A^2 - m_N c_2 \right) \,, \tag{4.398}$$

at the scale $\mu = m_N$ and the fourth-order contributions are one-loop graphs with exactly one insertion from $\mathcal{L}^{(2)}$. Here, $\bar{e}_1$ is a particular combination of the LECs e_{115} and e_{116} from the fourth-order Lagrangian $\mathcal{L}^{(4)}$; see Ref. [63]. Plugging in the physical values of the various masses and coupling constants together with the LECs c_i from Ref. [70] and setting the value of the fourth-order LEC at $\bar{e}_1(m_N) = 12.6\,\text{GeV}^{-3}$ leads to the following:[25]

$$m_N = \underbrace{869.5 \text{ MeV}}_{O(p^0)} + \underbrace{86.5 \text{ MeV}}_{O(p^2)} - \underbrace{15.4 \text{ MeV}}_{O(p^3)} - \underbrace{2.3 \text{ MeV}}_{O(p^4)} = 938.3 \text{ MeV} \,, \tag{4.399}$$

which is a rather well-converging series. Note here that the isospin limit has been defined via the charged particle masses. (We will come back to this subtle issue in the next sections.) Even higher-order corrections are given in Refs. [122, 123], and isospin-breaking effects on $O(p^4)$ are discussed in Ref. [124]. We note that all strange quark effects are absorbed in the nucleon mass in the chiral limit and in the pertinent LECs, as discussed in Section 4.9.4. This equation further allows us to make contact with the trace anomaly discussed in Section 3.13. Namely, apart from the strange quark effects, the nucleon mass in the chiral limit, $\overset{\circ}{m}$, subsumes all gluonic effects that are only implicit in the chiral EFT. We will return to this issue when discussing the so-called σ-terms. Finally, we note that current lattice QCD results show essentially a linear behavior of the nucleon mass with M_π, $m_N(M_\pi) = 800 \text{ MeV} + M_\pi$, the so-called

---

[25] The LEC $\bar{e}_1$ also appears in the fourth correction to $\sigma_{\pi N}$. Demanding that the sigma-term at this order reproduce the empirical determination [121] leads to this value.

"ruler plot"; see, for example, Ref. [125]. This is deviating from the fourth-order prediction already at pion masses of about 300 MeV, which is somewhat unexpected, as for such a value of the pion mass the higher-order corrections are still moderate.

We now turn to the calculation of the ground state octet baryon masses to third order, thus including the leading loop effects from pions, kaons and the eta. The pertinent three-flavor Lagrangians read as follows [126]:

$$\mathcal{L}^{(1)} = \langle \bar{B}\left(i\slashed{D} - m_0\right)B\rangle + \frac{D}{2}\langle \bar{B}\gamma^\mu \gamma_5 \{u_\mu, B\}\rangle + \frac{F}{2}\langle \bar{B}\gamma^\mu \gamma_5 [u_\mu, B]\rangle\,,$$

$$\mathcal{L}^{(2)} = b_D \langle \bar{B}\{\chi_+, B\}\rangle + b_F \langle \bar{B}[\chi_+, B]\rangle + b_0 \langle \bar{B}B\rangle \langle \chi_+ \rangle + \dots\,, \tag{4.400}$$

where B collects the baryon octet fields $(n, p, \Sigma^\pm, \Sigma^0, \Lambda, \Xi^-, \Xi^0)$,

$$B(x) = \frac{1}{\sqrt{2}}\sum_{a=1}^{8}\lambda^a B^a(x) = \begin{pmatrix} \frac{1}{\sqrt{2}}\Sigma^0 + \frac{1}{\sqrt{6}}\Lambda & \Sigma^+ & p \\ \Sigma^- & -\frac{1}{\sqrt{2}}\Sigma^0 + \frac{1}{\sqrt{6}}\Lambda & n \\ \Xi^- & \Xi^0 & -\frac{2}{\sqrt{6}}\Lambda \end{pmatrix}. \tag{4.401}$$

The matrix B transforms as $B \mapsto hBh^\dagger$ under chiral $SU(3)_L \times SU(3)_R$, with h the compensator field[26] (cf. Section 4.6). At leading order, we have 3 LECs, namely the baryon octet mass in the chiral limit, m_0 (which is different from its $SU(2)$-counterpart $\mathring{m}$), and the two axial-vector coupling constants D and F, respectively,[27] subject to the matching constraint $D + F = g_A$.[28] For the dimension-two Lagrangian, we have only displayed the symmetry breaking terms parameterized in terms of the LECs b_0, b_D, b_F. To third order, the chiral expansion of the baryon masses takes the form

$$m_B = m_0 + \gamma_B^D b_D + \gamma_B^F b_F - 2b_0(M_\pi^2 + 2M_K^2) - \frac{1}{24\pi F_\phi^2}\left[\alpha_B^\pi M_\pi^3 + \alpha_B^K M_K^3 + \alpha_B^\eta M_\eta^3\right], \tag{4.402}$$

with

$$\gamma_N^D = -4M_K^2\,, \quad \gamma_N^F = 4M_K^2 - 4M_\pi^2\,, \quad \gamma_\Sigma^D = -4M_\pi^2\,, \quad \gamma_\Sigma^F = 0\,,$$

$$\gamma_\Lambda^D = -\frac{16}{3}M_K^2 + \frac{4}{3}M_\pi^2\,, \quad \gamma_\Lambda^F = 0\,, \quad \gamma_\Xi^D = -4M_K^2\,, \quad \gamma_\Xi^F = -4M_K^2 + 4M_\pi^2\,,$$

$$\alpha_N^\pi = \frac{9}{4}(D+F)^2\,, \quad \alpha_N^K = \frac{1}{2}(5D^2 - 6DF + 9F^2)\,, \quad \alpha_N^\eta = \frac{1}{4}(D-3F)^2\,,$$

$$\alpha_\Sigma^\pi = D^2 + 6F^2\,, \quad \alpha_\Sigma^K = 3(D^2 + F^2)\,, \quad \alpha_\Sigma^\eta = D^2\,, \tag{4.403}$$

$$\alpha_\Lambda^\pi = 3D^2\,, \quad \alpha_\Lambda^K = D^2 + 9F^2\,, \quad \alpha_\Lambda^\eta = D^2\,,$$

$$\alpha_\Xi^\pi = \frac{9}{4}(D-F)^2\,, \quad \alpha_\Xi^K = \frac{1}{2}(5D^2 + 6DF + 9F^2)\,, \quad \alpha_\Xi^\eta = \frac{1}{4}(D+3F)^2\,.$$

[26] The baryon field transforms differently in the case of $N_f = 2, 3$ flavors. Namely, for $N_f = 2$, it belongs to the fundamental representation, transforming as $\Psi \mapsto h\Psi$. In case of $N_f = 3$, eight baryon fields form a basis of the adjoint representation.

[27] Note that symbol F also denotes the pion decay constant in the two-flavor chiral limit, but these two quantities cannot be mistaken.

[28] Strictly speaking, this refers to the couplings in the chiral limit. At any given order in the chiral expansion, it is, however, legitimate to equate the leading-order values with the physical ones, as done here.

Here, $F_\phi = (F_\pi + F_K + F_\eta)/3 \simeq 108\,\text{MeV}$ is the average pseudoscalar decay constant (to this order, one could equally well use F_π). Further $N = (n, p)$, $\Sigma = (\Sigma^-, \Sigma^0, \Sigma^+)$ and $\Xi = (\Xi^-, \Xi^0)$ denote the nucleon doublet, the Σ triplet and the Ξ doublet, in order. The chiral expansion to second order is interesting, as the LEC b_0 contributes equally to all masses, and its contribution can thus be absorbed into m_0. Being left with m_0 and two $O(p^2)$ LECs on one side and four masses on other side, one obtains the famous Gell–Mann–Okubo relation [127, 128],

$$2(m_N + m_\Xi) = m_\Sigma + 3m_\Lambda \,. \tag{4.404}$$

Using the octet-baryon masses in the isospin limit,[29] $m_N = 938.9\,\text{MeV}$, $m_\Sigma = 1193.2\,\text{MeV}$, $m_\Lambda = 1115.7\,\text{MeV}$ and $m_\Xi = 1318.3\,\text{MeV}$, we realize that this relation is fulfilled in nature to better than 1%. However, the loop corrections not only upset this relation but also severely question the convergence of this expansion, as, for example, $(M_K/M_\pi)^3 \simeq 45$. Various remedies to this situation exist, like going to fourth order in the various formulations of baryon ChPT, for example, one finds an improved convergence in the EOMS scheme compared to the HBChPT approach, as well as including the decuplet states, whose loop contribution tend to cancel the ones from the octet; see, for example, Refs. [129–135]. Such type of cancellations naturally arise at large-N_C; see, for example, Ref. [136].

We now turn to the so-called σ-terms, which are nothing but the expectation values of the quark mass terms in a given baryon state, already discussed in Section 3.16.2. Of particular interest is the pion–nucleon σ-term, $\sigma_{\pi N}$. It parameterizes the scalar couplings of the nucleon to the light up- and down-quarks. It also plays a key role in the search for physics beyond the Standard Model, such as direct-detection searches for dark matter; see, for example, Refs. [137–139], but also other searches that are sensitive to the scalar current coupling to nucleons; see, for example, Refs. [140, 141]. The term $\sigma_{\pi N}$ is defined as the expectation value of the light flavor (u, d) QCD quark mass term in the nucleon,

$$\sigma_{\pi N} = \frac{\hat{m}}{2m_N} \langle N | \bar{u}u + \bar{d}d | N \rangle \,, \tag{4.405}$$

where u and d are the up- and down-quark fields, respectively, and $|N\rangle$ is a properly normalized nucleon state. There is also a σ-term related to the strange quark field

$$\sigma_s = \frac{m_s}{2m_N} \langle N | \bar{s}s | N \rangle \,, \tag{4.406}$$

with the strange quark mass m_s. One can define another expression that characterizes the scalar nucleon structure, namely the strangeness content of the nucleon y. It is defined by

$$y = \frac{2\langle N | \bar{s}s | N \rangle}{\langle N | \bar{u}u + \bar{d}d | N \rangle} = \frac{2\hat{m}}{m_s} \frac{\sigma_s}{\sigma_{\pi N}} \,. \tag{4.407}$$

---

[29] For a given multiplet, this is nothing but the sum of the masses divided by the number of states.

In order to calculate the strangeness content, the πN σ-term usually is rewritten the following way:

$$\sigma_{\pi N} = \frac{\sigma_0}{1 - y} \,, \tag{4.408}$$

where σ_0 is given by

$$\sigma_0 = \frac{\hat{m}}{2m_N} \langle N | \bar{u}u + \bar{d}d - 2\bar{s}s | N \rangle \,. \tag{4.409}$$

The σ_0 allows us to deduce the strangeness content of the nucleon. For example, if y is equal to zero, $\sigma_{\pi N}$ and σ_0 are identical and the nucleon has a pure u- and d-quark content.

One way to calculate $\sigma_{\pi N}$ in ChPT is based on utilizing the Feynman–Hellmann theorem [142, 143]; see also Section 3.16.2. First, from Eq. (3.110), it is seen that the nucleon mass in the chiral limit is directly related to the trace anomaly:

$$2\overset{\circ}{m}^2 = -\frac{\beta(g_r)}{g_r^3} \lim_{m_u, m_d \to 0} \langle N | [\text{tr}_c (F_{\mu\nu} F^{\mu\nu})]^r | N \rangle \,. \tag{4.410}$$

Here, the relation is displayed for the two-flavor case. Further, from Eq. (3.388) we obtain:[30]

$$\sigma_{\pi N} = \hat{m} \left(\frac{\partial m_N}{\partial \hat{m}} \right) \,, \tag{4.411}$$

which is known as the Feynman–Hellmann theorem for the pion–nucleon σ-term. This is an important result in QCD. The effect of the explicit chiral symmetry breaking on the nucleon mass due to the light quarks is thus entirely given in terms of $\sigma_{\pi N}$. Now, using the expression (4.398), we straightforwardly obtain at lowest order

$$\sigma_{\pi N} = -4c_1 M_\pi^2 + \cdots . \tag{4.412}$$

Higher-order terms in the quark mass expansion can be obtained likewise, given, for example, in Ref. [124].

A similar calculation in the three-flavor case, where the nucleon mass is differentiated with respect to the strange quark mass m_s, leads to σ_s:

$$\sigma_s = m_s \left(\frac{\partial m_N}{\partial m_s} \right) \,. \tag{4.413}$$

Thus, at the leading order in quark masses, the quantities $\sigma_{\pi N}$ and σ_s characterize the shift of the nucleon mass due to the nonzero quark masses. Since, in the Standard Model, the quark masses emerge through the spontaneous chiral symmetry breaking via the Higgs field, then, loosely speaking, the preceding quantities measure the contribution of the Higgs mechanism to the nucleon mass. From Eq. (4.399) we may, however, conclude that in the two-flavor case this contribution is very moderate, and

---

[30] Note that a certain care is needed in differentiating Eq. (3.309), defining the trace anomaly, with respect to the quark masses, since the gluonic operator is not multiplicatively renormalizable; see, e.g., Ref. [144]. In particular, it mixes with the operator $[\bar{\psi} \mathcal{M}_r \psi]^r$ and thus implicitly depends on the quark masses. We have avoided this problem here, invoking the Feynman–Hellmann theorem directly.

the large part of the nucleon mass emerges through the gluon field energy, encoded in the trace anomaly. The three-flavor case is considered in what follows.

From the preceding discussion it is seen that the pion–nucleon σ-term can be calculated, if we know the nucleon mass as a function of the quark masses, which can be done using either ChPT or lattice QCD. Moreover, using dispersion relations for pion–nucleon scattering and input from ChPT, it is possible to evaluate $\sigma_{\pi N}$ from the available experimental data of πN scattering. Owing to the fact that the σ-terms are some of the most important parameters of the low-energy QCD in the one-nucleon sector, such studies have been carried out since the early days of current algebra. Here we briefly summarize the present status of knowledge about $\sigma_{\pi N}$, σ_s and σ_0. For a long time, the value of $\sigma_{\pi N} = (45 \pm 5)\,\text{MeV}$ from Ref. [145] has been widely accepted. This result was obtained from a thorough analysis of πN scattering data, which was available at that time [91, 146], with the use of dispersion relations. Recently, a new dispersion-theoretical analysis of the problem, based on the so-called Roy–Steiner equations, has been carried out (see the next section for more details). This analysis is using a larger data set that, apart from the more recent partial-wave analysis [147] of the πN scattering experiments, includes the latest, very accurate measurements of pionic hydrogen and deuterium, and also employs isospin corrections in the representation of the pion–nucleon scattering amplitude inside the unphysical region (where the sigma-term is defined). As a result, this analysis yields a substantially larger value, $\sigma_{\pi N} = (59.1 \pm 3.5)\,\text{MeV}$ [121]; for more details, see the review [124]. Moreover, excluding the constraints from the pionic atom measurements and fitting a representation based on the Roy–Steiner equations to the low-energy pion–nucleon scattering data base leads to a consistent but less precise value of $\sigma_{\pi N} = (58 \pm 5)\,\text{MeV}$ [148]. Present-day lattice QCD determination tends to give smaller values for this quantity.[31] A comparison of the dispersion-theoretical approach to lattice data is given in Ref. [150], which also discusses the intimate link between $\sigma_{\pi N}$ and the S-wave pion–nucleon scattering lengths (which is a consequence of chiral symmetry).

The strangeness σ-term can be determined from lattice QCD; a recent determination gives $\sigma_s = 39.8(5.5)\,\text{MeV}$ [151], consistent with earlier results displaying values centered around 50 MeV, but with a large spread; see, for example, Refs. [152, 153]. Adding these numbers, we find that the Higgs (the quark mass term) contribution to the nucleon mass is about 100 MeV, just 10%.

We end this section with a few remarks on σ_0. If we use the second-order meson-baryon Lagrangian given in Eq. (4.400), we can express these quantities entirely in terms of hadron masses:

$$\sigma_0 = 2M_\pi^2(b_D - 3b_F) = \frac{1}{2}\left(\frac{M_\pi^2}{M_K^2 - M_\pi^2}\right)(m_\Xi + m_\Sigma - 2m_N) \simeq 27\,\text{MeV}\,. \tag{4.414}$$

---

[31] Note, however, that, as claimed recently, the neglect of the πN, $\pi\pi N$ intermediate states in the fit of the three-point correlator may result in a significant underestimate of the σ-term, extracted on the lattice [149]. In our opinion, this result merely shows that, at present, not all systematic uncertainties are under full control in the lattice calculations of the σ-term.

However, at third and fourth order there are, similar to the case of the baryon masses, large corrections so that a very precise determination of this quantity is difficult. State-of-the-art fourth-order calculations including also the decuplet lead to central values between 50 and 60 MeV [135], consistent with a small strangeness contribution to the nucleon mass.

4.11.5 More on the Pion-Nucleon σ-Term

In this section, we take a closer look at the relation between the pion–nucleon scattering amplitude and the pion–nucleon σ-term that was alluded to in the preceding section. We start with some basic definitions of pion–nucleon scattering (see also Section 4.5.3).

Consider the process of elastic pion–nucleon (πN) scattering (for more details on the kinematics and conventions, see Refs. [91, 154]):

$$\pi^i(q) + N(p,s) \rightarrow \pi^j(q') + N(p',s') \,, \tag{4.415}$$

with pion isospin labels i, j, and we use the conventional Mandelstam variables $s = (p+q)^2, t = (q-q')^2, u = (p-q')^2$, which fulfill $s+t+u = 2m_N^2 + 2M_\pi^2$. From here on, we will identify the masses of the nucleon (m_N) and the pion (M_π) with the charged-particle masses. The following definitions are needed:

$$s = W^2, \quad v = \frac{s-u}{4m_N} \,, \tag{4.416}$$

with W the s-channel center-of-mass energy, and the threshold of the reaction is located at $\sqrt{s_0} = W_0 = m_N + M_\pi \simeq 1078 \,\mathrm{MeV}$.

The scattering amplitude may be expressed in terms of Lorentz-invariant amplitudes A, B and D according to

$$T^{ji}(s,t) = \delta^{ji}T^+(s,t) + \frac{1}{2}[\tau^j, \tau^i]T^-(s,t),$$

$$T^I(s,t) = \bar{u}(p',s')\left\{A^I(s,t) + \frac{\slashed{q}' + \slashed{q}}{2}B^I(s,t)\right\}u(p,s)$$

$$= \bar{u}(p',s')\left\{D^I(s,t) - \frac{[\slashed{q}', \slashed{q}]}{4m_N}B^I(s,t)\right\}u(p,s),$$

$$D^I(s,t) = A^I(s,t) + vB^I(s,t), \tag{4.417}$$

where $I = \pm$ is the total isospin, τ^i denotes the Pauli matrices, and the spinors are normalized as $\bar{u}u = 2m_N$. Here, $I = +$ refers to the isoscalar and $I = -$ to the isovector amplitude, respectively. These are related to the amplitudes with total isospin $1/2$ and $3/2$ via $T^+ - T^- = T^{3/2}$ and $T^+ + 2T^- = T^{1/2}$. Under crossing $s \leftrightarrow u$, the properties of the amplitudes become most transparent when they are written as functions of v and t:

$$A^\pm(v,t) = \pm A^\pm(-v,t), \quad B^\pm(v,t) = \mp B^\pm(-v,t), \quad D^\pm(v,t) = \pm D^\pm(-v,t). \tag{4.418}$$

Furthermore, isospin symmetry leaves only two independent amplitudes that are needed to describe all eight πN scattering reactions, characterized by total s-channel isospin $I_s \in \{1/2, 3/2\}$,

$$\mathcal{A}(\pi^+ p \to \pi^+ p) = \mathcal{A}(\pi^- n \to \pi^- n) = \mathcal{A}^+ - \mathcal{A}^- = \mathcal{A}^{3/2},$$

$$\mathcal{A}(\pi^- p \to \pi^- p) = \mathcal{A}(\pi^+ n \to \pi^+ n) = \mathcal{A}^+ + \mathcal{A}^- = \frac{1}{3}(2\mathcal{A}^{1/2} + \mathcal{A}^{3/2}),$$

$$\mathcal{A}(\pi^- p \to \pi^0 n) = \mathcal{A}(\pi^+ n \to \pi^0 p) = -\sqrt{2}\mathcal{A}^- = -\frac{\sqrt{2}}{3}(\mathcal{A}^{1/2} - \mathcal{A}^{3/2}),$$

$$\mathcal{A}(\pi^0 p \to \pi^0 p) = \mathcal{A}(\pi^0 n \to \pi^0 n) = \mathcal{A}^+ = \frac{1}{3}(\mathcal{A}^{1/2} + 2\mathcal{A}^{3/2}), \qquad (4.419)$$

where $\mathcal{A} \in \{A, B, D\}$. In the discussion of $\sigma_{\pi N}$, we also need the so-called pseudovector-Born-term-subtracted amplitudes,

$$\bar{A}^+(s,t) = A^+(s,t) - \frac{g_{\pi N}^2}{m_N}, \quad \bar{B}^+(s,t) = B^+(s,t) - g_{\pi N}^2 \left[\frac{1}{m_N^2 - s} - \frac{1}{m_N^2 - u} \right],$$

$$\bar{A}^-(s,t) = A^-(s,t), \quad \bar{B}^-(s,t) = B^-(s,t) - g_{\pi N}^2 \left[\frac{1}{m_N^2 - s} + \frac{1}{m_N^2 - u} \right] + \frac{g_{\pi N}^2}{2m_N^2},$$

$$(4.420)$$

with $g_{\pi N}$ the pion–nucleon coupling constant, which to leading order obeys the Goldberger–Treiman relation, $g_{\pi N} = g_A m_N / F_\pi$ [155]. We will use the most recent and precise value, $g_{\pi N}^2/(4\pi) = 13.7 \pm 0.2$ [156]. This subtraction is necessary, since the Born terms (which are the terms calculated with the LO Lagrangian) are ill-defined at the so-called Cheng–Dashen point ($s = m_N^2, t = 2M_\pi^2$); see below.

Separating factors of ν that are required by crossing symmetry, these amplitudes permit the expansions

$$\bar{A}^+(\nu,t) = \sum_{m,n=0}^\infty a_{mn}^+ \nu^{2m} t^n, \qquad \bar{B}^+(\nu,t) = \sum_{m,n=0}^\infty b_{mn}^+ \nu^{2m+1} t^n,$$

$$\bar{A}^-(\nu,t) = \sum_{m,n=0}^\infty a_{mn}^- \nu^{2m+1} t^n, \qquad \bar{B}^-(\nu,t) = \sum_{m,n=0}^\infty b_{mn}^- \nu^{2m} t^n, \qquad (4.421)$$

and similarly for $\bar{D}^\pm = \bar{A}^\pm + \nu \bar{B}^\pm$. The expansions around $\nu = t = 0$ are called subthreshold expansions, and the corresponding coefficients $a_{mn}^\pm$, $b_{mn}^\pm$ are called subthreshold parameters. These fulfill the relations

$$d_{mn}^+ = a_{mn}^+ + b_{m-1,n}^+, \qquad d_{mn}^- = a_{mn}^- + b_{mn}^-, \qquad d_{0n}^+ = a_{0n}^+. \qquad (4.422)$$

The central quantity for the following discussions is the Cheng–Dashen low-energy theorem (LET). The LET relates the Born-term-subtracted isoscalar amplitude $\bar{D}^+(\nu,t)$ evaluated at the Cheng–Dashen point ($\nu = 0$, $t = 2M_\pi^2$) (i.e., in the unphysical region) to the scalar form factor of the nucleon [157, 158]:

$$\sigma(t) = \frac{1}{2m_N} \langle p', s' | \hat{m}(\bar{u}u + \bar{d}d) | p, s \rangle, \quad \hat{m} = \frac{1}{2}(m_u + m_d), \qquad (4.423)$$

evaluated at momentum transfer $t = (p' - p)^2 = 2M_\pi^2$,

$$\bar{D}^+(0, 2M_\pi^2) = \sigma(2M_\pi^2) + \Delta_R, \qquad (4.424)$$

where Δ_R represents higher-order corrections in the chiral expansion. These corrections are expected to be very small. The nonanalytic terms agree at order $O(p^4)$ [103, 159], so that, based on the $SU(2)$ expansion parameter, the remaining effect would scale as $(M_\pi^2/m_N^2)\sigma_{\pi N} \sim 1\,\mathrm{MeV}$. Here, we will use the estimate [159],

$$|\Delta_R| \lesssim 2\,\mathrm{MeV}\,, \tag{4.425}$$

derived from resonance saturation for the $O(p^4)$ LECs.

To make better contact to the dispersion-theoretical representation of the πN scattering amplitudes, the relation (4.424) is often rewritten as

$$\sigma_{\pi N} = \sigma(0) = \Sigma_d + \Delta_D - \Delta_\sigma - \Delta_R, \tag{4.426}$$

with correction terms

$$\Delta_\sigma = \sigma(2M_\pi^2) - \sigma_{\pi N}, \quad \Delta_D = \bar{D}^+(0, 2M_\pi^2) - \Sigma_d, \quad \Sigma_d = F_\pi^2\left(d_{00}^+ + 2M_\pi^2 d_{01}^+\right)\,, \tag{4.427}$$

in terms of the subthreshold parameters d_{00}^+ and d_{01}^+. Here, Δ_σ measures the curvature in the scalar form factor, while Δ_D parameterizes contributions to the πN amplitude beyond the first two terms in the subthreshold expansion. As shown in Ref. [160], although these corrections are large individually due to strong rescattering in the isospin-0 $\pi\pi$ S-wave, they cancel to a large extent in the difference. For the numerical analysis we will use [161]

$$\Delta_D - \Delta_\sigma = (-1.8 \pm 0.2)\,\mathrm{MeV}\,. \tag{4.428}$$

The crucial remaining challenge thus consists of determining the subthreshold parameters to sufficient accuracy. In addition, isospin-breaking corrections to the LET become relevant when it comes to a precision determination of $\sigma_{\pi N}$. The corresponding RS analysis,[32] reported in Ref. [121], produced the following results:[33]

$$d_{00}^+ = -1.36(3)\,M_\pi^{-1}\,, \quad d_{01}^+ = 1.16(3)\,M_\pi^{-3}\,. \tag{4.429}$$

Furthermore, isospin-breaking in the Cheng–Dashen LET shifts $\sigma_{\pi N}$ by $3.0\,\mathrm{MeV}$, leading to

$$\sigma_{\pi N} = (59.1 \pm 1.9_{\mathrm{RS}} \pm 3.0_{\mathrm{LET}})\,\mathrm{MeV} = (59.1 \pm 3.5)\,\mathrm{MeV}\,, \tag{4.430}$$

where the first error is due to the uncertainty in the RS analysis, and the second one due to the isospin-breaking corrections.

One of the main results of the RS approach is a robust correlation between the σ-term and the S-wave πN scattering lengths,

$$\sigma_{\pi N} = (59.1 \pm 3.1)\,\mathrm{MeV} + \sum_{I_s} c_{I_s}\left(a^{I_s} - \bar{a}^{I_s}\right),$$

$$c_{1/2} = 0.242\,\mathrm{MeV}\cdot 10^3 M_\pi\,, \quad c_{3/2} = 0.874\,\mathrm{MeV}\cdot 10^3 M_\pi\,, \tag{4.431}$$

---

[32] The RS equations are a framework that maintains analyticity, unitarity and crossing symmetry of the scattering amplitude within a partial-wave expansion. For early works, see Refs. [162–166].

[33] We remark that this is the first ever dispersion-theoretical analysis of πN scattering that gives uncertainties on the phase shifts and the resulting subthreshold parameters.

where the sum extends over the two s-channel isospin channels and $a^{Is} - \bar{a}^{Is}$ measures the deviation of the scattering lengths from their reference values extracted from pionic atoms: $\bar{a}^{1/2} = (169.8 \pm 2.0) \cdot 10^{-3} M_\pi^{-1}$ and $a^{3/2} = (-86.3 \pm 1.8) \cdot 10^{-3} M_\pi^{-1}$ [167–170]. This is an important result and can serve as a check on lattice QCD determinations of $\sigma_{\pi N}$ [150]. Note also that one recovers the old value of $\sigma_{\pi N} = 45\,\text{MeV}$ from Eq. (4.431), when one uses the pion–nucleon scattering amplitudes from the Karlsruhe–Helsinki group (KH80) together with the then accepted (larger) value of $g_{\pi N}$ [91, 146]. For more details, see the review [124].

Finally, from the so-determined σ-term we can readily deduce the scalar nucleon couplings (that, e.g., are important in dark matter particle (WIMP) scattering of nuclei):

$$\langle N | m_q \bar{q}q | N \rangle = f_q^N m_N , \quad N = n, p , \quad q = u, d, s . \tag{4.432}$$

Including isospin corrections and using $m_u/m_d = 0.46$, this gives $f_u^p - (20.8 \pm 1.5) \cdot 10^{-3}$, $f_d^p = (41.1 \pm 2.8) \cdot 10^{-3}$, $f_u^n = (18.9 \pm 1.4) \cdot 10^{-3}$, and $f_d^p - (45.1 \pm 2.7) \cdot 10^{-3}$. Furthermore, we obtain for the combinations of couplings relevant for Higgs-mediated interactions

$$\sum_{q=u,\dots,t} f_q^N = \frac{2}{9} + \frac{7}{9} \left(f_u^N + f_d^N + f_s^N \right) = 0.305 \pm 0.009 , \tag{4.433}$$

where we used f_s^N from lattice QCD, as described in the preceding section.

4.12 Theoretical Uncertainties

Any theoretical calculation comes with an uncertainty, which must be quantified to make the resulting pre- or postdiction meaningful. In Sections 1.2.5 and 4.9.5 we had briefly discussed the uncertainty estimation based on naive dimensional analysis (NDA), which allows for a rough estimate, but can also be misleading in certain cases, as, for example, LECs are not always of natural size. We will take the discussed chiral expansion for the baryon masses and σ-terms to refine the machinery to assess the theoretical uncertainty for any calculation in the framework of ChPT (or any similar EFT).

First, it is important to realize that there are various sources of uncertainties in the framework of ChPT. These are:

1. The *systematic* uncertainty due to truncation of the chiral expansion at a given order.
2. The uncertainty in the estimation of the LECs contributing to a given process.
3. The uncertainties in the experimental data used to determine the LECs.

The last two issues are, of course, intimately connected, and we have encountered this already in Section 4.9.5. The uncertainty in the pion charge radius, $\langle (r_\pi^V)^2 \rangle = (0.4416 \pm 0.0004)\,\text{fm}$ [171], leads to an uncertainty in the determination of the LEC L_9,

$L_9^r(M_\rho) = (6.78 \pm 0.01) \cdot 10^{-3}$. Similarly, the quark mass dependence of the pion mass and the pion decay constant measured on the lattice allow one to determine the LECs L_3 and L_4, respectively [153]. Such types of uncertainties are usually readily quantified. However, the situation can be more messy when more LECs enter one observable, which is, in particular, the case for two-loop and higher loop calculations; see, for example, Ref. [172]. While the lower-order LECs can be determined to some precision, this becomes very difficult at higher orders, and thus the uncertainty in the LECs that is not related to any experimental error needs to be accounted for.

More difficult is the error due to the neglect of higher orders. Denote by Q the set of small parameters, like p/Λ_χ and M_ϕ/Λ_χ, with M_ϕ the pertinent Goldstone boson mass. If such a calculation is performed to order Q^n, in NDA the error would simply be given by cQ^{n+1}, with c a number of $O(1)$. As we had already seen, in the formulation of ChPT with nucleons, such naive estimates are upset by the closeness of the $\Delta(1232)$ resonance, which leads to the enhancements of certain LECs, and thus the number c in such an estimate can be sizably larger than $O(1)$. This can be overcome by the conservative method proposed in Ref. [173]. Consider an observable X that is calculated up to $O(p^n)$. Its truncation uncertainty is given by

$$\Delta X^{(n)} = \max \left(|X^{n_{LO}}| Q^{n-n_{LO}+1}, \left\{ \left| X^k - X^j \right| Q^{n-j} \right\} \right), \qquad (4.434)$$

where n_{LO} denotes the order of the leading-order result with $n_{LO} \leq j < k \leq n$ and Q the EFT expansion parameter,

$$Q = \max \left(\frac{p}{\Lambda_\chi}, \frac{M_\phi}{\Lambda_\chi} \right) , \quad Q = \max \left(\frac{p}{\Lambda_\chi}, \frac{M_\phi}{\Lambda_\chi}, \frac{\Delta}{\Lambda_\chi} \right) , \qquad (4.435)$$

where the first choice of Q is generic for ChPT or baryon ChPT, whereas the second choice refers to baryon ChPT with the inclusion of the decuplet, and Δ is the average octet-decuplet splitting. In the two-flavor case, one identifies M_ϕ with the pion mass, whereas in the three-flavor formulation, one takes either M_K or M_η. Note that for small external momenta p, the Goldstone boson mass corrections dominate the so-calculated uncertainty. Two remarks are in necessary: first, the method is conservative, as it utilizes the largest relative error between orders that occurs in the chiral expansion. Second, it does not provide a statistical interpretation. However, as shown in Ref. [174] in the framework of a Bayesian analysis (see below), such a procedure emerges from one class of naturalness priors considered, and that all such priors result in consistent quantitative predictions for 68% degree-of-belief intervals; see also Refs. [175, 176].

There exist various statistical methods to get a handle on theoretical uncertainties. A popular approach for extracting LECs and calculating the uncertainties from higher orders is based on Bayes' theorem; see, for example, Refs. [177, 178] for an introduction. The Bayesian approach is particularly useful when, besides experimental data, one wishes to implement some additional physical information in the fit, in order to make the latter more robust. For example, one could try to impose the constraints on the LECs, requiring that they are of natural size, or that their values do not upset agreement with other experiments and/or lattice calculations. Here we closely follow a formulation suitable for the discussion of EFTs, as pioneered in Ref. [179]. The Bayes' theorem states:[34]

$$\mathrm{pr}(\mathbf{a}|D,I) = \frac{\mathrm{pr}(D|\mathbf{a},I)\,\mathrm{pr}(\mathbf{a}|I)}{\mathrm{pr}(D|I)}. \tag{4.436}$$

Here, D is a given set of observables or data, including errors, $\mathbf{a}$ is a vector containing the LECs and I contains the information on the EFT under consideration, like the order of the expansion. Statistical quantities are (a) the posterior probability distribution function (pdf), $\mathrm{pr}(\mathbf{a}|D,I)$. It represents the probability that the parameters take particular values, encoded in the vector $\mathbf{a}$, given the data and the EFT under consideration. (b) $\mathrm{pr}(D|\mathbf{a},I)$ is the likelihood pdf, giving the probability of reproducing data D in the theory I, given the LECs $\mathbf{a}$. For normally distributed errors on independent measurements, it is given by $\exp(-\chi^2/2)$, with χ^2 the usual chi-squared function. (c) $\mathrm{pr}(\mathbf{a}|I)$ is the so-called prior pdf, which represents the knowledge we have about the parameters prior to the measurements (like, e.g., that the LECs are of natural size), and (d) $\mathrm{pr}(D|I)$ is called the evidence (which is often used only for normalization). Eq. (4.436) is a mathematical statement about how the knowledge of $\mathbf{a}$ is improved when new data become available. Consequently, if we have new data that are numerous and/or very accurate, the posterior pdf will not depend on the specific choice of prior pdf any more.

Let us be more specific on the prior knowledge collected in I. If one assumes the LECs to be of natural size, a typical prior would take the following form (here I is given in terms of N_χ, which is a number of independent LECs in a given order of the chiral expansion, and a parameter R, which encodes naturalness for the EFT under consideration):

$$\mathrm{pr}(\mathbf{a}|N_\chi,R) = \left(\frac{1}{\sqrt{2\pi}R}\right)^{N_\chi+1} \exp\left(-\frac{\mathbf{a}^2}{2R^2}\right). \tag{4.437}$$

The precise value of R depends, of course, on the EFT one is considering. Note that Eq. (4.437) is a multivariate Gaussian distribution with mean $\mu=0$ and standard deviation R, so it can be written as

$$\mathrm{pr}(\mathbf{a}|N_\chi,R) = \left(\prod_{i=0}^{N_\chi}\frac{1}{\sqrt{2\pi}R}\right) \exp\left(-\frac{\chi^2_{\mathrm{prior}}}{2}\right), \tag{4.438}$$

with

$$\chi^2_{\mathrm{prior}} = \sum_{i=0}^{N_\chi}\frac{a_i^2}{R^2}. \tag{4.439}$$

Hence, if the data take a Gaussian form, one can introduce the so-called "augmented χ^2,"

$$\chi^2_{\mathrm{aug}} = \chi^2 + \chi^2_{\mathrm{prior}}, \tag{4.440}$$

---

[34] In general, $\mathrm{pr}(A|B)$ denotes the conditional probability density of A given B.

and determine the expectation values of the parameters a_i by solving a least-square problem, with χ^2 replaced by χ^2_{aug}. This allows us to use the standard methods developed for least-square fits. For more details, see Ref. [179].

The formula Eq. (4.436) is most useful in EFT applications when marginalization is employed. Marginalization eliminates parameters that are not considered useful; for example, if one wishes to extract the LECs $\overset{\circ}{m}$ and c_1 from the nucleon mass formula Eq. (4.398) by a fit to lattice QCD data in a certain interval of pion masses, then the (combination of) LECs k_1 and k_2 are marginal and would be eliminated by integration. In general, the marginalization prescription is

$$\text{pr}(A|C) = \int dB \, \text{pr}(A, B|C) \, . \tag{4.441}$$

Note that marginalization with respect to B resembles integrating out the heavy degrees of freedom from the effective theory. For the case at hand, we split the vector $\mathbf{a}$ into a subset of relevant and marginal parameters, $\mathbf{a} := (\mathbf{a}_{\text{rel}}, \mathbf{a}_{\text{marg}})$, and integrate over the marginal ones. Similarly, the knowledge of R, which corresponds to naturalness of the LECs, is often not very precise, which amounts to marginalization in this parameter by integrating it between some minimal and maximal value, say between $R_{\text{min}} = 0.1$ and $R_{\text{max}} = 10$ for the aforementioned extraction of the two leading LECs contributing to the nucleon mass. We refer to Refs. [174, 179] for details and further reading.

The advantage of this Bayesian approach is that it provides one with a statistical interpretation of the calculated uncertainty. This is shown for general classes of prior pdfs in Ref. [174]. Furthermore, performing Bayesian fits while treating the ratio of small and hard scales as an independent parameter allows us to further pin down this hard scale more precisely than by estimates that stem from resonance saturation or from using dimensional arguments like $\Lambda \simeq 4\pi F_\pi$. In fact, analyzing the nucleon mass $m_N(M_\pi/\Lambda)$ in this way leads to a reduced hard scale of about $500\,\text{MeV}$ [179] as compared to the naive estimate of $\Lambda \simeq 1\,\text{GeV}$. Furthermore, a Bayesian analysis allows us to find correlations between LECs (or the corresponding operators) that point toward an over-complete basis. In the case of nucleon–nucleon scattering, this was indeed found at high orders in the expansion (N3LO) [180] that is consistent with arguments based on unitary transformation and renormalizability of the chiral two-nucleon potential [181].

We remark that the Bayesian approach is not the only one. Another standard method to assess theoretical uncertainties is the so-called frequentist approach. A very detailed introduction can, for example, be found in Ref. [182].

4.13 The Meaning of Low-Energy Theorems

We have already encountered low-energy theorems (LETs) at various points, such as the Cheng–Dashen LET that connects the pion–nucleon scattering amplitude to the σ-term (cf. Eq. (4.424)), or Weinberg's prediction for the S-wave $\pi\pi$ scatterings lengths (cf. Eq. (4.233)). These were derived in pre-ChPT times. Here, we want to address the following question: what is a low-energy theorem? We essentially follow Ref. [82].

Before addressing this in the framework of ChPT, the EFT of QCD, let us step back and discuss electromagnetism, which provides us with a well-known example of a LET. Consider the scattering of very soft photons on the proton, that is, the Compton scattering process $\gamma(k_1) + p(p_1) \to \gamma(k_2) + p(p_2)$, and denote by $\varepsilon\,(\varepsilon')$ the polarization vector of the incoming (outgoing) photon. The transition matrix element T, normalized to $d\sigma/d\Omega = |T|^2$, can be expanded in a Taylor series in the small parameter $\delta = |\mathbf{k}_1|/m_N$, with m_N the nucleon mass. In the forward direction and in a gauge where the polarization vectors have only space components, T is given by

$$T = c_0\,\boldsymbol{\varepsilon}' \cdot \boldsymbol{\varepsilon} + i c_1\,\delta\,\boldsymbol{\sigma} \cdot (\boldsymbol{\varepsilon}' \times \boldsymbol{\varepsilon}) + O(\delta^2)\,. \tag{4.442}$$

The parameter δ can be made arbitrarily small in the laboratory so that the first two terms in the Taylor expansion Eq. (4.442) dominate. To be precise, the first one proportional to c_0 gives the low-energy limit for the spin-averaged Compton amplitude, while the second ($\sim c_1$) is of pure spin-flip type and can directly be detected in polarized photon–proton scattering. The pertinent LETs fix the values of c_0 and c_1 in terms of measurable quantities [183, 184],

$$c_0 = -\frac{Z^2 e^2}{4\pi m_N}\,, \quad c_1 = -\frac{Z^2 e^2 \kappa_p^2}{8\pi m_N} \tag{4.443}$$

with $Z = 1$ the charge of the proton (in units of the elementary charge) and $\kappa_p = 1.793$ its anomalous magnetic moment. To arrive at Eq. (4.443), we only make use of gauge invariance and the fact that the T-matrix can be written in terms of a time-ordered product of two conserved vector currents sandwiched between proton states. The derivation proceeds by showing that for small enough photon energies the matrix element is determined by the electromagnetic form factor of the proton at $q^2 = 0$ [183, 184]. Two remarks are appropriate here: first, the predictions for the coefficients $c_{0,1}$ are intuitively clear, as a very low energetic photon cannot resolve the inner structure of the proton but sees only its bulk properties, and second, this LET holds to all orders in non-photonic momenta and masses.

Applying similar methods to the axial-vector currents in the strong interactions leads to the so-called "soft-pion" (also called low-energy) theorems, where the (almost) massless pion takes the role of the low-energy photon just discussed. The early derivations of the LETs like, for example, Eq. (4.233) were based on current algebra that was later shown to be equivalent to tree-level (leading-order) calculations based on the chiral effective Lagrangian, as is reviewed in detail in the book [185]; see also Section 3.17. As we have seen, using Chiral Perturbation Theory, we can now give a more general meaning of a LET:

A Low-Energy Theorem of $O(p^n)$ is a general prediction of ChPT to $O(p^n)$.

Here, by a general prediction we mean a strict consequence of the Standard Model depending on some LECs like $F_\pi, M_\pi, m_N, g_A, \kappa_p, \dots$, but without any model assumption about these parameters. This definition contains a precise prescription on how to obtain higher-order corrections to leading-order LETs and it should therefore be generally adopted for hadronic processes at low energies. More precisely, by a result to $O(p^n)$ we imply the use of the power counting in small momenta and pion masses; cf. Eqs. (4.262, 4.368).

First, let us revisit the LET for pion–pion scattering in view of this. As discussed in Section 4.8.4, the leading-order $O(p^2)$ LET for the S-wave, isospin zero pion–pion scattering length reads as follows:[35]

$$a_0^0 = \frac{7M_\pi^2}{32\pi F_\pi^2} = 0.16 \, . \tag{4.444}$$

The refined LET at one-loop, that is, at $O(p^4)$ reads [37]:

$$a_0^0 = \frac{7M_\pi^2}{32\pi F_\pi^2} \left\{ 1 + \frac{M_\pi^2}{3} \langle r^2 \rangle_S^\pi - \frac{M_\pi^2}{672\pi^2 F_\pi^2} (15\bar{l}_3 - 353) \right\} + \frac{25}{4} M_\pi^4 (a_2^0 + 2a_2^2)$$

$$= 0.20 \pm 0.01 \, , \tag{4.445}$$

with $\bar{l}_3$ the LEC defined at a scale $\mu = M_\pi$, $\langle r^2 \rangle_S^\pi \simeq 0.6 \, \text{fm}^2$ the scalar radius of the pion [186] and a_2^0, a_2^2 the D-wave scattering lengths. We note that this LET has by now been replaced by the even more precise prediction based on Roy equations combined with the two-loop representation of the $\pi\pi$ scattering amplitude; see Ref. [50] (as discussed in Section 4.9.1).

It is instructive to reconsider the LET for soft photon scattering of a proton. To be specific, let us work in the framework of HBChPT [80]. The spin-averaged Compton amplitude in the forward direction (in the Coulomb gauge $\varepsilon \cdot v = 0$) reads as follows:[36]

$$\frac{e^2}{m_N} \varepsilon^\mu \varepsilon^\nu T_{\mu\nu}(v,k) = e^2 \left[\varepsilon^2 U(\omega) + (\varepsilon \cdot k)^2 V(\omega) \right] , \tag{4.446}$$

with $\omega = v \cdot k$ (k is the photon momentum) and

$$T_{\mu\nu}(v,k) = \frac{i}{4} \sum_s \int d^4x \, e^{ik \cdot x} \langle p,s | T \, j_\mu(x) j_\nu(0) | p,s \rangle \, , \tag{4.447}$$

with j_μ the electromagnetic current and $p^\mu = m_N v^\mu$. All dynamical information is contained in the functions $U(\omega)$ and $V(\omega)$. Let us concentrate on $U(\omega)$ here, as in the Thomson limit ($|\mathbf{k}_1| = |\mathbf{k}_2| \to 0$) only $U(0)$ contributes to the amplitude. In the forward direction, the only quantities with nonzero chiral dimension are ω and M_π. In order to make this dependence explicit, we write $U(\omega, M_\pi)$ instead of $U(\omega)$. According to the power counting, the chiral expansion of $U(\omega, M_\pi)$ can be written as

---

[35] Note that here we display the angular momentum $l = 0$ as superscript, because we need to differentiate from the D-wave scattering lengths with $l = 2$ in what follows.

[36] We are using a trick due to Bernabeu and Tarrach [331] here. Instead of calculating the full Compton amplitude for general kinematics, enforcing the gauge condition $\varepsilon \cdot k = 0$ and then going to the forward limit $t = 0$, we directly take the forward limit and keep the structure $\varepsilon \cdot k$ in the evaluation of the Feynman integrals to read the magnetic polarizability from the terms proportional to $(\varepsilon \cdot k)^2$. Alternatively, one can calculate the full unpolarized Compton amplitude as we have described. This leads to the same result, and that type of calculation is given in Ref. [83].

$$U(\omega, M_\pi) = \sum_{k \geq -1} \omega^k f_k(\omega/M_\pi) ; \qquad (4.448)$$

see Ref. [82] for more details. We can now illuminate the difference and the interplay between the soft-photon limit and the low-energy expansion of ChPT. Let us consider first the leading terms in the chiral expansion, Eq. (4.448):

$$U(\omega, M_\pi) = \frac{1}{\omega} f_{-1}(\omega/M_\pi) + f_0(\omega/M_\pi) + O(p^3) . \qquad (4.449)$$

Eq. (4.368) tells us that only tree diagrams can contribute to the first two terms. However, the relevant tree diagrams shown in Fig. 4.14 do not contain pion lines. Consequently, the functions f_{-1}, f_0 cannot depend on M_π and are therefore constants. Since the soft-photon theorem [183, 184] requires $U(0, M_\pi)$ to be finite, f_{-1} must actually vanish, and the chiral expansion of $U(\omega, M_\pi)$ can be written as

$$U(\omega, M_\pi) = f_0 + \sum_{k \geq 1} \omega^k f_k(\omega/M_\pi) . \qquad (4.450)$$

But the soft-photon theorem yields additional information. Since the Compton amplitude is independent of M_π in the Thomson limit and since, due to gauge invariance, there is no term linear in ω in the spin-averaged amplitude, we find

$$\lim_{\omega \to 0} \omega^{n-1} f_n(\omega/M_\pi) = 0 \qquad (n \geq 1) , \qquad (4.451)$$

implying in particular that the constant f_0 describes the Thomson limit:

$$U(0, M_\pi) = f_0 . \qquad (4.452)$$

Let us now verify these results by explicit calculation. In the Coulomb gauge, there is no direct photon–nucleon coupling from the lowest-order effective Lagrangian $\mathcal{L}_{\pi N}^{(1)}$ since it is proportional to $\varepsilon \cdot v$. Consequently, the Born diagrams a,b in Fig. 4.14 vanish so that indeed $f_{-1} = 0$. On the other hand, the expansion of the relativistic Dirac Lagrangian leads to terms of the type $D^2/2m_N$ and $(v \cdot D)^2/2m_N$, where D_μ is a covariant derivative. Notice that although these terms belong to $\mathcal{L}_{\pi N}^{(2)}$, they do not contain novel LECs since they are of purely kinematical origin. These terms lead to a Feynman insertion (Fig. 4.14c) of the form

$$i \frac{e^2}{m_N} \frac{1}{2} (1 + \tau_3) \left[\varepsilon^2 - (\varepsilon \cdot v)^2 \right] = i \frac{e^2 Z^2}{m_N} \varepsilon^2 , \qquad (4.453)$$

producing the desired result $f_0 = Z^2/m_N$, the Thomson limit.

At the next order in the chiral expansion, $O(p^3)$, the function $f_1(\omega/M_\pi)$ is given by the finite sum of nine one-loop diagrams [80, 90]. According to Eq. (4.451), f_1 vanishes for $\omega \to 0$. The term linear in ω/M_π yields the leading contribution to the sum of the electric and magnetic polarizabilities of the nucleon, defined by the second-order Taylor coefficient in the expansion of $U(\omega, M_\pi)$ in ω:

$$f_1(\omega/M_\pi) = -\frac{11 g_A^2 \omega}{192 \pi F_\pi^2 M_\pi} + O(\omega^2) . \qquad (4.454)$$

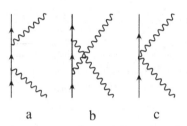

a   b   c

Figure 4.14 Tree diagrams of $O(p)$ (a,b) and $O(p^2)$ (c) for Compton scattering in HBChPT. Full (wavy) lines stand for nucleons (photons).

The $1/M_\pi$ behavior should not come as a surprise – in the chiral limit the pion cloud becomes long-ranged (instead of being Yukawa-suppressed) so that the polarizabilities explode. This behavior is specific to the leading contribution of $O(p^3)$. In fact, from the general form Eq. (4.450) we immediately derive that the contribution of $O(p^n)$ ($D_L = n - 2$) to the polarizabilities is of the form $c_n M_\pi^{n-4}$ ($n \geq 3$), where c_n is a constant that may be zero. One can perform a similar analysis for the amplitude $V(\omega)$ and for the spin–flip amplitude. We do not discuss these amplitudes here but refer the reader to Ref. [80] for details; see also Exercise 26.

A nice example of the power of ChPT is the much discussed LET for the electric dipole amplitude E_{0+} for neutral pion photoproduction of the nucleon at threshold. To be specific, consider the reaction $\gamma(k) + p(p_1) \to \pi^0(q) + p(p_2)$. The electric dipole amplitude at threshold is related to the cross section in the center-of-mass frame through; see, for example, Ref. [187],

$$\frac{|\mathbf{k}|}{|\mathbf{q}|} \frac{d\sigma}{d\Omega}\bigg|_{|\mathbf{q}|\to 0} = (E_{0+})^2 \,, \tag{4.455}$$

with $\mathbf{k}$ and $\mathbf{q}$ the photon and the pion momentum, in order. In the usual conventions, E_{0+} has physical dimension -1 and it can therefore be written as

$$E_{0+} = K \frac{eg_A}{F_\pi} A \left(\frac{M_\pi}{m_p} \right) \,, \tag{4.456}$$

where m_p is the proton mass and K is a kinematical factor. The dimensionless amplitude A can be expressed as a power series in M_π. The original LET due to de Baenst [188] and Vainsthein and Zakharov [189] was quite intriguing, as it gave two terms in the expansion of A, and thus of E_{0+},

$$E_{0+,\text{thr}} = -\frac{eg_A}{8\pi F_\pi} \frac{M_\pi}{m_p} \left[1 - \frac{M_\pi}{2m_p} (3 + \kappa_p) \right] + O(M_\pi^3) \,, \tag{4.457}$$

which, using the language of chiral Lagrangians, corresponds to a tree-level computation with insertions from $\mathcal{L}_{\pi N}^{(1)}$ and $\mathcal{L}_{\pi N}^{(2)}$. However, an additional assumption has to be made, namely that the expansion of the amplitudes in the variables ν, ν_B

Figure 4.15   The triangle graph in the process of the neutral pion photoproduction off nucleons. The solid, dashed and wiggly lines denote nucleons, pions and photons, respectively.

(linear combinations of s and u) is analytic in these variables.[37] However, as shown by an explicit one-loop calculations in Ref. [89], this is violated by the so-called triangle graph (see Fig. 4.15), leading to

$$E_{0+,\text{thr}} = -\frac{eg_A}{8\pi F_\pi}\frac{M_\pi}{m_p}\left[1 - \frac{M_\pi}{2m_p}(3+\kappa_p) - \frac{M_\pi m_p}{16F_\pi^2}\right] + O(M_\pi^3), \qquad (4.458)$$

which is the correct LET at $O(p^3)$. The appearance of this correction $\sim M_\pi^2$ can be understood as follows. The general form of the leading loop correction to the electric dipole amplitude can be written as

$$E_{0+}^{\text{loop}} = -\frac{eg_A M_\pi^2}{(4\pi F_\pi)^3}\left[f(\mu) - f(-\mu)\right],$$

$$f(\mu) = \int_0^1 dx \int_0^1 dy \frac{2xy}{\mu^2 + y^2 - 2xy\mu}, \qquad (4.459)$$

with $\mu = M_\pi/m_p$. The standard assumption of smoothness, or that $f(\mu)$ is analytic as μ goes to zero, means that $f(\mu) = f(-\mu)$ as $\mu \to 0$, so that E_{0+}^{loop} would start at $O(M_\pi^3)$. But as $f(0) = \int_0^1 dy(1/y)$ diverges, one instead has

$$\lim_{\mu \to 0^+}\left[f(\mu) - f(-\mu)\right] = \frac{\pi^2}{2} \neq 0. \qquad (4.460)$$

Therefore, E_{0+}^{loop} indeed starts contributing at $O(M_\pi^2)$. The physical origin of this behavior can be traced back to the triangle diagram: it has an IR singularity $\sim \log(\mu)$ in the chiral limit, so that the smoothness (analyticity) assumption is explicitly violated. For a more detailed discussion, see Refs. [82, 89]. We also remark that the convergence in this quantity is rather slow; for more details of the corresponding $O(p^4)$ calculations; see, for example, Refs. [190, 191].

4.14 Including Electromagnetic and Weak Interactions

4.14.1 Mesonic Sector

In the Standard Model, quarks interact with photons, as well as with $W^\pm$- and Z-bosons. At low energy, when quarks and gluons form hadrons, these interactions

---

[37] We remark that the authors of Ref. [189] were aware of this assumption. They did indeed check it by calculating the so-called rescattering diagram; that, however, does not lead to any nonanalyticity in the amplitudes.

translate into interactions between hadrons and the electromagnetic and weak gauge bosons as well as leptons. Then, as in pure QCD, using the underlying symmetries, we can construct an effective field theory that describes all these interactions. In this section we demonstrate how this can be achieved.

There are several reasons to look for such an extension of ChPT. First and foremost, the processes, which proceed through electromagnetic and weak interactions, provide important information about the strong interactions. In such a case one speaks about electromagnetic and weak *probes*. For example, at the quark level, the neutron β-decay is described by the elementary process $d \to ue^-\bar{\nu}_e$, mediated by a charged weak boson. The quarks, however, are not free, but bound in a nucleon and experience strong interactions that modify the decay amplitude. This is described by the matrix element of the vector and axial-vector currents between nucleon states. Moreover, since the momentum transfer in the decay process is very small, only the values of the pertinent form factors at $q^2 = (p' - p)^2 \simeq 0$ are relevant (here, p, p' denote the initial and final momenta of the nucleon, respectively). Further, Ward identities imply that the vector form factor is not renormalized at $q^2 = 0$, unlike the axial-vector form factor, given in Eq. (4.120). Namely, the constant $g_A = G_A(q^2 = 0)$, which appears in the amplitude for neutron β-decay, is unity for free quarks, whereas its present experimental value is $g_A \simeq 1.28$. Hence, the deviation of this constant from unity describes the effect of the strong interactions on this particular weak decay. An accurate calculation of these effects plays a crucial role, for example, for the precise determination of the Cabibbo–Kobayashi–Maskawa (CKM) matrix elements [192, 193], which should obey unitarity constraints in the Standard Model and, hence, the check of these constraints provides an important test for the physics beyond the Standard Model. Another example of an electroweak probe is the study of the structure of hadrons by measuring the electromagnetic form factors in scattering experiments of charged particles. The Fourier transform of these form factors characterizes the spatial distribution of charge inside a hadron and thus allows us to determine the characteristic size of the latter.

A further reason for the inclusion of the electroweak sector into the effective theory of QCD at low energy is that the (small) effects from electromagnetic corrections to the strong processes are no more negligible at present experimental precision. Examples are provided by the determination of the pion decay constant F_π from leptonic decays of charged pions (see, e.g., Refs. [194–196]) as well as by the precision analysis of πN scattering data. It should be pointed out that, even though some pronounced effects like the Coulomb enhancement of charged particle scattering amplitudes at threshold [197–199] are universal and can be evaluated without referring to the internal dynamics of hadrons, others are not. In order to evaluate these, the knowledge of the complicated interplay of the strong and the electroweak interactions at low energy is essential. A unified effective field theory of strong and effective interactions enables one to address these calculations systematically within a perturbative framework. Note that one deals here with a double expansion: Along with the usual expansion parameter p/Λ from the strong sector, another small expansion parameter is provided by the electromagnetic fine structure constant $\alpha \simeq 1/137$. In what follows, we shall see how these two expansions can be merged consistently.

In order to explain the basic idea beyond the construction of the EFT with the electroweak interactions, we shall start with a minimal extension of QCD, adding the interactions of (charged) quarks with photons only. The leptonic sector as well as weak interactions are completely ignored for the moment. The fermion sector of QCD takes the form

$$\mathcal{L}_F = \bar{\psi} i \gamma_\mu (D^\mu - ieQ\mathscr{A}^\mu) \psi + \text{terms with the sources } s, p, v_\mu, a_\mu, \qquad (4.461)$$

where D^μ contains the gluon field and Q denotes the quark charge matrix

$$Q = \begin{pmatrix} \frac{2}{3} & 0 & 0 \\ 0 & -\frac{1}{3} & 0 \\ 0 & 0 & -\frac{1}{3} \end{pmatrix}. \qquad (4.462)$$

Here, one encounters exactly the same situation as in the case of nonzero quark masses. The piece of the Lagrangian that contains the interaction with the electromagnetic field is not invariant under chiral transformations. The reason for this is simple, as the matrix Q does not commute with all generators of $SU(N_f)$. Moreover, as in the case of the quark mass matrix, this problem is cured by introducing a *spurion* field. Let us rewrite the electromagnetic part of the preceding Lagrangian in the following manner:

$$\mathcal{L}_F^{\text{em}} = \bar{\psi}_L i \gamma_\mu (\partial^\mu - ieQ_L \mathscr{A}^\mu) \psi_L + \bar{\psi}_R i \gamma_\mu (\partial^\mu - ieQ_R \mathscr{A}^\mu) \psi_R, \qquad (4.463)$$

where $Q_L(x), Q_R(x)$ are spurion fields, which transform according to

$$Q_{L,R}(x) \mapsto g_{L,R}(x) Q_{L,R}(x) g_{L,R}^\dagger(x), \qquad g_{L,R}(x) \in SU(N_f)_{L,R}. \qquad (4.464)$$

This transformation law renders the Lagrangian exactly invariant under $SU(N_f)_L \times SU(N_f)_R$. The original Lagrangian is reproduced when $Q_L(x) = Q_R(x) = Q$.

The rest follows the familiar path. Consider first the purely mesonic three-flavor sector [200]. (The $SU(2)$ case is considered in Ref. [38]; see also [201].) In the beginning, one constructs appropriate building blocks for the effective Lagrangian. The covariant derivative of the pion matrix and the quantity χ is now given by

$$D_\mu U = \partial_\mu U - i(v_\mu + eQ_R \mathscr{A}_\mu + a_\mu) U + iU(v_\mu + eQ_L \mathscr{A}_\mu - a_\mu),$$

$$D_\mu \chi = \partial_\mu \chi - i(v_\mu + eQ_R \mathscr{A}_\mu + a_\mu) \chi + i\chi(v_\mu + eQ_L \mathscr{A}_\mu - a_\mu). \qquad (4.465)$$

From the preceding equation it is clear that, for consistency, the electric charge e has to be counted as $O(p)$ in the chiral counting.[38] It is also seen that in the presence of electromagnetism, the vector source gets shifted $v_\mu \to v_\mu + eQ\mathscr{A}_\mu$. The quantities r_μ, l_μ, as well as the vector and axial-vector tensors are modified accordingly.

Further, the covariant derivative of the spurion fields is given by:

$$D_\mu^L Q_L = \partial_\mu Q_L - i[l_\mu, Q_L], \qquad D_\mu^R Q_R = \partial_\mu Q_R - i[r_\mu, Q_R]. \qquad (4.466)$$

---

[38] We count the photon field $\mathscr{A}_\mu$ as $O(1)$. Since this field is always accompanied by a factor e, there is some arbitrariness in these assignments. All what matters in the end is that $e\mathscr{A}_\mu = O(p)$.

Now, we can proceed with writing the effective Lagrangian. At lowest order, only the following operators are available:

$$\mathcal{L}^{(2)} = \frac{F_0^2}{4} \langle D^\mu U D_\mu U^\dagger + \chi^\dagger U + U^\dagger \chi \rangle + C_0 e^2 \langle Q_R U Q_L U^\dagger \rangle - \frac{1}{4} \mathscr{F}_{\mu\nu} \mathscr{F}^{\mu\nu} + \mathcal{L}_{\text{g.f.}},$$

(4.467)

where $\mathcal{L}_{\text{g.f.}}$ denotes the standard gauge-fixing term. At this order, only one new LEC, $C_0 = Z_0 F_0^4$, appears, responsible for the mass splitting between the charged and neutral members of the same isospin multiplets (cf. Eq. (4.214)). In the two-flavor case, the pertinent LECs are denoted by C, Z. Setting $Q_R = Q_L = Q$, elementary calculations yield

$$M_{\pi^\pm}^2 = 2\hat{m}B_0 + 2e^2 F_0^2 Z_0,$$

$$M_{\pi^0}^2 = 2\hat{m}B_0 - \frac{4}{3}(m_s - \hat{m})B_0 \frac{\sin^2 \varepsilon}{\cos 2\varepsilon},$$

$$M_{K^\pm}^2 = (m_u + m_s)B_0 + 2e^2 F_0^2 Z_0,$$

$$M_{K^0}^2 = M_{\bar{K}^0}^2 = (m_d + m_s)B_0,$$

$$M_\eta^2 = \frac{2}{3}(\hat{m} + 2m_s)B_0 + \frac{4}{3}(m_s - \hat{m})B_0 \frac{\sin^2 \varepsilon}{\cos 2\varepsilon}.$$

(4.468)

From the preceding expressions it can be concluded that the mass splitting in isospin multiplets arises from two distinct sources. They are on the one hand due to the quark mass difference $m_u \neq m_d$, and on the other hand are generated by electromagnetic effects proportional to e^2. Intuitively, you might think that the latter correction arises from the "hard" photon loop, in which the characteristic photon momentum is on the order of the hard QCD scale Λ. In reality, however, the situation is more complicated, because the separation between the "electromagnetic" and "non-electromagnetic" interactions is convention-dependent. This can be seen most easily by noticing that, in fact, the difference $m_d - m_u$ is partly of electromagnetic origin. We shall dwell on this issue in more detail later. Here, we adopt a commonly shared convention, identifying the "strong" and "electromagnetic" isospin breaking with different terms in the expressions just given. Following this identification, we immediately verify that the so-called Dashen theorem holds exactly for the charged-to-neutral pion and kaon electromagnetic mass differences at $O(p^2)$:

$$(\Delta M_\pi^2)^{(\text{em})} = (\Delta M_K^2)^{(\text{e.m.})} = 2e^2 F_0^2 Z_0.$$

(4.469)

This exact relation, called Dashen's theorem, is violated by the higher-order terms; see, for example, Refs. [202, 203].

In order to calculate the Goldstone boson masses at $O(p^4)$, one starts with the Lagrangian at this order. Apart from the strong $O(p^4)$ Lagrangian given in Eqs. (4.195), (4.196) and (4.197), there are terms that are proportional to e^2 and e^4. For the first time, these terms were listed in Ref. [200]:

$$\mathcal{L}^{(e^2 p^2, e^4)} = e^2 K_1 F_0^2 \langle D_\mu U D^\mu U^\dagger \rangle \langle Q^2 \rangle + e^2 K_2 F_0^2 \langle D_\mu U D^\mu U^\dagger \rangle \langle QUQU^\dagger \rangle$$

$$+ e^2 K_3 F_0^2 \left(\langle D_\mu U^\dagger QU \rangle \langle D^\mu U^\dagger QU \rangle + \langle D_\mu U Q U^\dagger \rangle \langle D^\mu U Q U^\dagger \rangle \right)$$

$$+ e^2 K_4 F_0^2 \langle D_\mu U^\dagger QU \rangle \langle D^\mu U Q U^\dagger \rangle$$

$$+ e^2 K_5 F_0^2 \langle (D_\mu U D^\mu U^\dagger + D_\mu U^\dagger D^\mu U) Q^2 \rangle$$

$$+ e^2 K_6 F_0^2 \langle D_\mu U^\dagger D^\mu U Q U^\dagger QU + D_\mu U D^\mu U^\dagger Q U Q U^\dagger \rangle$$

$$+ e^2 K_7 F_0^2 \langle \chi^\dagger U + U^\dagger \chi \rangle \langle Q^2 \rangle + e^2 K_8 F_0^2 \langle \chi^\dagger U + U^\dagger \chi \rangle \langle QUQU^\dagger \rangle$$

$$+ e^2 K_9 F_0^2 \langle (\chi^\dagger U + U^\dagger \chi + \chi U^\dagger + U \chi^\dagger) Q^2 \rangle$$

$$+ e^2 K_{10} F_0^2 \langle (\chi^\dagger U + U^\dagger \chi) Q U^\dagger QU + (\chi U^\dagger + U \chi^\dagger) Q U Q U^\dagger \rangle$$

$$+ e^2 K_{11} F_0^2 \langle (\chi^\dagger U - U^\dagger \chi) Q U^\dagger QU + (\chi U^\dagger - U \chi^\dagger) Q U Q U^\dagger \rangle$$

$$+ e^2 K_{12} F_0^2 \langle D_\mu U^\dagger [D^\mu Q, Q] U + D_\mu U [D^\mu Q, Q] U^\dagger \rangle$$

$$+ e^2 K_{13} F_0^2 \langle D^\mu Q U D_\mu Q U^\dagger \rangle + e^2 K_{14} F_0^2 \langle D_\mu Q D^\mu Q + D_\mu Q D^\mu Q \rangle$$

$$+ e^4 K_{15} F_0^4 (\langle QUQU^\dagger \rangle)^2 + e^4 K_{16} F_0^4 \langle QUQU^\dagger \rangle \langle Q^2 \rangle + e^4 K_{17} F_0^4 \langle Q^2 \rangle^2 .$$

$$(4.470)$$

In the two-flavor case, where the strong part is given by Eqs. (4.198), (4.199) and (4.200), the pertinent electromagnetic operators were first given in Ref. [38]:

$$\mathcal{L}^{(e^2 p^2, e^4)} = e^2 k_1 F^2 \langle D_\mu U^\dagger D^\mu U \rangle \langle Q^2 \rangle + e^2 k_2 F^2 \langle D_\mu U^\dagger D^\mu U \rangle \langle QUQU^\dagger \rangle$$

$$+ e^2 k_3 F^2 \left(\langle D_\mu U^\dagger QU \rangle \langle D^\mu U^\dagger QU \rangle + \langle D_\mu U Q U^\dagger \rangle \langle D^\mu U Q U^\dagger \rangle \right)$$

$$+ e^2 k_4 F^2 \langle D_\mu U^\dagger QU \rangle \langle D^\mu U Q U^\dagger \rangle$$

$$+ e^2 k_5 F^2 \langle \chi^\dagger U + U^\dagger \chi \rangle \langle Q^2 \rangle + e^2 k_6 F^2 \langle \chi^\dagger U + U^\dagger \chi \rangle \langle QUQU^\dagger \rangle$$

$$+ e^2 k_7 F^2 \langle (\chi U^\dagger + U \chi^\dagger + \chi^\dagger U + U^\dagger \chi) Q \rangle \langle Q \rangle$$

$$+ e^2 k_8 F^2 \langle (\chi^\dagger U - U^\dagger \chi) Q U^\dagger QU + (\chi U^\dagger - U \chi^\dagger) Q U Q U^\dagger \rangle$$

$$+ e^2 k_9 F^2 \langle D_\mu U^\dagger [D^\mu Q, Q] U + D_\mu U [D^\mu Q, Q] U^\dagger \rangle$$

$$+ e^2 k_{10} F^2 \langle D^\mu Q U D_\mu Q U^\dagger \rangle + e^2 k_{11} F^2 \langle D_\mu Q D^\mu Q + D_\mu Q D^\mu Q \rangle$$

$$+ e^4 k_{12} F^4 \langle Q^2 \rangle^2 + e^4 k_{13} F^4 \langle QUQU^\dagger \rangle \langle Q^2 \rangle + e^4 k_{14} F^4 (\langle QUQU^\dagger \rangle)^2 .$$

$$(4.471)$$

As in the strong sector, the LECs K_i and k_i are UV divergent:

$$K_i = \Sigma_i \lambda + K_i^r(\mu), \qquad k_i = \sigma_i \lambda + k_i^r(\mu), \qquad (4.472)$$

$$M_{\pi^\pm},\ M_{\pi^0}$$

$$l_i,\ k_i$$

Figure 4.16   Pion self-energy at $O(p^4)$. Dashed and wiggly lines denote pions and photons, respectively. At this order, the pion mass difference in the loops already matters.

where the Σ_i, σ_i can be chosen such that the one-loop effective action is UV-finite at $O(p^4)$ [38].

The self-energy of the pion at $O(p^4)$ (we remind the reader that this includes terms of $e^2 p^2$ and e^4 as well since $e = O(p)$) is given by the diagrams shown in Fig. 4.16. Note that, at this order, one already needs to take into account the fact that the masses of the charged and neutral particles from the same isospin multiplet running in the loops are different. Here, for illustration, we display the full result for the charged and neutral pion masses into two-flavor ChPT, derived at next-to-leading-order in Ref. [38]:

$$M_{\pi^0}^2 = 2\hat{m}B \left\{ 1 + \frac{2M_{\pi^0}^2}{F^2} l_3^r(\mu) + e^2 K_{\pi^0}^r(\mu) + \frac{M_{\pi^\pm}^2}{16\pi^2 F^2} \ln \frac{M_{\pi^\pm}^2}{\mu^2} - \frac{M_{\pi^0}^2}{32\pi^2 F^2} \ln \frac{M_{\pi^0}^2}{\mu^2} \right\}$$

$$- \frac{2B^2}{F^2}(m_d - m_u)^2 l_7 - \frac{4}{3} B(m_d - m_u)e^2 k_7 \,,$$

$$M_{\pi^\pm}^2 = 2\hat{m}B \left\{ 1 + \frac{e^2}{4\pi^2} + \frac{2M_{\pi^0}^2}{F^2} l_3^r(\mu) + e^2 K_{\pi^\pm}^r(\mu) + \frac{M_{\pi^0}^2}{32\pi^2 F^2} \ln \frac{M_{\pi^0}^2}{\mu^2} \right\}$$

$$+ 2e^2 F^2 \left\{ Z\left(1 + \frac{e^2}{4\pi^2}\right) + e^2 K_{\pi^\pm}^{\prime\prime r}(\mu) - (3 + 4Z)\frac{M_{\pi^\pm}^2}{32\pi^2 F^2} \ln \frac{M_{\pi^\pm}^2}{\mu^2} \right\}$$

$$- \frac{4}{3} B(m_d - m_u)e^2 k_7 \,, \tag{4.473}$$

where $K_{\pi^0}^r$, $K_{\pi^\pm}^r$ and $K_{\pi^\pm}^{\prime\prime r}$ are particular linear combinations of the original LECs:

$$K_{\pi^0}^r = -\frac{4}{9}\left(5k_1^r + 5k_2^r - 5k_5^r - 5k_6^r - k_7\right) + 4k_3^r - 2k_4^r \,,$$

$$K_{\pi^\pm}^r = -\frac{4}{9}\left(5k_1^r + 5k_2^r - 5k_5^r - 23k_6^r - k_7 - 18k_8^r\right),$$

$$K_{\pi^\pm}^{\prime\prime r} = -\frac{5}{9}\left(4Z(k_1^r + k_2^r) - k_{13}^r - 2k_{14}^r\right). \tag{4.474}$$

The numerical values of the electromagnetic LECs k_i, as well as their three-flavor counterparts K_i are very poorly known; see, for example, Ref. [60]. As a rule of thumb, we may use the NDA estimate, already discussed in the case of the strong LECs:

$$|k_i^r(\mu)| \le \frac{1}{16\pi^2} \,. \tag{4.475}$$

As mentioned before, such an assignment renders the results scale-dependent, because the $k_i^r(\mu)$ depend (logarithmically) on μ. Usually, μ is chosen to be on the order of the heavy scale, say, the ρ-meson mass. In a more elaborate approach, one could try to fit these LECs to experimental data. Using resonance saturation is yet another alternative. The best known example of this is the so-called DGMLY sum rule for the charged-to-neutral pion mass difference [204], which can be translated into a sum rule for the LEC Z (or Z_0 in the $SU(3)$ case). This sum rule, which is exact in the chiral limit and under the assumption of the dominance of the lowest-lying vector- and axial-vector mesons, states:

$$(\Delta M_\pi^2)^{(em)} = \frac{3e^2}{16\pi^2} \frac{M_A^2 M_V^2}{M_A^2 - M_V^2} \ln \frac{M_A^2}{M_V^2} . \tag{4.476}$$

One can also use the resonance saturation to determine the $O(p^4)$ electromagnetic LECs [205, 206]. The extended Nambu-Jona-Lasinio model has been also utilized for this purpose [207]. Further, matching conditions between the $SU(2)$ and $SU(3)$ LECs can be derived like in the strong sector [208]. Note also that, based on the preceding estimates for the electromagnetic LECs, there have been controversial claims about the magnitude of the violation of Dashen's theorem at $O(p^4)$ [202, 203, 205, 207]. This violation is conveniently parameterized by the quantity

$$\varepsilon = \frac{(\Delta M_K^2)^{(em)} - (\Delta M_\pi^2)^{(em)}}{\Delta M_\pi^2} . \tag{4.477}$$

The phenomenological estimates give a large spread for this parameter $\varepsilon = 0.7(5)$, cited in the recent review of the Flavor Lattice Averaging Group (FLAG) [153]. (Note that $\varepsilon = 0$ corresponds to no violation.) Ultimately, the issue has to be resolved by calculations in lattice QCD+QED, where some results are already available. Ref. [153] quotes $\varepsilon = 0.79(7)$ for $N_f = 2 + 1 + 1$ and $\varepsilon = 0.73(0.17)$ for $N_f = 2 + 1$ flavors. This already represents a significant improvement as compared to the accuracy of phenomenological determinations but still requires refinement.

The just described EFT methods can also be used in the case of the observables, different from the Goldstone boson masses. A nice example is provided by the calculation of the electromagnetic effects in the leptonic decay constants [87, 209, 210], which reproduces the long-range corrections from the photon loops, present in the pre-ChPT calculations [194–196], and systematically characterizes the short-range part of the corrections in terms of the LECs. We would like to briefly discuss here only one subtlety, which is addressed in detail in Ref. [210]. Namely, as we already know (see Section 4.8.3), in the absence of electromagnetic corrections, the pion decay constant, F_π, can be read from the two-point function of axial-vector currents:

$$i \int d^4x e^{ipx} \langle 0|TA_\mu(x)A_\nu^\dagger(0)|0\rangle = B(p^2)p_\mu p_\nu + C(p^2)g_{\mu\nu} , \tag{4.478}$$

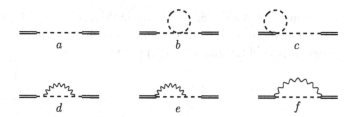

Figure 4.17 Loop corrections to the two-point function of the axial-vector current. The double, dashed and wiggly lines denote axial-vector current, pions and photons, respectively. Contributions from local operators with the corresponding LECs are not shown.

where

$$B(p^2) = \frac{2F_\pi^2}{M_\pi^2 - p^2} + \text{regular terms at } p^2 \to M_\pi^2. \qquad (4.479)$$

Here, $A_\mu(x) = (-A_\mu^1(x) - iA_\mu^2(x))/\sqrt{2}$.

Turning on the electromagnetic interactions $e \neq 0$, one can calculate this two-point function at $O(p^4)$. The pertinent diagrams are given in Fig. 4.17. The diagram 4.17d is of particular interest since, due to the fact that the photon is massless, it leads to an infrared singularity. More precisely, there is a branch cut starting at $p^2 = M_\pi^2$ in the function $B(p^2)$ instead of a pole. In order to circumvent this problem, one can give a small mass $m_\gamma \neq 0$ to the photon. The result for F_π at $e \neq 0$ is then divergent as $m_\gamma \to 0$. In Ref. [210] it was proposed to use dimensional regularization to tame both ultraviolet and infrared divergences. The photon stays massless, and gauge invariance is automatically obeyed (of course, the final answer does not depend on the particular regularization used).

In order to demonstrate how this approach works, let us consider the sum of the diagrams 4.17a+4.17d:

$$B_{\mu\nu}^{(a+d)} = \frac{2F^2}{M^2 - p^2} \left(1 + \frac{e^2 M^2 J(p^2)}{M^2 - p^2}\right), \quad M^2 = 2\hat{m}B + O(p^4), \qquad (4.480)$$

and

$$J(p^2) = \int \frac{d^D k}{(2\pi)^D i} \frac{1}{(M^2 - (p-k)^2)(-k^2)}$$

$$= \frac{\Gamma(2 - D/2)}{(4\pi)^{D/2}} \int_0^1 dx\, x^{D/2-2}(1 - z + zx)^{D/2-2}, \qquad (4.481)$$

where $z = p^2/M^2$. The result to this order in e^2 can be rewritten as

$$B_{\mu\nu}^{(a+d)} = \frac{2F^2}{M_1^2 - p^2}(1 + e^2 \underline{J}(p^2)), \quad J(p^2) = J(M^2) + (1-z)\underline{J}(p^2), \qquad (4.482)$$

and

$$M_1^2 = M^2(1 - e^2 J(M^2)). \qquad (4.483)$$

Two limits still have to be performed in this result: $D \to 4$ and $p^2 \to M^2$. Carrying out the integration explicitly, it is seen that in the vicinity of $z = 1$ and $D = 4$, the most singular part of the quantity $\underline{J}(p^2)$ behaves as

$$\underline{J}(p^2) = (\lambda + \kappa)\left[(1-z)^{D-4} - 1\right] + \cdots, \quad \kappa = \frac{1}{32\pi^2}\left[\ln\frac{M^2}{\mu^2} - 1\right], \quad (4.484)$$

where λ is a divergent quantity defined in Eq. (4.256). It is now evident that the result depends on the order of the limiting procedures:

1. If $D \to 4$ first, the first derivative of $\underline{J}(p^2)$ is logarithmically divergent at $z \to 1$:

$$\underline{J}(p^2) \to \frac{1}{16\pi^2}\ln(1-z), \qquad D = 4. \quad (4.485)$$

In this case, the quantity $B(p^2)$ develops a branch cut instead of a pole:

$$B(p^2) = \frac{2F^2}{M_1^2(1 - p^2/M_1^2)^{1-e^2/16\pi^2}} + \cdots. \quad (4.486)$$

2. The limit $z \to 1$ is regular for $D > 4$:

$$\underline{J}(M^2) = -\lambda - \kappa. \quad (4.487)$$

However, this expression diverges as $D \to 4$. In this case, the function $B(p^2)$ has a pole, but the residue of this pole is divergent. This is an infrared divergence, which, like the ultraviolet divergence, manifests itself as a pole in dimensional regularization. Introducing a photon mass, one would obtain a divergence that is proportional to $\ln m_\gamma^2$, instead of Eq. (4.487).

Since both the ultraviolet and infrared divergences look very similar in dimensional regularization, a question naturally arises as to how these can be distinguished. We remind the reader that only the ultraviolet divergences can be removed by renormalization. The infrared divergences are, in general, non-polynomial in momenta and should cancel automatically in all observable quantities [211].

An easy answer to the preceding question is provided by the observation that the infrared divergences always emerge in a particular kinematic configuration, where external particles go on shell. In our example, this corresponds to $z = 1$. On the contrary, the ultraviolet divergences are polynomial in momenta. From the expression given above we see that $\underline{J}(p^2)$ is finite at $D \to 4$ everywhere except at $z = 1$. Consequently, the divergence that emerges at $z = 1$ is of purely infrared origin.

Taking into account all these subtleties, a straightforward calculation of $B(p^2)$ in dimensional regularization can be performed. It is perfectly legitimate (if carried out consistently everywhere) to stick to the second option in the ordering of the limits: first $p^2 \to M^2$ and then $D \to 4$. In this case, the two-point function has a pole even at $e \neq 0$. The quantity $B(p^2)$ at order p^4 is given by

$$B(p^2) = \frac{2\tilde{F}^2}{M_{\pi^\pm}^2 - p^2} + \cdots, \quad (4.488)$$

where $M_{\pi^\pm}^2$ is the physical pion mass, given by Eq. (4.473), and

$$\tilde{F}^2 = F^2 \left(1 + a + e^2 b + O(M^4, e^4, e^2 M^2)\right),$$

$$a = -\frac{1}{16\pi^2 F^2} \left(M^2 \ln\frac{M^2}{\mu^2} + M_{\pi^\pm}^2 \ln\frac{M_{\pi^\pm}^2}{\mu^2}\right) + \frac{2M^2}{F^2} l_4^r,$$

$$b = (6 - 2\xi)\lambda + \frac{1}{16\pi^2}\left(3\ln\frac{M_{\pi^\pm}^2}{\mu^2} - 4\right) + \frac{20}{9}(k_1^r + k_2^r) + 4k_9^r, \qquad (4.489)$$

where ξ denotes the gauge parameter of the photon field, which enters in the gauge-fixing part of the Lagrangian $\mathcal{L}_{\text{g.f.}} = -\frac{1}{2\xi}(\partial^\mu \mathscr{A}_\mu)^2$. One sees now that the counterpart of the quantity F_π for $e \neq 0$, defined through the residue of two axial-vector currents, is both infrared-divergent and gauge-dependent. However, note that for $e \neq 0$ the quantity $\tilde{F}$ is not a physical observable. In order to see this, we have to return back to the effective weak Fermi Lagrangian that describes the process $\pi \to \ell\bar{\nu}_\ell$:

$$\mathcal{L}_{\text{Fermi}} = \frac{G_F}{\sqrt{2}}\bar{\ell}\gamma_\mu(1 - \gamma^5)\nu_\ell \bar{u}\gamma^\mu(1 - \gamma^5)d + \text{h.c.} + \cdots, \qquad (4.490)$$

where G_F is the weak Fermi constant. Taking the matrix element of this Lagrangian between the state containing a lepton–antilepton pair, and the one-pion state, we obtain the amplitude of the process one is looking for at $O(G_F)$. All terms of order G_F^2 are neglected. Further, in pure QCD, this matrix element factorizes, that is, the leptons and the quarks/gluons do not talk to each other in the limit $e \to 0$. The lepton part is easily calculated, and we are left with the matrix element of the bilinear quark current between the vacuum and the one-pion state:

$$\mathcal{M}(\pi \to \ell\bar{\nu}_\ell) \propto G_F \bar{u}(p_\ell)\gamma^\mu(1 - \gamma^5)v(p_{\bar{\nu}_\ell})\langle 0|\bar{u}\gamma^\mu(1 - \gamma^5)d|\pi^-\rangle. \qquad (4.491)$$

Since QCD conserves parity, only the axial-vector part of the matrix element contributes, and we finally arrive at the expression of the amplitude $\mathcal{M}(\pi \to \ell\bar{\nu}_\ell)$, which is proportional to the constant F_π modulo the kinematical factors. Thus, F_π at $e = 0$ is an observable.

The situation changes when $e \neq 0$. It is seen that now in the physical weak decay amplitude $\pi \to \ell\bar{\nu}_\ell$ there are diagrams, which do not contribute to $\tilde{F}$. These are the diagrams, where the photon field hooks on to the external lepton line. The pertinent diagrams are shown in Fig. 4.18. Moreover, in order to obtain an infrared-finite quantity, it is well known that the one-photon inner Bremsstrahlung rate $\pi \to \ell\bar{\nu}_\ell\gamma$ should be added to the original expression without the photon: $\Gamma(\pi \to \ell\bar{\nu}_\ell(\gamma)) = \Gamma(\pi \to \ell\bar{\nu}_\ell) + \Gamma(\pi \to \ell\bar{\nu}_\ell\gamma)$. It can be then shown by explicit calculations (see in what follows) that only the final result is both infrared-finite and gauge-invariant. Thus, at $e \neq 0$, the quantity $\tilde{F}$ is not a physical observable anymore, and no wonder that it is both gauge-dependent and infrared-singular.

The preceding fact has one more implication. In pure QCD, the non-singlet axial-vector current $A_\mu^i(x)$ is scale-independent. This follows from the fact that chiral symmetry in QCD is broken softly by the dimension-three quark mass term, and the

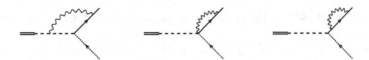

Figure 4.18   Diagrams that represent photons hooking on to the external lepton line. The double, dashed, solid, dotted and wiggly lines denote axial-vector currents, pions, leptons, antineutrinos and photons, respectively. Such diagrams are obviously not present in the two-point function of the axial-vector currents.

vector- and axial-vector charges still are generators of the $SU(N_f)_L \times SU(N_f)_R$ chiral algebra (see, e.g., Refs. [212–214]). In the presence of electromagnetic interactions the scale-independence does not hold anymore, because the matrix Q does not commute with all flavor generators. As a result, $\tilde{F}$, which is given by the matrix element of the axial-vector current, should in general depend on the QCD scale and thus is not an observable.[39] In our case, the QCD scale dependence enters through the LEC k_9^r, which depends on the QCD renormalization scale *as well as* on μ. This dependence can be visualized in explicit one-loop calculations, which were carried out in the linear σ-model [25], and has been discussed also in Ref. [206]. The bottom line is that, in order to be an observable, a quantity should be scale-independent, and the residue at the pole of the two-point function of axial currents does not qualify. Hence, a counterpart of F_π should be defined from an observable $\Gamma(\pi \to \ell\bar{\nu}_\ell(\gamma))$, calculated in an effective theory that includes *both* hadrons and leptons.

Here we shall briefly discuss the construction of such theory, following Ref. [87]. Along with the electromagnetic charge matrix Q, the left-handed weak charge matrix Q_L^w is introduced:

$$Q_L^w = -2\sqrt{2}G_F \begin{pmatrix} 0 & V_{ud} & V_{us} \\ 0 & 0 & 0 \\ 0 & 0 & 0 \end{pmatrix}. \tag{4.492}$$

Here, V_{ud}, V_{us} denote certain elements of the CKM matrix. Further, in order to retain chiral symmetry, the matrix Q_L^w is also promoted to a spurion field, which transforms according to

$$Q_L^w(x) \mapsto g_L(x)Q_L^w(x)g_L^\dagger(x), \qquad g_L(x) \in SU(N_f)_L. \tag{4.493}$$

As usual, after all invariant terms of the effective Lagrangian at a given order are written, one sets $Q_L^w(x) \to Q_L^w$.

The lowest-order three-flavor effective Lagrangian with leptons takes the following form [87]:

$$\mathcal{L}^{\mathrm{LO}} = \frac{F_0^2}{4}\langle u_\mu u^\mu + \chi_+\rangle + e^2 F_0^2 Z_0 \langle \mathcal{Q}_L \mathcal{Q}_R\rangle - \frac{1}{4}\mathcal{F}_{\mu\nu}\mathcal{F}^{\mu\nu} + \mathcal{L}_{\mathrm{g.f.}}$$
$$+ \sum_\ell \bar{\ell}(i\slashed{\partial} + e\slashed{A} - m_\ell)\ell + \sum_\ell \bar{\nu}_{\ell L}i\slashed{\partial}\nu_{\ell L}, \tag{4.494}$$

---

[39] Note that this scale should not be confused with the scale of the effective theory, which here is denoted by μ. Of course, $\tilde{F}$ does not depend on μ. We have already seen an example of two different scales in Section 1.6.

where

$$Q_L = uQu^\dagger, \qquad Q_R = u^\dagger Qu, \qquad Q_L^w = uQ_L^w u^\dagger. \tag{4.495}$$

Further,

$$u_\mu = i(u^\dagger(\partial_\mu - ir_\mu)u - u(\partial_\mu - il_\mu)u^\dagger),$$

$$l_\mu = v_\mu - a_\mu - eQ\mathscr{A}_\mu + \sum_\ell (\bar\ell\gamma_\mu v_{\ell L} Q_L^W + \bar{v}_{\ell L}\gamma_\mu \ell Q_L^{W\dagger}),$$

$$r_\mu = v_\mu + a_\mu - eQ\mathscr{A}_\mu. \tag{4.496}$$

Now, we are in a position to write down the Lagrangian at $O(p^4)$. Besides the strong part, given in Eqs. (4.195), (4.196) and (4.197), and the electromagnetic part, displayed in Eq. (4.470), additional terms should be taken into account:

$$\mathcal{L}_{\text{lept}}^{\text{NLO}} = e^2 F_0^2 \sum_\ell \Big(X_1 \bar\ell\gamma_\mu v_{\ell L} \langle u^\mu \{Q_R, Q_L^w\}\rangle + X_2 \bar\ell\gamma_\mu v_{\ell L} \langle u^\mu [Q_R, Q_L^w]\rangle$$

$$+ X_3 m_\ell \bar\ell v_{\ell L} \langle Q_L^w Q_R\rangle + iX_4 \ell\gamma_\mu v_{\ell L} \langle Q_L^w \hat{\nabla}^\mu Q_L\rangle + iX_5 \ell\gamma_\mu v_{\ell L}\langle Q_L^w \hat\nabla^\mu Q_R\rangle + \text{h.c.}\Big)$$

$$+ e^2 \sum_\ell \Big(\bar\ell(i\slashed\partial + e\slashed{\mathscr{A}})\ell + X_7 m_\ell \bar\ell\ell\Big) + e^2 X_8 \mathscr{F}_{\mu\nu}\mathscr{F}^{\mu\nu}, \tag{4.497}$$

where the X_i ($i = 1,\ldots,5$) are additional electroweak LECs, and

$$\hat\nabla_\mu Q_L = u(D_\mu^L Q)u^\dagger, \qquad \hat\nabla_\mu Q_R = u^\dagger(D_\mu^R Q)u. \tag{4.498}$$

The total rate of the process $\pi \to \ell\bar{v}_\ell$, calculated with this Lagrangian, is, of course, ultraviolet finite, but still contains infrared divergences. As already mentioned, in order to get an infrared-finite result, we have to add to the rate the contribution of the inner Bremsstrahlung, $\pi \to \ell\bar{v}_\ell\gamma$, to it. The sum of these two up to the terms of order e^4 is given by [87] (see also Ref. [209]):

$$\Gamma(\pi \to \ell\bar{v}_\ell(\gamma)) = \frac{G_F^2 |V_{ud}|^2 F_0^2 m_\ell^2 M_{\pi^\pm}}{4\pi}(1 - z_{\pi\ell})^2$$

$$\times \left\{ 1 + \frac{8}{F_0^2}\left[(M_\pi^2 + 2M_K^2)L_4^r + M_\pi^2 L_5^r\right]\right.$$

$$- \frac{1}{32\pi^2 F_0^2}\left[2M_{\pi^\pm}^2 \ln\frac{M_{\pi^\pm}^2}{\mu^2} + 2M_{\pi^0}^2 \ln\frac{M_{\pi^0}^2}{\mu^2} + M_{K^\pm}^2 \ln\frac{M_{K^\pm}^2}{\mu^2} + M_{K^0}^2 \ln\frac{M_{K^0}^2}{\mu^2}\right]$$

$$\left. + \frac{e^2}{16\pi^2}\left[3\ln\frac{M_\pi^2}{\mu^2} + H(z_{\pi\ell})\right] + e^2 E^r\right\}, \tag{4.499}$$

where $z_{\pi\ell} = m_\ell^2/M_{\pi^\pm}^2$ and $H(z)$ is a known function

$$H(z) = \frac{23}{2} - \frac{3}{1-z} + 11\ln z - \frac{2\ln z}{1-z} - \frac{3\ln z}{(1-z)^2} - 8\ln(1-z)$$

$$- \frac{4(1+z)}{1-z}\ln z\ln(1-z) + \frac{8(1+z)}{1-z}\int_0^{1-z} dt\,\frac{\ln(1-t)}{t}. \tag{4.500}$$

Further,

$$M_\pi^2 = 2\hat{m}B_0, \quad M_K^2 = (m_s + \hat{m})B_0,$$

$$E^r = \frac{8}{3}(K_1^r + K_2^r) + \frac{20}{9}(K_5^r + K_6^r) + 4K_{12}^r - \frac{4}{3}X_1 - 4X_2^r + 4X_3^r - X_6^r.$$

$$(4.501)$$

In order to identify the first-order QED corrections to the decay rate, it is convenient to rewrite this result in a standard form pioneered by Marciano and Sirlin [195]:

$$\Gamma(\pi \to \ell\bar{\nu}_\ell(\gamma)) = \frac{G_\mu^2 |V_{ud}|^2 F_\pi^2 M_{\pi^\pm} m_\ell^2}{4\pi}(1 - z_{\pi\ell}^2)$$

$$\times \left\{ 1 + \frac{\alpha}{\pi}\left[\ln\frac{M_Z^2}{M_\rho^2} + \frac{3}{2}\ln\frac{M_\rho}{M_{\pi^+}} + F(\sqrt{z_{\pi\ell}}) - C_1 \right] \right\}, \quad (4.502)$$

where $F(z) = H(z^2)/4 - 1/2$, and the quantity G_μ is, up to tiny radiative corrections of order $\alpha m_\ell^2/M_W^2$, equal to G_F. The structure-dependent correction C_1 is then given by a linear combination of LECs:

$$C_1 = -4\pi^2 E^r - \frac{1}{2} + \ln\frac{M_Z^2}{M_\rho^2} + \frac{Z_0}{4}\left(3 + 2\ln\frac{M_\pi^2}{M_\rho^2} + \ln\frac{M_K^2}{M_\rho^2} \right). \quad (4.503)$$

Here, it is also worth mentioning that in the preceding formulae the quantity F_π is *defined* through the reference masses $M_\pi = M_{\pi^\pm}$ and $M_K = M_{K^\pm}$. In other words, it is the pion decay constant in pure QCD with $m_u = m_d = \hat{m}$, in which the quark masses are tuned so that the pion and kaon masses are equal to M_π and M_K, respectively. In such a theory, the quantity F_π at $O(p^4)$ is given by

$$F_\pi = F_0\left\{ 1 + \frac{4}{F_0^2}\left[(M_\pi^2 + 2M_K^2)L_4^r + M_\pi^2 L_5^r \right] \right.$$

$$\left. - \frac{1}{32\pi^2 F_0^2}\left[2M_\pi^2 \ln\frac{M_\pi^2}{\mu^2} + M_K^2 \ln\frac{M_K^2}{\mu^2} \right] \right\}. \quad (4.504)$$

The ambiguity in the definition of a theory in the presence of electromagnetic interactions is inherent for all calculations of this sort. We shall discuss this ambiguity in more detail in the next section.

The equations (4.502) and (4.503) solve our problem completely. They allow us to extract the "purified" quantity F_π from the experimental data. The $O(\alpha)$ corrections in Eq. (4.502) are infrared-finite and do not depend on the QCD scale, as the dependence in the LECs K_i cancels with a similar dependence arising from the new electroweak LECs X_i [25].

4.14.2 Baryonic Sector

Including the electroweak interactions into the one-baryon sector proceeds along a similar path. Here, we skip the details common to the mesonic sector and briefly review only a couple of interesting applications. For simplicity, we restrict ourselves again to two-flavor ChPT with virtual photons. In the relativistic formulation, the covariant derivative of the nucleon field is given by

$$D_\mu \Psi = (\partial_\mu + \Gamma_\mu)\Psi, \qquad \Gamma_\mu = \frac{1}{2}[u^\dagger, \partial_\mu u] - \frac{i}{2}u^\dagger R_\mu u - \frac{i}{2}u L_\mu u^\dagger,$$

$$\begin{pmatrix} R_\mu \\ L_\mu \end{pmatrix} = v_\mu \pm a_\mu + Q_N \mathscr{A}_\mu. \tag{4.505}$$

Here, $Q_N = \mathrm{diag}(1,0)$ is the nucleon charge matrix (in units of the elementary charge e). In particular, the presence of $\mathscr{A}_\mu$ in the covariant derivative gives rise to the minimal coupling of the proton to the photon field. The $O(p^2)$ electromagnetic Lagrangian is given by

$$\mathcal{L}^{(e^2)} = e^2 \bar{\Psi}\left(f_1 \langle \hat{Q}_+^2 - Q_-^2 \rangle + f_2 \langle Q_+ \rangle \hat{Q}_+ + f_3 \langle \hat{Q}_+^2 + Q_-^2 \rangle\right)\Psi, \tag{4.506}$$

with

$$Q_\pm = \frac{1}{2}(u Q_N u^\dagger \pm u^\dagger Q_N u), \qquad \hat{Q}_+ = Q_+ - \frac{1}{2}\langle Q_+ \rangle. \tag{4.507}$$

At order p^2, there are only three additional electromagnetic LECs f_1, f_2, f_3, which are ultraviolet-finite. The photon loop starts to contribute at $O(p^3)$, and the pertinent third-order LECs are divergent. Note that the HB version of the $O(p^2)$ Lagrangian was first given in Ref. [215].

A large amount of work has been done to apply this framework for systematic studies of isospin-breaking corrections in the one-nucleon sector at low energy. From these applications, we mention only the calculation of the proton-neutron mass difference at lowest order, and isospin breaking in the pion–nucleon scattering lengths, which plays a crucial role in the determination of the precise value of the pion–nucleon σ-term discussed in Section 4.11.4. More information can be found in the original papers, which are cited in the literature guide at the end of this chapter.

4.14.3 The Proton–Neutron Mass Difference

Using Eq. (4.506) together with the strong Lagrangian at the same order, given in Eq. (4.325) and Table 4.2, it is straightforward to evaluate the proton and neutron masses at $O(p^2)$, where the loop diagrams still do not contribute:

$$m_p = \mathring{m} - 4c_1 B(m_u + m_d) + 2c_5 B(m_d - m_u) - \frac{F^2 e^2}{2}(f_1 + f_2 + f_3),$$

$$m_n = \mathring{m} - 4c_1 B(m_u + m_d) - 2c_5 B(m_d - m_u) - \frac{F^2 e^2}{2}(f_1 - f_2 + f_3). \tag{4.508}$$

Note that the proton-neutron mass difference contains terms where the quark mass difference enters as well as an electromagnetic contribution:

$$m_p - m_n = 4c_5(m_d - m_u) - F^2 e^2 f_2. \tag{4.509}$$

It is customary to call these two terms "strong" and "electromagnetic" mass difference at lowest order in ChPT. Such an identification, however, comes with a grain of salt due to the convention-dependence of this sort of splitting that was already mentioned. However, as seen from the preceding expression, this ambiguity disappears in the chiral limit, where the whole effect is of electromagnetic origin.[40] Consequently, the quantity f_2 does not suffer from this ambiguity, whereas higher terms in the expansion do.

Cottingham [216] derived a beautiful formula that gives the electromagnetic mass splitting of the proton and the neutron in terms of an integral over the doubly virtual, spin-averaged forward Compton scattering amplitude off nucleons, which is defined as

$$T^{\mu\nu}(p,q) = \frac{i}{4} \sum_s \int d^4x \, e^{iqx} \langle p,s|T j^\mu(x) j^\nu(0)|p,s\rangle. \tag{4.510}$$

Here, $j^\mu(x)$ denotes the electromagnetic current and $|p,s\rangle$ is a one-nucleon state. Further, using Lorentz invariance, it can be seen that the tensor $T^{\mu\nu}(p,q)$ can be written in terms of two scalar functions $T_{1,2}(\nu,q^2)$, where $\nu = p \cdot q/m_N$:

$$T^{\mu\nu}(p,q) = K_1^{\mu\nu}(p,q)T_1(\nu,q^2) + K_2^{\mu\nu}(p,q)T_2(\nu,q^2),$$

$$K_1^{\mu\nu}(p,q) = q^\mu q^\nu - g^{\mu\nu}q^2,$$

$$K_2^{\mu\nu}(p,q) = \frac{1}{m_N^2}\left((p^\mu q^\nu + p^\nu q^\mu)(p\cdot q) - g^{\mu\nu}(p\cdot q)^2 - p^\mu p^\nu q^2\right). \tag{4.511}$$

The contribution of the photon loop to the nucleon self-energy at $O(e^2)$ is shown in Fig. 4.19a. The shaded ellipse in this figure corresponds to the forward Compton scattering amplitude. The electromagnetic mass splitting is given by the difference of the proton and neutron self-energies. Closing the photon loop, we obtain

$$\delta m_N^{(\text{em})} = -\frac{e^2}{2m_N} \int \frac{d^4q}{(2\pi)^4 i} D_\Lambda(q^2) g_{\mu\nu}(T_p^{\mu\nu}(p,q) - T_n^{\mu\nu}(p,q)) + \delta m_{\text{ct}}. \tag{4.512}$$

Here, $p^2 = m_N^2$ and $D_\Lambda(q^2)$ denotes the regularized photon propagator taken, for example, in the form

$$D_\Lambda(q^2) = \frac{1}{-q^2}\frac{\Lambda^2}{\Lambda^2 - q^2}, \tag{4.513}$$

where Λ is the cutoff mass (note that the original integral diverges without regularization). Further, δm_{ct} is a counterterm.

Next, let us see where the counterterm arises from. In QCD with virtual photons, all observable quantities at $O(e^2)$ can be made finite by the renormalization of the strong coupling constant g and quark masses. This means that, at this order, only the matrix elements of the operators $G_{\mu\nu}G^{\mu\nu}$ and $\sum_{f=u,d} m_f Q_f^2 \bar{\psi}_f \psi_f$, emerging from the Lagrangian after renormalization, contribute to the right-hand side of Eq. (4.512). Since isoscalar operators cannot be present in the proton-neutron mass difference,

---

[40] Note that, due to chiral symmetry, quark masses are multiplicatively renormalizable in QCD with virtual photons.

one is left with the isovector operator $\bar{u}u - \bar{d}d$. At lowest order in the strong coupling constant, this leads to

$$\delta m_{ct} = -\frac{3\alpha C}{8\pi m_N} \ln \frac{\Lambda^2}{\mu^2}, \qquad C = \frac{4m_u - m_d}{18} \sum_s \langle p, s | \bar{u}u - \bar{d}d | p, s \rangle. \qquad (4.514)$$

The matrix element of the operator $\bar{u}u - \bar{d}d$ also enters the definition of the proton-neutron mass difference in pure QCD (i.e., without virtual photons):

$$\delta m_{QCD} = \frac{m_u - m_d}{4m_N} \sum_s \langle p, s | \bar{u}u - \bar{d}d | p, s \rangle (1 + O(m_d - m_u)). \qquad (4.515)$$

A crude estimate $\delta m_{QCD} \simeq -2\,\mathrm{MeV}$ leads to a tiny value $C \simeq 6 \cdot 10^{-4}\,\mathrm{GeV}^2$. The (logarithmic) divergence of the loop integral cancels against the logarithm in δm_{ct}, so that the limit $\Lambda \to \infty$ can be performed in the sum.[41]

Note now that the ambiguity, which was mentioned earlier, becomes clear using Eq. (4.514). The full answer depends on the scale μ, which here should be interpreted as the "matching scale." At this scale, the strong coupling constant and quark masses in QCD+QED and in pure QCD are set equal (this will be discussed in more detail in the next section). However, as anticipated, the μ-dependence disappears in the chiral limit. Moreover, for physical quark masses, the numerical value of the counterterm turns out to be tiny and negligible for practical purposes (see, e.g., Ref. [217, 218]).

In what follows, we briefly comment on the evaluation of the integral, which turns out to be a challenging exercise. In the rest-frame of the nucleon we may perform a Wick rotation, $q_0 = iQ_4$, $q^2 = -Q^2$. Carrying out the summation over Lorentz indices, we may express the mass shift in terms of the invariant amplitudes:

$$\delta m_N^{(em)} = \frac{e^2}{2m_N} \int \frac{d^4Q}{(2\pi)^4} D_\Lambda(-Q^2)\Phi(Q_4, Q^2) + \delta m_{ct},$$

$$\Phi(Q_4, Q^2) = 3Q^2 T_1(iQ_4, -Q^2) + (2Q_4^2 + Q^2)T_2(iQ_4, -Q^2). \qquad (4.516)$$

Here $T_{1,2} = T_{1,2}^p - T_{1,2}^n$ denotes the difference of the proton and neutron amplitudes.

In the next step, we may write dispersion relations for the amplitudes T_i. According to Ref. [217, 218], it is useful to consider the linear combinations $\bar{T} = T_1 + T_2/2$ and T_2 instead of the original set of amplitudes. Taking into account the leading Regge behavior, it can be shown that T_2 obeys an unsubtracted dispersion relation, whereas the dispersion relation for T_1 and thus for $\bar{T}$ requires one subtraction (see, e.g., Ref. [219]):

$$\bar{T}(\nu, -Q^2) = \bar{T}(iQ/2, -Q^2) + \frac{2}{\pi}(Q^2 + 4\nu^2) \int \frac{\nu' d\nu' \, \mathrm{Im}\, \bar{T}(\nu', -Q^2)}{(Q^2 + 4\nu'^2)(\nu'^2 - \nu^2 + i\varepsilon)},$$

$$T_2(\nu, -Q^2) = \frac{2}{\pi} \int \frac{\nu' d\nu' \, \mathrm{Im}\, T_2(\nu', -Q^2)}{\nu'^2 - \nu^2 + i\varepsilon}. \qquad (4.517)$$

It is worth mentioning that here, following Ref. [217, 218], we have performed a subtraction at $\nu = iQ/2$ instead of $\nu = 0$ used in other studies. As can be shown,

---

[41] The logarithm in Eq. (4.514) at higher orders in g gets modified, but this does not change the argument.

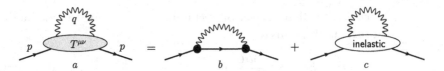

Figure 4.19 The photon loop contribution to the self-energy of the proton and the neutron: (a) the shaded ellipse depicts the full Compton scattering amplitude. This contribution can be split into (b) the Born term with the one-nucleon intermediate state and (c) the inelastic contribution with multiparticle intermediate states. The filled circles in (b) correspond to the nucleon electromagnetic form factors.

this helps to suppress the contribution to the integral from the regions where the integrand is poorly known. Substituting the dispersion relation into Eq. (4.516), we finally obtain the expression of the electromagnetic mass splitting in terms of the imaginary parts and the subtraction function $\bar{T}(iQ/2, -Q^2)$.[42] The ultraviolet divergence in this formulation emerges from the integral over the subtraction function only.

Crucial in the calculations is the fact that the imaginary parts of the Compton scattering amplitudes are proportional to the experimentally measured (transverse and longitudinal) electroproduction cross sections of the nucleon and, thus, the integral can be evaluated using known experimental input. The stumbling point is represented by the subtraction function. In the literature, there are two different approaches to the problem. The approach in Refs. [220, 221] is based on *modeling* the behavior of this function, using constraints at small and large values of Q^2. This approach, however, inherently contains a systematic uncertainty, which is very hard to control. Another approach, pioneered in Ref. [219] and further developed in Refs. [217, 218, 222], is based on the so-called Reggeon dominance hypothesis (or, the absence of the so-called fixed poles), which assumes a certain asymptotic behavior of the amplitudes in the variable v at fixed Q^2. This assumption enables us to *calculate* the subtraction function in terms of the imaginary parts and thus in terms of the observable cross sections as well.

Evaluating the integrals, it is seen that by far the largest contribution to the mass splitting is given by the so-called Born term, which corresponds to the diagram shown in Fig. 4.19b. The vertices in this diagram are given by the nucleon electromagnetic form factors dominated by the ρ-meson, so this contribution conceptually corresponds to the resonance saturation. The remaining contribution to the mass splitting, coming from the multiparticle intermediate states, as well as the counterterm contribution, turn out rather small. The most recent result for the mass splitting, quoted in Ref. [217, 218], amounts to $\delta m_N^{(\mathrm{em})} = (0.58 \pm 0.16)\,\mathrm{MeV}$. Given a rather large spread in numerical values for this quantity, which exists in the literature at present, an independent check of the preceding result on the lattice would be welcome.

---

[42] Traditionally, the mass shift is expressed in terms of $\mathrm{Im}\,T_{1,2}(v, -Q^2)$, and the subtraction function is given by $T_1(0, -Q^2)$. This, however, does not affect the main part of the argumentation given in what follows.

4.14.4 Isospin-Breaking in Pion-Nucleon Scattering

After discussing the virtual photon effects in the nucleon masses, let us turn to the very challenging task of evaluating the isospin-breaking corrections in the pion–nucleon scattering amplitudes in the vicinity of the threshold. This issue has been extensively addressed in the past both within the systematic EFT approach as well as using different methods. The details of the calculations can be found in the quoted references. Here, we would like to concentrate on the essentials only.

It is well known that, if isospin is conserved, the elastic πN scattering amplitude in all channels are expressed in terms of two independent amplitudes, corresponding to the scattering in the channels with total isospin $I = \frac{1}{2}, \frac{3}{2}$; see Eq. (4.417). If isospin violation is turned on, due to the quark mass difference $m_d - m_u$ as well as photons, this property is lost. Thus, we might ask:

1. In the isospin-symmetric world, there exist linear relations between the amplitudes in different physical channels. Can one determine how large the violation of these relations is, when isospin-breaking is turned on?

2. Consider the amplitude in one particular physical channel. Imagine the world with no isospin-breaking, that is, $e = 0$ and $m_u = m_d$. The same physical amplitude can be evaluated in this idealized world as well. How large are the corrections, when the isospin-breaking effects are turned on?

It is clear that these two questions are related but not identical. In particular, the answer to the first question can in principle be given in terms of observable characteristics, evaluated in the real world with $e \neq 0$ and $m_u \neq m_d$. On the contrary, the answer to the second question implies the introduction of the idealized, isospin-symmetric world and is thus convention-dependent. For example, this answer will depend on the value of the pion mass in the idealized world, which can be chosen to be equal to $M_{\pi^\pm}$ or M_{π^0}.

In the discussion here, we closely follow Refs. [65, 223]; see the literature guide for more references on the subject. When the virtual photons are taken into account, there emerges an additional subtlety that is not present in case of the mass. Namely, the amplitudes, in general, are not regular in the vicinity of the threshold. For a generic pion–nucleon amplitude, using Lorentz invariance, one could write (cf. Eq. (4.417))

$$T(\pi N \to \pi' N') = \bar{u}(p', s') \left(D(s, t) - \frac{1}{2(m' + m)} [\slashed{q}', \slashed{q}] B(s, t) \right) u(p, s). \quad (4.518)$$

Here, (p', s') and (p, s) are the momenta and spins of the outgoing and incoming nucleons and m', m denote their masses. The momenta q', q are those of the outgoing and incoming pions, and s, t, u are the usual Mandelstam variables.

Consider, for instance, the amplitude $\pi^- p \to \pi^- p$. The behavior of the scalar amplitudes D and B in the vicinity of the charged particle threshold $s = (m_p + M_{\pi^\pm})^2$ contains singularities that are produced by the diagrams shown in Fig. 4.20. Consider, for instance, the amplitude D (the amplitude B can be discussed in a similar way). The

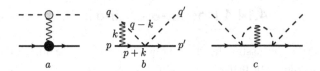

Figure 4.20   The diagrams that produce singularities at threshold: (a) one-photon exchange diagram, (b) the triangle diagram with incoming particles, and (c) exchange of a photon inside the hadronic loop. In case (b) there exists a similar diagram with outgoing particles (not shown here). The total contribution for (b) acquires the factor 2. Solid and dashed lines denote nucleons and pions, respectively, wiggly lines are photons, and the shaded/filled circles depict the pertinent form factors.

most singular part of this amplitude at $O(e^2)$ is given by the one-photon exchange diagram shown in Fig. 4.20a (such a diagram exists only in the case of the charged particle elastic scattering). To all chiral orders, this diagram is given in terms of the pion and proton electromagnetic form factors. One can subtract the most singular part, defining

$$\bar{D}(s,t) = D(s,t) - \frac{e^2 F_\pi^V(t) F_1(t)(s-u)}{2m_p t}. \qquad (4.519)$$

Here, $F_\pi^V(t)$ and $F_1(t)$ denote the pion electromagnetic form factor and the Dirac form factor for proton, respectively. F_π^V is given in Eq. (4.315), and F_1 is defined via the nucleon matrix element of the electromagnetic current,

$$\langle p',s'|j_\mu(0)|p,s\rangle = \bar{u}(p',s')\left[\gamma_\mu F_1(t) + i\frac{F_2(t)}{2m}\sigma_{\mu\nu}q^\nu\right]u(p,s), \qquad (4.520)$$

where $q_\mu = (p'-p)_\mu$, $t = (p'-p)^2$, m is the nucleon mass (m_p or m_n) and F_2 is the so-called Pauli form factor. For the proton, we have $F_1(0) = 1$. The remainder $\bar{D}$ is still singular at threshold, albeit the singularity is milder.

Next, let us consider the diagram shown in Fig. 4.20b and call its contribution $T^{(1)}$ (the details of the calculations of this diagram can be found in Ref. [65]):

$$T^{(1)} = \frac{e^2}{2F^2}\int\frac{d^D k}{(2\pi)^D i}\frac{\bar{u}(p',s')(\slashed{q}+\slashed{q}'-\slashed{k})(m_p+\slashed{p}+\slashed{k})(2\slashed{q}-\slashed{k})u(p,s)}{(-k^2)(m_p^2-(p+k)^2)(M_{\pi^\pm}^2-(q-k)^2)}. \qquad (4.521)$$

After simplifying the denominator, a contribution to the invariant amplitude $\bar{D}$ can be singled out; see Fig. 4.20b:

$$D^{(1)}(s,t) = \frac{e^2(s-u)}{8m_p F^2}\left(\frac{J_1}{2M_{\pi^\pm}^2} + 4J_2 + 4(s-m_p^2)J_\gamma(s)\right) + O(p^4). \qquad (4.522)$$

Here, $J_{1,2}$ are irrelevant (ultraviolet-divergent) constants, given by the following loop integrals:

$$J_1 = \int\frac{d^D k}{(2\pi)^D i}\frac{1}{M_{\pi^\pm}^2-k^2} = 2M_{\pi^\pm}^2\left(\lambda + \frac{1}{32\pi^2}\ln\frac{M_{\pi^\pm}^2}{\mu^2}\right),$$

$$J_2 = \int\frac{d^D k}{(2\pi)^D i}\frac{1}{(-k^2)(M_{\pi^\pm}^2-(q-k)^2)} = -2\left(\lambda + \frac{1}{32\pi^2}\left(\frac{M_{\pi^\pm}^2}{\mu^2}-1\right)\right). \qquad (4.523)$$

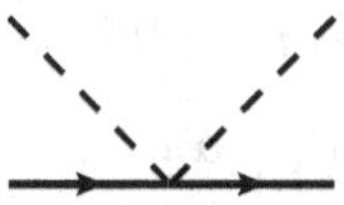

Figure 4.21 The Weinberg–Tomozawa vertex, contributing as $1/(4F^2)\,\bar{u}(p',s')(\not{q}' + \not{q})u(p,s)$ to the scattering amplitude of the process $\pi^- p \to \pi^- p$. Solid and dashed lines denote nucleons and pions, respectively.

The singularity is contained in the quantity $J_\gamma(s)$, which is given by

$$
J_\gamma(s) = \int \frac{d^D k}{(2\pi)^D i} \frac{1}{(-k^2)(m_p^2 - (p-k)^2)(M_{\pi^\pm}^2 - (q-k)^2)}
$$

$$
= \left(\frac{1}{m_p^2}\lambda + \frac{1}{32\pi^2 m_p^2} \left(\ln \frac{m_p^2}{\mu^2} + 1 \right) \right) \int_0^1 \frac{dx}{R} + \frac{1}{32\pi^2 m_p^2} \int_0^1 \frac{dx \ln R}{R},
$$

$$(4.524)$$

where

$$
R = \frac{1}{m_p^2} \left(xm_p^2 + (1-x)M_{\pi^\pm}^2 - x(1-x)s \right).
$$

$$(4.525)$$

Note that the preceding divergence is completely of infrared origin since, according to dimensional counting, $J_\gamma(s)$ is primitively convergent in the ultraviolet region.

The integrals over the Feynman parameter x can be explicitly performed in the vicinity of the threshold. The result is given by

$$
D^{(1)}(s,t) = \frac{\alpha M_\pi}{2F^2} \left(\frac{\pi \alpha \mu_c}{|\mathbf{p}|} + 2i\theta_C(|\mathbf{p}|) + C + \cdots \right) + O(p^4),
$$

$$(4.526)$$

where

$$
\mu_c = \frac{m_p M_{\pi^\pm}}{m_p + M_{\pi^\pm}}
$$

$$(4.527)$$

is the reduced mass, $|\mathbf{p}|$ is the magnitude of the relative momentum of the incoming or outgoing pion–nucleon pair, and the ellipses stand for terms that vanish at threshold. Further, $\theta_C(|\mathbf{p}|)$ denotes the so-called Coulomb phase, which is both infrared divergent and singular as $|\mathbf{p}| \to 0$:

$$
\theta_C(|\mathbf{p}|) = \frac{\mu_c}{|\mathbf{p}|} \mu^{D-4} \left(\frac{1}{D-4} - \frac{1}{2}(\Gamma'(1) + \ln 4\pi) + \ln \frac{2|\mathbf{p}|}{\mu} \right).
$$

$$(4.528)$$

The constant C in Eq. (4.526) is also infrared-divergent, albeit this singularity is canceled (at threshold), when all diagrams are added up. This is no more the case above threshold (i.e., the $O(|\mathbf{p}|)$ terms, shown by the ellipsis in Eq. (4.526). In this case, following the well-known path, in order to arrive at an infrared-finite result, we have to add the elastic and Bremsstrahlung cross sections.

Adding now the expression in Eq. (4.526) to the lowest-order result obtained by using the Weinberg–Tomozawa (WT) $O(p)$ strong Lagrangian, shown in Fig. 4.21, at $O(\alpha)$ we get

$$D^{\mathrm{WT}}(s,t) + D^{(1)}(s,t)$$

$$= \frac{M_\pi}{2F^2} \left(1 + \frac{\alpha\pi\alpha\mu_c}{|\mathbf{p}|} + 2i\alpha\theta_C(|\mathbf{p}|) + \alpha C + \cdots \right) + O(p^4)$$

$$= \exp(2i\alpha\theta_C(|\mathbf{p}|)) \left(1 + \frac{\alpha\pi\alpha\mu_c}{|\mathbf{p}|} \right) \frac{M_\pi}{2F^2} (1 + \alpha C + \cdots) + O(p^4). \tag{4.529}$$

As is well known, such a behavior persists in higher orders as well; in fact, the factor $|\mathbf{p}|^{-1}$ and the Coulomb phase, due to soft virtual photons, are universal and emerge in any QFT describing charged particles (see, e.g., Ref. [224]). The former factor is, of course, the well-known Coulomb singularity [197–199]. Note also that in Chapter 2 we have already derived the same result in the nonrelativistic EFT, which proves that this factor indeed emerges at all orders in the chiral expansion.

Last, we turn to the two-loop diagram shown in Fig. 4.20c. This diagram has also been treated in the nonrelativistic EFT, and we only quote the result here. The real part of the diagram has a singularity that is proportional to $\ln|\mathbf{p}|$. The imaginary part at threshold tends to a constant (we remind the reader that the imaginary part of the one-loop diagram, obtained by deleting the photon line, vanishes at threshold linearly with $|\mathbf{p}|$). All other diagrams are regular at threshold. Consequently, quite generally, we may write:

$$\mathrm{Re}\left(\exp(-2i\alpha\theta_C(|\mathbf{p}|))\bar{D}(s,t) \right)\bigg|_{|\mathbf{p}|\to 0} = \frac{\alpha D_{-1}}{|\mathbf{p}|} + \alpha D_0 \ln\frac{|\mathbf{p}|}{\mu_c} + \mathcal{T} + \cdots, \tag{4.530}$$

where the ellipsis stands for the terms that vanish at $|\mathbf{p}| \to 0$. Here, we stress once more that the behavior, shown in the preceding equation, is universal at $O(e^2)$, it holds to all orders in the chiral expansion. The quantity $\mathcal{T}$, the so-called threshold amplitude, is free of infrared singularities. Note also that the choice of the scale of the logarithm is a convention, and a change of this convention changes $\mathcal{T}$ by an additive constant, proportional to α.

After considering in detail the singularity structure of the amplitudes, we may return to the main question, which was stated in the beginning, and ask, how big are the isospin-breaking corrections to the scattering amplitudes? It is clear that, due to the singular behavior of the amplitudes at threshold, an answer to this question makes sense only for energies sufficiently far from threshold. This excludes, for example, scattering lengths, which are some of the most important characteristics of the low-energy scattering. In order to circumvent this difficulty, one might try to *identify* a quantity, which has a regular perturbative expansion in the isospin-breaking parameters α and $m_d - m_u$, and which in the isospin-symmetric world coincides with the scattering lengths, defined in a conventional manner. The threshold amplitude $\mathcal{T}$ from the preceding equation, as well as its counterparts for different physical channels of the πN scattering represent a convenient generalization for the purely strong scattering amplitudes in ChPT. Equivalently, one might also use the conventional definition of the so-called Coulomb-modified scattering length from nonrelativistic scattering theory [225]. The advantage of the present definition is that it is defined in a purely perturbative manner, without splitting photons into the non-perturbative "Coulomb photons" and the rest, which is treated perturbatively.

The isospin-breaking corrections to the πN scattering have been extensively studied both in the HB approach (see, e.g., Refs. [226–228]) as well as in the infrared regularization [65, 223, 229]. For illustrative purposes, we shall consider here only a few simple cases. More references to the work done, as well as to the calculations in other systems (e.g., the πK scattering) are given at the end of this chapter. One could start from the so-called triangle ratio, which exactly vanishes in the isospin limit and thus might be expected to serve as a convenient parameter that characterizes the degree of the isospin breaking:

$$R = 2\frac{f_{\ell,\pm}^{\pi^+ p \to \pi^+ p} - f_{\ell,\pm}^{\pi^- p \to \pi^- p} - \sqrt{2}f_{\ell,\pm}^{\pi^- p \to \pi^0 n}}{f_{\ell,\pm}^{\pi^+ p \to \pi^+ p} - f_{\ell,\pm}^{\pi^- p \to \pi^- p} + \sqrt{2}f_{\ell,\pm}^{\pi^- p \to \pi^0 n}}, \tag{4.531}$$

where $f_{\ell,\pm}(s)$ denote the partial-wave amplitudes for individual processes, which are defined in a standard manner (see, e.g., Ref. [223]). However, as we already know, the preceding amplitudes are, in general, infrared-divergent and contain threshold singularities, when $e \neq 0$. For this reason, an alternative definition for this ratio was introduced in Ref. [223], which includes, in particular, the Bremsstrahlung contributions. If we further restrict ourselves to the threshold only, it would be much simpler to use the threshold amplitudes defined in Eq. (4.530) instead, since the Bremsstrahlung contribution vanishes at threshold. Moreover, we carry out calculations up to and including $O(p^2)$ only, where the Coulomb singularities are still absent and the threshold amplitude $\mathcal{T}$ coincides with the invariant amplitude D. So, in this case, one could consider the ratio

$$R = 2\frac{D^{\pi^+ p \to \pi^+ p} - D^{\pi^- p \to \pi^- p} - \sqrt{2}D^{\pi^- p \to \pi^0 n}}{D^{\pi^+ p \to \pi^+ p} - D^{\pi^- p \to \pi^- p} + \sqrt{2}D^{\pi^- p \to \pi^0 n}}, \tag{4.532}$$

where all amplitudes are evaluated at the charged threshold.

Using the Lagrangian up to and including order p^2 for the evaluation of the tree diagrams (the Weinberg–Tomozawa contribution, nucleon pole diagrams, and the contributions coming from the LECs c_i, f_i), we get

$$\begin{aligned}
T_{\pi^+ p \to \pi^+ p} = \bar{u}(p',s')\Bigg\{ &-\frac{\slashed{q}' + \slashed{q}}{4F^2} - \frac{\overset{\circ}{g}_A^2}{2F^2} \slashed{q} \frac{1}{m_n + \slashed{p}' - \slashed{q}} \slashed{q}' - 4c_1\frac{2B\hat{m}}{F^2} \\
&+\frac{c_2}{2\overset{\circ}{m}^2 F^2}(q_\mu q'_\nu + q_\nu q'_\mu)(p'^\mu p'^\nu + p^\mu p^\nu) + 2c_3 q^\mu q'_\mu \\
&-\frac{ic_4}{F^2}\sigma^{\mu\nu}(q'_\mu q_\nu - q'_\nu q_\mu) - 2e^2 f_1 - \frac{e^2}{2}f_2 \Bigg\}u(p,s), \tag{4.533}
\end{aligned}$$

$$
T_{\pi^- p \to \pi^- p} = \bar{u}(p', s') \left\{ \frac{\slashed{q}' + \slashed{q}}{4F^2} - \frac{\overset{\circ}{g}_A^2}{2F^2} \slashed{q}' \frac{1}{m_n + \slashed{p}' + \slashed{q}'} \slashed{q} - 4c_1 \frac{2B\hat{m}}{F^2} \right.
$$

$$
+ \frac{c_2}{2\overset{\circ}{m}^2 F^2} (q_\mu q_\nu' + q_\nu q_\mu')(p'^\mu p'^\nu + p^\mu p^\nu) + 2c_3 q^\mu q_\mu'
$$

$$
\left. + \frac{ic_4}{F^2} \sigma^{\mu\nu} (q_\mu' q_\nu - q_\nu' q_\mu) - 2e^2 f_1 - \frac{e^2}{2} f_2 \right\} u(p, s), \tag{4.534}
$$

$$
T_{\pi^- p \to \pi^0 n} = \bar{u}(p', s') \left\{ -\frac{\slashed{q}' + \slashed{q}}{2\sqrt{2}F^2} + \frac{\overset{\circ}{g}_A^2}{2\sqrt{2}F^2} \slashed{q}' \frac{1}{m_n + \slashed{p}' + \slashed{q}'} \slashed{q} \right.
$$

$$
- \frac{\overset{\circ}{g}_A^2}{2\sqrt{2}F^2} \slashed{q} \frac{1}{m_p + \slashed{p}' - \slashed{q}} \slashed{q}' + \frac{ic_4\sqrt{2}}{F^2} \sigma^{\mu\nu} (q_\nu' q_\mu - q_\mu' q_\nu)
$$

$$
\left. + c_5 \frac{\sqrt{2}B(m_d - m_u)}{F^2} + \frac{e^2}{2\sqrt{2}} f_2 \right\} u(p, s). \tag{4.535}
$$

Evaluating these quantities at threshold, it is important to pay attention to the kinematics since, when isospin in broken, the $\pi^0 n$ pair has a nonzero relative momentum at the charged particle threshold. Assuming that $\alpha, (m_d - m_u) = O(\delta)$, where δ denotes a generic small parameter, one can expand the result in powers of this parameter, neglecting terms of order $p\delta$ and higher. The only nonvanishing contribution at this order comes from the Weinberg–Tomozawa term and from the contributions of the LECs c_5, f_2. The result at threshold takes the form

$$
D^{\pi^+ p \to \pi^+ p} - D^{\pi^- p \to \pi^- p} - \sqrt{2} D^{\pi^- p \to \pi^0 n}
$$

$$
= \frac{1}{2F^2} \left(m_p - m_n - 4c_5 B(m_d - m_u) - e^2 f_2 \right) + O(p\delta). \tag{4.536}
$$

Recalling now the lowest-order expression (4.509) for the proton-neutron mass difference, we finally obtain

$$
D^{\pi^+ p \to \pi^+ p} - D^{\pi^- p \to \pi^- p} - \sqrt{2} D^{\pi^- p \to \pi^0 n} = -\frac{e^2 f_2}{F^2} + O(p\delta). \tag{4.537}
$$

This means that, at threshold, the numerator can be expressed in terms of the electromagnetic mass difference between the proton and the neutron, which was already considered. On the other hand, the denominator at leading order is given by

$$
D^{\pi^+ p \to \pi^+ p} - D^{\pi^- p \to \pi^- p} + \sqrt{2} D^{\pi^- p \to \pi^0 n} = -\frac{2M_{\pi^+}}{F^2} + O(p^2). \tag{4.538}
$$

This means that the triangle relation at threshold is obeyed at the percent accuracy.

As mentioned before, we may also ask a question, what are the isospin-breaking corrections to individual amplitudes? The answer to this question depends on the prescription, which is used to define the isospin-symmetric world. In order to explain this, consider an expression of a single amplitude at threshold, here $\pi^- p \to \pi^- p$:

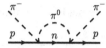

Figure 4.22   The one-loop diagram that is responsible for the cusp effect in the scattering process $\pi^- p \to \pi^- p$. Neutral particles are running in the loop.

$$D^{\pi^- p \to \pi^- p} = \frac{M_{\pi^\pm}}{2F^2} - \frac{\overset{\circ}{g}_A^2}{2F^2} \frac{M_{\pi^\pm}^2}{m_n + m_p + M_{\pi^\pm}} - \frac{4c_1 M_{\pi^0}^2}{F^2}$$

$$+ \frac{c_2}{2\overset{\circ}{m}^2 F^2} m_p^2 M_{\pi^\pm}^2 + c_3 M_{\pi^\pm}^2 - 2e^2 f_1 - \frac{e^2}{2} f_2. \qquad (4.539)$$

Defining now the isospin-symmetric world in which the pion and nucleon masses are taken equal to the charged pion and the proton masses, respectively, and replacing $F, \overset{\circ}{m}, \overset{\circ}{g}_A$ by their physical values, we may rewrite Eq. (4.539) at leading order as

$$D^{\pi^- p \to \pi^- p} = \frac{M_{\pi^\pm}}{2F_\pi^2} \left(1 - g_A^2 \frac{M_{\pi^\pm}}{2m_p + M_{\pi^\pm}} - (8c_1 - c_2 - 2c_3) M_{\pi^\pm} \right)$$

$$+ \Delta D^{\pi^- p \to \pi^- p}, \qquad (4.540)$$

where

$$\Delta D^{\pi^- p \to \pi^- p} = 8c_1 \frac{\Delta M_\pi^2}{F^2} - 2e^2 f_1 - \frac{e^2}{2} f_2. \qquad (4.541)$$

We may convince oneself that this expression will change, if another convention is used to define the isospin-symmetric world. We also would like to mention that the issue of the isospin-breaking corrections is crucial for the accuracy of the determination of the pion–nucleon σ-term, which was discussed previously. Namely, from pionic hydrogen experiments we extract a very accurate value of the threshold amplitude $D^{\pi^- p \to \pi^- p}$. It is, however, necessary to "purify" this result with respect to the isospin-breaking corrections and to extract the scattering length in the isospin-symmetric world, which will be further used in the Roy–Steiner equations [124]. It turns out, that the bulk of the systematic uncertainty is introduced by the LEC f_1, which is very poorly known. For a detailed discussion of isospin-breaking in the pion–nucleon scattering lengths, see Ref. [229].

Last but not least, we mention an effect, which is absent at tree level and arises first at one loop. The effect is caused by the so-called rescattering diagram shown in Fig. 4.22, where the masses of the particles, running in the loop, are not the same as the masses of the external particles (an example: elastic scattering $\pi^- p \to \pi^- p$, with the neutral pion and the neutron running in the loop). This diagram corresponds to the following integral:

$$J = \int \frac{d^D k}{(2\pi)^D i} \frac{\bar{u}(p's')(\slashed{q}' + \slashed{k})(m_n + \slashed{p}' + \slashed{q}' - \slashed{k})(\slashed{q} + \slashed{k}) u(p,s)}{(m_n^2 - (p' + q' - k)^2)(M_{\pi^0}^2 - k^2)}. \qquad (4.542)$$

For simplicity, we may neglect the numerator and consider only the scalar integral:

$$J_0(s) = \int \frac{d^D k}{(2\pi)^D i} \frac{1}{(m_n^2 - (p' + q' - k)^2)(M_{\pi^0}^2 - k^2)}$$

$$= -2\lambda - \frac{1}{16\pi^2} \left(1 + \int_0^1 dx \ln \frac{x m_n^2 + (1-x) M_{\pi^0}^2 - x(1-x)s - i\varepsilon}{\mu^2} \right).$$

$$(4.543)$$

This will not change anything in the argument. The imaginary part of the preceding expression is given by

$$\text{Im} J_0(s) = \frac{\lambda^{1/2}(s + i\varepsilon, m_n^2, M_{\pi^0}^2)}{16\pi s}, \qquad (4.544)$$

where $\lambda(x, y, z)$ is the triangle function. It is now seen that the isospin corrections to the imaginary part at threshold are of order $\sqrt{\delta}$ and not of order δ. Moreover, if the masses of the internal particles are higher than the masses of external particles (e.g., in the scattering $\pi^0 n \to \pi^0 n$), the preceding expression becomes purely imaginary and thus contributes to the real part of the amplitude, which in this case receives corrections of order $\sqrt{\delta}$. Then, the scattering cross section above threshold develops a cusp-like structure at $s = (m_p + M_{\pi^\pm})^2$. We have already studied similar structures within the nonrelativistic EFT; see Section 2.8. Note also that the cusp effect is not restricted to the case of broken isospin only – it emerges always, at the energies where the higher-mass threshold opens. This is most beautifully seen in the electric dipole amplitude E_{0+} in the threshold region of the photoproduction reaction $\gamma p \to \pi^0 p$ (where the initial pion is substituted by a photon); see Fig. 4.23. For a detailed discussion of threshold cusps in other hadronic reactions and further kinematical singularities, see Ref. [230].

4.15 Splitting of the Strong and the Electromagnetic Interactions

In previous sections, we have discussed the calculation of isospin-breaking corrections on several occasions. It was also mentioned that in some cases such corrections are convention-dependent. In this section, we would like to address this issue in more detail and explain what exactly is meant with "convention-dependence." To be specific, here we exclusively concentrate on the framework that includes QCD + virtual photons only. Thus, isospin breaking emerges from two sources, $m_u \neq m_d$ and $e \neq 0$. These two effects are intertwined, since the quark mass difference is partly due to electromagnetic effects. Respectively, one speaks of "strong" and "electromagnetic" breaking; even so, such a naming comes with a grain of salt.

Also, as already mentioned, there are two different interpretations of the isospin-breaking corrections that should be clearly distinguished:

1. In the isospin limit, there exist relations between observable quantities due to sym-metry. In the real world, these relations are broken, and the magnitude of the

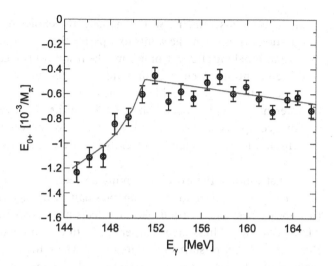

Figure 4.23   Cusp effect in the electric dipole amplitude E_{0+} for $\gamma p \to \pi^0 p$ at the opening of the $\pi^+ n$ threshold at $E_\gamma = 151.8\,\text{MeV}$, with E_γ the photon energy in the laboratory [191]. The data are from Ref. [231].

breaking characterizes the strength of the isospin-violating effects. In some cases, this criterion can be implemented straightforwardly. For example, the proton and neutron masses, as well as the masses of the charged and neutral pions, are the same in the isospin symmetry limit. Hence, $(m_p - m_n)/m_p$ and $(M_{\pi^\pm} - M_{\pi^0})/M_{\pi^\pm}$ serve as appropriate parameters that characterize isospin violation. In other cases, the identification of such a parameter can be less straightforward. For example, scattering amplitudes with charged particles, which were considered in the previous section, obey certain linear relations in the isospin limit. Namely, the so-called triangle ratio vanishes in this limit. However, since the scattering amplitudes are ill-defined quantities at $e \neq 0$ due to infrared problems, one cannot use the triangle ratio as it stands to study the isospin-breaking effects. Instead, one needs to choose an observable that resembles the triangle ratio but is well-defined even for $e \neq 0$. One possible choice, adopted in the previous section, was to use threshold amplitudes, which are purified from Coulomb singularities. Another choice is described in Ref. [229], and others are possible. Of course, there exists freedom in choosing a particular observable for such a study but, when the choice is done, the prediction made within a given full theory with $m_u \neq m_d$ and $e \neq 0$ is unambiguous. Note also that the observables considered before vanish in the isospin limit. Here we do not consider this case anymore.

2. A different situation emerges if we ask, what is an isospin-breaking correction to a particular observable, which does not vanish in the isospin limit? For example, let us introduce a common nucleon mass m_N in the isospin limit $m_u = m_d = \hat{m}$ and $e = 0$. One might ask whether the differences $m_p - m_N$ and $m_n - m_N$ can be evaluated systematically. It is clear that the answer to this question is convention-dependent, that is, the quantities $m_p - m_N$ and $m_n - m_N$ depend on the definition

of the isospin-symmetric world (namely, the choice of its parameters). The same question arises when one wants to separate the "strong" and the "electromagnetic" isospin-breaking effects. Comparing the results of two calculations, we have to bear this convention dependence in mind if, at the end, one does not want to compare apples with oranges. In addition, one is dealing with the isospin-breaking effects in different settings: in perturbative QCD, in the effective theory and on the lattice. For a comparison of results, it is important to know how the choice of a particular convention can be made consistently in different settings.

In what follows, the isospin-breaking corrections will be considered exclusively at order $O(\delta)$, where δ stands for a generic small parameter, either $m_d - m_u$ or $\alpha = e^2/4\pi$ (α appears here since virtual photon effects are always quadratic in e). No attempt will be done to address higher orders in δ. In our discussion, we shall mainly follow Refs. [25, 76]. Let us start from pure QCD, where the only parameters are the renormalized quark masses and the strong coupling constant. They run according to the RG equations, given in Section 3.4:

$$\mu \frac{dm_{u,d}(\mu)}{d\mu} = -\gamma_m(g(\mu))m_{u,d}(\mu), \qquad \mu \frac{dg(\mu)}{d\mu} = \beta(g(\mu)). \qquad (4.545)$$

Here, we are using the mass-independent $\overline{\text{MS}}$ scheme, in which $\gamma_m(g)$ depends only on the dimensionless coupling constant and is the same for all quark flavors f. The quark mass ratios are therefore RG-invariant. The isospin-symmetric world in this case can be defined as follows: in QCD, at some scale μ_1, the strong coupling constant and the quark masses are equal to $g(\mu_1)$ and $m_u(\mu_1)$, $m_d(\mu_1)$, respectively. (For simplicity, we consider here the two-flavor case, albeit the argument can be easily adapted to the case of any number of flavors). Then, in the isospin-symmetric world, the common quark mass and strong coupling constant at a scale $\mu = \mu_1$ are set equal to $\hat{m}(\mu_1) = \frac{1}{2}(m_u(\mu_1) + m_d(\mu_1))$ and $g(\mu_1)$, respectively, whereas at a different scale, they can be found from the RG equation. All observables can be evaluated both in the full theory with $m_u \neq m_d$ and in the isospin-symmetric world defined above (at least in principle). Consequently, the size of the isospin-breaking correction to any observable becomes uniquely calculable.

Let us now discuss the determination of the parameters of the theory from data. All observables in the full theory are functions of the parameters $g(\mu), m_u(\mu), m_d(\mu)$ and the running scale μ. Fixing $\mu = \mu_1$, we may adjust these three parameters to any three measured observables, which can be explicitly calculated in terms of these parameters. Then, the same measurement defines the isospin-symmetric world as well, and the magnitude of the isospin-breaking corrections to the observables does not depend on the choice of μ_1. Exactly the same line of reasoning applies to the determination of these parameters in lattice QCD. Here, initially, given the lattice parameters L/a, T/a are fixed, the parameters of the theory are the bare coupling g^0, the bare quark masses (in lattice units) am_u^0, am_d^0, and the scale given by the inverse lattice spacing $1/a$ in physical units. One can fit all these parameters to some measured observables. At the next step, we may use perturbation theory to relate these bare constants with the renormalized constants in the $\overline{\text{MS}}$ scheme at a scale $\mu = \mu_1$ (already in physical units). Again, the splitting of all observables into the isospin-symmetric piece and the rest does not depend on the choice of the matching scale μ_1, since the condition $m_d(\mu) - m_u(\mu) = 0$ is scale-invariant.

The implications for the effective low-energy theory are straightforward. The quark masses in any low-energy expansion enter only through the combination Bm_u, Bm_d, which do not depend on the QCD scale. (Recall that the quantity B is related to the vacuum expectation value of the renormalized operator $(\bar\psi\psi)$ in the chiral limit, whose RG running is governed by the same anomalous dimension $\gamma_m(g)$ as the running of quark masses in Eq. (4.545), but with an opposite sign.) The LECs do not depend on the quark masses by definition and, since they can be uniquely related to physical observables, they do not depend on the QCD scale μ as well. (This should not be mistaken for the scale of the effective theory μ_{eff}, on which the LECs obviously *do* depend.) Thus, in general, the LECs can be written as

$$l_i^r(\mu_{\text{eff}}) = (\Lambda_{\text{QCD}})^d f_i\left(\frac{\Lambda_{\text{QCD}}}{\mu_{\text{eff}}}\right), \tag{4.546}$$

where d denotes the mass dimension of a generic LEC l_i^r and f_i is dimensionless. From this, one sees that the LECs in the isospin-symmetric world (those who survive there) are the same as in QCD with $e = 0$ but $m_u \neq m_d$, because Λ_{QCD} is the same in both worlds, according to the matching condition.

The issue becomes much more subtle when the electromagnetic interactions are turned on. The RG functions in this case depend on two couplings: g and e. To lowest order in the electromagnetic coupling, we may write

$$\beta(g,e) = \beta(g) + e^2\beta_1(g) + O(e^4),$$

$$\gamma_m^{(f)}(g,e) = \gamma_m(g) + e^2\gamma_1^{(f)}(g) + O(e^4). \tag{4.547}$$

Here, $\beta(g)$ and $\gamma_m(g)$ are the same as in Eq. (4.545), and the quantity $\gamma_1^{(f)}(g)$ depends on the quark flavor. Note that e^2 also runs with its own RG equation, but due to the Ward identities, the running starts at $O(e^4)$ and will be neglected here.

Suppose that one acts exactly in the same manner as in the previous case. In the full theory with $m_u \neq m_d$ and $e \neq 0$ we may extract the values of the strong constant g and the quark masses at the matching scale $\mu = \mu_1$. This can be done, using experimental data as an input. Details are not relevant here. Next, one would like to define a world with no electromagnetic interactions but with quark masses still different, that is, $m_u \neq m_d$ and $e = 0$. We have seen that we can easily relate this world with the isospin-symmetric world, where the quark mass difference also vanishes. The first step, namely switching off the electromagnetic interactions, is, however, a subtle one. In order to see this, note that the RG running of the parameters is different for $e \neq 0$ and $e = 0$. So, for example, taking $\bar{m}_u(\mu)$ to be the u-quark mass at $e = 0$, and imposing the matching condition $\bar{m}_u(\mu_1) = m_u(\mu_1)$, one can find $\bar{m}_u(\mu)$ for any μ by using the RG running, Eq. (4.545). Shifting now the matching scale $\mu_1 \to \mu_1'$, it is seen that the

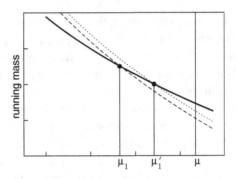

Figure 4.24 The running quark mass $m_u(\mu)$ in the full theory (solid line) and $\bar{m}_u(\mu)$ in the theory with $e = 0$ (dashed and dotted lines). It is seen that, if the matching point changes, $\mu_1 \to \mu_1'$, then the value of $\bar{m}_u(\mu)$ at arbitrary μ also changes.

value of $\bar{m}_u(\mu)$ also changes, because the RG running of $\bar{m}_u(\mu)$ and $m_u(\mu)$ is not the same. This statement is visualized in Fig. 4.24. Hence, the QCD parameters $\bar{g}(\mu, \mu_1)$, $\bar{m}_u(\mu, \mu_1)$, $\bar{m}_d(\mu, \mu_1)$, apart from depending on the running scale μ, implicitly depend on the matching scale μ_1, rendering the predictions convention-dependent.

A nice illustration to the splitting ambiguity is provided by the Cottingham formula, which was considered in Section 4.14.3. In particular, it was pointed out that the scale in Eq. (4.514) should be identified in fact with the matching scale μ_1. Below, we shall give a simple argument in favor of such an interpretation in the approximation, where we neglect strong interactions altogether and consider the electromagnetic interactions with point protons (a QED-like theory with protons). To order e^2, the proton mass shift is described by a one-loop self-energy diagram:

$$\Sigma(p) = -e^2 \int^\Lambda \frac{d^4k}{(2\pi)^4 i} \, \gamma_\mu \frac{1}{m - \not{p} + \not{k}} \gamma^\mu \frac{1}{-k^2}, \qquad (4.548)$$

where we, in order to facilitate the comparison with the Cottingham formula, used cutoff regularization, and m is the proton mass. Performing the mass-shell limit $\not{p} \to m$ and carrying out standard calculations, we get

$$\Sigma(p)\Big|_{\not{p} \to m} = -\frac{3e^2 m}{16\pi^2} \left(\ln \frac{\Lambda^2}{\mu^2} - \ln \frac{m^2}{\mu^2} \right) - \frac{e^2}{8\pi^2}. \qquad (4.549)$$

Here, μ is the running scale. The proton propagator is given by

$$S(p) = \frac{1}{m - \not{p} + \Sigma(p)}. \qquad (4.550)$$

Introducing the renormalized mass at order e^2,

$$m = m_r(\mu) + \frac{3e^2 m_r(\mu)}{16\pi^2} \ln \frac{\Lambda^2}{\mu^2}, \qquad (4.551)$$

for the physical mass at the same order we get

$$m_P = m_r(\mu) + \frac{3e^2 m_r(\mu)}{16\pi^2} \ln \frac{m_r^2(\mu)}{\mu^2} - \frac{e^2}{8\pi^2}. \tag{4.552}$$

Up to this point, the calculations were done in a standard way. Denote now the mass of proton "in the absence of electromagnetic interactions" by $\bar{m}$. According to the Cottingham formula, the physical mass is equal to $\bar{m}$ plus the closed photon loop (regularized by a cutoff Λ) plus the counterterm contribution,

$$\delta m_{\text{ct}} = \frac{3e^2 \bar{m}}{16\pi^2} \ln \frac{\Lambda^2}{\mu_1^2}, \tag{4.553}$$

where μ_1 is some unspecified scale at this moment. The contribution of the photon cloud is given by the integral in Eq. (4.548), the calculations are done exactly in the same way and the result looks very similar:

$$m_P = \bar{m} + \frac{3e^2 \bar{m}}{16\pi^2} \ln \frac{\bar{m}^2}{\mu_1^2} - \frac{e^2}{8\pi^2}. \tag{4.554}$$

As can be seen immediately, $\bar{m} = m_r(\mu_1)$ in this approximation; μ_1 indeed coincides with the matching scale, introduced in the present section. It remains to be seen whether this interpretation can be extended to higher orders without modification.

This splitting ambiguity has implications for the low-energy constants of the effective field theory of QCD+QED. The quark masses that appear in the chiral expansion are $\bar{m}_u(\mu, \mu_1)$, $\bar{m}_d(\mu, \mu_1)$ and the LECs depend on $\bar{g}(\mu, \mu_1)$. Hence, there is a $O(e^2)$ ambiguity in strong LECs, whereas for the electromagnetic LECs the ambiguity is of order one.

To summarize, when the electromagnetic interactions are switched off in QCD+QED, an inherent ambiguity arises, which was not present in the case of "purely strong" isospin-breaking corrections due to $m_d - m_u$. Namely, the concept of "switching off" the em interactions is scale-dependent, because the parameters run with different RG equations, when $e = 0$ and $e \neq 0$. We have introduced a matching scale μ_1, which is a useful and conceptually a clean way to parameterize the discussed ambiguity. However, it is obviously not the only one, and not the easiest one to impose, say, in lattice calculations. In order to use a particular splitting convention, all what is needed is to fix the values of the parameters in the purely strong theory at some scale $\mu = \mu_1$. It can be done directly by matching these parameters (as was done here), or matching some observables, which depend on these parameters. For example, the FLAG review (2019) [153] lists a few options, which have already been implemented in lattice calculations: namely, the so-called Dashen scheme, where the quark masses are tuned so that the flavor-diagonal mesons have the same mass in the full theory and in the symmetric world; the so-called $\Delta M^2 = 0$ scheme, which also uses the flavor-diagonal meson masses and is equivalent to the Dashen scheme; the $m_{\Sigma^+} - m_{\Sigma^-} = 0$ scheme, where the symmetric point is defined through the vanishing of the Σ^+ and Σ^- mass difference. Further, in the calculations of isospin-breaking corrections to the scattering amplitudes in the effective theory, which is in particular needed to extract precise values of the "purely hadronic" scattering lengths from the experimental data on the spectrum of hadronic atoms, the masses of the particles in the isospin multiplets (pions, nucleons and so on) have been routinely identified with the masses of *charged* particles (to cite only a few; see, for example, Refs. [65, 208, 229, 232]).[43] All these prescriptions are valid ones per se. However, one cannot directly compare results obtained in different prescriptions. For example, the μ_1-scheme, which we have discussed at length, is close to Dashen's scheme, but not exactly identical. This follows from the fact that isospin-breaking corrections to the neutral pion mass are of order $e^2 p$ and are small, but not completely negligible. At present, relating the existing splitting schemes remains a challenging problem, which still awaits the final solution.

4.16 Wess–Zumino–Witten Effective Action

The low-energy EFTs that we have constructed so far have not contained anomalies. In other words, the Green's functions obtained in such effective theories obey non-anomalous Ward identities. In order to be a consistent low-energy theory of QCD, we have, however, to include anomalies. Here we shall demonstrate how this goal can be achieved.

As mentioned in Section 3.7.4, in the presence of anomalies the set of the external sources s, p, v_μ, a_μ should be enlarged with the pseudoscalar source θ, which is coupled to the gluonic operator $\varepsilon_{\mu\nu\alpha\beta} \text{tr}_c(F^{\mu\nu}F^{\alpha\beta})$. The Fujikawa determinant, which emerges after the transformation of the fermionic measure, consists of two parts; see Eq. (3.198). The part containing the gluon fields is then compensated by the transformation of the source θ, given in Eq. (3.207). The remainder, which depends on the external sources only, renders the generating functional non-invariant with respect to the transformations from the chiral $U(N_f)_L \times U(N_f)_R$ group; see Eq. (3.208). Note also that, since this part does not depend on the gluon fields, Eq. (3.208) represents the variation of the QCD generating functional to all orders in the strong coupling constant.

Our aim here is to construct an effective theory that reproduces the anomalous term contained in Eq. (3.208). It should be stressed that the source of the anomaly is different in QCD and in the EFT. In QCD, the Lagrangian is explicitly invariant under chiral transformations and the anomaly is produced by a closed fermion (quark) loop. In the effective theory, we either have no fermions at all (purely bosonic ChPT), or fermion loop contributions are absent (HB, IR and EOMS formulations of ChPT with baryons). Hence, the anomalies can emerge, if and only if the effective action is not invariant under chiral transformations. The pertinent non-invariant term goes under the name of the Wess–Zumino–Witten (WZW) effective action [233, 234]. According to Eq. (3.208), under infinitesimal chiral $U(N_f)_L \times U(N_f)_R$ transformations, the variation of the WZW effective action is described by

---

[43] Note, by the way, that the corrections, introduced by the change of the prescription, are not that small, especially for the $\pi\pi$ scattering lengths, which are proportional to the square of the pion mass at lowest order. Using neutral pion mass instead of the charged one amounts up to an almost 8% effect!

$$D(\alpha, \beta) S_{\mathrm{WZW}}(U, v_\mu, a_\mu) = -\frac{N_c}{16\pi^2} \int d^4 x \, \mathrm{tr}_f \left(\beta(x) \Omega(x) \right). \tag{4.555}$$

Here, $\Omega(x)$ is a functional of the sources $v_\mu(x), a_\mu(x)$, given by Eq. (3.199), and $\alpha(x), \beta(x)$ denote the parameters of the vector and axial-vector transformations.[44] Note also that S_{WZW} is a functional, containing both the field U as well as the external sources, whereas the r.h.s. of the preceding equation contains the external sources only. Further, in Eq. (4.555), the differential operator $D(\alpha, \beta)$ describes the infinitesimal transformations for a generic functional $f(U, s, p, v_\mu, a_\mu, \theta)$:

$$D(\alpha, \beta) f = \delta U \frac{\delta f}{\delta U} + \delta s \frac{\delta f}{\delta s} + \delta p \frac{\delta f}{\delta p} + \delta v_\mu \frac{\delta f}{\delta v_\mu} + \delta a_\mu \frac{\delta f}{\delta a_\mu} + \delta \theta \frac{\delta f}{\delta \theta}, \tag{4.556}$$

where $\delta s, \ldots, \delta \theta$ are given in Eq. (3.207) and U transforms like $s + ip$.

From the group property, it is straightforward to conclude that S_{WZW} should obey the so-called Wess–Zumino consistency condition. Performing two subsequent infinitesimal transformations with parameters (α, β) and (α', β'), respectively, we get

$$[D(\alpha, \beta), D(\alpha', \beta')] S_{\mathrm{WZW}}(U, v_\mu, a_\mu) = D(\alpha'', \beta'') S_{\mathrm{WZW}}(U, v_\mu, a_\mu), \tag{4.557}$$

where (α'', β'') are expressed through (α, β) and (α', β'). In order to make this dependence explicit, it is convenient to introduce the parameters for the left- and right-handed chiral rotations:

$$\omega_R = \alpha + \beta, \qquad \omega_L = \alpha - \beta, \tag{4.558}$$

and similarly for (ω_R', ω_L'), (ω_R'', ω_L''). Next, since the left- and right-handed rotations decouple, one finally gets

$$\omega_{R,L}'' = i[\omega_{R,L}, \omega_{R,L}']. \tag{4.559}$$

Furthermore, Eq. (4.555) can be considered as a first-order differential equation that determines the functional S_{WZW}, and the consistency condition ensures its integrability. The boundary condition, which is needed to define the functional S_{WZW} unambiguously, can be chosen as follows:

$$S_{\mathrm{WZW}}(\mathbb{1}, v_\mu, a_\mu) = 0. \tag{4.560}$$

A method for obtaining the explicit form for the functional S_{WZW} from the known form of infinitesimal variation with respect to the chiral transformations, which we shall mainly follow, was first outlined in Ref. [233]. In Ref. [234], a compact form of the effective action, given in the form of a five-dimensional integral over a local Lagrange density, has been given. Different important aspects of the problem have been addressed in Refs. [235–238] as well. The finite chiral transformations are created by the exponential of the operator $D(\alpha, \beta)$:

---

[44] We would like to stress here once more that the right-hand side of Eq. (4.555) is not uniquely defined. Using the freedom in the choice of the renormalization prescription, any polynomial, containing only external sources, can be added to S_{WZW} without changing the physical content of the theory. The non-trivial part of the statement is that one cannot enforce the r.h.s. of this equation to vanish by any choice of the subtraction polynomial. Here we shall stick to the prescription in which there is no anomaly in the vector current and do not consider other schemes explicitly.

$$e^{D(\alpha,\beta)}U = g_R U g_L^\dagger,$$

$$e^{D(\alpha,\beta)}(v_\mu + a_\mu) = g_R(v_\mu + a_\mu)g_R^\dagger + ig_R \partial_\mu g_R^\dagger,$$

$$e^{D(\alpha,\beta)}(v_\mu - a_\mu) = g_L(v_\mu - a_\mu)g_L^\dagger + ig_L \partial_\mu g_L^\dagger. \qquad (4.561)$$

We can now choose a pair of the matrices α_U, β_U so that

$$e^{i(\alpha_U + \beta_U)}U e^{-i(\alpha_U - \beta_U)} = \mathbb{1}. \qquad (4.562)$$

The choice is obviously not unique, but the Wess–Zumino consistency condition guarantees that the final result does not depend on this choice. A convenient choice is given by

$$\alpha_U = 0, \qquad U = e^{-2i\beta_U}. \qquad (4.563)$$

Here we shall stick to this choice and suppress the argument α in the operator $D(\alpha,\beta) \to D(\beta_U)$. The boundary condition then gives

$$e^{D(\beta_U)}S_{\text{WZW}}(U, v_\mu, a_\mu) = 0. \qquad (4.564)$$

Expanding the exponent in a Taylor series, we get

$$\left(1 + \sum_{n=1}^{\infty} \frac{1}{n!}(D(\beta_U))^n\right)S_{\text{WZW}}(U, v_\mu, a_\mu) = 0. \qquad (4.565)$$

Taking into account Eq. (4.555), from the preceding equation we obtain

$$S_{\text{WZW}}(U, v_\mu, a_\mu) = \sum_{n=1}^{\infty} \frac{1}{n!}\frac{N_c}{16\pi^2}\int d^4x\, \text{tr}_f(\beta_U(x)D(\beta_U)^{n-1}\Omega(x)). \qquad (4.566)$$

Next, let us use the fact that the operator D is linear in β_U, that is, $D(t\beta_U) = tD(\beta_U)$. Then,

$$S_{\text{WZW}}(U, v_\mu, a_\mu) = \frac{N_c}{16\pi^2}\int_0^1 dt \int d^4x\, \text{tr}_f(\beta_U(x)e^{D(t\beta_U)}\Omega(x))$$

$$= \frac{N_c}{16\pi^2}\int_0^1 dt \int d^4x\, \text{tr}_f(\beta_U(x)\Omega(v_\mu^t, a_\mu^t)). \qquad (4.567)$$

Indeed, expanding the r.h.s. of the preceding equation in a Taylor series in t, and integrating over this variable term by term, one reproduces Eq. (4.566). The quantities v_μ^t, a_μ^t are defined through

$$v_\mu^t + a_\mu^t = r_\mu^t = u_t^\dagger(i\partial_\mu + r_\mu)u_t,$$

$$v_\mu^t - a_\mu^t = l_\mu^t = u_t(i\partial_\mu + l_\mu)u_t^\dagger. \qquad (4.568)$$

Here, $U_t = u_t^2$ and the matrix-valued field $u_t(x) = u(t,x) = \exp(-i\beta_U(x)t)$, defined in a five-dimensional domain with the coordinates (t,x), smoothly interpolates between the values $u(0,x) = \mathbb{1}$ and $u(1,x) = \exp(-i\beta_U(x)) = (U(x))^{1/2}$. Albeit the result was derived for the case of a linear interpolation, the integration path in the variable t can be deformed continuously, provided the endpoints at $t = 0, 1$ are fixed. Defining now

$$D_\mu U_t = \partial_\mu U_t - i r_\mu U_t + i U_t l_\mu ,$$

$$\hat{L}_\mu = D_\mu U_t U_t^\dagger , \quad \hat{R}_\mu = U_t^\dagger D_\mu U_t , \tag{4.569}$$

and using Eq. (4.568), one obtains:

$$\text{tr}_f(\beta_U(x)\Omega(v_\mu^t, a_\mu^t)) = \varepsilon^{\mu\nu\alpha\beta} \text{tr}_f\left(\beta_U(x)\left\{\frac{1}{3}(r_{\alpha\beta} r_{\mu\nu} + l_{\alpha\beta} l_{\mu\nu})\right.\right.$$

$$+ \frac{1}{6}(U_t^\dagger r_{\alpha\beta} U_t l_{\mu\nu} + U_t l_{\alpha\beta} U_t^\dagger r_{\mu\nu}) - \frac{2}{3}\hat{R}_\alpha \hat{R}_\beta \hat{R}_\mu \hat{R}_\nu \tag{4.570}$$

$$\left.\left. - \frac{i}{3}(\hat{L}_\mu \hat{L}_\nu r_{\alpha\beta} + \hat{R}_\mu \hat{R}_\nu l_{\alpha\beta} + r_{\alpha\beta} \hat{L}_\mu \hat{L}_\nu + l_{\alpha\beta} \hat{R}_\mu \hat{R}_\nu + \hat{L}_\mu r_{\alpha\beta} \hat{L}_\nu + \hat{R}_\mu l_{\alpha\beta} \hat{R}_\nu)\right\}\right).$$

Taking into account the fact that $2i\beta_U = -U_t^\dagger \partial_t U_t = U_t \partial_t U_t^\dagger$, the Wess–Zumino effective action, Eq. (4.567), can be rewritten as a sum of two terms. The first one contains only pion fields and is given in terms of an integral over the five-dimensional compact domain,[45] $S_4 \times [0, 1]$. It can be shown by explicit calculation that the integrand in the remainder, which depends on the external fields r_μ, l_μ, is a total derivative in the variable t that leaves the integral over S_4 only. Hence,

$$S_{\text{WZW}}(U, v_\mu, a_\mu)$$

$$= \frac{iN_c}{240\pi^2}\int d^5x\, \varepsilon^{abcde} \text{tr}_f\left(U_t^\dagger \partial_a U_t\, U_t^\dagger \partial_b U_t\, U_t^\dagger \partial_c U_t\, U_t^\dagger \partial_d U_t\, U_t^\dagger \partial_e U_t\right)$$

$$+ \frac{iN_c}{48\pi^2}\int d^4x\, \varepsilon^{\mu\nu\alpha\beta} \text{tr}_f(Z_{\mu\nu\alpha\beta}). \tag{4.571}$$

Here, the indices $a, b, \ldots$ label vectors in the five-dimensional space and ε^{abcde} is the totally antisymmetric tensor in this space. It is now clear that the result does not depend on the choice of coordinates on the five-dimensional domain, provided the values of the field U at the boundaries $t = 0$ and $t = 1$ are fixed. Further, the second term in this expression is given by (see, e.g., Ref. [236])

$$Z_{\mu\nu\alpha\beta} = -(r_\mu R_\nu R_\alpha R_\beta + l_\mu L_\nu L_\alpha L_\beta) + \frac{1}{2}(r_\mu R_\nu r_\alpha R_\beta - l_\mu L_\nu l_\alpha L_\beta)$$

$$- r_\mu U^\dagger l_\nu U R_\alpha R_\beta + l_\mu U r_\nu U^\dagger L_\alpha L_\beta + \partial_\mu r_\nu U^\dagger l_\alpha U R_\beta + \partial_\mu l_\nu U r_\alpha U^\dagger L_\beta$$

$$+ (r_\mu \partial_\nu r_\alpha + \partial_\mu r_\nu r_\alpha + r_\mu r_\nu r_\alpha + r_\mu U^\dagger l_\nu U r_\alpha)R_\beta$$

$$+ (l_\mu \partial_\nu l_\alpha + \partial_\mu l_\nu l_\alpha + l_\mu l_\nu l_\alpha + l_\mu U r_\nu U^\dagger l_\alpha)L_\beta$$

$$+ r_\mu U^\dagger l_\nu \partial_\alpha l_\beta U + r_\mu U^\dagger \partial_\nu l_\alpha l_\beta U + \partial_\mu r_\nu r_\alpha U^\dagger l_\beta U + r_\mu \partial_\nu r_\alpha U^\dagger l_\beta U$$

---

[45] Since the fields are assumed to vanish at infinity in the four-dimensional space, $\mathbb{R}^4$ can be compactified to the four-dimensional sphere S_4.

$$+ r_\mu r_\nu r_\alpha U^\dagger l_\beta U + r_\mu U^\dagger l_\nu l_\alpha l_\beta U + \frac{1}{2} r_\mu U^\dagger l_\nu U r_\alpha U^\dagger l_\beta U$$

$$- (l_\mu r_\nu - r_\mu l_\nu)(\partial_\alpha l_\beta + \partial_\alpha r_\beta) - l_\mu l_\nu l_\alpha r_\beta + r_\mu r_\nu r_\alpha l_\beta + \frac{1}{2} r_\mu l_\nu r_\alpha l_\beta . \qquad (4.572)$$

Here, we have introduced the notations:

$$L_\mu = \partial_\mu U U^\dagger , \qquad R_\mu = U^\dagger \partial_\mu U . \qquad (4.573)$$

Equations (4.571) and (4.572) solve our problem completely, as they provide an action functional, whose variation with respect to chiral transformations reproduces the anomalous part of the fermion determinant in QCD, given in Eq. (4.555). As a result, the WZW effective Lagrangian describes those processes, which proceed via the QCD anomaly. A prominent example of such a process is given by the decay $\pi^0 \to 2\gamma$, which has been already considered in Chapter 3. We shall demonstrate now how this result can be reproduced in the EFT framework. For simplicity, let us work in the two-flavor case and assume $r_\mu = l_\mu = eQ\mathscr{A}_\mu$, where $Q = \mathrm{diag}(\frac{2}{3}, -\frac{1}{3})$ is the charge matrix. Further, take $U(x) = \exp(i\pi^0(x)\tau^3/F)$ that leads to $R_\mu(x) = L_\mu(x) = i\partial_\mu \pi^0(x)\tau_3/F + \dots$. It is easy to see that in this case the expression for the WZW effective action simplifies to

$$S_{\mathrm{WZW}} = -\frac{N_c e^2}{96\pi^2 F} \int d^4x \, \varepsilon^{\mu\nu\alpha\beta} \mathscr{F}_{\mu\nu} \mathscr{F}_{\alpha\beta} \pi^0(x) + O(\pi^2) . \qquad (4.574)$$

For the decay amplitude, one obtains

$$\mathcal{M}(\pi^0 \to 2\gamma) = \frac{e^2 N_c}{12\pi^2 F} \varepsilon_{\mu\nu\alpha\beta} k_1^\mu \varepsilon^*(k_1)^\nu k_2^\alpha \varepsilon^*(k_2)^\beta , \qquad (4.575)$$

which coincides with the well-known result. Here, k_1, k_2 and $\varepsilon(k_1), \varepsilon(k_2)$ are the momenta and the polarization vectors of the outgoing photons, respectively.

It can be seen that the first term in Eq. (4.571), which is given by the five-dimensional integral, vanishes in the two-flavor case. In the case of three flavors, it describes processes with an odd number of the Goldstone bosons. Indeed, expanding in the Goldstone boson fields in the absence of the external sources, the WZW effective action takes the form

$$S_{\mathrm{WZW}} = -\frac{N_c}{240\pi^2 F_0^5} \int d^4x \, \varepsilon^{\mu\nu\alpha\beta} \mathrm{tr}_f(\Phi \partial_\mu \Phi \partial_\nu \Phi \partial_\alpha \Phi \partial_\beta \Phi) + O(\Phi^6) , \qquad (4.576)$$

where $\Phi = \phi^i \lambda^i$ is the matrix field of the Goldstone boson octet. For instance, the preceding Lagrangian describes the transition $K^+ K^- \to \pi^+ \pi^- \pi^0$, which was not present in the mesonic Lagrangian of ChPT considered in previous sections. The latter Lagrangian, namely, contains only the even number of boson fields at any order in chiral expansion and thus features a symmetry $\Phi(x) \to -\Phi(x)$, sometimes called "intrinsic parity," which is *not* a symmetry of QCD (see, e.g., Ref. [239]). Indeed, in QCD the ϕ-meson can decay in both $K^+ K^-$ and 3π. The introduction of the WZW Lagrangian lifts this symmetry.

In conclusion, we mention a few important points without providing further details:

• By construction, the variation of the functional S_{WZW} yields the expression of the non-Abelian anomaly, which depends on the external sources v_μ, a_μ, but not on the field U. The question naturally arises, if there are different structures with the same property. The answer is fortunately no. There is only one such term [20, 21].

- The (non)renormalization of the anomaly in the effective field theory has been studied, for example, in Refs. [240–242]. It was found that inserting the vertices generated from the WZW Lagrangian into the chiral loops yields a result that is chirally invariant to all orders. The non-invariant contribution to the effective action emerges exclusively at tree level. Clearly, this property is an EFT counterpart of the non-renormalization of the axial anomaly in QCD, which was already discussed in Chapter 3.

- In the derivation just given, the prefactor in front of the WZW effective action was fixed from the beginning by the anomaly. In a suitable normalization, it is given by the number of colors N_c. As first mentioned by Witten in Ref. [234], one could drop any reference to the anomaly whatsoever and try to construct a chirally invariant action functional that does not obey the symmetry under $\Phi \to -\Phi$. There exists no local Lagrangian in four dimensions with this property. We can, however, write down the action functional with the required properties in the form of an five-dimensional integral. The result is nothing but the WZW effective action with an undefined overall factor in front of it. Next, it turns out that, in order to be uniquely defined, the overall factor must be quantized and equal to an odd integer number.[46] (This statement can be derived in full analogy with Dirac's quantization condition of the magnetic charge for a monopole [243].) Needless to say, the preceding conclusion is consistent with our findings, where the prefactor was given by the integer number N_c.

4.17 The η'-Meson and Large-N_c ChPT

It can be shown that QCD in the limit of the large number of colors, $N_c \to \infty$, undergoes drastic simplifications [245–248]. This enables one to qualitatively understand its spectrum and the gross features of hadron–hadron interactions in this limit.[47] In particular, one can verify that the leading-order contributions to the Green's functions that describe hadronic processes stem from the so-called *planar* gluon diagrams, whereas diagrams with virtual quark loops are suppressed by powers of $1/N_c$. The low-energy spectrum contains a tower of weakly interacting mesons with all values of spin, with masses that approach a smooth limit as $N_c \to \infty$. The mesonic interactions scale as inverse powers of N_c. The OZI rule (see Section 3.18) is exact in the limit $N_c \to \infty$. Baryons consist of N_c quarks, and their masses scale linearly with N_c. Moreover, strong arguments were given in favor of the conjecture that in the limit of a large number of

---

[46] Obviously, N_c should be odd – otherwise, colorless baryons, which belong to the totally antisymmetric representation of the color $SU(N_c)$ group, would be bosons, not fermions [244].

[47] More precisely, the limit of a large number of colors implies $N_c \to \infty$ and the strong coupling constant $g \to 0$ with $\bar{g}^2 = N_c g^2$ fixed. It can be shown that the β-function, describing the running of the reduced coupling $\bar{g}$, possesses a nontrivial limit.

colors, baryons can be described as (semi-)classical localized soliton-like solutions of the mesonic EOM [248], as known from the time-honored Skyrme model [249].

In the context of the problems that we are considering, the large-N_c limit of QCD is interesting for several reasons. First, as already mentioned in Section 3.18, since the $U(1)_A$ anomaly is produced by a quark loop, it is subleading and vanishes in the large-N_c limit. Hence, the η' boson, whose mass has a contribution from the anomaly, becomes light in this limit as well. Moreover, compelling arguments can be used to justify that in the limit of large N_c, the pattern of the spontaneous symmetry breaking in QCD is given by $U(N_f)_L \times U(N_f)_R \to U(N_f)_V$ [250]. Thus, an adequate effective theory of large-N_c QCD should feature nine Goldstone bosons and should not exclude the η', as has been done so far.

Further, up to now, the dependence of the LECs on the parameter N_c has been obscure, albeit different scaling of these quantities with N_c might have important phenomenological consequences, since some of these (or linear combinations thereof) turn out to be suppressed in the large-N_c limit. Here we shall address this issue briefly.

Last but not least, up to now, we have considered the chiral expansion separately from the expansion in $1/N_c$. However, one might ask whether there exists a consistent way to merge these two expansions, and what we can learn from this merger. We shall address this question as well.

Constructing the EFT of QCD in the large-N_c limit has been discussed in the literature in sufficient detail [37, 237, 251–254]. In what follows, we shall mainly follow the argumentation of Refs. [37, 237]. We start with the generating functional of QCD $Z(s, p, v, a, \theta)$ defined in Section 3.10 which, being expanded in a Taylor series in the external sources, produces Green's functions of various quark bilinears as well as of the winding number density $\omega(x) = \varepsilon^{\mu\nu\alpha\beta} \mathrm{tr}_c(F_{\mu\nu}F_{\alpha\beta})/(32\pi^2)$. A generic Green's function with n_Q quark bilinears $j_k(x_k) = \bar{\psi}(x_k)\Gamma_k\psi(x_k)$, where the matrices Γ_k carry Lorentz and flavor indices, and n_ω insertions of the winding number density is given by:

$$G_{n_Q, n_\omega} = \langle 0 | T j_1(x_1) \cdots j_{n_Q}(x_{n_Q}) \omega(y_1) \cdots \omega(y_{n_\omega}) | 0 \rangle. \qquad (4.577)$$

Let us count the leading power in N_c for each type of such a Green's function. We start from the Green's functions that do not contain the field ω at all. A generic diagram contributing to this Green's function at leading order is depicted in Fig. 4.25a. It consists of a single quark loop, to which all bilinears are attached, and any number of gluon lines that are located in a plane (the so-called planar diagrams [246, 248]). Using the double-line notation, it can be straightforwardly shown that all such diagrams contribute at the same order in N_c. Hence, it suffices to consider a diagram without any gluon lines. Taking into account the fact that the color factor in each vertex, which describes the coupling of $j_k(x_k)$ to the quark line, is given by $\delta_{\alpha\beta}$, where α, β stand for the color indices of the quarks in the fundamental representation, it is seen that one color trace finally survives, yielding a factor N_c. This result does not depend on the number of the external legs n_Q, provided all legs are attached to a single quark loop. Any additional quark loop leads to a suppression factor $1/N_c$.

Consider now insertions of the operator ω, first without any $j_k(x_k)$. Rescaling the field G_μ in order to ensure that the kinetic term of gluons has the usual normalization

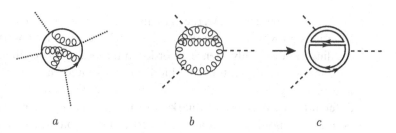

The leading contributions to the Green's functions of the quark bilinears and gluonic operators: (a) the operators j_k attached to a single quark line; (b) the gluonic operators ω attached to a gluon loop; (c) the same diagram in the double-line notation. Any number of planar gluons can be attached to each diagram. The solid, dotted, dashed and wiggly lines stand for quarks, the operators j_k, ω, and gluons, respectively.

will introduce the factor $g^2 = O(N_c^{-1})$ in front of ω. In other words, each ω counts like $1/N_c$. Further, since ω is a color-singlet made of gluon fields, its wave function will carry the factor δ_{AB}, where $A, B = 1, \ldots, N_c^2 - 1$ enumerates the generators of $SU(N_c)$. A single color trace in the gluon loop, shown in Fig. 4.25b, yields the factor $N_c^2 - 1 \simeq N_c^2$. The double-line notation, which leads to the two quark loops in Fig. 4.25c, yields the same color factor N_c^2, of course. Putting all factors together, it is seen that the diagram with n_ω insertions of ω only contributes at order $N_c^{2-n_\omega}$.

Finally, let us consider a generic Green's function G_{n_Q, n_ω} in Eq. (4.577). In order to insert the operators $j_k(x_k)$ into the diagram that already contains insertions of the operators ω, one should add at least one quark loop and attach all operators $j_k(x_k)$ to this loop. Adding more quark loops yields subleading contributions. Taking all factors into account gives the final result:

$$G_{n_Q, n_\omega} = O(N_c^{2-\ell-n_\omega}), \qquad (4.578)$$

where ℓ denotes the number of quark loops. Consequently, the effective action of QCD can be written in the following form:

$$S_{\text{eff}} = N_c^2 S_0(\vartheta) + N_c S_1(s, p, v, a, \vartheta) + N_c^0 S_2(s, p, v, a, \vartheta) + \cdots, \qquad (4.579)$$

where $\vartheta = \theta/N_c$ and S_ℓ collects contributions with $\ell = 0, 1, \ldots$ quark loops. Note that the scaling $\theta = O(N_c)$ and $s, p, v, a = O(1)$, displayed in Eq. (4.579), is consistent with the large-N_c scaling of the operators ω and j_k, respectively.

The following remark is needed. The first term in Eq. (4.579) does not contain quark loops at all and looks the same as in pure gluodynamics. However, this term depends on the parameter θ only through the ratio θ/N_c and thus seems to contradict the periodicity of the generating functional, which was already mentioned in Section 3.7.4. Namely, Z should be a periodic function of θ. A solution of this apparent paradox was offered in Ref. [255]; see also Ref. [237]. It has been argued that, even though they look superficially similar, the leading-order contributions in QCD and in pure gluodynamics differ. In particular, in gluodynamics, the path integral is performed over gluon configurations with fractional winding numbers ν (only $N_c\nu$ should be integer).

On the other hand, in QCD, the path integral is performed only over those gluonic contributions on which the quarks can live, and these involve contributions with integer winding numbers only. Hence, in order to reproduce the leading-order term in QCD, the sum over configurations in gluodynamics should be restricted to integer winding numbers only. The result then shows the required periodicity in θ modulo 2π. Stated differently, the presence of quarks affects the value of the path integral through the boundary conditions even at the lowest order, despite the fact that quark loops are not explicitly present in the expression. A more detailed discussion of this issue can be found in Refs. [237, 255].

As already noted, adding a quark loop introduces a factor $1/N_c$ in the effective action; see Eq. (4.579). Next, since the gluons carry no flavor, ℓ quark loops correspond to ℓ flavor traces involving the sources s, p, v, a. This enables one to establish the large-N_c counting rules for various LECs in the effective Lagrangians. Therefore, the N_c counting rules can be obtained by simply counting the number of the flavor traces that are present in an operator that multiplies a given LEC. This method can be also applied to the effective Lagrangians considered in the previous sections. Recall that these Lagrangians do not treat the η' particle explicitly (regarding it as infinitely heavy) but still, the large-N_c arguments apply in this case as well. The analysis was performed in Ref. [37], and we state here only the pertinent results. The sole subtle point in this analysis is that, since the η' is integrated out, it may contribute to some of the LECs. Owing to the fact that $M_{\eta'}^2 = O(N_c^{-1})$ in the chiral limit, it will upset the naive counting rules for these LECs.

We start from the $O(p^2)$ LECs. Noting that the wave function of a meson contains a color factor $N_c^{-1/2}\delta_{\alpha\beta}$ since it has to be normalized to unity, we get that F_0 (three-flavor case) and F (two-flavor case) both should be of order $N_c^{1/2}$. Similarly, B_0 and B are of order one. The preceding counting rules obviously agree with the result obtained by counting the flavor traces in the Lagrangian. Passing to the $O(p^4)$ LECs, it must first be noticed that the LEC $L_7 = O(N_c^2)$ (respectively, l_7 in the two-flavor case) plays a special role, as it gets contribution from the exchange of the η', which would be the ninth Goldstone boson as $N_c \to \infty$. The other $SU(2)$ LECs are all leading order in $1/N_c$ and scale as $O(N_c)$. The situation is different in $SU(3)$. There, the LECs $L_1, L_2, L_3, L_5, L_8, L_9, L_{10}$ all scale as $O(N_c)$; however, the LECs (or combinations thereof) $2L_1 - L_2, L_4, L_6$, which contain an additional flavor trace, are $O(1)$ and thus suppressed. This pattern was verified in the original analysis of Ref. [37] and is also reflected in Table 4.1 in Section 4.9.5 (see, however, Refs. [60, 256] for a critical discussion on this issue). In particular, it has been argued that the counting of the LEC L_7 at $O(N_c^2)$ is inconsistent since, in this case, the $O(p^4)$ contribution to the pion mass, proportional to $L_7(m_d - m_u)^2$, becomes dominant in the large-N_c limit and upsets chiral counting. This (seeming) paradox has been further addressed in Ref. [257], where it was pointed out that the limit $N_c \to \infty$ cannot be performed at fixed quark masses in the $SU(3)$-version of ChPT, since the η' meson also becomes light in this limit. On the other hand, the $1/N_c$ counting of the LECs is still valid, because these do not depend on the quark masses. Finally, note that the large-N_c counting for the meson-baryon LECs is discussed, for example, in Refs. [136, 258, 259].

Next, we shall demonstrate how the η' field can be explicitly included in the EFT framework. In a first step, no large-N_c arguments are invoked at all. All nine Goldstone fields can be collected in the unitary matrix

$$U(x) = \exp\left(\frac{i}{3}\phi^0(x)\right)\exp(i\Phi(x)), \qquad \Phi(x) = \lambda^i\phi^i(x), \qquad (4.580)$$

which transforms under $U(3)_L \times U(3)_R$ as $U \mapsto g_R U g_L^\dagger$. The field $\phi^0(x)$ describes the singlet component.

It is immediately seen that the combination $\phi^0 + \theta$ is invariant under chiral $U(3)_L \times U(3)_R$ transformations. Consequently, already at order p^0, one can write down an invariant Lagrangian,

$$\mathcal{L}^{(0)} = -V_0(\phi^0 + \theta), \qquad (4.581)$$

where the potential V_0 is an arbitrary function of its argument.

To order p^2, the most general Lagrangian has the form

$$\mathcal{L}^{(2)} = V_1\langle D_\mu U^\dagger D^\mu U\rangle + V_2\langle(s-ip)U\rangle + V_2^*\langle U^\dagger(s+ip)\rangle$$
$$+ V_3 D_\mu\phi^0 D^\mu\theta + V_4 D_\mu\theta D^\mu\theta + V_5 D_\mu\phi^0 D^\mu\phi^0. \qquad (4.582)$$

Here, the covariant derivatives are defined as

$$D_\mu U = \partial_\mu U - i(v_\mu + a_\mu)U + iU(v_\mu + a_\mu),$$
$$D_\mu\theta = \partial_\mu\theta + 2\langle a_\mu\rangle,$$
$$D_\mu\phi^0 = \partial_\mu\phi^0 - 2\langle a_\mu\rangle, \qquad (4.583)$$

and the potentials V_i are functions of $\phi^0 + \theta$. Parity invariance implies $V_i(\alpha) = V_i^*(-\alpha)$. V_0, V_1, V_3 and V_4 are real, whereas V_2 can be complex. Further, the term with V_5 can be eliminated by using the field transformations, so one could set $V_5 = 0$ from the beginning.

Expanding the preceding Lagrangian up to the second order in the fields, we may read off the expressions for the masses at lowest order. For our purpose, it is sufficient to consider the terms with ϕ^0 and ϕ^8 only. Assuming that the vacuum angle $\bar\theta$, defined in Section 3.10, vanishes and $m_u = m_d = \hat m$, the pertinent quadratic terms in the expansion of $\mathcal{L}^{(0)} + \mathcal{L}^{(2)}$ take the form

$$\mathcal{L}^{(0)} + \mathcal{L}^{(2)} = -\frac{1}{2}V_0''(0)\phi^0\phi^0 + V_1(0)\left(\frac{1}{3}(\partial\phi^0)^2 + 2(\partial\phi^8)^2\right)$$

$$- V_2(0)\left(\frac{1}{9}\langle\mathcal{M}\rangle\phi^0\phi^0 + \frac{4}{3\sqrt{3}}(\hat m - m_s)\phi^0\phi^8 + \frac{2}{3}(2m_s + \hat m)\phi^8\phi^8\right)$$

$$+ \langle\mathcal{M}\rangle V_2''(0)\phi^0\phi^0 + 2iV_2'(0)\phi^0\left(\frac{1}{3}\langle\mathcal{M}\rangle\phi^0 + \frac{2}{\sqrt{3}}(\hat m - m_s)\phi^8\right) + \cdots.$$

$$(4.584)$$

Introducing the notations,

$$F_0^2 M_0^2 = 6V_0''(0) + \frac{4}{3}\langle\mathcal{M}\rangle(V_2(0) - 6iV_2'(0) - 9V_2''(0)),$$

$$M_8^2 = \frac{2}{3}B_0(2m_s + \hat{m}), \quad \gamma = 1 - \frac{3iV_2'(0)}{V_2(0)},$$

$$F_0^2 = 4V_1(0), \quad F_0^2 B_0 = 2V_2(0), \tag{4.585}$$

we get

$$\mathcal{L}^{(0)} + \mathcal{L}^{(2)} = \frac{F_0^2}{2}\left(\frac{1}{6}(\partial\phi^0)^2 + (\partial\phi^8)^2\right) - \frac{M_0^2 F_0^2}{12}\phi^0\phi^0 - \frac{F_0^2 M_8^2}{2}\phi^8\phi^8$$

$$- \frac{4V_2(0)\gamma}{3\sqrt{3}}(\hat{m} - m_s)\phi^0\phi^8 + \cdots. \tag{4.586}$$

The preceding quadratic form can be diagonalized by introducing the physical η, η' fields,

$$F_0\phi^8 = \eta\cos\tilde{\delta} + \eta'\sin\tilde{\delta},$$

$$\frac{1}{\sqrt{6}}F_0\phi^0 = -\eta\sin\tilde{\delta} + \eta'\cos\tilde{\delta}, \tag{4.587}$$

where the angle $\tilde{\delta}$, which describes the octet-singlet mixing, is given by

$$\tan 2\tilde{\delta} = -\frac{4\sqrt{2}\gamma}{3(M_0^2 - M_8^2)}B_0(m_s - \hat{m}). \tag{4.588}$$

The masses of the η and η' at lowest order can be read off from the diagonalized Lagrangian:

$$M_{\eta'}^2 = M_0^2\cos^2\tilde{\delta} + M_8^2\sin^2\tilde{\delta} - \frac{4\sqrt{2}\gamma}{3(M_0^2 - M_8^2)}B_0(m_s - \hat{m})\sin\tilde{\delta}\cos\tilde{\delta},$$

$$M_\eta^2 = M_0^2\sin^2\tilde{\delta} + M_8^2\cos^2\tilde{\delta} + \frac{4\sqrt{2}\gamma}{3(M_0^2 - M_8^2)}B_0(m_s - \hat{m})\sin\tilde{\delta}\cos\tilde{\delta}. \tag{4.589}$$

Higher orders can be considered in a standard manner. In general, there are two types of terms, which are of the so-called natural/unnatural parity type. The latter contain one tensor $\varepsilon^{\mu\nu\alpha\beta}$. (Since the product of two such tensors can be expressed through a linear combination of $g^{\mu\nu}$ tensors, the number of antisymmetric tensors can be either zero or one.) The terms with unnatural parity arise first at $O(p^4)$. One of such terms is, of course, the WZW effective action, which was discussed in the previous section, but there are other, non-anomalous terms as well. At $O(p^4)$ there emerge six such non-anomalous terms:

$$\tilde{\mathcal{L}} = i\tilde{V}_1\langle\tilde{R}^{\mu\nu}D_\mu U D_\nu U^\dagger + \tilde{L}^{\mu\nu}D_\mu U^\dagger D_\nu U\rangle + \tilde{V}_2\langle\tilde{R}^{\mu\nu}U L_{\mu\nu}U^\dagger\rangle$$

$$+ \tilde{V}_3\langle\tilde{L}^{\mu\nu}L_{\mu\nu} + \tilde{R}^{\mu\nu}R_{\mu\nu}\rangle + \tilde{V}_4 i D_\mu\theta\langle\tilde{R}^{\mu\nu}D_\nu U U^\dagger - \tilde{L}^{\mu\nu}U^\dagger D_\nu U\rangle$$

$$+ \tilde{V}_5\left(\langle\tilde{L}^{\mu\nu}\rangle\langle L_{\mu\nu}\rangle + \langle\tilde{R}^{\mu\nu}\rangle\langle R_{\mu\nu}\rangle\right) + \tilde{V}_6\langle\tilde{R}^{\mu\nu}\rangle\langle L_{\mu\nu}\rangle, \tag{4.590}$$

where $\tilde{F}^{\mu\nu} = \frac{1}{2}\varepsilon^{\mu\nu\alpha\beta}F_{\alpha\beta}$ is the dual tensor. Due to the parity conservation, all potentials should be odd functions of ϕ^0, except $\tilde{V}_4$, which is even. Further, at this order, there are 52 independent potentials in the natural parity sector, which are listed, for example, in Ref. [251]. The extension to higher orders is straightforward.

Applying now the large N_c arguments to the preceding framework, the scaling rules for the various potentials can be established. For example, for the potentials from Eqs. (4.581) and (4.582), in the neighborhood of the origin the following rules hold:[48]

$$V_0(\alpha) = N_c^2 v_0(\alpha/N_c),$$

$$V_i(\alpha) = N_c v_i(\alpha/N_c), \quad i = 1, 2,$$

$$V_j(\alpha) = N_c^0 v_j(\alpha/N_c), \quad j = 3, 4. \tag{4.591}$$

This means that, expanding the potentials in Taylor series,

$$V_n(\alpha) = \sum_{k=0}^{\infty} V_{n,k}\alpha^k, \tag{4.592}$$

we have the following:[49]

$$V_{0,k} = O(N_c^{2-k}), \qquad V_{1,k}, V_{2,k} = O(N_c^{1-k}), \qquad V_{3,k}, V_{4,k} = O(N_c^{-k}). \tag{4.593}$$

At this point, it is convenient to introduce a *common counting* for these two otherwise unrelated expansions in the effective theory, the chiral expansion and the expansion in powers of $1/N_c$. In particular, it can be verified that

$$p = O(\sqrt{\delta}), \qquad 1/N_c = O(\delta), \tag{4.594}$$

where δ denotes a generic small parameter, leads to a consistent power-counting scheme (see, e.g., Ref. [237]). In this case, for example, the mixing angle $\tilde{\delta}$ is a quantity of order unity, and the masses of the η, η'-mesons at lowest order are given by a sum of two contributions, of order N_c^{-1} and $\mathcal{M}$, respectively, which are both $O(\delta)$ in the new counting. On the other hand, if one considers the chiral limit for a fixed N_c, then the octet Goldstone boson masses vanish, whereas the mass of η' approaches a finite limit determined by the topological susceptibility. In this limit, the heavy η' particle can be integrated out from the theory, leading to the $SU(3)$ meson Lagrangian that was already considered in the previous sections. In this way, we may also match the effective couplings in both theories – with and without η' – and verify the large-N_c scaling of various LECs in the effective theory without η', which were already mentioned in this section.

Last but not least, following Ref. [237], we shall consider the fate of the Kaplan–Manohar transformations [260] in the large-N_c limit. As shown in the latter paper, the standard $SU(3)$ Lagrangian of ChPT up to and including order p^4 is invariant under the transformation of the quark masses $m_u \to m_u + \lambda m_d m_s$ (with λ a real parameter) plus two other transformations, obtained from this by a cyclic permutation of u, d, s,

---

[48] V_0 describes the leading $O(N_c^2)$ term in the effective action, and V_1, V_2 contain one flavor trace and are thus suppressed by $1/N_c$. Finally, since $\theta, \phi_0 = O(N_c)$, V_3, V_4 must be of order one.

[49] There is a small loophole in this argument, related to the freedom of the choice of field variables. In Ref. [237] it has been shown that we can use the following scaling rules despite this ambiguity.

provided that certain $O(p^4)$ LECs are also adjusted.[50] In a compact form, the Kaplan–Manohar transformations can be given as

$$\mathcal{M} \rightarrow \mathcal{M} + \lambda (\mathcal{M}^\dagger)^{-1} \det \mathcal{M}^\dagger. \tag{4.595}$$

Considering the effective Lagrangian, the quark matrix $\mathcal{M}$ should be replaced by the matrix with the external sources $m(x) = s(x) + ip(x)$ in the preceding transformation. The generalization in case of the $U(3)$ theory takes the form

$$m(x) \rightarrow m(x) + \lambda e^{-i\theta} (m(x)^\dagger)^{-1} \det m(x)^\dagger. \tag{4.596}$$

In the $O(p^2)$ Lagrangian, $m(x)$ enters only through the terms that multiply the potentials V_2 and V_2^*. For example,

$$V_2 \langle U^\dagger m \rangle \rightarrow V_2 \langle U^\dagger m \rangle + \lambda V_2 e^{-i\theta} \langle U^\dagger (m^\dagger)^{-1} \rangle \det m^\dagger, \tag{4.597}$$

and, similarly, for the term with V_2^*. Using now the matrix identity,

$$\langle A^{-1} \rangle = \frac{\langle A \rangle^2 - \langle A^2 \rangle}{2 \det A}, \tag{4.598}$$

and taking into account that $\det U = \exp(i\phi^0)$, we obtain

$$V_2 \langle U^\dagger m \rangle \rightarrow V_2 \langle U^\dagger m \rangle + \frac{\lambda}{2} V_2 e^{-i(\phi^0 + \theta)} \left(\langle m^\dagger U \rangle^2 - \langle m^\dagger U m^\dagger U \rangle \right). \tag{4.599}$$

The pertinent terms in the $O(p^4)$ Lagrangian, which contain the matrix m, take the form

$$\mathcal{L}^{(4)} = \cdots + W_1 \langle U^\dagger m U^\dagger m \rangle + W_2 \langle U^\dagger m \rangle^2 + \text{h.c.} + \cdots, \tag{4.600}$$

where W_1, W_2 denote the $O(p^4)$ potentials. In the limit of a large $M_{\eta'}$, these become certain linear combinations of the LECs L_6, L_7, L_8. The Lagrangian is invariant, if the potentials are changed as

$$W_1 \rightarrow W_1 + \frac{\lambda}{2} V_2 e^{-i(\phi^0 + \theta)},$$

$$W_2 \rightarrow W_2 - \frac{\lambda}{2} V_2 e^{-i(\phi^0 + \theta)}. \tag{4.601}$$

By the way, it is now clear that the factor $\exp(-i\theta)$ in Eq. (4.596) guarantees that the transformed potential is also a function of $\phi^0 + \theta$. Again, in the limit of the large $M_{\eta'}$, this boils down to a certain transformation of the LECs L_6, L_7, L_8.

Note now that, if we consider only the chiral expansion for a fixed N_c, Eq. (4.601) represents a perfectly valid transformations of the potentials. However, this is no more so, if one treats $1/N_c$ and p on an equal footing. In the latter scheme, the potentials are polynomials in $\phi^0 + \theta$, whereas the transformed potentials are not, obviously, due to the exponential factor. Consequently, the transformation parameter λ should be zero to all orders in the expansion in $1/N_c$.

---

[50] We remark that this is an accidental symmetry broken at $O(p^6)$.

We end this section with a short remark on the effective Lagrangian including baryons. Here, the extension to $U(3)_L \times U(3)_R$ has mostly been considered to investigate the effects on the CP-violating θ-term; see, for example, Refs. [261–265]. The corresponding effective Lagrangians to one-loop accuracy are constructed and given in Refs. [266, 267].

4.18 Invariance of the ChPT Lagrangian

The symmetries, which the effective theories inherit from the underlying QCD, are encoded in the Ward identities obeyed by the Green's functions. These identities are equivalent to the invariance of the effective action with respect to local chiral transformations. In the presence of anomalies, this statement is replaced by an equally restrictive statement, which determines the variation of the effective action through a finite local polynomial of the external sources. Further, in our construction, the most general effective action was written as a path integral containing the most general invariant local Lagrangian with Goldstone boson fields. Our argument, however, contains a loophole. Consider, for simplicity, the case with no anomalies. Strictly speaking, from the invariance of the effective action it does not follow that the Lagrangian is also invariant. The latter may change by a total derivative as a result of local chiral transformations. No proof has been given up to now that there are no such noninvariant terms in the chiral Lagrangian, which lead to an invariant action functional, if the fields vanish sufficiently fast at infinity. It is clear that the presence of such terms would invalidate the construction scheme of the effective theories, which we have been using so far, because it was based on the assumption that the chiral Lagrangians to all orders are invariant. Following the discussion in Ref. [20] (see also Ref. [21]), we are going to close this gap now.

The effective Lagrangian, which appears in the generating functional of ChPT, depends on the Goldstone boson fields collected in the matrix U (which we shall call "pions" for brevity), the vector and axial-vector sources v_μ, a_μ and the scalar and the pseudo-scalar sources s, p that form the matrix χ. For a moment, assume that there are no anomalies (these will be considered later). Then, the problem can be formulated as follows. Suppose that the effective action $W_{\text{eff}}(s, p, v, a)$ defined via

$$e^{iW_{\text{eff}}(s,p,v,a)} = \int dU \exp\left\{i \int d^4x \mathcal{L}(U, s, p, v, a)\right\}, \tag{4.602}$$

is invariant under chiral transformations. Does this entail that the Lagrangian must be chirally invariant, too?

The answer to the preceding question is "yes" (in four space-time dimensions), and the proof is given in several steps. First of all, note that the effective potential at the leading order is given by the leading-order action functional $S^{(2)} = \int d^4x \mathcal{L}^{(2)}$, evaluated at the solution $U = U_c$ of the classical EOM. Loops do not contribute at this order. Thus, our problem boils down to a purely classical one: does the invariance of the classical action at $O(p^2)$ with respect to the local chiral transformations imply the invariance of the Lagrangian, or not?

Suppose that one was able to prove this statement. Let us now use induction to carry out the proof to all orders. Namely, at the next step, one assumes that Lagrangians up to and including $O(p^{2n})$ are invariant. So will be the loops, generated by the vertices entering these Lagrangians. Consider now the effective action at $O(p^{2n+2})$. Except for the loops, there is only one new contribution, and this is given by the classical action functional $S^{(2n+2)} = \int d^4x \, \mathcal{L}^{(2n+2)}$, again evaluated at the solution of the classical EOM. So, the problem is reduced to the classical one again, at the higher orders as well. The preceding discussion makes clear that all one needs to prove is that the invariance of the full *classical* action functional with respect to the local chiral transformation implies the invariance of the classical Lagrangian.

The full classical action $S(U,s,p,v,a)$ is invariant under the local chiral transformations. Here we use the same notations as in Section 4.6. Let $g \in G$ denote a generic element of the symmetry group G. Then, the symmetry of the action functional implies

$$S(U^g, s^g, p^g, v^g, a^g) = S(U,s,p,v,a). \tag{4.603}$$

Here, the transformed quantities are denoted by the superscript g. Further, one can always find an element g that transforms the matrix U to unity (such an element, of course, will depend on the pion fields). Assuming that g has this property, we get

$$S(U,s,p,v,a) = S(\mathbb{1}, s^g, p^g, v^g, a^g). \tag{4.604}$$

This implies on the Lagrangian level that the following difference is a total divergence:

$$\mathcal{L}(U,s,p,v,a) - \mathcal{L}(\mathbb{1}, s^g, p^g, v^g, a^g) = \partial_\mu \omega^\mu(U,s,p,v,a), \tag{4.605}$$

and

$$\partial_\mu \omega^\mu(\mathbb{1}, s, p, v, a) = 0. \tag{4.606}$$

Defining the new Lagrangian

$$\mathcal{L}(U,s,p,v,a) \to \mathcal{L}(U,s,p,v,a) - \partial_\mu \omega^\mu(U,s,p,v,a), \tag{4.607}$$

one concludes that it is now invariant under the action of the group element g, which brings U to the unit matrix.

Further, the action functional $S(\mathbb{1}, s, p, v, a)$ is invariant with respect to the diagonal subgroup H only:

$$S(\mathbb{1}, s, p, v, a) = S(\mathbb{1}, s^h, p^h, v^h, a^h), \quad h \in H. \tag{4.608}$$

Conversely, constructing the most general classical action, which has this property, and bringing the field U back to its original form, one arrives at the most general action functional that is invariant under the full group G. In order to see this, consider the effective action invariant under the full group:

$$S(U, f) = S(\tilde{g}U, \tilde{g}f), \quad \tilde{g} \in G. \tag{4.609}$$

Here, f is a cumulative notation for s, p, v, a. Let us now choose $g, g' \in G$ so that $gU = \mathbb{1}$ and $g'\tilde{g}U = \mathbb{1}$. Then, one easily obtains

$$S(\mathbb{1}, f) = S(\mathbb{1}, g'\tilde{g}g^{-1}f). \tag{4.610}$$

Using the notations from Section 4.6, we take $U = g^{-1} = \exp(i\xi A)$ and $U' = \tilde{g}U = g'^{-1} = \exp(i\xi'A)$. Hence, $\tilde{g}g^{-1} = \tilde{g}\exp(i\xi A) = \exp(i\xi'A)\exp(i\eta'V)$ and $g'\tilde{g}g^{-1} = \exp(i\eta'V)$. In other words, the object $h = g'\tilde{g}g^{-1}$ belongs to the diagonal group H for any choice of $\tilde{g} \in G$. Conversely, it is clear that for any $h \in H$ the pertinent element $\tilde{g} \in G$ can be found. Thus, our problem is finally reduced to the case of a vanishing pion field and to transformations from the diagonal subgroup only.

Next, taking $g_R = g_L = h$ in Eq. (3.74), it is seen that

$$v_\mu = \frac{1}{2}(r_\mu + l_\mu) \mapsto ih\partial_\mu h^\dagger + hv_\mu h^\dagger,$$

$$a_\mu = \frac{1}{2}(r_\mu - l_\mu) \mapsto ha_\mu h^\dagger,$$

$$\chi = (s + ip) \mapsto h\chi h^\dagger.$$

$$\chi^\dagger = (s - ip) \mapsto h\chi^\dagger h^\dagger. \tag{4.611}$$

This means that the sources a_μ and $\chi, \chi^\dagger$ transform covariantly. The same is true for the functional derivatives:

$$A^\mu(\chi, \chi^\dagger, v, a; x) = \frac{\delta S(\mathbb{1}, s, p, v, a)}{\delta a_\mu(x)} \mapsto hA^\mu(\chi, \chi^\dagger, v, a; x)h^\dagger,$$

$$m(\chi, \chi^\dagger, v, a; x) = \frac{\delta S(\mathbb{1}, s, p, v, a)}{\delta \chi^\dagger(x)} \mapsto hm(\chi, \chi^\dagger, v, a; x)h^\dagger,$$

$$m^\dagger(\chi, \chi^\dagger, v, a; x) = \frac{\delta S(\mathbb{1}, s, p, v, a)}{\delta \chi(x)} \mapsto hm(\chi, \chi^\dagger, v, a; x)h^\dagger. \tag{4.612}$$

Moreover,

$$S(\mathbb{1}, s, p, v, a) = S(\mathbb{1}, 0, 0, v, 0) + \int_0^1 dt \frac{d}{dt}S(\mathbb{1}, ts, tp, v, ta)$$

$$= S(\mathbb{1}, 0, 0, v, 0) + \int_0^1 dt \int d^4x\, \text{tr}\Big(a_\mu(x)A^\mu(t\chi, t\chi^\dagger, v, ta; x)$$

$$+ \chi^\dagger(x)m(t\chi, t\chi^\dagger, v, ta; x) + m^\dagger(t\chi, t\chi^\dagger, v, ta; x)\chi(x)\Big). \tag{4.613}$$

The action functional defines the Lagrangian up to a total derivative. One can use this freedom and define the effective Lagrangian as

$$\mathcal{L}(\mathbb{1}, s, p, v, a) = \mathcal{L}(\mathbb{1}, 0, 0, v, 0) + \int_0^1 dt\, \text{tr}\Big(a_\mu(x)A^\mu(t\chi, t\chi^\dagger, v, ta; x)$$

$$+ \chi^\dagger(x)m(t\chi, t\chi^\dagger, v, ta; x) + m^\dagger(t\chi, t\chi^\dagger, v, ta; x)\chi(x)\Big). \tag{4.614}$$

The additional terms are then explicitly gauge invariant.

Thus, our problem has been finally reduced to the case of the action functional that contains the vector source only. Does the action functional of a vector field, invariant under the diagonal subgroup H, necessarily imply the gauge invariance of the Lagrangian? The answer to this question, given in Refs. [20, 21], depends on the number of the space-time dimensions. Namely, it was shown that the effective Lagrangian is invariant for all dimensions except $D = 3$, where the so-called Chern–Simons term occurs. This term in the Lagrangian is itself not gauge-invariant, but its variation with respect to the gauge transformations is a total derivative.

This completes the discussion in the case with no anomalies. The extension to the case with anomalies can be done straightforwardly. Namely, since the WZW effective action reproduces all anomalies in QCD, the effective action $(S - S_{WZW})$ will be gauge-invariant, and we immediately arrive at the situation already considered. Here, we would like to reiterate that doing loops with the WZW Lagrangian does not lead to new contributions to the anomaly, and the path integral measure of the effective theory is explicitly invariant, unlike the fermionic measure in QCD. Thus, the WZW term is the sole non-invariant part of the effective Lagrangian.

4.19 Partially Quenched ChPT

It is important to realize that the mathematical methods that have led to the systematic formulation of the EFT of QCD can also be used in the context of other problems that exhibit the same gross features. Namely, the symmetry group should be broken to some subgroup that leads to the emergence of (pseudo-)Goldstone bosons. The low-energy dynamics of a system is then dominated by these (almost) massless degrees of freedom, whose Lagrangian can be written down by using the guidelines already considered.

In this section, we shall consider the EFT of (partially) quenched QCD on the lattice. The symmetries of the (partially) quenched theory were briefly considered in Section 3.20. There are several reasons why such a theory is interesting. First of all, it allows one to study the quark mass dependence of the observables as measured on the lattice (a dependence on both the valence and sea quark masses). The (partially) quenched QCD has some peculiar properties (e.g., the existence of the double poles in the two-point functions or unconventional chiral logarithms), which are absent in full QCD. Such effects can be observed on the lattice, and one can use (partially) quenched ChPT to understand these properties in the continuum setting. Further, as shown, for example, in Ref. [268], the partially quenched simulations provide a very convenient framework for extracting the low-energy constants in full QCD, because here we have more mass parameters (the valence and sea quark masses) that can be varied independently from each other. Moreover, the partially quenched effective field theory can be used for estimating the contributions from the so-called disconnected diagrams (see, e.g., Ref. [269]). Last but not least, the partially quenched effective theory can be used to study the finite-volume effects, arising from the so-called partial twisting on the lattice. Here we shall briefly review some of these applications.

As was argued in Section 3.20, the chiral symmetry breaking pattern in a partially quenched QCD is the same as in QCD. Essentially, this happens because the sea quark sector of partially quenched QCD is identical to full QCD – that is, the Green's functions of the currents built of sea quarks only are the same as the Green's functions in QCD. On the contrary, the statement that fully quenched QCD exhibits the same symmetry breaking pattern is an additional assumption.

We collect the Goldstone bosons emerging from the chiral symmetry breaking in the matrix $U(x)$, where

$$U(x) = \exp\left(\frac{\sqrt{2}i}{F_0}\Phi(x)\right), \qquad \Phi(x) = \phi^\alpha(x)T^\alpha, \qquad (4.615)$$

and the normalization of the generators T^α, defined in Section 3.20, is given by $\text{str}(T^\alpha T^\beta) = \pm\delta^{\alpha\beta}$. The matrix field Φ can be written as

$$\Phi(x) = \begin{pmatrix} \phi(x) & \eta(x) \\ \eta^\dagger(x) & \tilde{\phi}(x) \end{pmatrix}, \qquad \phi^\dagger(x) = \phi(x), \qquad \tilde{\phi}^\dagger(x) = \tilde{\phi}(x). \qquad (4.616)$$

Here, $\phi(x), \tilde{\phi}(x)$ and $\eta(x)$ are $(N_v + N_s) \times (N_v + N_s)$, $N_v \times N_v$ and $N_v \times (N_v + N_s)$ matrix fields, respectively. The field $\phi(x)$ contains the mesons made up of valence and sea quarks/antiquarks (all possible combinations), $\tilde{\phi}(x)$ collects the mesons consisting of a ghost and an antighost, and $\eta(x)$ represents the mesons that contain exactly one ghost or one antighost (the other quark/antiquark belongs to the valence or sea sectors). It is now seen that the spinless fields contained in the matrix η have fermion statistics; they are ghosts as well.

The matrix $\Phi(x)$ has the zero supertrace $\text{str}(\Phi) = \text{tr}(\phi) - \text{tr}(\tilde{\phi}) = 0$. It transforms according to

$$U \mapsto g_R U g_L^\dagger, \qquad g_L, g_R \in SU(N_v + N_s | N_v)_{L,R}. \qquad (4.617)$$

The transformation laws for the external sources s, p, v, a are defined in a standard manner, only the symmetry group is now different. The covariant derivative of the field U is also given by a well-known expression. We do not copy all these expressions here. The lowest-order chiral Lagrangian has the familiar form

$$\mathcal{L}^{(2)} = \frac{F_0^2}{4}\,\text{str}(D_\mu U D^\mu U^\dagger + \chi U^\dagger + U\chi^\dagger). \qquad (4.618)$$

The higher-order terms can be constructed following the well-known procedure. There is, however, one subtlety. In standard ChPT, we have used certain matrix identities for the $SU(3)$ matrices in order to reduce the number of the terms in the Lagrangian. The remaining minimal set of the LECs are those that can be uniquely determined from the experimental input. These identities do not hold for the present symmetry group anymore. An example at $O(p^4)$ is provided by the term $\langle L_\mu L_\nu L_\mu L_\nu \rangle$, where L_μ is defined in Eq. (4.573). If $U \in SU(3)$, this can be expressed in terms of $\langle L_\mu L_\mu L_\nu L_\nu \rangle$, $\langle L_\mu L_\nu \rangle \langle L_\mu L_\nu \rangle$ and $\langle L_\mu L_\mu \rangle^2$. This is no more possible for U transforming under the group $SU(N_v + N_s|N_v)$. Consequently, the effective Lagrangian in partially quenched ChPT will contain an additional term

$$\delta\mathcal{L} = C\Big(\text{str}(L_\mu L_\nu L_\mu L_\nu) + 2\,\text{str}(L_\mu L_\mu L_\nu L_\nu) - \frac{1}{2}\,(\text{str}(L_\mu L_\mu))^2$$

$$- \,\text{str}(L_\mu L_\nu)\text{str}(L_\mu L_\nu)\Big). \tag{4.619}$$

This linear combination is chosen in such a way that it vanishes in the $SU(3)$ case.

The question arises as to how the LECs in partially quenched ChPT are related to those in conventional ChPT? A very important property of the partially quenched theory is that it *includes* standard ChPT. Indeed, consider the generating functional in which the external sources couple only to the sea quarks (other sources are set to zero from the beginning). Integrating out the valence and ghost quarks produces determinants that exactly cancel each other, resulting in unquenched QCD, in which the quark masses are taken equal to the sea quark masses. Now take the meson field $\Phi(x)$ in Eq. (4.615) and eliminate all fields that describe mesons with at least one valence or ghost quark. Then, the partially quenched effective Lagrangian should reduce to the standard one for such fields. This leads us to a very important conclusion: since the LECs do not depend on the quark masses, *they should be exactly the same* in the partially quenched theory and in the standard one. This circumstance gives predictive power to the partially quenched theories and enables one to extract the values of the LECs in the full QCD from partially quenched simulations. Notably, there are also exceptions, which are displayed, for example, in Eq. (4.619), as the operator multiplying the constant C vanishes for such fields, yielding no restriction on the numerical value of the latter. The LECs like C emerge only in the partially quenched theory and are not present in full QCD. The number of such additional LECs is, however, limited.

Next, let us investigate the spectrum of the partially quenched free field theory. For this reason, we set all external sources to zero except the scalar source. This yields the quark mass matrix:

$$s = \mathcal{M} = \begin{pmatrix} \mathcal{M}_q & \\ & \mathcal{M}_v \end{pmatrix}, \qquad \mathcal{M}_q = \begin{pmatrix} \mathcal{M}_v & \\ & \mathcal{M}_s \end{pmatrix}, \tag{4.620}$$

and we remind the reader that the masses of the valence and ghost quarks are equal. What remains now is to expand the lowest-order chiral Lagrangian up to the second order in the fields:

$$\mathcal{L}^{(2)} = \frac{1}{2}\text{tr}\Big(\partial_\mu\phi\partial^\mu\phi + \partial_\mu\eta\partial^\mu\eta^\dagger \Big) - \frac{1}{2}\text{tr}\Big(\partial_\mu\eta^\dagger\partial^\mu\eta + \partial_\mu\tilde{\phi}\partial^\mu\tilde{\phi} \Big)$$

$$- B_0\text{tr}\Big((\phi^2 + \eta\eta^\dagger)\mathcal{M}_q \Big) + B_0\text{tr}\Big((\eta^\dagger\eta + \tilde{\phi}^2)\mathcal{M}_v \Big) + \cdots. \tag{4.621}$$

The masses of the quarks are $\mathcal{M}_v = \mathrm{diag}(m_u, m_d, m_s)$ and $\mathcal{M}_s = \mathrm{diag}(m'_u, m'_d, m'_s)$. The matrices $\phi, \eta, \tilde{\phi}$ have the following structure:

$$\phi = \begin{pmatrix} \phi_{vv} & \phi_{vs} \\ \phi_{sv} & \phi_{ss} \end{pmatrix}, \qquad \eta = \begin{pmatrix} \eta_{vg} \\ \eta_{sg} \end{pmatrix}, \qquad \tilde{\phi} = \tilde{\phi}_{gg}. \tag{4.622}$$

The vanishing of the supertrace, $\mathrm{str}(\Phi) = 0$, results in a constraint on the diagonal fields of the matrices ϕ_{vv}, ϕ_{ss} and $\tilde{\phi}_{gg}$; that is, one of the fields is redundant and can be expressed through the others. One could explicitly take this constraint into account and diagonalize the quadratic part of the Lagrangian straightforwardly. A more convenient way of doing this is the following. Let us add a singlet scalar field (the partially quenched η') to the Lagrangian $\Phi(x) \to \Phi(x) + (\mathbb{1}/\sqrt{3})\Phi_0(x)$. The modified Lagrangian is

$$\mathcal{L}^{(2)} = \frac{F_0^2}{4} \mathrm{str}(D_\mu U D^\mu U^\dagger + \chi U^\dagger + U \chi^\dagger) + \frac{m_0^2}{6} (\mathrm{str}(\Phi))^2. \tag{4.623}$$

If $m_0^2 \to \infty$, the singlet meson, described by the field Φ_0, disappears from the theory, leaving us with the constraint $\mathrm{str}(\Phi) = 0$. One could also add the kinetic term for Φ_0 to the preceding Lagrangian, but in the limit $m_0^2 \to \infty$ this term will not matter.

Let us now calculate the two-point function in this theory. We start from the non-diagonal mesons that are described by the fields, which do not appear on the diagonals of the matrices $\phi, \tilde{\phi}$. The last term of $\mathcal{L}^{(2)}$ does not contribute for such mesons, and we can immediately read off the propagator:

$$i \int d^4 x\, e^{-ipx} \langle 0|T\Phi_{ij}(x)\Phi_{ji}^\dagger(0)|0\rangle = \frac{\pm 1}{M_{ij}^2 - p^2}, \qquad i \neq j. \tag{4.624}$$

Here, the minus sign emerges only for the gg mesons. Also, $M_{ij}^2 = B_0(m_i + m_j)$ with $m_i = (m_u, m_d, m_s)$ for the valence and ghost quarks and with $m_i = (m'_u, m'_d, m'_s)$ for the sea quarks. The emergence of a wrong sign in the ghost–ghost meson propagator already signals that partially quenched theory is a sick theory: the vacuum at $\Phi = \Phi_0 = 0$ is not stable. In Ref. [270], it has been, however, shown that a rigorous treatment in perturbation theory is possible and leads to results identical to the naive treatment considered.

Next, we turn to the propagator of the diagonal mesons. Expanding the last term of $\mathcal{L}^{(2)}$ gives:

$$\frac{m_0^2}{6} (\mathrm{str}(\Phi))^2 = \frac{m_0^2}{6} \Phi_{ii} \varepsilon_i \varepsilon_j \Phi_{jj}, \tag{4.625}$$

where $\varepsilon_i = +1$ for valence and sea quarks, and $\varepsilon_i = -1$ for ghosts. The propagator can then be read from the quadratic term in the Lagrangian:

$$i \int d^4 x\, e^{-ipx} \langle 0|T\Phi_{ii}(x)\Phi_{jj}^\dagger(0)|0\rangle = G_{ij}(p) - \frac{m_0^2}{3} G_{ik}(p)\varepsilon_k \varepsilon_m G_{mj}(p) + \cdots, \tag{4.626}$$

where

$$G_{ij}(p) = \frac{\varepsilon_i \delta_{ij}}{M_{ii}^2 - p^2}. \tag{4.627}$$

Further, we take into account that

$$S \doteq \varepsilon_i G_{ij}(p)\varepsilon_j = \sum_i \frac{\varepsilon_i}{M_{ii}^2 - p^2} = \sum_{q=u,d,s} \frac{1}{M_q^2 - p^2}, \quad M_q^2 = 2B_0 m_a'. \quad (4.628)$$

The last equality follows from the fact that the contributions from the valence and ghost quarks cancel in the sum.

Using Eq. (4.628), we see that Eq. (4.626) is reduced to a geometric series that can be readily summed:

$$i \int d^4x e^{-ipx} \langle 0|T\Phi_{ii}(x)\Phi_{jj}^\dagger(0)|0\rangle = G_{ij}(p) - \frac{m_0^2/3}{(M_{ii}^2 - p^2)(M_{jj}^2 - p^2)} \frac{1}{1 + m_0^2 S/3}.$$
$$(4.629)$$

Introducing the physical meson masses, according to (see also Eq. (4.214))

$$M_{\pi^0}^2 = 2\hat{m}B_0 - \frac{4}{3}B_0(m_s - \hat{m})\frac{\sin^2\varepsilon}{\cos 2\varepsilon} + O(m_0^{-2}),$$

$$M_\eta^2 = \frac{2}{3}(\hat{m} + 2m_s)B_0 + \frac{4}{3}B_0(m_s - \hat{m})\frac{\sin^2\varepsilon}{\cos 2\varepsilon} + O(m_0^{-2}),$$

$$M_{\eta'}^2 = m_0^2 + \frac{2}{3}(2\hat{m} + m_s)B_0 + O(m_0^{-2}), \quad (4.630)$$

we get (up to the terms of order m_0^{-2})

$$\left(1 + \frac{m_0^2}{3}S\right)^{-1} = \frac{(M_u^2 - p^2)(M_d^2 - p^2)(M_s^2 - p^2)}{(M_{\pi^0}^2 - p^2)(M_\eta^2 - p^2)(M_{\eta'}^2 - p^2)}, \quad (4.631)$$

and, hence,

$$i \int d^4x e^{-ipx} \langle 0|T\Phi_{ii}(x)\Phi_{jj}^\dagger(0)|0\rangle = G_{ij}(p)$$

$$-\frac{m_0^2}{3} \frac{(M_u^2 - p^2)(M_d^2 - p^2)(M_s^2 - p^2)}{(M_{ii}^2 - p^2)(M_{jj}^2 - p^2)(M_{\pi^0}^2 - p^2)(M_\eta^2 - p^2)(M_{\eta'}^2 - p^2)}. \quad (4.632)$$

Performing the limit $m_0^2 \to \infty$ in the last term results in

$$\lim_{m_0^2 \to \infty} i \int d^4x e^{-ipx} \langle 0|T\Phi_{ii}(x)\Phi_{jj}^\dagger(0)|0\rangle = G_{ij}(p)$$

$$-\frac{1}{3} \frac{(M_u^2 - p^2)(M_d^2 - p^2)(M_s^2 - p^2)}{(M_{ii}^2 - p^2)(M_{jj}^2 - p^2)(M_{\pi^0}^2 - p^2)(M_\eta^2 - p^2)}. \quad (4.633)$$

It is seen that the propagator has a more complicated structure than in "ordinary" field theory. For example, it has a double pole in the valence and ghost sectors. This double pole leads to the emergence of nonconventional chiral logarithms. In order to see this, let us, for simplicity, assume exact $SU(3)$ symmetry with $m_u = m_d = m_s = m$

and $m'_u = m'_d = m'_s = m'$. Then, the propagator, say, in the valence quark sector takes the form

$$\lim_{m_0^2 \to \infty} i \int d^4x e^{-ipx} \langle 0|T\Phi_{ii}(x)\Phi^\dagger_{jj}(0)|0\rangle = \frac{1}{2B_0m - p^2} - \frac{1}{3}\frac{2B_0m' - p^2}{(2B_0m - p^2)^2}. \tag{4.634}$$

The propagator in the ghost sector differs from the preceding expression by a sign in the first term, and the propagator in the sea quark sector is obtained through the replacement $m \to m'$, leading to the cancellation of the double pole. The unconventional chiral logarithms emerge exactly from the double pole. Indeed, consider the one-loop correction to the Goldstone boson mass, which is described by a tadpole diagram in Fig. 4.9a. We have

$$\int \frac{d^Dk}{(2\pi)^Di} \frac{2B_0m' - p^2}{(2B_0m - p^2)^2} = -2B_0(m' - 2m)\left(2\lambda + \frac{1}{16\pi^2}\ln\frac{2B_0m}{\mu^2}\right)$$

$$- 2B_0(m' - m)\frac{1}{16\pi^2}. \tag{4.635}$$

Here, the ultraviolet divergence is contained in λ. The full expression for the valence pion mass in partially quenched ChPT takes the form [271, 272]:

$$(M_\pi^2)^{PQ} = 2B_0m\left(1 + \frac{1}{24\pi^2F_0^2}2B_0\left((2m - m')\ln\frac{2B_0m}{\mu^2} + (m - m')\right)\right.$$

$$+ \frac{16}{F_0^2}2B_0\left((2L_8^r - L_5^r)m + 3(2L_6^r - L_4^r)m'\right)\bigg\}. \tag{4.636}$$

It can be seen that the preceding expression contains an infrared divergence in the limit $m \to 0$, and m' stays finite. Hence, the chiral limit in partially quenched theory implies a simultaneous vanishing of the valence and sea quark masses, with their ratio kept fixed.

Next, one should note that the behavior of the fully quenched theory, where the sea quark sector is absent from the beginning, is very different from that of the partially quenched theory. As we have seen in Section 3.20, the symmetry-breaking pattern in these two theories is very different. Namely, the field Φ_0 that corresponds to the conventional η'-meson, stays massless in the chiral limit in the fully quenched theory, and the divergence of the pertinent axial-vector current does not contain an anomaly. Consequently, the Φ_0 cannot be integrated out in a fully quenched theory. One can see this directly from the expression of the two-point function in Eq. (4.632). Discarding the sea quark contributions, which are not there in the fully quenched theory, one obtains for the two-point function in the valence quark sector

$$i \int d^4x e^{-ipx}\langle 0|T\Phi_{ii}(x)\Phi^\dagger_{ii}(0)|0\rangle = \frac{1}{M_{ii}^2 - p^2} - \frac{m_0^2/3}{(M_{ii}^2 - p^2)^2}. \tag{4.637}$$

Here, for simplicity, we consider a diagonal propagator with $i = j$. It is now immediately seen that the limit $m_0^2 \to \infty$ cannot be performed anymore, and the Φ_0 is there to stay in the quenched theory. A systematic inclusion of the Φ_0 into the theory proceeds pretty much along the lines already described in Section 4.17: we introduce arbitrary

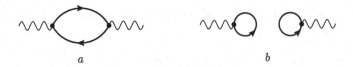

a b

Figure 4.26   Connected (a) and disconnected (b) contributions to the hadronic vacuum polarization tensor on the lattice. Any number of gluons can be exchanged between the quark lines (not shown explicitly). Solid and wiggly lines denote quarks/antiquarks and photons, respectively.

potentials depending on Φ_0 and perform an expansion in the derivatives of the fields. We do not repeat this construction here.

Next, we illustrate how the partially quenched ChPT can be used for the evaluation of the so-called disconnected contributions. To this end, we consider the example of the hadronic vacuum polarization in a theory with two degenerate flavors; see also Ref. [269]. The two-point function of the electromagnetic currents is defined as

$$\Pi_{\mu\nu}(p) = i \int d^4x e^{-ipx} \langle 0 | T j_\mu^{\text{em}}(x) j_\nu^{\text{em}}(0) | 0 \rangle ,$$

$$j_\mu^{\text{em}}(x) = \frac{2}{3} \bar{u}(x) \gamma_\mu u(x) - \frac{1}{3} \bar{d}(x) \gamma_\mu d(x) . \tag{4.638}$$

We may calculate the Euclidean version of the preceding expression in lattice QCD. Two types of Wick contractions contribute to it, as shown in Fig. 4.26. In the connected contributions, displayed in Fig. 4.26a, the quark/antiquark lines run through the diagram, whereas in the disconnected contributions in Fig. 4.26b, the quarks/antiquarks from the initial and final states annihilate, exchanging gluons. Usually, the latter are more difficult to calculate, leading to a larger statistical error. In many calculations, one drops the disconnected part by hand. In this case, however, it is very useful to give arguments in favor of the conjecture that the neglected contributions are small, otherwise one introduces an uncontrollable systematic error in the calculations.

One finds an elegant solution to the preceding problem using the partially quenched effective theory. To this end, note that in the initial expression for the two-point function all quarks are valence quarks (by construction). However, in this theory, nothing forbids considering correlators with different quark species. Replacing one valence quark with a sea quark, we may consider the T-product $T(j_\mu^{vs}(x) j_\nu^{vs\dagger}(0))$, where $j_\mu^{vs}(x) = \frac{2}{3} \bar{u}_v(x) \gamma_\mu u_s(x) - \frac{1}{3} \bar{d}_v(x) \gamma_\mu d_s(x)$. If the valence and sea quark masses are equal, then it is easy to see that the new two-point function will be given by exactly the same connected diagrams (the disconnected ones will be absent). Thus, calculating the two-point function of the valence-sea currents in the effective theory, one is separating the contribution of the connected diagrams. Acting in this manner, it was shown in Ref. [269] that the disconnected contribution amounts exactly to one tenth of the connected one at next-to-leading-order (all LECs cancel in the final expression). This method is widely used at present in the calculations of other observables as well. In the literature guide at the end of this chapter some references will be given.

Last but not least, we briefly mention the use of the partially quenched effective theories for the evaluation of the finite-volume effects emerging from the twisted boundary conditions [273, 274]. In lattice calculations, quark fields are subject to certain boundary conditions. One speaks about twisted boundary conditions, if the boundary conditions in the space directions are chosen in the following manner:

$$\psi_f(x + e_i L) = \exp(i\theta_i^f)\psi_f(x), \quad f = u, d, s, \quad i = 1, 2, 3. \tag{4.639}$$

These boundary conditions should be supplemented with the one in the timelike direction (here, usually antiperiodic boundary conditions are chosen). In the preceding expression, e_i are unit vectors in the direction of the spatial axes, and the three-vector θ^f describes the twisting angles in various directions for a quark with a flavor f. The crucial point here is that one can use the effective theory to study the artifacts that arise from twisting, because the boundary conditions for the mesons can be straightforwardly derived from those of the quarks. For a meson with an ff' quark–antiquark pair, the twisting angle will be $\theta^f - \theta^{f'}$ [275]. Consequently, imposing appropriate boundary conditions on the meson fields in conventional ChPT, we may study these finite-volume effects in perturbation theory. Moreover, partially quenched effective theories in the continuum allow us to investigate the effect of the so-called partial twisting, when the valence/ghost and sea quarks are twisted differently (e.g., one is using the periodic boundary conditions $\theta^f = 0$ for the sea quarks, which is a rather popular choice). This can be achieved by imposing different boundary conditions of the meson fields from different sectors. We shall consider this issue in more detail in Chapter 5.

4.20 The θ-Term, Strong CP Violation and Axions

4.20.1 θ-Vacua

As we already know, a chiral rotation of the fermion fields in the QCD generating functional leads to the anomaly. The conservation of the singlet axial-vector current is violated by a term that is proportional to the winding number density of the gluon field. The additional term in the Lagrangian is a total derivative (note that in this subsection we are working in the Euclidean metric):

$$-\frac{1}{16\pi^2} \text{tr}_c(F_{\mu\nu}\tilde{F}_{\mu\nu}) = \partial_\mu K_\mu, \tag{4.640}$$

where

$$K_\mu = -\frac{1}{8\pi^2} \varepsilon_{\mu\nu\alpha\beta} \text{tr}\left(G_\nu\left(\partial_\alpha G_\beta + \frac{2}{3} G_\alpha G_\beta\right)\right). \tag{4.641}$$

Note also that the current K_μ is not gauge-invariant, but its divergence is.

Naively, one might think that adding a total derivative to the Lagrangian should not have observable consequences. This conclusion is, however, wrong. The crucial point here is that in the path integral there exist gluon field configurations, which lead to a finite Euclidean action and, at the same time, for which the integral over $\partial_\mu K_\mu$ does not vanish, since the components of K_μ do not approach zero sufficiently fast at infinity. Instead, we have

$$Q = \int d^4x \, \partial_\mu K_\mu = \left(\lim_{t \to +\infty} - \lim_{t \to -\infty} \right) \int d^3\mathbf{x} \, K_0(\mathbf{x}, t) . \tag{4.642}$$

At $t \to \pm\infty$, the field configuration tends to the classical vacuum, in which the field tensor $F_{\mu\nu}$ vanishes. From this, we get $\varepsilon_{\mu\nu\alpha\beta}\partial_\alpha G_\beta = -\varepsilon_{\mu\nu\alpha\beta}G_\alpha G_\beta$, and

$$Q = \frac{1}{24\pi^2} \left(\lim_{t \to +\infty} - \lim_{t \to -\infty} \right) \int d^3\mathbf{x} \, \varepsilon_{0\nu\alpha\beta} \, \mathrm{tr}(G_\nu G_\alpha G_\beta) . \tag{4.643}$$

Let us now study these vacuum field configurations in more detail. We shall see that, instead of a single vacuum configuration, there exists a countable set of classical vacua, labeled by an integer number, the so-called winding number. In the following, it will be convenient to use the Hamilton gauge

$$G_0(x) = 0 . \tag{4.644}$$

In this gauge, the gauge freedom is reduced to the time-independent transformations

$$G_\mu(x) \mapsto \Lambda(\mathbf{x}) G_\mu(x) \Lambda^\dagger(\mathbf{x}) - \partial_\mu \Lambda(\mathbf{x}) \Lambda^\dagger(\mathbf{x}) , \qquad \Lambda(\mathbf{x}) \in SU(3) . \tag{4.645}$$

Since $F_{\mu\nu}(x) = 0$, the corresponding gluon field is a pure gauge, both in gluodynamics and in full QCD (the presence of quarks does not play a role in the subsequent discussion). Further, owing to the gauge condition, the gauge parameters should not depend on the Euclidean time t. Thus, the vacuum field configurations are described by

$$G_i(x) = \partial_i \Omega(\mathbf{x}) \Omega^\dagger(\mathbf{x}) , \qquad \Omega(\mathbf{x}) = \exp(i\alpha(\mathbf{x})) . \tag{4.646}$$

Moreover, it is seen that, for the configurations that give a nonzero contribution to the path integral, we may limit ourselves to the gauge transformations that obey the constraint

$$\lim_{|\mathbf{x}| \to \infty} \Omega(\mathbf{x}) = 1 . \tag{4.647}$$

The origin of this constraint is easy to understand. Consider the contour depicted in Fig. 4.27. On the horizontal and vertical axes, we plot the spatial distance and the Euclidean time, respectively (for convenience, only one spatial dimension is shown). The elongations of the contour in both directions, R and T, are taken to be very large, so that the gluon field can be assumed to be a pure gauge on the contour. We may now use the remaining gauge freedom in the Hamilton gauge to ensure that $\Omega(\mathbf{x}, -T/2) = 1$ for all $-R/2 \leq |\mathbf{x}| \leq R/2$. Further, owing to the gauge condition $G_0(\mathbf{x}, t) = 0$, the quantity $\Omega(\mathbf{x}, t)$ does not depend on t, if $|\mathbf{x}| = \pm R/2$. This finally gives $\Omega(\mathbf{x}, T/2) = 1$ for $|\mathbf{x}| = \pm R/2$.

The boundary condition, imposed on the function $\Omega(\mathbf{x})$, is crucial for the topological classification of the gluon field configurations. Namely, the condition, displayed

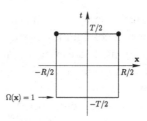

Figure 4.27 A large closed contour in Euclidean space. Using time-independent gauge transformations, it is possible to set $\Omega(\mathbf{x}) = 1$ on the bottom line of the contour. Further, due to gauge condition, $\Omega(\mathbf{x})$ is time-independent along the vertical paths at $|\mathbf{x}| = \pm R/2$ that describe the evolution in Euclidean time from $t = -T/2$ to $t = T/2$. Hence, $\Omega(\mathbf{x}) = 1$ at both ends of the upper line, shown by the small filled circles.

in Eq. (4.647), effectively compactifies the three-dimensional space into a three-dimensional sphere, $\mathbb{R}^3 \to S_3$, since all points at the infinity can be identified with a single point. The function $\Omega(\mathbf{x})$, which provides a mapping of a sphere S_3 into $SU(3)$, is single-valued at this point. Since the homotopy group $\pi_3(SU(3)) = \mathbb{Z}$ is nontrivial, all functions that describe this mapping fall into different sectors. These sectors are characterized by an integer number, which is referred to as the winding number, or the Pontryagin index:

$$N = \frac{1}{24\pi^2} \int d^3\mathbf{x}\, \varepsilon_{ijk} \mathrm{tr}\left[(\Omega\partial_i\Omega^\dagger)(\Omega\partial_j\Omega^\dagger)(\Omega\partial_k\Omega^\dagger)\right]. \tag{4.648}$$

It is also clear that the mappings belonging to different sectors cannot be continuously transformed into each other by gauge transformations. Indeed, the index N can take only integer values, which rules out a continuous change.

To get more familiar with this, we explicitly display few examples for such mappings that belong to different sectors. The most trivial example is, of course, $\Omega(\mathbf{x}) = 1$ everywhere, corresponding to $N = 0$. A representative mapping, corresponding to $N = 1$, is given by

$$\Omega_1(\mathbf{x}) = \exp(i\alpha_1(\mathbf{x}))\exp(i\pi\lambda^3), \qquad \alpha_1(\mathbf{x}) = \frac{\pi}{(\mathbf{x}^2+\rho^2)^{1/2}}\sum_{a=1}^{3} x^a\lambda^a, \tag{4.649}$$

with ρ is a real parameter (also referred to as the "instanton size"), and the λ^a denote the Gell–Mann matrices. Writing

$$\Omega_1(\mathbf{x}) = \begin{pmatrix} \tilde{\Omega}_1(\mathbf{x}) & 0 \\ 0 & 1 \end{pmatrix},$$

$$\tilde{\Omega}_1(\mathbf{x}) = -\cos\varphi - i\frac{\boldsymbol{\sigma}\mathbf{x}}{|\mathbf{x}|}\sin\varphi, \qquad \varphi = \frac{\pi|\mathbf{x}|}{(\mathbf{x}^2+\rho^2)^{1/2}}, \tag{4.650}$$

where the $\boldsymbol{\sigma}$ are the Pauli matrices and evaluating Eq. (4.648) explicitly, we indeed arrive at $N = 1$.

Note that all classical configurations in a given sector with $N = 0$ and $N = 1$ can be obtained by the gauge transformations from the particular configurations given.

These gauge transformations should, however, itself have the zero winding number, that the transformed configurations stay in a given sector. We refer to these as *small* gauge transformations. There exist also *large* transformations, which connect different sectors. For example, the matrix $\Omega_1(\mathbf{x})$, given in Eq. (4.649), is a gauge transformation by itself, corresponding to winding number $N = 1$. Applied to a matrix from the sector $N = 0$, a matrix from the sector with $N = 1$ is obtained. In general, $\Omega_1(\mathbf{x})$ relates the sectors N and $N + 1$. Further, $\left[\Omega_1(\mathbf{x})\right]^N$ belongs to a sector with the winding number N. The expression in Eq. (4.648) is invariant only with respect to small gauge transformations.

Finally, comparing the Eq. (4.648) with Eq. (4.643) and taking into account that for $t \to \pm\infty$, the gluon fields become time-independent pure gauge configurations, we directly obtain

$$Q = N_+ - N_- , \tag{4.651}$$

where $N_\pm$ is the winding number of the vacuum configurations, calculated at $t \to \pm\infty$.

To summarize, the *classical* vacuum field configurations in QCD fall into different classes labeled by integer values N. All these configurations have the same energy, because the classical Hamiltonian is invariant with respect to all gauge transformations (small as well as large). The fields G_μ interpolate between different vacua, emerging at $t \to -\infty$ and $t \to +\infty$, and the winding number of the field configuration corresponding to this is given by Eq. (4.651).

In the next step, one constructs quantum vacuum states in the vicinity of each classical vacuum configuration with a given N in a standard manner. The degrees of freedom, which should be quantized, are given by the space components G_i, the conjugate momenta and the $\Pi_i = (2/g^2)E_i$, where $E_i = F_{0i}$ is the (chromo)electric field. Here, g denotes the strong coupling constant. Classically, the fields should obey the Gauß law:

$$D_i E_i = \partial_i E_i + [G_i, E_i] = 0 . \tag{4.652}$$

In quantum field theory, this condition is rather imposed on the states. The physical states $|\text{phys}\rangle$ should satisfy the condition:

$$D_i E_i |\text{phys}\rangle = 0 . \tag{4.653}$$

Consider now an infinitesimal gauge transformation $\Lambda(\mathbf{x}) = \exp(i\lambda(\mathbf{x})) = 1 + i\lambda(\mathbf{x}) + O(\lambda^2)$. The gauge field transforms as

$$G_i \mapsto G_i - i(\partial_i \lambda + [G_i, \lambda]) = G_i - iD_i\lambda . \tag{4.654}$$

Consequently, the operator of a gauge transformation is given by

$$U(\lambda) = \exp\left(i \int d^3\mathbf{x}\, \mathrm{tr}(D_i\lambda(\mathbf{x})\Pi_i(\mathbf{x})) \right). \tag{4.655}$$

Assume now that $\lambda(\mathbf{x})$ describes a small gauge transformation. Then, we may integrate by parts (the surface terms vanish), and we get

$$U(\lambda) = \exp\left(-\frac{2i}{g^2} \int d^3\mathbf{x}\,\mathrm{tr}(\lambda(\mathbf{x})D_i E_i(\mathbf{x}))\right). \tag{4.656}$$

In the case of large gauge transformations, this does not work, since the surface terms do not vanish.

Applying now the operator $U(\lambda)$ on the states and using Gauß' law, it is immediately seen that the physical states are invariant under small gauge transformations:

$$U(\lambda)|\mathrm{phys}\rangle = |\mathrm{phys}\rangle. \tag{4.657}$$

Again, this statement does not hold for large gauge transformations.

Now, we can construct the vacuum states corresponding to the gluon field configurations in different topological sectors:

$$|N\rangle = \int d\lambda\, U(\lambda)|\Omega_N\rangle. \tag{4.658}$$

Here, the state $|\Omega_N\rangle$ is built around *a particular field configuration* from the sector with the winding number N (we have seen examples of such configurations, say, in Eq. (4.649)), and $d\lambda$ is the invariant group integration measure (Haar measure) [276, 277]. The integral runs over small gauge transformations. Applying now the operator of small transformations to both sides of the preceding equation, and using the invariance of the measure, it is immediately seen that Eq. (4.657) holds for the states $|N\rangle$. Note that these states are not invariant with respect to large gauge transformations. More precisely, if the transformation described by $\lambda(\mathbf{x})$ carries winding number N', then

$$U(\lambda)|N\rangle = |N+N'\rangle. \tag{4.659}$$

All this closely resembles the one-dimensional quantum-mechanical problem of a particle, moving in a periodic potential $V(r)$ (see Fig. 4.28); see, for example, Refs. [278, 279] for more details. The minima of the potential are labeled by an integer number N, and the large gauge transformations correspond to the translation operators. For example, the translation operator by one unit, U_1, can be constructed by using the matrix $\Omega_1(\mathbf{x})$, given in Eq. (4.649).

The crucial point in the following discussion is that the states $|N\rangle$ do not correspond to the true quantum-mechanical vacuum state. The reason is that there exist gluon field configurations with nonzero winding number, which connect the states with different N. And, since the Euclidean action of these configurations is finite (albeit their contribution to the path integral is exponentially suppressed at small g), tunneling between these states is possible. Then, as we shall see, the true vacuum state is a linear superposition of all states $|N\rangle$. In order to prove this, note that the translation operator U_1 commutes with the total Hamiltonian of the system, H, due to gauge invariance, and thus can be diagonalized simultaneously with the latter. Further, since U_1 is unitary,

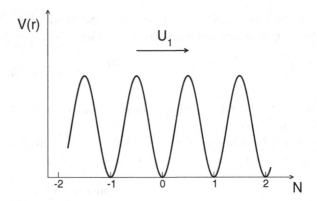

Figure 4.28 The one-dimensional periodic potential. The minima of this potential correspond to the classical vacua with different N. The "translation" operator U_1 transforms $|N\rangle$ into $|N+1\rangle$.

its eigenvalues must have a form $\exp\{-i\theta\}$, where $0 \leq \theta < 2\pi$. Thus, the true vacuum should simultaneously obey

$$H|\theta\rangle = 0, \qquad U_1|\theta\rangle = e^{-i\theta}|\theta\rangle. \tag{4.660}$$

Here, we used a notation that allows one to explicitly distinguish the vacua that correspond to the different values of θ. From this, it is seen that $|\theta\rangle$ is an analog of the Bloch wave function in solid-state physics:

$$|\theta\rangle = \sum_N e^{iN\theta}|N\rangle. \tag{4.661}$$

In particular, it can be directly checked that the preceding wave function indeed acquires a factor $\exp\{-i\theta\}$ under translations.

As was already mentioned, the different states $|N\rangle$ are connected by vacuum tunneling and are thus not orthogonal. Conversely, it can be seen that the states with different θ are orthogonal:

$$\langle\theta|\theta'\rangle = \sum_{N,N'} \langle N|N'\rangle e^{-iN\theta+iN'\theta'} = \sum_N e^{-iN(\theta-\theta')} \sum_Q \langle N|N+Q\rangle e^{iQ\theta'}. \tag{4.662}$$

Due to the "translational" invariance, the matrix element $\langle N+Q|N\rangle$ does not depend on N; thus, the two summations in the preceding equation can be carried out independently, yielding a factor $2\pi\delta(\theta - \theta')$ in front of the expression. Thus, the vectors $|\theta\rangle$ for different values of θ are orthogonal.

Moreover, it can be seen that the matrix elements of any gauge-invariant operator O between two states with different θ vanish:

$$\langle\theta|O|\theta'\rangle = \sum_{N,N'} \langle N|O|N'\rangle e^{-iN\theta+iN'\theta'} = \sum_N e^{-iN(\theta-\theta')} \sum_Q \langle N|O|N+Q\rangle e^{iQ\theta'}.$$

$$\tag{4.663}$$

Note now that the gauge invariance implies $[O,U_1] = 0$. Thus,

$$\langle N|[O,U_1]|N+Q\rangle = \langle N|O|N+Q+1\rangle - \langle N-1|O|N+Q\rangle = 0. \tag{4.664}$$

In other words, the matrix element of the operator O does not depend on N due to the "translational" invariance, which again leads to the emergence of the factor $2\pi\delta(\theta - \theta')$.

To summarize, the true vacuum of QCD is the linear superposition of all vacua with different winding numbers N. This vacuum is characterized by a parameter θ, and the sectors with different θ do not talk to each other. Thus, θ is a parameter of the theory, along with the quark masses and the strong coupling constant. The worlds with different values of θ are different, and we live in a world where this parameter takes one particular value.[51] The Green's functions are given by the matrix elements of the T-product of the field operators, evaluated in a vacuum with a given value of θ (i.e., they all depend on θ). For example, the matrix element of the evolution operator is given by

$$\langle\theta|e^{-HT}|\theta'\rangle = \sum_{N,N'} \langle N|e^{-HT}|N'\rangle e^{-iN\theta + iN'\theta'}$$

$$= 2\pi\delta(\theta - \theta') \sum_Q e^{iQ\theta} \langle N|e^{-HT}|N+Q\rangle. \qquad (4.665)$$

The matrix element on the r.h.s. of the preceding equation can be written down in a path integral formalism, integrating over all fields. The gluon fields are subject to a special boundary condition. They interpolate between vacua $|N+Q\rangle$ and $|N\rangle$, that is, they should carry winding number $-Q$. Recalling the definition of the winding number density, given in Eqs. (4.640) and (4.642), we get

$$e^{iQ\theta} \langle N|e^{-HT}|N+Q\rangle = \int (dG_\mu)_{-Q} d\psi d\bar\psi \exp\left(-S + i\frac{\theta}{16\pi^2} \int d^4x \, \text{tr}_c(F_{\mu\nu}\tilde{F}_{\mu\nu})\right).$$
$$(4.666)$$

Here, the subscript "$-Q$" for the gluon measure indicates that only configurations with a winding number $-Q$ are included. Now, note that the sum over all Q merely removes this constraint and we finally get

$$\langle\theta|e^{-HT}|\theta'\rangle = 2\pi\delta(\theta - \theta') \int dG_\mu d\psi d\bar\psi \exp\left(-S + i\frac{\theta}{16\pi^2} \int d^4x \, \text{tr}_c(F_{\mu\nu}\tilde{F}_{\mu\nu})\right).$$
$$(4.667)$$

Here, the integral over G_μ runs over all gluon configurations. As seen from the preceding expression, the parameter θ has migrated from the vacuum state to the Lagrangian: In order to evaluate the evolution operator (and Green's functions) in the vacuum with

[51] At this place, it is informative to discuss the relation with the energy levels in a crystal. As it is well known, in a periodic potential the energy levels exhibit a band structure. Energies in a band can be labeled by a continuous parameter θ, which is the counterpart of our parameter carrying the same name. However, in contrast to our case, we may experimentally observe the *whole band,* whereas we always refer to a *a single value* of θ. The difference arises because the band structure in solid-state physics emerges due to the translational invariance in space, and an average observer in a laboratory is not translationally invariant. In our case, different minima are transformed into each other with the help of large gauge transformations. Since no gauge non-invariant observers or devices are available, the superposition of states corresponding to different values of θ does not occur in experiment.

a particular value of the θ parameter, it suffices to add a term to the Lagrangian, which is proportional to the winding number density. In Minkowski space, the additional Lagrangian is

$$\mathcal{L}_\theta = \frac{\theta}{16\pi^2} \, \text{tr}_c(F_{\mu\nu}\tilde{F}^{\mu\nu}). \tag{4.668}$$

Using the language of the external sources, we may say that the Green's functions in a particular θ-vacuum are obtained by a Taylor expansion of the generating functional in powers of $(\theta(x) - \theta)$, where $\theta(x)$ denotes a source coupled to the winding number density (up to now, the expansion was done in the vicinity of $\theta(x) = 0$). The question now is; what is the value of θ in nature?

4.20.2 The Strong CP Problem

The θ-term, which was introduced in the previous subsection, breaks the P- and CP-symmetry of QCD,[52] unless it is equal to 0 or π. We have shown that this term naturally emerges due to the nontrivial topological structure of the QCD vacuum, even though the original Lagrangian did not violate discrete symmetries. The CP breaking leads to observable effects, for example, to a nonvanishing electric dipole moment of elementary particles. However, as we know, the strong interactions in nature *do respect* discrete symmetries (at least, at present, there exists no experimental evidence of the opposite). This experimental fact can be translated into the constraint on the parameter θ. The real challenge is to explain the origin of this constraint.

In order to obtain the constraint, we first note that the θ-parameter emerging in the Lagrangian is not independent from the quark mass matrix, which in the Standard Model arises from the Yukawa couplings of the Higgs to the chiral fermions and is neither diagonal nor real. The most general quark mass term can be written in the following manner:

$$\mathcal{L}_m = -\bar{\psi}_R \mathcal{M} \psi_L - \bar{\psi}_L \mathcal{M}^\dagger \psi_R. \tag{4.669}$$

The mass matrix can be always decomposed as

$$\mathcal{M} = U \mathcal{M}_d V^\dagger, \tag{4.670}$$

where U and V are unitary matrices, and $\mathcal{M}_d$ is the diagonal matrix with real entries, the quark masses. One can now independently rotate the right- and left-handed fermion fields by the $SU(3)$-part of the unitary matrices U and V, as well as apply $U(1)_V$ transformations. All these transformations are non-anomalous in the absence of the external sources. The matrix $\mathcal{M}$ takes the diagonal form:

$$\mathcal{M} = \text{diag}(m_u e^{-2i\theta_u}, m_d e^{-2i\theta_d}, m_s e^{-2i\theta_s}). \tag{4.671}$$

In general, $SU(3)$ transformations change $\theta_u, \theta_d, \theta_s$. Their sum, however, does not change under this transformation: $\theta_u + \theta_d + \theta_s = -\beta = \text{const}$. At the end, it takes

[52] This is most easily seen by writing it in terms of chromo-electric and chromo-magnetic fields, $\mathcal{L}_\theta \sim E^a \cdot B^a$, which is a product of a vector and an axial-vector.

$U(1)_A$ axial-vector transformation $g_R = g_L^\dagger = e^{i\beta/3}$ to remove a common phase in the mass matrix. However, as we know, this transformation is anomalous and alters the parameter θ by

$$\theta \to \bar{\theta} = \theta + 2\beta = \theta - 2(\theta_u + \theta_d + \theta_s) = \theta + \arg(\det(\mathcal{M})). \qquad (4.672)$$

In the resulting Lagrangian, the quark mass term is diagonal and contains only real masses. Hence, the parameter $\bar{\theta}$ (rather than the original θ) measures the strong CP breaking. Note also that, when the effect of the anomaly was neglected, we have assumed a diagonal and real quark mass matrix from the beginning. Introducing a completely general mass matrix affects the coupling of the anomalous term only. However, as we shall see, in the construction of the effective Lagrangian it will be convenient to make a reverse transformation, eliminating the θ-term completely in favor of complex quark masses.

As the parameter $\bar{\theta}$ is the sole one that describes the CP violation, the experimental data impose constraints on it. For example, the present experimental value of the electric dipole moment of the neutron is given by [280]:

$$d_n = (0.0 \pm 1.1_{\text{stat.}} \pm 0.2_{syst.}) \cdot 10^{-26} e\,\text{cm}. \qquad (4.673)$$

This results in the constraint $\bar{\theta} \leq 10^{-10}$; see in Section 4.20.4. The smallness of this quantity comes as a surprise and needs to be explained. From the naturalness argument, it could be anything between 0 and 2π. Moreover, even if it is fine-tuned to be very small in the beginning, there is no a priori guarantee that it does not get destabilized by electroweak radiative corrections at higher orders. This problem is usually referred to as the *strong CP problem*.[53]

The strong CP problem does not arise if one of the quarks is massless. The reason for this can be easily seen from the preceding discussion. Indeed, suppose that the u-quark is massless. Since the only constraint on the phases in the quark mass matrix is $\theta_u + \theta_d + \theta_s = $ const, one is free to assume $\theta_d = \theta_s = 0$. Further, we may completely rotate the anomalous term away, at the expense of a nonzero θ_u. Note now that in the mass matrix, the factor $\exp\{-2i\theta_u\}$ is multiplied by m_u and thus vanishes in the limit of the massless u-quark. Hence, there is no observable CP-violation in this limit. Unfortunately, this simple explanation does not work, since the hypothesis of a massless quark is not phenomenologically viable. The quark mass ratios are rather well established, both in phenomenology and on the lattice, and none of them turns out to be zero within errors. At present, the most natural and robust explanation of the strong CP-problem is given by axions. We shall consider this issue in the next subsection.

Last but not least, we would like to mention that the θ-term is not the only mechanism to break the CP invariance in nature. Beyond QCD, in the Standard Model the symmetry is broken by the complex phase in the Cabibbo–Cobayashi–Maskawa

---

[53] The choice $\bar{\theta} \simeq 0$ seems to be supported by the available experimental evidence, whereas an alternative choice $\bar{\theta} \simeq \pi$ leads to a different structure of the low-energy Goldstone boson spectrum in QCD [262] (see also [281, 282] for more details concerning the low-energy effective theory of QCD at $\theta = \pi$). In addition, we do not consider here a possibility of a *spontaneous* breaking of the CP symmetry, which may occur at $\theta = \pi$ [283, 284].

(CKM) mixing matrix. This breaking is, however, very small. For example, in the neutron electric dipole moment this would amount to a contribution, which is a few orders of magnitude smaller than the present experimental bound. A large effect could be caused by the mechanisms and phenomena that are not present in the Standard Model, making the measurement of the effect possible at a lower experimental precision. Such mechanisms can be of different nature and, as usual, can be characterized by a string of effective operators with an increasing mass dimension. Then, the experimental study of the relative size of the effects, caused by the different operators, may provide useful constraints on the various extensions of the Standard Model, which are considered at present. This is a large and vibrant field of research, which will not be further discussed here.

4.20.3 Peccei–Quinn Symmetry and the Axion

As mentioned in the preceding section, in the absence of massless quarks, the *CP*-violating parameter θ is observable (from now on, we shall use the notation θ instead of $\bar{\theta}$ for this parameter, and hope that this does not lead to any confusion). Peccei and Quinn [285, 286] have proposed a mechanism for the solution of the strong *CP* problem, which allows θ to be eaten up by a dynamical field a, called axion. In what follows, we shall briefly explain the main idea beyond this mechanism. Consider a toy model with the Lagrangian:

$$\mathcal{L}_{PQ} = \frac{1}{2g^2}\mathrm{tr}_c(F_{\mu\nu}F^{\mu\nu}) + \bar{\psi}i\gamma_\mu D^\mu \psi - (\bar{\psi}_L Y \psi_R \phi + \mathrm{h.c.})$$

$$- \frac{\theta}{16\pi^2}\mathrm{tr}_c(F_{\mu\nu}\tilde{F}^{\mu\nu}) + \partial_\mu\phi^*\partial^\mu\phi - m_0^2\phi^*\phi - \lambda(\phi^*\phi)^2. \qquad (4.674)$$

Here, ϕ is the complex scalar field (Higgs boson), and the matrix $Y = \mathrm{diag}(y_u, y_d, y_s)$ contains Yukawa couplings of the quarks to the Higgs. The quarks are initially massless and acquire masses through spontaneous symmetry breaking. Classically, the theory is invariant under the Peccei–Quinn symmetry:

$$U(1)_{PQ}: \qquad \phi \mapsto e^{2i\alpha}\phi, \qquad \psi \mapsto e^{-i\alpha\gamma_5}\psi. \qquad (4.675)$$

The field ϕ can be written down in terms of two real fields:

$$\phi(x) = \frac{f_a}{\sqrt{2}}\rho(x)e^{ia(x)/f_a}, \qquad (4.676)$$

where $f_a^2 = -m_0^2/\lambda$, with f_a the axion decay constant. The Lagrangian that depends on the Higgs field can be rewritten as:

$$\mathcal{L}_\phi = -f_a(\rho\,\bar{\psi}_L Y \psi_R e^{ia/f_a} + \mathrm{h.c.}) + \frac{f_a^2}{2}(\partial\rho)^2 + \frac{1}{2}(\partial a)^2 - \frac{m_0^2 f_a^2}{2}\rho^2 - \frac{\lambda f_a^4}{4}\rho^4.$$

$$(4.677)$$

It is now seen that, if $m_0^2 < 0$, a spontaneous breaking of the Peccei–Quinn symmetry occurs. The vacuum expectation value of the field ρ is equal to 1, and its mass is $-2m_0^2$.

Assuming that its mass is much larger than the characteristic scale of the strong interactions (e.g., the spontaneous symmetry breaking occurs at the electroweak scale), we
may integrate out the ρ-field from the theory, leaving only the light field a (the axion).
Thus, the initial Lagrangian is reduced to

$$\mathcal{L}_{PQ} = \frac{1}{2g^2} \text{tr}_c(F_{\mu\nu} F^{\mu\nu}) - \frac{\theta}{16\pi^2} \text{tr}_c(F_{\mu\nu} \tilde{F}^{\mu\nu}) + \bar{\psi} i \gamma_\mu D^\mu \psi$$

$$- \frac{f_a}{\sqrt{2}} \left(\bar{\psi}_L Y \psi_R e^{ia/f_a} + \text{h.c.} \right) + \frac{1}{2} (\partial a)^2 . \tag{4.678}$$

It is seen that the quark mass matrix $\mathcal{M} = f_a Y / \sqrt{2}$ becomes nonzero as a result of the
spontaneous symmetry breaking.

Next, we may rotate away the axion field from the interaction term with the quarks,
considering a $U(1)$ transformation of the fermion fields:

$$\psi \mapsto \exp\left(-i \frac{a}{2f_a N_f} \gamma^5 \right) \psi , \tag{4.679}$$

with $N_f = 3$ the number of flavors. As we know, this transformation is anomalous and
shifts the parameter θ. The transformed Lagrangian takes the form

$$\mathcal{L}_{PQ} = \frac{1}{2g^2} \text{tr}_c(F_{\mu\nu} F^{\mu\nu}) - \frac{1}{16\pi^2} \left(\theta + \frac{a}{f_a} \right) \text{tr}_c(F_{\mu\nu} \tilde{F}^{\mu\nu}) + \bar{\psi} i \gamma_\mu D^\mu \psi$$

$$- \frac{f_a}{\sqrt{2}} \left(\bar{\psi}_L Y \psi_R + \text{h.c.} \right) + \frac{1}{2} (\partial a)^2 . \tag{4.680}$$

Now, it is seen that the parameter θ can be eliminated by a linear shift of the axion
field $a \to a - \theta f_a$. The question, however, is, whether the shifted axion field develops a
nonzero vacuum expectation value. Peccei and Quinn have shown that the minimum
of the potential of the shifted field is located at $a = 0$. (Due to the periodicity of the
effective action, the minima are, in fact, at $a/f_a = 2\pi n$, where n is an integer number).
We rely here on a simple intuitive argument by Vafa and Witten [287] to show that this
is indeed the case. Let us calculate the effective potential of the shifted field, integrating
out the fermion fields in the Euclidean generating functional:

$$\exp\left(-\int d^4x V(a) \right) = \int dG_\mu \prod_{f=u,d,s} \det(\slashed{D} + m_f)$$

$$\times \exp\left(-\int d^4x \left[-\frac{1}{2g^2} \text{tr}(F_{\mu\nu} F_{\mu\nu}) + \frac{ia}{16\pi^2 f_a} \text{tr}(F_{\mu\nu} \tilde{F}_{\mu\nu}) \right] \right) . \tag{4.681}$$

The determinant is a positive quantity:

$$\det(\slashed{D} + m_f) = m_f^{N_0} \prod_n (i\lambda_n + m_f)(-i\lambda_n + m_f) = m_f^{N_0} \prod_n (\lambda_n^2 + m_f^2) , \tag{4.682}$$

where $N_0 = N_+ + N_-$ is the total number of the zero modes, and we have used the fact
that the nonzero eigenvalues always come in pairs; see Section 3.11. Further, the first
term in the action is real and the second is purely imaginary. Hence:

$$\exp\left(-\int d^4x V(a)\right) = \int dG_\mu \prod_{f=u,d,s} \det(\slashed{D} + m_f)$$

$$\times \exp\left(-\int d^4x\left[-\frac{1}{2g^2}\,\text{tr}(F_{\mu\nu}F_{\mu\nu})\right]\right)\cos\left(\frac{1}{16\pi^2 f_a}\int d^4x\, a\left[\text{tr}(F_{\mu\nu}\tilde{F}_{\mu\nu})\right]\right)$$

$$\leq \int dG_\mu \prod_{f=u,d,s} \det(\slashed{D} + m_f)\exp\left(-\int d^4x\left[-\frac{1}{2g^2}\,\text{tr}(F_{\mu\nu}F_{\mu\nu})\right]\right)$$

$$= \exp\left(-\int d^4x V(0)\right). \tag{4.683}$$

This gives $V(a) \geq V(0)$, so the minimum is indeed at $a = \langle a \rangle = 0$.

Moreover, even if the axion is a Goldstone boson, it acquires mass due to the anomaly (here, we switch back to Minkowski space):

$$m_a^2 = \frac{\partial^2 V(a)}{\partial a^2}\bigg|_{a=\langle a\rangle}$$

$$= \left(\frac{1}{16\pi^2 f_a}\right)^2 \int d^4x \langle \text{tr}_c(F_{\mu\nu}(x)\tilde{F}^{\mu\nu}(x))\text{tr}_c(F_{\mu\nu}(0)\tilde{F}^{\mu\nu}(0))\rangle\bigg|_{a=\langle a\rangle}. \tag{4.684}$$

Note that m_a^2 is inversely proportional to the axion coupling (decay constant) squared, f_a^2. Similarity to the Witten–Veneziano formula, considered in Section 3.18, is seen by a bare eye.

The inclusion of the axion in realistic models proceeds along a similar path. Here we merely list a few options. It takes two Higgs fields to introduce the axion in the Standard Model (Weinberg–Wilczek (WW) axion [288, 289]). In this scenario, both the axion mass and the coupling constant are tied to the electroweak scale. However, the fact that axions have not been discovered yet puts severe constraints on the axion mass and coupling constant, so that the WW axion is ruled out by phenomenology. One can weaken the bounds, if one unties the axion from the electroweak scale. For example, in Ref. [290, 291] (KSVZ axion) it was proposed to decouple the scalar field from the known quarks and consider the coupling to (hypothetical) heavy quarks. Another possibility is discussed in Ref. [292, 293] (the ZDFS axion). Here, a scalar field, which is a singlet with respect to the Standard Model symmetries, is added to the Lagrangian. We do not want to further proceed with the discussion of alternative scenarios, as well as other aspects of the physics of axions, and refer the interested reader, for example, to a nice review article [294], as well as the additional literature at the end of this chapter. It should only be noted that axions (or axion-like particles, ALPs) represent very robust candidates for dark matter and might contribute to the solution of that problem.

4.20.4 ChPT at a Nonzero θ and in the Presence of Axions

Here, a primary focus is to study the interactions of axions with the observed particles at low energy, namely hadrons, nuclei, as well as photons. In this section we briefly consider the framework, which enables one to address this problem in a systematic manner.

Let us first address the simpler question and assume that there are no axions, but the parameter θ is different from zero. One might ask, how the physical observables depend on the parameter θ. This question is directly answered by noting that the θ-term can be rotated away at the expense of a modification of the quark mass matrix, where the parameter θ now appears. Writing down the effective mesonic Lagrangian for this theory, at the lowest order we arrive at the following:[54]

$$\mathcal{L}^{(2)} = \frac{F_0^2}{4} \langle D_\mu U D^\mu U^\dagger + \chi_\theta U^\dagger + U \chi_\theta^\dagger \rangle, \tag{4.685}$$

where, in the absence of the external scalar and pseudoscalar sources,

$$\chi_\theta = 2B_0 \mathrm{diag}(m_u e^{-2i\theta_u}, m_d e^{-2i\theta_d}, m_s e^{-2i\theta_s})$$

$$\theta = -2(\theta_u + \theta_d + \theta_s). \tag{4.686}$$

It is important to stress that the LECs F_0 and B_0 are the same as at $\theta = 0$. The nonvanishing θ is present in the quark mass matrix *only*.

We expand the field U in the vicinity of the vacuum field configuration U_0, which minimizes the potential energy. This configuration takes the form

$$U_0 = \mathrm{diag}(e^{i\varphi_u}, e^{i\varphi_d}, e^{i\varphi_s}). \tag{4.687}$$

Since the common phase is already included in θ, it is possible to assume that $U \in SU(3)$ and hence $\varphi_u + \varphi_d + \varphi_s = 0$. The potential is then given by

$$V(\theta) = -F_0^2 B_0 \sum_{f=u,d,s} m_f \cos\phi_f, \qquad \phi_f = -\varphi_f - 2\theta_f. \tag{4.688}$$

Next, we have to find a minimum of this potential with respect to the angles, whose sum remains constant. In order to do this, one could express, say, ϕ_s in terms of ϕ_u, ϕ_d and find the minimum of the potential with respect to the independent variables ϕ_u, ϕ_d. The solution takes the following form:

$$m_u \sin\phi_u = m_d \sin\phi_d = m_s \sin\phi_s = 0, \qquad \phi_u + \phi_d + \phi_s = \theta. \tag{4.689}$$

In order to read off the masses of the Goldstone bosons for nonzero values of θ, it suffices to write down $U = U_0^{1/2} \tilde{U} U_0^{1/2}$, where $\tilde{U} = \exp\{i\Phi/F_0\}$ and $\Phi = \lambda^a \phi_a$ is the

---

[54] To carry out calculations, we may use either the $U(N_f)$ or the $SU(N_f)$ version of ChPT. If one is not interested in the properties of η' and works at momentum scales much smaller than the η' mass, both approaches are equivalent, as we already know. Indeed, the η' meson can be integrated out from the theory completely. Bearing this in mind, in what follows we shall opt for the $SU(N_f)$ version.

matrix containing the Goldstone boson octet fields. Expanding up to the second order in the fields and using the physical basis, we get (see, e.g., Ref. [295])

$$M_{\pi^\pm}^2 = B_0(m_u \cos \phi_u + m_d \cos \phi_d),$$

$$M_{K^\pm}^2 = B_0(m_u \cos \phi_u + m_s \cos \phi_s),$$

$$M_{K^0}^2 = M_{\bar{K}^0}^2 = B_0(m_d \cos \phi_d + m_s \cos \phi_s),$$

$$M_{\pi^0}^2 = B_0(m_u \cos \phi_u + m_d \cos \phi_d) - \xi,$$

$$M_\eta^2 = \frac{1}{3} B_0(m_u \cos \phi_u + m_d \cos \phi_d + 4m_s \cos \phi_s) + \xi, \qquad (4.690)$$

where

$$\xi = \frac{2}{3} B_0 (2m_s \cos \phi_s - m_u \cos \phi_u - m_d \cos \phi_d) \frac{\sin \varepsilon_\theta}{\cos 2\varepsilon_\theta}, \qquad (4.691)$$

and ε_θ is the $\pi^0 - \eta$ mixing angle for the nonzero value of θ:

$$\tan 2\varepsilon_\theta = \frac{\sqrt{3}(m_d \cos \phi_d - m_u \cos \phi_u)}{2m_s \cos \phi_s - m_u \cos \phi_u - m_d \cos \phi_d}. \qquad (4.692)$$

Needless to say, when $\theta = 0$, then $\phi_u = \phi_d = \phi_s = 0$ and these expressions agree with the ones derived earlier. For $\theta \neq 0$, ϕ_1, ϕ_2, ϕ_3 cannot be, in general, analytically expressed through θ. For small values of θ, the following relation holds:

$$m' = m_u \sin \phi_u = m_d \sin \phi_d = m_s \sin \phi_s = \frac{m_u m_d m_s}{m_u m_d + m_d m_s + m_s m_u} \theta + O(\theta^2). \qquad (4.693)$$

In the preceding simple example, it was demonstrated how the calculations of physical observables can be carried out for nonzero values of theta. The technique that was used here is similar to the one discussed in detail earlier in this chapter. The LECs are the same, the explicit dependence on θ emerges in the quark mass matrix only. The method can be readily extended to other observables and to the higher chiral orders, too. What is more interesting, such calculations can be extended to the sectors that contain one or more nucleons. In this way, for example, one can study the θ-dependence of the electromagnetic form factor of a nucleon and thus get a tool that enables one to "translate" the experimental constraints on the neutron dipole moment into constraints on the value of the parameter θ.

Our discussion here mainly follows Ref. [263] despite the fact that from the beginning we stick to the $SU(3)$-version of ChPT. In order to write down the Lagrangian in the one-baryon sector, we define the matrices u_R, u_L, which transform under $U(3)_L \times U(3)_R$ as

$$u_L \mapsto h u_L g_L^\dagger = g_L u_L h^\dagger, \qquad u_R \mapsto h u_R g_R^\dagger = g_R u_R h^\dagger, \qquad (4.694)$$

so that $U = u_R u_L^\dagger$ transforms as $U \mapsto g_R U g_L^\dagger$. The configuration that minimizes the vacuum is obtained at

$$u_R = u_L^\dagger = U_0^{1/2}. \qquad (4.695)$$

Consequently, we may define

$$u_R = U_0^{1/2}\tilde{u}_R, \qquad u_L = U_0^{-1/2}\tilde{u}_L. \tag{4.696}$$

At the end, we may fix the gauge as $\tilde{u}_R = \tilde{u}_L^\dagger = \tilde{u}$ and $\tilde{U} = \tilde{u}\tilde{u}$, where $\tilde{U}$ contains only physical Goldstone boson fields (the vacuum phase is removed).

The leading-order low-energy Lagrangians in the one-nucleon sector are given in Eq. (4.400). The building blocks, containing the derivative of the meson fields, are

$$\Gamma_\mu = \frac{1}{2}\left(u_R^\dagger(\partial_\mu - ir_\mu)u_R + u_L^\dagger(\partial_\mu - il_\mu)u_L\right),$$

$$u_\mu = i(u_R^\dagger(\partial_\mu - ir_\mu)u_R - u_L^\dagger(\partial_\mu - il_\mu)u_L). \tag{4.697}$$

It is straightforward to ensure that the phase U_0 cancels in these expressions. Hence, there are no CP-breaking interactions emerging from these terms. For the term containing the quark mass matrix, the situation is, however, different. We may write

$$\chi_\pm = u_R^\dagger \chi u_L \pm u_L^\dagger \chi^\dagger u_R. \tag{4.698}$$

Let us put the external sources s and p to zero, so that $\chi = \chi_\theta$, where χ_θ is given by Eq. (4.686). Taking out the phase factor U_0 and fixing the gauge $\tilde{u}_R = \tilde{u}_L^\dagger = \tilde{u}$, we get

$$\chi_\pm = \tilde{\chi}_\pm + im'(\tilde{U}^\dagger \mp \tilde{U}), \tag{4.699}$$

where m' is given by Eq. (4.693).

Using the preceding equation, one can derive the CP-breaking vertices at $O(p^2)$:

$$\mathcal{L}' = im'b_D\langle\bar{B}\{(\tilde{U}^\dagger - \tilde{U}),B\}\rangle + im'b_F\langle\bar{B}[(\tilde{U}^\dagger - \tilde{U}),B]\rangle + im'b_0\langle\bar{B}B\rangle\langle\tilde{U}^\dagger - \tilde{U}\rangle$$

$$= \frac{m'b_D}{F_0}\langle\bar{B}\{\Phi,B\}\rangle + \frac{m'b_F}{F_0}\langle\bar{B}[\Phi,B]\rangle + O(\Phi^2). \tag{4.700}$$

The electric dipole moment of the neutron emerges first at order $O(p^4)$ (we remind the reader that e actually counts as p). At this order, in the Lagrangian there are terms of two types:

i) The terms with the pseudoscalar couplings of the baryons. An example of such a term is given by

$$\mathcal{L}_5 = -c_0\langle\chi_-\rangle\langle\bar{B}\gamma_5 B\rangle. \tag{4.701}$$

Formally, this operator is of order p^2. However, since γ_5 connects the large and small components of the Dirac spinor, it counts at order p on the mass shell. Hence, such operators effectively start to contribute at $O(p^3)$. According to Eq. (4.699), the rotation of χ_- produces the shift:

$$\mathcal{L}_5 \to \mathcal{L}_5 - im'c_0\langle U^\dagger + U\rangle\langle\bar{B}\gamma_5 B\rangle = -6im'c_0\langle\bar{B}\gamma_5 B\rangle + O(\Phi). \tag{4.702}$$

Combining the term, proportional to m', with the mass term, it can be seen that it can be rotated away by an appropriate transformation of the baryon field:

$$-m_0\langle\bar{B}B\rangle - 6im'c_0\langle\bar{B}\gamma_5 B\rangle = -m_0\langle\bar{B}'B'\rangle + O(\theta^2), \tag{4.703}$$

where

$$B' = e^{3im'c_0\gamma_5} B. \tag{4.704}$$

The $O(p)$ meson–nucleon Lagrangian is invariant with respect to this field redefinition. However, it produces new CP-violating vertices in the $O(p^2)$ Lagrangian, which initially contains, for example, operators like $\langle \bar{B}\sigma^{\mu\nu}F_{\mu\nu}^+ B\rangle$. These vertices contribute to the anomalous magnetic moment at $O(p^4)$.

ii) There exists a string of terms in the $O(p^4)$ Lagrangian that produce CP-violating vertices. For example, using Eq. (4.699) in the operator $\langle \bar{B}\sigma^{\mu\nu}\gamma_5\{\chi_-, F_{\mu\nu}^+\}B\rangle$ produces a vertex that, at the tree level, contributes to the electric dipole moment at the accuracy we are working.

Putting things together, we may now carry out the calculation of the electromagnetic form factor of the nucleon, taking into account all CP-violating vertices that are proportional to the vacuum angle θ. From Lorentz covariance and $U(1)$ gauge invariance, the matrix element of the electromagnetic current between the one-nucleon states is given by

$$\langle p's'|j^\mu(0)|ps\rangle = \bar{u}(p's')\left(\gamma^\mu F_1(q^2) + \frac{i}{2m_N}\sigma^{\mu\nu}q_\nu F_2(q^2)\right.$$

$$\left. + i(\gamma^\mu q^2 - 2m_N q^\mu)\gamma_5 F_A(q^2) + \frac{1}{2m_N}\sigma^{\mu\nu}q_\nu \gamma_5 F_3(q^2)\right)u(ps), \tag{4.705}$$

where $q^2 = (p'-p)^2$, $F_1(q^2)$ and $F_2(q^2)$ are the P- and CP-conserving Dirac and Pauli form factors, respectively, while $F_A(q^2)$ denotes the P-violating anapole form factor. $F_3(q^2)$ is the CP-violating electric dipole form factor. The electric dipole moment is defined as

$$d_n = \frac{1}{2m_N}F_3(0). \tag{4.706}$$

There are two types of contributions to the quantity d_n at $O(p^4)$. These are shown in Fig. 4.29. First, there are loop graphs, which contain one CP-violating vertex from the $O(p^2)$ Lagrangian. Second, there are CP-violating vertices emerging at $O(p^4)$ that contribute at the tree level. Adding everything together, the following result was obtained in Ref. [263]:

$$d_n = \frac{em'}{4\pi^2 F_\pi^2}\left((b_D + b_F)(D+F)\ln\frac{M_\pi^2}{\mu^2} - (b_D - b_F)(D-F)\ln\frac{M_K^2}{\mu^2} + c^r(\mu)\right), \tag{4.707}$$

where $c^r(\mu)$ denotes a particular linear combination of the renormalized $O(p^4)$ LECs at the scale μ. We remind the reader that these are the same LECs that appear in the CP-conserving sector of the theory. Estimating the contribution coming from these LECs, the final result $d_n = (3.3 \pm 1.8) \times 10^{-16}\theta\, e \cdot$cm is quoted in Ref. [263]. This was updated

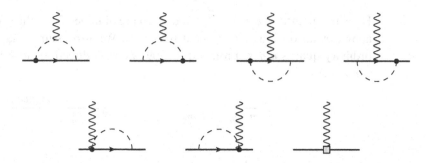

Feynman diagrams that contribute to the electric dipole moment of the neutron at $O(p^4)$. The filled dots and a square are the CP-violating vertices at $O(p^2)$ and $O(p^4)$, respectively, solid/dashed lines denote the baryons/mesons, and the wiggly lines represent photons.

in Ref. [296], leading to $d_n \simeq 1.1 \times 10^{-16}\theta\, e \cdot \mathrm{cm}$, which produces a more stringent constraint on the value of the parameter θ, extracted from the experimental bound on d_n. In the latter calculation, the $SU(2)$ formulation was employed, which is more precise than the $SU(3)$ framework utilized in Ref. [263]. Additionally, the relevant LECs are better constrained in the newer investigation. We do not enter into a more detailed discussion here, noting that the preceding example clearly demonstrates how to carry out calculations in ChPT in the presence of a nonzero θ-term.

As discussed in Section 4.20.3, most probably, the smallness of the parameter θ in nature is explained by the existence of axions. It is clear that the technique, which has been used so far, can be applied to the study of the axion interactions with matter as well. As we have already seen, originally the leading dependence on the axion field in the QCD Lagrangian emerges through the coupling to the winding number density; see Eq. (4.680). It is possible to rotate the axion field back to the quark Lagrangian, where it appears in the quark mass matrix. The construction of the effective theory proceeds then along the known path. Both $SU(3)$ and $U(3)$ formulations can be used. We shall not enter in the details here, except a demonstration of the axion mass calculation at the lowest order. As it is known, in the presence of axion, the vacuum energy $V(\theta)$ becomes dependent on the axion field $\theta \to \theta + a/f_a$, and the minimum is at $\theta + \langle a\rangle/f_a = 0$. Shifting $a \to a - \langle a\rangle$, and expanding the potential in powers of the new field a, we get, by definition,

$$V(a) = \frac{m_a^2}{2}\, a^2 + O(a^4). \tag{4.708}$$

Further, using the Eq. (4.688) and Eq. (4.693), we have

$$V(a) = -F_0^2 B_0 \sum_f m_f \cos\phi_f(a) \simeq \mathrm{const} + \frac{1}{2} F_0^2 B_0 [m'(a)]^2 \sum_f \frac{1}{m_f}$$

$$= \frac{F_0^2 B_0}{2 f_a^2}\, \frac{m_u m_d m_s}{m_u m_d + m_d m_s + m_s m_u}\, a^2. \tag{4.709}$$

Here, the quantities $\phi_f(a)$, $m'(a)$ and so on are obtained from the expressions, derived above, by an obvious replacement $\theta \to a/f_a$. We also assume that axions couple to ordinary quarks u, d, s. From Eq. (4.709) we may directly read off the value of the axion mass:

$$m_a^2 = \frac{F_0^2 B_0}{f_a^2} \frac{m_u m_d m_s}{m_u m_d + m_d m_s + m_s m_u} = \frac{F_0^2 M_\pi^2}{2 f_a^2 \hat{m}} \frac{m_u m_d m_s}{m_u m_d + m_d m_s + m_s m_u},$$
(4.710)

where $\hat{m} = (m_u + m_d)/2$.

Such calculations can be readily carried out at higher orders. For example, in Ref. [295], these were extended to order p^4. The result for the axion mass is given by

$$m_a = 5.89(1)\,\mu\text{eV} \cdot \frac{10^{12}\,\text{GeV}}{f_a}.$$
(4.711)

Moreover, other observables of interest can be systematically evaluated by using the same method (say, the axion coupling to two photons). We, however, feel that it is time to stop and refer the reader to a vast amount of available literature on the subject. A few references can be found in the literature guide at the end of the chapter.

4.21 Literature Guide

The description of the nonelementary compounds (bound states) in quantum field theory has been addressed, starting from the 1950s [10–12, 14–16]. These findings have been largely summarized in Ref. [13], where a few subtle issues are also addressed. The description of hadrons as bound states of quarks and gluons dates back to potential models. An adequate field-theoretical approach to the problem is based on the use of the Dyson–Schwinger [297, 298] and Bethe–Salpeter equations [299]. It has been, in particular, shown that, unlike potential models, chiral symmetry can be incorporated in this approach in a consistent manner. For more information on this subject, we refer the reader to the review article [300].

The linear σ-model was proposed by Gell–Mann and Levy [22]. We recommend two textbooks, the ones by Itzykson and Zuber [301] as well as the one by De Alfaro, Fubini, Furlan and Rosetti [40], which contain all necessary information. The inclusion of the baryons is a subtle issue, and different scenarios are possible. We recommend Ref. [28] for a very clear and concise discussion of the topic. Alternative scenarios have been studied, for example, in Refs. [302–305], which deal with the so-called parity doubling linear σ-models. The three-quark baryon currents have been constructed in Ref. [27] (see also [306]), where the chiral properties of different currents are also discussed.

The nonlinear realization of chiral symmetry was considered in Ref. [36]. An elegant mathematical formulation, which is given in the present book, stems from Refs. [34, 35].

Chiral Perturbation Theory for mesons has been introduced in the seminal papers by Weinberg [18] and Gasser and Leutwyler [19]. In the one-nucleon sector, the relativistic formulation was first discussed by Gasser, Sainio and Svarc [62], whereas different formulations, which respect the power counting, were considered in Refs. [74, 100, 110]. A breakthrough in the two-nucleon sector occurred when Weinberg suggested to extend the formalism by applying the power counting to the two-nucleon interaction potential rather than to the scattering amplitude, which is obtained by solving the Lippmann–Schwinger equation non-perturbatively [7, 8]. The inclusion of the electromagnetic and weak interactions has been considered in Refs. [87, 200]. Here, we certainly cannot adequately reflect even a small part of a huge work done within a framework of Chiral Perturbation theory, and only refer to the book by Scherer and Schindler [307], as well as to a couple of review articles [69, 83, 308, 309] for more information and additional references.

The mesonic Lagrangian that reproduces anomalies of QCD in the external fields was constructed by Wess and Zumino [233]. In the paper by Witten [234], an intricate connection between anomalies and topology is discovered, which leads to the quantization of the coefficient of the anomalous term. Chiral perturbation theory for the Goldstone boson nonet is considered in Ref. [37], and the chiral counting in the large-N_c limit is systematically addressed in Ref. [237]. The solution of the $U(1)_A$ problem and its relation to the anomaly was discussed, for example, in the papers by 't Hooft [310, 311]. This problem was reconsidered in the light of the $1/N_c$ expansion by Witten [247].

As already mentioned, there are two approaches to construct the low-energy effective field theory of partially quenched QCD, namely the one based on the graded symmetry [268, 270, 312, 313] and the one using the so-called replica method [314, 315]. One of the applications of such effective theories is the estimate of the contribution of the disconnected contributions in lattice calculations. In addition to the papers, which were already cited in the text, we give here a list of a few recent publications on this subject [316–319].

In the papers by Jackiw and Rebbi [320], and by Callan, Dashen and Gross [321], it has been realized that the vacuum of QCD has a complicated structure that can be attributed to the so-called instantons. As a result, the new parameter θ emerges in QCD, and the CP invariance is broken if θ is not equal to 0 or π. A vast literature emerged on this subject. We recommend here a few textbooks that give a very thorough discussion of this issue [278, 322, 323]. A lot of useful information about instantons can be found in the lectures by Coleman [324]. The relation of the parameter θ with the electric dipole moment of the neutron was first given in Ref. [261], where the calculation was performed in the MIT bag model. Seiler and Stamatescu [325] discuss the introduction of the θ parameter on the lattice. The axions have been invented by Peccei and Quinn [285, 286]. In the References section, you can find our suggestions for further reading, in addition to the work that has been already cited in the text [326–330]. These papers address different aspects of the problem, related to the physics of axions.

References

[1] M. Neubert, "Heavy quark symmetry," Phys. Rept. **245** (1994) 259.

[2] A. V. Manohar and M. B. Wise, "Heavy quark physics," Camb. Monogr. Part. Phys. Nucl. Phys. Cosmol. **10** (2000) 1.

[3] T. Mannel, Effective Field Theories in Flavour Physics, Springer Tracts in Modern Physics, Springer-Verlag (2004).

[4] G. Burdman and J. F. Donoghue, "Union of chiral and heavy quark symmetries," Phys. Lett. B **280** (1992) 287.

[5] M. B. Wise, "Chiral perturbation theory for hadrons containing a heavy quark," Phys. Rev. D **45** (1992) R2188.

[6] T. M. Yan, H. Y. Cheng, C. Y. Cheung et al. "Heavy quark symmetry and chiral dynamics," Phys. Rev. D **46** (1992) 1148 Erratum: [Phys. Rev. D **55** (1997) 5851].

[7] S. Weinberg, "Nuclear forces from chiral Lagrangians," Phys. Lett. B **251** (1990) 288.

[8] S. Weinberg, "Effective chiral Lagrangians for nucleon-pion interactions and nuclear forces," Nucl. Phys. B **363** (1991) 3.

[9] T. A. Lähde and U.-G. Meißner, "Nuclear Lattice effective field theory : An introduction," Lect. Notes Phys. **957** (2019) 1.

[10] S. Mandelstam, "Dynamical variables in the Bethe-Salpeter formalism," Proc. Roy. Soc. Lond. A **233** (1955) 248.

[11] K. Nishijima, "Formulation of field theories of composite particles," Phys. Rev. **111** (1959) 995.

[12] R. Blankenbecler, "A contraction rule for composite particles," Nucl. Phys. B **14** (1959) 97.

[13] K. Huang and H. A. Weldon, "Bound state wave functions and bound state scattering in relativistic field theory," Phys. Rev. D **11** (1975) 257.

[14] R. Haag, "Quantum field theories with composite particles and asymptotic conditions," Phys. Rev. **112** (1958) 669.

[15] S. S. Schweber and G. Sudarshan, "Asymptotic field operators in quantum field theory," Annals Phys. **19** (1962) 351.

[16] A. Klein and C. Zemach, "Many-body problem in quantum field theory," Phys. Rev. **108** (1957) 126.

[17] J. D. Bjorken and S. D. Drell, II: Relativistic Quantum Fields, McGraw-Hill (1965).

[18] S. Weinberg, "Phenomenological Lagrangians," Physica A **96** (1979) 327.

[19] J. Gasser and H. Leutwyler, "Chiral perturbation theory to one loop," Annals Phys. **158** (1984) 142.

[20] H. Leutwyler, "On the foundations of chiral perturbation theory," Annals Phys. **235** (1994) 165.

[21] E. D'Hoker and S. Weinberg, "General effective actions," Phys. Rev. D **50** (1994) R6050.

[22] M. Gell–Mann and M. Levy, "The axial vector current in beta decay," Nuovo Cim. **16** (1960) 705.

[23] I. Caprini, G. Colangelo and H. Leutwyler, "Mass and width of the lowest resonance in QCD," Phys. Rev. Lett. **96** (2006) 132001.

[24] J. Gasser and A. Zepeda, "Approaching the chiral limit in QCD," Nucl. Phys. B **174** (1980) 445.

[25] J. Gasser, A. Rusetsky and I. Scimemi, "Electromagnetic corrections in hadronic processes," Eur. Phys. J. C **32** (2003) 97.

[26] A. Nyffeler and A. Schenk, "Effective field theory of the linear $O(N)$ sigma model," Annals Phys. **241** (1995) 301.

[27] B. L. Ioffe, "Calculation of baryon masses in quantum chromodynamics," Nucl. Phys. B **188** (1981) 317 Erratum: [Nucl. Phys. B **191** (1981) 591].

[28] H. Georgi, Weak Interactions and Modern Particle Theory, Benjamin/Cummings (1984).

[29] V. Bernard, L. Elouadrhiri and U.-G. Meißner, "Axial structure of the nucleon: Topical review," J. Phys. G **28** (2002) R1.

[30] M. L. Goldberger and S. B. Treiman, "Form-factors in beta decay and muon capture," Phys. Rev. **111** (1958) 354.

[31] V. Baru, C. Hanhart, M. Hoferichter et al. "Precision calculation of the π^- deuteron scattering length and its impact on threshold π N scattering," Phys. Lett. B **694** (2011) 473.

[32] S. Weinberg, "Pion scattering lengths," Phys. Rev. Lett. **17** (1966) 616.

[33] Y. Tomozawa, "Axial vector coupling renormalization and the meson baryon scattering lengths," Nuovo Cim. A **46** (1966) 707.

[34] S. R. Coleman, J. Wess and B. Zumino, "Structure of phenomenological Lagrangians. 1.," Phys. Rev. **177** (1969) 2239.

[35] C. G. Callan, Jr., S. R. Coleman, J. Wess and B. Zumino, "Structure of phenomenological Lagrangians. 2.," Phys. Rev. **177** (1969) 2247.

[36] S. Weinberg, "Nonlinear realizations of chiral symmetry," Phys. Rev. **166** (1968) 1568.

[37] J. Gasser and H. Leutwyler, "Chiral perturbation theory: expansions in the mass of the strange quark," Nucl. Phys. B **250** (1985) 465.

[38] M. Knecht and R. Urech, "Virtual photons in low-energy $\pi\pi$ scattering," Nucl. Phys. B **519** (1998) 329.

[39] J. Bijnens, G. Colangelo and G. Ecker, "The mesonic chiral Lagrangian of order p^6," JHEP **9902** (1999) 020.

[40] C. De Alfaro, S. Fubini, S. Furlan and G. Rossetti, Currents in Hadronic Physics, North Holland & American Elsevier (1973).

[41] J. F. Donoghue, "Light quark masses and chiral symmetry," Ann. Rev. Nucl. Part. Sci. **39** (1989) 1.

[42] H. Leutwyler, "The ratios of the light quark masses," Phys. Lett. B **378** (1996) 313.

[43] P. Chang and F. Gursey, "Unified formulation of effective nonlinear pion-nucleon Lagrangians," Phys. Rev. **164** (1967) 1752.

[44] J. S. Schwinger, "Chiral dynamics," Phys. Lett. **24B** (1967) 473.

[45] H. Georgi, "Effective field theory," Ann. Rev. Nucl. Part. Sci. **43** (1993) 209.

[46] H. Lehmann, "Chiral invariance and effective range expansion for pion pion scattering," Phys. Lett. **41B** (1972) 529.

[47] J. Stern, H. Sazdjian and N. H. Fuchs, "What $\pi - \pi$ scattering tells us about chiral perturbation theory," Phys. Rev. D **47** (1993) 3814.

[48] G. Colangelo, J. Gasser and H. Leutwyler, "$\pi\pi$ scattering," Nucl. Phys. B **603** (2001) 125.

[49] V. Bernard and U.-G. Meißner, "Chiral perturbation theory," Ann. Rev. Nucl. Part. Sci. **57** (2007) 33.

[50] G. Colangelo, J. Gasser and H. Leutwyler, "The $\pi\pi$ S-wave scattering lengths," Phys. Lett. B **488** (2000) 261.

[51] J. R. Batley et al., "Determination of the S-wave $\pi\pi$ scattering lengths from a study of $K^{\pm} \to \pi^{\pm}\pi^0\pi^0$ decays," Eur. Phys. J. C **64** (2009) 589.

[52] J. R. Batley et al., "Precise tests of low energy QCD from K_{e4} decay properties," Eur. Phys. J. C **70** (2010) 635.

[53] G. Colangelo, J. Gasser and H. Leutwyler, "The quark condensate from K_{e4} decays," Phys. Rev. Lett. **86** (2001) 5008.

[54] U.-G. Meißner, "The pion cloud of the nucleon: Facts and popular fantasies," AIP Conf. Proc. **904** (2007) 142.

[55] A. Manohar and H. Georgi, "Chiral quarks and the nonrelativistic quark model," Nucl. Phys. B **234** (1984) 189.

[56] J. Bijnens, G. Colangelo, G. Ecker, J. Gasser and M. E. Sainio, "Pion-pion scattering at low energy," Nucl. Phys. B **508** (1997) 263 Erratum: [Nucl. Phys. B **517** (1998) 639].

[57] G. Ecker, J. Gasser, A. Pich and E. de Rafael, "The role of resonances in chiral perturbation theory," Nucl. Phys. B **321** (1989) 311.

[58] J. F. Donoghue, C. Ramirez and G. Valencia, "The spectrum of QCD and chiral Lagrangians of the strong and weak interactions," Phys. Rev. D **39** (1989) 1947.

[59] J. J. Sakurai, "Theory of strong interactions," Annals Phys. **11** (1960) 1.

[60] J. Bijnens and G. Ecker, "Mesonic low-energy constants," Ann. Rev. Nucl. Part. Sci. **64** (2014) 149.

[61] H. Pagels, "Departures from chiral symmetry: A review," Phys. Rept. **16** (1975) 219.

[62] J. Gasser, M. E. Sainio and A. Svarc, "Nucleons with chiral loops," Nucl. Phys. B **307** (1988) 779.

[63] N. Fettes, U.-G. Meißner, M. Mojžiš and S. Steininger, "The chiral effective pion nucleon Lagrangian of order p^4," Annals Phys. **283** (2000) 273 Erratum: [Annals Phys. **288** (2001) 249].

[64] N. Fettes, U.-G. Meißner and S. Steininger, "Pion - nucleon scattering in chiral perturbation theory. 1. Isospin symmetric case," Nucl. Phys. A **640** (1998) 199.

[65] J. Gasser, M. A. Ivanov, E. Lipartia, M. Mojžiš and A. Rusetsky, "Ground state energy of pionic hydrogen to one loop," Eur. Phys. J. C **26** (2002) 13.

[66] M. Frink and U.-G. Meißner, "On the chiral effective meson-baryon Lagrangian at third order," Eur. Phys. J. A **29** (2006) 255.

[67] J. A. Oller, M. Verbeni and J. Prades, "Meson-baryon effective chiral lagrangians to $O(q^3)$," JHEP **0609** (2006) 079.

[68] V. Bernard, "Chiral Perturbation Theory and baryon properties," Prog. Part. Nucl. Phys. **60** (2008) 82.

[69] V. Bernard, N. Kaiser and U.-G. Meißner, "Aspects of chiral pion - nucleon physics," Nucl. Phys. A **615** (1997) 483.

[70] M. Hoferichter, J. Ruiz de Elvira, B. Kubis and U.-G. Meißner, "Matching pion-nucleon Roy-Steiner equations to chiral perturbation theory," Phys. Rev. Lett. **115** (2015) 192301.

[71] M. Beneke and V. A. Smirnov, "Asymptotic expansion of Feynman integrals near threshold," Nucl. Phys. B **522** (1998) 321.

[72] R. F. Mohr, R. J. Furnstahl, R. J. Perry, K. G. Wilson and H.-W. Hammer, "Precise numerical results for limit cycles in the quantum three-body problem," Annals Phys. **321** (2006) 225.

[73] H. Georgi, "An effective field theory for heavy quarks at low-energies," Phys. Lett. B **240** (1990) 447.

[74] E. E. Jenkins and A. V. Manohar, "Baryon chiral perturbation theory using a heavy fermion Lagrangian," Phys. Lett. B **255** (1991) 558.

[75] J. Gasser, "Hadron masses and sigma commutator in the light of chiral perturbation theory," Annals Phys. **136** (1981) 62.

[76] J. Gasser and H. Leutwyler, "Quark masses," Phys. Rept. **87** (1982) 77.

[77] M. E. Luke and A. V. Manohar, "Reparametrization invariance constraints on heavy particle effective field theories," Phys. Lett. B **286** (1992) 348.

[78] L. L. Foldy and S. A. Wouthuysen, "On the Dirac theory of spin 1/2 particle and its nonrelativistic limit," Phys. Rev. **78** (1950) 29.

[79] T. Mannel, W. Roberts and Z. Ryzak, "A derivation of the heavy quark effective Lagrangian from QCD," Nucl. Phys. B **368** (1992) 204.

[80] V. Bernard, N. Kaiser, J. Kambor and U.-G. Meißner, "Chiral structure of the nucleon," Nucl. Phys. B **388** (1992) 315.

[81] G. Ecker, "Chiral invariant renormalization of the pion - nucleon interaction," Phys. Lett. B **336** (1994) 508.

[82] G. Ecker and U.-G. Meißner, "What is a low-energy theorem?," Comments Nucl. Part. Phys. **21** (1995) 347.

[83] V. Bernard, N. Kaiser and U.-G. Meißner, "Chiral dynamics in nucleons and nuclei," Int. J. Mod. Phys. E **4** (1995) 193.

[84] U.-G. Meißner, G. Müller and S. Steininger, "Renormalization of the chiral pion - nucleon Lagrangian beyond next-to-leading order," Annals Phys. **279** (2000) 1.

[85] H. Neufeld, J. Gasser and G. Ecker, "The one loop functional as a Berezinian," Phys. Lett. B **438** (1998) 106.

[86] H. Neufeld, "The super heat kernel expansion and the renormalization of the pion - nucleon interaction," Eur. Phys. J. C **7** (1999) 355.

[87] M. Knecht, H. Neufeld, H. Rupertsberger and P. Talavera, "Chiral perturbation theory with virtual photons and leptons," Eur. Phys. J. C **12** (2000) 469.

[88] P. Langacker and H. Pagels, "Applications of chiral perturbation theory: mass formulas and the decay $\eta \to 3\pi$," Phys. Rev. D **10** (1974) 2904.

[89] V. Bernard, N. Kaiser, J. Gasser and U.-G. Meißner, "Neutral pion photoproduction at threshold," Phys. Lett. B **268** (1991) 291.

[90] V. Bernard, N. Kaiser and U.-G. Meißner, "Chiral expansion of the nucleon's electromagnetic polarizabilities," Phys. Rev. Lett. **67** (1991) 1515.

[91] G. Höhler, "Pion–Nukleon-Streuung: Methoden und Ergebnisse," in Landolt-Börnstein, **9b2**, ed. H. Schopper, Springer Verlag, 1983.

[92] V. Bernard, N. Kaiser and U.-G. Meißner, "Nucleon electroweak form-factors: Analysis of their spectral functions," Nucl. Phys. A **611** (1996) 429.

[93] N. Kaiser, "Spectral functions of isoscalar scalar and isovector electromagnetic form-factors of the nucleon at two loop order," Phys. Rev. C **68** (2003) 025202.

[94] V. Bernard, N. Kaiser, U.-G. Meißner and A. Schmidt, "Aspects of nucleon Compton scattering," Z. Phys. A **348** (1994) 317.

[95] E. E. Jenkins and A. V. Manohar, "Chiral corrections to the baryon axial currents," Phys. Lett. B **259** (1991) 353.

[96] T. R. Hemmert, B. R. Holstein and J. Kambor, "Chiral Lagrangians and $\Delta(1232)$ interactions: formalism," J. Phys. G **24** (1998) 1831.

[97] V. Bernard, H. W. Fearing, T. R. Hemmert and U.-G. Meißner, "The form-factors of the nucleon at small momentum transfer," Nucl. Phys. A **635** (1998) 121 Erratum: [Nucl. Phys. A **642** (1998) 563].

[98] H. B. Tang, "A New approach to chiral perturbation theory for matter fields," hep-ph/9607436.

[99] P. J. Ellis and H. B. Tang, "Pion nucleon scattering in a new approach to chiral perturbation theory," Phys. Rev. C **57** (1998) 3356.

[100] T. Becher and H. Leutwyler, "Baryon chiral perturbation theory in manifestly Lorentz invariant form," Eur. Phys. J. C **9** (1999) 643.

[101] B. Kubis and U.-G. Meißner, "Low-energy analysis of the nucleon electromagnetic form-factors," Nucl. Phys. A **679** (2001) 698.

[102] M. Procura, T. R. Hemmert and W. Weise, "Nucleon mass, sigma term and lattice QCD," Phys. Rev. D **69** (2004) 034505.

[103] T. Becher and H. Leutwyler, "Low energy analysis of $\pi N \to \pi N$," JHEP **0106** (2001) 017.

[104] V. Bernard, T. R. Hemmert and U.-G. Meißner, "Spin structure of the nucleon at low-energies," Phys. Rev. D **67** (2003) 076008.

[105] J. L. Goity, D. Lehmann, G. Prezeau and J. Saez, "Regularization for effective field theory with two heavy particles," Phys. Lett. B **504** (2001) 21.

[106] V. Bernard, T. R. Hemmert and U.-G. Meißner, "Infrared regularization with spin 3/2 fields," Phys. Lett. B **565** (2003) 137.

[107] P. C. Bruns and U.-G. Meißner, "Infrared regularization for spin-1 fields," Eur. Phys. J. C **40** (2005) 97.

[108] P. C. Bruns and U.-G. Meißner, "Infrared regularization with vector mesons and baryons," Eur. Phys. J. C **58** (2008) 407.

[109] M. R. Schindler, J. Gegelia and S. Scherer, "Infrared and extended on mass shell renormalization of two loop diagrams," Nucl. Phys. B **682** (2004) 367.

[110] T. Fuchs, J. Gegelia, G. Japaridze and S. Scherer, "Renormalization of relativistic baryon chiral perturbation theory and power counting," Phys. Rev. D **68** (2003) 056005.

[111] J. Gegelia, G. S. Japaridze and K. S. Turashvili, "Calculation of loop integrals by dimensional counting," Theor. Math. Phys. **101** (1994) 1313 [Teor. Mat. Fiz. **101** (1994) 225].

[112] J. Gegelia and G. Japaridze, "Matching heavy particle approach to relativistic theory," Phys. Rev. D **60** (1999) 114038.

[113] M. R. Schindler, J. Gegelia and S. Scherer, "Infrared regularization of baryon chiral perturbation theory reformulated," Phys. Lett. B **586** (2004) 258.

[114] M. L. Du, F. K. Guo and U.-G. Meißner, "One-loop renormalization of the chiral Lagrangian for spinless matter fields in the SU(N) fundamental representation," J. Phys. G **44** (2017) 014001.

[115] M. L. Du, F. K. Guo and U.-G. Meißner, "Subtraction of power counting breaking terms in chiral perturbation theory: spinless matter fields," JHEP **1610** (2016) 122.

[116] D. Djukanovic, M. R. Schindler, J. Gegelia, G. Japaridze and S. Scherer, "Universality of the rho-meson coupling in effective field theory," Phys. Rev. Lett. **93** (2004) 122002.

[117] C. Hacker, N. Wies, J. Gegelia and S. Scherer, "Including the $\Delta(1232)$ resonance in baryon chiral perturbation theory," Phys. Rev. C **72** (2005) 055203.

[118] D. Djukanovic, J. Gegelia and S. Scherer, "Chiral structure of the Roper resonance using complex-mass scheme," Phys. Lett. B **690** (2010) 123.

[119] S. Steininger, U.-G. Meißner and N. Fettes, "On wave function renormalization and related aspects in heavy fermion effective field theories," JHEP **9809** (1998) 008.

[120] J. Kambor and M. Mojžiš, "Field redefinitions and wave function renormalization to $O(p^4)$ in heavy baryon chiral perturbation theory," JHEP **9904** (1999) 031.

[121] M. Hoferichter, J. Ruiz de Elvira, B. Kubis and U.-G. Meißner, "High-precision determination of the pion-nucleon σ term from Roy-Steiner equations," Phys. Rev. Lett. **115** (2015) 092301.

[122] J. A. McGovern and M. C. Birse, "On the absence of fifth order contributions to the nucleon mass in heavy baryon chiral perturbation theory," Phys. Lett. B **446** (1999) 300.

[123] M. R. Schindler, D. Djukanovic, J. Gegelia and S. Scherer, "Chiral expansion of the nucleon mass to $O(q^6)$," Phys. Lett. B **649** (2007) 390.

[124] M. Hoferichter, J. Ruiz de Elvira, B. Kubis and U.-G. Meißner, "Roy–Steiner-equation analysis of pion–nucleon scattering," Phys. Rept. **625** (2016) 1.

[125] A. Walker-Loud, "Nuclear physics review," PoS LATTICE **2013** (2014) 013.

[126] A. Krause, "Baryon matrix elements of the vector current in chiral perturbation theory," Helv. Phys. Acta **63** (1990) 3.

[127] M. Gell–Mann, "Symmetries of baryons and mesons," Phys. Rev. **125** (1962) 1067.

[128] S. Okubo, "Note on unitary symmetry in strong interactions," Prog. Theor. Phys. **27** (1962) 949.

[129] E. E. Jenkins and A. V. Manohar, "The sigma term and $m_s^{3/2}$ corrections to the proton mass," Phys. Lett. B **281** (1992) 336.

[130] V. Bernard, N. Kaiser and U.-G. Meißner, "Critical analysis of baryon masses and sigma terms in heavy baryon chiral perturbation theory," Z. Phys. C **60** (1993) 111.

[131] B. Borasoy and U.-G. Meißner, "Chiral expansion of baryon masses and σ-terms," Annals Phys. **254** (1997) 192.

[132] J. Martin Camalich, L. S. Geng and M. J. Vicente Vacas, "The lowest-lying baryon masses in covariant $SU(3)$-flavor chiral perturbation theory," Phys. Rev. D **82** (2010) 074504.

[133] X.-L. Ren, L. S. Geng, J. Martin Camalich, J. Meng and H. Toki, "Octet baryon masses in next-to-next-to-next-to-leading order covariant baryon chiral perturbation theory," JHEP **1212** (2012) 073.

[134] M. F. M. Lutz, Y. Heo and X. Y. Guo, "On the convergence of the chiral expansion for the baryon ground-state masses," Nucl. Phys. A **977** (2018) 146.

[135] D. Severt, U.-G. Meißner and J. Gegelia, "Flavor decomposition of the pion-nucleon σ-term," JHEP **1903** (2019) 202.

[136] R. F. Dashen, E. E. Jenkins and A. V. Manohar, "The $1/N_c$ expansion for baryons," Phys. Rev. D **49** (1994) 4713 Erratum: [Phys. Rev. D **51** (1995) 2489].

[137] A. Bottino, F. Donato, N. Fornengo and S. Scopel, "Implications for relic neutralinos of the theoretical uncertainties in the neutralino nucleon cross-section," Astropart. Phys. **13** (2000) 215.

[138] J. R. Ellis, K. A. Olive and C. Savage, "Hadronic uncertainties in the elastic scattering of supersymmetric dark matter," Phys. Rev. D **77** (2008) 065026.

[139] A. Crivellin, M. Hoferichter and M. Procura, "Accurate evaluation of hadronic uncertainties in spin-independent WIMP-nucleon scattering: disentangling two- and three-flavor effects," Phys. Rev. D **89** (2014) 054021.

[140] A. Crivellin, M. Hoferichter and M. Procura, "Improved predictions for $\mu \to e$ conversion in nuclei and Higgs-induced lepton flavor violation," Phys. Rev. D **89** (2014) 093024.

[141] J. de Vries, E. Mereghetti, C. Y. Seng and A. Walker-Loud, "Lattice QCD spectroscopy for hadronic CP violation," Phys. Lett. B **766** (2017) 254.

[142] H. Hellmann, "Einführung in die Quantenchemie," Deuticke Verlag (1937).

[143] R. P. Feynman, "Forces in molecules," Phys. Rev. **56** (1939) 340.

[144] R. Tarrach, "The renormalization of FF," Nucl. Phys. B **196** (1982) 45.

[145] J. Gasser, H. Leutwyler and M. E. Sainio, "Sigma term update," Phys. Lett. B **253** (1991) 252.

[146] R. Koch and E. Pietarinen, "Low-energy πN partial wave analysis," Nucl. Phys. A **336** (1980) 331.

[147] M. M. Pavan, I. I. Strakovsky, R. L. Workman and R. A. Arndt, "The pion-nucleon σ-term is definitely large: Results from a G.W.U. analysis of πN scattering data," PiN Newslett. **16** (2002) 110. [arXiv:hep-ph/0111066 [hep-ph]].

[148] J. Ruiz de Elvira, M. Hoferichter, B. Kubis and U.-G. Meißner, "Extracting the σ-term from low-energy pion-nucleon scattering," J. Phys. G **45** (2018) 024001.

[149] R. Gupta et al., "The nucleon sigma term from lattice QCD" [arXiv:2105.12095 [hep-lat]].

[150] M. Hoferichter, J. Ruiz de Elvira, B. Kubis and U.-G. Meißner, "Remarks on the pion–nucleon σ-term," Phys. Lett. B **760** (2016) 74.

[151] C. Alexandrou, S. Bacchio, M. Constantinou et al. "The nucleon axial, tensor and scalar charges and σ-terms in lattice QCD," Phys. Rev. D **102** (2020) 054517.

[152] P. Junnarkar and A. Walker-Loud, "Scalar strange content of the nucleon from lattice QCD," Phys. Rev. D **87** (2013) 114510.

[153] S. Aoki et al. [Flavour Lattice Averaging Group], "FLAG Review 2019," Eur. Phys. J. C **80** (2020) 113.

[154] C. Ditsche, M. Hoferichter, B. Kubis and U.-G. Meißner, "Roy-Steiner equations for pion-nucleon scattering," JHEP **1206** (2012) 043.

[155] M. L. Goldberger and S. B. Treiman, "Conserved currents in the theory of Fermi interactions," Phys. Rev. **110** (1958) 1478.

[156] V. Baru, C. Hanhart, M. Hoferichter et al., "Precision calculation of threshold $\pi^- d$ scattering, πN scattering lengths, and the GMO sum rule," Nucl. Phys. A **872** (2011) 69.

[157] T. P. Cheng and R. F. Dashen, "Is $SU(2) \times SU(2)$ a better symmetry than $SU(3)$?," Phys. Rev. Lett. **26** (1971) 594.

[158] L. S. Brown, W. J. Pardee and R. D. Peccei, "Adler-Weisberger theorem reexamined," Phys. Rev. D **4** (1971) 2801.

[159] V. Bernard, N. Kaiser and U.-G. Meißner, "On the analysis of the pion - nucleon sigma term: the size of the remainder at the Cheng-Dashen point," Phys. Lett. B **389** (1996) 144.

[160] J. Gasser, H. Leutwyler and M. E. Sainio, "Form-factor of the sigma term," Phys. Lett. B **253** (1991) 260.

[161] M. Hoferichter, C. Ditsche, B. Kubis and U.-G. Meißner, "Dispersive analysis of the scalar form factor of the nucleon," JHEP **1206** (2012) 063.

[162] J. Baacke and F. Steiner, "πn partial wave relations from fixed-t dispersion relations," Fortsch. Phys. **18** (1970) 67.

[163] F. Steiner, "On the generalized πn potential - a new representation from fixed-t dispersion relations," Fortsch. Phys. **18** (1970) 43.

[164] F. Steiner, "Partial wave crossing relations for meson-baryon scattering," Fortsch. Phys. **19** (1971) 115.

[165] S. M. Roy, "Exact integral equation for pion pion scattering involving only physical region partial waves," Phys. Lett. B **36** (1971) 353.

[166] G. E. Hite and F. Steiner, " New dispersion relations and their application to partial-wave amplitudes," Nuovo Cim. A **18** (1973) 237.

[167] D. Gotta et al., "Pionic hydrogen," Lect. Notes Phys. **745** (2008) 165.

[168] T. Strauch et al., "Pionic deuterium," Eur. Phys. J. A **47** (2011) 88.

[169] M. Hennebach et al., "Hadronic shift in pionic hydrogen," Eur. Phys. J. A **50** (2014) 190.

[170] A. Hirtl, D. F. Anagnostopoulos, D. S. Covita et al. "Redetermination of the strong-interaction width in pionic hydrogen," Eur. Phys. J. A **57** (2021) 70.

[171] C. Hanhart, S. Holz, B. Kubis et al. "The branching ratio $\omega \to \pi^+\pi^-$ revisited," Eur. Phys. J. C **77** (2017) 98 Erratum: [Eur. Phys. J. C **78** (2018) 450].

[172] J. Bijnens and N. Hermansson Truedsson, "The pion mass and decay constant at three loops in two-flavour Chiral Perturbation Theory," JHEP **1711** (2017) 181.

[173] E. Epelbaum, H. Krebs and U.-G. Meißner, "Improved chiral nucleon-nucleon potential up to next-to-next-to-next-to-leading order," Eur. Phys. J. A **51** (2015) 53.

[174] R. J. Furnstahl, N. Klco, D. R. Phillips and S. Wesolowski, "Quantifying truncation errors in effective field theory," Phys. Rev. C **92** (2015) 024005.

[175] J. A. Melendez, S. Wesolowski and R. J. Furnstahl, "Bayesian truncation errors in chiral effective field theory: nucleon-nucleon observables," Phys. Rev. C **96** (2017) 024003.

[176] E. Epelbaum, "High-precision nuclear forces: where do we stand?," PoS CD **2018** (2019) 006.

[177] D. J. C. MacKay, "Information theory, inference, and learning algorithms," Cambridge University Press (1995).

[178] D. S. Sivia, "Data analysis: A Bayesian tutorial," Oxford University Press (1996).

[179] M. R. Schindler and D. R. Phillips, "Bayesian methods for parameter estimation in effective field theories," Annals Phys. **324** (2009) 682 Erratum: [Annals Phys. **324** (2009) 2051].

[180] S. Wesolowski, R. J. Furnstahl, J. A. Melendez and D. R. Phillips, "Exploring Bayesian parameter estimation for chiral effective field theory using nucleon–nucleon phase shifts," J. Phys. G **46** (2019) 045102.

[181] P. Reinert, H. Krebs and E. Epelbaum, "Semilocal momentum-space regularized chiral two-nucleon potentials up to fifth order," Eur. Phys. J. A **54** (2018) 86.

[182] J. Charles, S. Descotes-Genon, V. Niess and L. Vale Silva, "Modeling theoretical uncertainties in phenomenological analyses for particle physics," Eur. Phys. J. C **77** (2017) 214.

[183] F. E. Low, "Scattering of light of very low frequency by systems of spin $1/2$," Phys. Rev. **96** (1954) 1428.

[184] M. Gell–Mann and M. L. Goldberger, "Scattering of low-energy photons by particles of spin $1/2$," Phys. Rev. **96** (1954) 1433.

[185] V. de Alfaro, S. Fubini, G. Furlan and C. Rossetti, "Currents in hadron physics," North-Holland Publishing Company (1973).

[186] J. F. Donoghue, J. Gasser and H. Leutwyler, "The decay of a light Higgs boson," Nucl. Phys. B **343** (1990) 341.

[187] F. A. Berends, A. Donnachie and D. L. Weaver, "Photoproduction and electro-production of pions. 1. Dispersion relation theory," Nucl. Phys. B **4** (1967) 1.

[188] P. De Baenst, "An improvement on the Kroll-Ruderman theorem," Nucl. Phys. B **24** (1970) 633.

[189] A. I. Vainshtein and V. I. Zakharov, "Low-energy theorems for photoproduction and electropion production at threshold," Nucl. Phys. B **36** (1972) 589.

[190] V. Bernard, N. Kaiser and U.-G. Meißner, "Neutral pion photoproduction off nucleons revisited," Z. Phys. C **70** (1996) 483.

[191] V. Bernard, B. Kubis and U.-G. Meißner, "The Fubini-Furlan-Rosetti sum rule and related aspects in light of covariant baryon chiral perturbation theory," Eur. Phys. J. A **25** (2005) 419. [nucl-th/0506023].

[192] M. Kobayashi and T. Maskawa, "*CP* violation in the renormalizable theory of weak interaction," Prog. Theor. Phys. **49** (1973) 652.

[193] N. Cabibbo, "Unitary symmetry and leptonic decays," Phys. Rev. Lett. **10** (1963) 531.

[194] A. Sirlin, "Current algebra formulation of radiative corrections in gauge theories and the universality of the weak interactions," Rev. Mod. Phys. **50** (1978) 573 [Erratum *ibid* **50** (1978) 905].

[195] W. J. Marciano and A. Sirlin, "Radiative corrections to π_{l_2} decays," Phys. Rev. Lett. **71** (1993) 3629.

[196] M. Finkemeier, "Radiative corrections to π_{l_2} and K_{l_2} decays," Phys. Lett. B **387** (1996) 391.

[197] A. Sommerfeld, "Atmobau und Spektralinien," Friedrich Vieweg & Sohn (1921).

[198] G. Gamow, "Zur Quantentheorie des Atomkernes," Z. Phys. **51** (1928) 204.

[199] A. D. Sakharov, "Interaction of an electron and positron in pair production," Sov. Phys. Usp. **34** (1991) 375.

[200] R. Urech, "Virtual photons in chiral perturbation theory," Nucl. Phys. B **433** (1995) 234.

[201] U.-G. Meißner, G. Müller and S. Steininger, "Virtual photons in $SU(2)$ chiral perturbation theory and electromagnetic corrections to $\pi\pi$ scattering," Phys. Lett. B **406** (1997) 154 Erratum: [Phys. Lett. B **407** (1997) 454].

[202] J. Bijnens, "Violations of Dashen's theorem," Phys. Lett. B **306** (1993) 343.

[203] J. F. Donoghue, B. R. Holstein and D. Wyler, "Electromagnetic selfenergies of pseudoscalar mesons and Dashen's theorem," Phys. Rev. D **47** (1993) 2089.

[204] T. Das, G. Guralnik, V. Mathur, F. Low and J. Young, "Electromagnetic mass difference of pions," Phys. Rev. Lett. **18** (1967) 759.

[205] R. Baur and R. Urech, "On the corrections to Dashen's theorem," Phys. Rev. D **53** (1996) 6552.

[206] B. Moussallam, "A sum rule approach to the violation of Dashen's theorem," Nucl. Phys. B **504** (1997) 381.

[207] J. Bijnens and J. Prades, "Electromagnetic corrections for pions and kaons: Masses and polarizabilities," Nucl. Phys. B **490** (1997) 239.

[208] J. Gasser, V. E. Lyubovitskij, A. Rusetsky and A. Gall, "Decays of the $\pi^+\pi^-$ atom," Phys. Rev. D **64** (2001) 016008.

[209] V. Cirigliano and H. Neufeld, "A note on isospin violation in $\pi_{l_2\gamma}$ decays," Phys. Lett. B **700** (2011) 7.

[210] J. Gasser and G. Zarnauskas, "On the pion decay constant," Phys. Lett. B **693** (2010) 122.

[211] F. Bloch and A. Nordsieck, "Note on the radiation field of the electron," Phys. Rev. **52** (1937) 54.

[212] J. C. Collins, "Renormalization," Cambridge University Press (1984).

[213] F. Yndurain, "The theory of quark and gluon interactions," Springer (2006).

[214] T. Muta, "Foundations of quantum chromodynamics: an introduction to perturbative methods in gauge theories," 3rd ed., World Scientific Lecture Notes in Physics, Band 78 (2008).

[215] U.-G. Meißner and S. Steininger, "Isospin violation in pion nucleon scattering," Phys. Lett. B **419** (1998) 403.

[216] W. Cottingham, "The neutron proton mass difference and electron scattering experiments," Annals Phys. **25** (1963) 424.

[217] J. Gasser, H. Leutwyler and A. Rusetsky, "On the mass difference between proton and neutron," Phys. Lett. B **814** (2021) 136087.

[218] J. Gasser, H. Leutwyler and A. Rusetsky, "Sum rule for the Compton amplitude and implications for the proton–neutron mass difference," Eur. Phys. J. C **80** (2020) 1121.

[219] J. Gasser and H. Leutwyler, "Implications of scaling for the proton-neutron mass difference," Nucl. Phys. B **94** (1975) 269.

[220] A. Walker-Loud, C. E. Carlson and G. A. Miller, "The electromagnetic self-energy contribution to $M_p - M_n$ and the isovector nucleon magnetic polarizability," Phys. Rev. Lett. **108** (2012) 232301.

[221] F. Erben, P. Shanahan, A. Thomas and R. Young, "Dispersive estimate of the electromagnetic charge symmetry violation in the octet baryon masses," Phys. Rev. C **90** (2014) 065205.

[222] J. Gasser, M. Hoferichter, H. Leutwyler and A. Rusetsky, "Cottingham formula and nucleon polarisabilities," Eur. Phys. J. C **75** (2015) 375.

[223] M. Hoferichter, B. Kubis and U.-G. Meißner, "Isospin violation in low-energy pion-nucleon scattering revisited," Nucl. Phys. A **833** (2010) 18.

[224] D. Yennie, S. C. Frautschi and H. Suura, "The infrared divergence phenomena and high-energy processes," Annals Phys. **13** (1961) 379.

[225] H. Bethe, "Theory of the effective range in nuclear scattering," Phys. Rev. **76** (1949) 38.

[226] N. Fettes, U.-G. Meißner and S. Steininger, "On the size of isospin violation in low-energy pion - nucleon scattering," Phys. Lett. B **451** (1999) 233.

[227] N. Fettes and U.-G. Meißner, "Towards an understanding of isospin violation in pion nucleon scattering," Phys. Rev. C **63** (2001) 045201.

[228] N. Fettes and U.-G. Meißner, "Complete analysis of pion nucleon scattering in chiral perturbation theory to third order," Nucl. Phys. A **693** (2001) 693.

[229] M. Hoferichter, B. Kubis and U.-G. Meißner, "Isospin breaking in the pion-nucleon scattering lengths," Phys. Lett. B **678** (2009) 65.

[230] F. K. Guo, X. H. Liu and S. Sakai, "Threshold cusps and triangle singularities in hadronic reactions," Prog. Part. Nucl. Phys. **112** (2020) 103757.

[231] A. Schmidt et al., "Test of low-energy theorems for $^1H(\vec{\gamma}, \pi^0)^1H$ in the threshold region," Phys. Rev. Lett. **87** (2001) 232501 Erratum: [Phys. Rev. Lett. **110** (2013) 039903].

[232] J. Gasser, V. Lyubovitskij and A. Rusetsky, "Hadronic atoms in QCD + QED," Phys. Rept. **456** (2008) 167.

[233] J. Wess and B. Zumino, "Consequences of anomalous Ward identities," Phys. Lett. B **37** (1971) 95.

[234] E. Witten, "Global aspects of current algebra," Nucl. Phys. B **223** (1983) 422.

[235] N. K. Pak and P. Rossi, "Gauged Goldstone boson effective action from direct integration of Bardeen anomaly," Nucl. Phys. B **250** (1985) 279.

[236] J. Manes, "Differential geometric construction of the gauged Wess-Zumino action," Nucl. Phys. B **250** (1985) 369.

[237] R. Kaiser and H. Leutwyler, "Large N_c in chiral perturbation theory," Eur. Phys. J. C **17** (2000) 623.

[238] R. Kaiser, "Anomalies and WZW term of two flavor QCD," Phys. Rev. D **63** (2001) 076010.

[239] U.-G. Meißner and I. Zahed, "Skyrmions in nuclear physics," Adv. Nucl. Phys. **17** (1986) 143.

[240] J. F. Donoghue and D. Wyler, "One loop renormalization of the Wess-Zumino-Witten anomaly Lagrangian," Nucl. Phys. B **316** (1989) 289.

[241] J. Bijnens, "The anomalous sector of the QCD effective Lagrangian," Nucl. Phys. B **367** (1991) 709.

[242] R. Akhoury and A. Alfakih, "Invariant background field method for chiral Lagrangians including Wess-Zumino terms," Annals Phys. **210** (1991) 81.

[243] P. A. M. Dirac, "Quantised singularities in the electromagnetic field," Proc. Roy. Soc. Lond. A **A133** (1931) 60.

[244] E. Witten, "Current algebra, baryons, and quark confinement," Nucl. Phys. B **223** (1983) 433.

[245] G. 't Hooft, "A two-dimensional model for mesons," Nucl. Phys. B **75** (1974) 461.

[246] G. 't Hooft, "A planar diagram theory for strong interactions," Nucl. Phys. B **72** (1974) 461.

[247] E. Witten, "Current algebra theorems for the $U(1)$ Goldstone boson," Nucl. Phys. B **156** (1979) 269.

[248] E. Witten, "Baryons in the $1/n$ expansion," Nucl. Phys. B **160** (1979) 57.

[249] T. Skyrme, "A nonlinear field theory," Proc. Roy. Soc. Lond. A **A260** (1961) 127.

[250] S. R. Coleman and E. Witten, "Chiral symmetry breakdown in large N chromodynamics," Phys. Rev. Lett. **45** (1980) 100.

[251] P. Herrera-Siklody, J. Latorre, P. Pascual and J. Taron, "Chiral effective Lagrangian in the large N_c limit: the Nonet case," Nucl. Phys. B **497** (1997) 345.

[252] H. Leutwyler, "Bounds on the light quark masses," Phys. Lett. B **374** (1996), 163–168.

[253] P. Herrera-Siklody, "Matching of $U_L(3) \times U_R(3)$ and $SU_L(3) \times SU_R(3)$ chiral perturbation theories," Phys. Lett. B **442** (1998) 359.

[254] T. Feldmann, "Quark structure of pseudoscalar mesons," Int. J. Mod. Phys. A **15** (2000) 159.

[255] H. Leutwyler and A. V. Smilga, "Spectrum of Dirac operator and role of winding number in QCD," Phys. Rev. D **46** (1992) 5607.

[256] S. Peris and E. de Rafael, "On the large N_c behavior of the L_7 coupling in χPT," Phys. Lett. B **348** (1995) 539.

[257] H. Leutwyler, "On the $1/N$ expansion in chiral perturbation theory," Nucl. Phys. B Proc. Suppl. **64** (1998) 223.

[258] R. F. Dashen, E. E. Jenkins and A. V. Manohar, "Spin flavor structure of large N_c baryons," Phys. Rev. D **51** (1995) 3697.

[259] E. E. Jenkins, "Chiral Lagrangian for baryons in the $1/n_c$ expansion," Phys. Rev. D **53** (1996) 2625.

[260] D. B. Kaplan and A. V. Manohar, "Current mass ratios of the light quarks," Phys. Rev. Lett. **56** (1986) 2004.

[261] V. Baluni, "CP violating effects in QCD," Phys. Rev. D **19** (1979) 2227.

[262] R. Crewther, P. Di Vecchia, G. Veneziano and E. Witten, "Chiral estimate of the electric dipole moment of the neutron in quantum chromodynamics," Phys. Lett. B **88** (1979) 123.

[263] A. Pich and E. de Rafael, "Strong CP violation in an effective chiral Lagrangian approach," Nucl. Phys. B **367** (1991) 313.

[264] B. Borasoy, "The electric dipole moment of the neutron in chiral perturbation theory," Phys. Rev. D **61** (2000) 114017.

[265] F. K. Guo and U.-G. Meißner, "Baryon electric dipole moments from strong CP violation," JHEP **12** (2012) 097.

[266] J. de Vries, E. Mereghetti, R. Timmermans and U. van Kolck, "The effective chiral Lagrangian from dimension-six parity and time-reversal violation," Annals Phys. **338** (2013) 50.

[267] J. Bsaisou, U.-G. Meißner, A. Nogga and A. Wirzba, "P- and T-violating Lagrangians in chiral effective field theory and nuclear electric dipole moments," Annals Phys. **359** (2015) 317.

[268] S. R. Sharpe and N. Shoresh, "Physical results from unphysical simulations," Phys. Rev. D **62** (2000) 094503.

[269] M. Della Morte and A. Jüttner, "Quark disconnected diagrams in chiral perturbation theory," JHEP **11** (2010) 154.

[270] S. R. Sharpe and N. Shoresh, "Partially quenched chiral perturbation theory without Φ_0," Phys. Rev. D **64** (2001) 114510.

[271] S. R. Sharpe, "Enhanced chiral logarithms in partially quenched QCD," Phys. Rev. D **56** (1997) 7052.

[272] M. F. L. Golterman and K. C. Leung, "Applications of partially quenched chiral perturbation theory," Phys. Rev. D **57** (1998) 5703.

[273] P. F. Bedaque, "Aharonov-Bohm effect and nucleon nucleon phase shifts on the lattice," Phys. Lett. B **593** (2004) 82.

[274] G.M. d. Divitiis, R. Petronzio and N. Tantalo, "On the discretization of physical momenta in lattice QCD," Phys. Lett. B **595** (2004) 408.

[275] C. T. Sachrajda and G. Villadoro, "Twisted boundary conditions in lattice simulations," Phys. Lett. B **609** (2005) 73.

[276] J. von Neumann, "Invariant measures," American Mathematical Society (1999).

[277] N. Bourbaki, "Elements of mathematics. Integration," Addison-Wesley (1975).

[278] R. Rajaraman, "Solitons and instantons: an introduction to solitons and instantons in quantum field theory," North Holland Publishing Company (1982).

[279] S. Coleman, "Aspects of symmetry," Cambridge University Press (1985).

[280] C. Abel et al. [nEDM], "Measurement of the permanent electric dipole moment of the neutron," Phys. Rev. Lett. **124** (2020) 081803.

[281] A. V. Smilga, "QCD at $\theta \sim \pi$," Phys. Rev. D **59** (1999) 114021.

[282] T. Vonk, F. K. Guo and U.-G. Meißner, "Aspects of the QCD θ-vacuum," JHEP **06** (2019) 106.

[283] R. F. Dashen, "Some features of chiral symmetry breaking," Phys. Rev. D **3** (1971) 1879.

[284] P. Di Vecchia and G. Veneziano, "Chiral dynamics in the large n limit," Nucl. Phys. B **171** (1980) 253.

[285] R. D. Peccei and H. R. Quinn, "*CP* conservation in the presence of instantons," Phys. Rev. Lett. **38** (1977) 1440.

[286] R. D. Peccei and H. R. Quinn, "Constraints imposed by *CP* conservation in the presence of instantons," Phys. Rev. D **16** (1977) 1791.

[287] C. Vafa and E. Witten, "Restrictions on symmetry breaking in vector-like gauge theories," Nucl. Phys. B **234** (1984) 173.

[288] S. Weinberg, "A new light boson?," Phys. Rev. Lett. **40** (1978) 223.

[289] F. Wilczek, "Problem of strong *P* and *T* invariance in the presence of instantons," Phys. Rev. Lett. **40** (1978) 279.

[290] J. E. Kim, "Weak interaction singlet and strong *CP* invariance," Phys. Rev. Lett. **43** (1979) 103.

[291] M. A. Shifman, A. I. Vainshtein and V. I. Zakharov, "Can confinement ensure natural *CP* invariance of strong interactions?," Nucl. Phys. B **166** (1980) 493.

[292] M. Dine, W. Fischler and M. Srednicki, "A simple solution to the strong *CP* problem with a harmless axion," Phys. Lett. B **104** (1981) 199.

[293] A. R. Zhitnitsky, "On possible suppression of the axion hadron interactions. (In Russian)," Sov. J. Nucl. Phys. **31** (1980) 260.

[294] J. E. Kim, "Light pseudoscalars, particle physics and cosmology," Phys. Rept. **150** (1987) 1.

[295] Z. Y. Lu, M. L. Du, F. K. Guo, U.-G. Meißner and T. Vonk, "QCD θ-vacuum energy and axion properties," JHEP **05** (2020) 001.

[296] K. Ottnad, B. Kubis, U.-G. Meißner and F. K. Guo, "New insights into the neutron electric dipole moment," Phys. Lett. B **687** (2010) 42.

[297] F. J. Dyson, "The *S*-matrix in quantum electrodynamics," Phys. Rev. **75** (1949) 1736.

[298] J. S. Schwinger, "On the Green's functions of quantized fields $1+2$," Proc. Nat. Acad. Sci. **37** (1951) 452.

[299] E. E. Salpeter and H. A. Bethe, "A relativistic equation for bound state problems," Phys. Rev. **84** (1951) 1232.

[300] C. D. Roberts and A. G. Williams, "Dyson-Schwinger equations and their application to hadronic physics," Prog. Part. Nucl. Phys. **33** (1994) 477.

[301] C. Itzykson and J. B. Zuber, "Quantum Field Theory," McGraw-Hill (1980) 705 pp. (International Series in Pure and Applied Physics).

[302] G. A. Christos, "Effective chiral Lagrangians with baryons: the mass splitting of the spin 1/2 parity partners," Z. Phys. C **21** (1983) 83.

[303] G. A. Christos, "Effective chiral Lagrangian with baryons," Phys. Rev. D **35** (1987) 330.

[304] C. E. Detar and T. Kunihiro, "Linear σ model with parity doubling," Phys. Rev. D **39** (1989) 2805.

[305] D. Jido, M. Oka and A. Hosaka, "Chiral symmetry of baryons," Prog. Theor. Phys. **106** (2001) 873.

[306] G. V. Efimov, M. A. Ivanov and V. E. Lyubovitskij, "Strong nucleon and Δ-isobar form factors in quark confinement model," Few Body Syst. **6** (1989) 17 [Acta Phys. Austriaca **6** (1989) no. 1, 17].

[307] S. Scherer and M. R. Schindler, "A primer for Chiral Perturbation Theory," Lect. Notes Phys. **830** (2012) 1.

[308] E. Epelbaum, H. W. Hammer and U.-G. Meißner, "Modern theory of nuclear forces," Rev. Mod. Phys. **81** (2009) 1773.

[309] H. W. Hammer, S. König and U. van Kolck, "Nuclear effective field theory: status and perspectives," Rev. Mod. Phys. **92** (2020) 025004.

[310] G. 't Hooft, "Symmetry breaking through Bell-Jackiw anomalies," Phys. Rev. Lett. **37** (1976) 8.

[311] G. 't Hooft, "Computation of the quantum effects due to a four-dimensional pseudoparticle," Phys. Rev. D **14** (1976) 3432.

[312] A. Morel, "Chiral logarithms in quenched QCD," J. Phys. (France) **48** (1987) 1111.

[313] C. W. Bernard and M. F. L. Golterman, "Partially quenched gauge theories and an application to staggered fermions," Phys. Rev. D **49** (1994) 486.

[314] P. H. Damgaard and K. Splittorff, "Partially quenched chiral perturbation theory and the replica method," Phys. Rev. D **62** (2000) 054509.

[315] P. H. Damgaard, "Partially quenched chiral condensates from the replica method," Phys. Lett. B **476** (2000) 465.

[316] N. R. Acharya, F. K. Guo, U.-G. Meißner and C. Y. Seng, "Connected and disconnected contractions in pion–pion scattering," Nucl. Phys. B **922** (2017) 480.

[317] F. K. Guo and C. Y. Seng, "Effective field theory in the study of long range nuclear parity violation on lattice," Eur. Phys. J. C **79** (2019) 22.

[318] N. R. Acharya, F. K. Guo, U.-G. Meißner and C. Y. Seng, "Constraints on disconnected contributions in $\pi\pi$ scattering," JHEP **04** (2019) 165.

[319] C. Kalmahalli Guruswamy, U.-G. Meißner and C. Y. Seng, "Contraction diagram analysis in pion-kaon scattering," Nucl. Phys. B **957** (2020) 115091.

[320] R. Jackiw and C. Rebbi, "Vacuum periodicity in a Yang-Mills quantum theory," Phys. Rev. Lett. **37** (1976) 172.

[321] C. G. Callan, Jr., R. F. Dashen and D. J. Gross, "The structure of the gauge theory vacuum," Phys. Lett. B **63** (1976) 334.

[322] V. A. Rubakov, "Classical theory of gauge fields," Princeton University (2002) 444 pp.

[323] A. V. Smilga, "Lectures on quantum chromodynamics," World Scientific (2001).

[324] S. R. Coleman, "The uses of instantons," Subnucl. Ser. **15** (1979) 805.

[325] E. Seiler and I. O. Stamatescu, "Lattice fermions and θ vacua," Phys. Rev. D **25** (1982) 2177.

[326] R. D. Peccei, "QCD, strong CP and axions," J. Korean Phys. Soc. **29** (1996) S199.

[327] G. Gabadadze and M. Shifman, "QCD vacuum and axions: what's happening?," Int. J. Mod. Phys. A **17** (2002) 3689.

[328] R. D. Peccei, "The strong *CP* problem and axions," Lect. Notes Phys. **741** (2008) 3.

[329] P. Sikivie, "Axion cosmology," Lect. Notes Phys. **741** (2008) 19.

[330] G. G. Raffelt, "Astrophysical axion bounds," Lect. Notes Phys. **741** (2008) 51.

[331] J. Bernabeu and R. Tarrach, "Long range potentials and the electromagnetic polarizabilities," Annals Phys. **102** (1976) 323.

Effective Theories in a Finite Volume

5.1 Introduction

Calculations on the lattice are always done at a finite lattice spacing a and in a finite volume V, as well as (often) with quark masses larger than the physical ones. The artifacts that arise from all these three sources can be treated with the help of different EFTs. For example, the effects emerging due to a finite lattice spacing are addressed using the Symanzik effective Lagrangian that explicitly includes the terms responsible for scaling violation, the so-called "improvement" [1, 2]. The results of simulations with unphysical quark masses can be extrapolated to the physical point by using chiral effective Lagrangians that were considered in Chapter 4. This chapter is exclusively devoted to the artifacts of the third type, which emerge when the system is confined to a finite box. Only at the very end, we briefly consider one example of using the Symanzik Lagrangian to study phase transitions in QCD at a finite lattice spacing.

Consider, for simplicity, a Euclidean lattice with a time elongation L_t and space elongation L (i.e., we take a cubic lattice in three spatial directions). So the volume is $V = L \times L \times L \times L_t$. Assume further that the time direction is taken very large (tending to infinity). The energy levels and various matrix elements that can be extracted on such a lattice still depend on L, and the limit $L \to \infty$ should be performed properly. A fundamental property of the finite volume effects is that their characteristic scale, given by $1/L$, is very different from the hadronic scale. The latter is of order of $\Lambda \simeq 1\,\mathrm{GeV}$. Moreover, the effective ultraviolet cutoff that is determined by the inverse lattice spacing should obey the condition $1/a \gg \Lambda$. This means that finite-volume artifacts can be treated independently from the others. For instance, considering finite-volume effects, we can work in the continuum, letting $a \to 0$. In case of QCD, the existence of disparate scales has further implications. As we know, ChPT is an EFT of QCD at the momenta below the hadronic scale. Since $1/L \ll \Lambda$, the effective Lagrangians of QCD in the infinite and in a finite volume *are the same*. Indeed, the low-energy constants in the Lagrangian, which encode the dynamics at the scale or order of Λ and higher, cannot be, by definition, affected by the infrared phenomena at the scales of order of $1/L$. Thus, the dependence on L emerges solely *through the loops produced by this Lagrangian*.

This is the key property that enables one to systematically use EFT methods to study the finite-volume effects in lattice QCD. The effective Lagrangian is a bridge connecting the finite and the infinite volume. In some cases, it suffices to merely assume that such effective Lagrangian exists. The actual form of this Lagrangian and the exact values of the low-energy constants do not matter; the connection between infinite and finite volume can be established directly for observables and is valid to all orders. In this chapter, we shall see examples of such relations. Furthermore, by the same token, the ChPT Lagrangian can be used to study the quark mass dependence of various observables in a finite volume, with the same values of the low-energy constants as in the infinite volume. Turning the argument around, the study of the quark mass dependence allows us to determine the values of these LECs from lattice calculations, especially for the so-called symmetry breakers. These are often difficult to extract from experiments, as discussed in Chapter 4.

Last but not least, note that ChPT is not the only effective theory of QCD that is used to study the finite-volume effects on the lattice. If L is sufficiently large, $1/L \ll M_\pi$, the characteristic momenta of particles become much smaller than the lightest particle mass. In this case, antiparticle degrees of freedom freeze out and nonrelativistic EFT represents the most natural framework to work with. We shall extensively treat this case in what follows.

5.2 The Mass of a Stable Particle in a Finite Volume

5.2.1 Preliminary Remarks: Scales and Regimes

As we know from Eq. (3.240), the momenta on a lattice are quantized. Focusing on three spatial dimensions and assuming periodic boundary conditions, we see that the three-momentum of a free particle in a cubic box with size L takes the values:

$$\mathbf{p} = \frac{2\pi}{L}\mathbf{n}, \qquad \mathbf{n} \in \mathbb{Z}^3. \tag{5.1}$$

Thus, characteristic momenta on a lattice are of order $1/L$. On the other hand, the characteristic hadronic scale Λ can be identified with $4\pi F_\pi$, the nucleon or the ρ-meson masses, each of them being of the order of $1\,\mathrm{GeV}$. If the box is small, the product ΛL is of order of one, so individual hadrons are largely distorted by the presence of the box, and a clear separation of the short-range effects is no more possible. Thus, in simulations, the condition $\Lambda L \gg 1$ should be fulfilled.

In QCD, there exists a second scale, which is set by the Goldstone boson mass M_π (for simplicity, we consider the two-flavor case here). In the infinite volume, M_π/Λ is a small parameter, controlling the convergence of the chiral expansion. The quantity $M_\pi L$ can, however, be small as well as large, leading to different EFTs (or, different regimes). If this quantity is of order unity, we may use the standard chiral expansion to calculate the observables in a finite volume. The only difference is that the integral over three-momenta in any given Feynman diagram is replaced by a sum over the discrete values of three-momenta, defined by Eq. (5.1). We shall see examples of such calculations later in this section. As we will further see, the finite-volume artifacts in such a case will be exponentially suppressed by a factor $\exp(-L\Delta)$, where Δ is generally of order of the lightest mass in the system (the pion mass, in our case).[1] Thus, as already mentioned, if the quantity $M_\pi L$ is of order one, the finite-volume artifacts are caused solely by the pion loops. In this case, one speaks about the p-regime.

There is, however, one important exception to this picture. In particular, in the case of some physical observables, the quantity Δ vanishes for a certain class of Feynman diagrams. Namely, as will be seen in what follows, the quantity Δ is determined by the singularities of the integrand of a particular diagram in the complex plane. It may then happen that some integrands are singular in the physical region. For example, the scattering amplitude of two pions above the elastic threshold possesses a cut, produced by the bubble diagrams corresponding to the S-wave rescattering. The topology of these graphs is the same as shown in Fig. 2.4. All other topologies that we could draw, for example, the loops in the t-, u- channels and so on, are non-singular just above the physical threshold and hence correspond to a finite Δ, of order of the pion mass.[2] Consequently, choosing $M_\pi L \gg 1$, we may clearly separate bubble diagrams from the rest. The finite-volume artifacts are exponentially suppressed in the latter but only power-law suppressed in the former. It is also clear that the non-relativistic EFT, considered in Chapter 2, provides the most natural description of the system in a finite volume for sufficiently large values of L, when the exponential corrections can be neglected. We shall encounter this situation in the study of scattering processes.

The opposite situation emerges in the chiral limit, in which $M_\pi L$ is a small quantity. In this case, which will be referred to as the δ-regime hereafter, the standard ChPT techniques cannot be applied anymore. The reason for this is very easy to understand [3]. Expanding the lowest-order chiral Lagrangian in the pion fields, we get for the Euclidean action functional:

$$S = \frac{1}{2} \int dx_4 \int_{-L/2}^{+L/2} d^3\mathbf{x} \left(-(\partial_4 \boldsymbol{\phi}(\mathbf{x}, x_4))^2 - (\nabla \boldsymbol{\phi}(\mathbf{x}, x_4))^2 + M^2 (\boldsymbol{\phi}(\mathbf{x}, x_4))^2 \right)$$

$$+ \text{ higher-order terms}, \tag{5.2}$$

where $M^2 = 2\hat{m}B$. Furthermore, since the pion field is subject to periodic boundary conditions in the three spatial dimensions

$$\boldsymbol{\phi}(\mathbf{x} + \mathbf{n}L, x_4) = \boldsymbol{\phi}(\mathbf{x}, x_4), \qquad \mathbf{n} \in \mathbb{Z}^3, \tag{5.3}$$

it can be expanded in a Fourier series over the normal modes:

$$\boldsymbol{\phi}(\mathbf{x}, x_4) = \sum_{\mathbf{n}} \exp\left(\frac{2\pi i \mathbf{n} \mathbf{x}}{L} \right) \boldsymbol{\phi}_{\mathbf{n}}(x_4). \tag{5.4}$$

---

[1] We assume that the lightest mass is nonzero. This automatically excludes the case with QCD plus QED, which features massless photons. EFTs can be used in this case as well, and yield finite-volume artifacts that are suppressed by powers of L instead of exponentials. We shall consider the EFTs with photons later in this chapter.

[2] Figure 2.4 refers to a nonrelativistic EFT, whereas the present discussion is carried out within the standard relativistic QFT. We believe that this cannot lead to any confusion. The *only* class of diagrams that is singular in the vicinity of threshold in the scattering process is the one that emerges both in the NREFT and in the relativistic theory.

The action functional can be rewritten as

$$S = \frac{L^3}{2} \sum_{\mathbf{n}} \int dx_4 \left(-(\dot{\boldsymbol{\phi}}_{\mathbf{n}}(x_4))^2 + \omega_{\mathbf{n}}^2 (\boldsymbol{\phi}_{\mathbf{n}}(x_4))^2 \right) + \cdots , \tag{5.5}$$

where

$$\omega_{\mathbf{n}}^2 = M^2 + \left(\frac{2\pi \mathbf{n}}{L} \right)^2 . \tag{5.6}$$

It is now seen that in the chiral limit, $M^2 \to 0$, the variable corresponding to the zero mode, $\mathbf{n} \to 0$, cannot be treated perturbatively, because the pertinent integral does not contain an exponential damping factor, which emerges due to the term $\omega_{\mathbf{n}}^2 (\boldsymbol{\phi}_{\mathbf{n}}(x_4))^2$. On the contrary, the nonzero modes can be still treated perturbatively. Turning the argument around, one concludes that spontaneous chiral symmetry breaking cannot take place in a finite volume (see a detailed discussion later). If the limit $M^2 \to 0$ is performed at a fixed L, then at some point the pion Compton wavelength becomes larger than the box size L, and the finite-volume effects start to be dominant and non-perturbative. For instance, it can be demonstrated that, in accordance with our statement about the absence of the spontaneous symmetry breaking in a finite volume, the quark condensate vanishes in the limit $\hat{m} \to 0$, $L \neq 0$ (see, e.g., Ref. [4]).

If the time elongation of the box is also taken finite, more regimes are possible. Note that in thermodynamics the Euclidean elongation of the hypercubic lattice is given by the inverse temperature $\beta = L_t$, so that an infinite time elongation corresponds to zero temperature. The Fourier series in case of a finite β can be written as

$$\phi(\mathbf{x}, x_4) = \sum_n \exp\left(\frac{2\pi i \mathbf{n} \mathbf{x}}{L} + \frac{2\pi i n_4 x_4}{L_t} \right) \phi_n , \qquad n_\mu = (n_4, \mathbf{n}) \in \mathbb{Z}^4 . \tag{5.7}$$

It is seen that, at sufficiently high temperatures $\beta \sim L$ or smaller, only the component with $(n_4, \mathbf{n}) = (0, \mathbf{0})$, which corresponds to a constant field in the space and time, should be treated non-perturbatively. We refer to this regime as to the ε-regime hereafter. A schematic picture that shows the different regimes is given in Fig. 5.1.

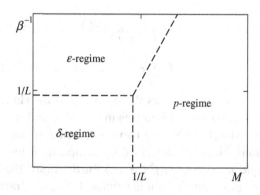

Figure 5.1   A schematic representation of different regimes on the (M, β^{-1}) plane; see the text for explanation. The figure is adopted from Ref. [3].

In this chapter, we shall explicitly address calculations in the different regimes. We start with the simplest case of the p-regime in what follows and evaluate the finite-volume correction to the pion mass at one loop.

5.2.2 Pion Mass in a Finite Volume at One Loop

In the following, the limits $L_t \to \infty$ and $a \to 0$ are assumed. The two-point function of a pion field $\phi^i(\mathbf{x}, x_4)$ on the Euclidean lattice is defined by

$$\delta^{ij} D(p_4, \mathbf{p}) = \int dx_4 \int_{-L/2}^{+L/2} d^3 x \, e^{-ipx} \langle 0 | T \phi^i(\mathbf{x}, x_4) \phi^j(0) | 0 \rangle. \tag{5.8}$$

Here, i, j are isospin indices and $px = p_4 x_4 + \mathbf{p}\mathbf{x}$. Note that the two-point function depends on the components of the four-momentum, and not on the variable $p_4^2 + \mathbf{p}^2$, because the cubic spatial lattice breaks the $SO(4)$ invariance. Furthermore, the free propagator is given by

$$D_0(p_4, \mathbf{p}) = \frac{1}{m^2 + p_4^2 + \mathbf{p}^2},$$

$$D_0(x_4, \mathbf{x}) = \frac{1}{L^3} \sum_{\mathbf{p}} \int \frac{dp_4}{2\pi} \frac{e^{ipx}}{m^2 + p_4^2 + \mathbf{p}^2} = \sum_{\mathbf{n} \in \mathbb{Z}^3} D_0^\infty(x_4, \mathbf{x} + \mathbf{n}L), \tag{5.9}$$

where D_0^∞ denotes the free Euclidean propagator in the infinite volume. It is seen that in a finite volume the propagator is modified. It contains an infinite sum of terms that emerge due to the imposition of periodic boundary conditions in the spatial direction.

Now, let us choose the rest-frame of a particle $\mathbf{p} = 0$ and insert a full set of eigenstates of the full Hamiltonian between two field operators in the full propagator. Assume further that $x_4 > 0$. In a finite volume, the spectrum of the Hamiltonian is discrete and in position space we get

$$\int_{-L/2}^{+L/2} d^3 x \, \langle 0 | T \phi^i(\mathbf{x}, x_4) \phi^j(0) | 0 \rangle$$

$$= \sum_n \int_{-L/2}^{+L/2} d^3 x \, e^{i\mathbf{p}_n \mathbf{x} - E_n(\mathbf{p}_n) x_4} \langle 0 | \phi^i(0) | n \rangle \langle n | \phi^j(0) | 0 \rangle$$

$$= L^3 \sum_n e^{-E_n x_4} \langle 0 | \phi^i(0) | n \rangle \langle n | \phi^j(0) | 0 \rangle. \tag{5.10}$$

Here, $E_n = E_n(\mathbf{0})$ denotes the spectrum of the Hamiltonian in the rest-frame. If x_4 is large, the first term dominates in the sum, and the argument of the exponential, E_n, can be extracted from the two-point function calculated on the lattice. (In these calculations, pions are described by composite fields constructed from quarks, like for example, $\phi^i(x) = \bar{\psi}(x) i \gamma^5 \tau^i \psi(x)$.) Furthermore, the lowest eigenvalue corresponds to the one-pion state in a finite volume. In the rest frame, the energy of this state is given by the pion mass M_L in a finite volume. Note also that, due to the absence of the $SO(4)$-invariance, we have to choose a particular reference frame in order to define a mass. The choice of the rest frame is both the most natural and commonly accepted one.

Performing the Fourier transform of the two-point function, we get:

$$\delta^{ij} D(p_4, \mathbf{0}) = \sum_n \int dx_4 \int_{-L/2}^{+L/2} d^3\mathbf{x}\, e^{-ip_4 x_4 - E_n x_4} \langle 0|\phi^i(0)|n\rangle \langle n|\phi^j(0)|0\rangle$$

$$= L^3 \sum_n \left\{ \langle 0|\phi^i(0)|n\rangle \frac{1}{E_n + ip_4} \langle n|\phi^j(0)|0\rangle \right.$$

$$\left. + \langle 0|\phi^j(0)|n\rangle \frac{1}{E_n - ip_4} \langle n|\phi^i(0)|0\rangle \right\}. \tag{5.11}$$

In other words, the full two-point function has poles at $p_4 = \mp iE_n$. The pole closest to the real axis corresponds to the pion mass in a finite volume.

Next, recall the definition of the self-energy part of the two-point function. In the absence of $O(4)$-invariance, the two-point function can be expressed through the self-energy part in the following manner:

$$D(p_4, \mathbf{p}) = \frac{1}{M^2 + p_4^2 + \mathbf{p}^2 - \Sigma(p_4, \mathbf{p})}. \tag{5.12}$$

Consequently, the quantity M_L is defined through the equation

$$M^2 - M_L^2 - \Sigma(iM_L, \mathbf{0}) = 0. \tag{5.13}$$

The calculation of the pion self-energy at one loop has been carried out in Section 4.9.2. Adapting Eq. (4.254) to a finite volume, we get

$$M_L^2 = M^2 + \frac{M^2}{2F^2} I_0^L + \frac{2M^4}{F^2} l_3. \tag{5.14}$$

Here, I_0^L denotes the tadpole loop in a finite volume, which is a counterpart of the infinite-volume quantity given in Eq. (4.252). As already mentioned, the spatial integration is replaced by a sum over discrete momenta in the finite volume. Furthermore, since the infinite-volume integral is divergent, dimensional regularization is implicit in the finite-volume expression as well. (We will explain later how the regularization can be implemented in this case.) Thus, the finite-volume tadpole takes the form

$$I_0^L = \frac{1}{L^d} \sum_{\mathbf{k}} \int \frac{dk_4}{2\pi} \frac{1}{M^2 + k_4^2 + \mathbf{k}^2}. \tag{5.15}$$

In the preceding expression, d denotes the number of space dimensions. In order to proceed with the calculation, it is convenient to use Poisson's formula, which in the case of one dimension reads

$$\sum_{n=-\infty}^{\infty} \delta(x - n) = \sum_{n=-\infty}^{\infty} e^{2\pi i x n}. \tag{5.16}$$

Using this, we may rewrite Eq. (5.15) in the following form:

$$I_0^L = \frac{1}{L^d} \sum_{\mathbf{n}} \int \frac{dk_4}{2\pi} \int d^d\mathbf{k}\, \delta^{(d)}\left(\mathbf{k} - \frac{2\pi}{L}\mathbf{n}\right) \frac{1}{M^2 + k_4^2 + \mathbf{k}^2}$$

$$= \sum_{\mathbf{n}} \int \frac{dk_4}{2\pi} \int \frac{d^d\mathbf{k}}{(2\pi)^d} \frac{e^{iL\mathbf{n}\mathbf{k}}}{M^2 + k_4^2 + \mathbf{k}^2} . \tag{5.17}$$

The term with $\mathbf{n} = \mathbf{0}$ coincides with the infinite-volume result I_0. Moreover, if $\mathbf{n} \neq \mathbf{0}$, each term in the integral is ultraviolet and infrared finite, so the limit $d \to 3$ can be readily performed there. Hence, one obtains $I_0^L = I_0 + \Delta I_0^L$, where

$$\Delta I_0^L = \sum_{\mathbf{n} \in \mathbb{Z}^3 \backslash \mathbf{0}} \int_0^\infty d\alpha \int \frac{d^4k}{(2\pi)^4} e^{iL\mathbf{n}\mathbf{k} - \alpha(M^2 + k_4^2 + \mathbf{k}^2)}$$

$$= \frac{1}{16\pi^2} \sum_{\mathbf{n} \in \mathbb{Z}^3 \backslash \mathbf{0}} \int_0^\infty \frac{d\alpha}{\alpha^2} \exp\left(-\alpha M^2 - \frac{L^2 \mathbf{n}^2}{4\alpha}\right) . \tag{5.18}$$

Using the known formula

$$\int_0^\infty d\alpha \, \alpha^{\nu-1} e^{-Z\alpha - Y/\alpha} = 2\left(\frac{Y}{Z}\right)^{\nu/2} K_\nu(2\sqrt{YZ}), \qquad K_{-\nu}(z) = K_\nu(z), \tag{5.19}$$

where $K_\nu(z)$ denotes the modified Bessel function of the second kind, we get

$$\Delta I_0^L = \sum_{\mathbf{n} \in \mathbb{Z}^3 \backslash \mathbf{0}} \frac{M}{4\pi^2 Ln} K_1(MLn), \qquad n = |\mathbf{n}| . \tag{5.20}$$

The asymptotic behavior of the modified Green's function for a large argument is given by

$$K_\nu(z) = \sqrt{\frac{\pi}{2z}} e^{-z} \left(1 + O(z^{-1})\right) . \tag{5.21}$$

Using this equation, we finally get

$$M_L^2 = M_\pi^2 + \sum_{\mathbf{n} \in \mathbb{Z}^3 \backslash \mathbf{0}} \frac{M_\pi^3}{8\pi^2 F_\pi^2 Ln} K_1(nM_\pi L)$$

$$= M_\pi^2 + \frac{M_\pi^4}{4F_\pi^2} \sum_{\mathbf{n} \in \mathbb{Z}^3 \backslash \mathbf{0}} \frac{e^{-nM_\pi L}}{(2\pi LnM_\pi)^{3/2}} + \cdots$$

$$= M_\pi^2 + \frac{3M_\pi^4}{2F_\pi^2} \frac{e^{-M_\pi L}}{(2\pi LM_\pi)^{3/2}} + O(e^{-\sqrt{2}M_\pi L}) . \tag{5.22}$$

Note that in the correction term we have replaced M^2, F by the physical quantities M_π^2, F_π. As anticipated, the finite-volume correction is exponentially suppressed with an argument proportional to $M_\pi L$. The leading contribution comes from the terms with $n = 1$.

The calculations of the finite-volume artifacts in other physical quantities, say, F_π, proceed along a similar path. The extension to higher orders is also straightforward. These calculations help to "purify" the "raw" lattice results, which are obtained on finite lattices and thus allow us to extract the observables in the infinite volume. We would like to repeat here that the key to the approach just described is that $1/L$ represents an infrared scale in the theory, that is, $\Lambda L \gg 1$. In other words, at the scale $1/L$ we deal already with hadrons and not with quarks and gluons. Hence, this inequality guarantees that ChPT with the same values of the coupling constants is the EFT of QCD both in the infinite and in a finite volume.

5.2.3 The Nucleon Mass in a Finite Volume

The study of the nucleon mass in a finite volume will allow us to gain more insight in the problem, namely, to clarify the relation of the quantity Δ, introduced in the beginning of this section, to the singularities of the pertinent Feynman diagrams. In addition, the breakdown of the counting rules will be discussed in the context of the finite-volume calculations. There is no need to carry out full-fledged calculations of the mass; it suffices to explain the procedure with the example of the pion–nucleon loop diagram shown in Fig. 4.12a that contributes to the nucleon mass at one loop. We work in the standard relativistic framework using dimensional regularization. The scalar loop diagram, which emerges here, is a finite-volume counterpart of a quantity defined in Eq. (4.330):

$$I_{\pi N}^L = \frac{1}{L^d} \sum_{\mathbf{k}} \int \frac{dk_4}{2\pi} \frac{1}{(M_\pi^2 + k^2)(m_N^2 + (p-k)^2)}, \qquad p = (p_4, \mathbf{0}). \qquad (5.23)$$

In order to evaluate this quantity, we use the Feynman trick and write

$$I_{\pi N}^L = \frac{1}{L^d} \sum_{\mathbf{k}} \int \frac{dk_4}{2\pi} \int_0^1 \frac{dx}{(g(x) + k_4^2 + \mathbf{k}^2)^2}, \qquad (5.24)$$

where

$$g(x) = (1-x)M_\pi^2 + xm_N^2 + x(1-x)p_4^2. \qquad (5.25)$$

Using Poisson's theorem, we can rewrite the integral in the following form:

$$I_{\pi N}^L = I_{\pi N} + \sum_{\mathbf{n} \in \mathbb{Z}^3 \setminus \mathbf{0}} \int_0^1 dx \int \frac{d^4k}{(2\pi)^4} \frac{e^{iL\mathbf{n}\mathbf{k}}}{(g(x) + k_4^2 + \mathbf{k}^2)^2}. \qquad (5.26)$$

As before, the infinite-volume part, $I_{\pi N}$, is obtained from the term with $\mathbf{n} = \mathbf{0}$. Now, we may again reduce this expression to the modified Bessel function with the argument that depends on x, and study the asymptotics of the explicit expression. We shall, however, choose a more intuitive approach. First, note that the leading exponential corrections emerge from the terms with $n = 1$. There are six such terms in the sum, all giving an equal contribution. We choose a representative vector $\mathbf{n} = (1,0,0)$, multiply this contribution by six and neglect all other (stronger suppressed) terms in the sum. Furthermore, performing the integral over the variables k_4 and $\mathbf{k}_\perp = (k_2, k_3)$, we get

$$I_{\pi N}^L = I_{\pi N} + \frac{3}{2} \int_0^1 dx \int_{-\infty}^\infty \frac{dk_1}{2\pi} \int \frac{d^2\mathbf{k}_\perp}{(2\pi)^2} \frac{e^{iLk_1}}{(g(x) + k_1^2 + \mathbf{k}_\perp^2)^{3/2}} + \cdots$$

$$= I_{\pi N} + \frac{3}{8\pi^2} \int_0^1 dx \int_{-\infty}^\infty dk_1 \frac{e^{iLk_1}}{(g(x) + k_1^2)^{1/2}} + \cdots$$

$$= I_{\pi N} + \frac{3}{8\pi^2} \int_0^1 dx \int_{-\infty}^\infty dk_1 \frac{e^{iL\sqrt{g(x)}k_1}}{(1 + k_1^2)^{1/2}} + \cdots. \qquad (5.27)$$

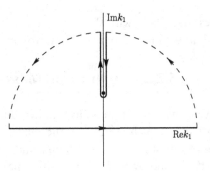

Deformation of the integration contour in the complex plane. The singularity of the integrand at $k_1 = i$ is indicated by a dot. The integral over the semicircle vanishes when the contour extends to infinity.

Here, the ellipses denote the terms with $n > 1$, which are exponentially suppressed as $L \to \infty$ with respect to the leading term.

The quantity $g(x)$ from Eq. (5.25), evaluated on the mass shell $p_4^2 = -m_N^2$, has a minimum at $x_0 = M_\pi^2/(2m_N^2)$ in the interval $0 \leq x \leq 1$. At this point, it takes the value $g_0 = g(x_0) = M_\pi^2 - M_\pi^4/(4m_N^2) = \Delta^2$, which is a positive quantity of order of the pion mass squared. It is now seen that in Eq. (5.27) it is possible to deform the integration contour in the complex plane as shown in Fig. 5.2. The integral over the large semicircle vanishes, and one is left with the integral over the imaginary axis from $k_1 = i$ (the position of the singularity of the denominator) to $i\infty$. Thus, for any positive g, we have

$$\int_{-\infty}^\infty dk_1 \frac{e^{iL\sqrt{g}k_1}}{(1+k_1^2)^{1/2}} = i \int_1^\infty dt\, e^{-L\sqrt{g}t} \left(\frac{1}{(1-t^2+i\varepsilon)^{1/2}} - \frac{1}{(1-t^2-i\varepsilon)^{1/2}} \right)$$

$$= 2 \int_1^\infty dt \frac{e^{-L\sqrt{g}t}}{(t^2-1)^{1/2}}$$

$$= 2\sqrt{2} e^{-L\sqrt{g}} \int_0^\infty du \frac{e^{-L\sqrt{g}u^2}}{(1+\frac{1}{2}u^2)^{1/2}}. \tag{5.28}$$

The asymptotic expansion for large L is obtained by expanding the denominator of the integrand in powers of u^2. This gives

$$\int_{-\infty}^\infty dk_1 \frac{e^{iL\sqrt{g}k_1}}{(1+k_1^2)^{1/2}} = 2\sqrt{\frac{\pi}{2\sqrt{g}L}} e^{-L\sqrt{g}} \left(1 + O(L^{-1/2})\right). \tag{5.29}$$

It now becomes clear that the rate of the exponential decay at large L is given by $\Delta = \min_x \sqrt{g(x)}$ (referred hereafter also as to a gap), which determines the distance to the nearest singularity of a Feynman diagram in the complex plane. As expected, this quantity is of the order of the pion mass.

A few final remarks are appropriate here. First, as previously discussed, the LECs of the ChPT Lagrangian must be the same in a finite and in the infinite volume. At this point, we want to quantify this statement. For example, assume that dynamics at a scale Λ involves some (unknown) particles with the mass of order Λ. Then, in a full theory, loops with these particles, which are approximated by the low-energy couplings in the effective theory, will also be present. Since these loops contain finite-volume corrections of order $e^{-\Lambda L}$, we arrive at the conclusion that the finite-volume artifacts in the LECs should be suppressed by the same factor.

Last but not least, the calculations here were carried out in a dimensionally regularized field theory, without invoking additional prescriptions to ensure the counting rules in baryon ChPT. What changes, if these are applied? It turns out that there is actually no change in the finite-volume part. (The result changes by the terms that contain exponentials of the type $e^{-m_N L}$ and thus are completely negligible.) Hence, only the infinite-volume part has to be modified in the ways described in Chapter 4, whereas the finite-volume terms automatically respect power counting. This result was, of course, expected from the beginning due to the decoupling of the finite-volume artifacts from the short-range physics.

5.2.4 Lüscher Formula for the Mass of a Stable Particle

From the Källén–Lehmann representation for the two-point function of stable particles, it is known that the inelastic threshold is located above the one-particle pole, with the distance between them being of order of the particle masses in the theory. For this reason, the particle masses always get exponentially suppressed contributions in a finite volume. We have seen this by considering particular Feynman graphs, but the result is valid to all orders in perturbation theory. The same statement applies to the physical quantities, which do not receive contributions from graphs with singularities. For example, the pion decay constant F_π or the electromagnetic form factor in the spacelike region receive only exponentially suppressed corrections in a finite volume. Using ChPT, we may evaluate these corrections order by order in the chiral expansion. It should be pointed out that, at a given order in the chiral expansion, the Feynman graphs contain all corrections in L. For example, the expression of the finite-volume pion mass in Eq. (5.22) in terms of Bessel functions describes the exact dependence on L and not only the leading exponential (the difference may be substantial for not-so-large values of L). On the other hand, Lüscher has derived a formula that gives only the leading exponential, but is valid to all orders in the perturbation theory [5]. (Note that the formalism of Ref. [5] is not limited to a specific form of a theory and is thus valid beyond ChPT.) We give this result in what follows without derivation, for the simplest case of a theory with a single scalar particle of mass m, described by a field φ. Further, let $T(p',q';p,q)$ be the on-shell scattering amplitude. The normalization is chosen so that the unitarity relation for the forward scattering amplitude takes the form

$$\operatorname{Im} T(p,q;p,q) = \sqrt{s(s-4m^2)}\,\sigma_{\text{tot}}(s)\,. \tag{5.30}$$

Here, $\sigma_{\text{tot}}(s)$ denotes the total cross section, and the kinematical variables are defined as

$$s = (p+q)^2, \qquad v = p \cdot q/m\,. \tag{5.31}$$

In general, the theory does not have a symmetry under $\varphi \to -\varphi$ and, hence, the φ^3 vertex does not vanish. Then, the forward scattering amplitude $F(v)$, which is a function of the variable v only, contains the one-particle pole at $v = \pm\frac{1}{2m}$. The three-particle coupling constant λ is related to the residue at this pole via

$$\lim_{v \to \pm\frac{1}{2}m} \left(v^2 - \frac{1}{4}m^2 \right) F(v) = \frac{1}{2}\lambda^2. \tag{5.32}$$

The Lüscher formula for the finite-volume mass shift is given by

$$m_L - m = -\frac{3}{16\pi m^2 L} \left(\lambda^2 e^{-\frac{\sqrt{3}}{2}mL} + \frac{m}{\pi} \int_{-\infty}^{\infty} dy\, e^{-L\sqrt{m^2+y^2}} F(iy) + O(e^{-\bar{m}L}) \right), \tag{5.33}$$

where $\bar{m} \geq \sqrt{3/2}m$. The proof given in Ref. [5] is based on abstract graph theory and is beautiful but rather voluminous. However, the idea behind the proof is very transparent. As we have already seen, the calculation of the diagrams in a finite volume boils down to the modification of the one-particle propagators described in Eq. (5.9). Using Poisson's formula, it can be shown that this modification amounts to the introduction of the exponential factors (cf. Eq. (5.17)):

$$D_0(x_4, \mathbf{x}) = \int \frac{d^4 p}{(2\pi)^4} \left(\sum_{\mathbf{n} \in \mathbb{Z}^3} e^{iL\mathbf{p}\mathbf{n}} \right) \frac{e^{ipx}}{m_4^2 + p_4^2 + \mathbf{p}^2}. \tag{5.34}$$

The propagator in the infinite volume is given by the term with $\mathbf{n} = \mathbf{0}$ in this sum. Note now that, in order to evaluate the leading exponential correction, it suffices to modify one propagator at once and retain only the terms with $|\mathbf{n}| = 1$. All possible topologies emerging in perturbation theory are shown in Fig. 5.3, which is adopted from Ref. [5]. The graphs in Fig. 5.3a give the first term in Eq. (5.33), whereas the "infinite-volume parts" of the diagrams in Fig. 5.3b,c sum up to give the forward amplitude $F(v)$, folded with the finite-volume propagator. This explains the origin of various terms in Eq. (5.33). In the case of the pion mass, the first term is absent, and the second term reproduces the leading exponential in Eq. (5.22), if $F(v)$, calculated at tree level, is substituted. Note also that it is possible to improve the accuracy of the Lüscher formula, relaxing the condition $n = 1$. Moreover, similar formulae can be derived for other quantities of interest, say, F_π. All this was done in Refs. [6, 7] where, in addition, a thorough numerical comparison of the results obtained in the chiral expansion and with the use of the Lüscher formula has been performed.

5.3 Scattering States: the Lüscher Equation

5.3.1 Maiani–Testa No-Go Theorem

As we have seen from Eq. (5.10), the mass of a stable particle can be extracted from the two-point function calculated on the lattice. Furthermore, since in massive theories there exists a nonzero mass gap between the one-particle pole and the continuum, the

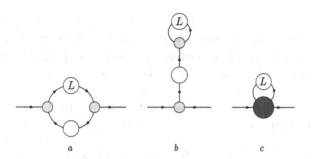

Figure 5.3 Self-energy contributions in a finite volume. The full propagators without and with the finite-volume modification are denoted by empty circles without and with the symbol L, respectively. The light shaded circles correspond to the φ^3 vertex, and the dark shaded circle in diagram c denotes the sum of all one-particle irreducible diagrams in the scattering amplitude. The figure is adopted from Ref. [5].

energy level corresponding to a single particle can be isolated from the rest. (The distance to the first excited level is equal to the mass gap in the infinite-volume limit.) The energy of a single-particle level exhibits a regular dependence on L and coincides with the infinite-volume particle mass in the large-L limit, up to exponentially suppressed contributions.

The same method can be applied in more complicated cases as well. Consider, for example, the calculation of the electromagnetic form factor of the pion. Define the field operators for $\pi^\pm$ moving with the three-momentum $\mathbf{p}$:

$$\phi_{\mathbf{p}}^\pm(x_4) = \int_{-L/2}^{+L/2} d^3\mathbf{x}\, e^{-i\mathbf{px}}\, \phi^\pm(\mathbf{x}, x_4). \tag{5.35}$$

The two-point function of these operators for the large values of the argument x_4 is given by

$$R_{\mathbf{p}}(x_4) = \langle 0|T\phi_{\mathbf{p}}^+(x_4)\phi_{\mathbf{p}}^-(0)|0\rangle = e^{-E_1(\mathbf{p})x_4}Z_{\mathbf{p}} + \cdots,$$

$$Z_{\mathbf{p}}^{1/2} = \langle 0|\phi_{\mathbf{p}}^+(0)|p\rangle. \tag{5.36}$$

Here, $E_1(\mathbf{p})$ denotes the energy of the one-particle state $|p\rangle$, which contains the π^+-meson moving with the three-momentum $\mathbf{p}$. The ellipses denote the contributions of the states with higher energy (these are exponentially suppressed as $x_4 \to \infty$).

Consider now the three-point function:

$$R_{\mathbf{p},\mathbf{q}}(x_4, y_4) = \langle 0|T\phi_{\mathbf{p}}^+(x_4)A_\mu(0)\phi_{\mathbf{q}}^-(y_4)|0\rangle, \tag{5.37}$$

where $A_\mu(0) = \bar\psi(0)\gamma_\mu Q\psi(0)$, where $Q = \text{diag}(\frac{2}{3}, -\frac{1}{3})$ is the electromagnetic charge. Further, consider the limit $x_4 \to +\infty$ and $y_4 \to -\infty$. It is straightforward to verify that the space-like form factor can be obtained from

$$\lim_{x_4 \to \infty, y_4 \to -\infty} R_{\mathbf{p},\mathbf{q}}(x_4, y_4)\left(\frac{R_{\mathbf{q}}(x_4)R_{\mathbf{p}}(-y_4)}{R_{\mathbf{q}}(-y_4)R_{\mathbf{p}}(x_4)R_{\mathbf{q}}(x_4 - y_4)R_{\mathbf{p}}(x_4 - y_4)}\right)^{1/2}$$

$$= \langle p|A_\mu(0)|q\rangle. \tag{5.38}$$

All what was said before about the mass of the stable particle applies to this matrix element as well. Since the one-pion state is separated from continuum by a mass gap, the matrix element exhibits a regular behavior in L. Furthermore, ChPT can be applied to evaluate finite-volume artifacts, which turn out to be exponentially suppressed by a factor $e^{-L\Delta}$, with Δ being of the order of the pion mass.

These statements, however, do not apply to the timelike form factor, as will be seen in what follows. In this case, one considers the following three-point function:

$$\tilde{R}_{\mathbf{p},\mathbf{q}}(x_4,y_4) = \langle 0|T\phi_{\mathbf{p}}^{+}(x_4)\phi_{\mathbf{q}}^{-}(y_4)A_\mu(0)|0\rangle. \tag{5.39}$$

Furthermore, since both pions belong now to the *out*-state, the limit $x_4,y_4 \to +\infty$ should be considered. Let us, for example, perform the limit $x_4 \to \infty$ first. We get

$$\tilde{R}_{\mathbf{p},\mathbf{q}}(x_4,y_4) = e^{-E_1(\mathbf{p})x_4}\langle 0|\phi_{\mathbf{p}}^{+}(0)|p\rangle F_{\mathbf{q}}(y_4) + \cdots, \tag{5.40}$$

where

$$F_{\mathbf{q}}(y_4) = \langle p|\phi_{\mathbf{q}}^{-}(y_4)A_\mu(0)|0\rangle$$

$$= \sum_n L^3 \delta_{\mathbf{p}+\mathbf{q},\mathbf{p}_n} e^{-(E_n(\mathbf{p}_n)-E_1(\mathbf{q}))y_4}\langle p|\phi_{\mathbf{q}}^{-}(0)|n\rangle\langle n|A_\mu(0)|0\rangle. \tag{5.41}$$

At large values of y_4 (but, still, $x_4 \gg y_4$), the state with the lowest energy will dominate in this sum. In our case, this is the $\pi^+\pi^-$ state with total three-momentum $\mathbf{p}_n = \mathbf{p} + \mathbf{q}$. For simplicity, we may further assume $\mathbf{p} = -\mathbf{q}$ and, hence, $\mathbf{p}_n = 0$. The key argument is that the energy of the intermediate state can be *smaller* than the energy of the *out*-state, which is equal to $2\sqrt{M_\pi^2+\mathbf{p}^2}$. Indeed, the intermediate state with the smallest energy would be the one containing two pions with zero momenta; the energy of such a state would be close to $2M_\pi$. This state is, however, excluded because the vector particle described by A_μ cannot decay into two pseudoscalars at rest. The configuration with back-to-back three momenta $\pm 2\pi/L$ is, however, allowed. If $|\mathbf{p}| > 2\pi/L$, one goes below the energy of the *out*-state, meaning that the three-point function $\tilde{R}$ is not dominated by the matrix element that we wish to extract. This statement gives the essence of the Maiani–Testa no-go theorem [8]. Note that the same statement applies to all matrix elements that have more than one stable particle in the *in*- or *out*-states. For example, the two-particle scattering amplitude falls into this category.

It should be pointed out that, in the preceding discussion, considering the limit of large L was essential. If $L \to \infty$, the energy of any intermediate state always collapses toward threshold, no matter what one does. However, for a finite L, we may reach a given value of energy, measuring sufficiently high-lying excited states, because the gap between the energy levels is finite. This circumstance plays a key role in the finite-volume Lüscher approach that we shall now consider.

5.3.2 The Lüscher Equation

In continuum field theory, the scattering amplitudes are related to the four-point Green's function through the LSZ formula. In order to ease the notation, we shall consider the case of a single spin-zero field with mass m here. As we have learned from the preceding discussion, letting the time arguments of the individual particles on the lattice tend to $\pm\infty$ independently is not a good idea, as contributions unrelated to the scattering amplitudes we are looking for dominate the sum over intermediate states.[3] Alternatively, we may consider the situation when the time separations between all particles in the *in*- and all particles in the *out*-state are kept finite separately. In its simplest form,[4] this boils down to studying the so-called equal-time four-point Green's function:

$$L^3 g(\mathbf{p}, \mathbf{q}; t) = \langle 0 | T \phi_{\mathbf{p}}(t) \phi_{-\mathbf{p}}(t) \phi_{\mathbf{q}}^\dagger(0) \phi_{-\mathbf{q}}^\dagger(0) | 0 \rangle. \tag{5.42}$$

The normalization factor L^3 is a matter of convenience. The choice of the momenta $\mathbf{p}, \mathbf{q}$ is irrelevant. We may for example, put them to zero, or take the smallest available momenta on the lattice. Now, for $t \to +\infty$, we get

$$L^3 g(\mathbf{p}, \mathbf{q}; t) = \sum_n e^{-E_n t} \langle 0 | \phi_{\mathbf{p}}(0) \phi_{-\mathbf{p}}(0) | n \rangle \langle n | \phi_{\mathbf{q}}^\dagger(0) \phi_{-\mathbf{q}}^\dagger(0) | 0 \rangle. \tag{5.43}$$

What one obtains in this limit are the energy levels of a system in a finite box in the two-particle sector (i.e., in the sector with quantum numbers of a two-particle state). We assume, for simplicity, that there is no contribution from the vacuum state in this sum (e.g. consider charged particles), or that the vacuum contribution, which is a t-independent constant, can be cleanly subtracted. Then, we may extract the energy levels from the function $g(\mathbf{p}, \mathbf{q}; t)$ calculated on the lattice. Both ground-state and excited-state energies can be resolved, using different techniques. For example, we can perform a multi-exponential fit to $g(\mathbf{p}, \mathbf{q}; t)$ or solve the generalized eigenvalue problem (GEVP) [11, 12]. We do not want to enter into these details here. Suppose that this is done and that the first few energy levels are known.

Our aim here is to demonstrate that the finite-volume spectrum E_n is uniquely determined by the on-shell scattering amplitudes in the infinite volume *evaluated at the same energy*. Reversing the argument, it will be possible to extract the on-shell amplitude from the measured energy spectrum on the lattice. Below, a detailed proof of the above statement is provided.

To start with, note that the Fourier transform of $g(\mathbf{p}, \mathbf{q}; t)$, defined as

$$g(\mathbf{p}, \mathbf{q}; P_4) = \int dt\, e^{-iP_4 t} g(\mathbf{p}, \mathbf{q}; t), \tag{5.44}$$

contains poles at $P_4 = \pm i E_n$. Furthermore, the equal-time Green's function can be related to the standard relativistic Green's function in the CM frame, according to

$$g(\mathbf{p}, \mathbf{q}; P_4) = \int \frac{dp_4}{2\pi} \frac{dq_4}{2\pi} G(\mathbf{p}, p_4, \mathbf{q}, q_4; P_4). \tag{5.45}$$

Now, using perturbation theory, we shall try to identify which diagrams lead to these

---

[3] However, if the Euclidean time separations are not taken to be infinite, we can still extract certain matrix elements, fitting the explicit time-dependence of appropriately chosen correlators; see, e.g., Ref. [8–10].

[4] For simplicity, we consider the center-of-mass frame, albeit very little effort is needed to reformulate the problem in an arbitrary moving frame.

Figure 5.4
Examples of two-particle irreducible and reducible diagrams in the $\lambda \phi^4$ theory up to and including $O(\lambda^2)$. The diagram b, which corresponds to the s-channel bubble, is reducible; it is possible to separate the initial and final particles by cutting just two internal lines. The other diagrams, including the t, u-channel bubbles shown in c,d, are irreducible.

poles as well as to determine their location. To this end, note that the relativistic Green's function can be written as

$$G(\mathbf{p}, p_4, \mathbf{q}, q_4; P_4) = \frac{2\pi\delta(p_4 - q_4)L^3 \delta_{\mathbf{pq}} + 2\pi\delta(p_4 + q_4)L^3 \delta_{\mathbf{p},-\mathbf{q}}}{(m^2 + (\frac{1}{2}P_4 + p_4)^2 + \mathbf{p}^2)(m^2 + (\frac{1}{2}P_4 - p_4)^2 + \mathbf{p}^2)}$$

$$+ \frac{1}{(m^2 + (\frac{1}{2}P_4 + p_4)^2 + \mathbf{p}^2)(m^2 + (\frac{1}{2}P_4 - p_4)^2 + \mathbf{p}^2)}$$

$$\times \frac{T(\mathbf{p}, p_4, \mathbf{q}, q_4; P_4)}{(m^2 + (\frac{1}{2}P_4 + q_4)^2 + \mathbf{q}^2)(m^2 + (\frac{1}{2}P_4 - q_4)^2 + \mathbf{q}^2)}. \qquad (5.46)$$

The first term in the preceding equation is the free Green's function $G_0(\mathbf{p}, p_4, \mathbf{q}, q_4; P_4)$, and T denotes the two-particle scattering amplitude, which satisfies the Bethe–Salpeter equation:

$$T(\mathbf{p}, p_4, \mathbf{q}, q_4; P_4) = K(\mathbf{p}, p_4, \mathbf{q}, q_4; P_4)$$

$$+ \frac{1}{2L^3} \sum_{\mathbf{k}} \int \frac{dk_4}{2\pi} \frac{K(\mathbf{p}, p_4, \mathbf{k}, k_4; P_4)T(\mathbf{k}, k_4, \mathbf{q}, q_4; P_4)}{(m^2 + (\frac{1}{2}P_4 + k_4)^2 + \mathbf{k}^2)(m^2 + (\frac{1}{2}P_4 - k_4)^2 + \mathbf{k}^2)}. \qquad (5.47)$$

Here, K denotes the kernel of the Bethe–Salpeter equation, which is defined as a sum of all two-particle irreducible graphs. These are the graphs in which the initial and final ends of the diagram cannot be made disconnected by cutting just two internal lines. Examples of both two-particle reducible as well as irreducible diagrams are given in Fig. 5.4. Note also that due to Bose symmetry the functions T and K are invariant with respect to $(\mathbf{p}, p_4) \rightarrow -(\mathbf{p}, p_4)$, and similarly for other momenta.

In order to derive an equation for the equal-time Green's function, we shall use the quasipotential method by Logunov and Tavkhelidze [13]. [5] First, for any quantity A, we introduce a useful notation that describes the "projection" on equal times:

$$\widetilde{A} = \int \frac{dp_4}{2\pi} \frac{dq_4}{2\pi} A(p_4, q_4). \qquad (5.48)$$

---

[5] Note that we shall do this in a somewhat cavalier manner, avoiding details in order to concentrate on the main message. For example, we never consider self-energy insertions in the external or internal lines explicitly, as well as the issues related to renormalization. All this can be done, and the result will remain unchanged.

Hence, the quantity g is obtained from the full Green's function G with the use of this "projection"; see Eq. (5.45). Furthermore, the free equal-time Green's function is defined by

$$g_0(\mathbf{p},\mathbf{q};P_4) = \int \frac{dp_4}{2\pi}\frac{dq_4}{2\pi}\frac{2\pi\delta(p_4-q_4)L^3\delta_{\mathbf{pq}}+2\pi\delta(p_4+q_4)L^3\delta_{\mathbf{p},-\mathbf{q}}}{(m^2+(\frac{1}{2}P_4+p_4)^2+\mathbf{p}^2)(m^2+(\frac{1}{2}P_4-p_4)^2+\mathbf{p}^2)}$$

$$= \frac{L^3\delta_{\mathbf{p},\mathbf{q}}+L^3\delta_{\mathbf{p},-\mathbf{q}}}{w(\mathbf{p})(P_4^2+4w^2(\mathbf{p}))}. \tag{5.49}$$

Next, "projecting" Eq. (5.46), we get

$$\widetilde{G} = \widetilde{G_0}+\frac{1}{4}\widetilde{G_0TG_0}. \tag{5.50}$$

Note that the factor $1/4$ arises due to Bose symmetry, as well as the factor $1/2$ in Eq. (5.47). Furthermore, it can be straightforwardly seen that the equal-time Green's function obeys the equation

$$g(\mathbf{p},\mathbf{q};P_4) = g_0(\mathbf{p},\mathbf{q};P_4)+\frac{1}{L^6}\sum_{\mathbf{k},\mathbf{l}}g_0(\mathbf{p},\mathbf{k};P_4)w(\mathbf{k},\mathbf{l};P_4)g(\mathbf{l},\mathbf{q};P_4), \tag{5.51}$$

where the quantity w is the quasipotential that can be expressed through the scattering amplitude in perturbation theory:

$$w = \frac{1}{4}\widetilde{G_0}^{-1}\widetilde{G_0TG_0}\widetilde{G}^{-1} = \frac{1}{4}\widetilde{G_0}^{-1}\widetilde{G_0TG_0}\widetilde{G_0}^{-1}\left(1-\frac{1}{4}\widetilde{G_0TG_0}\widetilde{G_0}^{-1}+\cdots\right). \tag{5.52}$$

The main idea behind this approach is that the quasipotential w is a regular function of P_4 in the scattering region below the inelastic threshold. This property, which is very plausible on grounds of unitarity, can be directly verified in perturbation theory. For simplicity, we shall check this within the $\lambda\phi^4$-theory with the Lagrangian

$$\mathcal{L} = \frac{1}{2}(\partial\phi)^2-\frac{m^2}{2}\phi^2-\frac{\lambda}{4}\phi^4. \tag{5.53}$$

Of course, for studying the singularities of Feynman diagrams, the exact form of the interaction vertices is irrelevant.

Up to and including $O(\lambda^2)$ terms, the scattering amplitude is given by

$$T(\mathbf{p},p_4,\mathbf{q},q_4;P_4) = -6\lambda+18\lambda^2(J_s+J_t+J_u)+O(\lambda^3). \tag{5.54}$$

Here,

$$J_{s,t,u} = \frac{1}{L^3}\sum_{\mathbf{k}}\int\frac{dk_4}{2\pi}\frac{1}{(m^2+(R_{s,t,u}-k)^2)(m^2+k^2)}, \tag{5.55}$$

and

$$R_s = (P_4,\mathbf{0}), \qquad R_t = (p_4-q_4,\mathbf{p}-\mathbf{q}), \qquad R_u = (p_4+q_4,\mathbf{p}+\mathbf{q}). \tag{5.56}$$

First, let us make sure that the t- and u-channel diagrams lead to a regular expression. To this end, let us concentrate on the t-channel piece. Carrying out the Cauchy integration over k_4, we get

$$J_t = \frac{1}{L^3} \sum_{\mathbf{k}} \frac{1}{4iw(\mathbf{k})w(\mathbf{p}-\mathbf{q}-\mathbf{k})} \left(\frac{1}{p_4 - q_4 - iw(\mathbf{p}-\mathbf{q}-\mathbf{k}) - iw(\mathbf{k})} \right.$$

$$\left. - \frac{1}{p_4 - q_4 + iw(\mathbf{p}-\mathbf{q}-\mathbf{k}) + iw(\mathbf{k})} \right). \tag{5.57}$$

This expression can be further used to calculate the contribution of the t-channel diagram to the quasipotential. We get

$$\frac{1}{4} \widetilde{G_0 T_t G_0} = 18\lambda^2 \int \frac{dp_4}{2\pi} \frac{dq_4}{2\pi} \frac{1}{(m^2 + (\frac{1}{2}P_4 + p_4)^2 + \mathbf{p}^2)(m^2 + (\frac{1}{2}P_4 - p_4)^2 + \mathbf{p}^2)}$$

$$\times \frac{J_t}{(m^2 + (\frac{1}{2}P_4 + q_4)^2 + \mathbf{q}^2)(m^2 + (\frac{1}{2}P_4 - q_4)^2 + \mathbf{q}^2)}. \tag{5.58}$$

Carrying out the Cauchy integration over p_4, q_4 is straightforward but very cumbersome. We are interested in the singularities that emerge near $P_4 = \pm 2im$. It turns out that the most singular terms emerge from the contributions of the poles contained in the external Green's functions G_0. These contributions contain the factor $(P_4^2 + 4w^2(\mathbf{p}))^{-1}(P_4^2 + 4w^2(\mathbf{q}))^{-1}$. There are no other singularities in the low-energy domain coming from the poles in the loop integral. Recalling now that, in order to arrive at the quasipotential w, we have to multiply $\frac{1}{4} \widetilde{G_0 T_t G_0}$ with the factors $\widetilde{G_0}^{-1}$ from the left and from the right, we immediately realize that the low-energy singularities in the quasipotential cancel. The same statement holds for the u-channel term as well.

In contrast, the s-channel term contains the singularity. Namely, since J_s does not depend on the momenta p, q, we may write

$$\frac{1}{4} \widetilde{G_0 T_s G_0} = \frac{1}{4} \widetilde{G_0} (18\lambda^2 J_s) \widetilde{G_0},$$

$$J_s = \frac{1}{L^3} \sum_{\mathbf{k}} \frac{1}{w(\mathbf{k})} \frac{1}{P_4^2 + 4w^2(\mathbf{k})}. \tag{5.59}$$

The factors $\widetilde{G_0}$ cancel against the factors $\widetilde{G_0}^{-1}$ in the expression of the quasipotential, but the singular factor J_s remains. In order to ensure its cancellation, it is necessary to consider the second-order term in Eq. (5.52), replacing the scattering amplitude with the tree-level expression $T = -6\lambda$. It can be seen that the obtained expression exactly cancels the term with J_s, leaving no low-energy singularities in the quasipotential whatsoever.

At the end of this rather lengthy discussion, we arrive at a remarkably simple result. The quasipotential, as defined by Eq. (5.52), does not possess any singularities in the low-energy region, more precisely in the vicinity of the elastic threshold. We have explicitly verified this statement at second order in perturbation theory in the $\lambda \phi^4$-theory, but it holds in general. Indeed, as we know from the study of the analytic properties of the scattering amplitudes in the infinite volume, the only singularities in the elastic region are coming from the two-particle intermediate states in the s-channel, as these can go on shell. These are exactly the states that are *not* included in the definition of the quasipotential. All the rest is included and is regular in the low-energy region. Furthermore, in a finite volume, the absence of low-energy singularities ensures that the finite-volume artifacts related to the quasipotential can be evaluated with the use of Poisson's formula and are exponentially suppressed for large values of L. Choosing mL very large, we could neglect these artifacts altogether and use the quasipotential calculated in the finite volume. This means that the quasipotential equation for the two-particle scattering amplitude, which is defined through the equal-time Green's function as

$$g(\mathbf{p},\mathbf{q},P_4) = g_0(\mathbf{p},\mathbf{q},P_4) + \frac{1}{L^6}\sum_{\mathbf{k},\mathbf{l}}g_0(\mathbf{p},\mathbf{k},P_4)t(\mathbf{k},\mathbf{l},P_4)g_0(\mathbf{l},\mathbf{q},P_4),\tag{5.60}$$

takes the form

$$t(\mathbf{p},\mathbf{q},P_4) = w(\mathbf{p},\mathbf{q},P_4) + \frac{1}{L^3}\sum_{\mathbf{k}}w(\mathbf{p},\mathbf{k},P_4)\frac{2}{w(\mathbf{k})(P_4^2+4w^2(\mathbf{k}))}t(\mathbf{k},\mathbf{q},P_4).\tag{5.61}$$

Note also that, changing the normalization of the scattering matrix and the quasipotential,

$$t(\mathbf{p},\mathbf{q},P_4) \to \frac{1}{2w(\mathbf{p})}t(\mathbf{p},\mathbf{q},P_4)\frac{1}{2w(\mathbf{q})},$$

$$w(\mathbf{p},\mathbf{q},P_4) \to \frac{1}{2w(\mathbf{p})}w(\mathbf{p},\mathbf{q},P_4)\frac{1}{2w(\mathbf{q})},\tag{5.62}$$

we arrive at the equation that formally resembles the nonrelativistic Lippmann–Schwinger equation

$$t(\mathbf{p},\mathbf{q},P_4) = w(\mathbf{p},\mathbf{q},P_4) + \frac{1}{L^3}\sum_{\mathbf{k}}w(\mathbf{p},\mathbf{k},P_4)\frac{1}{(\mathbf{k}^2-q_0^2)}t(\mathbf{k},\mathbf{q},P_4),\tag{5.63}$$

where q_0 is given by the relativistic expression [5]

$$q_0^2 = -\frac{P_4^2}{4} - m^2.\tag{5.64}$$

This finite-volume equation should be solved exactly, since the Green's function contains simple poles at the free-particle energies and, thus, iterations will explode in the vicinity of these energies. As we shall see, the poles in the exact solution are, as expected, shifted from the free-particle values, and the location of the poles in the equal-time Green's function coincides with those in the scattering amplitude. Furthermore, according to Eq. (5.44), the poles in the upper plane, $P_4 = iE_n$, are responsible for the damping exponentials $e^{-E_n t}$ in the Euclidean equal-time Green's function. Thus, the location of the poles determines the spectrum of the full theory in the finite volume.

Here, an equation that determines the position of these poles, the famous Lüscher equation [14], will be derived. To this end, we first consider the partial-wave expansion of the quasipotential and the scattering amplitude. The quasipotential is a quantity defined in the infinite volume, where rotational invariance holds. Thus,

$$w(\mathbf{p},\mathbf{q};iE_n) = 4\pi\sum_{\ell m}\mathscr{Y}_{\ell m}(\mathbf{p})w_\ell(p,q)\mathscr{Y}_{\ell m}^*(\mathbf{q}). \tag{5.65}$$

Here, $\mathscr{Y}_{\ell m}(\mathbf{p}) = p^\ell Y_{\ell m}(\hat{p})$ are the solid harmonics, where $p = |\mathbf{p}|$; the unit vector in the direction of $\mathbf{p}$ is given by $\hat{p} = \mathbf{p}/p$; and the $Y_{\ell m}(\hat{p})$ denote the spherical harmonics. The dependence on the total energy E_n is not explicitly displayed in $w_\ell(p,q)$.

The quantity t is, however, defined in a finite cubic box. In the absence of rotational invariance, its partial-wave expansion takes the form

$$t(\mathbf{p},\mathbf{q};iE_n) = 4\pi\sum_{\ell m}\sum_{\ell'm'}\mathscr{Y}_{\ell m}(\mathbf{p})t_{\ell m,\ell'm'}(p,q)\mathscr{Y}_{\ell'm'}^*(\mathbf{q}). \tag{5.66}$$

Substituting this expansion in Eq. (5.63), we get

$$t_{\ell m,\ell'm'}(p,q) = \delta_{\ell\ell'}\delta_{mm'}w_\ell(p,q) + \frac{4\pi}{L^3}\sum_{\mathbf{k}}\sum_{\ell''m''}\frac{\mathscr{Y}_{\ell m}^*(\mathbf{k})\mathscr{Y}_{\ell''m''}(\mathbf{k})}{\mathbf{k}^2 - q_0^2}$$

$$\times\left\{w_\ell(p,k)t_{\ell''m'',\ell'm'}(k,q) - \left(\frac{q_0}{k}\right)^{\ell''}w_\ell(p,q_0)t_{\ell''m'',\ell'm'}(q_0,q)\right\}$$

$$+ w_\ell(p,q_0)\frac{4\pi}{L^3}\sum_{\mathbf{k}}\sum_{\ell''m''}\left(\frac{q_0}{k}\right)^{\ell''}\frac{\mathscr{Y}_{\ell m}^*(\mathbf{k})\mathscr{Y}_{\ell''m''}(\mathbf{k})}{\mathbf{k}^2 - q_0^2}t_{\ell''m'',\ell'm'}(q_0,q). \tag{5.67}$$

Now, the first sum is not singular at $k^2 = q_0^2$ (the numerator also vanishes at this point) and thus, the summation can be replaced by integration. The effects that emerge due to such a replacement are exponentially suppressed and are negligible, as long as $mL \gg 1$. This statement, which has already been used on many occasions, is known under the name of the regular summation theorem [5]. Note also that we are free to choose the prescription for circumventing the singularity of the integrand in the curly brackets, because the whole integrand is not singular. We make use of this freedom and shall consistently apply the principal-value prescription everywhere. Thus, we obtain

$$t_{\ell m,\ell'm'}(p,q) = \delta_{\ell\ell'}\delta_{mm'}\left(w_\ell(p,q) + \text{P.V.}\int\frac{k^{2(1+\ell)}dk}{2\pi^2}\frac{w_\ell(p,k)t_{\ell m,\ell'm'}(k,q)}{\mathbf{k}^2 - q_0^2}\right)$$

$$+ 4\pi\sum_{\ell''m''}w_\ell(p,q_0)H_{\ell m,\ell''m''}(q_0)t_{\ell''m'',\ell'm'}(q_0,q), \tag{5.68}$$

where

$$H_{\ell m,\ell''m''}(q_0) = \left\{\frac{1}{L^3}\sum_{\mathbf{k}} - \text{P.V.}\int\frac{d^3\mathbf{k}}{(2\pi)^3}\right\}\left(\frac{q_0}{k}\right)^{\ell''}\frac{\mathscr{Y}_{\ell m}^*(\mathbf{k})\mathscr{Y}_{\ell''m''}(\mathbf{k})}{\mathbf{k}^2 - q_0^2}. \tag{5.69}$$

Defining

$$r_\ell(p,q) = w_\ell(p,q) + \text{P.V.}\int\frac{k^{2(1+\ell)}dk}{2\pi^2}\frac{w_\ell(p,k)r_\ell(k,q)}{\mathbf{k}^2 - q_0^2}, \tag{5.70}$$

we can rewrite Eq. (5.68) in a simpler form. Taking, in addition, $p = q = q_0$, we obtain

$$
t_{\ell m, \ell' m'}(q_0, q_0) = \delta_{\ell\ell'} \delta_{mm'} r_\ell(q_0, q_0)
$$

$$
+ 4\pi \sum_{\ell'' m''} r_\ell(q_0, q_0) H_{\ell m, \ell'' m''}(q_0) t_{\ell'' m'', \ell' m'}(q_0, q_0). \tag{5.71}
$$

Note that, unlike the quasipotential, the infinite-volume quantity $r_\ell(q_0, q_0)$ taken on the mass shell is related to an observable quantity, the scattering phase shift [13]. In the normalization chosen in Eq. (5.70) this relation takes the form

$$
r_\ell(q_0, q_0) = 4\pi q_0^{-(2\ell+1)} \tan \delta_\ell(q_0). \tag{5.72}
$$

Furthermore, using the following relation for the solid harmonics:

$$
\mathscr{Y}_{\ell m}^*(\mathbf{k}) \mathscr{Y}_{\ell'' m''}(\mathbf{k}) = \frac{1}{\sqrt{4\pi}} \sum_{j=|\ell-\ell''|}^{\ell+\ell''} \sum_{s=-j}^{j} i^{-\ell-\ell''+j} k^{\ell+\ell''-j} C_{\ell m, js, \ell'' m''} \mathscr{Y}_{js}^*(\mathbf{k}), \tag{5.73}
$$

where

$$
C_{\ell m, js, \ell'' m''} = (-1)^{m''} i^{\ell+\ell''-j} \sqrt{(2\ell+1)(2\ell''+1)(2j+1)}
$$

$$
\times \begin{pmatrix} \ell & j & \ell'' \\ m & s & -m'' \end{pmatrix} \begin{pmatrix} \ell & j & \ell'' \\ 0 & 0 & 0 \end{pmatrix}, \tag{5.74}
$$

we may simplify the expression of the matrix H in Eq. (5.75). To this end, note first that, owing to the properties of the Wigner $3j$-symbols that appear in Eq. (5.74) (see Appendix A for their definition), the integer $\ell + \ell'' - j$ must be even. Then, the factor $(q_0/k)^{\ell''} k^{\ell+\ell''-j}$ can be replaced by $q_0^{\ell+\ell''-j}$. The difference between these two expressions has the form $(k^2 - q_0^2) \times$ (polynomial in k^2), which, according to the regular summation theorem, gives only exponentially suppressed contributions. Thus, we may write

$$
H_{\ell m, \ell'' m''}(q_0) = \frac{i^{\ell-\ell''}}{16\pi^2} q_0^{\ell+\ell''+1} \mathscr{M}_{\ell m, \ell'' m''}(q_0), \tag{5.75}
$$

where

$$
\mathscr{M}_{\ell m, \ell'' m''}(q_0) = \frac{(-1)^\ell}{\pi^{3/2}} \sum_{js} C_{\ell m, js, \ell'' m''} \left[Z_{js}(1; \eta^2) \right]^*. \tag{5.76}
$$

Here, $\eta = 2\pi q_0/L$ and

$$
Z_{js}(1; \eta^2) = \left\{ \sum_{\mathbf{n} \in \mathbb{Z}^3} - \text{P.V.} \int d^3 \mathbf{n} \right\} \frac{\mathscr{Y}_{js}(\mathbf{n})}{\mathbf{n}^2 - \eta^2}, \tag{5.77}
$$

which is referred to as Lüscher's zeta function. Note that both the sum and the integral diverge on the upper limit, whereas their difference is finite. For actual calculations, it is useful to employ a cutoff:

$$
Z_{js}(1; \eta^2) = \lim_{N \to \infty} \left\{ \sum_{\mathbf{n} \in \mathbb{Z}^3}^{N} \frac{\mathscr{Y}_{js}(\mathbf{n})}{\mathbf{n}^2 - \eta^2} - \delta_{\ell 0} \delta_{m0} \sqrt{4\pi} N \right\}. \tag{5.78}
$$

Since $Z_{js}(1;\eta^2)$ is ultraviolet-finite, the final result does not depend on the regularization used.

Equation (5.71) is a linear matrix equation in the space of angular-momentum indices ℓ, m. In order to solve it, we should cut off the angular momenta at some maximal value $\ell = \ell_{max}$, assuming that all phase shifts above ℓ_{max} are zero (such an assumption is natural in the low-energy region). The position of the poles in the finite-volume scattering amplitude t are determined by the zeros of the determinant of the linear equations. Rewriting Eq. (5.71) as

$$\sum_{\ell m, \ell'' m''} \tilde{D}_{\ell m, \ell'' m''}(q^0) t_{\ell'' m'', \ell' m'}(q_0, q_0) = \delta_{\ell \ell'} \delta_{mm'} r_\ell(q_0, q_0), \qquad (5.79)$$

where

$$D_{\ell m, \ell'' m''}(q^0) = (-iq_0)^{-\ell} \left(\delta_{\ell \ell''} \delta_{mm''} - \tan \delta_\ell(q_0) \mathcal{M}_{\ell m, \ell'' m''}(q_0) \right) (-iq_0)^{\ell''}, \qquad (5.80)$$

we finally arrive at the *Lüscher equation* that relates the finite-volume energy levels to the infinite-volume phase shifts $\delta_\ell(q_0)$:

$$\det \mathcal{D}(q_0) = 0,$$

$$\mathcal{D}_{\ell m, \ell'' m''}(q^0) = \delta_{\ell \ell''} \delta_{mm''} - \tan \delta_\ell(q_0) \mathcal{M}_{\ell m, \ell'' m''}(q_0). \qquad (5.81)$$

Given the phase shifts, the Lüscher equation enables one to predict the energy levels in a finite volume. The inverse problem is, however, more interesting. Given the finite-volume energy levels measured on the lattice, we could solve this equation and extract the infinite-volume phase shifts from these measurements, thus finally achieving the goal that was stated in the beginning. In this respect, note that the function $\mathcal{M}_{\ell m, \ell'' m''}(q_0)$ in Eq. (5.81) does not depend on the interaction and is fully determined by the geometry of the box. Measuring the energy levels on the lattice is equivalent to measuring q_0, and evaluating the quantity $\mathcal{M}_{\ell m, \ell'' m''}(q_0)$, we get linear equations for determining the values of the phase shifts *at the same value of q_0*.

At the end of this rather lengthy and detailed derivation, it is useful to briefly summarize, emphasizing the essential points. Some additional remarks are also needed:

- In the scattering amplitude, one deals with the finite-volume artifacts in two different cases. First, there are so-called polarization effects that emerge from the diagrams with no singularities in the physical region. These effects are always exponentially suppressed with a factor $e^{-L\Delta}$, with the quantity Δ being of order of the particle mass m. In the second case, one considers effects emerging from the singular diagrams. There exists two types of such effects. As we shall see later, corrections to the finite-volume energy levels exhibit a power-law dependence on L (modulo logarithms). On the other hand, the dependence of scattering and decay matrix elements on L is irregular, being described by the Lüscher zeta function.

The equal-time (quasipotential) formalism, which was used in this section, enables one to effectively separate these two cases. The quasipotential contains only polarization effects. Choosing $mL \gg 1$, we may neglect these exponentially suppressed contributions altogether.

- It is very interesting to address the situation with an unnaturally large scattering length a. The question is whether in this case the smallness of the polarization effects is controlled by the ratio of a/L rather than mL. In order to answer this question, we recall that the polarization effects we are referring to emerge from the t- and u-channel diagrams occurring in the quasipotential. These diagrams do not have singularities close to the physical region, even if the physical amplitude develops a pole nearby the threshold, as implied by an unnaturally large value of the scattering length. In other words, the quantity Δ is of order of m in these diagrams even in case of existence of a shallow bound state. So, whether the scattering length is unnaturally large or not, the Lüscher approach is applicable, as long as $mL \gg 1$. Note that, in general, there are exponentially suppressed contributions with an unnaturally small argument. For example, in the energy shift of the two-particle ground state, there exists a contribution that goes with a factor $e^{-\gamma L}$, where $\gamma \ll m$ denotes the binding momentum of the shallow bound state. Such contributions, however, emerge from the s-channel diagrams and are readily reproduced within the Lüscher approach. We shall see examples of such calculations in what follows.

- The *two-body quantization condition* or, equivalently, the Lüscher equation (5.81), gives the relation of the finite-volume energy levels with the S-matrix elements in the infinite volume that are parameterized by the scattering phases. The latter are observables. All unobservable quantities used in the proof (like the quasipotential, or the Lagrangian) have disappeared from the final relation, which is therefore valid without any reference to a particular model or to the details of short-range dynamics between the particles. In fact, we are dealing here with one more example of decoupling at different scales. Namely, when $mL \gg 1$, the particles are very far from each other (we remind the reader that the characteristic range of interactions is given by $1/m$), and the short-range details cannot matter in the final result.

- In a finite volume, rotational invariance is broken. This leads to partial-wave mixing, which is explicit in Eq. (5.81). Tractable equations are obtained, if and only if a cutoff on the orbital angular momentum is imposed. A remnant of the rotational invariance (the octahedral symmetry group, or different little groups thereof in case of moving frames) allows us to partially diagonalize the Lüscher equation. We shall consider this question in what follows.

- The Lüscher equation, as it stands, is valid in a region of elastic scattering only. There is, however, a straightforward generalization to the case of *coupled two-particle channels*. In this case, the quantity $\mathscr{D}$ in Eq. (5.81) becomes, in addition, a matrix in the channel space; for example, Refs. [15–23].

5.4 Moving Frames, Angular Momentum and Spin

5.4.1 NREFT in a Finite Volume

A large part of the proof given in the previous section was devoted to the separation of the polarization effects from the power-law corrections that is achieved by invoking the quasipotential approach. The idea behind presenting such a detailed proof was mainly pedagogical, as it allows one to gain insight into the behavior of relativistic field theories in a finite volume. However, as discussed in Chapter 2, the same goal can be accomplished with substantially less effort by using suitable non-relativistic EFTs. In these theories, particle number is conserved in each vertex, and the backward propagation in time is absent. As a result, only the s-channel bubbles (exactly those that were considered in the derivation of the Lüscher equation) are treated explicitly in two-particle scattering, whereas all other effects are implicitly included in the low-energy effective couplings. Hence, instead of a long detour with the equal-time Green's functions, we may merely assume that the low-energy couplings in NREFT contain only exponentially suppressed contributions that can be safely neglected in the limit $mL \gg 1$. Such an approach, which was first proposed in Ref. [24], enables one to re-derive the Lüscher equation with surprising ease.

On the other hand, the Lüscher equation, which has been derived originally in the CM frame for identical spinless particles, was generalized to arbitrary moving frames, in which the incoming/outgoing particles have a nonzero total three-momentum, as well as for the case of nonidentical particles, particles with spin and elongated boxes; see, for example, Refs. [21, 25–33]. In this section, we consider all these generalizations, albeit some in more detail than others. We shall use the NREFT approach since, as already stated, it represents a very economic way to address the problem. As we already know, there will be no problem with implementing explicitly relativistic kinematics. The modified NREFT considered in Section 2.7 provides a convenient tool for this.

Our derivation of the Lüscher equation closely follows the one given in Ref. [34]. As an example, we consider nonidentical spinless particles in arbitrary moving frames.[6] In this case, the Lippmann–Schwinger equation in the infinite volume can be written down in the following form:

$$t(\mathbf{p}_1, \mathbf{p}_2; \mathbf{q}_1, \mathbf{q}_2) = w(\mathbf{p}_1, \mathbf{p}_2; \mathbf{q}_1, \mathbf{q}_2) + \int \frac{d^d \mathbf{k}_1}{(2\pi)^3 2w_1(\mathbf{k}_1)} \frac{d^d \mathbf{k}_2}{(2\pi)^3 2w_2(\mathbf{k}_2)}$$

$$\times (2\pi)^d \delta^{(d)}(\mathbf{p}_1 + \mathbf{p}_2 - \mathbf{k}_1 - \mathbf{k}_2) \frac{w(\mathbf{p}_1, \mathbf{p}_2; \mathbf{q}_1, \mathbf{q}_2) t(\mathbf{p}_1, \mathbf{p}_2; \mathbf{q}_1, \mathbf{q}_2)}{w_1(\mathbf{k}_1) + w_2(\mathbf{k}_2) - w_1(\mathbf{q}_1) - w_2(\mathbf{q}_2)}. \quad (5.82)$$

Here, $w_{1,2}(\mathbf{p}) = \sqrt{m_{1,2}^2 + \mathbf{p}^2}$ and $m_{1,2}$ are the masses of the particles. Dimensional regularization is used throughout. Furthermore,

$$(2\pi)^d \delta^{(d)}(\mathbf{p}_1 + \mathbf{p}_2 - \mathbf{q}_1 - \mathbf{q}_2) w(\mathbf{p}_1, \mathbf{p}_2; \mathbf{q}_1, \mathbf{q}_2) = -\langle \mathbf{p}_1, \mathbf{p}_2 | H_I | \mathbf{q}_1, \mathbf{q}_2 \rangle, \quad (5.83)$$

where H_I is the interaction Hamiltonian, obtained from the interaction Lagrangian $\mathcal{L}^{(0)} + \mathcal{L}^{(2)} + \ldots$, see Eq. (2.105). The first few terms in this expansion are given by

$$w(\mathbf{p}_1, \mathbf{p}_2; \mathbf{q}_1, \mathbf{q}_2) = C_0 + C_1(p_1 p_2 + q_1 q_2 - 2m_1 m_2) + C_2(p_2 - p_1)(q_2 - q_1) + \cdots. \quad (5.84)$$

---

[6] This situation can be realized on the lattice by assigning the momenta $\mathbf{p}_1, \mathbf{p}_2$ and $\mathbf{q}_1, \mathbf{q}_2$ with $\mathbf{p}_1 + \mathbf{p}_2 = \mathbf{q}_1 + \mathbf{q}_2 \neq 0$ to the outgoing and the incoming particles, respectively. Evaluating energy levels in different moving frames is very useful, as it provides many additional data points that can be used, e.g., in a global fit of the infinite-volume phase shifts to lattice data.

All momenta are assumed to be on shell in this expansion.

We further define the center of mass and relative four-momenta of the incoming and outgoing particles, according to Eqs. (2.109) and (2.115). In order to perform a partial-wave expansion of the potential, we should boost it to the rest-frame. The result is given by

$$w(\mathbf{p}_1, \mathbf{p}_2; \mathbf{q}_1, \mathbf{q}_2) = 4\pi \sum_{\ell m} \mathscr{Y}_{\ell m}(\mathbf{p}^*) w_\ell(p^*, q^*) \mathscr{Y}_{\ell m}^*(\mathbf{q}^*). \tag{5.85}$$

Here, $\mathbf{p}^*, \mathbf{q}^*$ are the relative three-momenta, boosted to the rest-frame:

$$\mathbf{p}^* = \mathbf{p} + \mathbf{P}\left[\left(\frac{1}{\sqrt{1-v^2}} - 1\right)\frac{\mathbf{p}\mathbf{P}}{\mathbf{P}^2} - \frac{v}{\sqrt{1-v^2}}\frac{p^0}{|\mathbf{P}|}\right], \qquad v = \frac{\mathbf{P}}{P^0},$$

$$\mathbf{q}^* = \mathbf{q} + \mathbf{Q}\left[\left(\frac{1}{\sqrt{1-v'^2}} - 1\right)\frac{\mathbf{q}\mathbf{Q}}{\mathbf{Q}^2} - \frac{v'}{\sqrt{1-v'^2}}\frac{q^0}{|\mathbf{Q}|}\right], \qquad v' = \frac{\mathbf{Q}}{Q^0}, \tag{5.86}$$

and p^*, q^* are the magnitudes of these momenta.

Because dimensional regularization is used in the infinite volume, on the energy shell we have (cf. with Eq. (5.72):[7]

$$w(p^*, p^*) = r(p^*, p^*) = r_\ell(s) = \frac{8\pi\sqrt{s}}{p^{*2\ell+1}}\tan\delta_\ell(s),$$

$$s = \left(\sqrt{m_1^2 + p^{*2}} + \sqrt{m_2^2 + p^{*2}}\right)^2. \tag{5.87}$$

The following steps are similar to the ones already used for the derivation of the Lüscher equation in the previous section. The partial-wave expansion of the scattering amplitude is given by

$$t(\mathbf{p}_1, \mathbf{p}_2; \mathbf{q}_1, \mathbf{q}_2) = 4\pi \sum_{\ell m} \sum_{\ell' m'} \mathscr{Y}_{\ell m}(\mathbf{p}^*) t_{\ell m, \ell' m'}(p^*, q^*; \mathbf{P}) \mathscr{Y}_{\ell' m'}^*(\mathbf{q}^*). \tag{5.88}$$

In the CM frame, this expression coincides with Eq. (5.66) (of course, the normalization of the amplitudes is different in the two cases). Substituting the partial-wave expansion in the finite-volume Lippmann–Schwinger equation, we get an expression that closely resembles Eq. (5.71):

$$t_{\ell m, \ell' m'}(p^*, p^*; \mathbf{P}) = \delta_{\ell \ell'} \delta_{mm'} r_\ell(s)$$

$$+ 4\pi \sum_{\ell'' m''} r_\ell(s) H_{\ell m, \ell'' m''}(s; \mathbf{P}) t_{\ell'' m'', \ell' m'}(p^*, p^*; \mathbf{P}). \tag{5.89}$$

Here,

$$H_{\ell m, \ell'' m''}(s; \mathbf{P}) = \frac{1}{L^3} \sum_{\mathbf{k}_1} \frac{\mathscr{Y}_{\ell m}^*(\mathbf{k}^*) \mathscr{Y}_{\ell'' m''}(\mathbf{k}^*)}{2w_1(\mathbf{k}_1) 2w_2(\mathbf{P} - \mathbf{k}_1)(w_1(\mathbf{k}_1) + w_2(\mathbf{P} - \mathbf{k}_1) - P^0)},$$

$$\tag{5.90}$$

and $\mathbf{k}^*$ is defined analogously, as in Eq. (5.86):

---

[7] Note that the relativistic normalization used here is different from Eq. (5.72).

$$\mathbf{k}^* = \mathbf{k} + \mathbf{K}\left[\left(\frac{1}{\sqrt{1-v''^2}}-1\right)\frac{\mathbf{k}\mathbf{K}}{\mathbf{K}^2}-\frac{v''}{\sqrt{1-v''^2}}\frac{k^0}{|\mathbf{K}|}\right], \qquad v'' = \frac{\mathbf{K}}{K^0}. \qquad (5.91)$$

In the preceding expressions, k^μ, K^μ are the relative and total four-momenta in the intermediate state. Further, on the energy shell $K^0 = P^0 = Q^0$, leading to

$$\mathbf{k}^* = (\gamma^{-1}k_\parallel, \mathbf{k}_\perp), \qquad \gamma = \frac{1}{\sqrt{1-v^2}}, \qquad (5.92)$$

where $k_\parallel, \mathbf{k}_\perp$ are the components of the three-vector $\mathbf{k}$ parallel and perpendicular to the total momentum $\mathbf{P}$, respectively,

$$k_\parallel = \frac{\mathbf{k}\mathbf{P}}{|\mathbf{P}|}, \qquad \mathbf{k}_\perp = \mathbf{k} - \mathbf{P}\frac{\mathbf{k}\mathbf{P}}{\mathbf{P}^2}. \qquad (5.93)$$

Using now the identity from Eq. (2.108), we may transform the expression for the matrix H into the following form:

$$H_{\ell m,\ell''m''}(s;\mathbf{P}) = \frac{1}{2P^0L^3}\sum_{\mathbf{k}=\mathbf{k}_1+\mu_1\mathbf{P}}\frac{\mathscr{Y}_{\ell m}^*(\mathbf{k}^*)\mathscr{Y}_{\ell''m''}(\mathbf{k}^*)}{\mathbf{k}^2 - \frac{(\mathbf{k}\mathbf{P})^2}{\mathbf{P}^2} - p^{*2}}. \qquad (5.94)$$

With the help of Eq. (5.73), we finally arrive at the expression

$$H_{\ell m,\ell''m''}(s;\mathbf{P}) = \frac{1}{2\sqrt{s}}\frac{i^{\ell-\ell''}}{16\pi^2}p^{*\ell+\ell''+1}\mathscr{M}_{\ell m,\ell''m''}(s;\mathbf{P}), \qquad (5.95)$$

where

$$\mathscr{M}_{\ell m,\ell''m''}(s;\mathbf{P}) = \frac{(-1)^\ell}{\pi^{3/2}\gamma}\sum_{js}C_{\ell m,js,\ell''m''}\left[Z_{js}^{\mathbf{d}}(1;s)\right]^*. \qquad (5.96)$$

Here, $\mathbf{d} = L\mathbf{P}/(2\pi)$, $\eta = p^*L/(2\pi)$ and

$$Z_{js}^{\mathbf{d}}(1;s) = \sum_{\mathbf{r}\in P_d}\frac{\mathscr{Y}_{js}(\mathbf{n})}{\mathbf{n}^2-\eta^2},$$

$$P_d = \left\{\mathbf{r}\in\mathbb{R}^3\,|\,r_\parallel = \gamma^{-1}(n_\parallel - \mu_1|\mathbf{d}|), \mathbf{r}_\perp = \mathbf{n}_\perp, \mathbf{n}\in\mathbb{Z}^3\right\}. \qquad (5.97)$$

In the preceding expression, dimensional regularization is implicitly assumed. Note that Z is a function of $\mathbf{P}, s$ and not merely of η^2 as in the rest-frame.

The Lüscher equation in an arbitrary moving frame takes the form:

$$\det\mathscr{D}(s,\mathbf{P}) = 0,$$

$$\mathscr{D}_{\ell m,\ell''m''}(s,\mathbf{P}) = \delta_{\ell\ell''}\delta_{mm''} - \tan\delta_\ell(p^*)\mathscr{M}_{\ell m,\ell''m''}(s,\mathbf{P}). \qquad (5.98)$$

To summarize, here we have derived the two-body quantization condition for non-equal mass particles in an arbitrary frame. The novel derivation is based on the nonrelativistic effective Lagrangian and makes a shortcut that simplifies life considerably. Namely, sorting the diagrams with/without low-energy singularities is now avoided. All contributions without singularities are included in the effective couplings from the beginning. This results in a much simpler bookkeeping of the diagrams, which allows us to achieve the same goal with less effort. This property of the NREFT is especially important in more complex cases, for example, in studying various matrix elements on the lattice, or in the derivation of the three-body quantization condition (see what follows). Note also that despite being branded "non-relativistic," the modified approach considered here operates with a relativistic invariant scattering two-body amplitude. All effects that emerge due to a non-covariant form of the two-body Green's function are exponentially suppressed for large values of L.

5.4.2 Symmetries in a Box

As already mentioned, rotational symmetry in a finite cubic box is broken. As a result, partial waves with different values of ℓ in the Lüscher equation mix, complicating the situation considerably. The bookkeeping becomes even more obscure in case of particles with spin. In what follows, we shall briefly address this situation and demonstrate that the use of the space-time symmetries enables one to partially diagonalize the Lüscher equation and to grossly reduce the dimensionality of the matrix entering this equation. We give only a brief sketch of the method here. For the details, a reader is referred to the original papers; see, for example, Refs. [14, 30, 31, 35, 36].

As we have already seen, the Lüscher equation for two spinless particles in an arbitrary moving frame is given by Eq. (5.98). This equation undergoes only minor modification for particles with spin. As an example, we give here the equation, corresponding to the scattering of spin-$\frac{1}{2}$ and spin-0 particles (such as pion–nucleon scattering) [31, 35]. Here, the relevant quantum numbers are the total angular momentum J, its projection on the z-axis μ and the orbital momentum $\ell = J \pm \frac{1}{2}$. The equation takes the form

$$\det \mathscr{D}(s, \mathbf{P}) = 0,$$

$$\mathscr{D}_{J\ell\mu, J''\ell''\mu''}(s, \mathbf{P}) = \delta_{JJ''} \delta_{\ell\ell''} \delta_{\mu\mu''} - \tan \delta_{J\ell}(p^*) \mathscr{M}_{J\ell\mu, J''\ell''\mu''}(s, \mathbf{P}), \qquad (5.99)$$

where the $\delta_{J\ell}$ denote the scattering phase shifts, and

$$\mathscr{M}_{J\ell\mu, J''\ell''\mu''}(s, \mathbf{P}) = \sum_{m\sigma, m'\sigma'} \langle \ell m \tfrac{1}{2} \sigma | J\mu \rangle \langle \ell''m'', \tfrac{1}{2}\sigma'' | J''\mu'' \rangle \mathscr{M}_{\ell m, \ell''m'}(s, \mathbf{P}). \qquad (5.100)$$

The quantities $\langle \cdots | \cdots \rangle$ denote Clebsch–Gordan coefficients (see Appendix A). The generalization to the case with a different spin is obvious (e.g. see Ref. [36], where the scattering of spin-1 particles is considered in all details).

As mentioned already, since the lattice breaks rotational invariance, the different partial waves in the Lüscher equation mix. However, the rotational invariance is not broken down completely. It is rather reduced to a finite-dimensional group of space transformations that leaves the lattice invariant. For example, in case of a cubic lattice and zero total momentum, the symmetry group consists of 24 rotations that transform a cube into itself, plus the space inversion, that is, 48 elements in total. These form the octahedral group O_h. In case of fermions, we have to consider the so-called double cover of the octahedral group, denoted by ^2O_h. This group consists of the rotations and inversion, supplemented by an element that we denote by J. This element corresponds to the rotation by an angle of 2π around any axis. Thus, the group ^2O_h consists in total of 96 elements g and $\bar{g} = Jg$, where g denotes any element of O_h. The double-valued representations of O_h correspond to those single-valued representations of ^2O_h, in which the matrix, corresponding to the element J, differs in sign from the matrix corresponding to the unit element I, that is, $T(J) = -T(I)$.[8]

In case of non-rest frames or elongated lattices (i.e., the lattices for which the different space dimensions L_1, L_2, L_3 are not all the same), the symmetry group is further reduced. Consider, for instance, the case of nonzero total momentum (the case of an elongated box can be treated similarly). Let $\mathbf{d}$ be an integer vector, which coincides in direction with the total three-momentum $\mathbf{P}$. The symmetry transformations, in addition, should leave this vector invariant:

$$g\mathbf{d} = \mathbf{d}. \tag{5.101}$$

In case of equal masses, the above condition is replaced by

$$g\mathbf{d} = \pm\mathbf{d}. \tag{5.102}$$

These conditions pick out certain elements g of O_h or ^2O_h, which form different subgroups of these groups, the so-called *little groups*, depending on the choice of the vector $\mathbf{d}$. Explicitly, these elements for different choices of $\mathbf{d}$ are listed in Ref. [31].

We do not aim here to give a complete introduction into point groups and their representations.[9] Rather, we will focus on a general line of reasoning only. If the rotational symmetry were in place, the Lüscher equation would be diagonalized in the basis of vectors that correspond to the different irreducible representations (irreps) of the rotation group. There are infinitely many irreps, labeled by $\ell = 0, 1, \ldots$ (boson-boson scattering) and by $J = \frac{1}{2}, \frac{3}{2}, \ldots, \ell = J \pm \frac{1}{2}$ (scattering of the spin-$\frac{1}{2}$ and spin-0 particles). The basis vectors in these irreps are $|\ell m\rangle$ with $-\ell \le m \le \ell$ and $|J\ell\mu\rangle$ with $-J \le \mu \le J$, respectively. On the lattice, the continuous rotation group is restricted to a discrete subgroup thereof, which has only a finite number of irreps. For example, the group O_h has two one-dimensional irreps A_1 and A_2, one two-dimensional irrep E and two three-dimensional irreps T_1, T_2. These representations are defined by matrices:

$$T^{\Gamma}(g): \quad g \in O_h, \quad \Gamma = A_1, A_2, E, T_1, T_2. \tag{5.103}$$

In case of the little groups, or the group ^2O_h, different irreps emerge. All these irreps and matrices $T^{\Gamma}(g)$ in these irreps are known. They are listed explicitly in Refs. [31, 35].

It is clear that, since all finite groups considered here are subgroups of the rotational group, all these irreps can be realized in a linear space that is formed by the basis vectors $|\ell m\rangle$ and $|J\ell\mu\rangle$, corresponding to a fixed ℓ and J, ℓ, respectively. However, in general, more than one irrep of a given finite group can be realized in each linear space, so that the basis vectors of these irreps will be given by particular linear combinations of the

---

[8] For illustrative purposes, note that the group $SU(2)$ represents a double cover of the rotation group $SO(3)$. Restricting the group parameters to a discrete set corresponding to all spatial symmetries of a cube the structure of the double cover ^2O_h is reproduced by the 2×2 matrices belonging to $SU(2)$.

[9] For a very brief and concise introduction we recommend the excellent textbook [37]; see also [38–40].

initial basis vectors $|\ell m\rangle$ and $|J\ell\mu\rangle$. Our aim here is to find the matrix that transforms the vectors $|\ell m\rangle$ and $|J\ell\mu\rangle$ into the basis vectors corresponding to a particular irrep Γ.

The solution of this problem is well known from group theory. In a given irrep Γ, the basis vectors are labeled by an index $\alpha = 1, \ldots, s_\Gamma$, where s_Γ denotes the dimension of the irrep Γ. We denote these basis vectors by $|\Gamma_t \alpha, \ell\rangle$ and $|\Gamma_t \alpha, J\ell\rangle$, respectively. Here, the index t labels different irreps corresponding to the same Γ and takes the values $t = 1, \ldots, N_\Gamma$, where N_Γ denotes the multiplicity of the irrep Γ. The projector on the basis vector is given by

$$\Pi_\alpha = \frac{s_\Gamma}{G} \sum_g \left(T^\Gamma_{\alpha\rho}(g)\right)^* g, \qquad (5.104)$$

where the sum runs over all elements of the (little) group, and G denotes a total number of elements in this sum. Furthermore, the index ρ is fixed (to avoid clutter of indices, we do not indicate it explicitly in the following). $T^\Gamma_{\alpha\beta}(g)$ denote the matrices of a given irrep.

Let us act now with this projector on a vector $|\ell m\rangle$ with a fixed m (for simplicity, we consider the bosonic case first). Using the relation

$$g|\ell m\rangle = \sum_{m'=-\ell}^{\ell} \mathscr{D}^\ell_{m'm}(g)|\ell m'\rangle, \qquad (5.105)$$

where $\mathscr{D}^\ell_{m'm}(g)$ denote the Wigner $\mathscr{D}$-matrices, corresponding to the transformation g [41], one finally arrives at the relation

$$\Pi_\alpha|\ell m\rangle = \sum_{m'=-\ell}^{\ell} C^\alpha_{mm'}|\ell m'\rangle, \qquad (5.106)$$

with some well-defined coefficients $C^\alpha_{mm'}$. We have now to repeat this procedure for all m from $-\ell$ to ℓ. If a given irrep Γ is not contained in the irrep of the rotation group, labeled by ℓ, we will get zero for all m. In general, the vectors $\Pi_\alpha|\ell m\rangle$ will be nonzero for some values of m, and the dimension of the linear space spanned by these vectors is equal to N_Γ. Finally, repeating this procedure to all α in a given irrep Γ and for all irreps Γ, we obtain all basis vectors we are looking for. These basis vectors are given by the linear combinations

$$|\Gamma_t \alpha, \ell\rangle = \sum_{m=-\ell}^{\ell} c^{\Gamma_t\alpha}_{\ell m}|\ell m\rangle, \qquad (5.107)$$

where the coefficients $c^{\Gamma_t\alpha}_{\ell m}$ can be easily read from the known values of $C^\alpha_{mm'}$. Furthermore, the fermion–boson case can be dealt with similarly. The transformation of the basis vectors of the rotation group irreps in this case is defined as

$$g|J\ell\mu\rangle = \sum_{\mu=-J}^{J} \mathscr{D}^J_{\mu'\mu}(g)|J\ell\mu'\rangle, \qquad (5.108)$$

and the basis vectors in a given irrep Γ_t are given by

$$|\Gamma_l \alpha, J\ell\rangle = \sum_{\mu=-J}^{J} c_{J\ell\mu}^{\Gamma_l \alpha} |J\ell\mu\rangle.$$ (5.109)

For low values of ℓ and J, the coefficients $c_{\ell m}^{\Gamma_l \alpha}$ and $c_{J\ell\mu}^{\Gamma_l \alpha}$ are listed explicitly in Refs. [31, 35].

The rest is now straightforward. The matrix elements of the operator $\mathcal{M}$ in the new basis are given by

$$\langle \Gamma_l \alpha, \ell | \mathcal{M} | \Gamma_{l'}' \alpha', \ell' \rangle = \sum_{m=-\ell}^{\ell} \sum_{m'=-\ell'}^{\ell'} \left(c_{\ell m}^{\Gamma_l \alpha} \right)^* c_{\ell' m'}^{\Gamma_{l'}' \alpha'} \mathcal{M}_{\ell m, \ell' m'},$$ (5.110)

and

$$\langle \Gamma_l \alpha, J\ell | \mathcal{M} | \Gamma_{l'}' \alpha', J'\ell' \rangle = \sum_{\mu=-J}^{J} \sum_{\mu'=-J'}^{J'} \left(c_{J\ell\mu}^{\Gamma_l \alpha} \right)^* c_{J'\ell'\mu'}^{\Gamma_{l'}' \alpha'} \mathcal{M}_{J\ell\mu, J'\ell'\mu'},$$ (5.111)

respectively. Furthermore, according to Schur's first lemma,

$$\langle \Gamma_l \alpha, \ell | \mathcal{M} | \Gamma_{l'}' \alpha', \ell' \rangle = \delta_{\Gamma\Gamma'} \delta_{\alpha\alpha'} \mathcal{M}_{\ell t, \ell' t'}^{\Gamma},$$

$$\langle \Gamma_l \alpha, J\ell | \mathcal{M} | \Gamma_{l'}' \alpha', J'\ell' \rangle = \delta_{\Gamma\Gamma'} \delta_{\alpha\alpha'} \mathcal{M}_{J\ell t, J'\ell' t'}^{\Gamma}.$$ (5.112)

This solves our problem. We see that the Lüscher equation is maximally diagonalized in a manner allowed by the remaining spatial symmetries. In practice this means a huge reduction of the dimensionality of the matrix $\mathcal{D}$ entering the Lüscher equation. For example, in the CM frame scattering of two identical bosons, cutting off the angular momenta at $\ell \leq 3$, we get a simple equation for the $\Gamma = A_1$ irrep, which does not even contain a determinant anymore:

$$1 - \tan\delta_\ell(q_0) w_{00} = 0, \qquad w_{\ell m} = \frac{1}{\pi^{3/2}} \eta^{-(\ell+1)} Z_{\ell m}(1, \eta^2).$$ (5.113)

Last but not least, it should be pointed out that the space-time symmetries enforce substantial restrictions on the $w_{\ell m}$. According to these, many of the $w_{\ell m}$ vanish. For example, in the CM frame, ℓ must be even. This can be checked immediately, changing the summation $\mathbf{n} \to -\mathbf{n}$ and recalling that $\mathscr{Y}_{\ell m}(-\mathbf{n}) = (-1)^\ell \mathscr{Y}_{\ell m}(\mathbf{n})$. Furthermore, in case of $\mathbf{d} = (0,0,1)$, the quantity w_{10} vanishes in the equal-mass case. This can be shown by applying a combined transformation $\mathbf{n} \to \mathbf{n} + \mathbf{d}$ and $\mathbf{n} \to -\mathbf{n}$. There are many more relations stemming from symmetry [31]. When using Schur's first lemma as done above, all these relations are taken into account, in addition.

5.4.3 Resonances

At low energies, where the higher partial waves can be neglected, the Lüscher equation takes the form given in Eq. (5.113). In the S-wave, for instance, it can be rewritten as

$$q_0 \cot\delta_0(q_0) = \frac{2}{\sqrt{\pi}L} Z_{00}(1; \eta^2).$$ (5.114)

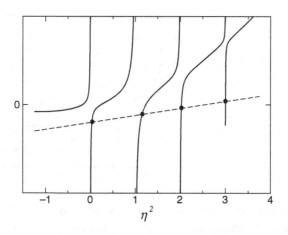

Figure 5.5   Schematic representation of the left- and right-hand sides of the Lüscher equation, Eq. (5.114). The Lüscher function on the right-hand side (solid line) has simple poles at the values of η^2 that correspond to the free-energy levels. On the contrary, the left-hand side (dashed line) is in general a continuous function. The points where these two lines cross each other (filled dots) determine the interacting energy levels.

Lüscher's zeta-function has poles at the energies that correspond to the free energy levels in the finite box. This function is plotted in Fig. 5.5. On the contrary, the left-hand side of Eq. (5.114) is, in general, a smooth function of the momentum q_0. The values of the points where the two curves cross determine the energy levels for a given value of L.

In the infinite volume, the presence of narrow resonances is one of the most spectacular features of the phase shift. It is interesting to know whether resonances have a similar impact of the finite-volume energy levels, which is immediately seen without solving the Lüscher equation first. This question is answered in Fig. 5.6, where the dependence on the energy levels on the box size L is shown. In order to obtain this picture, we have used a simple resonance parameterization of the phase shift:

$$q_0 \cot \delta_0(q_0) = \frac{q_R^2 - q_0^2}{\gamma_R}, \qquad (5.115)$$

where q_R^2 is the relative momentum corresponding to the resonance excitation, $E_R = q_R^2/m$, and γ_R is proportional to the width.[10] In the figure, curves corresponding to different values of γ_R are shown. We see that when γ_R decreases, a particular structure of the energy levels, the so-called avoided level crossing, sets in. Such a structure does not come as a surprise. Indeed, assume that $\gamma_R \to 0$. As we know, a very narrow resonance can be, to a good approximation, described by an s-channel exchange of a fictitious particle of a mass equal to the resonance energy. When $\gamma_R \to 0$, this particle becomes stable. Then, the finite-volume spectrum consists of a (meta)stable particle, which gives an L-independent horizontal line plus free two-particle levels decreasing with L (because we have assumed no interaction except the resonance exchange). If

---

[10] Strictly speaking, this is only true in the limit of a narrow resonance. We shall consider a general case in what follows.

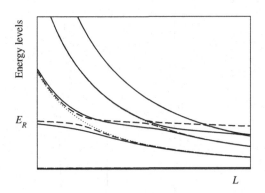

Figure 5.6 Avoided level crossing in the vicinity of the resonance energy E_R. For smaller values of the width (dashed lines), this structure in the energy levels is more pronounced. Dotted lines denote free energy levels.

γ_R is very small but nonzero, the picture is slightly modified, the energy levels merge and the crossing is no more present. Still, it strongly resembles the original picture. To summarize, avoided level crossing is a salient feature of the lattice spectrum in a finite volume that provides a very clear signature of the presence of a resonance in elastic scattering.[11]

This, however, is not the whole story. As it is known, resonances correspond to the singularities of the S-matrix elements in the complex plane. So, strictly speaking, q_R^2 and γ_R do not exactly correspond to the resonance energy and width. In order to determine these, we should perform an analytic continuation in the complex plane. Most easily, this is achieved by using a parameterization for the phase shift, which is compatible with the general properties of the amplitudes in field theory, like analyticity and unitarity. For example, in the vicinity of a threshold, we might consider the effective-range expansion:

$$q_0 \cot \delta_0(q_0) = -\frac{1}{a} + \frac{1}{2} r q_0^2 + \cdots . \tag{5.116}$$

The effective range expansion parameters $a, r, \ldots$ can be determined from a fit to the lattice energy levels. In the next step, we should determine the pole position(s) on the second Riemann sheet[12], solving the following equation:

$$-\frac{1}{a} + \frac{1}{2} r \left(\frac{s}{4} - m^2 \right) + \cdots - \sqrt{m^2 - \frac{s}{4}} = 0, \tag{5.117}$$

with $a, r, \ldots$ determined on the lattice. As expected, in the limit of small γ_R the previous result is readily reproduced.

5.4.4 Bound States

The leading finite-volume artifacts for shallow bound states can also be evaluated with the use of the Lüscher approach. If the energies are below threshold, the corresponding

---

[11] Note that, in the multichannel case, a very similar structure can emerge in the presence of an inelastic two-body threshold [17–19].

[12] This implies a change of sign in front of the square root in Eq. (5.117).

momentum becomes complex and we can set $q_0 = i\gamma_L$. In the infinite volume limit, $\gamma_L \to \gamma$. It is further assumed that $\gamma \ll m$. There are only exponentially suppressed corrections to the binding energy (because a bound state is stable). However, these corrections are of two types. The Lüscher equation allows one to reproduce the corrections that go like $e^{-\gamma L}$, but not like e^{-mL}. Hence, the condition $\gamma \ll m$ is needed in order to keep these two effects clearly apart.

The derivation of the energy level shift of the shallow bound state is an easy task. Using Poisson's summation formula, which is valid below threshold, it can be shown that

$$\frac{2}{\sqrt{\pi}L} Z_{00}(1;\eta^2) = \frac{1}{\pi L} \sum_{\mathbf{n} \in \mathbb{Z}^3 \setminus 0} \int d^3\mathbf{k} \frac{e^{2\pi i \mathbf{n} \mathbf{k}}}{\mathbf{k}^2 + (L\gamma_L/(2\pi))^2}$$

$$= \frac{6}{L} e^{-L\gamma_L} + O(e^{-\sqrt{2}L\gamma_L}). \tag{5.118}$$

The leading contribution is provided by the terms with $|\mathbf{n}| = 1$.

Using the effective-range expansion, the Lüscher equation in the vicinity of the elastic threshold takes the following form:[13]

$$-\frac{1}{a} - \frac{1}{2} r\gamma_L^2 + \cdots + \gamma_L = \frac{6}{L} e^{-L\gamma_L} + \cdots. \tag{5.119}$$

On the other hand, the infinite-volume quantity γ obeys the equation

$$-\frac{1}{a} - \frac{1}{2} r\gamma^2 + \gamma = 0. \tag{5.120}$$

Solving Eq. (5.119) iteratively, one obtains

$$\gamma_L = \gamma + \frac{6}{L(1 - r\gamma)} e^{-L\gamma} + \cdots. \tag{5.121}$$

Finally, the energy shift of the shallow bound state in a finite volume is determined by the following expression:

$$E_L = \frac{\gamma_L^2}{m} = \frac{\gamma^2}{m} \left(1 + \frac{12}{\gamma L(1 - r\gamma)} e^{-L\gamma} + \cdots \right). \tag{5.122}$$

For the first time, this expression was derived in Ref. [42]; see also Ref. [43]. The energy shift of the bound states in higher partial waves has been derived in Ref. [44].

5.4.5 Threshold Expansion

The free energy levels in a finite volume are given by $E_n = 2\sqrt{m^2 + (2\pi\mathbf{n}/L)^2}$ and behave at large L as $E_n = 2m + O(L^{-2})$. Furthermore, in the presence of interactions, these

---

[13] Note that, in fact, one makes two types of expansion here: an expansion in the finite-volume exponentials and the effective-range expansion. These two expansions are not correlated with each other. However, it is assumed that $|\gamma_L - \gamma| \ll \gamma$.

levels are subject to a shift. As we shall see, these shifts are of order of L^{-3}. Hence, perturbation theory is applicable, because the shift is smaller than the distance between the neighboring levels.

In this section, we shall derive the expression of the ground-state energy shift up to and including terms of $O(L^{-5})$. To this end, let us Laurent-expand the zeta-function in the vicinity of this level corresponding to $\eta^2 = 0$. If the simple pole at $\eta^2 = 0$ is subtracted, the rest can be Taylor-expanded, because the distance to the nearest singularity is much larger than the shift of η^2 we are looking for. To the order we are working, we may write

$$Z_{00}(1;\eta^2) = \frac{1}{\sqrt{4\pi}} \left(-\frac{1}{\eta^2} + I + \eta^2 J + O(\eta^4) \right),$$

$$I = \sum_{\mathbf{n}\in\mathbb{Z}^3\backslash\mathbf{0}} \frac{1}{\mathbf{n}^2} \simeq -8.9136,$$

$$J = \sum_{\mathbf{n}\in\mathbb{Z}^3\backslash\mathbf{0}} \frac{1}{\mathbf{n}^4} \simeq 16.5323. \tag{5.123}$$

A comment is needed only in case of the first sum, which is (formally) ultraviolet-divergent. Note, however, that it is (implicitly) regularized by subtracting the infinite-volume counterpart thereof that is displayed, for example, in Eq. (5.77). In practice, this is done in the following manner. We subtract and add an expression evaluated at some point $\eta^2 = -\mu^2$ below threshold:

$$I = \sum_{\mathbf{n}\in\mathbb{Z}^3\backslash\mathbf{0}} \left(\frac{1}{\mathbf{n}^2} - \frac{1}{\mathbf{n}^2+\mu^2} \right) + \sum_{\mathbf{n}\in\mathbb{Z}^3} \int d^3\mathbf{k}\, \frac{e^{2\pi i\mathbf{n}\mathbf{k}}}{\mathbf{k}^2+\mu^2} - \frac{1}{\mu^2}. \tag{5.124}$$

The first sum here is explicitly convergent. In the second sum, the (dimensionally regularized) infinite-volume part $\int d^d\mathbf{k}/(\mathbf{k}^2+\mu^2)$ should be subtracted, rendering this term also convergent. Evaluating these sums numerically, we arrive at the result given in Eq. (5.123). This final result does not depend on μ, of course.

Furthermore, the equation for η^2 can be solved iteratively. To this end, we use the effective-range expansion $q_0 \cot \delta_0(q_0) = -1/a + O(q_0^2)$ in the Lüscher equation and obtain

$$\eta^2 = \frac{a}{\pi L} \left(1 + \frac{a}{\pi L} I + \frac{a}{\pi L} \eta^2 J + \cdots \right)^{-1}$$

$$= \frac{a}{\pi L} \left(1 - \frac{a}{\pi L} I + \left(\frac{a}{\pi L} \right)^2 (I^2 - J) + O(L^{-3}) \right). \tag{5.125}$$

Finally, the (nonrelativistic) energy shift of the ground state is given by

$$\Delta E = \frac{4\pi^2}{mL^2} \eta^2 = \frac{4\pi a}{mL^3} \left(1 - \frac{a}{\pi L} I + \left(\frac{a}{\pi L} \right)^2 (I^2 - J) + O(L^{-3}) \right). \tag{5.126}$$

This expression, which in the context of lattice calculations was first derived in Ref. [45], is very convenient for the extraction of the hadronic scattering lengths from the measured ground-state energy. In what follows, we shall provide an alternative derivation of this formula, which can be straightforwardly extended to many-particle systems.

Two concluding remarks are appropriate here. First, the shift of the two-particle energy levels decreases as L^{-3} for large L, in contrast to the single-particle levels, which contain only exponentially suppressed artifacts. As we know, such a power-law behavior is caused by the singularities of the Feynman diagrams for two-particle scattering, which emerge in the physical region. Second, in contrast to the shallow bound-state case, here the effective-range expansion is correlated with the $1/L$ expansion, so that each power of momentum counts like $1/L$. For this reason, for example, a contribution from the effective range r appears first at order L^{-6}. Hence, we did not consider this contribution explicitly. Moreover, there are relativistic corrections at $O(L^{-6})$. We shall address these questions later, when an alternative derivation will be considered.

5.5 Twisted Boundary Conditions

The components of three-momenta on the lattice take integer values in units of $2\pi/L$. For typical lattices used nowadays, this quantity is rather large. This leads to a large spacing between energy levels, so that only few of them can fit the "allowed" interval below the multiparticle thresholds. On the other hand, even if one were able to work with very large lattices, disentangling many close-by levels would constitute a serious challenge. Furthermore, since the Lüscher equation enables one to extract the scattering phase *exactly at the energy of the measured level,* it is clear that we will be able to measure phases only at a few values of the argument, which are separated by large intervals. A detailed scan in energy remains elusive.

Note further that the problem cannot be reduced alone to the measurement of the energies and phase shifts. For example, calculating matrix elements (say, the pion electromagnetic form factor), we have to fix the momenta of the initial and final particles. As a result, the momentum transfer in the form factor is also quantized. Considering moving frames helps to alleviate the severity of this problem only partially. Moreover, for such high values of the lattice momenta, the stochastic noise in the measured correlators increases significantly. This puts certain limits on the use of this method.

In Refs. [46–48], a novel approach was proposed. It is based on the so-called twisted boundary conditions. The quark fields in this case satisfy the following condition on the boundaries of the spatial box:

$$\psi(\mathbf{x} + \mathbf{e}_a L, x_4) = e^{i\theta e_a} \psi(\mathbf{x}, x_4), \qquad \bar{\psi}(\mathbf{x} + \mathbf{e}_a L, x_4) = e^{-i\theta e_a} \bar{\psi}(\mathbf{x}, x_4). \qquad (5.127)$$

Here, the $\mathbf{e}_{1,2,3}$ denote unit vectors in the direction of the lattice axes, and $\boldsymbol{\theta}$ is the so-called twist angle. The components of this three-vector lie in the interval $0 \leq \theta_i < 2\pi$. The twist angle can be different for different fermion species. Thus, $\boldsymbol{\theta}$ represents a diagonal matrix in flavor space, with the entries $\boldsymbol{\theta} = \mathrm{diag}(\boldsymbol{\theta}_u, \boldsymbol{\theta}_d, \boldsymbol{\theta}_s)$. If one wishes, we might impose a restriction $\theta_u + \theta_d + \theta_s = 0$. This would merely eliminate a common phase of the three quark fields.

Note that the twisted boundary conditions are compatible with the general principles of quantum field theory on the lattice. Requiring that the fields are single-valued after the rotation on the torus (this corresponds to $\mathbf{x} \to \mathbf{x} + \mathbf{e}_a L$) is not necessary. It suffices if the observables are single-valued or, equivalently, the action is single-valued. It is straightforward to check that the QCD action indeed remains single-valued after imposing periodic boundary conditions.

For the further discussion, it will be useful to redefine the quark fields as

$$\psi(\mathbf{x}, x_4) = V(x)\tilde{\psi}(\mathbf{x}, x_4), \qquad V(x) = \exp\left(i\frac{\boldsymbol{\theta}\mathbf{x}}{L}\right). \tag{5.128}$$

It is easy to check that the redefined the quark fields $\tilde{\psi}(x)$ should obey the usual periodic boundary conditions.

The Euclidean fermion Lagrangian in terms of the redefined fields can be rewritten as

$$\mathcal{L}_F = \bar{\tilde{\psi}}(x)\left(\gamma_\mu D_\mu + V^\dagger(x)\gamma_\mu \partial_\mu V(x) - \mathcal{M}\right)\tilde{\psi}(x)$$

$$= \bar{\tilde{\psi}}(x)\left(\gamma_\mu D_\mu + i\gamma_\mu B_\mu - \mathcal{M}\right)\tilde{\psi}(x), \tag{5.129}$$

where

$$B_\mu = \frac{1}{L}(0, \boldsymbol{\theta}). \tag{5.130}$$

Thus, imposing twisted boundary conditions is mathematically equivalent to switching on a constant Abelian background field B_μ. These two descriptions are related by a gauge transformation.

In order to demonstrate that imposing the twisted boundary conditions enables one to "inject" a three-momentum into a quark line, we consider the free quark propagator. Since the redefined fields obey periodic boundary conditions in Euclidean space, we have

$$\langle 0|T\,\tilde{\psi}(x)\bar{\tilde{\psi}}(y)|0\rangle = \int \frac{dk_4}{2\pi}\frac{1}{L^3}\sum_{\mathbf{k}}\frac{e^{ik(x-y)}}{\mathcal{M} + i\gamma_\mu(k_\mu + B_\mu)}. \tag{5.131}$$

Here, the summation runs over $\mathbf{k} = 2\pi\mathbf{n}/L$, $\mathbf{n} \in \mathbb{Z}^3$. Further, taking into account that

$$\langle 0|T\,\psi(x)\bar{\psi}(y)|0\rangle = \exp(iBx)\langle 0|T\,\tilde{\psi}(x)\bar{\tilde{\psi}}(y)|0\rangle \exp(-iBy)$$

$$= \int \frac{dk_4}{2\pi}\frac{1}{L^3}\sum_{\mathbf{k}}\frac{e^{i(k+B)(x-y)}}{\mathcal{M} + i\gamma_\mu(k_\mu + B_\mu)}, \tag{5.132}$$

in the expression of the propagator k_μ is shifted to $k_\mu + B_\mu$.

Here, we are primarily interested in imposing the twisted boundary conditions at the level of effective hadronic Lagrangians. For instance, it is clear that, if a meson consists of a quark with flavor f and an antiquark with flavor f', the boundary condition for the meson fields reads

$$\phi_{ff'}(\mathbf{x} + \mathbf{e}_a L, x_4) = e^{i(\boldsymbol{\theta}_f - \boldsymbol{\theta}_{f'})\,\mathbf{e}_a} \phi_{ff'}(\mathbf{x}, x_4). \tag{5.133}$$

The twisted boundary condition for a baryon field, composed of three quark fields with arbitrary flavor, can be written down similarly. Furthermore, evaluating the two-point function of a (composite) meson field on the lattice and measuring the ground-state energy shift, we will get

$$E_0 = \sqrt{M_L^2 + \frac{(\boldsymbol{\theta}_f - \boldsymbol{\theta}_{f'})^2}{L^2}}, \tag{5.134}$$

where M_L denotes the finite-volume meson mass as measured by imposing periodic boundary conditions. We see that indeed the three-momentum, injected in a meson by imposing twisted boundary conditions, is equal to $\mathbf{p}_\theta = (\boldsymbol{\theta}_f - \boldsymbol{\theta}_{f'})/L$. This means that the neutral mesons like π^0 or η do not change momenta, whereas the charged mesons do. This fact will lead in particular to a change of the structure of energy levels. For example, in the isospin-symmetric case, the two-particle thresholds corresponding to the $2\pi^0$ and $\pi^+\pi^-$ states coincide when periodic boundary conditions are imposed. This is no more the case with twisted boundary conditions, where these two thresholds are split, leading effectively to a two-channel problem for the extraction of the phase shifts.

In the construction of the effective Lagrangian, we have used the meson matrix $U(x)$, which transformed under chiral transformations in a particularly simple form. The boundary conditions on the meson field matrix $U(x)$ are written as

$$U(\mathbf{x} + \mathbf{e}_a L, x_4) = \exp\left(i\frac{\boldsymbol{\theta}\mathbf{x}}{L}\right) U(\mathbf{x}, x_4) \exp\left(-i\frac{\boldsymbol{\theta}\mathbf{x}}{L}\right). \tag{5.135}$$

Furthermore, like the quark fields, it is possible to redefine the meson matrix and the field χ that contains the external scalar and pseudoscalar sources:

$$U(x) = V(x)\tilde{U}(x)V(x)^\dagger, \qquad \chi(x) = V(x)\tilde{\chi}(x)V(x)^\dagger. \tag{5.136}$$

The covariant derivative transforms as

$$D_\mu U = \partial_\mu U + i[v_\mu, U] + i\{a_\mu, U\} = V(\tilde{D}_\mu U)V^\dagger,$$

$$\tilde{D}_\mu U = \partial_\mu \tilde{U} + i[(V^\dagger v_\mu V + B_\mu), \tilde{U}] + i\{(V^\dagger a_\mu V), \tilde{U}\}. \tag{5.137}$$

Here, v_μ, a_μ denote the external vector and axial-vector sources. It is clear that any meson four-momentum that appears in the derivative vertex will be shifted by a constant value B_μ, similarly to the propagator. The derivatives acting on the baryon fields will be modified correspondingly.

The calculation of the finite-volume artifacts can now be carried out in a straightforward fashion. As an example, we reconsider the calculation of the finite-volume corrections to the pion mass, already performed in Section 5.2.2 (for simplicity, we again work in the theory with two flavors). There are both charged and neutral pions running in the loop, and now one needs to distinguish them. The tadpole diagram in Eq. (5.15) remains the same in case of a neutral pion loop. It is still given by the expression in Eq. (5.20). The charged pion loop, however, gets modified, since now the momenta $\mathbf{k}$ run over a different set. Denoting $\boldsymbol{\theta} = \boldsymbol{\theta}_u - \boldsymbol{\theta}_d$, instead of Eq. (5.17) we obtain

$$I_+^L = \frac{1}{L^3} \sum_{\mathbf{k}=(2\pi\mathbf{n}+\boldsymbol{\theta})/L,\ \mathbf{n}\in\mathbb{Z}^3} \int \frac{dk_4}{2\pi} \frac{1}{M^2+k_4^2+\mathbf{k}^2}$$

$$= \sum_{\mathbf{n}\in\mathbb{Z}^3} \int \frac{dk_4}{2\pi} \int \frac{d^d\mathbf{k}}{(2\pi)^d} \frac{e^{iL\mathbf{n}\mathbf{k}}e^{-i\mathbf{n}\boldsymbol{\theta}}}{M^2+k_4^2+\mathbf{k}^2}. \tag{5.138}$$

The infinite-volume term is again obtained at $\mathbf{n}=0$. The leading correction to the mass emerges from the six terms in the sum with $|\mathbf{n}|=1$. Dropping the sub-leading corrections, one readily gets (cf. with Eq. (5.20)):

$$\Delta I_+^L = \frac{M}{2\pi^2 L}(\cos\theta_1 + \cos\theta_2 + \cos\theta_3)K_1(ML) + \cdots. \tag{5.139}$$

Here, the θ_i are the spatial components of the vector $\boldsymbol{\theta}$.

What remains is to redo the combinatorics in the calculation of the pion self-energy. Here, we must separate the loops with neutral and charged pions. This is a straightforward exercise in applying Wick's theorem, and the result for the self-energies of the $\pi^\pm, \pi^0$ takes the form

$$\pi^\pm: \qquad M_L^2 = M^2 + \frac{M^2}{2F^2}I_0^L + \frac{2M^4}{F^2}l_3,$$

$$\pi^0: \qquad M_L^2 = M^2 + \frac{M^2}{2F^2}(2I_+^L - I_0^L) + \frac{2M^4}{F^2}l_3. \tag{5.140}$$

The leading exponential correction to the charged and neutral pion masses is therefore

$$\Delta M_{\pi^\pm}^2 = \frac{3M_\pi^4}{2F_\pi^2} \frac{e^{-LM_\pi}}{(2\pi LM_\pi)^{3/2}},$$

$$\Delta M_{\pi^0}^2 = \frac{3M_\pi^4}{2F_\pi^2} \frac{e^{-LM_\pi}}{(2\pi LM_\pi)^{3/2}}\left(\frac{2}{3}(\cos\theta_1 + \cos\theta_2 + \cos\theta_3) - 1\right). \tag{5.141}$$

Expressions of this type, and also the expressions for the finite-volume shift of the charged and neutral pion decay constants, were first obtained in Ref. [47]. Similar formulae can be worked out for other observables on the lattice, which do not receive contributions from singular diagrams in the physical region, say, the spacelike form factors. Examples of such calculations are given, for instance, in Ref. [49].

Application of the twisted boundary conditions to scattering processes opens quite a few new opportunities. Below, we shall only mention these without going into details. The expression for the Lüscher zeta-function is modified. For example, in the CM frame,

$$Z_{js}^\theta(1;\eta^2) = \left\{\sum_{\mathbf{n}\in\mathbb{Z}^3} -\text{P.V.}\int d^3\mathbf{n}\right\} \frac{\mathcal{Y}_{js}(\mathbf{n}+\frac{\boldsymbol{\theta}}{2\pi})}{(\mathbf{n}+\frac{\boldsymbol{\theta}}{2\pi})^2 - \eta^2}, \tag{5.142}$$

the (continuous) dependence of the zeta-function on the twist angle can be exploited. For example, in Ref. [50], it has been shown that choosing the vector $\boldsymbol{\theta}$ in a particular manner, it is possible to make the leading exponential corrections to the binding energy of the deuteron vanish. The systematic uncertainty in the determination of the energy on the lattice will be drastically reduced in this case. Moreover, studying the θ-dependence of shallow bound states in a finite volume, we might judge about the nature of these bound states, for example, whether they are the weakly bound hadronic molecules or represent tightly bound compounds of quarks and gluons. The difference between the two alternatives is determined by the typical size of these states, which is given by the inverse of the binding momentum γ. Consequently, hadronic molecules exhibit a much stronger volume-dependence than the tight compounds, for which γ is of the order of the pion mass or higher. This qualitative criterion can be translated into a quantitative one that boils down to measuring the bound-state wave function renormalization constant Z on the lattice and analyzing how well Weinberg's criterion for the nonelementary particles $Z = 0$ [51] is fulfilled. In order to carry out this test, however, measurements at different values of L should be performed that make this enterprise rather expensive. As shown in Ref. [52], the same goal can be achieved by studying the θ-dependence of the binding energy in the simulations with a single value of L.

Last but not least, imposing twisted boundary conditions allows one to "scan" the energy continuously in scattering processes [46]. This is particularly useful for resolving those structures in the phase shifts that emerge at relatively small energy scales. For example, in the coupled-channel $\pi\pi$ and $K\bar{K}$ scattering, considered in Ref. [17], the narrow resonance $f_0(980)$ very close to the $K\bar{K}$ threshold shows up. Choosing periodic boundary conditions, an energy level very close to $2M_K$ emerges in the scattering spectrum. Extracting the parameters of the resonance in this case is difficult, because the presence of the nearby threshold has the same impact on the energy levels as a resonance. Both result in an avoided level crossing. Imposing twisted boundary conditions on the s-quark, we may continuously shift the energy level away from $2M_K$ and eventually extract the resonance parameters (see also Refs. [18, 19], where a similar problem has been addressed).

The new possibilities provided by the use of twisted boundary conditions do not come for free, however. In order to carry out (fully) twisted calculations, we have to first generate lattice configurations for each value of the twist angle separately. This is a rather expensive procedure. A cheap solution, referred to as partial twisting, is to use the configurations produced at zero twist angle, and to twist only those quark fields that appear in the Green's functions explicitly. This situation is demonstrated in Fig. 5.7, which describes pion–pion scattering in terms of quark and gluon diagrams (left panel). The valence quarks that are present in the initial or final state are twisted. However, the quarks that are present in closed loop are subject to periodic boundary conditions. Since the effect from the choice of the boundary conditions should anyway disappear as $L \to \infty$, the results obtained in these simulations should also converge to the correct answer in this limit. Two questions, however, remain. The first question concerns the observables that do not contain contributions from the singular diagrams (like the stable particle masses). How fast is the convergence? Can we evaluate the artifacts at finite values of L? The second question concerning the scattering observables

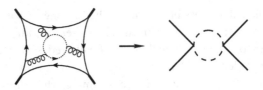

Figure 5.7 Left panel: Meson–meson scattering with partially twisted boundary conditions. The valence quarks (solid lines) obey twisted boundary conditions, whereas the sea quarks (dotted line) are subject to the periodic boundary condition. In the language of the EFT, this corresponds to the loop diagram shown in the right panel. A meson consisting of a valence quark and a sea antiquark, depicted by the dashed line, runs in the loop (right panel).

is more difficult. Namely, as we already know, a certain finite-volume formalism is needed to extract these observables from the lattice data (like the framework based on the Lüscher equation), and, therefore, the infinite-volume limit can be performed only after such a extraction is carried out. Introducing full twisting modifies the formalism only through the modification of the Lüscher zeta-function given in Eq. (5.142). How should one modify the framework in case of partial twisting?

The answer to all these questions can be found with the use of the EFT considered in Section 4.19 (we can include baryons in the partially quenched effective theory as well; see, for example, Ref. [53]). Imposing partially twisted boundary conditions is equivalent to twisting valence and ghost quarks, whereas the sea quarks obey periodic boundary conditions,

$$\psi_{v,g}(\mathbf{x} + \mathbf{e}_a L, x_4) = \exp\left(i\frac{\boldsymbol{\theta}\mathbf{x}}{L}\right)\psi_{v,g}(\mathbf{x}, x_4),$$

$$\psi_s(\mathbf{x} + \mathbf{e}_a L, x_4) = \psi_s(\mathbf{x}, x_4). \tag{5.143}$$

The masses of all quarks are taken equal. Furthermore, in the effective theory, all mesons consist of one quark and one antiquark of either type, and the baryons consist of three quarks. The boundary conditions on the hadron fields follow from the boundary conditions on the quarks. This allows us to use the EFT to straightforwardly evaluate exponentially suppressed finite-volume artifacts in particle masses, decay constants and so on.

The answer to the second question is, however, more complicated and closely related to the unitarity in the different sectors of the partially twisted theory. In case of scattering observables, the finite-volume artifacts are only power-law suppressed. Are all such effects captured by the (twisted) Lüscher equation in case of partial twisting? In the case of the nucleon–nucleon scattering, an elegant argument has been given in Ref. [54]. In this case, isospin conservation and quark number conservation in the valence quark sector forbid the coupling of a state with two valence nucleons (i.e., nucleons that contain only valence quarks) with any two-nucleon state that contains other nucleons. In other words, two-particle reducible diagrams can only contain two valence nucleons in the intermediate state. Other nucleons as well as mesons can emerge in the irreducible parts of the diagrams, as shown in Fig. 5.8. These, however, are at least one pion mass higher in energy, and the finite-volume corrections emerging from such diagrams to the

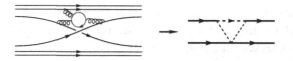

Figure 5.8   An example of a diagram quark diagram in the nucleon–nucleon scattering (left panel) that leads to intermediate states with non-valence nucleons and pions in the effective theory (right panel). The dotted lines correspond to the sea/ghost quarks and the dashed lines denote non-valence hadrons. The intermediate state energies in the diagram on the right panel are at least one pion mass away from the two-nucleon threshold. Therefore, the contribution from this diagram will be exponentially suppressed in the elastic region.

kernel of the Bethe–Salpeter equation (or, alternatively, to the couplings of the non-relativistic effective Lagrangian) are suppressed by a factor e^{-LM_π}. Hence, up to such exponentially suppressed contributions, the Lüscher equations in the fully twisted and the partially twisted theories are the same.

As pointed out in Ref. [54] itself, the argument is, in general, invalidated in the presence of mesons, since the valence quark and antiquark may annihilate and produce a sea or ghost quark–antiquark pair. As a result, other types of meson pairs besides the mesons with only valence quarks can be present in the intermediate states of the two-particle reducible diagrams, rendering the original Lüscher equation incomplete; see Fig. 5.7. We may say that imposing partially twisted boundary condition awakens mesons with ghost and sea quarks, there are more intermediate states now than in the case of full twisting. The modified Lüscher equation can also be derived. To this end, we have to write down a nonrelativistic effective theory that, apart from the valence mesons, contains mesons with a different quark content. The scattering amplitudes in this theory obey the multichannel Lippmann–Schwinger equation in the elastic region, as the states containing pairs of mesons with different quark content are coupled with each other. The Lüscher equation can be obtained by considering this Lippmann–Schwinger equation in a finite volume. The zeta-functions, corresponding to the different intermediate states, will differ according to the particular boundary condition imposed on the particles. For more details, we refer the reader to Ref. [55].

As mentioned, in general, such a Lüscher-type equation is of minimal interest, since the scattering amplitudes of mesons with different quark content (except the valence-valence amplitudes) are unobservable. However, as shown in Ref. [55], in certain cases it is possible to extract valuable information even from this equation. Namely, the graded symmetry of the original theory imposes stringent constraints. It relates the matrix elements of the *potential* of the Lippmann–Schwinger equation in different channels.[14] If the same symmetry for the zeta functions would also hold, the multichannel Lüscher equation would factorize and reduce to the standard Lüscher equation (with twisting). This does not happen in general but can be made to happen for a particular choice of the twist angles for different quark flavors [55]. As a rule of thumb, if a valence quark present in the initial/final state cannot annihilate (there is no antiquark with the same

---

[14] Note that the potentials can be evaluated in the infinite volume, because all finite-volume corrections there are exponentially suppressed.

flavor in the initial/final state), then the partial twisting of this quark is equivalent to the full twisting. This is no more the case if a quark is allowed to annihilate. This rule can be checked on a case-to-case basis [52, 55]. To the best of our knowledge, no rigorous proof is yet available that covers all possible cases.

To summarize, using twisted boundary conditions provides us with a very powerful tool in lattice calculations. It allows us to "inject" an external momentum into particles that makes it possible to continuously "scan" intervals in kinematic variables. Since the full twisting is rather expensive, partial twisting is often considered, which amounts to calculating the fermion determinant on the periodic configurations. It can be argued that, in many cases, the Lüscher equation in the partially twisted case corresponds to the fully twisted one, modulo exponentially suppressed corrections. This allows one to use partial twisting for the extraction of scattering amplitudes.

5.6 Two-Particle Decays

In two-particle scattering, the s-channel loops have singularities in the physical region. Similar diagrams that describe final-state interactions emerge in the study of two-particle decays as well. Furthermore, the presence of such diagrams leads to an irregular behavior of the measured observables in a finite volume. Hence, a specially designed finite-volume framework is needed to extract appropriate quantities from lattice calculations that possess a smooth infinite-volume limit. In previous sections we have demonstrated how this can be achieved for the phase shift in elastic scattering. As will be seen in what follows, a very similar method can be applied also to the two-particle decays in a finite volume.

First, note that decays to be studied on the lattice can be categorized into two classes. The decays that fall into the first class proceed through interactions that are of different origin than the strong interaction. The latter are responsible for the formation of hadrons themselves. A standard example for such decays is provided by the weak decay $K \to \pi\pi$. Both kaons and pions are stable states in QCD. The decays occur through the exchange of the $W^{\pm}$ bosons within the Standard Model, which may lead

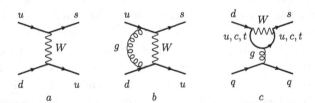

Figure 5.9   Representative diagrams in the Standard Model that contribute to the $\Delta S = 1$ effective Hamiltonian. a: Tree diagram that gives rise to the effective Fermi Hamiltonian displayed in Eq. (5.144); b,c: A gluon loop and the penguin diagram that lead to the renormalization of the Fermi Hamiltonian and operator mixing. The symbol $q = u, d, c, s, t, b$ denotes one of the six quarks.

to $s \to u$ transitions; see Fig. 5.9. At the energies much smaller than the masses of the $W^{\pm}, Z$ bosons, these decays are described by the effective $\Delta S = 1$ Hamiltonian. At tree level (see diagram 5.9a), the effective Hamiltonian contains only one operator, which reduces to the effective Fermi Hamiltonian with charged quark currents:

$$\mathcal{H}_F^0 = \frac{G_F}{\sqrt{2}} V_{us}^* V_{ud} \left(\bar{s} \gamma_\mu (1 - \gamma^5) u \right) \left(\bar{u} \gamma^\mu (1 - \gamma^5) d \right). \tag{5.144}$$

Here, G_F denotes the Fermi constant, $V_{ff'}$ are the elements of the Cabibbo-Kobayashi-Maskawa matrix, and the superscript "0" indicates that this Hamiltonian is obtained at lowest order in perturbation theory. At higher orders, the operator $O_1 = \left(\bar{s} \gamma_\mu (1 - \gamma^5) u \right) \left(\bar{u} \gamma^\mu (1 - \gamma^5) d \right)$ mixes with other operators through RG equations [56, 57]; see also chapter 18 of the textbook [58] for a very clear and concise discussion of this issue. Some diagrams that lead to such mixing are displayed in Fig. 5.9. The full expression of the $\Delta S = 1$ weak effective Hamiltonian is given by

$$\mathcal{H}_F = \sum_i c_i O_i, \tag{5.145}$$

where the c_i denote the pertinent Wilson coefficients and the sum runs over all linearly independent set of operators listed, for example, in Ref. [57].

Furthermore, the weak interactions do not preserve isospin; in other words, the preceding Hamiltonian is not a scalar with respect to isospin transformations. Treating isospin breaking at leading order, it is possible to write down the operators O_i in form of a linear combination of irreducible tensor operators, that transform according to the different irreducible representations of isospin. Only the values $I = \frac{1}{2}, \frac{3}{2}$ are allowed. Hence, the weak Hamiltonian can induce transitions with both $\Delta I = \frac{1}{2}, \frac{3}{2}$ in $K \to 2\pi$ decays, and these transitions are described by two independent amplitudes.

Since the Fermi constant is very small, we may restrict oneself to first order in perturbation theory:

$$\langle \pi(p_1) \pi(p_2); out | K(P) \rangle = -i (2\pi)^4 \delta^{(4)} (p_1 + p_2 - P) \langle \pi(p_1) \pi(p_2); out | \mathcal{H}_F(0) | K(P) \rangle. \tag{5.146}$$

In the preceding equation, the *in*- and *out*-states are eigenstates of the pure QCD Hamiltonian, that is, the kaon state is considered to be a stable particle (the quantity $P^2 = M_K^2$ is real). We could also include the decays proceeding through some type of symmetry violation into this category. A nice example for this is the decay $\omega \to \pi^+ \pi^-$ that occurs only if isospin is broken, and the decay amplitude is proportional to $m_d - m_u$. Because of the smallness of the isospin-breaking effects, we could treat the symmetry-breaking part of the QCD Hamiltonian, which is proportional to the operator $O_m = (m_d - m_u)(\bar{u}u - \bar{d}d)$, at first order in perturbation theory, calculating the decay amplitude by using a formula similar to Eq. (5.146). For this case, the unperturbed eigenstates in the formula refer to QCD with no isospin breaking, $m_u = m_d = \hat{m}$. The pole, corresponding to the ω-meson in such a theory, lies again on the real axis. On the contrary, the second category includes the decays of QCD resonances, where the pole in P^2 shifts into the unphysical Riemann sheets of a complex plane. The pole occurs at $P^2 \to s_R = m_R^2 - i m_R \Gamma_R$ where, by definition, m_R and Γ_R are the resonance mass and width, respectively. Using Fermi's golden rule, Eq. (5.146) is, strictly speaking, no more possible in this case. A good example for such decays is provided by $\rho \to 2\pi$, where the mass and width of the resonance are equal to approximately 775 MeV and 145 MeV, respectively. The formalism, which will be considered in what follows, can be applied, with small modifications, to all types of the decays. For simplicity, however, we shall concentrate on the decays of the first category and discuss the example of a kaon decay in all details.

Below, in order to avoid unnecessary complications, it will be further assumed that isospin symmetry in QCD holds exactly. Then, using the tables of the Clebsch–Gordan coefficients, the states $|\pi^+\pi^-\rangle$, $|\pi^0\pi^0\rangle$ and $|\pi^+\pi^0\rangle$ can be represented as linear combinations of the states with total isospin $I = 0, 2$. (We remind the reader that the kaons have spin 0 and that $I = 1$ is absent in the S-wave.) Recalling now that the effective Hamiltonian contains $\Delta I = \frac{1}{2}, \frac{3}{2}$ pieces with arbitrary coefficients, and using the Wigner–Eckart theorem, we finally arrive at the following representation of the decay amplitudes in the different channels:

$$-A(K^0 \to \pi^+\pi^-) = A_0 e^{i\delta_0} + \frac{1}{\sqrt{2}} A_2 e^{i\delta_2},$$

$$A(K^0 \to \pi^0\pi^0) = A_0 e^{i\delta_0} - \sqrt{2}A_2 e^{i\delta_2},$$

$$-A(K^+ \to \pi^+\pi^0) = \frac{3}{2} A_2 e^{i\delta_2}. \tag{5.147}$$

Here, A_0, A_2 are two independent real constants that correspond to the transitions with $\Delta I = \frac{1}{2}, \frac{3}{2}$, respectively. Further, δ_0, δ_2 are the phases of the pertinent transition amplitudes. According to Watson's theorem, these coincide with the $\pi\pi$ scattering phase shifts with the total isospin $I = 0, 2$, evaluated at the CM energy $s = M_K^2$.

Up to now, the discussion was carried out in the infinite volume. Let us see now whether the quantities A_0, A_2 can be extracted on the lattice. (We know already that the phases can be measured by using the Lüscher framework.) In order to make the discussion more transparent, we limit ourselves to the decay $K^+ \to \pi^+\pi^0$, where only the $I = 2$ component contributes. Further, we restrict ourselves to kaon decays in the CM frame, that is, $\mathbf{P}_K = \mathbf{0}$ and $\mathbf{p}_1 = -\mathbf{p}_2 = \mathbf{p}$. In order to evaluate the decay matrix element on the lattice, let us consider the following Euclidean Green's function:

$$G(t',t) = \langle 0|T\pi_{\mathbf{p}}^+(t')\pi_{-\mathbf{p}}^0(t')H_F\big(K_{\mathbf{0}}^+(t)\big)^\dagger|0\rangle, \tag{5.148}$$

where all meson fields are composed of quark fields, and the definition of the meson fields carrying different three-momenta is given in Eq. (5.35). Furthermore, the weak Hamiltonian is expressed through the Hamiltonian density, according to

$$H_F = \int_{-L/2}^{+L/2} d^3\mathbf{x}\,\mathcal{H}_F(\mathbf{x},0). \tag{5.149}$$

In addition, we consider the following Green's functions:

$$g(t',t) = \langle 0|T\pi_{\mathbf{p}}^+(t')\pi_{-\mathbf{p}}^0(t')\big(\pi_{\mathbf{p}}^+(t)\big)^\dagger\big(\pi_{-\mathbf{p}}^0(t)\big)^\dagger|0\rangle,$$

$$g_K(t',t) = \langle 0|TK_{\mathbf{0}}^+(t')\big(K_{\mathbf{0}}^+(t)\big)^\dagger|0\rangle, \tag{5.150}$$

and study the limit $t' \to +\infty$, $t \to -\infty$, inserting a full set of eigenvectors between the operators. It should be noted that these eigenvectors correspond to states in pure QCD, that is, without weak interactions. Thus, we obtain

$$G(t',t) \to \sum_{n=2\pi,\dots} \sum_{m=K,\dots} e^{-E_n t' - E_m |t|} \langle 0 | \pi_{\mathbf{p}}^+(0) \pi_{-\mathbf{p}}^0(0) | n \rangle$$

$$\times \langle n | H_F | m \rangle \langle m | \left(K_{\mathbf{0}}^+(0) \right)^\dagger | 0 \rangle,$$

$$g(t',t) \to \sum_{n=2\pi,\dots} e^{-E_n(t'-t)} \langle 0 | \pi_{\mathbf{p}}^+(0) \pi_{-\mathbf{p}}^0(0) | n \rangle \langle n | \left(\pi_{\mathbf{p}}^+(0) \right)^\dagger \left(\pi_{-\mathbf{p}}^0(0) \right)^\dagger | 0 \rangle,$$

$$g_K(t',t) \to \sum_{m=K,\dots} e^{-E_m(t'-t)} \langle 0 | K_{\mathbf{0}}^+(0) | m \rangle \langle m | \left(K_{\mathbf{0}}^+(0) \right)^\dagger | 0 \rangle. \tag{5.151}$$

The eigenvectors with the lowest energies in the various sums are the two-pion states and the kaon states, respectively. Singling those out and considering the limit of large Euclidean times, we get

$$\lim_{t' \to +\infty, t \to -\infty} G(t',t) \left(\frac{g(0,t) g_K(t',0)}{g(t',0) g_K(0,t) g(t',t) g_K(t',t)} \right)^{1/2}$$

$$= \langle \pi^+ \pi^0 | H_F | K \rangle_L. \tag{5.152}$$

Moreover, it is implicitly assumed that the size of the box L is tuned so that the energy of the state $|\pi^+ \pi^0\rangle$ is exactly equal to M_K. Note, however, that the state $|\pi^+ \pi^0\rangle$ does *not* describe two pions that move freely, it is just an eigenstate of the full QCD Hamiltonian with the quantum numbers of the two-pion state. The equation (5.152) represents, in fact, the definition of the finite-volume matrix element $\langle \pi^+ \pi^0 | \mathcal{H}_F(0) | K \rangle_L$, which, in the infinite volume, corresponds to the decay matrix element. The main question that arises here is, whether the right-hand side of Eq. (5.152) has a smooth limit $L \to \infty$ and, if yes, whether in this limit it coincides with the matrix element we are looking for.

It is immediately seen that there are serious reasons to doubt the preceding conjecture. For example, it is known that the infinite-volume matrix element is complex, whereas any quantity that is extracted from calculations in a finite volume is real for all values of L[15]. Moreover, as will be seen later, the finite-volume matrix element has an irregular behavior in L. Hence, in analogy to the two-particle elastic scattering, we have first to identify the quantities, which exhibit a regular (exponentially suppressed) dependence on L. As in the former case, the nonrelativistic EFT provides us with a very efficient tool that allows us to achieve the goal with remarkable ease.

---

[15] In order to understand why the imaginary part disappears in a finite volume, let us recall that in the infinite volume it emerges due to the $i\varepsilon$ prescription in the momentum-space propagators. In a finite volume, we have to deal with two limits: $\varepsilon \to 0$ and $L \to \infty$. The order of these two limits matters. Conventional lattice calculations imply the extraction of *real* energy levels in a *finite volume*, i.e., performing first the limit $\varepsilon \to 0$ and then $L \to \infty$. It can be verified that in this case the resulting amplitude is real, possessing an infinite tower of poles, which correspond to the finite-volume energy levels along the positive real energy axis instead of an unitary cut. The latter emerges, if the limits are performed in an reversed order. In this case, distinct poles come close and merge in a cut. For more detailed discussion of this issue, as well as for alternative methods for the extraction of the scattering observables, we refer the reader to Refs. [59, 60].

Figure 5.10 String of diagrams in the nonrelativistic EFT that contribute to the kaon decay $K \to 2\pi$. The double and single lines correspond to kaons and pions, respectively.

In order to obtain the relativistic kinematics of pions right from the beginning, we shall use the covariant version of the NREFT framework considered in Section 2.7. The Lagrangian that describes the $K^+ \to \pi^+ \pi^0$ decay can be written down as follows (cf. Eqs. (2.105), (2.106) and (2.106)):

$$
\mathcal{L} = \sum_{i=\pm,0} \pi_i^\dagger 2W \left(i\partial_t - W \right) \pi_i + K_+^\dagger 2W_K \left(i\partial_t - W_K \right) K_+
$$

$$
+ C_0 \pi_+^\dagger \pi_0^\dagger \pi_+ \pi_0 + C_1 \left((\pi_+^\mu)^\dagger (\pi_{0\mu})^\dagger \pi_+ \pi_0 - M^2 \pi_+^\dagger \pi_0^\dagger \pi_+ \pi_0 + \text{h.c.} \right) + \cdots
$$

$$
+ g_K \left(K_+^\dagger \pi_+ \pi_0 + \text{h.c.} \right). \tag{5.153}
$$

Here, $W = \sqrt{M^2 - \nabla^2}$ and $W_K = \sqrt{M_K^2 - \nabla^2}$, where M, M_K denote the pion and kaon masses, respectively, and the quantities π_+^μ, π_0^μ are defined analogously to Eq. (2.124). Further, the coupling g_K describes the weak decay of the kaon into the two-pion pair. This coupling is proportional to the Fermi coupling constant G_F. Consequently, one needs to calculate the amplitude to first order in g_K only.

A short comment is appropriate here. In this Lagrangian, we did not include derivative terms in the Lagrangian that describes $K \to 2\pi$ transitions. In fact, these are not needed. The decay amplitude that we are after is evaluated at $s = P^2 = M_K^2$. All possible derivative terms in the S-wave could be obtained by expanding the two-particle irreducible decay amplitude in a Taylor series in $s - M_K^2$. All these terms would, however, give a vanishing contribution on the kaon mass shell. In principle, the same arguments apply to the string of constants $C_0, C_1, \ldots$. Re-expanding the two-particle irreducible amplitude (the potential) in the vicinity of $s = M_K^2$ instead of $s = 4M^2$, we need to retain only the first term. However, in order to avoid a confusion in comparison with the formulae from Section 2.7, we have decided to still consider the expansion at the two-pion threshold and retain all terms.

In the infinite volume, the matrix element $\langle \pi^+(p_1) \pi^0(p_2); out | \mathcal{H}_F(0) | K(P) \rangle$ is given by a string of bubble diagrams shown in Fig. 5.10. The geometric series can be easily summed up, yielding

$$
-\langle \pi^+(p_1) \pi^0(p_2); out | \mathcal{H}_F(0) | K(P) \rangle = g_K \left\{ 1 + V(s) I(s) + \ldots \right\}
$$

$$
= \frac{g_K V^{-1}(s)}{V^{-1}(s) - I(s)}, \tag{5.154}
$$

where $V(s)$ is given by the matrix element of the interaction Lagrangian between free two-particle states,

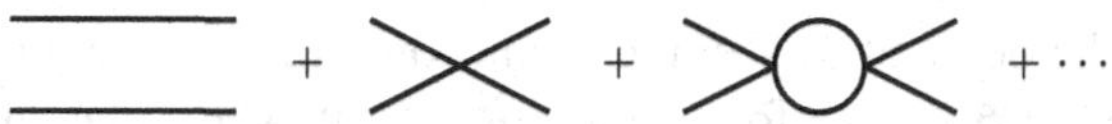

Figure 5.11   String of diagrams in the nonrelativistic EFT, which contribute to the four-pion Green's function given in Eq. (5.160).

$$V(s) = C_0 + C_1(s - 4M^2) + \cdots = \frac{8\pi\sqrt{s}}{p(s)} \tan \delta_2(s), \qquad (5.155)$$

and $I(s)$ is the loop function, calculated in Eq. (2.113):

$$I(s) = \frac{ip(s)}{8\pi\sqrt{s}}, \qquad p(s) = \sqrt{\frac{s}{4} - M^2}. \qquad (5.156)$$

Substituting these expressions into the matrix element and setting $s = M_K^2$, we get

$$-\langle \pi^+(p_1)\pi^0(p_2); out|\mathcal{H}_F(0)|K(P)\rangle = g_K \cos\delta_2(M_K^2) e^{i\delta_2(M_K^2)}. \qquad (5.157)$$

Comparing this with Eq. (5.147), we finally obtain the matching condition for the sole still undefined constant g_K:

$$\frac{3}{2}A_2 = g_K \cos\delta_2(M_K^2). \qquad (5.158)$$

Let us now carry out the calculation of the analogous matrix element in a finite volume. It is crucial that the Lagrangian, which will be used in these calculations, is the same (up to exponentially suppressed corrections). The Euclidean propagator for the pion fields for $i = \pm, 0$ is given by

$$i\langle 0|T\pi_i(x)\pi_i^\dagger(y)|0\rangle = \int \frac{d^4k}{(2\pi)^4} \frac{e^{ik(x-y)}}{2w(\mathbf{k})(k_4 - iw(\mathbf{k}))}, \qquad w(\mathbf{k}) = \sqrt{M^2 + \mathbf{k}^2}, \qquad (5.159)$$

and the propagator for the kaon can be written down in a similar fashion.

We start by evaluating the Green's function $g(t',t)$, defined in Eq. (5.150). This Green's function is defined by the diagrams shown in Fig. 5.11. Summing up the bubbles, we have

$$g(t',t) = L^3 \int \frac{dP_4}{2\pi} e^{iP_4(t'-t)} \left\{ \frac{-iL^3}{4w^2(\mathbf{p})(P_4 - 2iw(\mathbf{p}))} \right.$$
$$\left. - \frac{T_{\pi\pi}(s)}{(4w^2(\mathbf{p})(P_4 - 2iw(\mathbf{p})))^2} \right\}, \qquad (5.160)$$

where $s = -P_4^2$ and $T_{\pi\pi}(s)$ is the $\pi\pi$ scattering amplitude in a finite volume:

$$T_{\pi\pi}(s) = \frac{8\pi\sqrt{s}}{p(s)\cot\delta_2(s) + p(s)\cot\phi(s)}, \qquad \tan\phi(s) = -\frac{\pi^{3/2}\eta}{Z_{00}(1;\eta^2)}. \qquad (5.161)$$

Here, $\eta = p(s)L/(2\pi)$, and $Z_{00}(1;\eta^2)$ denotes the Lüscher zeta-function. Note also that we have neglected the partial-wave mixing here. In the CM frame, considered here, the S-wave couples first with the G-wave (this follows from the fact that the A_1 irrep of the octahedral group contains only the representations of the rotation group with $L = 0, 4, 6, \ldots$; see, for example, Ref. [14]). At the mass of the K-meson, this contribution is of academic interest only.

As we have mentioned, in the lattice calculations we tune the parameter L such that one of the energy levels exactly coincides with the kaon mass M_K. Then, the amplitude $T_{\pi\pi}(s)$ has a pole at $P_4 = iM_K$. In the vicinity of the pole, it behaves as

$$T_{\pi\pi}(s) = \frac{32\pi \sin^2 \delta_2(M_K^2)}{\delta_2'(M_K^2) + \phi'(M_K^2)} \frac{1}{M_K + iP_4} + \text{regular terms}, \tag{5.162}$$

where the derivative is taken with respect to the variable $p = p(s)$. Performing the contour integral over the variable P_4, we obtain, for large Euclidean time separations,[16]

$$g(t',t) \to L^3 \frac{32\pi \sin^2 \delta_2(M_K^2)}{\delta_2'(M_K^2) + \phi'(M_K^2)} \frac{e^{-M_K(t'-t)}}{\left(4w^2(\mathbf{p})(M_K - 2w(\mathbf{p}))\right)^2}. \tag{5.163}$$

The asymptotic behavior of the function $g_K(t',t)$ from Eq. (5.150) can be easily established. The leading contribution is given in this case from the one kaon state. The result is

$$g_K(t',t) \to L^3 \frac{e^{-M_K(t'-t)}}{2M_K}. \tag{5.164}$$

Finally, we consider the Green's function $G(t',t)$ from Eq. (5.148). The diagrams contributing to this Green's function are the same as depicted in Fig. 5.10. Again, summing up the s-channel bubbles, we get

$$G(t',t) = -g_K L^3 \int \frac{dP_4}{2\pi} \frac{1}{4w^2(\mathbf{p})(P_4 - 2iw(\mathbf{p}))} \frac{e^{iP_4 t'} p(s) \cot \delta_2(s)}{p(s) \cot \delta_2(s) + p(s) \cot \phi(s)}$$

$$\times \int \frac{dQ_4}{2\pi} e^{-iQ_4 t} \frac{1}{2M_K(Q_4 - iM_K)}. \tag{5.165}$$

Performing the contour integration in the variables P_4, Q_4, we readily obtain the leading asymptotic behavior at large Euclidean times:

$$G(t',t) \to g_K L^3 \frac{p(M_K^2) e^{-M_K(t'-t)}}{2M_K^2 w^2(\mathbf{p})(M_K - 2w(\mathbf{p}))} \frac{\sin \delta_2(M_K^2) \cos \delta_2(M_K^2)}{\delta_2'(M_K^2) + \phi'(M_K^2)}. \tag{5.166}$$

It remains to substitute everything in Eq. (5.152). One obtains

$$\left| \langle \pi^+ \pi^0 | H_F | K \rangle_L \right| = \frac{g_K \cos \delta_2(M_K^2)}{\left| 4\pi p (\delta_2'(M_K^2) + \phi'(M_K^2)) \right|^{1/2}} \left(\frac{p(M_K^2)}{M_K} \right)^{3/2}. \tag{5.167}$$

Now, using Eq. (5.158) and taking into account the fact that

$$p(\delta_2'(M_K^2) + \phi'(M_K^2)) = p \frac{d}{dp} \delta_2 + \eta \frac{d}{d\eta} \phi, \qquad \eta = \frac{pL}{2\pi}, \tag{5.168}$$

---

[16] Note that the "free particle" poles at $P_4 = 2iw(\mathbf{p})$ always cancel in the integrand.

we arrive finally at the celebrated Lellouch–Lüscher formula [61]:

$$\left| p\frac{d}{dp}\delta_2 + \eta\frac{d}{d\eta}\phi \right| \left(\langle \pi^+\pi^0|H_F|K\rangle_L \right)^2 = \frac{p^3}{4\pi M_K^3}\left(\frac{3}{2}A_2\right)^2. \qquad (5.169)$$

Note also that the denominator in case of the decay into two *identical* particles contains the factor 8π instead of 4π, cf. with Ref. [61].

The Lellouch–Lüscher formula relates the infinite- and finite-volume matrix elements through the universal Lellouch–Lüscher (LL) factor that does not depend on the kaon interaction vertex. In fact, as we see from the preceding formula, it contains the derivative of the phase shift δ_2 and describes the final-state interactions in a finite volume. The LL factor also contains a singular geometric function $\phi(s)$, which exhibits an irregular behavior in L at a given value of the kinematic variable s. Once the LL factor is removed from the finite-volume matrix element, the remainder exhibits exponentially suppressed corrections only and converges to the correct limit.

Similar tools can be straightforwardly applied to study various generalizations of the Lellouch–Lüscher formalism. For example, we mention the extraction of the matrix elements in moving frames [28, 62] and in case of multichannel systems [63]. Note also that in Ref. [64] a comprehensive lattice study of $K \to 2\pi$ decays in various channels has been carried out. The methods, similar to those considered in the present section, can be applied to study of the form factors in the timelike region [65] as well. Moreover, as we have mentioned, the effective field theory framework can be used to study the matrix elements of the QCD resonances, where the analytic continuation into the complex energy planes is required [34, 66–69]. In all these cases, the effective Lagrangian plays the role of a "bridge" between the finite and infinite volumes, because the effective couplings are the same. The explicit form of the Lagrangian is unimportant. At the end of the day, everything is expressed in terms of observables, and any reference to the Lagrangian disappears. We have seen an example of this in the Lellouch–Lüscher formula.

5.7 Multiparticle Systems: Perturbative Expansion for the Energy Levels

In Section 5.4.4, we have derived a formula for the shift of the two-particle ground state up to and including $O(L^{-5})$, starting from the Lüscher equation. Here we consider another, much simpler derivation of this result, which can be straightforwardly generalized to more complex cases, say, for multiparticle systems or for excited levels. This method, which in the present context was first used in Ref. [70],[17] approaches the problem by using standard Rayleigh–Schrödinger perturbation theory in quantum mechanics. Note also that here we will use the original, non-covariant version of the NREFT that allows one to arrive at simpler expressions in the sums over momenta of the intermediate states. The price to pay for this is that the relativistic effects should

---

[17] See also Refs. [71–74], where the further developments of the method are discussed.

be taken special care of, namely they will be included perturbatively, that is, order by order in the expansion in powers of $1/L$.

In order to simplify the discussion, we consider the case of a single nonrelativistic scalar particle with mass M. The Lagrangian is given by

$$\mathcal{L} = \phi^\dagger \left(i\partial_t - M + \frac{\nabla^2}{2M} + \frac{\nabla^4}{8M^3} + \cdots \right) \phi$$

$$+ C_0 \phi^\dagger \phi^\dagger \phi \phi + C_1 \left((\phi^\dagger \overset{\leftrightarrow}{\nabla}{}^2 \phi^\dagger)(\phi\phi) + \text{h.c.} \right) + C_2 \left((\phi^\dagger \phi^\dagger)\nabla^2(\phi\phi) + \text{h.c.} \right) + \cdots$$

$$+ D_0 \phi^\dagger \phi^\dagger \phi^\dagger \phi \phi \phi + \cdots . \tag{5.170}$$

In this expression, we display explicitly all terms, which contribute to the N-particle energy levels up to and including $O(L^{-6})$. Since the typical scale of three-momenta on the lattice is given by $2\pi/L$, derivative vertices give contributions to the energy that are suppressed by powers of L. This enables one to truncate the effective Lagrangian at a given order. The same power counting based on dimensional arguments holds also for the many-particle vertices. For example, the six-particle vertex (three-body force) proportional to the coupling D_0 starts to contribute at $O(L^{-6})$ to the N-particle ground-state energies for all $N \geq 3$. To this order, there are still no contributions from the four-particle force, and so on.

The matching of the couplings C_0, C_1, C_2, D_0 has been considered in great detail in Section 2. The two-particle couplings are related to the parameters of the effective-range expansion, namely the S-wave scattering length a_0 and the effective range r_0; see Eq. (2.87). These expressions already include the relativistic corrections up to the order one is working at. Furthermore, the coupling D_0 can be also related to the threshold 3-particle scattering amplitude. This relation has, however, a more complicated form and is discussed, for example, in Refs. [73, 74].

The three-momenta of free particles in a box take the discrete values $\mathbf{p}_i = 2\pi\mathbf{n}_i/L$, where $i = 1, \ldots N$. For simplicity, we shall consider the CM frame $\mathbf{p}_1 + \cdots + \mathbf{p}_N = \mathbf{0}$, albeit the formalism can be applied in moving frames as well. The energy of an unperturbed state is given by $E_n = \frac{1}{2M}(\mathbf{p}_1^2 + \cdots + \mathbf{p}_N^2)$. All energy levels except the ground state $\mathbf{n}_i = \mathbf{0}$ are degenerate, that is, there exist multiple choices of the components of the integer vectors $\mathbf{n}_i$, which lead to the same energy. As we know, these sets can be classified into the different irreps of the octahedral group. There is only one irrep A_1 for the ground state, but the degenerate excited states contain, in general, more than one irrep. When the interactions are turned on, the energy levels are shifted and split, because the shift is the same only for the state vectors belonging to a given irrep. In what follows, we shall give a formulation of the perturbative expansion, which can be directly used in the case of degenerate energy levels as well (we mainly follow Ref. [74] here).

We start by splitting the Hamiltonian of the system into the free and interaction parts:

$$\mathbf{H} = \int_{-L/2}^{+L/2} d^3\mathbf{x} (\mathcal{H}_0 + \mathcal{H}_I), \tag{5.171}$$

where

$$\mathcal{H}_0 = \phi^\dagger \left(M - \frac{\nabla^2}{2M} \right) \phi \,, \tag{5.172}$$

and

$$\mathcal{H}_I = -C_0 \phi^\dagger \phi^\dagger \phi \phi - C_1 \big((\phi^\dagger \overset{\leftrightarrow}{\nabla}^2 \phi^\dagger)(\phi\phi) + \text{h.c.} \big) - C_2 \big((\phi^\dagger \phi^\dagger) \nabla^2 (\phi\phi) + \text{h.c.} \big)$$

$$- \phi^\dagger \frac{\nabla^4}{8M^3} \phi - D_0 \phi^\dagger \phi^\dagger \phi^\dagger \phi \phi \phi \,. \tag{5.173}$$

Note again that the kinetic energy operator is entirely given by $\nabla^2/(2M)$. The field operators here are free fields and can be expanded in a Fourier series in creation/annihilation operators,

$$\phi(\mathbf{x},t) = \frac{1}{L^3} \sum_{\mathbf{p}} \exp(-iE_{\mathbf{p}}t + i\mathbf{p}\mathbf{x}) a_{\mathbf{p}} \,,$$

$$\phi^\dagger(\mathbf{x},t) = \frac{1}{L^3} \sum_{\mathbf{p}} \exp(iE_{\mathbf{p}}t - i\mathbf{p}\mathbf{x}) a_{\mathbf{p}}^\dagger \,, \tag{5.174}$$

where $E_{\mathbf{p}} = M + \mathbf{p}^2/2M$. These operators obey the commutation relations:

$$[a(\mathbf{p}), a^\dagger(\mathbf{q})] = L^3 \delta_{\mathbf{pq}} \,. \tag{5.175}$$

The properly normalized N-particle states are given by

$$|\mathbf{p}_1, \cdots, \mathbf{p}_N\rangle = \frac{1}{\sqrt{N!} L^{3N/2}} a^\dagger(\mathbf{p}_1) \cdots a^\dagger(\mathbf{p}_N) |0\rangle \,. \tag{5.176}$$

These states are the eigenstates of the free Hamiltonian:

$$\mathbf{H}_0 |\mathbf{p}_1, \cdots, \mathbf{p}_N\rangle = (E_{\mathbf{p}_1} + \cdots + E_{\mathbf{p}_N}) |\mathbf{p}_1, \cdots, \mathbf{p}_N\rangle \,. \tag{5.177}$$

The closure relation takes the form

$$\sum_{\mathbf{p}_1 \cdots \mathbf{p}_N} |\mathbf{p}_1, \cdots, \mathbf{p}_N\rangle \langle \mathbf{p}_1, \cdots, \mathbf{p}_N| = 1 \,. \tag{5.178}$$

The T-matrix is defined by the Lippmann–Schwinger equation:

$$\mathbf{T}(E) = \mathbf{H}_I + \mathbf{H}_I \mathbf{G}_0(E) \mathbf{T}(E) \,, \qquad \mathbf{G}_0(E) = \frac{1}{E - \mathbf{H}_0} \,. \tag{5.179}$$

Let us now sandwich the Lippmann–Schwinger equation in between N-particle states. Introducing a shorthand notation for the set of momenta $p \doteq (\mathbf{p}_1, \ldots, \mathbf{p}_N)$, we get

$$\langle p | \mathbf{T}(E) | q \rangle = \langle p | \mathbf{H}_I | q \rangle + \sum_k \langle p | \mathbf{H}_I | k \rangle \frac{1}{E - E_k} \langle k | \mathbf{T}(E) | q \rangle \,. \tag{5.180}$$

Suppose that one wishes to calculate the shift of some unperturbed level. Unless it is the ground state, this level is degenerate. For illustration, let us consider a two-particle system in the CM frame $|\mathbf{p}, -\mathbf{p}\rangle$. In the first excited state, the momentum $\mathbf{p}$ can take six different values $(1,0,0), (0,1,0), (0,0,1), (-1,0,0), (0,-1,0)$ and $(0,0,-1)$ (in units of

$2\pi/L$). All these momenta are related to each other through the transformation of the octahedral group and correspond to the same energy:

$$E_1 = \frac{1}{M}\left(\frac{2\pi}{L}\right)^2 (1^2 + 0^2 + 0^2) = \frac{4\pi^2}{ML^2}. \tag{5.181}$$

Note that higher excited states may contain states that are *not* related to each other through the transformations of the octahedral group. For example, since $3^2 + 0^2 + 0^2 = 2^2 + 2^2 + 1^2 = 9$, the state with the energy $E_9 = 36\pi^2/ML^2$ will contain the vectors with the components $(3,0,0)$ and $(2,2,1)$, not related by the octahedral symmetry. The sets of vectors related to each other by the transformation of the octahedral group is termed *shells* [75]. In this example, the state with the energy E_9 consists of two shells.

Let now $|\alpha\rangle$ denote all states that belong to a given state with the back-to-back momentum k_0. Separating these states in the Lippmann–Schwinger equation (5.180), we arrive at a system of two equations:

$$\langle p|\mathbf{T}(E)|q\rangle = \langle p|\mathbf{\Omega}(E)|q\rangle + \sum_\alpha \langle p|\mathbf{\Omega}(E)|\alpha\rangle \frac{1}{E - E_{k_0}} \langle\alpha|\mathbf{T}(E)|q\rangle,$$

$$\langle p|\mathbf{\Omega}(E)|q\rangle = \langle p|\mathbf{H}_I|q\rangle + \sum_{k \neq k_0} \langle p|\mathbf{H}_I|k\rangle \frac{1}{E - E_k} \langle k|\mathbf{\Omega}(E)|p\rangle. \tag{5.182}$$

Let us now choose $p = q = k_0$ in the first equation. We can again use Greek letters to label the vectors in this state. Then, the first equation can be rewritten as

$$\langle\beta|\mathbf{T}(E)|\gamma\rangle = \langle\beta|\mathbf{\Omega}(E)|\gamma\rangle + \sum_\alpha \langle\beta|\mathbf{\Omega}(E)|\alpha\rangle \frac{1}{E - E_{k_0}} \langle\alpha|\mathbf{T}(E)|\gamma\rangle. \tag{5.183}$$

This is a finite-dimensional system of linear equations in the space spanned by the vectors belonging to the state k_0. Introducing the matrix $\mathbf{\Omega}(E)$ with the matrix elements

$$\mathbf{\Omega}_{\beta\gamma}(E) = \langle\beta|\mathbf{\Omega}(E)|\gamma\rangle, \tag{5.184}$$

it is immediately seen that the linear system of equations becomes singular, if

$$\det\big((E - E_{k_0})\mathbb{1} - \mathbf{\Omega}(E)\big) = 0. \tag{5.185}$$

The matrix elements of $\mathbf{T}(E)$ diverge, if E approaches one of the roots of this equation. This will be termed a *quantization condition*, or the secular equation hereafter. In other words, the roots of the secular equation determine the finite-volume spectrum of the full Hamiltonian. Note also that, albeit we have considered the two-particle case for illustration, this equation has the same form for N particles.

A remark is necessary at this point. The symmetries of the theory (both the spatial symmetry as well as internal symmetries) may help to partially diagonalize the matrix that enters the quantization condition. This may happen, if the basis vectors are chosen so that all of them belong to only one invariant subspace of the symmetry group that corresponds to a given irrep. A standard method for constructing these basis vectors, which is based on the use of projection operators, was already considered above for the two-particle case and can be straightforwardly used for any number of particles. For simplicity, we assume that this is already done and that the matrix has a block-diagonal form. Then, the determinant becomes a product of determinants, one for each irrep, and it suffices to consider any one of them. It is also clear that the unperturbed energy level, corresponding to the state k_0, will be split into several levels when the interaction is turned on. The number of the resulting levels coincides with the number of the invariant subspaces in the state k_0.[18]

Next, consider a perturbative solution of the secular equation corresponding to a given irrep. Since the energy shift $E - E_{k_0}$ is much smaller than the distance between the different unperturbed levels, we may Taylor-expand the matrix $\mathbf{\Omega}(E)$. (We remind the reader that, in the definition of this matrix given by Eq. (5.182), the term with $k = k_0$ is absent.) Hence, the secular equation takes the form

$$\det\left((E - E_{k_0})\mathbb{1} - \mathbf{\Omega}(E_{k_0}) - \frac{1}{1!}(E - E_{k_0})\mathbf{\Omega}'(E_{k_0}) \right.$$

$$\left. - \frac{1}{2!}(E - E_{k_0})^2 \mathbf{\Omega}''(E_{k_0}) + \cdots \right) = 0. \tag{5.186}$$

This equation can be readily solved by iteration. For illustration, let us consider the case of a 1×1 matrix (nondegenerate perturbation theory). At the lowest order, we have $E - E_{k_0} = \mathbf{\Omega}(E_{k_0})$ (the matrix $\mathbf{\Omega}$ has no indices in this case). Substituting this expression back into the equation, we get a corrected expression:

$$E - E_{k_0} = \mathbf{\Omega}(E_{k_0}) + \mathbf{\Omega}(E_{k_0})\mathbf{\Omega}'(E_{k_0}) + \frac{1}{2!}(\mathbf{\Omega}(E_{k_0}))^2 \mathbf{\Omega}''(E_{k_0}) + \cdots, \tag{5.187}$$

and so on. On the other hand, the matrix $\langle k_0 | \mathbf{\Omega}(E_{k_0}) | k_0 \rangle$ in perturbation theory is given by the series

$$\langle k_0 | \mathbf{\Omega}(E_{k_0}) | k_0 \rangle = V_{k_0 k_0} + \sum_{k \neq k_0} \frac{V_{k_0 k} V_{k k_0}}{E_{k_0} - E_k}$$

$$+ \sum_{k \neq k_0} \sum_{p \neq k_0} \frac{V_{k_0 k} V_{k p} V_{p k_0}}{(E_{k_0} - E_k)(E_{k_0} - E_p)} + \cdots, \tag{5.188}$$

where we have introduced a shorthand notation,

$$V_{pq} = \langle p | \mathbf{H}_I | q \rangle, \tag{5.189}$$

and the vector $|k_0\rangle$ denotes the basis vector of the one-dimensional invariant subspace under consideration.

Using Eqs. (5.187) and (5.188), it is then an easy task to write down a perturbative expansion of the energy shift $E - E_{k_0}$. The terms of order V, V^2, V^3 and V^4 are given by

$$\Delta E_1 = V_{k_0 k_0},$$

$$\Delta E_2 = \sum_{k \neq k_0} \frac{V_{k_0 k} V_{k k_0}}{E_{k_0} - E_k},$$

[18] If the multiplicity of a given irrep in the state k_0 is larger than one, each copy of the same irrep leads, in general, to a different energy shift.

$$\Delta E_3 = \sum_{k \neq k_0} \sum_{p \neq k_0} \frac{V_{k_0 k} V_{k p} V_{p k_0}}{(E_{k_0} - E_k)(E_{k_0} - E_p)} - V_{k_0 k_0} \sum_{k \neq k_0} \frac{V_{k_0 k} V_{k k_0}}{(E_{k_0} - E_k)^2},$$

$$\Delta E_4 = \sum_{k \neq k_0} \sum_{p \neq k_0} \sum_{q \neq k_0} \frac{V_{k_0 k} V_{k p} V_{p q} V_{q k_0}}{(E_{k_0} - E_k)(E_{k_0} - E_p)(E_{k_0} - E_q)}$$

$$- \sum_{k \neq k_0} \sum_{p \neq k_0} \frac{V_{k_0 k} V_{k k_0} V_{k_0 p} V_{p k_0}}{(E_{k_0} - E_k)^2 (E_{k_0} - E_p)} + (V_{k_0 k_0})^2 \sum_{k \neq k_0} \frac{V_{k_0 k} V_{k k_0}}{(E_{k_0} - E_k)^3}$$

$$- 2 V_{k_0 k_0} \sum_{k \neq k_0} \sum_{p \neq k_0} \frac{V_{k_0 k} V_{k p} V_{p k_0}}{(E_{k_0} - E_k)^2 (E_{k_0} - E_p)}. \tag{5.190}$$

It is seen that in this way the standard formulae for the Rayleigh–Schrödinger perturbation theory (nondegenerate case) are reproduced.

Consider next briefly the degenerate case. We remind the reader that the matrix $\mathbf{\Omega}(E)$ is Hermitean. Therefore, it is possible to find a unitary matrix U that diagonalizes the matrix $\mathbf{\Omega}(E_{k_0})$:

$$U \mathbf{\Omega}(E_{k_0}) U^\dagger = \Lambda = \mathrm{diag}(\lambda_1, \cdots, \lambda_r). \tag{5.191}$$

Then, the secular equation can be rewritten as

$$\det\big((E - E_{k_0})\mathbb{1} - \Lambda - F(E)\big) = 0, \tag{5.192}$$

where

$$F(E) = U\left((E - E_{k_0})\mathbf{\Omega}'(E_{k_0}) + \frac{1}{2}(E - E_{k_0})^2 \mathbf{\Omega}''(E_{k_0}) + \cdots\right) U^\dagger. \tag{5.193}$$

At lowest order, the secular equation has r roots: $E_1 = E_{k_0} + \lambda_1, \ldots, E_r = E_{k_0} + \lambda_r$. Here, for simplicity, we assume that all these roots are nondegenerate. In order to calculate perturbative corrections to a given root $E_s = E_{k_0} + \lambda_s$ with $1 \leq s \leq r$, note that the determinant is the linear function of $(E - E_s)$. Next, we shall renumber the rows and columns in the matrix in Eq. (5.192) so that $(E - E_s)$ appears in the upper left corner,

$$(E - E_{k_0})\mathbb{1} - \Lambda - F(E) = \begin{pmatrix} E - E_s - F_{ss}(E) & -F_{sa}(E) \\ -F_{bs}(E) & (E - E_a)\delta_{ba} - F_{ba}(E) \end{pmatrix}. \tag{5.194}$$

Here, the indices a, b run from 2 to r.

Using now a well-known formula for the determinant of a matrix:

$$\det \begin{pmatrix} A & B \\ C & D \end{pmatrix} = \det(D)\det\big(A - BD^{-1}C\big), \tag{5.195}$$

the secular equation in the vicinity of $E = E_r$ can be rewritten as

$$E - E_s = F_{ss}(E) + \sum_{a,b=2}^{r} F_{sb}(E) D_{ba}^{-1}(E) F_{as}(E),$$

$$D_{ba}(E) = (E - E_a)\delta_{ba} - F_{ba}(E). \tag{5.196}$$

This equation can be solved by iteration, similarly to the nondegenerate case. To this end, one substitutes $E = E_s$ in the r.h.s. of the preceding equation and evaluates the correction. Simultaneously, the expansion of the matrix $F(E)$ in terms of the potential matrix V should be carried out. The iterations stop when the result at a given order in V does not change anymore. This pretty straightforward procedure allows to write down the perturbative series in case of the degenerate unperturbed levels up to a given order in V. However, the pertinent explicit formulae are rather voluminous even at lowest orders, and we refrain from displaying them here. The only assumption made during the derivation was that the degeneracy is completely removed already at the first order in perturbation theory. Should this not be the case, the solution requires further modification along very similar lines. We will, however, not pursue this issue further.

What remains now is to use the explicit form of the potential, which can be calculated by sandwiching Eq. (5.173) between free states, and to calculate the energy shift at the desired order. Below, we shall briefly consider examples of such calculations and concentrate on the ground-state shift only, where the unperturbed level is not degenerate. In the two-particle case, we have

$$
V_{pq} = \langle \mathbf{p}_1, \mathbf{p}_2 | \mathbf{H}_I | \mathbf{q}_1, \mathbf{q}_2 \rangle
$$

$$
= \left\{ -\frac{2}{L^3} \left(C_0 - C_1 (\mathbf{p}^2 + \mathbf{q}^2) \right) - \frac{\mathbf{p}^4}{2M^3} \delta_{\mathbf{pq}} \right\} \delta_{\mathbf{p}_1 + \mathbf{p}_2, \, \mathbf{q}_1 + \mathbf{q}_2}. \tag{5.197}
$$

Here, $\mathbf{p} = \frac{1}{2} (\mathbf{p}_1 - \mathbf{p}_2)$ and $\mathbf{q} = \frac{1}{2} (\mathbf{q}_1 - \mathbf{q}_2)$. In the CM frame, the contribution from the term with C_2 vanishes.

For the ground state with $|k_0\rangle = |\mathbf{0}, \mathbf{0}\rangle$, we have

$$
\Delta E_1 = V_{k_0 k_0} = -\frac{2C_0}{L^3} = \frac{4\pi a_0}{ML^3}. \tag{5.198}
$$

The last equation immediately follows from the matching condition. It is also seen that this is the only contribution to the ΔE_1, all other two-body matrix elements contain either an initial or a final momentum that vanishes in the ground state.

At second order, the contribution from the relativistic correction term still vanishes for the same reason. Hence, the second-order correction to the ground-state energy is given by

$$
\Delta E_2 = \frac{4}{L^6} \sum_{\mathbf{k} \neq 0} \frac{(C_0 - C_1 \mathbf{k}^2)^2}{0 - \mathbf{k}^2 / M}. \tag{5.199}
$$

The summation momenta are $\mathbf{k} = 2\pi \mathbf{n}/L$. Thus, ΔE_2 contains contributions of order L^{-4}, L^{-6} and L^{-8}:

$$
\Delta E_2 = -\frac{MC_0^2}{\pi^2 L^4} \sum_{\mathbf{n} \in \mathbb{Z}^3 \setminus \mathbf{0}} \frac{1}{\mathbf{n}^2} + \frac{8MC_0 C_1}{L^6} \sum_{\mathbf{n} \in \mathbb{Z}^3 \setminus \mathbf{0}} 1 - \frac{16\pi^2 MC_1^2}{L^8} \sum_{\mathbf{n} \in \mathbb{Z}^3 \setminus \mathbf{0}} \mathbf{n}^2. \tag{5.200}
$$

The first sum in the above expression coincides with the quantity I defined in Eq. (5.123). The second and third sums are also formally divergent and should be defined within some regularization. Note that the second sum can be rewritten as

$$\sum_{\mathbf{n}\in\mathbb{Z}^3\setminus 0} 1 = -1 + \sum_{\mathbf{n}\in\mathbb{Z}^3} 1. \qquad (5.201)$$

Since the summand is a regular function, the sum on the r.h.s. of this equation should be equal to the corresponding integral, up to exponentially suppressed corrections that are neglected here. Hence, regularizing the sum is equivalent to regularizing the integral. For consistency, we should use the same regularization both in the infinite and in a finite volume, in the matching as well as in the energy level calculation. If dimensional regularization is used everywhere, then this (no-scale) integral vanishes identically, and the above expression becomes equal to -1. By the same token, it can be shown that the last sum in Eq. (5.200) vanishes. Using the matching condition for C_0, C_1, one finally obtains

$$\Delta E_2 = -\frac{4a_0^2}{ML^4} I + \frac{8\pi^2 a_0^3}{ML^6}\left(r_0 - \frac{1}{a_0 M^2}\right). \qquad (5.202)$$

The general pattern of the calculations becomes crystal-clear now. It is easily seen that going to higher orders in perturbation theory gives each time one additional power of $1/L$. Indeed, each momentum counts as $1/L$, the leading term in the potential scales as $1/L^3$ and a factor L^2 appears in every additional energy denominator.[19] Below, we display (without derivation) the full expression for the ground-state energy shift of the two-body system at order L^{-6}:

$$\Delta E = \frac{4\pi a_0}{ML^3}\left\{1 - \frac{a_0}{\pi L} I + \left(\frac{a_0}{\pi L}\right)^2 (I^2 - J) - \left(\frac{a_0}{\pi L}\right)^3 (I^3 - 3IJ + K)\right.$$

$$\left. -\frac{\pi a_0}{M^2 L^3} + \frac{2\pi a_0^2 r_0}{L^3}\right\}, \qquad (5.203)$$

where I, J are defined in Eq. (5.123) and K is given by

$$K = \sum_{\mathbf{n}\in\mathbb{Z}^3\setminus 0} \frac{1}{\mathbf{n}^6} \simeq 8.40192. \qquad (5.204)$$

Note that the expression for the energy shift contains both the correction due to the effective range term as well as the relativistic correction (the term proportional to $1/M^2$ inside the curly brackets). Both these effects emerge first at $O(L^{-6})$.

It turns out that the same method can be straightforwardly applied to the calculation of the energy levels of N particles. The combinatorics becomes trickier but still, a result for a generic $N \geq 3$ can be written down compactly in the case of the ground state:

$$\Delta E_N = \frac{N!}{2!(N-2)!} \frac{4\pi a_0}{ML^3}\left\{1 - \frac{a_0}{\pi L} I + \left(\frac{a_0}{\pi L}\right)^2 (I^2 + (2N-5)J)\right.$$

$$\left. -\left(\frac{a_0}{\pi L}\right)^3 (I^3 + (2N-7)IJ + (5N^2 - 41N + 63)K + 8(N-2)(2Q^r + R^r))\right.$$

---

[19]   At this stage, it is seen why the terms with only two derivatives are retained in the two-body Lagrangian. The operators with more derivatives do not contribute up to and including $O(L^{-6})$.

$$+ (4N - 9)\frac{\pi a_0}{M^2 L^3} + (4N - 6)\frac{\pi a_0^2 r_0}{L^3}\Bigg\}$$

$$+ \frac{N!}{3!(N-3)!}\left\{\frac{64\pi a_0^4}{ML^6}(3\sqrt{3} - 4\pi)\ln(\mu L) - \frac{6D_0^r(\mu)}{L^6}\right\}. \tag{5.205}$$

As usual, μ is the scale of dimensional regularization. Some comments to this formula are needed. Unlike the two-body case, here one encounters logarithmically divergent sums of two different types:[20]

$$Q = \frac{1}{L^{2d}}\frac{1}{(2\pi)^{2(d-3)}}\sum_{\mathbf{p},\mathbf{q}\neq 0}\frac{1}{\mathbf{p}^2\,\mathbf{q}^2(\mathbf{p}^2 + \mathbf{q}^2 + (\mathbf{p}+\mathbf{q})^2)}\,,$$

$$R = \frac{1}{L^{2d}}\frac{1}{(2\pi)^{2(d-3)}}\sum_{\mathbf{p},\mathbf{q}\neq 0}\frac{1}{\mathbf{p}^4(\mathbf{p}^2 + \mathbf{q}^2 + (\mathbf{p}+\mathbf{q})^2)}\,. \tag{5.206}$$

Here, we have anticipated the use of dimensional regularization and utilized the only available scale at our disposal, L, to adjust the dimension of the sums to their counterparts in the infinite volume. In order to carry out dimensional regularization in practice, let us first rescale the momenta according to $\mathbf{p} = 2\pi\mathbf{n}/L$, $\mathbf{q} = 2\pi\mathbf{m}/L$. The first sum, for example, can be rewritten as

$$Q = \frac{1}{L^{2(d-3)}(2\pi)^{2d}}\sum_{\mathbf{n},\mathbf{m}\in\mathbb{Z}^3\setminus 0}\frac{1}{2\mathbf{n}^2\mathbf{m}^2(\mathbf{n}^2 + \mathbf{m}^2 + \mathbf{nm})}\,. \tag{5.207}$$

We rewrite the summand as

$$\frac{(1 - \delta_{\mathbf{n}0})(1 - \delta_{\mathbf{m}0})}{2\mathbf{n}^2\mathbf{m}^2(\mathbf{n}^2 + \mathbf{m}^2 + \mathbf{nm})} = \frac{1}{2(\mathbf{n}^2 + 1)(\mathbf{m}^2 + 1)(\mathbf{n}^2 + \mathbf{m}^2 + \mathbf{nm} + 1)}$$

$$+ \left(\frac{(1 - \delta_{\mathbf{n}0})(1 - \delta_{\mathbf{m}0})}{2\mathbf{n}^2\,\mathbf{m}^2(\mathbf{n}^2 + \mathbf{m}^2 + \mathbf{nm})} - \frac{1}{2(\mathbf{n}^2 + 1)(\mathbf{m}^2 + 1)(\mathbf{n}^2 + \mathbf{m}^2 + \mathbf{nm} + 1)}\right). \tag{5.208}$$

The sum over the expression in brackets is convergent. In the first term, using Poisson's formula the sum can be replaced by an integral. Hence,

$$Q = \frac{1}{L^{2(d-3)}}\sum_{\mathbf{n},\mathbf{m}\in\mathbb{Z}^3}\int\frac{d^d\mathbf{p}\,d^d\mathbf{q}}{(2\pi)^{2d}}\frac{\delta^{2\pi i\mathbf{p}\mathbf{n} + 2\pi i\mathbf{q}\mathbf{m}}}{2(\mathbf{p}^2 + 1)(\mathbf{q}^2 + 1)(\mathbf{p}^2 + \mathbf{q}^2 + \mathbf{pq} + 1)}$$

$$+ \frac{1}{L^{2(d-3)}(2\pi)^{2d}}\sum_{\mathbf{n},\mathbf{m}\in\mathbb{Z}^3}\left(\frac{(1 - \delta_{\mathbf{n}0})(1 - \delta_{\mathbf{m}0})}{2\mathbf{n}^2\mathbf{m}^2(\mathbf{n}^2 + \mathbf{m}^2 + \mathbf{nm})}\right.$$

$$\left. - \frac{1}{2(\mathbf{n}^2 + 1)(\mathbf{m}^2 + 1)(\mathbf{n}^2 + \mathbf{m}^2 + \mathbf{nm} + 1)}\right)$$

$$= \mu^{2(d-3)}\left\{\frac{1}{48\pi^2}\left(\ln(\mu L) - \frac{1}{2(d-3)}\right) + \frac{1}{(2\pi)^6}Q^r\right\}. \tag{5.209}$$

Here, $Q^r \simeq -102.1556$ is the renormalized value of the sum.

---

[20] The choice of the numerical prefactor in the dimensionally regularized sums is a priori arbitrary. The given choice is consistent with the choice made in the infinite volume.

The second sum can be calculated analogously, and the result is given by

$$R = \mu^{2(d-3)} \left\{ -\frac{\sqrt{3}}{32\pi^3} \left(\ln(\mu L) - \frac{1}{2(d-3)} \right) + \frac{1}{(2\pi)^6} R^r \right\}, \qquad (5.210)$$

with $R^r \simeq 19.1869$.

The ultraviolet divergences are canceled by defining the renormalized coupling constant,

$$D_0 = \mu^{2(d-3)} \left\{ -\frac{16\pi a_0^4}{3M} \frac{1}{d-3} (3\sqrt{3} - 4\pi) + D_0^r \right\}. \qquad (5.211)$$

The μ-dependence in $D_0^r = D_0^r(\mu)$ cancels against the μ-dependence in the logarithmic term in the final result (as it should). There is only one three-body coupling constant at this order. Hence, the renormalization of this single constant, carried out in the infinite volume, should render all N-body energy levels finite. Finally, note that the same method can be used without any modification to calculate the energies of the excited states.

To summarize, it is seen that the perturbative approach based on the non-relativistic effective Hamiltonian enables one to evaluate the shift of the finite-volume energy levels in a much more straightforward and systematic fashion than carrying out the threshold expansion of the quantization condition. The only restriction of the method is that it works away from resonances, where shifts are small and can be evaluated using perturbation theory. It should be also mentioned that the explicit formulae for the shifts can be used to extract the fundamental parameters of the theory, for instance, the two-body effective-range exchange parameters, or the three-body coupling D_0^r, from measurements on the lattice. At present, this method is widely used.

Finally, it should be pointed out that the leading-order formula for the two-particle shift (5.203) bears a close analogy to the DGBT formula, considered in Section 2.12. Indeed, the normalized wave function of two particles in a cubic box is given by $\Psi(\mathbf{x}) = L^{-3/2} e^{i \mathbf{p} \mathbf{x}}$. (Here, $\mathbf{p}$ and $\mathbf{x}$ denote the relative momentum and relative distance, respectively.) Consequently, $|\Psi(0)|^2 = L^{-3}$ and ΔE is proportional to a_0/L^3. This similarity is not an accident. Indeed, in both situations, we have an exactly solvable long-range potential (Coulomb or box), which creates a discrete spectrum (in case of the Coulomb potential, we refer to the bound-state spectrum), and a short-range potential, which leads to a displacement of the energy levels. At the first order in perturbation theory, where the details of the long-range potential do not matter, the expressions for the displacement are identical.

5.8 Three-Particle Quantization Condition

As just discussed, perturbative calculations of the multiparticle energy levels in a cubic box can be carried out in a rather straightforward manner. However, obtaining an exact quantization condition, which can also be used in the vicinity of the resonance energies, as well as for the study of the shallow bound states, has proven to be much more difficult. The first attempt in this direction was made in Ref. [76]. There, it was demonstrated that the finite-volume energy levels are determined solely by the three-body S-matrix elements and do not depend on the details of the short-range interactions. This result has sparked activities in the field and, within a few years, three alternative formulations of the three-body quantization condition have become available in the literature [77–81]. Note that all these three approaches are based on very similar premises. In analogy to the two-particle case, the irregular dependence on L is produced here by the so-called three-particle reducible diagrams, that is, the diagrams in which the initial and final states can be made disconnected by cutting exactly three internal lines. The rest belongs to the class of the irreducible diagrams and exhibits only exponentially suppressed corrections in a finite volume. Writing down integral equations for the full three-body amplitude, and replacing the integral with a sum, we again arrive at the quantization condition. The energy levels are obtained by demanding that the determinant of this system of linear equations vanishes. Albeit differing in details, all three approaches follow the pattern described above. It is not our aim here to carry out a comparison of these approaches. In what follows we merely describe one of them, which is based on nonrelativistic effective Lagrangians, and in which the effectiveness of the EFT method is most clearly visible. The infinite-volume version of this approach was already discussed at length in Section 2.9. In this approach, all irreducible diagrams are already hidden in the effective couplings of the low-energy Lagrangian. This fact, as well as the use of the particle-dimer language, make the bookkeeping of the diagrams here remarkably easy. For simplicity, we shall use non-relativistic kinematics first. As we know, the infinite-volume scattering amplitude obeys Eq. (2.166), where the kernel is given by Eq. (2.167) at lowest order. In a finite volume, this equation can be rewritten as follows:[21]

$$\mathcal{M}_L(\mathbf{p}, \mathbf{q}; P^0) = Z(\mathbf{p}, \mathbf{q}; P^0) + \frac{1}{L^3} \sum_{\mathbf{k}}^{\Lambda} Z(\mathbf{p}, \mathbf{k}; P^0) \tau_L(\mathbf{k}; P^0 - E_{\mathbf{k}}) \mathcal{M}_L(\mathbf{k}, \mathbf{q}; P^0).$$

(5.212)

Here, the "driving term" Z is the same as in the infinite volume and, at lowest order, is given again by Eq. (2.167). The quantity τ, which is related to the particle-dimer propagator, however, gets modified. Instead of

$$8\pi \tau^{-1}(\mathbf{P}; P^0) = q_0 \cot \delta(q_0) - \frac{4\pi}{M} \int \frac{d^d \mathbf{q}}{(2\pi)^3} \frac{1}{E_{\mathbf{q}} + E_{\mathbf{P}-\mathbf{q}} - P^0},$$

(5.213)

(cf. Eqs. (2.157) and (2.150)) we have

$$8\pi \tau_L^{-1}(\mathbf{P}; P^0) = q_0 \cot \delta(q_0) - \frac{4\pi}{ML^3} \sum_{\mathbf{q}} \frac{1}{E_{\mathbf{q}} + E_{\mathbf{P}-\mathbf{q}} - P^0}.$$

(5.214)

---

[21] For simplicity, we use the Minkowski metric in the following derivation.

Here, $\delta(q_0)$ denotes the S-wave phase shift. Furthermore, since we are using non-relativistic kinematics, q_0 is given by

$$q_0 = \sqrt{MP^0 - 2M^2 - \frac{\mathbf{P}^2}{4}}.\qquad(5.215)$$

Hence,

$$8\pi\tau_L^{-1}(\mathbf{k};P^0 - E_\mathbf{k}) = q_0 \cot\delta(q_0) - \frac{4\pi}{L^3}\sum_\mathbf{q}\frac{1}{\mathbf{k}^2 + \mathbf{q}^2 + \mathbf{kq} - M(P^0 - 3M)}.$$

$$(5.216)$$

For consistency with the infinite-volume formalism, it is assumed that the above sum is regularized by using dimensional regularization. This can be done in analogy, for example, with Eq. (5.124).

As in the two-particle case, the finite-volume energy levels are determined by the position of the poles of the three-particle scattering amplitude, which is expressed through $\mathcal{M}_L$ via the finite-volume counterpart of Eq. (2.163). The poles emerge, if the system of linear equations that determines $\mathcal{M}_L$ becomes singular, or, stated differently, when the determinant of this system of equations vanishes:

$$\det(\tau_L^{-1}(P^0) - Z(P^0)) = 0.\qquad(5.217)$$

The matrices in this expression are defined in the space of the spectator momenta:

$$\langle\mathbf{p}|\tau_L^{-1}(P^0)|\mathbf{q}\rangle = L^3\delta_{\mathbf{pq}}\tau_L^{-1}(\mathbf{p};P^0 - E_\mathbf{p}),$$

$$\langle\mathbf{p}|Z(P^0)|\mathbf{q}\rangle = Z(\mathbf{p},\mathbf{q};P^0).\qquad(5.218)$$

Including higher partial waves, this matrix will depend on additional indices ℓm and $\ell'm'$, corresponding to the angular momentum and its projection. In order to ease the notation, we do not display these indices explicitly.

Let us discuss briefly the workflow during the extraction of the three-particle observables from lattice data. During this extraction, the two-particle effective-range expansion parameters that enter the phase shift are considered as input. They can be fitted to the two-particle data in the same simulations separately. As a result of the fit, we may determine three-body couplings (the nonderivative coupling h_0, as well as higher-order derivative couplings). These couplings are not observables as they depend on the cutoff Λ. However, due to their short-range nature, all finite-volume corrections to these quantities are exponentially suppressed. Thus, we can take them the same in a finite and in the infinite volume. Moreover, using power-counting arguments, we can always truncate the derivative expansion in the three-body Lagrangian, retaining only a finite number of unknown couplings that will be fitted to the data on the finite-volume energy levels.

Note also that the infinite-volume spectrum of three particles has a complex structure. In particular, along with the three-particle continuum, there are particle-dimer scattering states (when stable dimers exist) and the three-body bound states (again, when these exist). The finite-volume spectrum contains counterparts of all these states,

albeit an unambiguous identification of a *particular* state with its infinite-volume counterpart is challenging (see, for example, the discussion in Ref. [75]). The quantization
condition allows one to treat all these finite-volume states in one go; see [75, 79, 82–84].
In particular, in the case of the three-particle shallow bound states (the Efimov states
considered in Section 2.9), the volume-dependence is again of exponential form,

$$\Delta E_3 = \text{const} \cdot (\kappa L)^{-3/2} \exp\left(-\frac{2\kappa L}{\sqrt{3}}\right), \tag{5.219}$$

where $\kappa = \sqrt{M E_B}$ is the bound-state momentum related to the three-particle binding
energy E_B (with $E_B \geq 0$). In case of shallow bound states, the quantity κ obeys the
condition $\kappa \ll M$ and hence, retaining the terms $\sim \exp(-\kappa L)$ while neglecting the ones
$\sim \exp(-ML)$ is justified.

At the end of the day, when all couplings are fitted to data, we may *calculate* the
infinite-volume observables (various S-matrix elements in the infinite volume) by using
the Faddeev equation (2.166) *with the same cutoff.* Thus, the extraction of the S-
matrix elements proceeds in two steps in the three-particle sector, in contrast to the
Lüscher equation that directly relates the two-body S-matrix elements to the energy
spectrum. All attempts to write analogous equations in the three-particle sector in three
dimensions have failed so far.

The use of the covariant version of the NREFT, which was discussed in Section 2.7,
is straightforward and does not lead to any noteworthy change, except the use of
relativistic kinematics and the different normalization of the one-particle states. The
three-particle quantization condition in this case takes the form

$$\det(A(P^0)) = 0,$$

$$\langle \mathbf{p} | A(P^0) | \mathbf{q} \rangle = L^3 \delta_{\mathbf{pq}} 2w(\mathbf{p}) \tau_L^{-1}(\mathbf{p}; P^0 - w(\mathbf{p})) - Z(\mathbf{p}, \mathbf{q}; P^0). \tag{5.220}$$

Here,

$$Z(\mathbf{p}, \mathbf{q}; P^0) = \frac{1}{2w(\mathbf{p}+\mathbf{q})(w(\mathbf{p})+w(\mathbf{q})+w(\mathbf{p}+\mathbf{q})-P^0)} + \frac{h_0}{f_0^2}, \tag{5.221}$$

and

$$\tau_L(\mathbf{p}; P^0) = \frac{16\pi\sqrt{s}}{p^* \cot \delta(p^*) - p^* \cot \phi^{\mathbf{d}}(s)}, \tag{5.222}$$

where

$$s = (P^0)^2 - \mathbf{P}^2, \qquad p^* = \sqrt{\frac{s}{4} - M^2}, \qquad \mathbf{d} = \frac{\mathbf{P}L}{2\pi}, \tag{5.223}$$

and

$$\cot \phi^{\mathbf{d}}(s) = -\frac{Z_{00}^{\mathbf{d}}(1; s)}{\pi^{3/2} \gamma \eta}, \qquad \eta = \frac{p^* L}{2\pi}. \tag{5.224}$$

An explicit expression for the function $Z_{\ell m}^{\mathbf{d}}(1; s)$ is given in Eq. (5.97). Again, in
order to ease notation, we have restricted ourselves to the lowest-order particle-dimer

coupling and retained only the S-wave dimer. Including the higher-order effects is straightforward and will not be considered here.

As a warning, note that, despite using relativistic expression for the one-particle energies, the framework in the given form is not relativistically invariant. (Here, of course, we refer to the infinite volume since, in a finite volume, Lorentz invariance is broken by the box.) Namely, the particle-dimer scattering amplitude, calculated for different values of the full center-of-mass momentum $\mathbf{P}$, is not the same, albeit the physical singularities of the amplitude are located at the right place in all reference frames. The reason for this is that the quantity Z is not Lorentz-invariant. This situation differs from the two-particle case, considered in Section 2.7. There, amending the prescription for the calculation of Feynman diagrams, it was possible to obtain explicitly invariant expressions at all orders.

As in the two-particle case, the quantization condition can be partially diagonalized, using octahedral symmetry [75]. To this end, let us recall the definition of the *shells*. These are sets of lattice momenta, which are related to each other by the transformations from the octahedral group O_h (we are considering here the CM system of three particles). In other words, each shell s is characterized by the *reference momentum* $\mathbf{p}_0(s)$, so that all momenta in this shall can be represented as $\mathbf{p}(s) = g\mathbf{p}_0(s)$ with $g \in O_h$. Further, let us define the *multiplicity* of a shell, $\vartheta(s)$, as the number of different vectors in a given shell s. Then, let us consider a system of linear equations

$$\frac{1}{L^3} \sum_{\mathbf{k}} \langle \mathbf{p}|A(P^0)|\mathbf{k}\rangle f(\mathbf{k}) = 0, \tag{5.225}$$

with the matrix element of $A(P^0)$ given by Eq. (5.220). The determinant of this system of linear equations vanishes exactly at the energies that satisfy the quantization condition.

Next, we introduce what can be considered the counterpart of introducing polar coordinates in a finite cubic volume. Namely, we rewrite the sum over all lattice momenta as

$$\frac{1}{L^3} \sum_{s} \sum_{g \in O_h} \frac{\vartheta(s)}{G} \langle \mathbf{p}|A(P^0)|g\mathbf{k}_0(s)\rangle f(g\mathbf{k}_0(s)) = 0, \tag{5.226}$$

where $G = 48$ is the number of elements in O_h. In the above expression, the sum runs over shells, as well as over all elements of O_h. The normalization factor $\vartheta(s)/G$ ensures that all different momenta are counted exactly once.

Furthermore, an arbitrary function $f(\mathbf{p})$ can be expanded as

$$f(\mathbf{p}) = f(g\mathbf{p}_0(s)) = \sum_{\Gamma} \sum_{\lambda\rho} T_{\lambda\rho}^{\Gamma}(g) f_{\rho\lambda}^{\Gamma}(\mathbf{p}_0(s)). \tag{5.227}$$

The inverse transformation is

$$f_{\lambda\rho}^{\Gamma}(\mathbf{p}_0(s)) = \frac{s_{\Gamma}}{G} \sum_{g \in O_h} \left(T_{\rho\lambda}^{\Gamma}(g)\right)^* f(g\mathbf{p}_0(s)). \tag{5.228}$$

Here, the $T_{\lambda\rho}^{\Gamma}(g)$ are matrices of the irreps of the group O_h, and s_Γ is the dimension of these irreps. The indices λ,ρ label the basis vectors in these irreps. The above transformations can be considered as the partial-wave expansion in a cubic volume. Using now Eq. (5.227) and projecting Eq. (5.226) onto a given irrep Γ, we get

$$\frac{s_\Gamma}{G} \sum_{g\in O_h} \left(T_{\rho\lambda}^{\Gamma}(g)\right)^* \frac{1}{L^3} \sum_s \sum_{g'\in O_h} \frac{\vartheta(s)}{G} \langle g\mathbf{p}_0(r)|A(P^0)|g'\mathbf{k}_0(s)\rangle$$

$$\times \sum_{\Gamma'}\sum_{\delta\sigma} T_{\delta\sigma}^{\Gamma'}(g') f_{\sigma\delta}^{\Gamma'}(\mathbf{k}_0(s)) = 0. \tag{5.229}$$

Next, let us take into account the invariance of the quantity $A(P^0)$ with respect to the transformations from O_h:

$$\langle g\mathbf{p}_0(r)|A(P^0)|g'\mathbf{k}_0(s)\rangle = \langle g'^{-1}g\mathbf{p}_0(r)|A(P^0)|\mathbf{k}_0(s)\rangle. \tag{5.230}$$

This allows us to simplify the sum over all elements of O_h:

$$\sum_{g,g'\in O_h} \left(T_{\rho\lambda}^{\Gamma}(g)\right)^* \langle g\mathbf{p}_0(r)|A(P^0)|g'\mathbf{k}_0(s)\rangle T_{\delta\sigma}^{\Gamma'}(g')$$

$$= \sum_{g,g'\in O_h} \left(T_{\rho\lambda}^{\Gamma}(g'g)\right)^* \langle g\mathbf{p}_0(r)|A(P^0)|\mathbf{k}_0(s)\rangle T_{\delta\sigma}^{\Gamma'}(g')$$

$$= \sum_{\omega}\sum_{g\in O_h} \left(T_{\omega\lambda}^{\Gamma}(g)\right)^* \langle g\,\mathbf{p}_0(r)|A(P^0)|\mathbf{k}_0(s)\rangle \sum_{g'\in O_h} \left(T_{\rho\omega}^{\Gamma}(g')\right)^* T_{\delta\sigma}^{\Gamma'}(g')$$

$$= \sum_{\omega}\sum_{g\in O_h} \left(T_{\omega\lambda}^{\Gamma}(g)\right)^* \langle g\,\mathbf{p}_0(r)|A(P^0)|\mathbf{k}_0(s)\rangle \frac{G}{s_\Gamma}\delta_{\Gamma\Gamma'}\delta_{\rho\delta}\delta_{\omega\sigma}. \tag{5.231}$$

In the last line, we have used the orthogonality of the matrices of the irreps. Substituting this expression into Eq. (5.229), we obtain

$$\frac{1}{L^3} \sum_s \frac{\vartheta(s)}{G} \sum_\sigma A_{\lambda\sigma}^{\Gamma}(r,s) f_{\sigma\rho}(s) = 0, \tag{5.232}$$

where

$$A_{\lambda\sigma}^{\Gamma}(r,s) = \sum_{g\in O_h} \left(T_{\sigma\lambda}^{\Gamma}(g)\right)^* \langle g\mathbf{p}_0(r)|A(P^0)|\mathbf{k}_0(s)\rangle. \tag{5.233}$$

Here, we already took into account the fact that the choice of the reference momentum is arbitrary and, hence, $A_{\lambda\sigma}^{\Gamma}(r,s)$ depends only on the shells rather than the individual momenta in these shells.

The equations (5.232) and (5.233) solve our problem completely. The matrix in the quantization condition takes a block-diagonal form, each block corresponding to different irreps:

$$\det A^{\Gamma} = 0. \tag{5.234}$$

In this expression, the matrix A^{Γ} is given by Eq. (5.233) with indices that label the shells as well as the vectors in a given irrep Γ. The indices λ,σ run from 1 to s_Γ.

5.9 Three-Particle Decays on the Lattice

In this section we shall derive a three-particle analog of the Lellouch–Lüscher equation, following exactly the same pattern as in the two-body case. The method used here is similar to the one from Ref. [85] (for an alternative approach, see Ref. [86]). Within this method, one carries out the calculation of the decay matrix element twice, namely in the infinite as well as in a finite volume. The bridge between both calculations is provided by the effective couplings in the Lagrangian, which are the same up to exponentially suppressed finite-volume artifacts.

In order to avoid unnecessary details, we shall consider here the decay of a spinless particle with a mass M into three identical particles with the mass m, where $M > 3m$. This resembles closely the decay $K \to 3\pi$ (e.g., there are nonidentical particles present in the final states of kaon decays). For brevity, we shall still refer to the heavy and light particles as to "kaon" and "pion," respectively. Furthermore, here we shall use the covariant version of the three-particle equations, just to ensure that the low-energy singularities of the three-particle decay amplitude are all at the correct place from the beginning.

In the particle-dimer picture, the Lagrangian that describes the decay of a kaon into three pions is given by

$$\mathcal{L} = K^\dagger 2W(i\partial_t - W)K + \phi^\dagger 2w(i\partial_t - w)\phi + \sigma d^\dagger d + \frac{f_0}{2}\left(d^\dagger \phi\phi + \text{h.c.}\right)$$

$$+ h_0 d^\dagger d\phi^\dagger \phi + g_0(K^\dagger d\phi + \text{h.c.}) + \cdots. \tag{5.235}$$

Here, K, ϕ and d denote the kaon, pion and dimer fields, respectively, $W = \sqrt{M^2 - \nabla^2}$ and $w = \sqrt{m^2 - \nabla^2}$ are the relativistic energies of the kaon and the pions, and the ellipses denote the higher-order terms. In the particle-dimer picture, the kaon decays into the particle-dimer pair first, and subsequently the dimer decays into a pion pair. Hence, the weak coupling g_0, which is present in the preceding equation, is a counterpart of the coupling g_K introduced in Eq. (5.153). Moreover, integrating out the (redundant) dimer field in the Lagrangian, we may relate the coupling g_0 to the coupling that describes the direct decay of a kaon into three pions. There is no need to do this pretty straightforward exercise here. Note, however, that there is an important difference to the two-particle decay. As already mentioned in Section 5.6, there is no need to consider other couplings that describe the decay $K \to 2\pi$ beyond the leading nonderivative coupling g_K, since the kinematics of the two-particle decay always fixes the magnitude of the relative momenta of the decay products in the CM frame as well as the orbital momentum. On the contrary, two relative momenta and two orbital momenta of the decay products in the three-particle decay are only partially fixed by the conservation of total energy and total orbital momentum. For this reason, the Lagrangian that describes the $K \to 3\pi$ should contain a bunch of derivative operators with increasing mass dimension, each coming with an independent coupling. Hence, it does not suffice to carry out one measurement on the lattice to determine the matrix element. Instead, carrying out a truncation at some order, one ends up with N independent couplings, and thus N independent measurements are needed to determine these. Consequently, we anticipate that the Lellouch–Lüscher factor in the three- (and more) particle case is not just one number. Instead, it is a $N \times N$ matrix that relates measurements on the lattice with the amplitudes in the infinite volume, evaluated for different configurations of the momenta of the decaying particles (see a more detailed discussion of this issue in what follows). Here, we shall restrict our derivation to the leading order, where there is only one coupling. The inclusion of the higher orders is straightforward.

In order to proceed with the derivation, let us define a composite operator constructed out of three pion fields,

$$\mathcal{O}(\{\mathbf{k}\}; x_4) = \prod_{i=1}^{3} \int d^3 \mathbf{x}_i e^{-i \mathbf{k}_i \mathbf{x}_i} \phi(\mathbf{x}_i, x_4). \tag{5.236}$$

In the free-field case, this operator creates three pions with momenta $\{\mathbf{k}\} = (\mathbf{k}_1, \mathbf{k}_2, \mathbf{k}_3)$, acting on the vacuum bra-vector $\langle 0|$ from the right. Obviously, the choice of the interpolating field does not play any role in the following; we just stick to the simplest and most natural one.

Now, we shall closely follow the steps of the derivation in the two-particle case; see Section 5.6. Namely, we first take $x_4 > y_4$ and consider the two-point function

$$\langle 0| \mathcal{O}(\{\mathbf{k}\}; x_4) \mathcal{O}^\dagger(\{\mathbf{k}\}; y_4) |0\rangle = \sum_n e^{-E_n(x_4 - y_4)} \big| \langle 0| \mathcal{O}(\{\mathbf{k}\}; 0) |n\rangle \big|^2. \tag{5.237}$$

Note that here, as usual, $|n\rangle$ denotes the eigenvectors of the interacting system in a finite volume, which have a nonzero overlap with the operator $\mathcal{O}(\{\mathbf{k}\}; 0)$. An identification of this state with the asymptotic state of three pions with definite momenta is not possible.

Furthermore, we calculate the same matrix element in NREFT. By analogy with Eq. (5.160),

$$\langle 0| \mathcal{O}(\{\mathbf{k}\}; x_4) \mathcal{O}^\dagger(\{\mathbf{k}\}; y_4) |0\rangle = \int \frac{dP_4}{2\pi} e^{iP_4(x_4 - y_4)}$$

$$\times \left\{ \frac{-i L^9 \left(1 + \delta_{\mathbf{k}_1 \mathbf{k}_2} + \delta_{\mathbf{k}_1 \mathbf{k}_3} + \delta_{\mathbf{k}_2 \mathbf{k}_3} + 2\delta_{\mathbf{k}_1 \mathbf{k}_2} \delta_{\mathbf{k}_2 \mathbf{k}_3}\right)}{8 w(\mathbf{k}_1) w(\mathbf{k}_2) w(\mathbf{k}_3)(P_4 - i(w(\mathbf{k}_1) + w(\mathbf{k}_2) + w(\mathbf{k}_3)))} \right.$$

$$\left. - \frac{L^3 T_L(\{\mathbf{k}\}, \{\mathbf{k}\}; -iP_4)}{\left(8 w(\mathbf{k}_1) w(\mathbf{k}_2) w(\mathbf{k}_3)(P_4 - i(w(\mathbf{k}_1) + w(\mathbf{k}_2) + w(\mathbf{k}_3)))\right)^2} \right\}. \tag{5.238}$$

The three-particle amplitude has simple poles at $P_4 = iE_n$:

$$T_L(\{\mathbf{p}\}, \{\mathbf{q}\}; -iP_4) = \frac{\Psi_L^{(n)}(\{\mathbf{p}\}) \Psi_L^{(n)}(\{\mathbf{q}\})}{E_n + iP_4} + \text{regular terms at } P_4 \to iE_n. \tag{5.239}$$

Here, $\Psi_L^{(n)}(\{\mathbf{p}\})$ denotes the three-particle wave function for the state $|n\rangle$. Furthermore, performing the contour integral, we get

$$\langle 0|\mathcal{O}(\{\mathbf{k}\};x_4)\mathcal{O}^\dagger(\{\mathbf{k}\};y_4)|0\rangle$$

$$= \sum_n \frac{L^3 e^{-E_n(x_4-y_4)}\left(\Psi_L^{(n)}(\{\mathbf{p}\})\right)^2}{\left(8w(\mathbf{k}_1)w(\mathbf{k}_2)w(\mathbf{k}_3)(E_n-w(\mathbf{k}_1)-w(\mathbf{k}_2)-w(\mathbf{k}_3))\right)^2}, \tag{5.240}$$

and, hence, the absolute value of the matrix element of the composite field operator $\mathcal{O}$ between the vacuum and a given eigenstate of the strong Hamiltonian is given by

$$\left|\langle 0|\mathcal{O}(\{\mathbf{k}\};0)|n\rangle\right| = L^{3/2} \frac{\left|\Psi_L^{(n)}(\{\mathbf{p}\})\right|}{\left|8w(\mathbf{k}_1)w(\mathbf{k}_2)w(\mathbf{k}_3)(E_n-w(\mathbf{k}_1)-w(\mathbf{k}_2)-w(\mathbf{k}_3))\right|}. \tag{5.241}$$

Next, we consider the decay of a kaon into three pions. In lattice QCD, to lowest order in the weak coupling, the matrix element that describes this decay is given by the expression $-\langle n|\mathcal{H}_F(0)|K\rangle_L$, where $\langle n|$ is a given three-pion eigenstate of the *purely strong* Hamiltonian, and the parameter L is tuned so that E_n exactly coincides with the mass of the kaon M. At the level of the effective Lagrangian, the decay $K \to 3\pi$ is described by the local vertex,

$$\mathcal{L}_{\text{decay}}(x) = J_K^\dagger(x)K(x) + \text{h.c.}, \tag{5.242}$$

where, using the lowest-order Lagrangian (5.235), we get

$$J_K^\dagger(x) = g_0 d^\dagger(x)\phi^\dagger(x). \tag{5.243}$$

Within the EFT, to lowest order in the weak decay parameter g_0, the decay amplitude is given by Fermi's golden rule

$$\langle n|\mathcal{L}_{\text{decay}}(0)|K\rangle_L = \langle n|J_K^\dagger(0)|0\rangle_L\langle 0|K(0)|K\rangle_L = \langle n|J_K^\dagger(0)|0\rangle_L. \tag{5.244}$$

The matrix element on the left-hand side of this equation can be extracted from the Euclidean two-point function:

$$\langle 0|\mathcal{O}(\{\mathbf{k}\};x_4)J_K^\dagger(0)|0\rangle = \sum_n e^{-E_n x_4}\langle 0|\mathcal{O}(\{\mathbf{k}\};0)|n\rangle\langle n|J_K^\dagger(0)|0\rangle. \tag{5.245}$$

Again, let us evaluate this two-point function, summing up the graphs in the effective theory:

$$\langle 0|\mathcal{O}(\{\mathbf{k}\};x_4)J_K^\dagger(0)|0\rangle$$

$$= -i \int \frac{dP_4}{2\pi} \frac{e^{iP_x x_4} F_L(\{\mathbf{k}\};-iP_4)}{8w(\mathbf{k}_1)w(\mathbf{k}_2)w(\mathbf{k}_3)(P_4-i(w(\mathbf{k}_1)+w(\mathbf{k}_2)+w(\mathbf{k}_3)))}, \tag{5.246}$$

where the quantity F_L, calculated with the use of the lowest-order Lagrangian (5.235), is given by (see Fig. 5.12)

$$F_L(\{\mathbf{k}\};-iP_4) = \frac{g_0}{f_0} \sum_{\alpha=1}^3 \tau_L(-\mathbf{k}_\alpha;-iP_4)$$

$$\times \left\{1 + \frac{1}{L^3}\sum_{\mathbf{q}}^\Lambda \mathcal{M}_L(-\mathbf{k}_\alpha,-\mathbf{q};-iP_4)\frac{1}{2w(\mathbf{q})}\tau_L(-\mathbf{q};-iP_4)\right\}. \tag{5.247}$$

Diagrams describing the decay $K \to 3\pi$ in NREFT. The wiggly line denotes the current $J_K^\dagger(0)$, and the filled circle represents the weak decay vertex. The single and double lines correspond to the pions and the dimer, respectively.

The particle-dimer scattering amplitude also has a pole, exactly at the same energy:

$$\mathcal{M}_L(\mathbf{p}, \mathbf{q}; P_4) = \frac{\psi_L^{(n)}(\mathbf{p})\, \psi_L^{(n)}(\mathbf{q})}{E_n + iP_4} + \text{regular terms at } P_4 \to iE_n. \qquad (5.248)$$

Here, $\psi_L^{(n)}(\mathbf{p})$ is the particle-dimer wave function.

Carrying out the Cauchy integration over P_4, we get

$$\langle 0 | \mathcal{O}(\{\mathbf{k}\}; x_4) J_K^\dagger(0) | 0 \rangle$$

$$= \sum_n \frac{e^{-E_n x_4}}{8w(\mathbf{k}_1) w(\mathbf{k}_2) w(\mathbf{k}_3)(w(\mathbf{k}_1) + w(\mathbf{k}_2) + w(\mathbf{k}_3) - E_n)}$$

$$\times \sum_{\alpha=1}^{3} \tau(-\mathbf{k}_\alpha; E_n) \psi_L^{(n)}(-\mathbf{k}_\alpha) \frac{g_0}{f_0} \frac{1}{L^3} \sum_{\mathbf{q}}^{\Lambda} \psi_L^{(n)}(-\mathbf{q}) \frac{1}{2w(\mathbf{q})} \tau_L(-\mathbf{q}; E_n). \qquad (5.249)$$

In order to proceed further, we have to establish the relation between the wave functions $\Psi_L^{(n)}(\{\mathbf{p}\})$ and $\psi_L^{(n)}(\mathbf{p})$. This can be achieved exactly in the same manner as in the infinite volume; see Section 2.9. To this end, we consider the relation between the three-particle scattering amplitude and the particle-dimer scattering amplitude. The former is obtained from the latter by attaching two particles to each incoming/outgoing dimer leg. The summation over all possible choices of incoming and outgoing particles as a spectator is performed (see Fig. 2.10). The relation takes the form (cf. Eq. (2.163))

$$T_L(\{\mathbf{p}\}, \{\mathbf{q}\}; -iP_4) = \sum_{\alpha,\beta=1}^{3} \left\{ \tau_L(-\mathbf{p}_\alpha; -iP_4) 2w(\mathbf{p}_\alpha) L^3 \delta_{\mathbf{p}_\alpha \mathbf{q}_\beta} \right.$$

$$\left. + \tau_L(-\mathbf{p}_\alpha; -iP_4) \mathcal{M}_L(-\mathbf{p}_\alpha, -\mathbf{q}_\beta; -iP_4) \tau_L(-\mathbf{q}_\alpha; -iP_4) \right\}. \qquad (5.250)$$

Extracting the pole at $P_4 = iE_n$ from both sides, we finally get

$$\Psi_L^{(n)}(\{\mathbf{p}\}) = \sum_{\alpha=1}^{3} \tau_L(-\mathbf{p}_\alpha; E_n) \psi_L^{(n)}(\mathbf{p}), \qquad (5.251)$$

where the sum runs over all choices of the spectator momentum $\mathbf{p}_\alpha$. Note also that the particle-dimer wave function obeys the homogeneous equation

$$\psi_L^{(n)}(\mathbf{p}) = \frac{1}{L^3} \sum_{\mathbf{k}}^{\Lambda} Z(\mathbf{p}, \mathbf{k}; E_n) \frac{\tau_L(\mathbf{k}; E_n)}{2w(\mathbf{k})} \psi_L^{(n)}(\mathbf{k}). \qquad (5.252)$$

In order to calculate the kaon decay matrix element, we also need to determine the normalization of the wave function. Since the wave function obeys Eq. (5.252), where Z and τ both depend on the energy in a nontrivial manner, it cannot be merely normalized to unity. The goal can be achieved along the same path as in the infinite volume; see Section 2.9. In a finite volume, the normalization is given by

$$\frac{1}{L^6} \sum_{\mathbf{p},\mathbf{k}}^{\Lambda} \psi_L^{(n)}(\mathbf{p}) \frac{\tau_L(\mathbf{p};E_n)}{2w(\mathbf{p})} \frac{dZ(\mathbf{p},\mathbf{k};E_n)}{dE_n} \frac{\tau_L(\mathbf{k};E_n)}{2w(\mathbf{p})} \psi_L^{(n)}(\mathbf{k};E_n)$$

$$+ \frac{1}{L^3} \sum_{\mathbf{p}}^{\Lambda} \psi_L^{(n)}(\mathbf{p}) \frac{1}{2w(\mathbf{p})} \frac{d\tau(\mathbf{p};E_n)}{dE_n} \psi_L^{(n)}(\mathbf{p}) = 1. \qquad (5.253)$$

Using now Eqs. (5.241), (5.249) and (5.251), we can give the explicit expression for the decay matrix element in a finite volume:

$$L^{3/2} \left| \langle n | J_K^\dagger(0) | 0 \rangle \right| = \left| \frac{g_0}{f_0} \frac{1}{L^3} \sum_{\mathbf{q}}^{\Lambda} \psi_L^{(n)}(-\mathbf{q}) \frac{1}{2w(\mathbf{q})} \tau_L(-\mathbf{q};E_n) \right|. \qquad (5.254)$$

On the other hand, we can calculate the decay matrix element in the infinite volume by using the same effective Lagrangian. Summing up the diagrams in perturbation theory, we arrive at the expression

$$\langle \mathbf{k}_1, \mathbf{k}_2, \mathbf{k}_3; out | J_K^\dagger(0) | 0 \rangle = \frac{g_0}{f_0} \sum_{\alpha=1}^{3} \tau(-\mathbf{k}_\alpha;M)$$

$$\times \left\{ 1 + \int^{\Lambda} \frac{d^3\mathbf{q}}{(2\pi)^3} \mathcal{M}(-\mathbf{k}_\alpha, -\mathbf{q};M) \frac{1}{2w(\mathbf{q})} \tau(-\mathbf{q};M) \right\}. \qquad (5.255)$$

Comparing Eqs (5.254) and (5.255), we may finally conclude that

$$\langle \mathbf{k}_1, \mathbf{k}_2, \mathbf{k}_3; out | J_K^\dagger(0) | 0 \rangle = \Phi_3(\{\mathbf{k}\}) L^{3/2} \langle n | J_K^\dagger(0) | 0 \rangle \qquad (5.256)$$

where the three-particle counterpart of the Lellouch–Lüscher factor is given by

$$\Phi_3(\{\mathbf{k}\}) = \pm \frac{\displaystyle\sum_{\alpha=1}^{3} \tau(-\mathbf{k}_\alpha;M) \left\{ 1 + \int^{\Lambda} \frac{d^3\mathbf{q}}{(2\pi)^3} \mathcal{M}(-\mathbf{k}_\alpha, -\mathbf{q};M) \frac{1}{2w(\mathbf{q})} \tau(-\mathbf{q};M) \right\}}{\displaystyle\frac{1}{L^3} \sum_{\mathbf{q}}^{\Lambda} \psi_L^{(n)}(-\mathbf{q}) \frac{1}{2w(\mathbf{q})} \tau_L(-\mathbf{q};E_n)}.$$

$$(5.257)$$

Here, as we have already mentioned, the box size L is tuned so that $E_n = M$. Note also that the preceding equation does not determine the sign of this factor.

A few concluding remarks are needed here:

- In this section, we have derived the three-particle analog of the Lellouch–Lüscher (LL) relation that relates the decay matrix element measured on the lattice with its infinite-volume counterpart. Unlike the two-body case, a closed expression for the pertinent LL factor cannot be provided to all orders. Here, the derivation has been carried out at the lowest order.

- At higher orders, the LL factor becomes, in general, a matrix. In order to understand its structure, note first that in the two-body case the matrix element can be
  extracted from any (ground or excited) level, whose energy is tuned to the kaon mass.
  On the contrary, in the three-particle case, the measurements for different levels n
  are independent measurements. Suppose the calculations are done at a given order
  in NREFT, where N independent couplings $g_0, \ldots, g_N$ describe the momentum-
  dependence of the decay amplitude at tree level (these are the counterparts of the
  couplings G_i, H_i, considered in Section 2.8). Hence, in the infinite volume, we need
  to measure the amplitude for N different configurations of momenta in order to
  resolve all these couplings separately. On the other hand, in a finite volume, we have
  to carry out the measurement of N different energy levels to achieve the same goal.
  Hence, the LL factor that connects the results of the lattice measurements to the
  infinite-volume amplitudes is a $N \times N$ matrix in this case.
- The LL factor depends on the couplings that describe two- and three-particle interactions in the final state through the quantities $\tau, \tau_L, \mathcal{M}$ and the particle-dimer wave
  function in a finite volume $\psi_L^{(n)}$. Thus, for example, we should first extract the three-
  body force from the fit of the quantization condition to the energy levels in order to
  be able to calculate the LL factor. On the other hand, the LL factor does not depend
  on the kaon decay couplings $g_0, \ldots,$ by definition.
- The LL factor that we have introduced here is complex. Its imaginary part stems
  from the infinite-volume particle-dimer amplitude $\mathcal{M}$ at the energy equal to the
  kaon mass. This is nothing but the three-particle analog of Watson's theorem.
  Note that the lattice calculations determine the infinite-volume amplitude up to an
  overall sign.

5.10 Photons in a Finite Volume

At present, lattice calculations for many hadronic observables have reached a level of
accuracy at which electromagnetic corrections to these observables become important.
The inclusion of the electromagnetic interaction in such calculations is a very interesting problem from different points of view. Here, we shall be exclusively concerned with
the finite-volume aspects of this problem.

The use of effective Lagrangians allows us to address the calculation of the finite-
volume artifacts in a systematic fashion. We shall further restrict ourselves to a single
application of the approach, namely, to the evaluation of the finite-volume corrections
to stable particle masses in the presence of the electromagnetic interaction. The reason
for such a choice is that these calculations nicely highlight a very fundamental property
of (effective) field theories, namely, a deep interconnection of the *locality* of a theory
with the *decoupling of the heavy degrees of freedom*. This can be considered to be the
main lesson to be learned from this example.

We shall start with the subtle issues that accompany the introduction of the dynamic
electromagnetic field in a finite volume. A straightforward imposition of periodic

boundary conditions on the photon field is in contradiction with Gauß' law in QED. Accordingly, the electric field has a divergence proportional to the charge density:

$$\nabla \mathbf{E}(\mathbf{x},t) = e\rho(\mathbf{x},t). \tag{5.258}$$

Consider, for instance, a unit static point charge, located at the origin, $\rho(\mathbf{x},t) = \delta^{(3)}(\mathbf{x})$. Integrating both sides of the above equation over the cubic box, and taking into account that $\mathbf{E}(\mathbf{x},t)$ obeys periodic boundary conditions in space, we get a vanishing result on the left-hand side. The integral on the right-hand side is, however, nonzero and equal to the total charge of the source. Hence, if the particle is not neutral, we arrive at the contradiction mentioned above.

The same problem appears in a different context, namely if one tries to write down the perturbation expansion in a finite volume. Consider, for instance, the photon propagator in the Feynman gauge (for simplicity, we take the temporal elongation of the box to be very large):

$$i\langle 0|T\mathscr{A}^{\mu}(x)\mathscr{A}^{\nu}(y)|0\rangle = \frac{1}{L^3}\sum_{\mathbf{k}}\int_{-\infty}^{+\infty}\frac{dk^0}{2\pi}\frac{g^{\mu\nu}e^{-ik(x-y)}}{k^2+i\varepsilon}. \tag{5.259}$$

In the infinite volume, the corresponding four-dimensional integral is mathematically well defined. This is different in a finite volume. It is the zero mode with $\mathbf{k} = 0$ in this sum that causes the problem, as the integral over k^0 develops a pinch-singularity in the complex plane.

Now we shall discuss one possible solution to these problems, following mainly Refs. [87, 88]. It turns out that both problems are caused by the zero-momentum mode of the dynamical photon field. Thus, we should eliminate this mode from the theory. To this end, let us consider the Fourier-transform of the vector-potential in position space:

$$\mathscr{A}^{\mu}(\mathbf{x},t) = \frac{1}{L^3}\sum_{\mathbf{k}}e^{i\mathbf{k}\mathbf{x}}\mathscr{A}^{\mu}(\mathbf{k},t). \tag{5.260}$$

Dropping the zero mode by hand, we arrive at the modified photon field:

$$\hat{\mathscr{A}}^{\mu}(\mathbf{x},t) = \frac{1}{L^3}\sum_{\mathbf{k}\neq\mathbf{0}}e^{i\mathbf{k}\mathbf{x}}\mathscr{A}^{\mu}(\mathbf{k},t). \tag{5.261}$$

The differential operator in the free photon Lagrangian becomes invertible for such fields, and the pertinent propagator takes the form

$$i\langle 0|T\hat{\mathscr{A}}^{\mu}(x)\hat{\mathscr{A}}^{\nu}(y)|0\rangle = \frac{1}{L^3}\sum_{\mathbf{k}\neq\mathbf{0}}\int_{-\infty}^{+\infty}\frac{dk^0}{2\pi}\frac{g^{\mu\nu}e^{-ik(x-y)}}{k^2+i\varepsilon}, \tag{5.262}$$

in which the contribution of the zero mode is obviously absent.

Note that, the removal of the zero mode can be described by the transformation:

$$\mathscr{A}^{\mu}(\mathbf{x},t) \mapsto \mathscr{A}^{\mu}(\mathbf{x},t) + \frac{1}{L^3}b^{\mu}(t), \qquad b^{\mu}(t) = -\mathscr{A}^{\mu}(\mathbf{k}=\mathbf{0},t). \tag{5.263}$$

One also has to modify the interaction Lagrangian, in order to ensure that the zero modes are not produced in the interaction processes of photons with matter fields and thus stay completely decoupled. For example, in spinor QED, the pertinent interaction Lagrangian is given by $\mathcal{L}_I(x) = e\mathscr{A}^\mu(x)j_\mu(x)$, where $j_\mu(x)$ denotes the electromagnetic current. Introducing the Fourier transform of this current,

$$j^\mu(\mathbf{x},t) = \frac{1}{L^3}\sum_{\mathbf{k}} e^{i\mathbf{k}\mathbf{x}} j^\mu(\mathbf{k},t),\tag{5.264}$$

the action functional can be rewritten as follows:

$$S_I = e\int_{-\infty}^{+\infty} dt \int_{-L/2}^{+L/2} d^3\mathbf{x}\,\mathscr{A}^\mu(\mathbf{x},t)j_\mu(\mathbf{x},t)$$

$$= e\int_{-\infty}^{+\infty} dt \frac{1}{L^3}\sum_{\mathbf{k}} \mathscr{A}^\mu(\mathbf{k},t)j_\mu(-\mathbf{k},t).\tag{5.265}$$

In order to eliminate the coupling with $\mathbf{k} = \mathbf{0}$, we have to modify the current $j^\mu(\mathbf{k},t) \to \hat{j}^\mu(\mathbf{k},t) = j^\mu(\mathbf{k},t) - \delta_{\mathbf{k0}}j^\mu(\mathbf{0},t)$. In position space, this corresponds to the modification

$$\hat{j}^\mu(\mathbf{x},t) = j^\mu(\mathbf{x},t) - \frac{1}{L^3}\int_{-L/2}^{+L/2} d^3\mathbf{y}\,j^\mu(\mathbf{y},t).\tag{5.266}$$

In particular, in case of a static point charge, which we previously considered, the modified current takes the form

$$\hat{j}^\mu(\mathbf{x},t) = eg^{\mu 0}\left(\delta^{(3)}(\mathbf{x}) - \frac{1}{L^3}\right).\tag{5.267}$$

Now, the Maxwell equation contains the current $\hat{j}^\mu$ instead of j^μ:

$$\partial_\nu \mathscr{F}^{\mu\nu}(\mathbf{x},t) = e\hat{j}^\mu(\mathbf{x},t),\tag{5.268}$$

and it can be straightforwardly checked that Gauß' law is obeyed. Hence, the elimination of the zero mode solves both problems.

Before we proceed further, the following brief remarks are in necessary:

- The proposed modification of the theory, which implies the removal of the zero mode, is nonlocal. This is best seen, for example, from Eq. (5.266). It is evident that the photon field interacts with the electromagnetic current, which is integrated over the whole cubic box. Of course, this nonlocality is accompanied by the coefficient $1/L^3$ and thus vanishes in the infinite volume limit. Alternatively, we might argue that dropping a single mode leads to an effect that disappears in the infinite volume limit, where the discrete modes condense to the continuum. While this is true in general, the elimination of the zero mode still leads to very important finite-volume artifacts. We shall address this issue in detail in the following.
- The prescription for the removal of the spatial zero mode leads to the formulation of the finite-volume QED known as the QED$_L$ [87, 89]. Note that there exist alternative formulations. For example, when both the space and time elongation of the lattice are considered finite, we might try to remove only the zero mode with $k_0 = 0$ and $\mathbf{k} = \mathbf{0}$ (see, for example, Ref. [90]). The corresponding theory, known under the

name QED$_{TL}$, does not, however, possess a bounded transfer matrix. The QED$_{SF}$ prescription [91], which is defined through the restriction of the $k^\mu = 0$ zero mode to a particular interval, does not possess a well-defined transfer matrix either. Further, in Ref. [92], it has been suggested to regularize QED by introducing a small photon mass and study the zero-mass limit of observables numerically. Last but not least, a very promising approach consists in using the so-called C^* boundary conditions instead of the periodic ones [93]. This method, originally invented by Kronfeld and Wiese in Refs. [94, 95], leads to a local theory, differently from theories where the zero mode is removed. However, in QED$_C$ (as this theory is referred to), charge and flavor conservation are violated in a finite volume. In what follows, we shall exclusively focus on QED$_L$, leaving out other realizations of the electromagnetic field on the lattice.

It is very instructive to consider the static Coulomb potential in a finite volume. In the infinite volume, an exchange of a single static photon gives rise to a potential that falls off as the inverse of the distance:[22]

$$V(r) = e^2 \int \frac{d^3\mathbf{k}}{(2\pi)^3} \frac{e^{i\mathbf{k}\mathbf{r}}}{\mathbf{k}^2} = \frac{\alpha}{r}. \tag{5.269}$$

Here, $\alpha = e^2/(4\pi)$ denotes the fine-structure constant. In QED$_L$, this expression is modified [96]:

$$V_L(\mathbf{r}) = \frac{e^2}{L^3} \sum_{\mathbf{k}\neq 0} \frac{e^{i\mathbf{k}\,\mathbf{r}}}{\mathbf{k}^2} = \frac{\alpha}{\pi L} \sum_{\mathbf{n}\in\mathbb{Z}^3\backslash\{0\}} \frac{\exp(2\pi i\mathbf{n}\mathbf{r}/L)}{\mathbf{n}^2}. \tag{5.270}$$

In order to calculate this sum numerically, we may write

$$V_L(\mathbf{r}) = \frac{\alpha}{\pi L} \left(\sum_{\mathbf{n}\in\mathbb{Z}^3\backslash\{0\}} \frac{\exp(2\pi i\mathbf{n}\mathbf{r}/L)e^{-\mathbf{n}^2}}{\mathbf{n}^2} - 1 \right.$$

$$\left. + \sum_{\mathbf{n}\in\mathbb{Z}^3} \frac{\exp(2\pi i\mathbf{n}\mathbf{r}/L)(1-e^{-\mathbf{n}^2})}{\mathbf{n}^2} \right). \tag{5.271}$$

Using the Poisson formula in the last sum, we get

$$V_L(\mathbf{r}) = \frac{\alpha}{\pi L} \left(\sum_{\mathbf{n}\in\mathbb{Z}^3\backslash\{0\}} \frac{\exp(2\pi i\mathbf{n}\mathbf{r}/L)e^{-\mathbf{n}^2}}{\mathbf{n}^2} - 1 \right.$$

$$\left. + \sum_{\mathbf{n}\in\mathbb{Z}^3} \int_0^1 dt \left(\frac{\pi}{t}\right)^{3/2} \exp\left(-\frac{\pi^2}{t}\left(\mathbf{n}-\frac{\mathbf{r}}{L}\right)^2\right) \right). \tag{5.272}$$

With this formula, we can evaluate the dependence of the Coulomb potential on $\mathbf{r}$ in a finite volume. This leads to the result shown in Fig. 5.13. Since the result is not rotationally invariant, we have chosen a particular configuration $\mathbf{r} = (0,0,r)$. As seen, for small r/L, V_L almost coincides with the infinite-volume potential. This is very different from

---

[22] Once again, here for brevity we refer to the potential energy of two charged particles as the "potential."

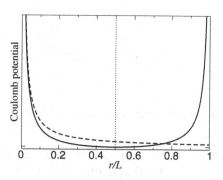

Figure 5.13 The Coulomb potential in a finite volume, given by Eq. (5.272), (solid line) versus the infinite-volume Coulomb potential (dashed line). The finite-volume potential is symmetric with respect to the dotted vertical line at $r/L = 0.5$.

the region $r \sim L$, where the finite-volume result diverges, whereas the infinite-volume result stays finite.

After this simple but instructive example, let us turn to our central problem and evaluate the finite-volume corrections to the particle masses, which are due to virtual photons. These masses will be described in the effective theory with the spatial zero mode discarded. The calculations are pretty straightforward. Since the photon has vanishing mass, one expects finite-volume artifacts that vanish as powers of $1/L$. Of course, exponentially vanishing artifacts will also exist. These, however, are subleading and will not be of concern here, as the condition $ML \gg 1$ is imposed (with M denoting the mass of the lightest massive particle in the theory). One further expects that the power-law artifacts can be readily evaluated in a finite-volume NREFT, since the difference between the relativistic and non-relativistic results should be suppressed exponentially. We shall, however, explicitly demonstrate that this expectation is incorrect. It will be further seen that, due to the nonlocal nature of the theory without zero modes, the particle and antiparticle degrees of freedom do not decouple completely, and the antiparticles also contribute to the power-law behavior.

In order to prove the preceding statement, we consider first the electromagnetic self-energy in the relativistic case. Calculations will be carried out for particles with positive unit charge in the two theories, namely scalar QED and spinor QED (here, we mainly follow Ref. [97]). The one-loop corrections to the scalar propagator are shown in Fig. 5.14a. Adding up these two diagrams, the self-energy of the scalar with mass M in Minkowski space takes the form

$$\Sigma_S(p^0, \mathbf{p}) = e^2 \int \frac{dk^0}{2\pi i} \frac{1}{L^3} \sum_{\mathbf{k} \neq 0} \frac{(2p+k)^2}{k^2(M^2 - (p+k)^2)} + 4e^2 \int \frac{dk^0}{2\pi i} \frac{1}{L^3} \sum_{\mathbf{k} \neq 0} \frac{1}{k^2}. \quad (5.273)$$

This expression is formally UV-divergent. However, as it is known, this divergence is contained in the infinite-volume part. Since here we are interested in the finite-volume corrections only, a somewhat cavalier treatment of the ultraviolet divergences is justified. (For example, we have already replaced $g^\mu_\mu = D \to 4$ in the coefficient of the second term.)

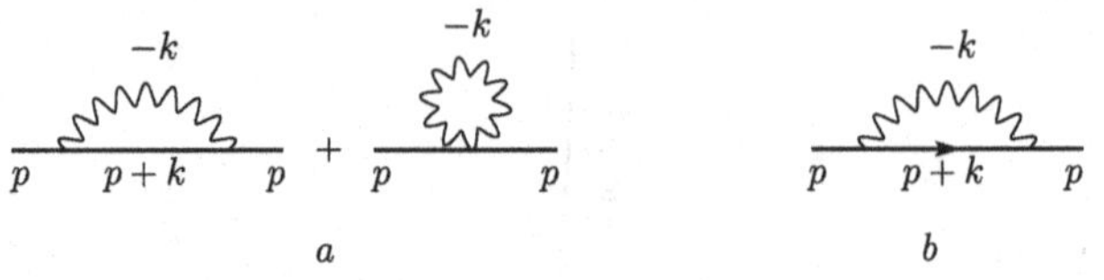

Figure 5.14 One loop correction to the particle self-energy in (a) scalar QED and (b) spinor QED.

The one-loop finite-volume correction to the particle mass is given by $\Delta M^2 = 2M\Delta M = -\Sigma_S(M,\mathbf{0}) + \Sigma_S^\infty(M,\mathbf{0})$. (We remind the reader that the finite-volume self-energy is not relativistically invariant, and the rest-frame is chosen to define the mass.) Carrying out the contour integration over the variable k^0, we get

$$\Delta M = e^2 \int \frac{dk^0}{2\pi i} \frac{1}{L^3} {\sum_{\mathbf{k}\neq 0}}' \frac{1}{k^2 + i\varepsilon} \left(\frac{2M + k^0}{2Mk^0 + k^2 + i\varepsilon} - \frac{3}{2M} \right)$$

$$= \frac{e^2}{L^3} {\sum_{\mathbf{k}\neq 0}}' \left(\frac{2M - |\mathbf{k}|}{4Mk^2} + \frac{3}{4M|\mathbf{k}|} - \frac{M - W(\mathbf{k})}{4MW(\mathbf{k})(M + W(\mathbf{k}))} \right), \tag{5.274}$$

where $W(\mathbf{k}) = \sqrt{M^2 + \mathbf{k}^2}$ and the notation $\sum'$ implicitly assumes that the infinite-volume contribution is subtracted. The crucial point here is that in the last term, we could extend the summation to $\mathbf{k} = \mathbf{0}$ as well, since the numerator vanishes at this point. Then, the Poisson formula can be used in this term and, since it does not contain massless propagators, the pertinent contribution to the mass shift turns out to be exponentially suppressed. The power-law behavior emerges from the first two terms and, at this order, we can write

$$\Delta M = \frac{\alpha}{2\pi L} c_2 + \frac{\alpha}{ML^2} c_1 + O(e^{-\Delta L}). \tag{5.275}$$

Here,

$$c_k = {\sum_{\mathbf{n} = \mathbb{Z}^3 \backslash 0}}' \frac{1}{|\mathbf{n}|^k}. \tag{5.276}$$

In particular, $c_0 = -1$, $c_1 \simeq -2.83730$ and $c_2 = I = \pi c_1 \simeq -8.91363$; see also Eq. (5.123).

In spinor QED, which describes a charged fermion of the mass m coupled to photons, the calculations are completely similar. Only the diagram shown in Fig. 5.14b contributes. The correction to the fermion mass at one loop is given by

$$\Delta m = 2e^2 \int \frac{dk^0}{2\pi i} \frac{1}{L^3} {\sum_{\mathbf{k}\neq 0}}' \frac{m - k^0}{(k^2 + i\varepsilon)(2mk^0 + k^2 + i\varepsilon)}$$

$$= \frac{2e^2}{L^3} {\sum_{\mathbf{k}\neq 0}}' \left(\frac{m + |\mathbf{k}|}{4m|\mathbf{k}|^2} - \frac{2m + w(\mathbf{k})}{4mw(\mathbf{k})(m + w(\mathbf{k}))} \right), \tag{5.277}$$

where $w(\mathbf{k}) = \sqrt{m^2 + \mathbf{k}^2}$. In contrast to the scalar case, the last term here contributes to the power-suppressed corrections. Indeed, substituting $\mathbf{k} = \mathbf{0}$ in the last term gives a nonzero contribution, since the numerator does not vanish anymore. The final result is given by

$$\Delta m = \frac{\alpha}{2\pi L} c_2 + \frac{\alpha}{mL^2} c_1 + \frac{3\pi\alpha}{m^2 L^3} + O(e^{-\Delta L}) \,. \tag{5.278}$$

On the other hand, according to the commonly shared wisdom, the terms that exhibit a power-law behavior in L should be reproducible in the nonrelativistic calculations. In order to carry out the latter, we have to repeat all steps outlined in Chapter 2, namely, we have to write the effective Lagrangian first, match the couplings to the relativistic theory, and finally remove the zero mode of the photon field and carry out calculations in a finite volume. These calculations, which are described in all detail, for example, in Ref. [96], lead to the same answer for the scalar and fail to reproduce the $O(L^{-3})$ term for the fermion. Fortunately, in order to verify this result, we do not need to repeat the whole calculation here. As discussed earlier, there exists a shortcut, which is based on the expansion of the integrand in the relativistic Feynman integrals (the so-called threshold expansion, or the expansion by regions).[23] To this end, we first write the relativistic particle propagator as

$$\frac{1}{2Mk^0 + k^2 + i\varepsilon} = \frac{1}{(M + k^0 - W(\mathbf{k}) + i\varepsilon)(M + k^0 + W(\mathbf{k}) - i\varepsilon)} \,. \tag{5.279}$$

The propagator in the case of the fermion field looks similar. Note now that the pole at $k^0 = W(\mathbf{k}) - M = O(\mathbf{k}^2)$ corresponds to a particle. The quantity k^0 is small as compared to the particle mass M in the vicinity of this pole, and we can expand

$$\frac{1}{M + k^0 + W(\mathbf{k}) - i\varepsilon} = \frac{1}{M + W(\mathbf{k})} \left(1 - \frac{k^0}{M + W(\mathbf{k})} + \frac{(k^0)^2}{(M + W(\mathbf{k}))^2} + \cdots \right) \,. \tag{5.280}$$

We should get the same result as in NREFT, if we substitute this expansion into Eq. (5.274) and integrates over k^0, using

$$\int \frac{dk^0}{2\pi i} \frac{1}{(k^2 + i\varepsilon)(M + k^0 - W(\mathbf{k}) + i\varepsilon)} = \frac{1}{-2|\mathbf{k}|(M - |\mathbf{k}| - W(\mathbf{k}) + i\varepsilon)}$$

$$\int \frac{dk^0}{2\pi i} \frac{k^0}{(k^2 + i\varepsilon)(M + k^0 - W(\mathbf{k}) + i\varepsilon)} = \frac{1}{2(M - |\mathbf{k}| - W(\mathbf{k}) + i\varepsilon)}$$

$$\int \frac{dk^0}{2\pi i} \frac{(k^0)^2}{(k^2 + i\varepsilon)(M + k^0 - W(\mathbf{k}) + i\varepsilon)}$$

$$= \frac{-|\mathbf{k}|}{2(M - |\mathbf{k}| - W(\mathbf{k}) + i\varepsilon)} + \int \frac{dk^0}{2\pi i} \frac{1}{(M + k^0 - W(\mathbf{k}) + i\varepsilon)}$$

$$= \frac{-|\mathbf{k}|}{(M - |\mathbf{k}| - W(\mathbf{k}) + i\varepsilon)} - \frac{1}{2} \,, \tag{5.281}$$

---

[23] Here, we follow the method used in Ref. [97].

Figure 5.15 Particle–antiparticle scattering through one-photon annihilation, shown in panel a, leads to a four-fermion vertex after integrating out the hard photon. Here, single and double lines denote the fermion and antifermion, respectively, and the wiggly line corresponds to the photon. A tadpole diagram with the antifermion in the loop, which contributes to the fermion self-energy in NREFT, is shown in panel b.

and so on. Note that, in order to arrive at the last result, we have split the integrand in the second term into the principal-value part and the δ-function, and assumed that the principal-value integral vanishes (which is the case for symmetric boundaries).[24]

Applying this for the scalar, we obtain

$$\Delta M_{\text{NREFT}} = \frac{e^2}{L^3} {\sum_{\mathbf{k}\neq 0}}' \left(\frac{2M-|\mathbf{k}|}{4M\mathbf{k}^2} + \frac{3}{4M|\mathbf{k}|} + O(|\mathbf{k}|) \right)$$

$$= \frac{\alpha}{2\pi L} c_2 + \frac{\alpha}{ML^2} c_1 + \cdots, \tag{5.282}$$

and, hence, the answer is the same as in the relativistic case.

Differently from the scalar case, employing the same method for a fermion yields

$$\Delta m = \frac{2e^2}{L^3} {\sum_{\mathbf{k}\neq 0}}' \left(\frac{m+|\mathbf{k}|}{4m|\mathbf{k}|^2} - \frac{3}{16m^2} + O(\mathbf{k}) \right)$$

$$= \frac{\alpha}{2\pi L} c_2 + \frac{\alpha}{mL^2} c_1 + \frac{3\pi\alpha}{2m^2L^3} + \cdots. \tag{5.283}$$

As we see, there is a factor-of-two difference with the $O(L^{-3})$ term in Eq. (5.278).

The essence of this contradiction has been clarified in Refs. [97, 98]; see also the discussion in Ref. [88]. As it is known, the NREFT Lagrangian contains four-fermion vertices that are obtained by integrating out hard protons; see Fig. 5.15a. In particular, the Lagrangian contains the term that describes a fermion–antifermion interaction:

$$\mathcal{L}_{4F} = -\frac{e^2}{4m^2} \left(\psi^\dagger \boldsymbol{\sigma}\sigma_2 \chi^* \right) \left(\chi^T \boldsymbol{\sigma}\sigma_2 \psi \right). \tag{5.284}$$

The contribution of the tadpole diagram, shown in Fig. 5.15b, is

$$\Delta m_{4F} = -\frac{3\pi\alpha}{m^2L^3} \int \frac{dk^0}{2\pi i} \sum_{\mathbf{k}\neq 0} \frac{1}{E_{\mathbf{k}} - k^0 - i\varepsilon}, \tag{5.285}$$

where $E_{\mathbf{k}} = m + \mathbf{k}^2/(2m)$. Hence,

$$\Delta m_{4F} = -\frac{3\pi\alpha}{2m^2L^3} \sum_{\mathbf{k}\neq 0} 1 = \frac{3\pi\alpha}{2m^2L^3}. \tag{5.286}$$

[24] More precisely, this means $\lim_{\Lambda\to\infty} \text{P.V.} \int_{-\Lambda}^{+\Lambda} \frac{dx}{x-x_0} = \lim_{\Lambda\to\infty} \ln\left|\frac{\Lambda-x_0}{\Lambda+x_0}\right| = 0.$

This contribution should be added to Eq. (5.283), reproducing the relativistic result given in Eq. (5.277). (In the scalar case, the tadpole contribution is absent at leading order, because the scalar and anti-scalar cannot scatter in S-wave via the one-photon annihilation [97].) However, such a solution comes at the cost of creating further problems. For instance, we could ask, why are the zero modes in the propagator of the *antiparticles (not the photons)* left out when calculating the tadpole contribution? Of course, we could argue here that the four-fermion vertex at $O(e^2)$ arises after integrating out a hard photon. In a finite volume, the photon has no zero mode and, consequently, the sum of three-momenta of a particle and an antiparticle cannot vanish. Since in the calculation of the particle mass its three-momentum is fixed to be zero, the sum over the three-momenta of the antiparticle in the tadpole runs only over non-zero values [98]. This would, however, mean that the matching between the relativistic and nonrelativistic theories should be performed in a finite volume (because, in the infinite volume, the photon field possesses all modes and the preceding argument is not valid). In other words, there is no complete decoupling of the long-range (manifested in the L-dependence) and the short-range effects that enter in the effective Lagrangian. The statement that the short-range effects are exponentially suppressed, with the characteristic scale appearing in the argument of the exponent, does not hold anymore. In fact, we have seen explicitly that the short-range effects are suppressed only by some power of L. This effect, which is a direct consequence of the nonlocal nature of the zero-mode-removed theory, is discussed in more detail in Refs. [96, 99], and we refer the interested reader to these articles. Moreover, note that the arguments that were applied during the matching of the relativistic theory (say, the ChPT with virtual photons) to the NREFT, are applicable, in principle, to the matching of QCD + QED to ChPT + QED as well. In this case, power-law-suppressed contributions to the physical observables, depending on the quark–gluon dynamics at the hadronic scale, should also emerge. A small parameter, controlling the expansion in the inverse powers of L in this case, will be $1/(\Lambda L)$ instead of $1/(M_\pi L)$. (Here, $\Lambda \simeq 1$ GeV denotes the hadronic scale in QCD.)

Last but not least, note that the above calculations did not take into account the internal structure of the particles. Most easily, these effects can be included within NREFT, where, as we know from Chapter 2, the anomalous magnetic moment κ, the charge radius $\sqrt{\langle r^2 \rangle}$ and so on, define the effective couplings in the Lagrangian. One expects that the nonlocal character of the finite-volume theory with photons does not affect the result at leading order, which is given by the diagrams shown in Fig. 5.16a,b. Here, we merely display this result in the case of spin-0 and spin-1/2 particles, which was first obtained in Ref. [96]:

$$\Delta M = \frac{\alpha}{2\pi L} c_2 + \frac{\alpha}{ML^2} c_1 + \frac{2\pi\alpha}{3L^3} \langle r^2 \rangle + O(L^{-4}),$$

$$\Delta m = \frac{\alpha}{2\pi L} c_2 + \frac{\alpha}{mL^2} c_1 + \frac{2\pi\alpha}{3L^3} \langle r^2 \rangle$$

$$+ \frac{\pi\alpha}{m^2 L^3} (3 + 2\kappa + \kappa^2) + O(L^{-4}). \tag{5.287}$$

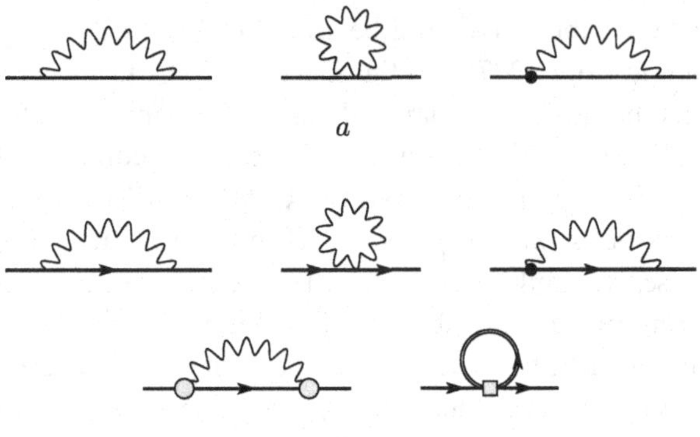

Figure 5.16 NREFT diagrams that contribute to the self-energy of a scalar (a) and a fermion (b). Single, double and wiggly lines denote the particles, antiparticles and photons, respectively. A filled black dot corresponds to the contribution proportional to the effective radius $\langle r^2 \rangle$, a filled light-gray dot stands for the vertex with the anomalous magnetic moment, and the light-gray square is the four-fermion vertex. The nonrelativistic propagators of the particles contain the kinetic energy $\mathbf{k}^2/(2M)$, and the different terms in the expansion of the propagator in the soft regime $k^0 \sim |\mathbf{k}|$ are not shown explicitly (for more details, we refer to Chapter 2 and to the original paper [96]). Note that crossed diagrams are not shown.

Relativistic calculations, carried out in Ref. [97], have confirmed this result. As can be seen, the leading and next-to-leading corrections are universal (i.e., these are the same as in the case of point particles). The internal structure shows up first at N^2LO in the expansion in $1/L$.

5.11 The Chiral Limit in a Finite Volume

5.11.1 The Chiral Limit versus the Thermodynamic Limit

In Section 3.14.5, we have already seen that the quark condensate in the chiral limit, evaluated for a given configuration of the background gluon field G_μ, is given by the expression

$$\langle 0|\bar\psi\psi|0\rangle_G = -\lim_{\mathcal{M}\to 0} 2m \int_0^\infty \frac{d\lambda\, \rho_G(\lambda)}{m^2 + \lambda^2}. \tag{5.288}$$

In this expression, ψ represents a light quark field of any flavor (no summation over flavors), m is the quark mass of that flavor and $\rho_G(\lambda)$ is the spectral density of the massless Euclidean Dirac operator in the background gluon field G. The chiral limit $\mathcal{M} \to 0$ is performed in a manner that the ratios of quark masses with different flavors $m_f/m_{f'}$ stay finite. In order to obtain the conventional quark condensate in QCD, we should average this formula over all gluon field configurations. Performing the limit, we finally arrive at the Banks–Casher formula; see Eq. (3.345), which relates the quark condensate in a background gluon field to the limiting value of the spectral density at the origin, $\rho(0)$, which is obtained by averaging $\rho_G(0)$ over gluon field configurations.

It is instructive to consider the fate of the Banks–Casher relation in a finite volume. For simplicity, let us consider a hypercubic box of size L. The volume of this box is given by $V_4 = L^4$. In this case, the eigenvalues are discrete, and the integral in Eq. (5.288) is replaced by a sum over these eigenvalues. Integrating over the box, we get

$$\frac{1}{V_4} \int_{-L/2}^{+L/2} d^4x \langle 0| \bar{\psi}(x)\psi(x)|0\rangle = - \lim_{\mathcal{M}\to 0} \frac{2m}{V_4} \sum_{\lambda_n \neq 0} \frac{1}{m^2 + \lambda_n^2}. \tag{5.289}$$

Here, $i\lambda_n$ are the eigenvalues of the operator $\gamma_\mu D_\mu$, and we have dropped the zero modes with $\lambda_n = 0$ from this sum. This approximation needs further comments. Indeed, as we have seen also in Section 3.14.5, the above relation can be equivalently expressed in the form

$$\frac{1}{V_4} \int_{-L/2}^{+L/2} d^4x \langle 0| \bar{\psi}(x)\psi(x)|0\rangle = -\frac{1}{V_4} \lim_{\mathcal{M}\to 0} \int_{-\infty}^{+\infty} \frac{d\lambda\, \rho_G(\lambda)}{m - i\lambda}, \tag{5.290}$$

where the spectral density $\rho_G(\lambda)$ is given by the sum over *all* eigenvalues:

$$\rho(\lambda) = \sum_{\lambda_n} \delta(\lambda - \lambda_n). \tag{5.291}$$

The eigenvalues $\lambda_n = \lambda_n(G)$ are functionals of the background gluon field G_μ. Furthermore, there are configurations of G_μ, for which exact zero modes $\lambda_n(G) = 0$ exist. Indeed, according to the Atiyah–Singer index theorem (see Section 3.11), the number of the right-handed minus left-handed zero modes of the Euclidean Dirac operator in the background field G_μ is equal to the winding number of this field. Hence, in the sector where G_μ carries a nonzero winding number ν, there exists at least ν zero modes. Suppose now that, for a given G_μ, there are $k(G) \neq 0$ zero modes. The contribution of these zero modes to the spectral function is given by $\rho_G^{(0)}(\lambda) = k(G)\delta(\lambda)$. The contribution to the full spectral function from the zero modes implies averaging this result over all gluon field configurations. This averaging procedure involves a weight factor given by the fermion determinant:

$$\det(\gamma_\mu D_\mu + \mathcal{M}) = (\det \mathcal{M})^k \det'(\gamma_\mu D_\mu + \mathcal{M}), \tag{5.292}$$

where $\det'$ denotes the determinant without zero modes. Suppose now that we have $N_f > 1$ quark flavors. Then, the determinant in Eq. (5.292) behaves as m^{N_f} in the chiral limit, with m being the mass of any light quark. Consequently, the zero-mode contribution to the spectral function from a given sector behaves like $k(G)m^{N_f k(G)}$ and, according to Eq. (5.290), the contribution to the quark condensate vanishes as $k(G)m^{N_f k(G)-1}$ in this limit. Consequently, for $N_f > 1$, it is justified to neglect the zero-mode contributions to the quark condensate. The case with $N_f = 1$ is physically different. As one knows, a spontaneous breaking of symmetry occurs only when $N_f > 1$. In the following, we shall exclusively concentrate on this case.

Our discussion here largely follows Ref. [100]. From Eq. (5.289), we clearly see that the chiral limit $\mathcal{M} \to 0$ and the thermodynamic limit $V_4 \to \infty$ are not interchangeable.

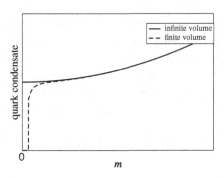

Figure 5.17 A schematic picture of the quark mass dependence of the condensate in an infinite and in a finite volume. As seen from this picture, the quark condensate can be measured on a finite lattice to a good accuracy, as long as the quark masses are taken small but nonzero (above the point where the curves start to diverge).

In order to arrive at a nonzero condensate, we have to perform the limit $V_4 \to \infty$ first. In this limit, the finite contribution to the condensate is provided by the *near-zero* modes, accumulating toward $\lambda \to 0$. The average level spacing is given by $\Delta\lambda \simeq (V_4\rho(\lambda))^{-1}$ and, owing to the Banks–Casher relation, is given by $\Delta\lambda = \pi(V_4|\langle 0|\bar{\psi}\psi|0\rangle|)^{-1}$ near the origin (in all these expressions, averaging over gluon configurations is implicit). Furthermore, if the quark mass is much larger than average distance between levels, that is,

$$V_4 m|\langle 0|\bar{\psi}\psi|0\rangle| \gg 1, \tag{5.293}$$

then the sum over the eigenvalues can be approximated by an integral, and we arrive at the Banks–Casher relation. We remark that reversing the order of limits and performing the limit $\mathcal{M} \to 0$ first gives a vanishing condensate, as immediately seen from Eq. (5.289). Hence, a nonzero quark condensate can be seen in a finite volume (on the lattice) as well, since, if V_4 is very large, the condition (5.293) excludes only tiny quark masses. A schematic picture that describes the dependence of the condensate on the quark mass in the infinite and in a finite volume is given in Fig. 5.17. Last but not least, note that, as shown in Ref. [100], if the condition (5.293) is fulfilled, the gluon configurations with the winding numbers $0, \pm 1, \pm 2, \ldots$ give comparable contributions to the formation of the condensate. This demonstrates the important role of such configurations for the spontaneous chiral symmetry breaking in QCD.

5.11.2 Chiral Limit of the Pion Mass in a Finite Volume

The problems that emerge, when the chiral limit is interchanged with the large volume limit, are also seen in the expression for the finite-volume correction to the pion mass, which was obtained in Section 5.2.2. Indeed, consider

$$M_L^2 = M_\pi^2 \left(1 + \frac{1}{32\pi^2 F^2}\Delta I_0^L + \cdots\right), \tag{5.294}$$

where the quantity ΔI_0^L is given by Eq. (5.18), it becomes clear that this expression does not make sense in the chiral limit, since the correction diverges[25] as M_π^{-1}. As we may conclude from Eq. (5.15), this divergence is generated by the term with $\mathbf{k} = \mathbf{0}$. This is in line with the discussion given in Section 5.2.1. There is no exponential damping for the mode with $\mathbf{k} = \mathbf{0}$ in the path integral in the chiral limit and, hence, this mode should be treated non-perturbatively. A criterion for a non-perturbative behavior can be directly deduced from Eq. (5.5). The fluctuations of fields here are of order of F, as the pion fields are given in terms of $U = \exp(i\phi/F)$. Hence, the damping factor in the Gaussian path integral from the mass term is given by $F^2 L^3 w_{\mathbf{n}}^2$. Thus, in case of a zero mode, a non-perturbative treatment is needed, if $F^2 L^3 M^2 \ll M$ (here, M is the pion mass at lowest order in the chiral expansion). For the nonzero modes, we have $w_{\mathbf{n}} = \sqrt{M^2 + (2\pi\mathbf{n}/L)^2}$. If $M \ll 2\pi/L$, these modes are always perturbative, when $FL \gg 1$.

At this stage, it is useful to introduce formal counting rules, which enable one to relegate the contributions from the zero as well as nonzero modes of the pion field to pertinent orders in the expansion. A convenient scheme was proposed in Ref. [3], where

$$L = O(\delta^{-1}), \qquad L_t = O(\delta^{-3}), \qquad M = O(\delta^3). \tag{5.295}$$

Here, L, L_t denote the spatial and temporal elongation of the box, respectively, and δ is a generic small parameter. This regime is called the δ-regime.

Having adopted the counting rules of the δ-regime, it becomes obvious that the leading correction to the finite-volume pion mass shift is provided by the modes with $\mathbf{k} = \mathbf{0}$. Indeed, this counting implies that the momentum components scale like $\mathbf{k} = O(\delta)$ and $k_4 = O(\delta^3)$. Hence, for example, the leading contribution to the tadpole given by Eq. (5.15), which comes from the $\mathbf{k} = \mathbf{0}$ mode, scales like $O(\delta^0)$, whereas the nonzero modes give a contribution at $O(\delta^4)$.

Freezing the nonzero modes of the pion field makes the pion matrix $U = \exp(i\lambda^i\phi^i/F)$ depend on time only [3]. In Minkowski space, the Lagrangian can then be written as

$$L(t) = \int_{-L/2}^{+L/2} d^3\mathbf{x}\, \frac{F^2}{4} \left\langle \dot{U}(t)\dot{U}^\dagger(t) + M^2\left[U(t) + U^\dagger(t)\right]\right\rangle$$

$$= \frac{F^2 L^3}{2}\left(g_{ij}(\phi)\dot{\phi}^i(t)\dot{\phi}^j(t) + M^2\,\mathrm{Re}\,\langle U(t)\rangle\right). \tag{5.296}$$

Here, the group metric for $SU(N_f)$ is

$$g_{ij}(\phi) = \frac{1}{2}\left\langle \frac{\partial U}{\partial\phi^i}\frac{\partial U}{\partial\phi^j}\right\rangle. \tag{5.297}$$

The canonical momentum variable for the pion field is defined as

$$p^i = \frac{\partial L}{\partial\dot{\phi}^i} = F^2 L^3 g_{ij}\dot{\phi}^j, \tag{5.298}$$

---

[25] In the limit $M_\pi \to 0$, the sum over $\mathbf{n}$ diverges. For large $\mathbf{n}$, we may replace summation by integration over $\mathbf{n}$. This leads to the integral $\int_0^\infty d\alpha\, e^{-\alpha M_\pi^2}/\alpha^{1/2}$, which behaves like M_π^{-1} in the limit of the vanishing pion mass.

and the Hamiltonian is given by the following expression:

$$H = \dot{\phi}^i \frac{\partial L}{\partial \dot{\phi}^i} - L = \frac{1}{2F^2L^3} p^i g^{ij}(\phi) p^j - \frac{F^2L^3}{2} M^2 \mathrm{Re}\,\langle U(t) \rangle. \tag{5.299}$$

The matrix g^{ij} obeys the relation $g^{ij} g_{jk} = \delta^i_k$.

Canonical quantization implies $p^i \to -i(\partial/\partial\phi_i)$. Note that, since $g^{ij}(\phi)$ depends on ϕ, there is an ambiguity here, related to the operator ordering. A self-adjoint quantum-mechanical Hamiltonian is written in terms of the Laplace–Beltrami operator Δ:

$$H_{QM} = -\frac{1}{2F^2L^3}\Delta - \frac{F^2L^3}{2}M^2\mathrm{Re}\,\langle U(t)\rangle,$$

$$\Delta = \frac{1}{\sqrt{g}}\frac{\partial}{\partial\phi^i}\left(g^{ij}(\phi)\frac{\partial}{\partial\phi^j}\right), \tag{5.300}$$

with g the absolute value of the determinant of g_{ij}. The energy levels of QCD in the box are given by the eigenvalues of the operator H_{QM},

$$H_{QM}\Psi_n(\phi) = E_n\Psi_n(\phi). \tag{5.301}$$

It is instructive to focus on the chiral limit, where an exact solution to the problem can be found. The first excited level contains the Goldstone bosons, which are massless in the infinite volume. Hence, the energy of this level equals the finite-volume shift of the pion mass. Taking into account that the corresponding eigenvalue of the Laplace–Beltrami operator is equal to $-2(N_f^2 - 1)/N_f$, we finally arrive at the formula for the leading-order mass shift of pions in the chiral limit [3]:

$$M_L = \frac{N_f^2 - 1}{N_f F^2 L^3} + \cdots. \tag{5.302}$$

Thus, "squeezing" the pion inside a box gives rise to a finite mass shift even in the chiral limit.

In conclusion, note that the hierarchies that emerge among the various scales that are present in the problem lead to different effective theories. This is manifested in different counting rules, even if the same effective Lagrangian is used in the calculations. We have just witnessed that, in the chiral limit, the zero-mode of the pion field with $\mathbf{k} = \mathbf{0}$ becomes dominant and should be treated non-perturbatively in the calculation of the finite-volume shift of the pion mass. Note also that in these calculations the time and space elongations of a box were chosen differently, namely $L_t \gg L$. This fact is reflected in the δ-counting, defined by Eq. (5.295), and is a natural choice in case of the calculation of a mass, which is extracted from the asymptotic behavior of the two-particle correlators at large Euclidean times on the lattice. In order to study QCD at *both* at a finite temperature and volume, a different counting was proposed in Ref. [4]:

$$L_t = O(\varepsilon^{-1}), \qquad L = O(\varepsilon^{-1}), \qquad M = O(\varepsilon^2). \tag{5.303}$$

Here, ε is a new generic small parameter and the temperature is given by $T = 1/L_t$. In this regime, called the ε-regime in the literature, the leading contributions are given by a mode with $k_\mu = 0$, which corresponds to a constant pion matrix U in the Euclidean space. Using ChPT in this regime enables us to study, for example, the dependence of the quark condensate on the volume and on the temperature simultaneously [4, 101, 102]. In particular, it has been shown that the condensate decreases with increasing temperature and approaches zero at some critical temperature. Note, however, that the calculation of the critical temperature cannot be done in ChPT, as at that point the corrections to the condensate are as large as the leading term.

5.11.3 Aoki Phase

We would like to discuss next an example that, strictly speaking, is not directly related to the finite-volume effects. Still, we decided to briefly discuss it here, as it nicely demonstrates that the phase structure of the theory on the lattice becomes much richer when one moves away from the continuum limit. In particular, it will be shown that isospin can be spontaneously broken and some of the pions may become exactly massless, even if the quark masses are nonzero. Of course, this will have some implications for the finite-volume artifacts as well, providing a hint at an interesting interplay of the finite-volume and finite-lattice-spacing effects in actual lattice calculations. However, the example serves primarily as a yet another beautiful illustration of the predictive power of EFTs in the context of various, very different physical problems on the lattice.

The effects of the finite lattice spacing (finite UV cutoff) can be treated within continuum field theory, using the language of the Symanzik effective Lagrangian [1, 2]. Below, we shall dwell on this issue only very briefly. In the simplest case of the one-component scalar φ^4-theory, the effective Lagrangian has the form

$$\mathcal{L}_{\text{eff}} = \mathcal{L} + a^2 \mathcal{L}_2 + a^4 \mathcal{L}_4 + \cdots, \tag{5.304}$$

where a is the finite lattice spacing and $\mathcal{L}$ is the original Lagrangian of the φ^4-theory.[26] The terms in Eq. (5.304) that vanish, when $a \to 0$, are responsible for mimicking the terms in the Green's functions that describe scaling violation; see Eq. (1.202). Note that, since Lorentz invariance is broken on the hypercubic lattice, $\mathcal{L}_2, \mathcal{L}_4, \ldots$ contain, in general, not only Lorenz-invariant operators, whereas $\mathcal{L}$ does, by definition. The coupling constants in front of the various operators in $\mathcal{L}_2, \mathcal{L}_4, \ldots$ can be tuned (in perturbation theory or otherwise) so that the scaling violating terms in all Green's functions of the original theory are reproduced at any given order in a.

Adapting ChPT to the case with a finite lattice spacing is a subtle issue, because chiral symmetry, which is a basic guiding principle in the construction of the ChPT Lagrangian, is broken down at a finite a, say, for the Wilson fermions. Therefore, it is convenient to use a two-step procedure; see, for example, Ref. [103]. First, we consider the QCD Lagrangian modified through a-dependent terms:

---

[26] Strictly speaking, the couplings in the Lagrangians $\mathcal{L}_2, \mathcal{L}_4, \ldots$ contain a residual dependence on a. This dependence is, however, only logarithmic, proportional to $\ln^k a\mu$, and does not affect any of our conclusions (here, μ denotes the renormalization scale). We shall consistently neglect this logarithmic dependence in the following.

$$\mathcal{L}_{QCD} \rightarrow \mathcal{L}_{QCD} + \sum_{k=1}^{\infty} a^k O_k, \tag{5.305}$$

where the operators O_k have the mass dimension $[O_k] = k + 4$. The list of operators of dimension 5 is given by [104]:

$$O_1^{(1)} = \bar{\psi} i \sigma_{\mu\nu} F_{\mu\nu} \psi, \qquad O_1^{(2)} = \bar{\psi}(\slashed{D} + \mathcal{M})^2 \psi, \qquad O_1^{(3)} = \bar{\psi}\mathcal{M}(\slashed{D} + \mathcal{M})\psi,$$

$$O_1^{(4)} = \bar{\psi}\mathcal{M}^2 \psi, \qquad O_1^{(5)} = \langle \mathcal{M} \rangle \mathrm{tr}_c (F_{\mu\nu}F_{\mu\nu}). \tag{5.306}$$

The operators $O_1^{(2)}, O_1^{(3)}$ can be eliminated by using the equations of motion. The operators $O_1^{(4)}, O_1^{(5)}$ can be safely ignored, if $a\mathcal{M} \ll 1$ is assumed (we shall always consider this regime). Consequently, at the lowest order, only $O_1^{(1)}$ needs to be considered. This operator breaks chiral symmetry explicitly. As usual, in order to write down the chiral effective Lagrangian, we use the spurion technique. Introducing the spurion field A, which transforms as $A \mapsto g_L A g_R^\dagger$, we rewrite $O_1^{(1)}$ in the following manner:

$$O_1^{(1)} \rightarrow \bar{\psi}_L i \sigma_{\mu\nu} F_{\mu\nu} A \psi_R + \bar{\psi}_R i \sigma_{\mu\nu} F_{\mu\nu} A^\dagger \psi_L. \tag{5.307}$$

Setting $A = 1$ in the end restores the original expression for $O_1^{(1)}$. Higher dimensional operators can be treated in a similar fashion.

Now, let us write down an effective chiral Lagrangian and then find the pion field configuration in the ground state, minimizing the effective potential with respect to the (constant) pion field. We shall further restrict ourselves to $N_f = 2$ and assume that $m_u = m_d = \hat{m}$. The construction of the effective Lagrangian is considered in detail in Section 4.7, and the full set of operators up to order p^4 is listed. We only need those operators that do not contain derivatives of the pion field U, as well as vector- and axial-vector sources $\sim v_\mu, a_\mu$. It can be verified that there is only one such operator at $O(p^2)$ and $O(p^4)$, $\langle \chi^\dagger U + \chi U^\dagger \rangle$ and $\langle \chi^\dagger U + \chi U^\dagger \rangle^2$, respectively. (We remind the reader that isospin symmetry $m_u = m_d$ is assumed.) In the presence of the spurion A, further operators of the same form will emerge, in which some (all) χ's are replaced by A's. Since both χ and A are proportional to the unit 2×2 matrix, we may finally write down the effective potential of the pion field:

$$V(U) = -\frac{c_1}{4} \langle U + U^\dagger \rangle + \frac{c_2}{16} \langle U + U^\dagger \rangle^2, \tag{5.308}$$

where the couplings are given by

$$c_1 = x_1 \hat{m}\Lambda^3 + x_2 a\Lambda^5, \qquad c_2 = y_1 \hat{m}^2 \Lambda^2 + y_2 \hat{m}a\Lambda^4 + y_3 a^2\Lambda^6, \tag{5.309}$$

where Λ denotes the hadronic scale, and the x_i, y_i are (dimensionless) numerical constants. Note that c_1 describes a discretization effect in the quark mass, which can be absorbed by redefining $\hat{m}' = \hat{m} - (x_2/x_1)a\Lambda^2$.

The behavior of the system critically depends on the hierarchy of two scales, $\hat{m}'$ and $a\Lambda^2$. If these two scales are of the same order of magnitude, then $c_2/c_1 = O(\hat{m}'/\Lambda)$ is a small quantity, and the second term in the potential $V(U)$ represents a small perturbation. The situation changes, however, dramatically, if we have $\hat{m}' \sim a^2\Lambda^3$. Then,

$c_1 = x_1 \hat{m}' \Lambda^3$ and $c_2 = y_3 a^2 \Lambda^6 + \ldots$ are of the same order of magnitude (the ellipses denote the subleading terms in this regime). As we shall see, the minimization of the potential allows for the possibility of spontaneous symmetry breaking for a certain choice of the parameters. Indeed, using the familiar σ-model parameterization for the pion field, $U = \sigma + i \tau \pi / F$, we find

$$V(U) = -c_1 \sigma + c_2 \sigma^2, \qquad |\sigma| = \left| \sqrt{1 - \frac{\pi^2}{F^2}} \right| \leq 1. \tag{5.310}$$

The constants c_1, c_2 are not known a a priori. Note that c_1 depends linearly on the quark mass $\hat{m}'$ and changes sign together with it.

- Assume first $c_2 < 0$. The curve that determines the dependence of $V(U)$ on σ is a parabola. On the interval $\sigma \in [-1, 1]$, it takes its minimal value either at $\sigma = 1$ or at $\sigma = -1$, depending on the sign of c_1. Expanding around either minimum, we get

$$V(U) = -|c_1| \left(1 - \frac{\pi^2}{F^2} \right)^{1/2} - |c_2| \left(1 - \frac{\pi^2}{F^2} \right)$$

$$= -(|c_1| + |c_2|) - \frac{1}{2} \pi^2 \frac{|c_1| + 2|c_2|}{F^2} + \cdots. \tag{5.311}$$

We see that isospin symmetry is not broken, and the pion at the lowest order has the mass $M_\pi^2 = (|c_1| + 2|c_2|)/F^2$.

- Assume now $c_2 > 0$. The function $V(U)$ has a minimum at $\sigma_0 = c_1/(2c_2)$. If $|\sigma_0| > 1$, the minimum of the potential is localized again at the ends of the interval, $\sigma = \pm 1$, depending on the sign of c_1. Expanding the potential in this case, we get

$$V(U) = -|c_1| \left(1 - \frac{\pi^2}{F^2} \right)^{1/2} + |c_2| \left(1 - \frac{\pi^2}{F^2} \right)$$

$$= -(|c_1| - |c_2|) - \frac{1}{2} \pi^2 \frac{|c_1| - 2|c_2|}{F^2} + \cdots, \tag{5.312}$$

and, hence, $M_\pi^2 = (|c_1| - 2|c_2|)/F^2$.

- However, when $|\sigma_0| < 1$, the minimum of the potential is located inside the interval and a phase transition occurs. This can be most easily seen from the expression of the pion mass in the preceding case. This quantity becomes negative, if $|c_1| < 2|c_2|$, indicating spontaneous breaking of the symmetry. Also, if $|\sigma| < 1$ at the minimum, the pion field should develop a vacuum expectation value, because $\sigma^2 + \pi^2/F^2 = 1$. Without loss of generality, it can be assumed that the condensate aligns to the third axis, that is, $\langle \pi^{1,2} \rangle = 0$ and $\langle \pi^3 \rangle = v \neq 0$. The expansion of the quantity σ in the shifted pion field $\pi^i \to \pi^i - v\delta^{i3}$ is given by

$$\sigma = \left(\sigma_0^2 + \frac{\pi^2 - 2v\pi^3}{F^2} \right)^{1/2} = \sigma_0 + \frac{\pi^2 - 2v\pi^3}{2F^2 \sigma_0^2} - \frac{v^2 (\pi^3)^2}{2F^4 \sigma_0^4} + \cdots, \tag{5.313}$$

where $\sigma_0^2 = 1 - v^2/F^2$. Substituting this expansion into the potential, we get

$$V(U) = -c_1\sigma_0 + c_2\sigma_0^2 + \frac{c_1 v^2}{2F^4\sigma_0^3}(\pi_3)^2 + \cdots. \tag{5.314}$$

It is immediately seen that the charged pions become massless after the phase transition, whereas the neutral pion stays massive:

$$M_{\pi^\pm}^2 = 0, \qquad M_{\pi^0}^2 = \frac{c_1 v^2}{F^4\sigma_0^3} = \frac{c_1}{F^2\sigma_0}\left(1 - \sigma_0^2\right). \tag{5.315}$$

Hence, in this new phase, which was conjectured by Aoki [105, 106], both parity and isospin symmetry are spontaneously broken. (The latter is broken down to the $U(1)$ subgroup that corresponds to rotations around the third axis.) The EFT-based treatment displayed here follows mainly the path outlined in Ref. [103]; see also Ref. [107].

As we see, the existence of the Aoki phase critically depends on the parameters of the theory (the effective couplings c_1, c_2) that cannot be predicted a priori. A signature of the phase transition is provided, for example, by a nonzero value of the isospin- and parity-breaking quark condensate $\langle \psi i \gamma^5 \tau^i \psi \rangle \neq 0$. This quantity can be measured on the lattice and indeed turns out to be nonzero at certain values of input parameters (see, e.g., Ref. [108]), thus confirming the existence of the Aoki phase.

5.11.4 Applications in Solid-State Physics

Effective field theories are widely used for the study of different phenomena in solid-state physics that are described by collective excitations (phonons, magnons, etc.). If, as a result of a phase transition, symmetry breaking occurs, exactly zero-mass particles emerge. This is reminiscent of the chiral limit in QCD. The δ- and ε-regimes are then particularly useful to study the properties of these systems at a finite temperature and volume.

In this section, we shall consider a formulation of a systematic finite-volume EFT framework for the $O(N)$ model in D dimensions; see, for example, Refs. [109–113]. In $D = 3$ space-time dimensions, this model describes the spin-$\frac{1}{2}$ antiferromagnetic Heisenberg model, whereas in $D = 4$ dimensions it is an effective theory of QCD with two massless quarks.

The lowest-order Lagrangian in Euclidean space is that of the nonlinear σ-model:

$$\mathcal{L}^{(2)} = \frac{F^2}{2}\partial_\mu \mathbf{S}\partial_\mu \mathbf{S} - \Sigma(\mathbf{HS}). \tag{5.316}$$

Here, $\mathbf{S} = (S_0, S_1, \ldots, S_{N-1})$ is the N-component unit vector, $\mathbf{S}^2 = 1$, and $\mathbf{H}$ denotes an external magnetic field. In the infinite volume, the fields obey familiar counting rules, $\mathbf{S} = O(p^0)$ and $\mathbf{H} = O(p^2)$. Only two LECs, F and Σ, enter the lowest-order Lagrangian.

The external magnetic field plays the role of the quark mass matrix here. In the presence of this field, the $O(N)$ symmetry is broken explicitly and the Goldstone boson acquires a mass. At lowest order in $H = |\mathbf{H}|$, the expression of this mass is given by

$$M_H^2 = \frac{\Sigma H}{F^2} . \tag{5.317}$$

Using the equations of motion and field redefinitions in order to eliminate redundant terms, it can be shown that at NLO only five additional LECs k_1, k_2, k_3, k_4, k_5 enter. The Lagrangian at NLO takes the form

$$\mathcal{L}^{(4)} = \frac{\Sigma}{F^2} k_1 (\mathbf{HS})(\partial_\mu \mathbf{S} \partial_\mu \mathbf{S}) - \frac{\Sigma^2}{F^4} k_2 (\mathbf{HS})^2 - \frac{\Sigma^2}{F^4} k_3 (\mathbf{HH})$$

$$+ \frac{k_4}{4} (\partial_\mu \mathbf{S} \partial_\mu \mathbf{S})^2 + \frac{k_5}{4} (\partial_\mu \mathbf{S} \partial_\nu \mathbf{S})^2 . \tag{5.318}$$

The full Lagrangian is given by $\mathcal{L} = \mathcal{L}^{(2)} + \mathcal{L}^{(4)} + \ldots$.

Considering the system in a finite volume implies, in general, a modification of the counting rules. The choice of a particular regime is dictated by the hierarchy between the spatial and temporal elongations of the box. Consider first the ε-regime where, for simplicity, we consider a hypercubic box with both spatial and temporal elongations equal to L. As one already knows, the component of the field $\mathbf{S}(x)$ with $k_\mu = 0$ in this regime should be treated non-perturbatively. This component corresponds to the net magnetization, defined in the following manner:

$$\mathbf{m} = \frac{1}{L^D} \int_{-L/2}^{+L/2} d^D x \, \mathbf{S}(x) . \tag{5.319}$$

The contribution to the action functional stemming from the net magnetization is given by

$$\delta S_m = - \int_{-L/2}^{+L/2} d^D x \, \Sigma(\mathbf{HS}) = -\Sigma L^D (\mathbf{Hm}) . \tag{5.320}$$

Consequently, in the limit $L \to \infty$, spin alignment occurs if the quantity $\Sigma H L^D$ does not vanish. The ε-regime implies the assumption that this product stays finite in this limit, which is equivalent to the assumption $H = O(L^{-D})$. Using now Eq. (5.317) leads to $M_H L = O(L^{-D/2+1})$. Thus, the product $M_H L$ stays small in the limit $L \to \infty$, as expected.

In order to carry out a systematic expansion of physical observables within a given EFT framework, we should first decouple the "slow" component of the field $\mathbf{S}(x)$, corresponding to $k_\mu = 0$, from the other, "fast" modes. To this end, it is convenient to use the Faddeev–Popov procedure [114] in the path integral. Introducing the unit multiplier,

$$\int d\mathbf{m} \, \delta \left(\mathbf{m} - \frac{1}{L^D} \int_{-L/2}^{+L/2} d^D x \, \mathbf{S}(x) \right) = 1, \tag{5.321}$$

into the partition function, we may further split the group integral over the direction of $\mathbf{m}$ and its magnitude:

$$\mathbf{m} = m \mathbf{e}, \qquad d\mathbf{m} = m^{N-1} dm \, d\mathbf{e}. \tag{5.322}$$

It will be convenient to replace the integral over the unit vector $\mathbf{e}$ by the integral over the $O(N)$ rotation group parameters. Defining

$$\mathbf{e} = \Omega^T \mathbf{n}, \qquad \mathbf{n} = (1, 0, \ldots, 0), \tag{5.323}$$

where $\Omega \in O(N)$, the integration measure $d\mathbf{e}$ in the path integral can be replaced by $d\Omega$. Performing further a change of the integration variable $\mathbf{S}(x) = \Omega^T \mathbf{R}(x)$, the partition function $Z = Z(H)$ can be rewritten in the following form:

$$Z = \int d\Omega \, d\mathbf{R} \, m^{N-1} dm \, \delta \left(m\mathbf{n} - \frac{1}{L^D} \int d^D x \mathbf{R}(x) \right) \exp\left(- \int d^D x \mathcal{L}(\mathbf{R}, \Omega H) \right). \tag{5.324}$$

Carrying out the integration over m gives

$$Z = \int d\Omega \, d\mathbf{R} \prod_{i=1}^{N-1} \delta \left(\frac{1}{L^D} \int d^D x R^i(x) \right) \exp\left(- \int d^D x \mathcal{L}(\mathbf{R}, \Omega H) \right.$$

$$\left. + (N-1) \ln \left(\frac{1}{L^D} \int d^D x R^0(x) \right) \right). \tag{5.325}$$

Hence, we may define

$$R^i(x) = \Pi^i(x), \quad i = 1, \cdots, N-1, \qquad R^0(x) = \sqrt{1 - \Pi^2(x)}. \tag{5.326}$$

Here, the field $\Pi(x)$ does not contain the zero-momentum mode $k_\mu = 0$, because of the constraint $\int d^D x \Pi(x) = 0$ enforced by the δ-function in the path integral. Hence, in the propagator of this field, the chiral limit $H \to 0$ can be readily performed.

Now, in order to set the counting rules for the field Π, let us note that, in order to obtain a non-trivial theory in the limit $L \to \infty$, the leading kinetic term of this field in the action functional should count as $O(1)$ in this limit. Since the derivative ∂_μ counts at $O(L^{-1})$, then $\Pi = O(L^{-D/2+1})$. This, together with $H = O(L^{-D})$, unambiguously sets the counting rules for the various terms in the Lagrangian. According to these rules, the effective action in Eq. (5.325) can be re-expanded, $S^{\text{eff}} = S_0^{\text{eff}} + S_1^{\text{eff}} + S_2^{\text{eff}} + \ldots$, where

$$S_0^{\text{eff}} = \frac{F^2}{2} \int d^D x \, \partial_\mu \Pi \partial_\mu \Pi - \Sigma H L^D \Omega^{00},$$

$$S_1^{\text{eff}} = \int d^D x \left(\frac{F^2}{2} (\Pi \partial_\mu \Pi)(\Pi \partial_\mu \Pi) + \frac{N-1}{2L^D} \Pi^2 + \frac{1}{2} \Sigma H \Omega^{00} \Pi^2 \right), \tag{5.327}$$

and so on. It can then be verified straightforwardly that $S_0^{\text{eff}}, S_1^{\text{eff}}, \ldots$ count as $O(1)$, $O(L^{2-D}), \ldots$. The quantity Ω^{00} denotes the "00" component of the rotation matrix $\Omega^{ij}, i, j = 0, \ldots N-1$.

Now, it is straightforward to set up a systematic framework for the calculation of the partition function at a given order in $1/L$. For example, at next-to-leading order $O(L^{2-D})$, only $S_0^{\text{eff}}, S_1^{\text{eff}}$ contribute. Expanding in a Taylor series, we get

$$Z = \int d\Omega \, e^{\Sigma H L^D \Omega^{00}} \int' [d\Pi] \exp\left(-\frac{F^2}{2} \int d^D x \, \partial_\mu \Pi \partial_\mu \Pi \right)$$

$$\times \left\{ 1 + \int d^D y \left(\frac{F^2}{2} (\mathbf{\Pi} \partial_\mu \mathbf{\Pi})(\mathbf{\Pi} \partial_\mu \mathbf{\Pi}) + \frac{N-1}{2L^D} \mathbf{\Pi}^2 + \frac{1}{2} \Sigma H \Omega^{00} \mathbf{\Pi}^2 \right) \right\}, \quad (5.328)$$

where $\int'$ denotes the path integral, where the zero-momentum modes of the field $\mathbf{\Pi}(x)$ are "chopped off" by the pertinent δ-function. The measure $[d\mathbf{\Pi}]$ contains the factor

$$[d\mathbf{\Pi}] = d\mathbf{\Pi} \exp \left(-\frac{L^D}{2} \int d^D x \ln(1 - \mathbf{\Pi}^2(x)) \right), \quad (5.329)$$

which is nothing but the Jacobian of the variable transformation from $\mathbf{R}$ to $\mathbf{\Pi}$. This H-independent factor does not contribute to the following expressions and will be ignored hereafter. (See also the following discussion.)

Next, we may use Wick's theorem for the calculation of the path integral over the field $\mathbf{\Pi}$. The result is given by

$$Z = \int d\Omega \, e^{\Sigma H L^D \Omega^{00}} \left\{ 1 + L^D \left(\frac{F^2}{2} \langle \Pi^i \Pi^j \rangle \langle \partial_\mu \Pi^i \partial_\mu \Pi^j \rangle \right. \right.$$

$$\left. + \frac{N-1}{2L^D} \langle \mathbf{\Pi} \mathbf{\Pi} \rangle + \frac{1}{2} \Sigma H \Omega^{00} \langle \mathbf{\Pi} \mathbf{\Pi} \rangle \right) \right\}$$

$$= \int d\Omega \exp \left\{ \Sigma H L^D \Omega^{00} + L^D \left(\frac{F^2}{2} \langle \Pi^i \Pi^j \rangle \langle \partial_\mu \Pi^i \partial_\mu \Pi^j \rangle \right. \right.$$

$$\left. \left. + \frac{N-1}{2L^D} \langle \mathbf{\Pi} \mathbf{\Pi} \rangle + \frac{1}{2} \Sigma H \Omega^{00} \langle \mathbf{\Pi} \mathbf{\Pi} \rangle \right) \right\}. \quad (5.330)$$

Here, the tadpoles are given by

$$\langle \Pi^i \Pi^j \rangle = \lim_{x \to y} \langle 0 | T \Pi^i(x) \Pi^j(y) | 0 \rangle = \delta^{ij} \frac{1}{F^2 L^D} \sum_{p \neq 0} \frac{1}{p^2} \quad (5.331)$$

and

$$\langle \partial_\mu \Pi^i \partial_\mu \Pi^j \rangle = \lim_{x \to y} \partial_\mu^x \partial_\mu^y \langle 0 | T \Pi^i(x) \Pi^j(y) | 0 \rangle. \quad (5.332)$$

Now, taking into account that $p_\mu = 2\pi n_\mu / L$, we may write

$$\frac{1}{L^D} \sum_{p \neq 0} \frac{1}{p^2} = \frac{4\pi^2}{L^{D-2}} \sum_{n \in \mathbb{Z}^D \setminus 0} \frac{1}{n^2} = -\frac{\beta_1}{L^{D-2}}, \quad (5.333)$$

where the numerical constant β_1 depends on D only.

Taking now into account the fact that the partition function $Z = Z(H)$ is normalized, $Z(0) = 1$, it is immediately seen that the H-independent terms in Eq. (5.330) do not contribute, since they can be included in the normalization. Hence,

$$Z = \int d\Omega \exp \left\{ \rho \Sigma H L^D \Omega^{00} \right\}, \quad (5.334)$$

where

$$\rho = 1 + \frac{(N-1)\beta_1}{2F^2 L^{D-2}}. \quad (5.335)$$

It remains to evaluate the integral over the $O(N)$ group. This can be done most clearly using Eq. (5.323) and performing the integral over $\mathbf{e}$. Introducing polar coordinates in the N-dimensional space,

$$e^0 = \cos\theta_1\,,$$

$$e^1 = \sin\theta_1\cos\theta_2\,,$$

$$\cdots$$

$$e^{N-2} = \sin\theta_1\cdots\sin\theta_{N-2}\cos\theta_{N-1}\,,$$

$$e^{N-1} = \sin\theta_1\cdots\sin\theta_{N-2}\sin\theta_{N-1}\,, \tag{5.336}$$

with $\Omega^{00} = \cos\theta_1$ and

$$d\mathbf{e} = \sin^{N-2}\theta_1\cdots\sin\theta_{N-2}d\theta_1\cdots d\theta_{N-1}\,, \tag{5.337}$$

we get

$$\int d\mathbf{e}\, e^{z\Omega^{00}} = \frac{\int_0^\pi d\theta_1 \sin^{N-2}\theta_1 e^{z\cos\theta}}{\int_0^\pi d\theta_1 \sin^{N-2}\theta_1} = \Gamma\left(\frac{N}{2}\right) Y_N(z)\,,$$

$$Y_N(z) = \left(\frac{2}{z}\right)^{(N-2)/2} I_{(N-2)/2}(z)\,, \tag{5.338}$$

where $I_\nu(z)$ is the modified Bessel function of the first kind. Finally,

$$Z = \Gamma\left(\frac{N}{2}\right) Y_N(\rho\Sigma HL^D)\,. \tag{5.339}$$

This expression for the partition function can be used to obtain the volume-dependence of the observables in the ε-regime. For example, the magnetization is defined as

$$\Sigma_H = \frac{1}{L^D}\frac{\partial}{\partial H}\ln Z = \rho\Sigma\frac{Y_N'(\rho\Sigma HL^D)}{Y_N(\rho\Sigma HL^D)}\,. \tag{5.340}$$

Thus, the size of Σ_H is governed by the dimensionless parameter $z = \rho\Sigma HL^D$. The derivative in the right-hand side of the preceding expression vanishes linearly for small z. Consequently, for a fixed L the magnetization vanishes linearly in H when $H \to 0$, in accordance with the vanishing of the finite-volume quark condensate in the chiral limit.

Below, we shall very briefly consider the EFT framework in the δ-regime, which is relevant for the analysis of lattice data in different solid-state simulations (such as the Hubbard model), performed at zero (very low) temperature. The Faddeev–Popov procedure can again be used to separate the "slow" and the "fast" modes. However, the magnetization is time-dependent in the δ-regime. This corresponds to eliminating the $\mathbf{k} = \mathbf{0}$ mode instead of $k_\mu = 0$. Hence, instead of Eq. (5.321), one introduces the following unit multiplier in the path integral:

$$\int d\mathbf{m}(t)\,\delta\left(\mathbf{m}(t) - \frac{1}{L^d}\int_{-L/2}^{+L/2} d^d\mathbf{x}\,\mathbf{S}(\mathbf{x},t)\right) = 1\,. \tag{5.341}$$

Here, $d = D - 1$ and L is the spatial elongation of the box (the temporal elongation is taken to infinity).

The further steps are reminiscent of the ones just done for the ε-regime, with the difference that now the $O(N)$-rotation depends on time. This means that $\Omega(t)$ should be elevated to a dynamical variable. Defining the transformed fields as follows:[27]

$$\mathbf{m}(t) = m(t)\mathbf{e}(t), \tag{5.342}$$

$$\mathbf{e}(t) = \Omega^T(t)\mathbf{n}, \qquad \mathbf{S}(\mathbf{x}, t) = \Omega^T(t)\mathbf{R}(\mathbf{x}, t), \tag{5.343}$$

substituting everything into the partition function and integrating over $m(t)$, one finally has

$$Z = \int d\mathbf{e}(t)\, d\mathbf{R} \prod_{i=1}^{N-1} \delta\left(\frac{1}{L^d}\int d^d\mathbf{x}\, R^i(\mathbf{x}, t)\right) \exp\left\{-\int dt \int d^d\mathbf{x}\, \mathcal{L}(\Omega^T(t)\mathbf{R}, H)\right.$$

$$\left. - (N-1)\delta(0)\int dt \int d^d\mathbf{x} \ln\left(\frac{1}{L^d}\int d^d\mathbf{x} \ln(R^0(\mathbf{x}, t))\right)\right\}. \tag{5.344}$$

The factor $\delta(0)$ arises from the integration over t in the *infinite* interval. The same factor arises in the Jacobian of variable transformation from $\mathbf{R}$ to $\mathbf{\Pi}$ (cf. Eq. (5.329)):

$$[d\mathbf{\Pi}] = d\mathbf{\Pi} \exp\left(-\frac{L^d}{2}\delta(0)\int dt \int d^d\mathbf{x} \ln(1 - \mathbf{\Pi}^2(\mathbf{x}))\right). \tag{5.345}$$

In dimensional regularization, $\delta(0) = 0$. We shall use this regularization in the following and thus neglect such terms. The partition function is then rewritten in the following form:

$$Z = \int d\mathbf{e}(t)\, d\mathbf{\Pi} \prod_{i=1}^{N-1} \delta\left(\int d^d\mathbf{x}\, \Pi^i(\mathbf{x}, t)\right) \exp\left\{-\int dt \int d^d\mathbf{x}\, \mathcal{L}(\Omega^T(t)\mathbf{R}, H)\right\}. \tag{5.346}$$

It is assumed that $\mathbf{R}$ and Ω in the action functional are expressed through $\mathbf{\Pi}$ and $\mathbf{e}$.

The counting rules in the δ-regime are different from those in the ε-regime. The "slow" field $\mathbf{e}(t)$ counts as $O(\delta^0)$, where $\delta \sim 1/L$ is a generic small parameter. Furthermore, the space and time derivatives are counted differently. It is convenient to assign the following rules (cf. Eq. (5.295)):

$$\partial_t = O(\delta^3), \qquad \partial_x = O(\delta). \tag{5.347}$$

Consider now the two-point function of the (massless) "fast" mode:

$$\langle 0|T\mathbf{\Pi}(\mathbf{x}, t)\mathbf{\Pi}(\mathbf{y}, t')|0\rangle = \int \frac{dE}{2\pi} \frac{1}{L^d} \sum_{\mathbf{k}}{}' \frac{e^{iE(t-t')+i\mathbf{k}(\mathbf{x}-\mathbf{y})}}{E^2 + \mathbf{k}^2}, \tag{5.348}$$

---

[27] A more complicated transformation, involving an additional matrix $\Sigma(t)$, is considered in Refs. [112, 113]. This leads to a simplified structure of the perturbative expansion that contains less terms but is, strictly speaking, not necessary. We could express $\Omega(t)$ in terms of $\mathbf{e}(t)$ explicitly and treat $\mathbf{e}(t)$, $\mathbf{R}(\mathbf{x}, t)$ as independent degrees of freedom.

with the $\mathbf{k} = \mathbf{0}$ contribution omitted in the sum. Taking into account that $E = O(\delta^3)$, we see that the right-hand side counts as $O(\delta^{d+1})$ and, hence,

$$\Pi = O(\delta^{(d+1)/2}). \tag{5.349}$$

The rest is relatively straightforward. The lowest-order term in the effective action emerges from the kinetic term of the field $\mathbf{R}$ and is given by

$$
\begin{aligned}
S_0^{\text{eff}} &= \frac{F^2}{2} \int dt \int d^d\mathbf{x}\, \partial_\mu (\Omega_{ij}^T(t) R^i(\mathbf{x},t)) \partial_\mu (\Omega_{ik}^T(t) R^k(\mathbf{x},t)) \\
&= \frac{F^2}{2} \int dt \int d^d\mathbf{x}\, \partial_\mu (\Omega_{i0}^T(t)(1 - \mathbf{\Pi}^2(\mathbf{x},t))^{1/2}) \partial_\mu (\Omega_{i0}^T(t)(1 - \mathbf{\Pi}^2(\mathbf{x},t))^{1/2}) + \cdots \\
&= \frac{F^2}{2} \int dt \int d^d\mathbf{x}\, (\dot{\Omega}_{i0}^T(t))^2 + \cdots = \frac{F^2 L^d}{2} \int dt\, \dot{\mathbf{e}}^2(t) + \cdots .
\end{aligned}
\tag{5.350}
$$

Here, the ellipses represent the higher-order terms in δ. Thus, at lowest order the action functional of the system describes the rigid quantum-mechanical rotor with the moment of inertia,

$$\Theta = F^2 L^d , \tag{5.351}$$

and the spectrum of the rotor is given by

$$E_n = \frac{1}{2\Theta} n(n+N-2), \qquad n = 0,1,\cdots . \tag{5.352}$$

The case of $N_f = 2$ in QCD corresponds to the $O(4)$ σ-model. Taking $N = 4$ and $n = 1$, it is seen that Eq. (5.302) is readily reproduced.

5.12 Literature Guide

The use of ChPT to study QCD at a finite temperature and volume started with the seminal papers by Gasser and Leutwyler [4, 101, 115]; see also [102]. The finite-volume corrections to the stable particle masses have been studied by Lüscher for generic relativistic field theories [5]. In particular, there it has been demonstrated that the leading exponentially suppressed contributions to the finite-volume mass shift can be summed up explicitly to all orders in perturbation theory and expressed through the integral over the infinite-volume elastic scattering amplitude in a particular kinematics. On the contrary, the ChPT framework enables us to calculate finite-volume artifacts order by order, but to take into account all subleading corrections in the volume. This alternative is explained very well in Refs. [6, 7].

The two-particle scattering in a finite volume has been first studied by Lüscher [14], who wrote down the two-particle quantization condition in the rest-frame and carried out (partial) diagonalization of this equation, using octahedral symmetry on the lattice. Various generalizations of the quantization condition have been considered and the diagonalization issue has been addressed in Refs. [21, 25–33]. The signatures of the resonances in a finite volume have been explored in Refs. [116, 117].

The volume-dependence of the $K \to 2\pi$ decay matrix element has been studied by Lellouch and Lüscher in Ref. [61]. The generalization of the approach to the moving frames, as well as multiple two-body channels has been performed in Refs. [28, 62, 63]. The papers [34, 65–69] address the calculation of the matrix elements between multiparticle states, including the case of the resonance matrix elements.

The study of the finite-volume shifts of multiparticle scattering states in perturbation theory has a decades-long history [118–121]. A systematic use of the nonrelativistic EFT for this purpose has been proposed in Ref. [70]. Later, this approach has been successfully used to study perturbative shifts both in the ground and excited states in Refs. [71, 73, 74]. Moreover, in Ref. [72], the long-range Coulomb force has been also included.

In Ref. [76], it was demonstrated that the three-particle spectrum in a finite volume is determined by the infinite-volume S-matrix elements only and, thus, does not depend on the sort-range details of interactions. This result had a stimulating effect on the activities in the field, where, within a short period, three different but essentially equivalent forms of the three-particle quantization condition have been proposed [77–81]. A full coverage of the latest developments cannot be given here. Rather, the reader who wants to get familiar with the subject is referred to a recent brief review [122], which contains an extended list of references. Note only that the three-particle analog of the Lellouch–Lüscher equation has been derived in Refs. [85, 86].

The formation of the quark condensate in a finite volume and the role of topologically nontrivial gluon fields has been studied in Ref. [100]. Related issues are discussed also in Ref. [123–128]. Ref. [129] is a useful review on the role of instantons in QCD.

The ε- and δ-regimes in QCD have been introduced in Refs. [3, 4, 101]. In particular, in Ref. [3] a calculation of the finite-volume pion mass in the chiral limit has been performed. A systematic perturbation theory in these regimes has been developed in Refs. [109–113]. It should also be mentioned that the separation of the zero mode has been carried out in the one-nucleon sector as well (in the δ-regime). The resulting quantum-mechanical Hamiltonian again coincides with the Hamiltonian of a rigid rotor [130].

The existence of a new phase, where isospin and parity are spontaneously broken, was first suggested in Refs. [105, 106]. The papers [103, 107, 131] address this problem systematically in the EFT setting; see also Ref. [132], where the finite-volume aspects of the problem are discussed. An analog of Aoki phase in case of staggered fermions has been studied in Ref. [133]. In the lectures [134, 135], many interesting applications of EFT in the context of lattice QCD are discussed, including the Symanzik effective Lagrangian and the formation of the Aoki phase.

References

[1] K. Symanzik, "Continuum limit and improved action in lattice theories. 1. principles and φ^4 theory," Nucl. Phys. B **226** (1983) 187.

[2] K. Symanzik, "Continuum limit and improved action in lattice theories. 2. $O(N)$ nonlinear sigma model in perturbation theory," Nucl. Phys. B **226** (1983) 205.

[3] H. Leutwyler, "Energy levels of light quarks confined to a box," Phys. Lett. B **189** (1987) 197.

[4] J. Gasser and H. Leutwyler, "Thermodynamics of chiral symmetry," Phys. Lett. B **188** (1987) 477.

[5] M. Lüscher, "Volume dependence of the energy spectrum in massive quantum field theories. 1. Stable particle states," Commun. Math. Phys. **104** (1986) 177.

[6] G. Colangelo and S. Dürr, "The pion mass in finite volume," Eur. Phys. J. C **33** (2004) 543.

[7] G. Colangelo, S. Dürr and C. Haefeli, "Finite volume effects for meson masses and decay constants," Nucl. Phys. B **721** (2005) 136.

[8] L. Maiani and M. Testa, "Final state interactions from Euclidean correlation functions," Phys. Lett. B **245** (1990) 585.

[9] C. Michael, "Particle decay in lattice gauge theory," Nucl. Phys. B **327** (1989) 515.

[10] M. Bruno and M. T. Hansen, "Variations on the Maiani-Testa approach and the inverse problem" [arXiv:2012.11488 [hep-lat]].

[11] C. Michael, "Adjoint sources in lattice gauge theory," Nucl. Phys. B **259** (1985) 58.

[12] M. Lüscher and U. Wolff, "How to calculate the elastic scattering matrix in two-dimensional quantum field theories by numerical simulation," Nucl. Phys. B **339** (1990) 222.

[13] A. A. Logunov and A. N. Tavkhelidze, "Quasioptical approach in quantum field theory," Nuovo Cim. **29** (1963) 380.

[14] M. Lüscher, "Two particle states on a torus and their relation to the scattering matrix," Nucl. Phys. B **354** (1991) 531.

[15] C. Liu, X. Feng and S. He, "Two particle states in a box and the S-matrix in multi-channel scattering," Int. J. Mod. Phys. A **21** (2006) 847.

[16] M. Lage, U.-G. Meißner and A. Rusetsky, "A method to measure the antikaon-nucleon scattering length in lattice QCD," Phys. Lett. B **681** (2009) 439.

[17] V. Bernard, M. Lage, U.-G. Meißner and A. Rusetsky, "Scalar mesons in a finite volume," JHEP **01** (2011) 019.

[18] M. Döring, U.-G. Meißner, E. Oset and A. Rusetsky, "Unitarized Chiral Perturbation Theory in a finite volume: scalar meson sector," Eur. Phys. J. A **47** (2011) 139.

[19] M. Döring, J. Haidenbauer, U.-G. Meißner and A. Rusetsky, "Dynamical coupled-channel approaches on a momentum lattice," Eur. Phys. J. A **47** (2011) 163.

[20] M. Döring and U.-G. Meißner, "Finite volume effects in pion-kaon scattering and reconstruction of the $\kappa(800)$ resonance," JHEP **01** (2012) 009.

[21] R. A. Briceño and Z. Davoudi, "Moving multichannel systems in a finite volume with application to proton-proton fusion," Phys. Rev. D **88** (2013) 094507.

[22] M. Döring, U.-G. Meißner, E. Oset and A. Rusetsky, "Scalar mesons moving in a finite volume and the role of partial wave mixing," Eur. Phys. J. A **48** (2012) 114.

[23] R. A. Briceño, "Two-particle multichannel systems in a finite volume with arbitrary spin," Phys. Rev. D **89** (2014) 074507.

[24] S. R. Beane, P. F. Bedaque, A. Parreno and M. J. Savage, "Exploring hyperons and hypernuclei with lattice QCD," Nucl. Phys. A **747** (2005) 55.

[25] K. Rummukainen and S. A. Gottlieb, "Resonance scattering phase shifts on a nonrest frame lattice," Nucl. Phys. B **450** (1995) 397.

[26] X. Li and C. Liu, "Two particle states in an asymmetric box," Phys. Lett. B **587** (2004) 100.

[27] X. Feng, X. Li and C. Liu, "Two particle states in an asymmetric box and the elastic scattering phases," Phys. Rev. D **70** (2004) 014505.

[28] C. H. Kim, C. T. Sachrajda and S. R. Sharpe, "Finite-volume effects for two-hadron states in moving frames," Nucl. Phys. B **727** (2005) 218.

[29] Z. Fu, "Rummukainen-Gottlieb's formula on two-particle system with different mass," Phys. Rev. D **85** (2012) 014506.

[30] L. Leskovec and S. Prelovsek, "Scattering phase shifts for two particles of different mass and non-zero total momentum in lattice QCD," Phys. Rev. D **85** (2012) 114507.

[31] M. Göckeler, R. Horsley, M. Lage et al. "Scattering phases for meson and baryon resonances on general moving-frame lattices," Phys. Rev. D **86** (2012) 094513.

[32] N. Li and C. Liu, "Generalized Lüscher formula in multichannel baryon-meson scattering," Phys. Rev. D **87** (2013) 014502.

[33] F. X. Lee and A. Alexandru, "Scattering phase-shift formulas for mesons and baryons in elongated boxes," Phys. Rev. D **96** (2017) 054508.

[34] V. Bernard, D. Hoja, U.-G. Meißner and A. Rusetsky, "Matrix elements of unstable states," JHEP **09** (2012) 023.

[35] V. Bernard, M. Lage, U.-G. Meißner and A. Rusetsky, "Resonance properties from the finite-volume energy spectrum," JHEP **08** (2008) 024.

[36] F. Romero-López, A. Rusetsky and C. Urbach, "Vector particle scattering on the lattice," Phys. Rev. D **98** (2018) 014503.

[37] J. P. Elliott and P. G. Dawber, Symmetry in Physics, Volume I: Principles and Simple Applications, Macmillan (1979).

[38] M. Tinkham, Group Theory and Quantum Mechanics, McGraw-Hill (1964).

[39] R. C. Johnson, "Angular momentum on a lattice," Phys. Lett. B **114** (1982) 147.

[40] S. l. Altmann and A. P. Cracknell, "Lattice Harmonics 1. Cubic Groups," Rev. Mod. Phys. **37** (1965) 19.

[41] D. A. Varshalovich, A. N. Moskalev and V. K. Khersonsky, Quantum Theory of Angular Momentum: Irreducible Tensors, Spherical Harmonics, Vector Coupling Coefficients, $3nj$ Symbols, World Scientific (1988).

[42] S. R. Beane, P. F. Bedaque, A. Parreno and M. J. Savage, "Two nucleons on a lattice," Phys. Lett. B **585** (2004) 106.

[43] S. Sasaki and T. Yamazaki, "Signatures of S-wave bound-state formation in finite volume," Phys. Rev. D **74** (2006) 114507.

[44] S. König, D. Lee and H.-W. Hammer, "Non-relativistic bound states in a finite volume," Annals Phys. **327** (2012) 1450.

[45] M. Lüscher, "Volume dependence of the energy spectrum in massive quantum field theories. 2. Scattering states," Commun. Math. Phys. **105** (1986) 153.

[46] P. F. Bedaque, "Aharonov-Bohm effect and nucleon nucleon phase shifts on the lattice," Phys. Lett. B **593** (2004) 82.

[47] C. T. Sachrajda and G. Villadoro, "Twisted boundary conditions in lattice simulations," Phys. Lett. B **609** (2005) 73.

[48] G. M. de Divitiis, R. Petronzio and N. Tantalo, "On the discretization of physical momenta in lattice QCD," Phys. Lett. B **595** (2004) 408.

[49] J. Bijnens and J. Relefors, "Masses, decay constants and electromagnetic form-factors with twisted boundary conditions," JHEP **05** (2014) 015.

[50] R. A. Briceño, Z. Davoudi, T. C. Luu and M. J. Savage, "Two-baryon systems with twisted boundary conditions," Phys. Rev. D **89** (2014) 074509.

[51] S. Weinberg, "Evidence that the deuteron is not an elementary particle," Phys. Rev. **137** (1965) B672.

[52] D. Agadjanov, F. K. Guo, G. Ríos and A. Rusetsky, "Bound states on the lattice with partially twisted boundary conditions," JHEP **01** (2015) 118.

[53] J. W. Chen and M. J. Savage, "Baryons in partially quenched chiral perturbation theory," Phys. Rev. D **65** (2002) 094001.

[54] P. F. Bedaque and J. W. Chen, "Twisted valence quarks and hadron interactions on the lattice," Phys. Lett. B **616** (2005) 208.

[55] D. Agadjanov, U.-G. Meißner and A. Rusetsky, "Partial twisting for scalar mesons," JHEP **01** (2014) 103.

[56] F. J. Gilman and M. B. Wise, "Effective Hamiltonian for $\Delta S = 1$ weak nonleptonic decays in the six quark model," Phys. Rev. D **20** (1979) 2392.

[57] G. Buchalla, A. J. Buras and M. E. Lautenbacher, "Weak decays beyond leading logarithms," Rev. Mod. Phys. **68** (1996) 1125.

[58] M. E. Peskin and D. V. Schroeder, An Introduction to Quantum Field Theory, Addison-Wesley (1995) 842 pp.

[59] U.-G. Meißner, K. Polejaeva and A. Rusetsky, "Extraction of the resonance parameters at finite times," Nucl. Phys. B **846** (2011) 1.

[60] D. Agadjanov, M. Döring, M. Mai, U.-G. Meißner and A. Rusetsky, "The optical potential on the lattice," JHEP **06** (2016) 043.

[61] L. Lellouch and M. Lüscher, "Weak transition matrix elements from finite volume correlation functions," Commun. Math. Phys. **219** (2001) 31.

[62] N. H. Christ, C. Kim and T. Yamazaki, "Finite volume corrections to the two-particle decay of states with non-zero momentum," Phys. Rev. D **72** (2005) 114506.

[63] M. T. Hansen and S. R. Sharpe, "Multiple-channel generalization of Lellouch-Lüscher formula," Phys. Rev. D **86** (2012) 016007.

[64] R. Abbott et al. [RBC and UKQCD], "Direct CP violation and the $\Delta I = 1/2$ rule in $K \to \pi\pi$ decay from the standard model," Phys. Rev. D **102** (2020) 054509.

[65] H. B. Meyer, "Lattice QCD and the timelike pion form factor," Phys. Rev. Lett. **107** (2011) 072002.

[66] R. A. Briceño and M. T. Hansen, "Multichannel $0 \to 2$ and $1 \to 2$ transition amplitudes for arbitrary spin particles in a finite volume," Phys. Rev. D **92** (2015) 074509.

[67] R. A. Briceño, M. T. Hansen and A. Walker-Loud, "Multichannel $1 \to 2$ transition amplitudes in a finite volume," Phys. Rev. D **91** (2015) 034501.

[68] A. Agadjanov, V. Bernard, U.-G. Meißner and A. Rusetsky, "The $B \to K^*$ form factors on the lattice," Nucl. Phys. B **910** (2016) 387.

[69] A. Agadjanov, V. Bernard, U.-G. Meißner and A. Rusetsky, "A framework for the calculation of the $\Delta N\gamma^*$ transition form factors on the lattice," Nucl. Phys. B **886** (2014) 1199.

[70] S. R. Beane, W. Detmold and M. J. Savage, "n-boson energies at finite volume and three-boson interactions," Phys. Rev. D **76** (2007) 074507.

[71] W. Detmold and M. J. Savage, "The energy of n identical bosons in a finite volume at $O(L^{-7})$," Phys. Rev. D **77** (2008) 057502.

[72] S. R. Beane, et al. "Charged multi-hadron systems in lattice QCD+QED," Phys. Rev. D **103** (2021) 054504.

[73] F. Romero-López, A. Rusetsky, N. Schlage and C. Urbach, "Relativistic N-particle energy shift in finite volume," JHEP **02** (2021) 060.

[74] F. Müller, A. Rusetsky and T. Yu, "Finite-volume energy shift of the three-pion ground state," Phys. Rev. D **103** (2021) 054506.

[75] M. Döring et al. "Three-body spectrum in a finite volume: the role of cubic symmetry," Phys. Rev. D **97** (2018) 114508.

[76] K. Polejaeva and A. Rusetsky, "Three particles in a finite volume," Eur. Phys. J. A **48** (2012) 67.

[77] M. T. Hansen and S. R. Sharpe, "Relativistic, model-independent, three-particle quantization condition," Phys. Rev. D **90** (2014) 116003.

[78] M. T. Hansen and S. R. Sharpe, "Expressing the three-particle finite-volume spectrum in terms of the three-to-three scattering amplitude," Phys. Rev. D **92** (2015) 114509.

[79] H. W. Hammer, J. Y. Pang and A. Rusetsky, "Three-particle quantization condition in a finite volume: 1. The role of the three-particle force," JHEP **09** (2017) 109.

[80] H. W. Hammer, J. Y. Pang and A. Rusetsky, "Three particle quantization condition in a finite volume: 2. general formalism and the analysis of data," JHEP **10** (2017) 115.

[81] M. Mai and M. Döring, "Three-body unitarity in the finite volume," Eur. Phys. J. A **53** (2017) 240.

[82] U.-G. Meißner, G. Ríos and A. Rusetsky, "Spectrum of three-body bound states in a finite volume," Phys. Rev. Lett. **114** (2015) 091602 [erratum: Phys. Rev. Lett. **117** (2016) 069902].

[83] M. T. Hansen and S. R. Sharpe, "Applying the relativistic quantization condition to a three-particle bound state in a periodic box," Phys. Rev. D **95** (2017) 034501.

[84] S. König and D. Lee, "Volume dependence of n-body bound states," Phys. Lett. B **779** (2018) 9.

[85] F. Müller and A. Rusetsky, "On the three-particle analog of the Lellouch-Lüscher formula," JHEP **21** (2020) 152.

[86] M. T. Hansen, F. Romero-López and S. R. Sharpe, "Decay amplitudes to three hadrons from finite-volume matrix elements," JHEP **04** (2021) 113.

[87] M. Hayakawa and S. Uno, "QED in finite volume and finite size scaling effect on electromagnetic properties of hadrons," Prog. Theor. Phys. **120** (2008) 413.

[88] Z. Davoudi, J. Harrison, A. Jüttner, A. Portelli and M. J. Savage, "Theoretical aspects of quantum electrodynamics in a finite volume with periodic boundary conditions," Phys. Rev. D **99** (2019) 034510.

[89] T. Blum, T. Doi, M. Hayakawa, T. Izubuchi and N. Yamada, "Determination of light quark masses from the electromagnetic splitting of pseudoscalar meson masses computed with two flavors of domain wall fermions," Phys. Rev. D **76** (2007) 114508.

[90] A. Duncan, E. Eichten and H. Thacker, "Electromagnetic splittings and light quark masses in lattice QCD," Phys. Rev. Lett. **76** (1996) 3894.

[91] M. Göckeler, R. Horsley, E. Laermann et al. "QED: A lattice investigation of the chiral phase transition and the nature of the continuum limit," Nucl. Phys. B **334** (1990) 527.

[92] M. G. Endres, A. Shindler, B. C. Tiburzi and A. Walker-Loud, "Massive photons: An infrared regularization scheme for lattice QCD+QED," Phys. Rev. Lett. **117** (2016) 072002.

[93] M. Hansen, B. Lucini, A. Patella and N. Tantalo, "Gauge invariant determination of charged hadron masses," JHEP **05** (2018) 146.

[94] A. S. Kronfeld and U. J. Wiese, "$SU(N)$ gauge theories with C-periodic boundary conditions. 1. Topological structure," Nucl. Phys. B **357** (1991) 521.

[95] A. S. Kronfeld and U. J. Wiese, "$SU(N)$ gauge theories with C-periodic boundary conditions. 2. Small volume dynamics," Nucl. Phys. B **401** (1993) 190.

[96] Z. Davoudi and M. J. Savage, "Finite-volume electromagnetic corrections to the masses of mesons, baryons and nuclei," Phys. Rev. D **90** (2014) 054503.

[97] J. W. Lee and B. C. Tiburzi, "Finite volume corrections to the electromagnetic mass of composite particles," Phys. Rev. D **93** (2016) 034012.

[98] Z. Fodor et al., "Quantum electrodynamics in finite volume and nonrelativistic effective field theories," Phys. Lett. B **755** (2016) 245.

[99] A. Patella, "QED corrections to hadronic observables," PoS **LATTICE2016** (2017) 020 [arXiv:1702.03857 [hep-lat]].

[100] H. Leutwyler and A. V. Smilga, "Spectrum of Dirac operator and role of winding number in QCD," Phys. Rev. D **46** (1992) 5607.

[101] J. Gasser and H. Leutwyler, "Light quarks at low temperatures," Phys. Lett. B **184** (1987) 83.

[102] P. Gerber and H. Leutwyler, "Hadrons below the chiral phase transition," Nucl. Phys. B **321** (1989) 387.

[103] S. R. Sharpe and R. L. Singleton, Jr, "Spontaneous flavor and parity breaking with Wilson fermions," Phys. Rev. D **58** (1998) 074501.

[104] B. Sheikholeslami and R. Wohlert, "Improved continuum limit lattice action for QCD with Wilson fermions," Nucl. Phys. B **259** (1985) 572.

[105] S. Aoki, "New phase structure for lattice QCD with Wilson fermions," Phys. Rev. D **30** (1984) 2653.

[106] S. Aoki, "A solution to the $U(1)$ problem on a lattice," Phys. Rev. Lett. **57** (1986) 3136.

[107] S. R. Sharpe, "On the consistency of the Aoki-phase," Phys. Rev. D **79** (2009) 054503.

[108] E. M. Ilgenfritz, W. Kerler, M. Müller-Preussker, A. Sternbeck and H. Stuben, "A numerical reinvestigation of the Aoki phase with $N_f = 2$ Wilson fermions at zero temperature," Phys. Rev. D **69** (2004) 074511.

[109] P. Hasenfratz and H. Leutwyler, "Goldstone boson related finite size effects in field theory and critical phenomena with $O(N)$ symmetry," Nucl. Phys. B **343** (1990) 241.

[110] P. Hasenfratz, "Perturbation theory and zero modes in $O(N)$ lattice σ models," Phys. Lett. B **141** (1984) 385.

[111] P. Hasenfratz, "The QCD rotator in the chiral limit," Nucl. Phys. B **828** (2010) 201.

[112] F. Niedermayer and C. Weiermann, "The rotator spectrum in the delta-regime of the $O(n)$ effective field theory in 3 and 4 dimensions," Nucl. Phys. B **842** (2011) 248.

[113] P. Hasenfratz and F. Niedermayer, "Finite size and temperature effects in the AF Heisenberg model," Z. Phys. B **92** (1993) 91.

[114] L. D. Faddeev and V. N. Popov, "Feynman diagrams for the Yang-Mills field," Phys. Lett. B **25** (1967) 29.

[115] J. Gasser and H. Leutwyler, "Spontaneously broken symmetries: effective Lagrangians at finite volume," Nucl. Phys. B **307** (1988) 763.

[116] U.-J. Wiese, "Identification of resonance parameters from the finite volume energy spectrum," Nucl. Phys. B Proc. Suppl. **9** (1989) 609.

[117] M. Lüscher, "Signatures of unstable particles in finite volume," Nucl. Phys. B **364** (1991) 237.

[118] T. D. Lee, K. Huang and C. N. Yang, "Eigenvalues and eigenfunctions of a Bose system of hard spheres and its low-temperature properties," Phys. Rev. **106** (1957) 1135.

[119] K. Huang and C. N. Yang, "Quantum-mechanical many-body problem with hard-sphere interaction," Phys. Rev. **105** (1957) 767.

[120] T. T. Wu, "Ground state of a Bose system of hard spheres," Phys. Rev. **115** (1959) 1390.

[121] S. Tan, "Three-boson problem at low energy and implications for dilute Bose-Einstein condensates," Phys. Rev. A **78** (2008) 013636.

[122] M. Mai, M. Döring and A. Rusetsky, "Multi-particle systems on the lattice and chiral extrapolations: a brief review," Eur. Phys. J. ST **230** (2021) 1623.

[123] A. V. Smilga, "QCD at $\theta \sim \pi$," Phys. Rev. D **59** (1999) 114021.

[124] A. V. Smilga, "Aspects of chiral symmetry" [arXiv:hep-ph/0010049 [hep-ph]].

[125] R. Brower, S. Chandrasekharan, J. W. Negele and U. J. Wiese, "QCD at fixed topology," Phys. Lett. B **560** (2003) 64.

[126] P. H. Damgaard and H. Fukaya, "The chiral condensate in a finite volume," JHEP **01** (2009) 052.

[127] D. Toublan and J. J. M. Verbaarschot, "Effective low-energy theories and QCD Dirac spectra," Ser. Adv. Quant. Many Body Theor. **3** (2000) 114.

[128] F. Niedermayer, "Exact chiral symmetry, topological charge and related topics," Nucl. Phys. B Proc. Suppl. **73** (1999) 105.

[129] T. Schäfer and E. V. Shuryak, "Instantons in QCD," Rev. Mod. Phys. **70** (1998) 323.

[130] N. D. Vlasii and U. J. Wiese, "Rotor spectra and Berry phases in the chiral limit of QCD on a torus," Phys. Rev. D **97** (2018) 114029.

[131] M. Creutz, "Wilson fermions at finite temperature" [arXiv:hep-lat/9608024 [hep-lat]].

[132] O. Bär, S. Necco and S. Schaefer, "The Epsilon regime with Wilson fermions," JHEP **03** (2009) 006.

[133] W. J. Lee and S. R. Sharpe, "Partial flavor symmetry restoration for chiral staggered fermions," Phys. Rev. D **60** (1999) 114503.

[134] M. Golterman, "Applications of chiral perturbation theory to lattice QCD" [arXiv:0912.4042 [hep-lat]].

[135] P. Weisz, "Renormalization and lattice artifacts" [arXiv:1004.3462 [hep-lat]].

We decided to collect all exercises in a separate chapter. The reason for this is that some of these exercises refer to concepts that are considered in different chapters. Moreover, some are rather unconventional and could be properly branded as small research projects. In cases where these were inspired by some paper, we display a reference to the original paper in the text of the exercise. The solutions are given online. If a reference is present, you can consult it for further hints.

1. Dimensional Transmutation in Quantum Mechanics

The purpose of this exercise is to demonstrate the essence of dimensional transmutation and reveal the role of an ultraviolet divergence in a simple, exactly solvable model. Consider a point-like interaction in *two-dimensional* quantum mechanics. The Schrödinger equation takes the form

$$\left(-\triangle - \lambda_0 \delta^{(2)}(\mathbf{x})\right)\psi(\mathbf{x}) = E\psi(\mathbf{x}). \tag{1}$$

Here, $\mathbf{x}$ is a vector in the two-dimensional plane.

a) Show that λ_0 should be dimensionless.
b) Write the Lippmann–Schwinger equation for the scattering amplitude in the two-dimensional space.
c) What is the mass dimension of the scattering amplitude? How should the on-shell amplitude behave at large energies based on dimensional arguments?
d) Next, solve the Lippmann–Schwinger equation. In this case, an explicit solution can be found. Note that the point-like potential is singular and leads to an ultraviolet divergence in momentum space. Use cutoff regularization to tame this divergence.
e) Show that a bound state solution with an *arbitrary* binding energy B is allowed. How does a dimensionful quantity emerge?
f) Express the on-shell amplitude in terms of B. Does the on-shell amplitude obey the expected dimensional scaling at high energies?

The interested reader is referred to the original article for more details:

C. B. Thorn, "Quark confinement in the infinite momentum frame," Phys. Rev. D **19** (1979) 639.

2. Dimensional Transmutation in Scalar QED

Consider massless scalar QED, which is described by the Lagrangian

$$\mathcal{L} = -\frac{1}{4}\mathscr{F}_{\mu\nu}\mathscr{F}^{\mu\nu} + (D_\mu\varphi)^* D^\mu\varphi - \frac{\lambda_0}{6}(\varphi^*\varphi)^2, \tag{1}$$

where

$$D_\mu\varphi = \partial_\mu\varphi + ie\mathscr{A}_\mu\varphi. \tag{2}$$

Both couplings e and λ_0 are dimensionless. Since we consider a massless theory, there is no dimensionful parameter whatsoever. However, as we shall see in what follows, dimensional transmutation happens, and the "photon" acquires mass as a result of spontaneous breaking of symmetry, which is caused by loop effects. There is obviously no breaking at tree level, because the scalar potential $V(\varphi) = (\lambda_0/6)(\varphi^*\varphi)^2$ has only one minimum at $\varphi = 0$.

a) First, get rid of the (unphysical) phase of the scalar field using

$$\varphi = \frac{1}{\sqrt{2}}\rho e^{i\theta}, \qquad \rho, \theta \text{ real}. \tag{3}$$

Further, show that the phase θ can be gauged away by making a gauge transformation of the photon field $\mathscr{A}_\mu$, so that the Lagrangian can be rewritten in terms of ρ and $\mathscr{A}_\mu$ alone:

$$\mathcal{L} = -\frac{1}{4}\mathscr{F}_{\mu\nu}\mathscr{F}^{\mu\nu} + \frac{1}{2}(\partial\rho)^2 + \frac{e^2}{2}\rho^2\mathscr{A}_\mu\mathscr{A}^\mu - \frac{\lambda_0}{4!}\rho^4. \tag{4}$$

b) Now, calculate the one-loop effective action for the field ρ by integrating out the field $\mathscr{A}_\mu$. Here, we have to take into account that $\mathscr{A}_\mu$ is a gauge field, so it contains unphysical degrees of freedom that can be eliminated by imposing a gauge condition. In our case, the simplest choice is to use the Lorentz gauge $\partial_\mu\mathscr{A}^\mu = 0$. Thus, the generating functional for the Green's functions of the field ρ can be written as

$$Z[j] = \int d\rho\, d\mathscr{A}_\mu\, \delta(\partial\mathscr{A}) \exp\left(i\int d^4x\,(\mathcal{L}(x) + j(x)\rho(x))\right). \tag{5}$$

For the calculations, it is convenient to use the following representation of the δ-function in the path-integral formalism

$$\delta(f) = \int d\xi \exp\left(i\int d^4x\,\xi(x)f(x)\right). \tag{6}$$

Here, $\xi(x)$ is some auxiliary variable, in which the path integration is performed. Taking $f(x) = \partial_\mu\mathscr{A}^\mu(x)$ in this formula, it is seen that the path integral in the variable $\mathscr{A}_\mu$ in Eq. (5) is Gaussian. Show that, shifting the integration variable

$$\mathscr{A}_\mu \to \mathscr{A}_\mu + D^{-1}\partial_\mu\xi, \tag{7}$$

the argument of the exponential in Eq. (5) is given by (up to a total derivative)

$$i \int d^4x \left(\frac{1}{2} \mathscr{A}_\mu D \mathscr{A}^\mu - \frac{1}{2} \xi \Box D^{-1} \xi - \frac{1}{2} \rho \Box \rho - \frac{\lambda_0}{4!} \rho^4 + j\rho \right). \tag{8}$$

Here, D denotes the operator:

$$D = \Box + e^2 \rho^2. \tag{9}$$

c) Perform the (Gaussian) integration in the variables $\mathscr{A}_\mu$ and ξ (remember that there are four variables $\mathscr{A}_\mu$ with $\mu = 0, 1, 2, 3$). The effective action for the field ρ is defined as follows:

$$Z[j] = \int d\rho \, e^{iS_{\text{eff}}[\rho, j]}. \tag{10}$$

Show that

$$e^{iS_{\text{eff}}[\rho, j]} = (\det(\Box))^{-1/2} (\det(D))^{-3/2} \exp\left(i \int d^4x \left(-\frac{1}{2} \rho \Box \rho - \frac{\lambda_0}{4!} \rho^4 + j\rho \right) \right). \tag{11}$$

The first factor in the Eq. (11) is an unimportant constant, so that

$$S_{\text{eff}}[\rho, j] = \int d^4x \left(-\frac{1}{2} \rho \Box \rho - \frac{\lambda_0}{4!} \rho^4 + j\rho \right) + \frac{3i}{2} \text{tr}(\ln D). \tag{12}$$

d) Assume now that the field $\rho(x) = \rho = \text{const}$. Put the external source $j = 0$ and calculate the effective potential $U(\rho) = -\mathcal{L}(\rho)$ (one does exactly the same to obtain the Higgs potential that leads to spontaneous symmetry breaking at tree level). Performing the Wick rotation, obtain in Euclidean space

$$U(\rho) = \frac{\lambda_0}{4!} \rho^4 + \frac{3}{2} \int \frac{d^4 k_E}{(2\pi)^4} \ln(k_E^2 + e^2 \rho^2). \tag{13}$$

e) The preceding one-loop effective potential is ultraviolet-divergent and needs regularization/renormalization. Do this integral in your favorite regularization and give the definition of the renormalized coupling. The following result is obtained in dimensional regularization, where we have replaced the logarithm by using the trick

$$\ln(x) = \lim_{\varepsilon \to 0} \frac{1}{\varepsilon} (x^\varepsilon - 1). \tag{14}$$

The renormalized coupling λ_r is defined using the $\overline{\text{MS}}$ scheme. The result is

$$U(\rho) = \rho^4 \left(\frac{\lambda_r}{4!} + \frac{3e^4}{64\pi^2} \left(-\frac{3}{2} + \ln \frac{e^2 \rho^2}{\mu^2} \right) \right). \tag{15}$$

Here, μ is the scale of dimensional regularization. You are free to use, for example, cutoff regularization instead. The constant $-3/2$ is convention-dependent. Obviously, it can be removed by a redefinition of the scale μ.

f) As we know, there is no spontaneous symmetry breaking at tree level. However, the one-loop effective potential leads to symmetry breaking at a nonzero vacuum expectation value of the field ρ:

$$\left.\frac{dU(\rho)}{d\rho}\right|_{\rho=v} = 0, \qquad v = \langle \rho \rangle. \tag{16}$$

Write the equation for the minimum of the potential and express the scale μ in terms of the vacuum expectation value v.

g) Since the symmetry is broken, the scalar and the vector field both acquire masses. Find these masses, using the known formulae for the case of the ordinary Higgs mechanism.

Let us briefly summarize what we have learned from this exercise:

- In the beginning, we only had dimensionless couplings e and λ_0.
- The one-loop effective potential is ultraviolet-divergent. We have to regularize and renormalize, and this leads to the appearance of the scale μ in the theory. The scale is unphysical, as no observable depends on it.
- We have spontaneous symmetry breaking generated by the one-loop potential. We could express μ in terms of the v (related to the observable mass of the vector meson) and the scale-dependent coupling $\lambda_r(\mu)$ – welcome to dimensional transmutation!
- The theory becomes massive – both the scalar and the vector particles acquire a mass due to dimensional transmutation, even if the original theory contained only dimensionless parameters!

For more details, please consult the original paper:

S. R. Coleman and E. J. Weinberg, "Radiative corrections as the origin of spontaneous symmetry breaking," Phys. Rev. D **7** (1973) 1888.

3. Matching of EFT Couplings

Let the original theory be described by the Lagrangian

$$\mathcal{L} = \frac{1}{2}(\partial\phi)^2 - \frac{m^2}{2}\phi^2 + \frac{1}{2}(\partial\psi)^2 - \frac{M^2}{2}\psi^2 - \lambda\phi^4 - \frac{g}{2}\phi^2\psi^2. \tag{1}$$

If $M \gg m$, the Lagrangian of the low-energy EFT only contains the field ϕ and is given by

$$\mathcal{L}_{\text{eff}} = \frac{1}{2}(\partial\phi)^2 - \frac{\tilde{m}^2}{2}\phi^2 - C_0\phi^4 - C_2\phi^2\square^2\phi^2 + \cdots. \tag{2}$$

a) Carry out the matching of $\tilde{m}, C_0, C_2$ to the underlying theory at tree level.
b) Carry out the matching at one-loop level, up to and including terms of $O(\lambda, g^2, M^{-4})$. Give the relations between the *renormalized* parameters.

Hint: This is most simply done in the path integral formalism. To this end, one integrates out the field ψ (this is a Gaussian integration!). In order to calculate the functional determinant that emerges as a result of this, we may use the formula $\det A = \exp(\ln \operatorname{tr} A)$, where A is an arbitrary operator. Then, we expand the logarithm in powers of g, up to and including order g^2. Finally, the loop functions are expanded as Taylor series in external momenta. The matching condition is then directly read from the effective Lagrangian.

4. Heat-Kernel Renormalization

In this exercise, we shall solve a problem similar to Exercise 3, but using a different technique. In this manner, you will become familiar with the heat-kernel method.

Consider the theory described by the Lagrangian

$$\mathcal{L}(\phi, \psi) = \frac{1}{2}(\partial\phi)^2 - \frac{m^2}{2}\phi^2 + \frac{1}{2}(\partial\psi)^2 - \frac{M^2}{2}\psi^2 - \lambda\phi^4 + \rho\phi^2\psi + g\phi^2\psi^2, \quad (1)$$

with a light field ϕ and a heavy field ψ. The goal of this exercise is to integrate out the heavy field and to arrive at the effective Lagrangian

$$\mathcal{L}_{\text{eff}}(\phi) = \frac{1}{2}(\partial\phi)^2 - \frac{m^2}{2}\phi^2 - \lambda_1\phi^4 - \lambda_2(\partial\phi)^4 + \cdots, \quad (2)$$

using the path-integral techniques. This procedure involves divergences, caused by the heavy particles running in the loops. These divergences are handled by using the heat-kernel method. At one loop, we shall establish a full set of operators with divergent coefficients and carry out renormalization in the original theory.

We shall use dimensional regularization and the $\overline{\text{MS}}$ renormalization scheme throughout the exercise. D is the number of space-time dimensions, with $D \to 4$ at the end.

a) Let us work in Euclidean space, $x_0 \to -ix_0$, in order to ensure the exponential damping of the argument of the path integral. Decompose $\psi = \psi_c + \eta$ into a classical solution, ψ_c, and a fluctuation, η. ψ_c fulfills the equation of motion. Show that the Lagrangian can be written as

$$\mathcal{L}(\phi, \psi) = \mathcal{L}(\phi, \psi_c) - \frac{1}{2}\eta(-\square + M^2 - 2g\phi^2)\eta, \quad \square = \partial_\mu\partial_\mu. \quad (3)$$

It particular, it is seen that there are no linear terms in η, since ψ_c is a solution of the classical EOM.

b) Show that the contribution to the generating functional due to heavy-particle loops is given by

$$e^{-\Gamma_{\text{loop}}} = \int d\eta\, e^{-\frac{1}{2}\int d^4x\, \eta(-\square + M^2 - 2g\phi^2)\eta}. \quad (4)$$

Perform the path integral explicitly and use $\det A = \exp(\operatorname{tr}\ln A)$ to arrive at

$$\Gamma_{\text{loop}} = \frac{1}{2}\text{tr}\ln(-\Box + M^2 - 2g\phi^2). \tag{5}$$

Hint: A path integral over a quadratic form, $\eta(x)\mathcal{D}\eta(x)$, with a differential operator, $\mathcal{D}$, can most conveniently be performed by expanding in eigenfunctions of $\mathcal{D}$:

$$\eta = \sum_n a_n\eta_n, \quad \mathcal{D}\eta_n = \lambda_n\eta_n, \quad \int d^D x\, \eta_n^*(x)\eta_m(x) = \delta_{nm}, \tag{6}$$

such that the functional integration collapses to a product of Gaussian integrals. The determinant $\det \mathcal{D} = \prod_n \lambda_n$ is formally defined as the product of the eigenvalues. You may drop any normalization constants, as these can be absorbed into the normalization of the generating functional.

c) Let $\mathcal{D} = -\Box + \sigma$, where $\sigma = M^2 - 2g\phi^2$. Define the *heat kernel:*[1]

$$K(x|\tau|y) = \langle x|e^{-\tau\mathcal{D}}|y\rangle. \tag{7}$$

Define also $\mathcal{D}_0 = -\Box + M^2$. Using the expansion in the basis of the eigenfunctions of the operators $\mathcal{D}$ and $\mathcal{D}_0$, show that

$$\text{tr}(\ln\mathcal{D} - \ln\mathcal{D}_0) = \int d^D x \int_0^\infty \frac{d\tau}{\tau} \left(\langle x|e^{-\tau\mathcal{D}}|x\rangle - \langle x|e^{-\tau\mathcal{D}_0}|x\rangle \right). \tag{8}$$

d) Show that K obeys the following equation:

$$\frac{\partial}{\partial\tau}K(x|\tau|y) = -\mathcal{D}K(x|\tau|y) \quad \text{for } x \neq y,$$

$$\lim_{\tau\to 0} K(x|\tau|y) = \delta^{(D)}(x - y). \tag{9}$$

e) The most singular part of the operator $\mathcal{D}$ is $-\Box$. Check that the solution with $\mathcal{D} = -\Box$ takes the form

$$K_\Box(x|\tau|y) = \langle x|e^{\tau\Box}|y\rangle = \frac{1}{(4\pi\tau)^{D/2}} \exp\left(-\frac{z^2}{4\tau}\right), \quad z = x - y. \tag{10}$$

f) Next, we can write

$$K(x|\tau|y) = \frac{1}{(4\pi\tau)^{D/2}} \exp\left(-\frac{z^2}{4\tau} - M^2\tau\right) H(x|\tau|y), \tag{11}$$

where $H(x|\tau|y)$ obeys the boundary condition: $H(x|0|x) = 1$. Show that $H(x|\tau|y)$ obeys the differential equation

$$\frac{\partial}{\partial\tau}H + \frac{1}{\tau}z_\mu\partial_\mu H - \partial_\mu\partial_\mu H + \sigma' H = 0, \tag{12}$$

where $\sigma' = \sigma - M^2 = -2g\phi^2$ and $\partial_\mu = \partial/\partial x^\mu$.

---

[1] Here, ∂_μ acts on the space-time argument of the following function.

g) Expand $H(x|\tau|y)$ in a Taylor series in the variable τ:

$$H(x|\tau|y) = \sum_{n=0}^{\infty} \tau^n H_n(x|y) \,. \tag{13}$$

The H_n are called *Seeley–DeWitt coefficients*. Show that the coefficients obey the following recurrence relations:

$$(n+1+z_\mu \partial_\mu) H_{n+1} + (-\partial_\mu \partial_\mu + \sigma') H_n = 0 \,, \qquad z_\mu \partial_\mu H_0 = 0 \,. \tag{14}$$

h) Show that the solution of the recurrence relations for the first three coefficients for $x = y$ is given by

$$H_0(x|x) = 1 \,, \quad H_1(x|x) = -\sigma'(x) \,,$$

$$H_2(x|x) = \frac{1}{2} \sigma'^2(x) - \frac{1}{6} [\partial_\mu, [\partial_\mu, \sigma'(x)]] \,. \tag{15}$$

i) Show that, using the preceding equations, we can write

$$\operatorname{tr} \ln \mathscr{D} = -\frac{1}{(4\pi)^{D/2}} \int d^D x \int_0^\infty d\tau e^{-M^2 \tau} \tau^{-D/2-1} \sum_{n=0}^{\infty} \tau^n H_n(x|x) \,. \tag{16}$$

Carry out the τ integration and prove that only the terms with $n = 0, 1, 2$ give divergent contributions as $D \to 4$:

$$\operatorname{tr} \ln \mathscr{D} \Big|_{\text{div}} = L \int d^4 x (M^4 - 4M^2 g \phi^2(x) + 4g^2 \phi^4(x)) \,, \tag{17}$$

where

$$L = \frac{\mu^{D-4}}{16\pi^2} \left(\frac{1}{D-4} - \frac{1}{2} (\Gamma'(1) + \ln 4\pi) \right) \,, \tag{18}$$

and μ denotes the scale of dimensional regularization.

j) The light field ϕ is still considered as *classical,* that is, there are no loops involving ϕ. Consequently, all couplings in the effective theory are finite at this level. From the preceding expression, read off the divergent part of the heavy particle loop that contributes to the effective Lagrangian:

$$\Gamma_{\text{eff,div}} = \int d^4 x \, \mathcal{L}_{\text{eff,div}} \,. \tag{19}$$

$$\mathcal{L}_{\text{eff,div}} = -2M^2 L g \phi^2 + 2g^2 L \phi^4 + \text{const} \,. \tag{20}$$

The constant term corresponds to an infinite vacuum energy and is irrelevant.

k) Show that these divergent terms can be removed by the renormalization of the light mass and coupling constant in the original theory, rendering the parameters of the effective theory finite at this order:

$$m^2 \to m^2 + 4g M^2 L \,, \qquad \lambda \to \lambda - 2g^2 L \,. \tag{21}$$

5. β-Function in the $O(N)$ ϕ^4-Model at $N \to \infty$

Consider the Lagrangian of the $O(N)$ ϕ^4-model,

$$\mathcal{L} = \frac{1}{2} \partial_\mu \phi^a \partial^\mu \phi^a - \frac{m_0^2}{2} \phi^a \phi^a - \frac{\lambda_0}{8N} (\phi^a \phi^a)^2, \tag{1}$$

where the vector ϕ^a, $a = 1, \ldots, N$ transforms under the $O(N)$ group. The theory is explicitly invariant under $O(N)$.

In perturbation theory, the β-function is given as a Taylor series in the quartic coupling. It turns out, however, that in the limit $N \to \infty$, the theory considerably simplifies. (Here, a similarity emerges with QCD in the limit of large N_c.) Namely, in this limit, it is possible to calculate the β-function to all orders in the coupling. We shall verify this statement below.

In order to simplify the bookkeeping of the powers of N, let us first introduce an auxiliary field χ by adding a piece to the Lagrangian

$$\mathcal{L} \to \mathcal{L} + \frac{N}{2\lambda_0} \left(\chi - \frac{\lambda_0}{2N} \phi^a \phi^a - m_0^2 \right)^2. \tag{2}$$

a) Use either the canonical or the path integral formalism to show that the additional term does not change the theory.

b) Explicitly rewrite the Lagrangian in terms of the χ, ϕ^a fields. What is the propagator of χ? What are the Feynman rules?

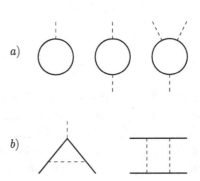

Various Green's functions of the ϕ and χ fields, denoted by solid and dashed lines, respectively. It can be shown that the one-loop diagrams with external χ legs only (a) give the leading $O(N)$ contribution, whereas the diagrams in (b) and (c) are $O(1)$.

c) Taking into account the fact that the factors of N in the Feynman diagrams arise from two sources:

- The propagator of χ is proportional to N^{-1};
- The closed loop of the ϕ^a fields is proportional to N, because of the trace over the index a,

show that the leading N contribution to the effective action $\sim N$ emerges in the *one-loop* graphs with any number of external χ fields. These diagrams are shown in Fig. 6.1a. Demonstrate that the diagrams in Figs. 6.1b,c are subleading in N.

d) In c) it was proven that, in order to get the effective action in the limit of large N, one-loop calculations are sufficient. Let us now carry out one-loop calculations in the saddle-point approximation. To this end, expand the fields in the generating functional around the classical solutions

$$\phi^a(x) = \phi_c^a(x) + h^a(x), \qquad \chi(x) = \chi_c(x) + \eta(x), \tag{3}$$

and integrate over the fluctuations $h^a(x)$, $\eta(x)$ in the path integral. For our purposes, it suffices to consider the case of constant classical fields $\phi_c^a(x) = \phi_c^a$ and $\chi_c(x) = \chi_c$ and to calculate the *effective potential*. Demonstrate that the one-loop effective potential takes the form

$$V_{\text{eff}} = -\frac{N}{2\lambda_0}\chi_c^2 + \frac{1}{2}\chi_c\phi_c^a\phi_c^a + \frac{Nm_0^2}{\lambda_0}\chi_c + \frac{N}{2}\int\frac{d^Dk}{(2\pi)^{Di}}\ln(k^2+\chi_c), \tag{4}$$

where the integral is performed in Euclidean space. Note that we have opted to tame the ultraviolet divergence in dimensional regularization.

e) Prove that in dimensional regularization (see also Exercise 2)

$$\frac{N}{2}\int\frac{d^Dk}{(2\pi)^{Di}}\ln(k^2+\chi_c) = \frac{N}{2}\chi_c^2L - \frac{N\chi_c^2}{64\pi^2}\left(\frac{3}{2}-\ln\frac{\chi_c}{\mu^2}\right), \tag{5}$$

$$L = \frac{\mu^{D-4}}{16\pi^2}\left(\frac{1}{D-4}-\frac{1}{2}(\Gamma'(1)+\ln 4\pi)\right). \tag{6}$$

f) From Eqs. (4) and (5), read off the renormalization of the quartic coupling in the $\overline{\text{MS}}$ scheme,

$$\frac{1}{\lambda_r(\mu)} = \frac{1}{\lambda_0} - L, \tag{7}$$

where $\lambda_r(\mu)$ denotes the renormalized coupling.

g) Derive the RG equation in the large-N limit, valid *to all orders in λ_r*:

$$\mu\frac{d\lambda_r(\mu)}{d\mu} = \frac{\lambda_r^2(\mu)}{16\pi^2} + O(N^{-1}). \tag{8}$$

Note: For more information, you may consult the following:

S. R. Coleman, R. Jackiw and H. D. Politzer, "Spontaneous symmetry breaking in the $O(N)$ model for large N," Phys. Rev. D **10** (1974) 2491.

6. Invariance under Field Redefinitions

This exercise serves multiple purposes. First, in a simple, exactly solvable field-theoretical model it will be seen that there exist several equivalent nonrelativistic reductions of a given relativistic theory. The Lagrangians corresponding to the different reductions can be related to each other by use of the EOM and field redefinitions and, hence, lead to the same S-matrix elements. Second, it will be verified that these equivalent theories have the same bound-state sector, even though the matching is performed in the scattering sector only. Last but not least, it will be demonstrated that, summing up $1/M$ corrections to the bound-state energies to all orders, it is possible to reproduce the relativistic result.

Consider the charged relativistic field H interacting with an external static Coulomb field $\phi(\mathbf{x})$. The Lagrangian takes the form

$$\mathcal{L}_{\text{rel}} = \partial_\mu H^\dagger \partial^\mu H - M^2 H^\dagger H + \lambda H^\dagger H \phi, \qquad \phi(\mathbf{x}) = \frac{1}{|\mathbf{x}|}. \tag{1}$$

The (bound-state) energy levels of this system are given by

$$E_n^2 = M^2 - \frac{\lambda^2}{4n^2}, \qquad n = 1, 2, \cdots. \tag{2}$$

This is nothing but the hydrogen atom spectrum that emerges after solving the Schrödinger equation with a static Coulomb potential.

The nonrelativistic Lagrangian that corresponds to a given relativistic theory is not fixed uniquely. Let h denote a nonrelativistic field, which is a positive-energy component of H. In what follows we mainly follow the discussion from

V. Antonelli, A. Gall, J. Gasser and A. Rusetsky, "Effective Lagrangians in bound state calculations," Annals Phys. **286** (2001) 108.

As shown in that paper, using different reduction techniques, we may obtain different nonrelativistic Lagrangians. For example,

$$\mathcal{L}_1 = h^\dagger \left(i\partial_t - M + \frac{\triangle}{2M} \right) h + \sum_{n=0}^\infty \frac{1}{(2M)^{2n+1}} h^\dagger \hat{O}_{2n+1} h,$$

$$\hat{O}_1 = \lambda \phi,$$

$$\hat{O}_3 = \triangle^2 + \lambda(\phi\triangle + \triangle\phi) + \lambda^2 \phi^2,$$

$$\hat{O}^5 = \frac{1}{2}\left(4\triangle^3 + \lambda(2\triangle\phi\triangle + 5\phi\triangle^2 + 5\triangle^2\phi) \right.$$

$$\left. + 2\lambda^2(2\phi^2\triangle + \triangle\phi^2 + 3\phi\triangle\phi) + 4\lambda^3\phi^3 \right) \quad \cdots, \tag{3}$$

and

$$\mathcal{L}_2 = h^\dagger \left(i\partial_t - (M^2 - \triangle - \lambda\phi)^{1/2} \right) h. \tag{4}$$

a) Expanding the square root in $\mathcal{L}_2$, show that a difference between the two Lagrangians emerges first at order $1/M^5$. Show that this difference $\delta\mathcal{L}$ is given by

$$\delta\mathcal{L} = \mathcal{L}_1 - \mathcal{L}_2 = \frac{\lambda}{64M^5} h^\dagger (\triangle^2\phi - 2\triangle\phi\triangle + \phi\triangle^2 - 2\lambda\triangle\phi^2 + 2\lambda\phi\triangle\phi)h + \cdots . \quad (5)$$

b) Summing up all $O(\lambda^0)$ insertions in the propagator of the field h, we arrive at the free propagator:

$$G^0(p_0, \mathbf{p}, \mathbf{k}) = \int dx^0 d^3\mathbf{x}\, d^3\mathbf{y}\, e^{ip_0(x_0-y_0)-i\,\mathbf{p}\mathbf{x}+i\mathbf{k}\mathbf{y}} i\langle 0|Th(x)h^\dagger(y)|0\rangle\bigg|_{O(\lambda^0)}$$

$$= \frac{(2\pi)^3 \delta^{(3)}(\mathbf{p}-\mathbf{k})}{w(\mathbf{p}) - p_0 - i\varepsilon} . \quad (6)$$

The full two-point function of the field h takes the form

$$G(p_0, \mathbf{p}, \mathbf{k}) = \int dx^0 d^3\mathbf{x}\, d^3\mathbf{y}\, e^{ip_0(x_0-y_0)-i\mathbf{p}\,\mathbf{x}+i\mathbf{k}\mathbf{y}} i\langle 0|Th(x)h^\dagger(y)|0\rangle$$

$$= \frac{(2\pi)^3 \delta^{(3)}(\mathbf{p}-\mathbf{k})}{w(\mathbf{p}) - p_0 - i\varepsilon} + \frac{(2w(\mathbf{p}))^{-1/2} T(\mathbf{p}, \mathbf{k}; p_0)(2w(\mathbf{k}))^{-1/2}}{(w(\mathbf{p}) - p_0 - i\varepsilon)(w(\mathbf{k}) - p_0 - i\varepsilon)} ,$$

$$w(\mathbf{p}) = \sqrt{M^2 + \mathbf{p}^2}, \quad (7)$$

with the Lagrangian $\mathcal{L}_1$ or $\mathcal{L}_2$. Show that the difference in the T-matrices due to $\delta\mathcal{L}$ is an off-shell effect. Namely, at $O(\lambda)$, you should find

$$T_1 - T_2 \propto \lambda \tilde{\phi}(\mathbf{p}-\mathbf{k})(\mathbf{p}^2 - \mathbf{k}^2)^2 + \cdots . \quad (8)$$

Here, $\tilde{\phi}$ is the Fourier transform of the Coulomb field. The preceding expression vanishes on the energy shell $\mathbf{p}^2 = \mathbf{k}^2$. Consequently, since we are matching only observables (S-matrix elements), we cannot distinguish between two nonrelativistic theories described by the Lagrangians $\mathcal{L}_1$ and $\mathcal{L}_2$. In general, there exist infinitely many nonrelativistic Lagrangians that coincide on-shell and differ off-shell. From the physical point of view, they are all equivalent.

c) Show that up to order $1/M^5$ these two theories can be connected by a field redefinition:

$$h \to (1+X)h, \qquad X = -\frac{1}{32M^4} (\lambda(\triangle\phi - \phi\triangle) - \lambda^2\phi^2). \quad (9)$$

You will need to use the EOM in addition to prove the equivalence.

d) Next, starting from the non-relativistic theory, we should be able to include relativistic corrections order by order and to arrive at the known relativistic result. The unperturbed solution corresponds to a nonrelativistic particle in an external Coulomb field. Using dimensional regularization to handle the ultraviolet divergences, the bound-state equation in momentum space is given by

$$\left(\bar{E}_n - M - \frac{\mathbf{p}^2}{2M}\right)\psi_n(\mathbf{p}) = -\frac{2\pi\lambda}{M}\int\frac{d^d\mathbf{k}}{(2\pi)^d}\frac{1}{|\mathbf{p}-\mathbf{k}|^2}\psi_n(\mathbf{k})\,, \tag{10}$$

where

$$\bar{E}_n = M - \frac{\lambda^2}{8Mn^2}\,, \quad \text{as } d \to 3\,. \tag{11}$$

Note that we will never need the explicit form of the wave functions.

Use the Lagrangian $\mathcal{L}_2$ and consider (dimensionally regularized) perturbation theory around this unperturbed solution. Show that the results for the binding energies agree to all orders in λ.

Hint: The unperturbed states in the Fock space are described by the vectors:

$$|n\rangle = \int\frac{d^d\mathbf{q}}{(2\pi)^d}\psi_n(\mathbf{q})|\mathbf{q}\rangle\,. \tag{12}$$

Our main goal is to show that, for any integer a,

$$\int d^dx\, h^\dagger(\triangle + \lambda\phi)^a h|n\rangle = (2M(\bar{E}_n - M))^a|n\rangle\,. \tag{13}$$

With use of the preceding formula, we may sum up relativistic corrections to all orders.

e) Show that the energy shift due to the $\delta\mathcal{L}$ vanishes at order $1/M^5$

$$\langle n|\delta\mathcal{L}|n\rangle = O(M^{-7})\,. \tag{14}$$

This demonstrates a very important property of field theory: If two theories are related by a field redefinition, the bound-state spectrum of these two theories coincides.

Hint: Note that in these calculations some of the terms of $O(\lambda^2)$ will be canceled by terms of $O(\lambda)$ after using the EOM.

7. Reparameterization Invariance in HQET

Heavy Quark Effective Theory (HQET) is used for the description of the mesons and baryons containing heavy c, b quarks along with the light ones. A nice introduction to the subject is given in the following references:

M. Neubert, "Heavy quark symmetry," Phys. Rept. **245** (1994) 259,

A. V. Manohar and M. B. Wise, "Heavy quark physics," Camb. Monogr. Part. Phys. Nucl. Phys. Cosmol. **10** (2000) 1,

T. Mannel, "Effective Field Theories in Flavour Physics," Springer Tracts in Modern Physics, Springer-Verlag Berlin Heidelberg (2004).

Although HQET is not a subject of the present book, we decided to include the following exercise that nicely demonstrates the power and beauty of the reparameterization invariance. To solve this exercise, no prior knowledge in HQET is required.

The mathematical structure of the heavy quark effective Lagrangian is similar to the Lagrangian of HBChPT considered in this book, and the steps to derive it are the same. One starts from QCD and splits the heavy quark fields into the positive- and negative-energy components. Integrating out the latter, we arrive at the Lagrangian that is built of the fields h_v (the positive-energy components of the quark fields) and the gluon field G_μ. As in the HBChPT Lagrangian, v_μ represents a timelike unit vector. Different terms in the Lagrangian are ordered according to their mass dimension, and the heavy quark mass m_Q plays the role of the hard scale.

The Lagrangian at $O(1/m_Q)$ reads

$$\mathcal{L} = \bar{h}_v(iv \cdot D)h_v + \frac{1}{2m_Q}\bar{h}_v(iD_\perp)^2 h_v - c_F \frac{g}{4m_Q}\bar{h}_v \sigma_{\alpha\beta}F^{\alpha\beta}h_v, \tag{1}$$

where

$$D_\perp^\mu = D^\mu - v^\mu v \cdot D, \quad [D^\mu, D^\nu] = gF^{\mu\nu}. \tag{2}$$

Here, g is the strong coupling constant, $D_\mu = \partial_\mu - igG_\mu^a T^a$ is the covariant derivative and $F_{\mu\nu} = -iF_{\mu\nu}^a T^a$ is the gluon field strength tensor. The quantity c_F is a constant, and T^a denotes the generators of the $SU(3)_c$ group. The last term in the Lagrangian describes the so-called chromomagnetic moment of a heavy quark. The terms with light quarks are not shown explicitly.

a) Show on general grounds that $\mathcal{L}$ must be invariant under the redefinition

$$v \to v + \frac{\varepsilon}{m_Q}, \quad h_v \to e^{i\varepsilon \cdot x}\left(1 + \frac{\slashed{\varepsilon}}{2m_Q}\right)h_v. \tag{3}$$

Here, ε_μ is a small parameter (four-vector) and $v \cdot \varepsilon = 0$. The requirement that the physics must be independent of the particular choice of v is referred to as reparameterization invariance of the theory.

b) Show that under the given transformation,

$$\bar{h}_v(iv \cdot D)h_v \to \bar{h}_v(iv \cdot D)h_v + \frac{1}{m_Q}\bar{h}_v(i\varepsilon \cdot D)h_v + O(1/m_Q^2), \tag{4}$$

$$\frac{1}{2m_Q}\bar{h}_v(iD_\perp)^2 h_v \to \frac{1}{2m_Q}\bar{h}_v(iD_\perp)^2 h_v - \frac{1}{m_Q}\bar{h}_v(i\varepsilon \cdot D)h_v + O(1/m_Q^2), \tag{5}$$

$$\frac{g}{4m_Q}\bar{h}_v \sigma_{\alpha\beta}G^{\alpha\beta}h_v \to \frac{g}{4m_Q}\bar{h}_v \sigma_{\alpha\beta}G^{\alpha\beta}h_v + O(1/m_Q^2). \tag{6}$$

This proves the invariance of $\mathcal{L}$ explicitly up to terms of $O(1/m_Q^2)$. What conclusion can you draw for the coefficients of the kinetic energy and magnetic moment operator?

8. The Gross–Neveu Model

The Gross–Neveu model is a model of fermions in two space-time dimensions that reveals a discrete chiral symmetry. It was introduced in the following paper:

D. J. Gross and A. Neveu, "Dynamical symmetry breaking in asymptotically free field theories," Phys. Rev. D **10** (1974) 3235.

The Lagrangian of the model is given by

$$\mathcal{L} = \bar{\psi}_a i \partial\!\!\!/ \psi_a + \frac{g^2}{2} (\bar{\psi}_a \psi_a)^2, \tag{1}$$

with $a = 1, \ldots, N$. The kinetic term of two-dimensional fermions is built from the matrices γ^μ that satisfy the two-dimensional Dirac algebra. These are 2×2 matrices

$$\gamma^0 = \sigma^2, \quad \gamma^1 = i\sigma^1, \tag{2}$$

where the σ^i are Pauli matrices. Define

$$\gamma^5 = \gamma^0 \gamma^1 = \sigma^3. \tag{3}$$

This matrix anticommutes with γ^μ.

a) Show that this theory is invariant under

$$\psi_a \to \gamma_5 \psi_a, \tag{4}$$

and that this symmetry forbids the appearance of a fermion mass.

b) Show that this theory is renormalizable in two dimensions (at the level of dimensional analysis).

c) Show that the path integral for this theory can be represented in the following form

$$\int d\psi d\bar{\psi} \exp\left(i \int d^2x \mathcal{L} \right)$$

$$= \int d\sigma d\psi d\bar{\psi} \exp\left(i \int d^2x \left(\bar{\psi}_a i \partial\!\!\!/ \psi_a - \sigma \bar{\psi}_a \psi_a - \frac{1}{2g^2} \sigma^2 \right) \right), \tag{5}$$

where $\sigma(x)$ is a new auxiliary scalar field with no kinetic energy terms. This trick goes under the name of *Hubbard–Stratonovich transformation* (though it really is nothing but a Gaussian quadrature).

d) Compute the leading correction to the effective potential for σ by integrating over the fermion fields. Renormalize by minimal subtraction.

e) Ignoring two-loop contributions, minimize this potential. Show that the σ-field acquires a vacuum expectation value that breaks the discrete chiral symmetry. Convince yourself that this result does not depend on the particular renormalization condition chosen.

9. Ward Identities in QED

The purpose of this exercise is to demonstrate how the technique described in Chapter 3 can be used to derive Ward identities in QED, leading to the well-known results.

The generating functional in QED in an arbitrary covariant gauge is given by

$$Z(J,\eta,\bar{\eta}) = \int d\mathscr{A}_\mu d\psi d\bar{\psi} \exp\left\{i \int d^4x \left(\bar{\psi}(i\slashed{D}-m)\psi - \frac{1}{4}\mathscr{F}_{\mu\nu}\mathscr{F}^{\mu\nu}\right.\right.$$

$$\left.\left. -\frac{1}{2\alpha}(\partial_\mu\mathscr{A}^\mu)^2 + J_\mu\mathscr{A}^\mu + \bar{\eta}\psi + \bar{\psi}\eta\right)\right\}. \tag{1}$$

Here, α denotes the gauge fixing parameter and $D_\mu = \partial_\mu + ie\mathscr{A}_\mu$. The quantities $J_\mu, \bar{\eta}, \eta$ are external sources, associated with the photon, electron and positron fields, respectively. The Green's functions are obtained by differentiating the generating functional with respect to the pertinent sources and putting the sources tò zero at the end.

In order to simplify the notation, in what follows we use:

$$S = \int d^4x \left(\bar{\psi}(i\slashed{D}-m)\psi - \frac{1}{4}\mathscr{F}_{\mu\nu}\mathscr{F}^{\mu\nu} - \frac{1}{2\alpha}(\partial_\mu\mathscr{A}^\mu)^2\right). \tag{2}$$

a) Perform infinitesimal gauge transformations in the path integral:

$$\psi(x) \to (1+ie\theta(x))\psi(x),$$

$$\bar{\psi}(x) \to \bar{\psi}(x)(1-ie\theta(x)),$$

$$\mathscr{A}_\mu(x) \to \mathscr{A}_\mu(x) - \partial_\mu\theta(x), \tag{3}$$

where $\theta(x)$ denotes the parameter of the infinitesimal gauge transformation. The integration measure is invariant under this transformation. Expand the integrand up to first order in θ and show that one obtains the following Ward identity:

$$\int d\mathscr{A}_\mu d\psi d\bar{\psi} \exp\left\{iS + i\int d^4x \left(J_\mu\mathscr{A}^\mu + \bar{\eta}\psi + \bar{\psi}\eta\right)\right\}$$

$$\times \left(\frac{1}{\alpha}\Box(\partial^\mu A_\mu(y)) + \partial^\mu J_\mu(y) + ie\bar{\eta}(y)\psi(y) - ie\bar{\psi}(y)\eta(y)\right) = 0. \tag{4}$$

b) Differentiate the above identity with respect to $J_\mu(z)$ and set $J_\mu = \eta = \bar{\eta} = 0$ at the end. Show that the resulting identity is given by

$$\int d\mathscr{A}_\mu d\psi d\bar{\psi}\, e^{iS} \left(-\frac{i}{\alpha}\mathscr{A}_\lambda(z)\Box_y(\partial_y^\mu\mathscr{A}_\mu(y)) - \partial_y^\mu\delta^{(4)}(z-y)g_{\lambda\mu}\right) = 0. \tag{5}$$

c) Use

$$\int d\mathscr{A}_\mu d\psi d\bar{\psi}\, e^{iS} = 1,$$

$$i\int d\mathscr{A}_\mu d\psi d\bar{\psi}\, e^{iS}\mathscr{A}_\lambda(z)\mathscr{A}_\mu(y) = D_{\lambda\mu}(z-y), \tag{6}$$

where $D_{\lambda\mu}(z-y)$ is the photon propagator $i\langle 0|T\mathscr{A}_\lambda(z)\mathscr{A}_\mu(y)|0\rangle$, and derive the Ward identity

$$-\frac{1}{\alpha}\Box_y(\partial_y^\mu D_{\lambda\mu}(z-y)) = \partial_y^\mu\delta^4(z-y)g_{\lambda\mu}. \tag{7}$$

d) Rewrite the Ward identity in momentum space and show that it has the solution

$$D_{\lambda\mu}(p) = \left(g_{\lambda\mu} - \frac{p_\lambda p_\mu}{p^2}\right)D(p^2) + \frac{\alpha p_\lambda p_\mu}{(p^2)^2}. \tag{8}$$

In other words, the longitudinal component of the photon propagator is not altered by radiative corrections.

e) Using the expression for the free photon propagator,

$$D_{\lambda\mu}^{(0)}(p) = \left(g_{\lambda\mu} - \frac{p_\lambda p_\mu}{p^2}\right)\frac{1}{p^2} + \frac{\alpha p_\lambda p_\mu}{(p^2)^2}, \tag{9}$$

show that the self-energy of the photon, which enters the Dyson–Schwinger equation,

$$D = D^{(0)} + D^{(0)}\Pi D, \tag{10}$$

is transverse,

$$\Pi_{\lambda\mu}(p) = (g_{\lambda\mu}p^2 - p_\lambda p_\mu)\Pi(p^2), \tag{11}$$

and

$$D(p^2) = \frac{1}{p^2(1 - \Pi(p^2))}. \tag{12}$$

This means that the radiative corrections do not alter the position of the pole at $p^2 = 0$: The photon stays massless in the presence of interactions thanks to gauge invariance.

f) Derive the Ward identity for the electron–positron–photon vertex by differentiating the generating functional with respect to the pertinent sources and compare it to the known result from your favorite textbook in Quantum Field Theory.

10. The Axial Current in Two Dimensions

In this exercise, we learn in a simple example how anomalies emerge from short-distance singularities. We consider massless electrodynamics with one fermion flavor, described by a field $\psi(x)$ in two space-time dimensions (Schwinger model). As we know, putting two fields "on top of each other" (at the same space-time point) can be dangerous, as this can provoke divergences in quantum field theories. Therefore, for the axial-vector current, we introduce the following gauge-invariant definition:

$$A^\mu = \bar\psi(x+\varepsilon)\gamma^\mu\gamma^5 \exp\left(-ie\int_x^{x+\varepsilon} dz_\nu \mathscr{A}^\nu(z)\right)\psi(x), \tag{1}$$

where ε^μ is an infinitesimal displacement vector and the exponential, containing the vector potential $\mathscr{A}^\mu$, is required to make this expression gauge invariant.

a) Applying the product rule will lead to three terms when differentiating the current. Show that to first order in ε^μ

$$\partial_\mu A^\mu = ie\bar\psi(x+\varepsilon)\gamma^\mu\gamma^5\varepsilon^\nu\mathscr{F}_{\nu\mu}(x)\psi(x), \tag{2}$$

where $\mathscr{F}_{\mu\nu}$ is the field strength tensor for the potential $\mathscr{A}_\mu$.

b) Then, show that

$$\psi(y)\bar{\psi}(z) = i \int \frac{d^2k}{(2\pi)^2} e^{-ik(y-z)} \frac{\not{k}}{k^2} + \cdots = \frac{-1}{4\pi} \not{\partial} \ln(-\mu^2(y-z)^2) + \cdots, \qquad (3)$$

where the ellipses represent the less singular terms and μ is an arbitrary scale. Note that the derivative acts on the difference $y - z$.

c) Using this result, show that

$$\bar{\psi}(x+\varepsilon)\Gamma\psi(x) = \frac{i\varepsilon_\mu}{2\pi\varepsilon^2} \text{Tr}(\gamma^\mu\Gamma) + \cdots, \qquad (4)$$

where Γ denotes any combination of γ-matrices (in our case, $\Gamma = \gamma^\mu\gamma^5$). The ellipses again denote the less singular terms in ε^μ.

d) Our definition of the limit $\varepsilon^\mu \to 0$ implies averaging over all directions in space-time. Using the expression for two dimensions, $\langle \varepsilon^\mu\varepsilon^\nu \rangle = \frac{1}{2}g^{\mu\nu}\varepsilon^2$, show that

$$\partial_\mu A^\mu = -\frac{e}{2\pi}\varepsilon^{\mu\nu}\mathscr{F}_{\mu\nu}. \qquad (5)$$

This is an axial anomaly in two dimensions. Here, $\varepsilon^{\mu\nu}$ denotes the totally antisymmetric tensor in two dimensions, $\varepsilon^{01} = -\varepsilon^{10} = 1$ and $\varepsilon^{00} = \varepsilon^{11} = 0$.

11. Zero Modes of the Dirac Operator

The purpose of this exercise is to gain a deeper understanding of the zero modes of the massless Euclidean Dirac operator, which are directly related to the anomaly via the Atiyah–Singer index theorem. To simplify the algebra, we consider an $SU(2)$ gauge theory, that is, $N_c = 2$. We shall demonstrate that for the particular Euclidean gluon field configurations that will be considered in what follows, the massless Dirac operator has a zero mode. These modes are the topologically nontrivial classical field configurations, the so-called instantons, which are given by

$$G_\mu^a(x) = \frac{2\eta_{\mu\nu}^a x_\nu}{x^2 + \rho^2}. \qquad (1)$$

Here, $a = 1, 2, 3$ is the color index of the gluon field in the adjoint representation of the $SU(2)$ group, ρ is an arbitrary parameter (length scale) characterizing the size of the instanton and the $\eta_{\mu\nu}^a$ are the so-called 't Hooft symbols:

$$\eta_{00}^a = 0, \qquad \eta_{ij}^a = \varepsilon_{aij}, \qquad \eta_{0i}^a = -\delta_{ai}, \qquad \eta_{i0}^a = \delta_{ai}, \qquad i, j = 1, 2, 3. \qquad (2)$$

Here, δ_{ij} and ε_{ijk} denote the Kronecker and Levi–Civita symbols, respectively.

a) Using the given expression for the gluon field, calculate the gluon field strength tensor and show that

$$F_{\mu\nu}(x) = -\frac{4\rho^2 \eta_{\mu\nu}^a}{(x^2+\rho^2)^2}. \tag{3}$$

b) The dual tensor is defined by

$$\tilde{F}_{\mu\nu} = \frac{1}{2}\varepsilon_{\mu\nu\alpha\beta}F_{\alpha\beta}. \tag{4}$$

Prove that the just defined one-instanton field configuration is self-dual, that is,

$$\tilde{F}_{\mu\nu} = F_{\mu\nu}. \tag{5}$$

c) The winding number (Pontryagin index) Q for any field configuration that leads to a finite action functional is defined as

$$Q = \frac{1}{16\pi^2}\int d^4x\, \mathrm{tr}\left[F_{\mu\nu}\tilde{F}_{\mu\nu}\right]. \tag{6}$$

It can be shown in general that this number should be an integer, as it is conserved. What is its value for the one-instanton configuration? The configurations that correspond to a nonzero winding number are said to be nontrivial.

d) Using the trivial identity,

$$\int d^4x\, \mathrm{tr}\left[(F_{\mu\nu}-\tilde{F}_{\mu\nu})(F_{\mu\nu}-\tilde{F}_{\mu\nu})\right] \geq 0, \tag{7}$$

show that the purely gluonic action functional obeys the bound:

$$S = \frac{1}{2g^2}\int d^4x\, \mathrm{tr}\left[F_{\mu\nu}F_{\mu\nu}\right] \geq \frac{8\pi^2 Q}{g^2}. \tag{8}$$

Moreover, it is clear that the self-dual instanton configuration is the solution of the equations of motion in the purely gluonic theory, since it minimizes the action functional in a sector with a given Q!

e) Now, consider the Euclidean Dirac equation in the one-instanton background field and check that

$$\psi_0(x) = \frac{\rho}{\pi(x^2+\rho^2)^{3/2}}\begin{pmatrix}\varepsilon_{m\sigma}\\ 0\end{pmatrix} \tag{9}$$

is the zero mode of this equation:

$$\gamma_\mu(\partial_\mu + G_\mu)\psi_0(x) = 0. \tag{10}$$

Here, $m = 1, 2$ are color and $\sigma = 1, 2$ spinor indices, respectively. The quantity $\varepsilon_{m\sigma}$ denotes an antisymmetric tensor in two-dimensional space.

Hint: Use the chiral representation of the Dirac γ-matrices:

$$\gamma_\mu = \begin{pmatrix}0 & -\sigma_\mu^\dagger\\ \sigma_\mu & 0\end{pmatrix}, \qquad \sigma_\mu = (i, \boldsymbol{\sigma}), \tag{11}$$

and

$$\gamma_5 = \gamma_0\gamma_1\gamma_2\gamma_3 = \begin{pmatrix}1 & 0\\ 0 & -1\end{pmatrix}. \tag{12}$$

The matrices $\sigma_\mu, \sigma_\mu^\dagger$ satisfy the relations

$$\sigma_\mu^\dagger \sigma_\nu + \sigma_\nu^\dagger \sigma_\mu = \sigma_\mu \sigma_\nu^\dagger + \sigma_\nu \sigma_\mu^\dagger = 2\delta_{\mu\nu},$$

$$\sigma_\mu^\dagger \sigma_\nu - \sigma_\nu^\dagger \sigma_\mu = 2i\eta_{\mu\nu}^a \sigma^a,$$

$$\sigma_\mu \sigma_\nu^\dagger - \sigma_\nu \sigma_\mu^\dagger = 2i\bar\eta_{\mu\nu}^a \sigma^a,$$

$$\sigma_\mu^\dagger \sigma_\nu \sigma_\mu^\dagger = -2\sigma_\nu^\dagger, \qquad \sigma_\mu \sigma_\nu^\dagger \sigma_\mu = -2\sigma_\nu,$$

$$\sigma_\mu^T = -\sigma_2 \sigma_\mu^\dagger \sigma_2, \tag{13}$$

where

$$\bar\eta_{00}^a = 0, \qquad \bar\eta_{ij}^a = \varepsilon_{aij}, \qquad \bar\eta_{0i}^a = \delta_{ai}, \qquad \bar\eta_{i0}^a = -\delta_{ai}, \qquad i,j = 1,2,3. \tag{14}$$

f) Show that $\psi_0(x)$ is a *right-handed* solution, that is,

$$\gamma_5 \psi_0(x) = \psi_0(x). \tag{15}$$

More information can be found in the following textbooks:

R. Rajaraman, Solitons and Instantons: An Introduction to Solitons and Instantons in Quantum Field Theory, North Holland Publishing Company, Amsterdam (1982).

A. V. Smilga, Lectures on Quantum Chromodynamics, World Scientific (2001).

12. Spontaneous Symmetry Breaking in a Model with a Matrix Field

Consider a field theory described in terms of the elements of a complex $N \times N$ matrix M by a Lagrangian,

$$\mathcal{L} = \mathrm{tr}(\partial_\mu M \partial^\mu M^\dagger) - k\,\mathrm{tr}(M^\dagger M) - g\,\mathrm{tr}(M^\dagger M M^\dagger M), \tag{1}$$

where tr denotes the matrix trace and $g > 0$.

a) Show that this theory is invariant under the symmetry group $U(N) \times U(N)/U(1)$ for transformations given by $M \to AMB^{-1}$ for $A, B \in U(N)$ and where $U(1)$ corresponds to $A = B = e^{i\theta}\mathbb{1}$. (Note that if H is a subgroup of G, then G/H is a group if H belongs to the center of G, i.e $hg = gh$ for all $h \in H$, $g \in G$.)

b) Show that if $k < 0$, spontaneous symmetry breaking occurs and that in the ground state, $M_0^\dagger M_0 = v^2 \mathbb{1}$ for some v. What is the unbroken symmetry group and how many Goldstone modes are there?

c) If $\mathcal{L} \to \mathcal{L} + \mathcal{L}'$, where

$$\mathcal{L}' = \lambda\left(\det(M) + \det(M^\dagger)\right), \tag{2}$$

what is the symmetry group and how many Goldstone modes are there now after spontaneous symmetry breaking? (Hint: Assume the ground state still satisfies $M_0^\dagger M_0 = v^2 \mathbb{1}$.)

13. The Veltman–Sutherland Theorem

In this exercise, one checks explicitly that the Veltman–Sutherland theorem, which predicts that the pion decay amplitude into two photons vanishes in the chiral limit, is violated by triangle graphs. These diagrams are responsible for the anomaly. The calculations are carried out in a simplified version of the σ-model, where the fermion field describes the proton (with charge $+1$), the neutral pseudoscalar field π corresponds to the π^0-meson and σ is a hypothetical scalar particle. The Lagrangian of this model is given by

$$\mathcal{L} = \bar{N}(i\partial\!\!\!/ + g(\sigma - i\pi\gamma^5))N$$

$$+ \frac{1}{2}((\partial\pi)^2 + (\partial\sigma)^2) - \frac{\mu^2}{2}(\pi^2 + \sigma^2) - \frac{\lambda}{4}(\pi^2 + \sigma^2)^2 + c\sigma. \qquad (1)$$

Note that the nucleon mass m is taken to vanish, in order to obey chiral symmetry. If the parameter $\mu^2 < 0$, spontaneous breaking of chiral symmetry occurs and the pion becomes massless (in the limit $c \to 0$), whereas the σ and the nucleon acquire mass.

a) Find the vacuum expectation value v of the field σ in tree approximation.
b) Find the masses of all particles in tree approximation.
c) What is the symmetry of the theory, and what is its breaking pattern? Using Noether's theorem, find the expression of the vector and axial-vector currents. Find the divergence of these currents that follows from the Noether theorem.
d) Using minimal substitution, equip the Lagrangian with photons and calculate the decay amplitude $\pi^0(q) \to \gamma(k_1)\gamma(k_2)$ at one loop. Does this amplitude vanish at $q^2 \to 0$?
e) Argue that the nonvanishing amplitude corresponds to the anomaly. To this end, use PCAC and relate the decay amplitude to the three-point function of two vector and one axial-vector currents. Furthermore, write the most general expression of this three-point function in terms of scalar amplitudes and argue that, according to the PCAC hypothesis, the pion decay amplitude should vanish in the chiral limit. How does this result change, if the anomalous term is taken into account in the divergence of the axial-vector current?
f) Calculate the decay width, express everything in terms of physical observables (F_π, M_π, etc.) and find its numerical value. Compare with the experimental value that can be found in the literature.

14. Vacuum Expectation Value of the σ beyond the Tree Level

The aim of the present exercise is to get familiar with the renormalization of the vacuum expectation value and, eventually, the Goldstone theorem beyond the tree level. The explicit one-loop calculations will be carried out within the $U(1)$ σ-model, whose Lagrangian takes the form

$$\mathcal{L} = \frac{1}{2}(\partial\sigma)^2 + \frac{1}{2}(\partial\pi)^2 - \frac{m_0^2}{2}(\sigma^2 + \pi^2) - \frac{\lambda}{4}(\sigma^2 + \pi^2)^2. \qquad (1)$$

a) Assume that the π-field has a vanishing vacuum expectation value. Minimize the tree-level effective potential for the σ field and find the vacuum expectation value $v = \langle 0|\sigma|0\rangle$ at tree level.

b) Calculate the one-loop effective potential for the σ-field by using the saddle-point technique. You may again assume that the vacuum expectation value of the π-field vanishes.

c) Minimize the one-loop effective potential for the σ-field and evaluate the vacuum expectation value v at one loop. Carry out the renormalization of v at one loop in the $\overline{\text{MS}}$ scheme.

15. Redundant Terms in the Chiral Lagrangian

In Section 4.9.3 we derived the equation of motion from the $O(p^2)$ Lagrangian of $N_f = 3$ flavor ChPT. This EOM has the form

$$O_{\text{EOM}}^{(2)} = (D_\mu D^\mu U)U^\dagger - U D_\mu D^\mu U^\dagger + U\chi^\dagger - \chi U^\dagger - \frac{1}{3}\langle U\chi^\dagger - \chi U^\dagger\rangle \mathbb{1} = 0. \quad (1)$$

Furthermore, the $O(p^4)$ Lagrangian is explicitly displayed in Section 4.7; see Eq. (4.196). The list of operators does not include, for example, the operator

$$O_4' = \langle (D_\mu D^\mu U)\chi^\dagger + \chi(D_\mu D^\mu U^\dagger)\rangle. \quad (2)$$

Show that this term is redundant, that is, can be eliminated with the use of the EOM in favor of terms that are already present in the Lagrangian.

Hint: Consider the following expression that vanishes for the $O(p^2)$ EOM:

$$\langle (\chi U^\dagger - U\chi^\dagger)O_{\text{EOM}}^{(2)}\rangle = 0. \quad (3)$$

16. $\pi\pi$ Scattering at Tree Level in the Nonlinear σ-Model

The Lagrangian of the nonlinear σ-model in the two-flavor case is given by

$$\mathcal{L} = \frac{F_\pi^2}{4}\langle \partial_\mu U \partial^\mu U^\dagger\rangle + \frac{F_\pi^2 M_\pi^2}{4}\langle U^\dagger + U\rangle. \quad (1)$$

The 2×2 unitary matrix U is expressed through the pion fields $\pi^i(x)$. Different parameterizations of U are possible in terms of the pion fields (note that we use F_π, M_π instead of F, M everywhere; at tree level this is allowed):

- σ-model parameterization

$$U = \sqrt{1 - \pi^2/F_\pi^2} + i\tau\pi/F_\pi. \quad (2)$$

- Exponential parameterization

$$U = \exp\left(i\boldsymbol{\tau}\boldsymbol{\pi}/F_\pi\right). \tag{3}$$

- General parameterization

$$U = \exp\left(ig(\boldsymbol{\pi}^2/F_\pi^2)\boldsymbol{\tau}\boldsymbol{\pi}/F_\pi\right), \tag{4}$$

where $g(x)$ is a regular function that obeys $g(0) = 1$.

The scattering amplitude $\pi^i(p_1)\pi^j(p_2) \to \pi^k(p_3)\pi^l(p_4)$ is given by

$$T^{ij,kl}(s,t,u) = \delta^{ij}\delta^{kl}A(s,t,u) + \delta^{ik}\delta^{jl}A(t,u,s) + \delta^{il}\delta^{jk}A(u,s,t), \tag{5}$$

where s,t,u are the Mandelstam variables $s = (p_1 + p_2)^2$, $t = (p_1 - p_3)^2$ and $u = (p_1 - p_4)^2$, with $s + t + u = 4M_\pi^2$.

a) Expanding the Lagrangian in pion fields up to the fourth order, verify that, using the partial integration and EOM, the Lagrangians in different parameterizations can be reduced to each other. Furthermore, evaluate the amplitude $A(s,t,u)$ at tree level in different parameterizations and verify explicitly that the answer does not depend on the parameterization used.

b) Two pions, each with isospin 1, can form states with total isospin $I = 0,1,2$. This quantity is conserved. Let us calculate the scattering amplitudes T_I, corresponding to different total isospin. To this end, first define the projectors onto different total isospin:[2]

$$P_0^{ij,kl} = \frac{1}{3}\delta^{ij}\delta^{kl},$$

$$P_1^{ij,kl} = \frac{1}{2}\left(\delta^{ik}\delta^{jl} - \delta^{il}\delta^{jk}\right),$$

$$P_2^{ij,kl} = \frac{1}{2}\left(\delta^{ik}\delta^{jl} + \delta^{il}\delta^{jk}\right) - \frac{1}{3}\delta^{ij}\delta^{kl}, \tag{6}$$

and prove that

$$P_I^{ij,rs}P_{I'}^{rs,kl} = \delta_{II'}P_I^{ij,kl}. \tag{7}$$

Express now the isospin amplitudes T_I in terms of the amplitudes A from Eq. (5) and give these explicitly.

c) The partial-wave expansion of the amplitudes takes the form

$$T^I(s,t) = 32\pi \sum_{\ell=0}^{\infty}(2\ell + 1)P_\ell(\cos\theta)T_\ell^I(s),$$

$$s = 4(M_\pi^2 + p^2), \qquad t = -2p^2(1 - \cos\theta), \tag{8}$$

---

[2] You can derive these formulae, using known values of the Clebsch–Gordan coefficients for the $SU(2)$ group.

and ℓ labels the angular momentum of the $\pi\pi$-system, $\ell = 0,1,2,\ldots$. Further, unitarity implies that

$$T_\ell^I = \frac{\sqrt{s}}{4ip} \left(e^{2i\delta_\ell^I(s)} - 1 \right), \tag{9}$$

where $\delta_\ell^I(s)$ denotes the (real) $\pi\pi$ scattering phase shifts.

In the S-wave ($\ell = 0$), the expansion of the phase shift near threshold takes the form (the well-known effective range expansion)

$$p \cot \delta_0^I(s) = -\frac{1}{a^I} + \frac{1}{2} r^I p^2 + O(p^4). \tag{10}$$

Express the a^I through the value of the amplitude at threshold. Using this expression, obtain the S-wave scattering lengths with total isospin $I = 0,2$. The S-wave scattering length with $I = 1$ vanishes. Why? Calculate the tree-level S-wave $\pi\pi$ scattering lengths.

17. Unitarity Bound in $\pi\pi$ Scattering

Below the first inelastic threshold, the unitarity condition for the $\pi\pi$ partial-wave scattering amplitudes with the definite isospin $I = 0,1,2$ reads

$$\operatorname{Im} T_\ell^I(s) = \left(\frac{s - 4M_\pi^2}{s} \right)^{1/2} |T_\ell^I(s)|^2. \tag{1}$$

a) How is Eq. (1) modified above the first inelastic threshold? What is the energy corresponding to this threshold?

b) Show that Eq. (1) leads to the inequality (unitarity bound)

$$\frac{s - 4M_\pi^2}{s} |T_\ell^I(s)|^2 \leq 1. \tag{2}$$

c) At which energy is this bound violated by the tree-level amplitude in the partial wave $\ell = I = 0$ calculated in Chiral Perturbation Theory? The expression of this amplitude is given by

$$T_0^0(s) = \frac{2s - M_\pi^2}{32\pi F_\pi^2}, \tag{3}$$

with $F_\pi \simeq 92.2$ MeV the weak pion decay constant.

18. Pion Mass Difference in Pure QCD

Using the invariance of QCD under charge conjugation, prove that the mass difference between the charged and neutral pions is of second order in $m_d - m_u$, that is,

$$M_{\pi^\pm}^2 - M_{\pi^0}^2 = O((m_d - m_u)^2). \tag{1}$$

The electromagnetic interactions are assumed to be switched off.

To this end,

a) Write the part of the Hamiltonian that is responsible for the mass splitting. How does this Hamiltonian transform under charge conjugation?
b) Write the mass shift of π^+, π^0, π^- mesons at first order in $m_d - m_u$. Use isospin symmetry to understand how these shifts are related to each other.
c) Use invariance under charge conjugation for the same purpose. Draw conclusions.

In addition,

d) At lowest order, the Goldstone boson masses are described by the part of the Lagrangian proportional to $\langle \chi^\dagger U + U^\dagger \chi \rangle$. Show that, in the case of $N_f = 3$ flavors, the mass differences in the isospin multiplets are expressed through the constants d^{ijk}. Verify that the absence of the linear mass splitting between the charged and neutral pions is related to the fact that $d^{3ij} = 0$ for $i, j = 1, 2, 3^3$.

19. The Decay $\eta \to \pi^+ \pi^- \pi^0$

Show that $\eta \to \pi^+ \pi^- \pi^0$ is forbidden when the isospin and C-parity are conserved, but is allowed when the isospin breaking is turned on. Determine the possible values of total isospin and angular momenta for the final pions, assuming parity conservation as well.

Note: There are two angular momenta: ℓ denotes the angular momentum of the $\pi^+ \pi^-$ pair, and ℓ' is the angular momentum of π^0 with respect to the center of mass of the $\pi^+ \pi^-$ pair.

a) The simplest argument, when the isospin is not broken, utilizes the G-parity. Give this argument.[4]
b) Let us now invent an argument that is not based on the conserved G-parity. Assume first that the isospin is conserved. Consider the $\pi^+ \pi^-$ pair in the final state. What should be the total isospin of this pair?
c) Consider the wave function of the $\pi^+ \pi^-$ system in the final state. What is the angular momentum ℓ of this pair?
d) For simplicity, instead of the process $\eta \to \pi^0 \pi^+ \pi^-$, we could consider the crossed process $\eta \pi^0 \to \pi^+ \pi^-$. What is the C-parity of the state $|\eta \pi^0\rangle$? What is the C-parity of the state $\pi^+ \pi^-$ with the antisymmetric wave function? Can the process $\eta \to \pi^0 \pi^+ \pi^-$ take place, if charge conjugation is a good symmetry?
e) Assume that parity is conserved. How does this restrict ℓ, ℓ'?

---

[3] In general, for any N, the traceless, Hermitean generators of the $SU(N)$ group T^a, $a = 1, \ldots, N_f^2$, fulfill the relation $\{T^a, T^b\} = \delta^{ab} \mathbb{1}_N/N + \sum_{c=1}^{N^2-1} d^{abc} T^c$ in the fundamental representation. However, due to its structure, in $SU(2)$, the term $\sim d^{abc}$ does not exist.

[4] $G = C \exp(i\pi I_2)$ parity is a combination of the charge conjugation C and the isospin rotation $\exp(i\pi I_2)$ around the second axis by π. It applies to the whole isospin multiplets with zero average charge.

f) Finally, assume that the isospin is not conserved in the decay vertex, but it is still assumed in the final-state interactions. Adjust your argumentation in this case. What can we say about the total isospin of the three pions and possible values of ℓ, ℓ' in this case?

20. The Amplitude of the $\eta \to 3\pi$ Decays

a) Using the lowest order $N_f = 3$ chiral Lagrangian, calculate the $\eta \to 3\pi$ decay amplitude and verify that it is proportional to the quark mass difference $m_d - m_u$.

b) We could single out this prefactor and calculate the rest in the isospin limit $m_u = m_d$. Show that in the rest-frame of η the amplitude for the decay $\eta \to \pi^+\pi^-\pi^0$ can be written in the form

$$A = A_0\left(1 + \sigma\left(\frac{3T_0}{Q} - 1\right)\right), \tag{1}$$

where T_0 is the kinetic energy of the π^0 and $Q = M_\eta - 3M_\pi$. The quantities A_0 and σ are independent of momenta. The latter is also termed *the slope*. Derive the expressions for A_0 and σ at this order.

c) Compare your results with experiment. You may use the experimental result for the slope, for example, from

> F. Ambrosino et al. [KLOE Collaboration], "Determination of $\eta \to \pi^+\pi^-\pi^0$ Dalitz plot slopes and asymmetries with the KLOE detector," JHEP **0805** (2008) 006.

d) Not only the quark mass difference, but also the electromagnetic corrections can cause isospin breaking. Check that at $O(p^2)$ the $\eta \to 3\pi$ decay proceeds solely through the strong isospin breaking (Sutherland theorem).

21. Calculation of F_π at One Loop

Consider the two-point function of the axial-vector current $A_\mu^i = \bar{\psi}\gamma_\mu\gamma_5\frac{1}{2}\tau^i\psi$:

$$\delta^{ik}D_{\mu\nu}(p) = i\int dx\, e^{ipx}\,\langle 0|TA_\mu^i(x)A_\nu^k(0)|0\rangle. \tag{1}$$

a) Show that, as $p^2 \to M_\pi^2$,

$$D_{\mu\nu}(p) \to \frac{p_\mu p_\nu F_\pi^2}{M_\pi^2 - p^2} + \text{regular terms}, \tag{2}$$

where the pion decay constant is defined from $\langle 0|A_\mu^i(0)|\pi^k(p)\rangle = ip_\mu F_\pi\delta^{ik}$ and $|\pi^k(p)\rangle$ denotes the one-pion state with the on-shell momentum p^μ and isospin label k.

b) Assume isospin symmetry and calculate the two-point function at one loop in the two-flavor ChPT. Relate F_π to the low-energy constant F at one loop.

22. Power Counting and Inelasticities

In this exercise, we want to understand how the power counting for pion–pion scattering is modified for energies beyond the elastic region. To be specific, consider the three-flavor case and let the S-matrix be $S = \exp(2i\delta - 2\phi)$, with ϕ the inelasticity. Consider S-wave scattering in the isospin zero channel.

a) Show that the unitarity relation for the T-matrix takes the form

$$\text{Im } \sigma T_0^0 = \left|\sigma T_0^0\right|^2 + \frac{1}{4}\left(1 - e^{-4\phi}\right) . \tag{1}$$

b) Show that $2n\pi$ final states generate terms of $O(p^{4n})$, whereas $K\bar{K}$ states are of order $O(p^4)$, so that

$$\phi = \begin{cases} 0, & 4M_\pi^2, s < 16M_\pi^2, \\ O(E^8), & 16M_\pi^2 < s < 4M_K^2, \\ O(E^4), & s > 4M_K^2. \end{cases} \tag{2}$$

c) Show that the absorptive parts $\text{Im } F$ in the scalar or vector form factor of the pion are $(\text{Im } F)_{2n\pi} = O(p^{4n-2})$ and $(\text{Im } F)_{K\bar{K}} = O(p^2)$ for $2n\pi$ and $K\bar{K}$ intermediate states, respectively. The vector form factor of the pion is given in Eq. (4.315), and the scalar form factor is defined by

$$\langle \pi^a(p')\pi^b(p)|\hat{m}(\bar{u}u + \bar{d}d)|0\rangle = \delta^{ab} F_\pi^S(s) , \tag{3}$$

with $s = (p' + p)^2$ and $F_\pi^S(0) = M_\pi^2$ at leading order. It is advisable to consider the normalized scalar form factor $F(s) = F_\pi^S(s)/F_\pi^S(0)$. Despite these counting rules, one sets $(\text{Im } F)_{K\bar{K}} = 0$ in the one-loop form factor calculation. Why?

Details can be found in the following:

J. Gasser and U.-G. Meißner, "Chiral expansion of pion form-factors beyond one loop," Nucl. Phys. B **357** (1991) 90.

23. Interaction of Pions with Heavy Kaons

The s-quark is much heavier than u, d quarks. Consequently, the convergence problem in ChPT with three quark flavors should be more severe than in its two-flavor counterpart.

In the following paper,

A. Roessl, "Pion kaon scattering near the threshold in chiral $SU(2)$ perturbation theory," Nucl. Phys. B **555** (1999) 507,

it was proposed to treat kaons as heavy particles (like nucleons), and to use two-flavor ChPT to study, for example, the process of pion–kaon scattering. The present exercise deals with the construction of the effective Lagrangian with heavy kaons along the path considered in this paper.

In the following, exact isospin symmetry is assumed. The (heavy) kaon fields come in the isospin doublets:

$$K = \begin{pmatrix} K^+ \\ K^0 \end{pmatrix}, \quad \begin{pmatrix} \bar{K}^0 \\ K^- \end{pmatrix}. \tag{1}$$

The $SU(2) \times SU(2)$ chiral transformations take the form

$$u \mapsto g_R u h^\dagger, \quad K \mapsto hK. \tag{2}$$

Here, $u = U^{1/2}$ collects the pion fields, and h is the compensator that belongs to the diagonal subgroup.

In analogy to the baryon case, the Lagrangian with pion and heavy kaon fields can be written as

$$\mathcal{L} = \mathcal{L}_{\pi\pi}^{(2)} + \mathcal{L}_{\pi\pi}^{(4)} + \cdots + \mathcal{L}_{\pi K}^{(1)} + \mathcal{L}_{\pi K}^{(2)} + \mathcal{L}_{\pi K}^{(3)} + \cdots. \tag{3}$$

Here, $\mathcal{L}_{\pi\pi}^{(2)}, \mathcal{L}_{\pi\pi}^{(4)}$ denote the conventional Lagrangians with pions only, and the rest describes interactions of pions with a single kaon. The terms describing processes with two and more kaons are not shown explicitly. The EOM and field redefinitions are used to arrive at the minimal set of operators.

a) Write down invariant building blocks for constructing the Lagrangian.
b) In analogy with baryon ChPT, set up the counting rules.
c) Assume, for simplicity, that the vector and axial-vector external sources are set to zero from the beginning (these anyway do not contribute to pion–kaon scattering). Construct $\mathcal{L}_{\pi K}^{(1)}, \mathcal{L}_{\pi K}^{(2)}, \mathcal{L}_{\pi K}^{(3)}$.
d) Calculate the πK scattering amplitude and the πK scattering lengths at leading order. Compare this to the result obtained in the $N_f = 3$ flavor QCD.
e) Consider the matching of the "heavy-kaon ChPT" to the $N_f = 3$ flavor QCD. Which requirements should be obeyed by different mass scales in the problem, in order to render such a matching meaningful?

24. The Berezinian

In this exercise, we shall extend the saddle-point technique to the case where fermions are present in the theory along with bosons. Consider the Euclidean Lagrangian:

$$\mathcal{L} = \frac{1}{2}(\partial\phi)^2 + \frac{M^2}{2}\phi^2 + \frac{\lambda}{4}\phi^4 + \bar{\psi}(\slashed{\partial} + m - g\phi)\psi, \tag{1}$$

where ϕ and ψ denote the boson and the fermion fields, respectively. The generating functional is given by

$$Z(j, \eta, \bar{\eta}) = \int d\phi\, d\psi\, d\bar{\psi} \exp\left(-\int d^4x\left(\mathcal{L} + j\phi + \bar{\eta}\psi + \bar{\psi}\eta\right)\right). \tag{2}$$

a) Write down the classical equations of motion for ϕ and ψ.

b) Expand around the solutions of the classical EOM

$$\phi = \phi_c + \xi, \qquad \psi = \psi_c + \lambda, \qquad \bar\psi = \bar\psi_c + \bar\lambda. \tag{3}$$

Next, introduce a vector containing both bosonic and fermionic entries,

$$\rho^T = (\xi, \lambda^T, \bar\lambda), \tag{4}$$

and show that up to and including terms of the second order in small fluctuations, the Euclidean action functional can be written as

$$\int d^4x (\mathcal{L} + j\phi + \bar\eta\psi + \bar\psi\eta) = S(\phi_c, \psi_c, \bar\psi_c) + \frac{1}{2} \int d^4x \rho^T K \rho + \cdots. \tag{5}$$

Here, K denotes the differential operator

$$K = \begin{pmatrix} A & \bar\Gamma & -\Gamma^T \\ -\bar\Gamma^T & 0 & -B^T \\ \Gamma & B & 0 \end{pmatrix}, \tag{6}$$

with

$$A = -\Box + M^2 + 3\lambda\phi_c^2, \qquad B = \not\partial + m - g\phi_c, \qquad \Gamma = -g\psi_c. \tag{7}$$

The (super)matrix K has both bosonic and fermionic entries and takes the form

$$K = \begin{pmatrix} a & \alpha \\ \beta & b \end{pmatrix}, \tag{8}$$

where a, b and α, β are bosonic and fermionic variables, respectively. For this super-matrix, we may define the supertrace and the superdeterminant, also referred to as the Berezinian:

$$\mathrm{str} K = \mathrm{tr} a - \mathrm{tr} b, \qquad \mathrm{sdet} K = \det(a - \alpha b^{-1}\beta)/\det b, \tag{9}$$

and

$$\mathrm{sdet} K = \exp(\mathrm{str} \ln K). \tag{10}$$

c) Performing the Gaussian integral, we get

$$Z(j, \eta, \bar\eta) = \frac{1}{2} \mathrm{sdet} K \exp\left(-S(\phi_c, \psi_c, \bar\psi_c)\right). \tag{11}$$

The one-loop divergences are all contained in the Berezinian. Defining

$$K_0 = \begin{pmatrix} A_0 & 0 & 0 \\ 0 & 0 & -B_0^T \\ 0 & B_0 & 0 \end{pmatrix}, \qquad A_0 = -\Box + M^2, \qquad B_0 = \not\partial + m, \tag{12}$$

and $K_1 = K - K_0$, we may write

$$\mathrm{sdet} K = \exp(\mathrm{str} \ln K_0 + \mathrm{str} \ln(1 + K_0^{-1} K_1))$$

$$= \mathrm{const} \exp\left(\mathrm{str}\left(K_0^{-1} K_1 - \frac{1}{2} K_0^{-1} K_1 K_0^{-1} K_1 + \cdots\right)\right). \tag{13}$$

Using this expansion, show that in dimensional regularization the divergent part of the one-loop effective action is given by

$$S_{\text{div}}(\phi_c, \psi_c, \bar{\psi}_c) = L \int d^4x \left(3\lambda \phi_c^2 + \frac{9\lambda}{2}\phi_c^4 \right.$$

$$+ 8m^3 g\phi_c - 12m^2 g^2 \phi_c^2 + 8mg^3 \phi_c^3 - 2g^4 \phi_c^4$$

$$\left. - 2g^2(\partial\phi_c)^2 + g^2\bar{\psi}_c(-\slashed{\partial} + 2m - 2g\phi_c)\psi_c \right), \tag{14}$$

where

$$L = \frac{\mu^{D-4}}{16\pi^2}\left(\frac{1}{D-4} + \Gamma'(1) + \ln 4\pi \right). \tag{15}$$

Here, D denotes the number of space-time dimensions.

For more details, we refer the reader to the original article:

> H. Neufeld, J. Gasser and G. Ecker, "The one loop functional as a Berezinian," Phys. Lett. B **438** (1998) 106.

25. First Encounter with the WZW Effective Action

The WZW effective action was constructed in Section 4.16. In the absence of external fields, it is given by the first term in Eq. (4.571):

$$S_{\text{WZW}} = \frac{iN_c}{240\pi^2} \int d^5x \, \varepsilon^{abcde} \text{tr}_f \left(U_t^\dagger \partial_a U_t \, U_t^\dagger \partial_b U_t \, U_t^\dagger \partial_c U_t \, U_t^\dagger \partial_d U_t \, U_t^\dagger \partial_e U_t \right). \tag{1}$$

Here, the unitary matrix U_t containing the pion field is defined on the five-dimensional space

$$x^a = (x^\mu, t), \qquad \mu = 0, \ldots, 3, \qquad 0 \le t \le 1. \tag{2}$$

The index a takes the values $a = 0, \ldots, 4$ and ε^{abcde} is the totally antisymmetric tensor in the five-dimensional space, with $\varepsilon_{01234} = -\varepsilon^{01234} = -1$.

A convenient parameterization in terms of the pion fields is given by

$$U_t(x^a) = \exp\left(i\frac{\Phi(x^\mu, t)}{F_0} \right). \tag{3}$$

Here, $\Phi(x^\mu, t) = \lambda^i \phi^i(x^\mu, t)$ denotes the matrix containing the eight Goldstone boson fields, defined on the five-dimensional space. The conventional Goldstone boson fields are obtained at $t = 1$, namely $\phi^i(x^\mu, t)\big|_{t=1} = \phi^i(x^\mu)$. Moreover, the field $\phi^i(x^\mu, t)$ is subject to the boundary condition $\phi^i(x^\mu, t)\big|_{t=0} = 0$. Accordingly, the conventional matrix U is obtained from U_t as $U = U_t\big|_{t=1}$ and obeys the boundary condition $U_t\big|_{t=0} = 1$.

Below, we assume that all quarks are massless and that there are no external fields.

a) Show that the S^0_{WZW} vanishes for $SU(2)$.

b) Combine this action with the conventional $O(p^2)$ effective action

$$S^{(2)} = \int d^4x \frac{F_0^2}{4} \langle \partial_\mu U \partial^\mu U^\dagger \rangle, \tag{4}$$

and show that the equations of motion are given by

$$\partial_\mu \left(\frac{F_0^2}{2} U \partial^\mu U^\dagger \right) + \frac{iN_c}{48\pi^2} \varepsilon^{\mu\nu\rho\sigma} U \partial_\mu U^\dagger U \partial_\nu U^\dagger U \partial_\rho U^\dagger U \partial_\sigma U^\dagger = 0. \tag{5}$$

Hint: Consider the variation of the field U_t:

$$U_t(x^a) \to U'_t(x^a) = \exp(i\Delta(x^\mu, t)) U_t(x^a). \tag{6}$$

Here, Δ denotes an infinitesimal parameter that determines this variation. For consistency, the boundary condition $\Delta(x^\mu, t)|_{t=0} = 0$ should hold.

First, show that

$$U_t^\dagger \partial_a U_t \to U_t^\dagger \partial_a U_t + U_t^\dagger \left(\partial_a \Delta(x^\mu, t) \right) U_t + O(\Delta^2). \tag{7}$$

Furthermore, the variation of the field U is determined through $\Delta(x^\mu, t)|_{t=1} \doteq \Delta(x^\mu)$.

Using the preceding expressions, we may readily verify that the variation of the effective Lagrangian is a total derivative, and that all other terms cancel explicitly. Furthermore, the surface terms in the directions $0, 1, 2, 3$ vanish, because the fields are assumed to vanish at spatial infinity. The surface terms in the variable t give, however, a nonvanishing contribution at $t = 1$ (the contribution at $t = 0$ vanishes, because of the boundary condition imposed on Δ). At the end, we arrive at the equation of motion written down in Eq. (5). Note also that the equation of motion is written solely in terms of *physical* Goldstone boson fields at $t = 1$. This is a consequence of the fact that the variation of the Lagrangian is a total derivative.

c) Show that the two terms in the EOM transform differently under separate

$$\mathbf{x} \mapsto -\mathbf{x}, \; x_0 \mapsto x_0, \; U \mapsto U \quad \text{and} \quad \mathbf{x} \mapsto \mathbf{x}, \; x_0 \mapsto x_0, \; U \mapsto U^\dagger \tag{8}$$

transformations. This in fact lifts the boson number conservation obtained in the nonlinear σ-model.

d) Expand U_t in terms of Goldstone boson fields and show that S_{WZW} at lowest order contains five Goldstone boson fields:

$$S_{WZW} = -\frac{N_c}{240\pi^2 F_0^5} \int d^4x \varepsilon^{\mu\nu\rho\sigma} \langle \Phi \partial_\mu \Phi \partial_\nu \Phi \partial_\rho \Phi \partial_\sigma \Phi \rangle + \dots, \tag{9}$$

where $\Phi = \Phi(x^\mu) = \Phi(x^\mu, t) \big|_{t=1}$. Note that, after expansion, the Lagrangian becomes a total derivative, and the integral over the auxiliary variable t can be explicitly carried out. Then, S_{WZW} is expressed solely in terms of the physical fields at $t = 1$.

For more details, consult the article:

E. Witten, "Global aspects of current algebra," Nucl. Phys. B **223** (1983) 422.

26. Nucleon Polarizabilities

In this exercise, we want to get a better understanding of nucleon chiral dynamics, exemplified through the so-called electric and magnetic polarizabilities. To be specific, consider the Compton tensor for photon scattering off a nucleon in forward direction (in the heavy baryon limit), $\gamma(k) + N(p,s) \to \gamma(k) + N(p,s)$,

$$T_{\mu\nu}(v,k) = \frac{i}{4} \sum_s \int d^4x\, e^{ikx} \langle p,s | T j_\mu(x) j_\nu(0) | p,s \rangle , \tag{1}$$

for photon-momentum k and nucleon four-velocity v, so that $p = mv$. Furthermore, $j_\mu(x)$ in the preceding expression denotes the electromagnetic current.

a) Show that

$$\Theta = \frac{e^2}{m} \varepsilon^\mu \varepsilon^\nu T_{\mu\nu}(v,k) = e^2 \left\{ \varepsilon^2 U(\omega) + (\varepsilon \cdot k)^2 V(\omega) \right\} , \tag{2}$$

with $p = mv$ the nucleon momentum, m the nucleon mass, $\omega = v \cdot k$, ε^μ (ε^ν) the polarization vector of the outgoing (incoming) photon and we use the gauge $\varepsilon \cdot v = 0$. In particular, under crossing $U(\omega) = U(-\omega)$ and $V(\omega) = V(-\omega)$.

b) Show that the Thomson limit is entirely generated by the seagull diagram with $U(0) = Z^2/m$, with $Z = 1(0)$ the charge of the proton (neutron). Why do the leading-order Born and crossed Born diagrams not contribute?

c) Show that at low energies, the spin-averaged Compton tensor in forward direction can be written as

$$\Theta = \left[\frac{e^2 Z^2}{m} - 4\pi\omega^2(\alpha + \beta) \right] \varepsilon^2 - 4\pi\beta(\varepsilon \cdot k)^2 + \dots, \tag{3}$$

which defines the polarizabilities via

$$\alpha + \beta = -\frac{e^2}{8\pi} U''(0) , \quad \beta = -\frac{e^2}{4\pi} V''(0) , \tag{4}$$

and the derivatives are taken with respect to the variable ω.

d) Show that in this gauge, the leading-order one-loop graphs at $O(p^3)$ all come with the prefactor g_A^2/F^2. In total, nine diagrams are left. Calculate these and verify:

$$U(\omega) = \frac{g_A^2 M_\pi}{8\pi F^2} \left[-\frac{3}{2} - \frac{1}{z} + \left(1 + \frac{1}{z}\right) \sqrt{1-z} + \frac{1}{\sqrt{z}} \arcsin\sqrt{z} \right] ,$$

$$V(\omega) = \frac{g_A^2}{16\pi F^2 M_\pi} \left[-\frac{4}{z^2} - \frac{9}{2z} + \left(\frac{4}{z^2} + \frac{5}{2z}\right) \sqrt{1-z} \right.$$

$$\left. + \left(1 + \frac{4}{z}\right) \frac{1}{\sqrt{z}} \arcsin\sqrt{z} \right] , \quad z = \frac{\omega^2}{M_\pi^2} , \tag{5}$$

for $|\omega| < M_\pi$, that is, $|z| < 1$. If $|z| > 1$, we have to perform an analytic continuation such that $\sqrt{1-z} \to -i\sqrt{z-1}$ and $\arcsin\sqrt{z} \to \pi/2 + i\ln(\sqrt{z} + \sqrt{z-1})$.

Figure 6.2 Two-loop contribution to the nucleon self-energy in the form of a sunset diagram. Solid and wiggly lines denote nucleons and photons, respectively.

e) Use these formulae to show that

$$\alpha_p = \alpha_n = 10\beta_p = 10\beta_n = \frac{5e^2 g_A^2}{384\pi^2 F^2 M_\pi} . \tag{6}$$

Why do these polarizabilities diverge in the chiral limit? At which order would you expect differences between the proton and the neutron polarizabilities? Why are the electric polarizabilities less affected by the omitted $\Delta(1232)$ resonance than the magnetic ones? How will the results be affected at fourth order?

Many details concerning these calculations can be found in the following references:

V. Bernard, N. Kaiser and U.-G. Meißner, "Chiral expansion of the nucleon's electromagnetic polarizabilities," Phys. Rev. Lett. **67** (1991) 1515.

V. Bernard, N. Kaiser, J. Kambor and U.-G. Meißner, "Chiral structure of the nucleon," Nucl. Phys. B **388** (1992) 315.

V. Bernard, N. Kaiser, U.-G. Meißner and A. Schmidt, "Aspects of nucleon Compton scattering," Z. Phys. A **348** (1994) 317.

27. Double Logs

In this exercise, we will encounter the appearance of so-called double logarithms that are generated in two-loop calculations. Using as always dimensional regularization and the modified $\overline{\text{MS}}$ scheme (or any mass-independent renormalization scheme), the two-loop divergences take the general form

$$k(D)\frac{\mu^{2\varepsilon}}{(4\pi)^4}\left[\frac{1}{\varepsilon^2} + \frac{2}{\varepsilon^2}\ln\frac{M_\pi}{\mu} + \ln^2\frac{M_\pi}{\mu} + \dots\right], \tag{1}$$

with μ the regularization scale, D the number of space-time dimensions, $\varepsilon = D - 4$ and $k(D)$ is a function of D that depends on the specific diagram. The last term in the square brackets is a double logarithm (double log). Consider the so-called sunset diagram depicted in Fig. 6.2 with external momentum p. Using the two-flavor heavy baryon formulation, show that:

a) The nucleon self-energy in HBChPT does not have a double log,
b) The nucleon Z-factor in HBChPT contains a double log.

For details, see the following:

J. A. McGovern and M. C. Birse, "On the absence of fifth order contributions to the nucleon mass in heavy baryon chiral perturbation theory," Phys. Lett. B **446** (1999) 300.

V. Bernard and U.-G. Meißner, "The nucleon axial-vector coupling beyond one loop," Phys. Lett. B **639** (2006) 278.

M. Buchler and G. Colangelo, "Renormalization group equations for effective field theories," Eur. Phys. J. C **32** (2003) 427.

28. Harmonic Trap

The following exercise is inspired by the paper that studies the dependence of the spectrum of trapped cold atoms on the atom–atom scattering length:

Th. Busch, B.-G. Englert, K. Rzazewski and M. Wilkens, "Two cold atoms in a harmonic trap," Foundation of Physics **28** (1998) 549.

Imagine that we have (nonrelativistic) atoms that are individually trapped by some external potential (e.g., an optical lattice). Near the minimum, this potential is well approximated by a harmonic oscillator:

$$V_{\text{osc}}(\mathbf{x}) = \frac{m\omega^2 \mathbf{x}^2}{2}, \tag{1}$$

where m denotes the atom mass and ω is the parameter characterizing the second derivative of the potential at the minimum.

In experiments, usually more than one atom is confined in a rather small volume that is defined by the parameter ω. These atoms interact with each other. Hence, the multiparticle spectrum will not be just a multiple of single-particle energies, but these will be shifted. Typically, the range of the atom–atom interaction is much smaller than the size of the trap and, therefore, these interactions can be approximated by a point-like four-particle vertex. At lowest order, this is a nonderivative vertex with a coupling that can be related to the scattering length. (It is assumed that dimensional regularization is used everywhere to handle the ultraviolet divergences.)

Consider now two trapped atoms and calculate the leading-order energy shift in the two-body scattering length a_0. To this end:

a) Write the effective Lagrangian that describes the problem.
b) Derive the scattering equations and carry out the separation between the CM and relative variables of two particles.
c) Derive the formula for the energy shift at leading order in a_0.
d) Discuss the dynamical scales that emerge in the theory and their relation to the parameters in the Lagrangian. Discuss the condition for applicability of the perturbation theory in the scattering length. Consider also the case of an unnaturally large scattering length (in the vicinity of a Feshbach resonance).

The following remarks are appropriate. First, despite a striking similarity to the problem of hadronic atoms, there is a subtle difference. Namely, in the latter case, the hadrons are bound by the long-range Coulomb potential acting among them. In contrast, in the case considered here, the harmonic potential acts on each atom individually. The atoms are trapped together merely because they are trapped in the same cell. This peculiarity shows up when one separates the CM motion from the relative one (you will see this explicitly). The resulting equations for the relative motion, however, are similar, and so are the results. Note also that the many-body problem on the lattice is conceptually closer to the case of a harmonic trap rather than to that of hadronic atoms.

Furthermore, note that implementing a harmonic trap on the lattice has been proposed in order to facilitate the extraction of the nucleon–nucleon phase shifts from the measured lattice spectrum. For more details, refer to the following:

T. Luu, M. J. Savage, A. Schwenk and J. P. Vary, "Nucleon-nucleon scattering in a harmonic potential," Phys. Rev. C **82** (2010) 034003.

Appendix Notations and Conventions

A.1 Units and Metric

Throughout this book, we work in natural units, defined by

$$\hbar = c = k_B = 1, \tag{A.1}$$

where $\hbar \equiv h/2\pi$ is the reduced Planck constant, c is the speed of light in vacuum, and k_B is the Boltzmann constant. Only in a few places are these constants explicitly displayed. The unit charge is called e, and it is assumed to be positive. Up-to-date values can be found at the Particle Data Group website, $\mathtt{http://pdg.lbl.gov/}$.

The natural length scale in QCD is given by the approximate size of the proton or by $1/\Lambda_{\mathrm{QCD}}$, that is,

$$1 \,\mathrm{fm} = 10^{-15} \,\mathrm{m} \,. \tag{A.2}$$

Energies are given in mega-electronvolts (MeV) or giga-electronvolts (GeV), with $1\,\mathrm{GeV} = 1000\,\mathrm{MeV}$. The conversion factor from energy to length in natural units is

$$\hbar c = 197.327 \,\mathrm{MeV} \cdot \mathrm{fm} = 0.197327 \,\mathrm{GeV} \cdot \mathrm{fm} \,. \tag{A.3}$$

The Compton wavelength λ_C of a particle with mass m is

$$\lambda_C = \frac{1}{m} \,. \tag{A.4}$$

The metric tensor in Minkowski space is

$$g_{\mu\nu} \equiv \mathrm{diag}(1, -1, -1, -1), \tag{A.5}$$

and in Euclidean space it reads

$$g_{\mu\nu} \equiv \delta_{\mu\nu} \equiv \mathrm{diag}(1, 1, 1, 1), \tag{A.6}$$

with $\mu, \nu = 0, 1, 2, 3$ (in Minkowski space) and $\mu, \nu = 1, 2, 3, 4$ (in Euclidean space). Further, the totally antisymmetric (Levi–Civita) tensor in four dimensions is defined as

$$\varepsilon^{\mu\nu\alpha\beta} = \begin{cases} +1, & \text{if } \mu\nu\alpha\beta \text{ is an even permutation of } 0123 \,, \\ -1, & \text{if } \mu\nu\alpha\beta \text{ is an odd permutation of } 0123 \,, \\ 0, & \text{otherwise} \,. \end{cases} \tag{A.7}$$

The scalar product in four dimensions is

$$a_\mu b^\mu = g_{\mu\nu} a^\mu b^\nu = ab, \quad \mu, \nu = 0,1,2,3, \tag{A.8}$$

where the Einstein summation convention $\sum_{\mu=0}^{3} a_\mu b^\mu \equiv a_\mu b^\mu$ is understood and in some cases, we also write $a_\mu b^\mu = a \cdot b$ for clarity.
The scalar product in three dimensions is written as

$$a_i b_i = \mathbf{ab}, \tag{A.9}$$

and in some cases we write $a_i b_i = \mathbf{a} \cdot \mathbf{b}$ for clarity. Here, the sum runs over the space indices $i = 1,2,3$ only.

A.2 Pauli Matrices

The Pauli matrices represent the $SU(2)$ spin or isospin algebra. They obey the following relations:

$$[\sigma^i, \sigma^j] = 2i\varepsilon^{ijk}\sigma^k, \tag{A.10}$$

and

$$\sigma^i \sigma^j = \delta^{ij} \mathbb{1}_2 + i\varepsilon^{ijk}\sigma^k, \tag{A.11}$$

for $i,j,k = 1,2,3$. Here, ε^{ijk} is the totally antisymmetric tensor,

$$\varepsilon_{ijk} = \varepsilon^{ijk} = \begin{cases} +1, & \text{if } ijk \text{ is an even permutation of } 123, \\ -1, & \text{if } ijk \text{ is an odd permutation of } 123 \\ 0, & \text{otherwise}, \end{cases} \tag{A.12}$$

and the Kronecker delta is

$$\delta^{ij} = \delta_{ij} = \begin{cases} +1, & \text{if } i = j, \\ 0, & \text{if } i \neq j. \end{cases} \tag{A.13}$$

The explicit representation of the σ matrices is

$$\sigma^1 \equiv \begin{pmatrix} 0 & 1 \\ 1 & 0 \end{pmatrix}, \quad \sigma^2 \equiv \begin{pmatrix} 0 & -i \\ i & 0 \end{pmatrix}, \quad \sigma^3 \equiv \begin{pmatrix} 1 & 0 \\ 0 & -1 \end{pmatrix}. \tag{A.14}$$

The pertinent 2×2 unit matrix is

$$\mathbb{1}_2 = \begin{pmatrix} 1 & 0 \\ 0 & 1 \end{pmatrix}. \tag{A.15}$$

A.3 Gell–Mann Matrices

The $SU(3)$ Lie algebra is given in terms of the Gell–Mann matrices λ^i ($i = 1, \ldots, 8$),

$$\left[\frac{1}{2}\lambda^i, \frac{1}{2}\lambda^j \right] = if^{ijk}\frac{1}{2}\lambda^k, \tag{A.16}$$

where the f^{ijk} are the totally antisymmetric structure constants. The nonvanishing f^{ijk} are

$$f^{123} = 1, \quad f^{147} = f^{246} = f^{257} = f^{345} = f^{516} = f^{637} = \frac{1}{2}, \quad f^{458} = f^{678} = \frac{\sqrt{3}}{2}, \tag{A.17}$$

and permutations thereof, $f^{\sigma(i)\sigma(j)\sigma(k)} = \varepsilon(\sigma)f^{ijk}$, where $\varepsilon(\sigma)$ is the signature of the permutation $\sigma \in S_3$.

The explicit form of the λ^i is

$$\lambda^1 = \begin{pmatrix} 0 & 1 & 0 \\ 1 & 0 & 0 \\ 0 & 0 & 0 \end{pmatrix} \quad \lambda^2 = \begin{pmatrix} 0 & -i & 0 \\ i & 0 & 0 \\ 0 & 0 & 0 \end{pmatrix} \quad \lambda^3 = \begin{pmatrix} 1 & 0 & 0 \\ 0 & -1 & 0 \\ 0 & 0 & 0 \end{pmatrix}$$

$$\lambda^4 = \begin{pmatrix} 0 & 0 & 1 \\ 0 & 0 & 0 \\ 1 & 0 & 0 \end{pmatrix} \quad \lambda^5 = \begin{pmatrix} 0 & 0 & -i \\ 0 & 0 & 0 \\ i & 0 & 0 \end{pmatrix} \quad \lambda^6 = \begin{pmatrix} 0 & 0 & 0 \\ 0 & 0 & 1 \\ 0 & 1 & 0 \end{pmatrix}$$

$$\lambda^7 = \begin{pmatrix} 0 & 0 & 0 \\ 0 & 0 & -i \\ 0 & i & 0 \end{pmatrix} \quad \lambda^8 = \frac{1}{\sqrt{3}}\begin{pmatrix} 1 & 0 & 0 \\ 0 & 1 & 0 \\ 0 & 0 & -2 \end{pmatrix} \quad . \tag{A.18}$$

In addition, the λ^i obey the relation

$$\left\{ \frac{1}{2}\lambda^i, \frac{1}{2}\lambda^j \right\} = \frac{1}{3}\delta^{ij} + d^{ijk}\frac{1}{2}\lambda^k, \tag{A.19}$$

where the nonvanishing totally symmetric d^{ijk} are given by

$$d^{118} = d^{228} = d^{338} = -d^{888} = \frac{1}{\sqrt{3}}, \quad d^{146} = d^{157} = d^{256} = d^{344} = d^{355} = \frac{1}{2},$$

$$d^{247} = d^{366} = d^{377} = -\frac{1}{2}, \quad d^{448} = d^{558} = d^{668} = d^{778} = \frac{1}{2\sqrt{3}}, \tag{A.20}$$

and permutations thereof.

A.4 Dirac Matrices

The 4×4 matrices γ^μ, where $\mu = 0, 1, 2, 3$, obey

$$\{\gamma^\mu, \gamma^\nu\} = \gamma^\mu \gamma^\nu + \gamma^\nu \gamma^\mu = 2g^{\mu\nu} , \quad \mu, \nu = 0, 1, 2, 3 . \tag{A.21}$$

Further, γ_5 is defined via

$$\gamma_5 = \gamma^5 = i\gamma^0 \gamma^1 \gamma^2 \gamma^3 , \tag{A.22}$$

with

$$\gamma_5^2 = \mathbb{1}_4 , \quad \{\gamma_5, \gamma^\mu\} = 0 . \tag{A.23}$$

Some useful properties of γ matrices are

$$\sigma^{\mu\nu} = \frac{i}{2}[\gamma^\mu, \gamma^\nu] , \quad \gamma^\mu \gamma^\nu = g^{\mu\nu} - i\sigma^{\mu\nu} ,$$

$$[\gamma^5, \sigma^{\mu\nu}] = 0 , \quad \gamma_5 \sigma^{\mu\nu} = \frac{i}{2}\varepsilon^{\mu\nu\rho\sigma}\sigma_{\rho\sigma} ,$$

$$\gamma^\mu \gamma_\mu = 4 , \quad \gamma^\mu \gamma^\nu \gamma_\mu = -2\gamma^\nu , \quad \gamma^\mu \gamma^\nu \gamma^\lambda \gamma_\mu = 4g^{\nu\lambda} . \tag{A.24}$$

Often needed are traces, such as

$$\mathrm{Tr}(\mathbb{1}_4) = 4 ,$$

$$\mathrm{Tr}(\gamma^\mu) = \mathrm{Tr}(\gamma^5) = \mathrm{Tr}(\gamma^5 \gamma^\mu) = \mathrm{Tr}(\sigma^{\mu\nu}) = \mathrm{Tr}(\gamma^\mu \gamma^\nu \gamma^5) = 0 ,$$

$$\mathrm{Tr}(\gamma^\mu \gamma^\nu) = 4g^{\mu\nu} , \quad \mathrm{Tr}(\gamma^\mu \gamma^\nu \gamma^\rho \gamma^\sigma) = 4(g^{\mu\nu}g^{\rho\sigma} + g^{\mu\sigma}g^{\nu\rho} - g^{\mu\rho}g^{\nu\sigma}) ,$$

$$\mathrm{Tr}(\gamma^5 \gamma^\mu \gamma^\nu \gamma^\rho \gamma^\sigma) = -4i\varepsilon^{\mu\nu\rho\sigma} = 4i\varepsilon_{\mu\nu\rho\sigma} . \tag{A.25}$$

The explicit form of the γ-matrices in the so-called Dirac representation is

$$\gamma^0 = \begin{pmatrix} \mathbb{1}_2 & 0 \\ 0 & -\mathbb{1}_2 \end{pmatrix} , \quad \boldsymbol{\gamma} = \begin{pmatrix} 0 & \boldsymbol{\sigma} \\ -\boldsymbol{\sigma} & 0 \end{pmatrix} , \quad \gamma^5 = \begin{pmatrix} 0 & \mathbb{1}_2 \\ \mathbb{1}_2 & 0 \end{pmatrix} . \tag{A.26}$$

The Dirac spinors are normalized as

$$\bar{u}(p, s')u(p, s) = 2m\delta_{s', s} , \tag{A.27}$$

with m the fermion mass.

A.5 Isospin and *SU*(3) Flavor Symmetry

Isospin symmetry was first postulated by Heisenberg, who noticed that the nucleon–nucleon interactions are approximately invariant under

$$N \mapsto N' = UN , \quad U \in SU(2) , \quad N = \begin{pmatrix} p \\ n \end{pmatrix} ; \tag{A.28}$$

that is, the proton and the neutron behave like spin-up and spin-down states in the abstract isospin space. The generators of the corresponding Lie algebra are the Pauli matrices. Also used in the discussion of nuclear interactions is charge symmetry, which is a rotation by $180°$ in isospin space.

In QCD, isospin symmetry refers to the light up and down quarks. For $m_u = m_d$, QCD is invariant under $SU(2)$ isospin transformations

$$q \mapsto q' = U q , \quad q = \begin{pmatrix} u \\ d \end{pmatrix} \quad U = \begin{pmatrix} a^* & b^* \\ -b & a \end{pmatrix} \quad |a|^2 + |b|^2 = 1 . \tag{A.29}$$

The two-flavor QCD quark mass term can thus be rewritten as

$$\mathcal{H}_{QCD}^{mass} = m_u \, \bar{u}u + m_d \, \bar{d}d$$
$$= \underbrace{\frac{1}{2}(m_u + m_d)(\bar{u}u + \bar{d}d)}_{I=0} + \underbrace{\frac{1}{2}(m_u - m_d)(\bar{u}u - \bar{d}d)}_{I=1} \tag{A.30}$$

in terms of an isoscalar ($I = 0$) and an isovector ($I = 1$) piece. The latter is responsible for strong isospin breaking. This breaking is parametrized in terms of $(m_d - m_u)/\Lambda_{QCD}$ and thus expected to be of the order of a few percent.

$SU(3)$ flavor symmetry was originally introduced by Gell–Mann and Zweig to bring order into the so-called "hadron zoo." In QCD, it refers to the invariance under $SU(3)_V$ transformations when $m_s = m_d = m_u$ (the $SU(3)$ limit). This symmetry is broken stronger than isospin, as $m_s \gg m_u, m_d$.

A.6 Spherical Harmonics

The spherical harmonics $Y_{\ell m}(\theta, \phi)$ are defined as

$$Y_{\ell m}(\theta, \phi) \equiv \sqrt{\frac{(2l+1)}{4\pi} \frac{(\ell - m)!}{(\ell + m)!}} \, P_{\ell,m}(\cos \theta) \exp(im\phi) . \tag{A.31}$$

Here, the $P_{\ell,m}(x)$ are the associated Legendre polynomials,

$$P_{\ell,m}(x) \equiv (-1)^m (1 - x^2)^{m/2} \frac{d^m}{dx^m} P_\ell(x), \tag{A.32}$$

where

$$P_\ell(x) \equiv \frac{1}{2^\ell \ell!} \frac{d^\ell}{dx^\ell} (x^2 - 1)^\ell \tag{A.33}$$

are the Legendre polynomials. The spherical harmonics are orthonormal functions,

$$\int d\Omega \, Y_{\ell m}^*(\Omega) Y_{\ell m'}(\Omega) = \delta_{\ell\ell'} \delta_{mm'}, \tag{A.34}$$

with $d\Omega \equiv \sin\theta \, d\theta \, d\phi$. Under complex conjugation,

$$Y_{\ell m}^*(\theta, \phi) = (-1)^m Y_{\ell,-m}(\theta, \phi) . \tag{A.35}$$

Note that these definitions correspond to the Condon–Shortley phase convention.

A.7 Clebsch–Gordan Coefficients

The Clebsch–Gordan (CG) coefficients for the coupling of two angular momenta (j_1, m_1) and (j_2, m_2) to total angular momentum (J, J_z) are

$$|JJ_z\rangle = \sum_{m_1=-j_1}^{j_1} \sum_{m_2=-j_2}^{j_2} |j_1 m_1, j_2 m_2\rangle \langle j_1 m_1, j_2 m_2 | JJ_z\rangle. \tag{A.36}$$

These satisfy

$$\sum_{J=|j_1-j_2|}^{j_1+j_2} \sum_{J_z=-J}^{J} \langle j_1 m_1, j_2 m_2 | JJ_z\rangle \langle j_1 m_1', j_2 m_2' | JJ_z\rangle = \delta_{m_1 m_1'} \delta_{m_2 m_2'},$$

$$\sum_{m_1, m_2} \langle j_1 m_1, j_2 m_2 | JJ_z\rangle \langle j_1 m_1, j_2 m_2 | J'J_z'\rangle = \delta_{JJ'} \delta_{J_z J_z'}. \tag{A.37}$$

The CG coefficients vanish unless $J_z = m_1 + m_2$. Some special cases of CG coefficients are

$$\langle j_1 m_1, j_2 m_2 | 00\rangle = \frac{(-1)^{j_1-m_1}}{\sqrt{2j_1+1}} \delta_{j_1 j_2} \delta_{m_1, -m_2}, \tag{A.38}$$

for $J = 0$, and

$$\langle j_1 j_1, j_2 j_2 | (j_1 + j_2)(j_1 + j_2)\rangle = 1, \tag{A.39}$$

for $J = j_1 + j_2$ and $J = J_z$. We also note the integral relation between the spherical harmonics and certain CG coefficients:

$$\int d\Omega \, Y_{\ell_1, m_1}^*(\Omega) Y_{\ell_2, m_2}^*(\Omega) Y_{\ell, m}(\Omega)$$

$$= \sqrt{\frac{(2\ell_1+1)(2\ell_2+1)}{4\pi(2\ell+1)}} \langle \ell_1 0, \ell_2 0 | \ell 0\rangle \langle \ell_1 m_1, \ell_2 m_2 | \ell m\rangle. \tag{A.40}$$

The Wigner $3j$ symbols can be related to the CG coefficients via

$$\langle j_1 m_1, j_2 m_2 | JJ_z\rangle = (-1)^{-j_1+j_2-J_z} \sqrt{2J+1} \begin{pmatrix} j_1 & j_2 & J \\ m_1 & m_2 & -J_z \end{pmatrix}. \tag{A.41}$$

The $3j$ symbol $\begin{pmatrix} j_1 & j_2 & j_3 \\ m_1 & m_2 & m_3 \end{pmatrix}$ vanishes except for

- $m_i \in \{-j_i, -j_i + 1, \ldots, j_i - 1, j_i\}$, $i = 1, 2, 3$,

- $m_1 + m_2 + m_3 = 0$,

- $|j_1 - j_2| \leq j_3 \leq j_1 + j_2$,

- $j_1 + j_2 + j_3 \in \mathbb{Z}$. \hfill (A.42)

A.8 Bessel Functions

In the scattering problem in Chapter 1, we encountered the spherical Bessel functions of the first and second kind. These can be obtained from

$$j_\ell(x) = (-x)^\ell \left(\frac{1}{x}\frac{d}{dx}\right)^\ell \frac{\sin x}{x}, \tag{A.43}$$

$$n_\ell(x) = -(-x)^\ell \left(\frac{1}{x}\frac{d}{dx}\right)^\ell \frac{\cos x}{x}, \tag{A.44}$$

with $\ell = 0,1,2,\dots$. The $n_\ell(x)$ are also called von Neumann functions.

In Chapter 5, we required the modified Bessel functions of the first and second kind, $I_\nu(z)$ and $K_\nu(z)$. Integral representation for these, valid for $\mathrm{Re}\,\nu > 1/2$, are

$$I_\nu(z) = \frac{\left(\frac{z}{2}\right)^\nu}{\Gamma\left(\nu+\frac{1}{2}\right)\Gamma\left(\frac{1}{2}\right)} \int_0^\pi d\theta \, e^{\pm z\cos\theta} \sin^{2\nu}\theta,$$

$$K_\nu(z) = \frac{\left(\frac{z}{2}\right)^\nu \Gamma\left(\frac{1}{2}\right)}{\Gamma\left(\nu+\frac{1}{2}\right)} \int_0^\infty dt \, e^{-z\cosh t} \sinh^{2\nu} t. \tag{A.45}$$

Here,

$$\Gamma(z) = \int_0^\infty dt \, t^{z-1} e^{-t} \tag{A.46}$$

is the Euler Gamma-function.

The asymptotic behavior for large positive z of the modified Bessel functions of the first and second kind reads

$$I_\nu(z) \to \frac{e^z}{\sqrt{2\pi z}},$$

$$K_\nu(z) \to \sqrt{\frac{\pi}{2z}} e^{-z}. \tag{A.47}$$

The series representation of these functions in the vicinity of $z = 0$ for the positive integer values of $\nu = n$ takes the form

$$I_n(z) = \sum_{k=0}^\infty \frac{1}{k!(n+k)!} \left(\frac{z}{2}\right)^{n+2k},$$

$$K_n(z) = \frac{1}{2}\sum_{k=0}^{n-1} (-1)^k \frac{(n-k-1)!}{k!} \left(\frac{z}{2}\right)^{2k-n}$$

$$+ (-1)^{n+1}\sum_{k=0}^\infty \frac{1}{k!(n+k)!} \left(\frac{z}{2}\right)^{n+2k} \left(\ln\frac{z}{2} - \frac{1}{2}\psi(k+1) - \frac{1}{2}\psi(n+k+1)\right). \tag{A.48}$$

Here, $\psi(z) = \Gamma'(z)/\Gamma(z)$ is the logarithmic derivative of the Gamma-function.

Index

Printed in the United States
by Baker & Taylor Publisher Services